17th International Workshop on Advanced Computing and Analysis Techniques in Physics Research (ACAT 2016)

Journal of Physics: Conference Series Volume 762

Valparaiso, Chile
18 – 22 January 2016

Editors:

Luis Salinas
Claudio Torres

ISBN: 978-1-5108-3275-6
ISSN: 1742-6588

Printed from e-media with permission by:

Curran Associates, Inc.
57 Morehouse Lane
Red Hook, NY 12571

Some format issues inherent in the e-media version may also appear in this print version.

Printed by Curran Associates, Inc. (2017)

For permission requests, please contact the Institute of Physics
at the address below.

Institute of Physics
Dirac House, Temple Back
Bristol BS1 6BE UK

Phone: 44 1 17 929 7481
Fax: 44 1 17 920 0979

techtracking@iop.org

Additional copies of this publication are available from:

Curran Associates, Inc.
57 Morehouse Lane
Red Hook, NY 12571 USA
Phone: 845-758-0400
Fax: 845-758-2633
Email: curran@proceedings.com
Web: www.proceedings.com

Table of contents

Volume 762

17th International Workshop on Advanced Computing and Analysis Techniques in Physics Research (ACAT2016)

18–22 January 2016, Valparaíso, Chile

Accepted papers received: 28 September 2016
Published online: 21 November 2016

Preface

Papers

T1 - Computing Technology for Physics Research

T2 - Data Analysis - Algorithms and Tools

Journal of Physics: Conference Series **762** (2016) 011001 doi:10.1088/1742-6596/762/1/011001

17th International Workshop on Advanced Computing and Analysis Techniques in Physics Research (ACAT 2016)

Preface

The 2016 version of the International Workshop on Advanced Computing and Analysis Techniques in Physics Research took place on January 18-22, 2016, at the Universidad Técnica Federico Santa María –UTFSM– in Valparaíso, Chile.

The present volume of IOP Conference Series is devoted to the selected scientific contributions presented at the workshop. In order to guarantee the scientific quality of the Proceedings all papers were thoroughly peer-reviewed by an ad-hoc Editorial Committee with the help of many careful reviewers.

The ACAT Workshop series has a long tradition starting in 1990 (Lyon, France), and takes place in intervals of a year and a half. Formerly these workshops were known under the name AIHENP (Artificial Intelligence for High Energy and Nuclear Physics). Each edition brings together experimental and theoretical physicists and computer scientists/experts, from particle and nuclear physics, astronomy and astrophysics in order to exchange knowledge and experience in computing and data analysis in physics. Three tracks cover the main topics: Computing technology: languages and system architectures. Data analysis: algorithms and tools. Theoretical Physics: techniques and methods. Although most contributions and discussions are related to particle physics and computing, other fields like condensed matter physics, earth physics, biophysics are often addressed in the hope to share our approaches and visions. It created a forum for exchanging ideas among fields, exploring and promoting cutting-edge computing technologies and debating hot topics.

The 17th edition, ACAT 2016, attacted over 100 participants from all over the world including 18 invited speakers who presented key topics on advanced computing and analysis techniques in physics. During the workshop, 57 talks and 27 posters were presented in the three tracks

ACAT 2016 was generously sponsored by:

- Comisión Nacional de Investigación Científica y Tecnológica – CONICYT, Chile;
- Centro Científico Tecnológico de Valparaíso – CCTVal, Valparaíso, Chile;
- Center for Mathematical Modeling –CMM– of the Universidad de Chile;
- National Laboratory for High Performance Computing – NLHPC;
- UTFSM Office for Scientific Research, Innovation and Graduate Studies; and
- Universidad del Bio-Bio, Chillán, Chile.

Many persons made possible the preparation of this Proceedings volume. We specially thank the track liaisons Niko Neufeld, Lorenzo Moneta and York Schröder for their work on the scientific program and the Proceedings preparation.

The Editors and the members of the International Advisory and Coordination Committee –IACC– of ACAT would also like to express their gratitude to all referees for their work on ensuring the quality of the published contributions in the Proceedings.

Our special thanks extend to the conference liaisons Jerome Lauret and Andrei Kataev, as well as to the conference advisors Andrej Arbuzov, David Britton, Gudrun Heinrich, Alexander Kryukov, Milos Lokajicek, Fons Rademakers, Tord Riemann, José Seixas, Liliana Teodorescu, and Gordon Watts, who worked with the local contacts and made this conference possible.

We also would like to thank the members of the ACAT 2016 staff at UTFSM, Alejandro Álvarez, Claudia Arancibia, Juan Carlos Garcés, Roberto León, Tamara Parkes and Benjamin Vergara, who helped us on the background.

Finally we want to warmly thank the Program Committee Chair Federico Carminati and the Conference Chair Denis Perret-Gallix for their global supervision.

Luis Salinas and Claudio Torres, Editors
Universidad Técnica Federico Santa María
Valparaíso, Chile

Local Organizing Committee

UTFSM Physics Dept., Valparaiso:

William Brooks

Hayk Hakobyan

Sergey Kuleshov

UTFSM Computer Sciences Dept., Valparaiso:

Roberto Leon

Luis Salinas, LOC Chair

Claudio Torres, LOC Co-Chair

Dept. of Basic Sciences, University of Bio-Bio, Chillán:

Igor Kondrashuk

York Schröder

Program Committee: Federico Carminati, Chair, CERN, Switzerland.

Track 1:
- Niko Neufeld (CERN, Switzerland); Liaison.
- Axel Naumann (CERN, Switzerland)
- Luis Salinas (UTFSM, Chile)

Track 2: • Lorenzo Moneta (CERN, Switzerlland); Liaison.
• Gordon Watts (U. Washington, USA)
• William Brooks (UTFSM, Chile)

Track 3 • York Schröder (U. Bio-Bio, Chile); Liaison.
• Alexander Poslavsky (IHEP, Protvino, Russia).
• Gionata Luisoni (CERN, Switzerland).

International Advisory and Coordination Committee

Andrej Arbuzov, BLTP JINR, Russia
Pushpalatha Bhat, Fermilab, USA
David Britton, Glasgow Univ., UK
Federico Carminati, SPC chair, CERN, Switzerland
Denis Oliveira Damazio, BNL, USA
Bruce Denby, U. Pierre et Marie Curie, France
Junpei Fujimoto, KEK, Japan
Clara Gaspar, CERN, Switzerland
Gudrun Heinrich, MPI-Munich, Germany
Andrei Kataev, INR, Russia
Alexander Kryukov, SINP, Russia
Jerome Lauret, BNL, USA
Milos Lokajicek, LOC Chair, FZU Prague, Czech Republic
Daniel Maitre, Durham, UK
Axel Naumann, CERN, Switzerland
Denis Perret-Gallix, IACC Chair, IN2P3/CNRS, France
Fons Rademakers, CERN, Switzerland
Tord Riemann, DESY, Germany
Grigory Rubtsov, INR, Russia
Luis Salinas, LOC Chair, UTFSM, Chile
Jose Seixas, Rio de Janeiro Federal Univ., Brazil
Liliana Teodorescu, Brunel Univ., UK
Gordon T. Watts, University of Washington, USA
Monique Werlen, LAPTH, France

CONICYT
Ministerio de Educación
Gobierno de Chile

La Ciencia nos cambia la vida
CONICYT

UNIVERSIDAD DE CHILE

EX UMBRA IN SOLEM
UNIVERSIDAD TECNICA FEDERICO SANTA MARIA

UNIVERSIDAD DEL BÍO-BÍO

CMM
Center for Mathematical Modeling

CCT Val
Centro Científico Tecnológico de Valparaíso

NLHPC
National Laboratory for High Performance Computing Chile

Peer review statement

All papers published in this volume of *Journal of Physics: Conference Series* have been peer reviewed through processes administered by the proceedings Editors. Reviews were conducted by expert referees to the professional and scientific standards expected of a proceedings journal published by IOP Publishing.

Journal of Physics: Conference Series **762** (2016) 012001 doi:10.1088/1742-6596/762/1/012001

A new type of smart pointer for data object reference both in memory and in root files

T Li[1] **and X T Huang**[1]

[1]School of Physics, Shandong University, No.27 Shanda South Road, Jinan, Shandong, 250100, China

liteng@hepg.sdu.edu.cn, huangxt@sdu.edu.cn

Abstract. Based on ROOT framework, we develop a new mechanism named as SmartRef for event data correlations in offline data processing of High Energy Physics experiments. SmartRef uses the Universally Unique Identifier in ROOT to handle correlations between event data objects, both in memory and in ROOT files. It also provides a lazy-loading functionality to speed up event selection processes and simplify data input system. We have applied SmartRef to the JUNO offline software and used it as a test-bed. The test results show that SmartRef provides a significant improvement in event selection processes.

1. Introduction

In offline data processing applications of High Energy Physics experiments, the event data model defines entities of the event data management, file input/output (I/O), even data storage and physics analysis. Therefore, the design of event data model plays an important role for functionalities and performances of the whole offline software.

From the physics analysis point of view, neutrino physics experiments are different from accelerator physics experiments. The formers such as Daya Bay [1] and JUNO [2, 3] have two special requirements. Firstly, time relations between signal events must be supported due to the working principle of the Liquid Scintillator detector [4]. Secondly, high efficient data access and storage capabilities must be implemented due to the rare neutrino signal events compared with the backgrounds.

In order to meet the requirements mentioned above, both pointers of C++ and TRef of ROOT are investigated. Plain C++ pointers or smart pointers within the standard C++ library can only handle the relationship of objects in memory, but the correlated objects cannot be saved into different files. TRef is provided by ROOT [5] as a mechanism for the correlation of TObjects, but does have some limitations. Firstly, the lazy-loading mechanism of TRef requires correlated objects to be saved within TTrees with the same entry number, and it is designed for object correlation within one event. If TRef is used for object correlations across different events, the TBranchRef, which holds extra information for the lazy-loading mechanism, takes too much space. Figure 1 shows that the size of TBranchRef rapidly increases as a function of the number of correlated objects saved into a TTree.

Journal of Physics: Conference Series **762** (2016) 012001 doi:10.1088/1742-6596/762/1/012001

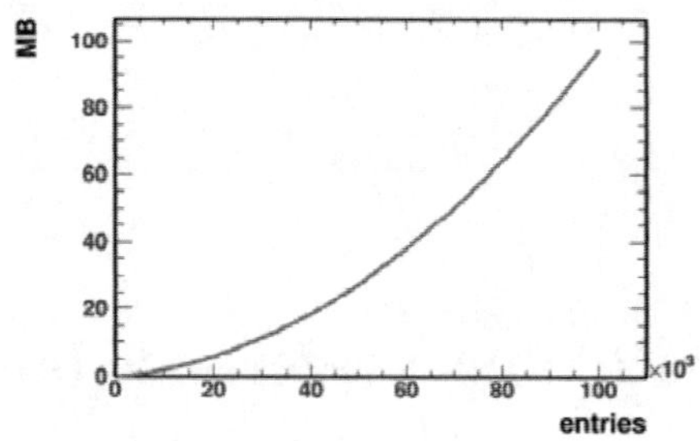

Figure 1. Space usage of TBranchRef

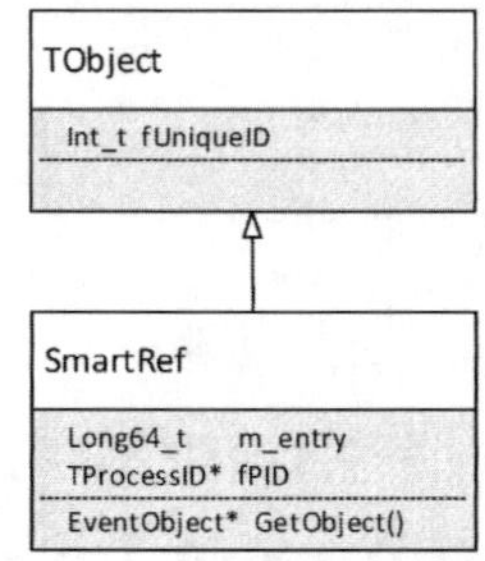

Figure 2. Design of SmartRef

2. SmartRef

2.1. Design of SmartRef

SmartRef is derived from TObject of ROOT and its key members are listed in Figure 2. The fUniqueID variable of type integer and the fPID, a pointer to the TProcessID object make use of the UUID (Universally Unique IDentifier) tool [6] in ROOT to build correlations between SmartRef and its referenced object (RO). The m_entry variable of type long integer records location of the RO when written into a ROOT file in term of TTree. The GetObject() function is designed to retrieve the RO and directly returns a pointer to it if the RO has been already loaded into memory. And if not, SmartRef firstly invokes the lazy-loading mechanism to load the RO from ROOT files and then returns a pointer to it.

2.2. Working principle

UUIDs are deployed to build correlations between SmartRef and its ROs. In a ROOT process, one or several instances of TProcessID are automatically created by the global TROOT, each of which keeps a UUID and one global integer. Associated with the TProcessID instance, an array of pointers is created simultaneously. When a SmartRef refers to a TObject, the value of the global integer is added by 1 and assigned to the fUniqueID member of both the SmartRef and the referenced TObject. Meanwhile, the fPID member of the SmartRef is set to the address of the current TProcessID and the RO is put into the n-th (n equals to the fUniqueID) slot of the array. The reference between SmartRef and its RO is constructed with the UUID and fUniqueID as showing in figure 3. In reverse, when users retrieve the RO, the SmartRef underlyingly queries its fPID and fUniqueID to get the array holding ROs and the position of the RO in the array.

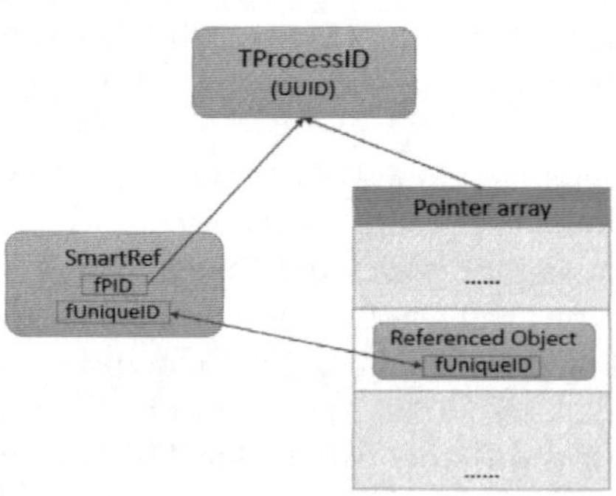

Figure 3. Working principle of SmartRef

When a SmartRef and its RO are written into a ROOT file, the corresponding TProcessID are also written into the file at the same time. When the ROOT file is opened, TProcessIDs are automatically loaded into memory, and the corresponding array is created simultaneously. Once a RO is loaded into memory, it is put into that array accordingly.

2.3. Lazy loading

For neutrino experiments like Daya Bay and JUNO, one signal event consists of a very large load of data. It is not necessary and inefficient to read the whole event into memory at one blow during the event selection process because only partial of the event data is enough to perform fast event selection. And it adds burden to I/O operations if more redundant information is read. In this case, the lazy-loading mechanism is very essential.

We design the SmartRef and an I/O manager module, InputElementKeeper, to achieve the lazy-loading functionality. During an output process, when a TTree holding ROs is created, a special table, TablePerTree, which keeps integer values, is also created. Whenever a RO is written into a TTree, its fUniqueID member is saved into the TablePerTree too. Before a ROOT file is closed, all objects of type TablePerTree are written into this file associated with the TTrees. At the same time, the corresponding UUID of the TProcessID are also written into the file as the meta-data.

When the input system is initializing, the InputElementKeeper module scans all input files, analyze the metadata of these files and then construct a table, which keeps the mapping relationship between UUIDs and files. When users access an object referenced by a SmartRef instance, the SmartRef instance firstly looks it up in the memory to check if the object has been loaded. If not, it will query the InputElementKeeper and search for the file that holds the RO. Once the file is found, InputElementKeeper will search the TablePerTree and load the RO into memory.

Comparing with the relatively large event data object, the metadata of files and TablePerTrees only take a small space, about 4 bytes per event object, and it takes O (1) time complexity to search the TablePerTree. So the efficiency of lazy loading is high, and the extra space needed is relatively small.

3. Application in JUNO

The SmartRef mechanism is successfully deployed in the offline software of JUNO, and it plays an important role in the event data model.

3.1. Design of JUNO event data model

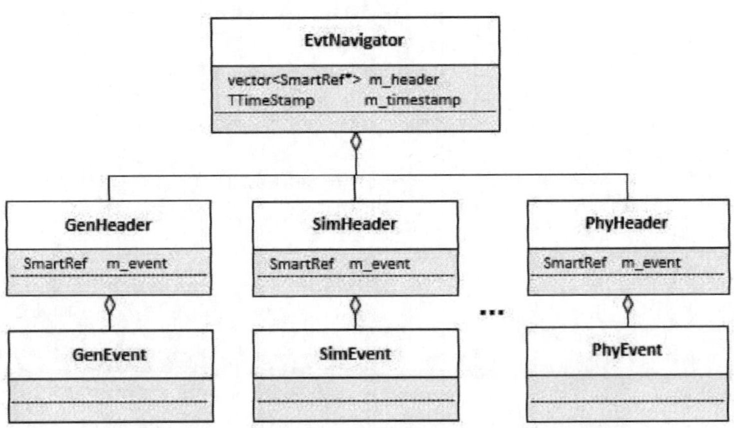

Figure 4. Design of JUNO Event Data Model

Figure 4 shows the design schema of JUNO event data model. All the event classes are derived from TObject directly or indirectly in order to take advantage of functionalities of data management provided by ROOT. To achieve quick event selection, the event data objects are divided into two levels: header and event. They keep the meta-information and full-information, respectively. In each header class, one or more SmartRefs are deployed to correlate its event objects. In this way, users are able to get the header of one event quickly for the meta-data and make decisions whether or not to load the whole event object to perform further analysis.

The EvtNavigator is another application of SmartRef. It keeps a list of SmartRef instances that refers event headers of different data processing stages (including generator, simulation, reconstruction, calibration, etc.) of one event and is used to easily navigate objects between different processing stages in one event. It turns out that SmartRef makes the correlation analyses of the inverse beta decay very conveniently and efficiently.

3.2. Performance test

To test the performance of SmartRef mechanism, we use JUNO offline software framework based on SNiPER [7] as a test-bed and generate two large statistical data samples with and without SmartRef, respectively. Figure 5 and Figure 6 show the comparison of the space usage of ROOT files and the time consuming of loading the data from ROOT files in two cases. The result shows that the extra space needed by SmartRef is very small, and the speed of lazy-loading and normal-loading is almost the same.

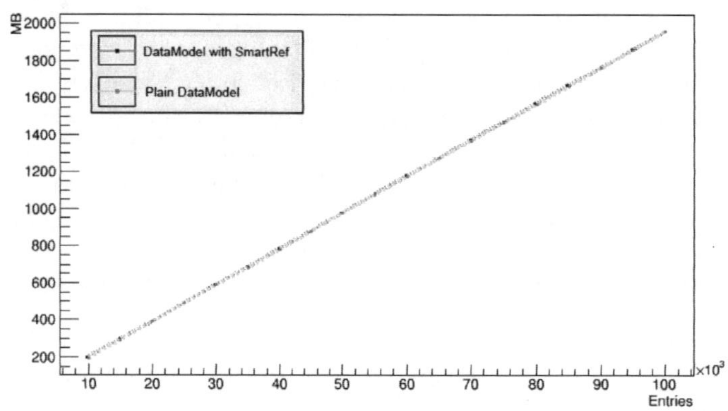

Figure 5. Comparison of space usage between plain data and data with SmartRef.

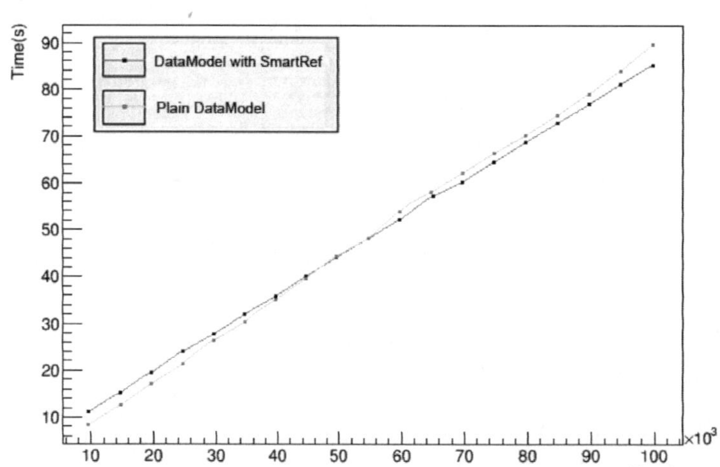

Figure 6. Comparison of the loading-speed between plain data and data with SmartRef.

We also test the event selection process by randomly selecting about 50% of the generated event data and comparing the time consuming of two cases. The result shows that the lazy-loading speeds up the event-selection process by a factor of 35%, as shown in Figure 7.

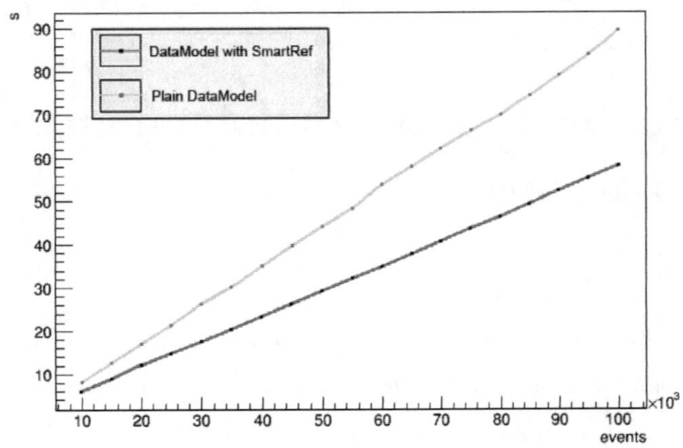

Figure 4. Comparison of time consuming of event-selection between plain data and data with SmartRef

4. Conclusion

We design and develop a smart pointer mechanism, called SmartRef, based on ROOT. The mechanism supports lazy-loading from persistent ROOT files and correlations between objects within one event and across events. The testing shows that it works very well in JUNO and might be used by other experiments too.

Acknowledgements

We acknowledge the supports from the Joint Large-Scale Scientific Facility Funds of the NSFC and CAS (U1532258), the Program for New Century Excellent Talents in University（NCET-13-0342）and the Shandong Natural Science Funds for Distinguished Young Scholar (JQ201402).

References

[1] F. P. An et al. (Daya Bay Collaboration), A Precision Measurement of the Neutrino Mixing Angle theta_13 using Reactor Antineutrinos at Daya Bay, arXiv:1202.6181
[2] Z. Djurcic et al. (JUNO Collaboration), JUNO Conceptual Design Report, arXiv:1508.07166
[3] F.P. An et al. (JUNO Collaboration), Neutrino Physics with JUNO, arXiv:1507.05613
[4] F. P. An et al. (Daya Bay Collaboration), A side-by-side comparison of Daya Bay antineutrino detectors, Nucl. Instrum. and Meth. A 685 78 (2012).
[5] R. Brun, F. Rademakers, ROOT — An object oriented data analysis framework, Nucl. Inst. Meth. in Phys. Res. A 389 (1997) 81–86.
[6] Balkić Z, Šoštarić D, Horvat G. Geohash and uuid identifier for multi-agent system, Technologies and Applications Lecture Notes in Computer Science Volume 7327, 2012, pp 290-298
[7] J. H. Zou et al. SNiPER: an offline software framework for non-collider physics experiments, J. Phys. Conf. Ser., 664(7): 072053 (2015)

A scalable architecture for online anomaly detection of WLCG batch jobs

E Kuehn, M Fischer, M Giffels, C Jung, A Petzold
Karlsruhe Institute of Technology, Steinbuch Centre for Computing,
Hermann-von-Helmholtz-Platz 1, 76344 Eggenstein-Leopoldshafen, Germany

E-mail: {eileen.kuehn, max.fischer, manuel.giffels, christopher.jung, andreas.petzold}@kit.edu

Abstract. For data centres it is increasingly important to monitor the network usage, and learn from network usage patterns. Especially configuration issues or misbehaving batch jobs preventing a smooth operation need to be detected as early as possible. At the GridKa data and computing centre we therefore operate a tool BPNetMon for monitoring traffic data and characteristics of WLCG batch jobs and pilots locally on different worker nodes. On the one hand local information itself are not sufficient to detect anomalies for several reasons, e.g. the underlying job distribution on a single worker node might change or there might be a local misconfiguration. On the other hand a centralised anomaly detection approach does not scale regarding network communication as well as computational costs. We therefore propose a scalable architecture based on concepts of a super-peer network.

1. Introduction

Misbehaving batch jobs can be a notable issue in batch systems as they can negatively affect other batch jobs. Especially negative effects on network traffic rates have potential to increase the walltime enormously. E.g., batch jobs report their results and therefore rely on the network. However, existing monitoring solutions do not provide network metrics distinguished by batch job. Therefore, a network traffic monitoring tool BPNetMon with process level resolution has been developed and put into operation at the GridKa data and computing centre [1]. It monitors information on UNIX process level with a focus on network traffic. By tracking the call hierarchy of single processes started by a batch job, information can be aggregated on batch job level.

Especially in the field of high energy physics the tool is useful to gain insights into the workflows of pilots [2]. Pilots form a virtual overlay batch system [3]. They load batch jobs, following *payloads*, from an external batch system queue. This happens independently from the underlying batch system. While this improves scheduling for the users, the information available to the underlying system is drastically reduced.

These information can also be monitored with BPNetMon. As all monitoring information are reported in a tree data structure, relevant information on payload level can be extracted by identifying the parent process. With access to details on payload level, we now want to introduce an online anomaly detection, that enables a near real time recognition of anomalous payloads. A working anomaly detection on payload level ensures an equal and undisturbed usage of resources for all users of the batch system.

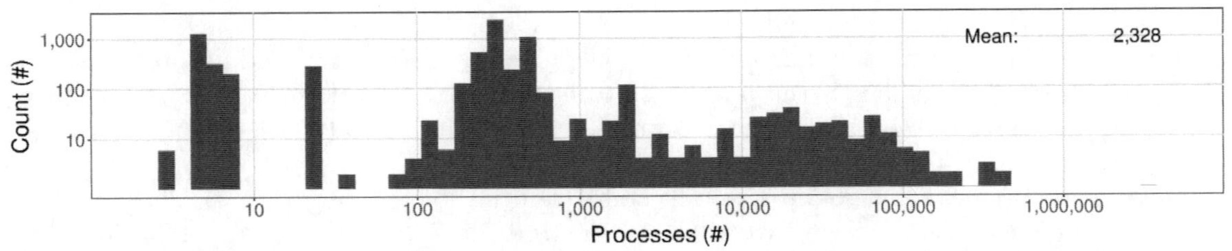

Figure 1. Number of processes to consider when analysing CMS payloads. The mean number of processes is around 2,300 whereas the maximum number recorded is up to 3,600,000 processes. This requires a good scalability. In addition, there is evidence for clusters of similar payloads based on their process count.

2. Problem description

The following discussions and assumptions are mainly based on monitoring data from the CMS collaboration [4]. In addition, only pilots are considered. Statistics are based on data that has been collected offline on one of the 600 worker nodes of the GridKa data and computing centre. The selected worker node offers 24 job slots. The data has been collected over a period of 12 months from July 2014 to June 2015.

2.1. Anomaly detection

The basis to detect anomalous batch jobs is knowledge about normal behaviour. As normal behaviour is unknown, it first needs to be learned. To define normal behaviour we consider majority of batch jobs as normal. Only a small fraction are outliers. If this holds true, normal behaviour can be defined based on natural breaks regarding the distribution of process count per payload. Given the dataset of payloads for CMS collaboration of one worker node, we therefore distinguish five groups of payloads describing normal behaviour (compare Figure 1). For each of those groups, the average payload is calculated. This average is called a *prototype* in the following.

A prototype is the representative for the whole group and can be used to measure the distance to a running batch job. Based on that distance, one can decide if the batch job belongs to that group, and therefore behaves normally. Otherwise it can be treated as an outlier.

2.2. Distance measurement for trees

As the batch job monitoring works on process level, we deal with tree data structures. To measure distances between trees there exist many solutions in literature, e.g. the tree edit distance [5]. The distance is given by the number of operations required to transform one tree T_1 with n nodes into another tree T_2 with m nodes. Operations permitted are the deletion, addition, and relabelling of nodes. State of the art algorithms of tree edit distance require $O(n^3)$ time and $O(nm)$ space resources [6].

However, these algorithms are designed to measure exact distances for static trees. Our use case requires an efficient solution that works on an event stream $E_{\text{job}} = \langle e_1, \ldots, e_n \rangle$ with bounded memory and CPU resources. The events given by E_{job} describe a tree $T_{\text{job}} = \langle T_1, \ldots, T_n \rangle$ of a batch job that is dynamic over time. Distances repeatedly need to be calculated to a given set of prototype trees $P = \{T_{\text{prototype},1}, \ldots, T_{\text{prototype},i}\}$. Complexity for exact distance measurements for i prototypes extends to $O(in^3)$ time and $O(inm)$ space requirements. Therefore an incremental distance measurement is preferred.

2.3. Exemplary demonstration

Let us consider to perform anomaly detection for an average payload with 2,300 processes for 5 groups of normal behaviour based on tree edit distance. We first need to measure distances to each of the 5 prototypes. We require resources for $5 \cdot 2,300 \cdot 2,300 = 26,450,000$ objects in average for one event at a time.

As there are up to 24 job slots per worker node, we need to consider up to $24 \cdot 26,450,000 = 634,800,000$ objects. In addition, we still need to store the data per process per payload and prototype. On average, each process requires 38 B of memory. This requires a total memory of 2.00 MB for payload as well as 0.42 MB for prototype storage and 2.36 GB for distance calculations.

Taking into account, that memory should not exceed 4.00 GB one out of the 24 payloads with 40,400 processes already exceeds the given limits. This is the case for 1.10 % of the payloads on the given worker node. Hence, it is practically impossible that limits are not exceeded when running on 600 worker nodes with tree edit distance.

2.4. Requirements

In summary, different aspects need to be considered to put a scalable architecture for online anomaly detection into practice. First, algorithmic complexity to determine distances between trees is of interest. So far, no sufficient scalable algorithm has been identified. Currently we are testing an incremental distance measurement approach that looks promising. It is based on lossy compression of trees and the usage of hash tables.

Second, the learning of prototypes needs to be considered. This can be done offline on a central node before distributing the prototypes to the different worker nodes. But it might also be considered to have different prototypes per worker node.

Most importantly the uninterrupted operation of the batch system is key. Both, a low communication overhead as well as low memory consumption and CPU load on worker nodes are crucial.

3. A scalable architecture for online anomaly detection

Currently the monitoring tool BPNetMon is deployed on two racks, each consisting of 32 worker nodes. Each worker node provides 24 job slots. The tool is operated as a sensor that transfers monitoring events to a central instance for further offline analysis and prototype learning. An online anomaly detection has not been deployed yet. Following, we will discuss feasible options for deployment and summarise advantages and disadvantages of the presented architectures.

3.1. Centralised systems

The initial situation enables a centralised system with one central server. The main task of the central server is the anomaly detection. Necessary calculations do therefore not interfere with batch jobs that are running on the worker nodes. We also benefit from data consistency, coherence, and a centralised prototype model. However, we expect very high CPU load as well as memory demands on that central server. The current setup already requires the parallel analysis of $64 \cdot 24 = 1,536$ payloads. Based on a tree edit distance, the distance calculations themselves require around 151 GB of memory. A scaling to the whole GridKa data and computing centre can therefore not be achieved.

Based on measurements for several months we simulated the network overhead for such a centralised system. We want to ensure that the tool generated traffic does not exceed 3.00 % of traffic that is induced by batch jobs and system relevant processes. The simulation shows that the tool generated traffic exceeds the threshold (compare Figure 2). Actually the simulation showed, that the threshold is exceeded in 10.44 % of all monitoring intervals. To ensure scalability, a preselection of data should be performed on the different worker nodes.

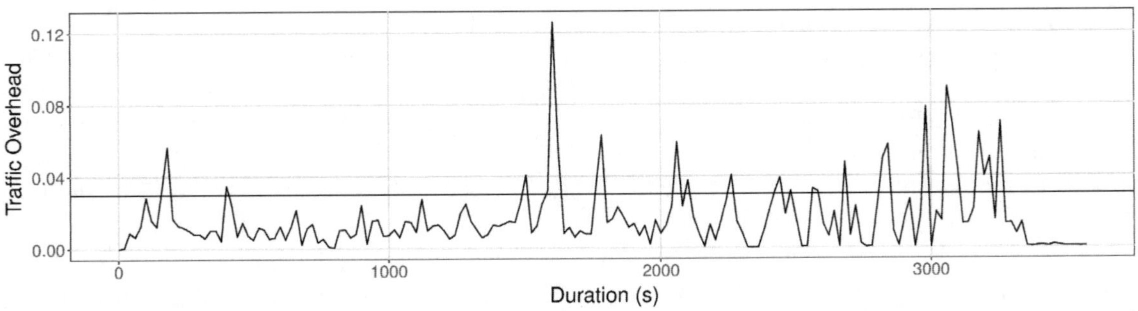

Figure 2. Simulated traffic overhead for a centralised system. The tool generated traffic should not exceed 3.00 % of batch job induced traffic. As shown, the generated traffic exceeds the threshold. In addition, the simulation showed that this is the case in 10.44 % of all monitored time intervals. To minimise this overhead, a preselection of data is required.

The setup consisted of 10 worker nodes and one central instance. The different hosts were connected by 1 Gbit/s full-duplex ethernet connections. Randomly a replay from log files was started to induce network activity for the different worker nodes. We first considered base traffic for batch jobs and system relevant processes. In addition we simulated the events produced by BPNetMon and transfered those to the central host.

3.2. Hierarchical systems
To tackle the disadvantages of centralised systems we can introduce additional layers of local servers. Each local server groups several worker nodes. Worker nodes still send monitoring information to their associated server. The local servers collect information from a group of worker nodes and perform a preselection of received data. Selected data are then send to the central server. The central server performs anomaly detection on a reduced set of data. Hence, a logical hierarchy is implemented that allows further scalability. As information flow is retained, data consistency, coherence, and a centralised prototype model are ensured.

A hierarchical system offers improved scalability over centralised system but still requires additional hardware for servers. However, chances for a failure of the whole anomaly detection process are reduced but the detection may still fail for groups of worker nodes.

3.3. Peer-to-peer systems
A more flexible architecture regarding the failure of the whole system or groups of worker nodes is a peer-to-peer system. In peer-to-peer systems any node can exchange data with any other node. Here, every node performs a local anomaly detection and reports its results.

While peer-to-peer systems enable horizontal scalability, we need to deal with high CPU load and memory demands on the worker nodes. In addition, the learning of prototypes needs to be handled as there is no central instance. Either a central server needs to take over calculation of prototypes and therefore needs to gather monitoring information from worker nodes (see 3.4), or different prototype models are used on each worker node.

When using differing prototypes, anomalies depend on their local context. The local context is given by the results from neighbouring worker nodes e.g. worker nodes of the same rack. One batch job does not necessarily have to be an anomaly inside the system, when it is different from currently known prototypes on one single worker node. It might be consistent with prototype models on differing worker nodes. Therefore, before results are reported, the local context can be considered for local anomaly detection.

3.4. Hybrid peer-to-peer systems
A hybrid peer-to-peer system additionally introduces a central server. As already mentioned for peer-to-peer systems, a central server can take over responsibility to calculate and distribute the prototype model. Besides this, a hybrid peer-to-peer system can be used to centralise anomaly detection on the central server. Worker nodes are then responsible for performing a preselection of data to reduce communication overhead. Again, the central server becomes a single point of failure and bottleneck of the system.

3.5. Super-peer systems
A final architecture to consider is a super-peer system. Super-peers act as a centralised server to a subset of worker nodes. Different super-peers are connected to each other as peers. In addition, the subsets of worker nodes are also connected to each other as peers. Worker nodes perform preselection of data and report to their super-peer. The super-peers perform the actual anomaly detection for their associated set of worker nodes.

Super-peers are a temporarily used resource and therefore an interesting option for opportunistic resources. The already established monitoring of batch jobs allows the recognition of unsaturated worker nodes. Those include worker nodes that wait until all batch jobs are finished to e.g. perform software updates. The resources of such a worker node can be temporarily utilised as a super-peer. As soon as that state is dropped, data is lost and the associated worker nodes need to report their data to another super-peer.

4. Conclusion
To perform an online anomaly detection on payload level, a scalable architecture is required. Particularly given that batch jobs need to run uninterrupted even though distance measurements and anomaly detection require high computational cost and memory demands.

The most promising architecture combines elements from a hierarchical system as well as a super-peer system. A temporary utilisation of unsaturated worker nodes with additional central server components as fallback instances when resources need to be dropped brings a high flexibility. Hence, data consistency and coherence can be guaranteed. In addition, a modular software design that allows the distribution of different algorithmic steps such as preselection and anomaly detection, enables a good utilisation of that flexibility. On this way, variations such as anomaly detection based on local contexts can be explored.

Acknowledgments
The authors would like to thank all people and institutions involved in the project Large Scale Data Management and Analysis, the German Helmholtz Association, and the Karlsruhe School of Elementary Particle and Astroparticle Physics for supporting and funding the work.

References
[1] Kuehn E, Fischer M, Jung C, Petzold A and Streit A 2014 Monitoring data streams at process level in scientific big data batch clusters *BDC '14: Proc. of the 2014 IEEE/ACM Int. Symposium on Big Data Computing* (IEEE Computer Society)
[2] Kuehn E, Fischer M, Giffels M, Jung C and Petzold A 2015 *J. Phys.: Conf. Ser.* **664** 052019
[3] Thain D, Tannenbaum T and Livny M 2005 *Concurr. Comput.: Pract. Exper.* **17** 323–356
[4] Chatrchyan S *et al.* (CMS) 2012 *Phys. Lett.* **B716** 30–61 (*Preprint* 1207.7235)
[5] Bille P 2005 *Theor. Comput. Sci.* **337** 217 – 239
[6] Pawlik M and Augsten N 2011 *Proc. VLDB Endow.* **5** 334–345

The Run-2 ATLAS Trigger System

A Ruiz Martínez on behalf of the ATLAS Collaboration
Department of Physics, Carleton University, Ottawa, ON, Canada
E-mail: aranzazu.ruiz.martinez@cern.ch

Abstract. The ATLAS trigger successfully collected collision data during the first run of the LHC between 2009-2013 at different centre-of-mass energies between 900 GeV and 8 TeV. The trigger system consists of a hardware Level-1 and a software-based high level trigger (HLT) that reduces the event rate from the design bunch-crossing rate of 40 MHz to an average recording rate of a few hundred Hz. In Run-2, the LHC will operate at centre-of-mass energies of 13 and 14 TeV and higher luminosity, resulting in up to five times higher rates of processes of interest. A brief review of the ATLAS trigger system upgrades that were implemented between Run-1 and Run-2, allowing to cope with the increased trigger rates while maintaining or even improving the efficiency to select physics processes of interest, will be given. This includes changes to the Level-1 calorimeter and muon trigger systems, the introduction of a new Level-1 topological trigger module and the merging of the previously two-level HLT system into a single event processing farm. A few examples will be shown, such as the impressive performance improvements in the HLT trigger algorithms used to identify leptons, hadrons and global event quantities like missing transverse energy. Finally, the status of the commissioning of the trigger system and its performance during the 2015 run will be presented.

1. Introduction

The trigger system is an essential component of the ATLAS [1] experiment at the LHC [2], since it is responsible for deciding whether or not to keep a given collision event for later study. In the LHC Run-1 (2009-2013), the ATLAS trigger system [3] operated efficiently at instantaneous luminosities of up to $8 \times 10^{33}\,\mathrm{cm}^{-2}\mathrm{s}^{-1}$ at different centre-of-mass energies between 900 GeV and 8 TeV and collected more than three billion events. In the LHC Run-2, which started in 2015, the increased collision energy to 13 TeV, higher luminosity and higher pile-up lead to an increase of the rates as compared to the Run-1 trigger selections by up to a factor five, exceeding the capabilities of the Run-1 trigger system. The LHC long shutdown (2013-2014) was therefore used to perform major upgrades in the different components of the trigger system.

2. The ATLAS Trigger System Upgrades for Run-2

The trigger system in Run-2 consists of a hardware-based first level trigger (Level-1) [4] and a software-based high level trigger (HLT) [5]. The Level-1 trigger uses custom electronics to determine Regions-of-Interest (RoIs) in the detector, taking as input coarse granularity calorimeter and muon detector information. The Level-1 trigger reduces the event rate from the LHC bunch crossing rate of approximately 30 MHz to 100 kHz. The decision time for a Level-1 accept is $2.5\,\mu$s. The RoIs formed at Level-1 are sent to the HLT in which sophisticated selection algorithms are run using full granularity detector information in either the RoI or the whole event. The HLT reduces the rate from the Level-1 output rate of 100 kHz to approximately

1 kHz on average within a processing time of about 200 ms. A schematic overview of the upgraded ATLAS trigger and data acquisition system is shown in Fig. 1.

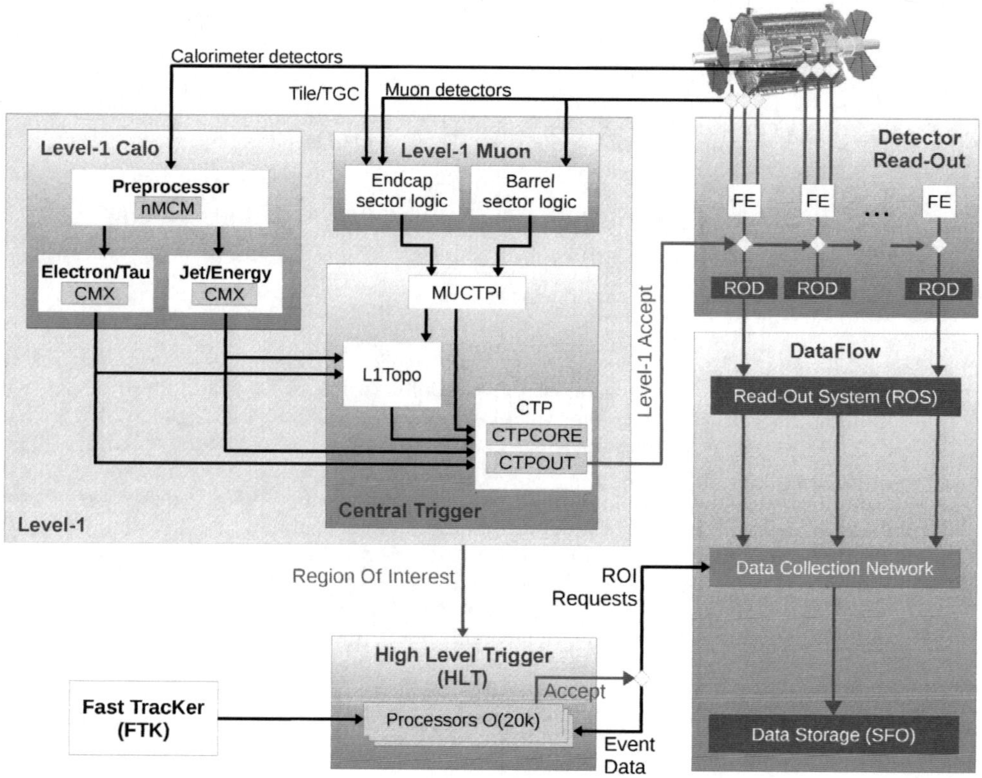

Figure 1. Schematic layout of the ATLAS trigger and data acquisition system in Run-2.

2.1. *Level-1 Trigger Upgrades*

Several upgrades have been introduced in the different components of the ATLAS Level-1 trigger system for Run-2 data taking. The upgrades, both in the Level-1 trigger hardware and in the detector readout, allowed to rise the maximum Level-1 trigger rate from 70 kHz in Run-1 to 100 kHz in Run-2.

The Level-1 Calorimeter trigger makes use of reduced granularity information from the electromagnetic and hadronic calorimeters to search for electrons, photons, taus and jets, as well as high total and missing transverse energy ($E_{\mathrm{T}}^{\mathrm{miss}}$). One of the main upgrades in the Level-1 Calorimeter trigger is the new Multi-Chip Modules (nMCM), based on field-programmable gate array (FPGA) technology, which replace the application-specific integrated circuits (ASICs) included in the modules used in Run-1. This new hardware allows the use of auto-correlation filters and a new bunch-by-bunch dynamic pedestal correction, meant to suppress pile-up effects. The effect of these corrections in linearising the $E_{\mathrm{T}}^{\mathrm{miss}}$ trigger rates as function of the instantaneous luminosity is illustrated in Fig. 2.

The Level-1 Muon trigger system, which consists of a barrel section and two endcap sections, provides fast trigger signals from the muon detectors for the Level-1 trigger decision. For Run-2, various improvements were added to the Level-1 Muon trigger. To suppress most of the fake muon triggers, the muon endcap trigger ($1.05 < |\eta| < 2.4$)[1] now requires a coincidence with hits from the innermost muon chambers (known as Forward Inner, FI), which as shown in Fig. 3 reduces the L1_MU15 rate by approximately 50% in the $1.3 < |\eta| < 1.9$ region with a very small signal efficiency loss of only ~2%. The L1_MU15 trigger requires that a candidate passed the 15 GeV threshold requirement of the Level-1 muon trigger system. In the future, a similar coincidence logic will also be applied using the outermost layer of the hadronic Tile Calorimeter for $1.05 < |\eta| < 1.3$. Moreover, the trigger coverage is expected to be improved by 4% in the barrel region due to the installation of new chambers in the feet region of the muon detector, which are currently being commissioned.

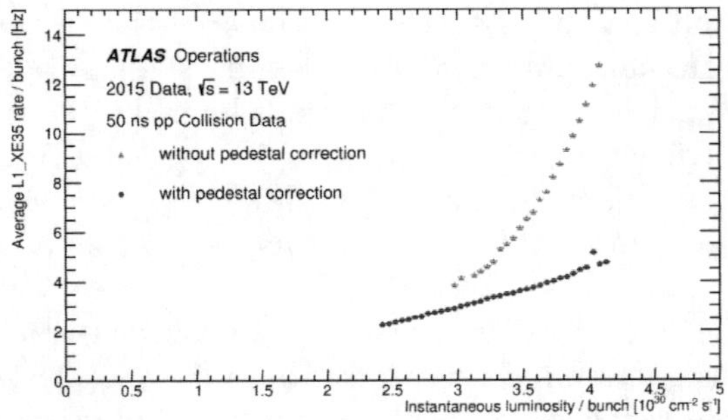

Figure 2. Trigger rate per bunch for an $E_{\mathrm{T}}^{\mathrm{miss}}$ trigger with a threshold at 35 GeV as function of the instantaneous bunch luminosity with and without Level-1 Calorimeter pedestal correction applied [6].

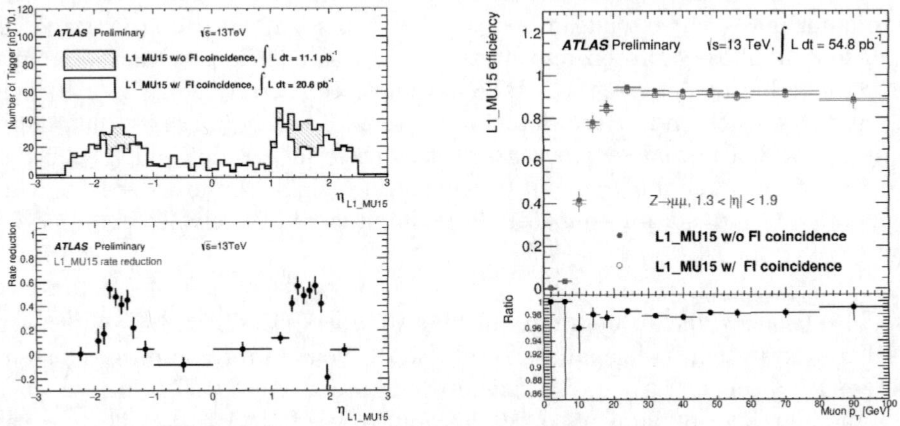

Figure 3. Distribution of the RoI η from the L1_MU15 trigger with and without the FI coincidence enabled (top left), and reduction of the trigger rate due to this coincidence (bottom left) [7]. On the right, the efficiency of the L1_MU15 trigger in the endcap region with and without the FI coincidence enabled [7].

[1] ATLAS uses a right-handed coordinate system with its origin at the nominal interaction point (IP) in the centre of the detector and the z-axis along the beam pipe. The x-axis points from the IP to the centre of the LHC ring, and the y-axis points upward. Cylindrical coordinates (r, ϕ) are used in the transverse plane, ϕ being the azimuthal angle around the z-axis. The pseudorapidity is defined in terms of the polar angle θ as $\eta = -\ln\tan(\theta/2)$.

A new topological trigger processor (L1Topo) has been deployed for Run-2 and is currently under commissioning. It takes as input the Level-1 trigger objects from the output merger modules (CMX) of the Level-1 Calorimeter trigger and the Level-1 Muon trigger (MuCTPi), which were both also upgraded. The new L1Topo system allows to apply topological selections at the Level-1 trigger combining kinematic information from multiple calorimeter and muon trigger objects, such as angular separation, invariant mass requirements, or global event quantities such as the sum of the transverse momenta of all Level-1 jet objects. The use of L1Topo significantly suppresses backgrounds for many trigger selections, in many cases by more than a factor of two, allowing to preserve the sensitivity for channels such as $B_s \to \mu\mu$ and $H \to \tau\tau$ at the current level with increasing luminosity.

The central trigger processor (CTP) forms the Level-1 trigger decision based on the information received from the L1Topo, Level-1 Calorimeter trigger and Level-1 Muon trigger, and distributes the Level-1 accept signal and LHC timing signals to the sub-detector readout systems via the Timing, Trigger and Control network. The main part of the CTP was replaced to increase the number of trigger inputs and trigger items, achieving more flexibility in the selection of events relevant for physics analyses. These improvements required the upgrade of the hardware or firmware of most CTP components, as well as a complete re-design of its software architecture.

The Fast TracKer (FTK) [8] project is a fast hardware-based tracking system designed to perform a global track reconstruction receiving input from the ATLAS silicon tracking detectors after each Level-1 trigger and provides full-event track information to the HLT. This can be used to improve many trigger selections requiring full-event tracking information, like b-jets, which would not be otherwise affordable at HLT. In addition the precise estimation of the pile-up conditions per event can be used to develop pile-up robust trigger strategies. With the Run-2 data taking ongoing, the first boards of FTK are being installed, already reading ATLAS data in parasitic mode. The full deployment and operation of the barrel region is expected for mid 2016, after which the integration within HLT and the extension to full coverage will follow.

The minimum bias trigger scintillators (MBTS) provided the primary trigger for selecting events from low luminosity proton-proton and lead-lead collisions with the smallest possible bias. Due to the radiation damage the MBTS have been replaced during the LHC long shutdown and are mounted at each end of the detector in front of the liquid-argon end-cap calorimeter cryostats at $z = \pm 3.56$ m and segmented into two rings in pseudorapidity ($2.07 < |\eta| < 2.76$ and $2.76 < |\eta| < 3.86$). The inner ring is segmented into eight azimuthal sectors while the outer ring is segmented into four azimuthal sectors, giving a total of twelve sectors per side.

2.2. High Level Trigger Upgrades

There have also been significant upgrades in the HLT during the LHC long shutdown. In Run-1 the ATLAS trigger system had separate Level-2 and Event Filter computer clusters. For Run-2, the system has been merged into a single event processing HLT farm, which reduces the complexity and allows for dynamic resource sharing between algorithms. This new arrangement reduces code and algorithm duplication and results in a more flexible HLT, reducing duplicated data-fetching.

Most of the trigger reconstruction algorithms were re-optimized during the shutdown to minimize differences between the HLT and the offline analysis selections, which in some cases, such as in the hadronic tau triggers, reduced inefficiencies by more than a factor two. The HLT tracking algorithms have also been prepared for the future inclusion of the FTK. The HLT processing performed within RoIs has been augmented for some triggers to also allow aggregation of the RoIs into a single object. This reduces the amount of CPU processing required for events with a large multiplicity of partially overlapping RoIs. The average output rate of the HLT has been increased from 400 Hz to 1 kHz, as imposed by data storage constraints.

Journal of Physics: Conference Series **762** (2016) 012003 doi:10.1088/1742-6596/762/1/012003

3. ATLAS Trigger System Commissioning and Performance in Run-2

The ATLAS trigger system has been successfully commissioned in the start-up phase of Run-2 with cosmic ray data and early $\sqrt{s} = 13\,\text{TeV}$ collision data. Most of the new trigger components have been used during the 2015 data taking period, allowing ATLAS to efficiently select events in Run-2. In this section, the efficiencies of the main physics triggers in the initial phase of Run-2 and comparisons to Monte Carlo (MC) simulations are discussed.

Photons are triggered using a cut-based identification criteria with different selections available (loose, medium and tight) [9]. The efficiency of several single photon triggers is shown in Fig. 4 with respect to photon candidates reconstructed using the full ATLAS reconstruction software passing a tight identification selection. The efficiency is measured as a function of the offline photon transverse energy for $|\eta| < 2.37$ excluding the transition region between the barrel and endcap electromagnetic calorimeters at $1.37 < |\eta| < 1.52$, using events recorded with a Level-1 trigger requiring an electromagnetic energy cluster with $E_T > 7\,\text{GeV}$.

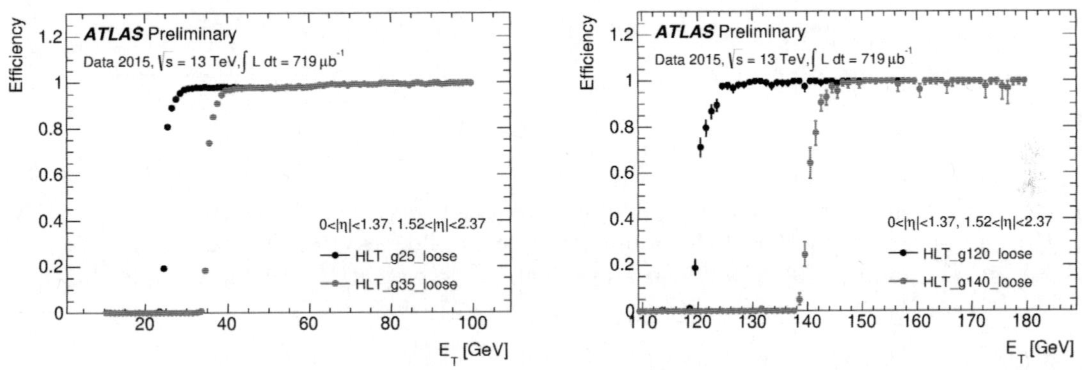

Figure 4. Efficiency for the single photon triggers requiring $E_T > 25\,\text{GeV}$ and $35\,\text{GeV}$ (left), and requiring $E_T > 120\,\text{GeV}$ and $140\,\text{GeV}$ (right) as a function of the E_T of the offline photon candidates [10]. Only statistical uncertainties are shown.

Electrons are selected at the HLT using a likelihood-based identification criteria which takes as input electromagnetic shower shape and tracking information with different selections available (lhloose, lhmedium and lhtight) [11]. Figure 5 shows the efficiency of the combined Level-1 and HLT single electron e12_lhloose_L1EM10VH and e24_lhmedium_L1EM20VH triggers as a function of the E_T of the offline electron candidates. These triggers require an electron candidate with $E_T > 12\,\text{GeV}$ satisfying the lhloose identification, and $E_T > 24\,\text{GeV}$ satisfying the lhmedium identification, respectively. They are seeded by the Level-1 triggers L1_EM10VH and L1_EM20VH, respectively, that apply an E_T dependent veto against energy deposited in the hadronic calorimeter behind the electron candidates electromagnetic cluster. The efficiency is measured with a tag-and-probe method using $Z \to ee$ decays and compared to the expectation from $Z \to ee$ MC simulation.

Muons are reconstructed at the HLT by combining the inner detector and muon spectrometer tracks. The absolute efficiency of the L1_MU15 trigger and absolute and relative efficiencies of the OR combination of mu20_iloose and mu50 HLT triggers are shown in Fig. 6 as a function of the p_T of the offline muon candidates in the barrel and endcap detector regions. The efficiency is computed with respect to the offline muon candidates requiring to pass a medium quality requirement defined in Ref. [12]. The L1_MU15 trigger requires that a candidate passed the 15 GeV threshold requirement of the Level-1 muon trigger system. The mu20_iloose trigger is seeded by the L1_MU15 trigger and is required to satisfy a 20 GeV HLT threshold and to pass a loose isolation selection computed using inner detector tracks reconstructed online by the HLT. The mu50 trigger is seeded by the L1_MU20 trigger and is required to satisfy a 50 GeV HLT threshold. The efficiency is measured with a tag-and-probe method using $Z \to \mu\mu$ candidates. The Level-1 trigger efficiency of $\sim$70% observed in the barrel region is mainly due to the detector geometrical acceptance.

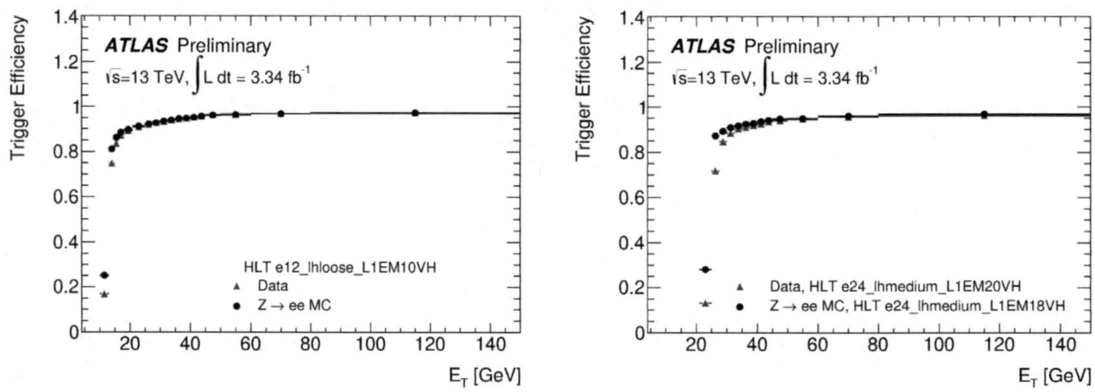

Figure 5. Efficiency for single electron triggers requiring $E_{\mathrm{T}} > 12\,\mathrm{GeV}$ (left) and $E_{\mathrm{T}} > 24\,\mathrm{GeV}$ (right) as a function of the E_{T} of the offline electron candidates. Results are shown for data and $Z \to ee$ MC simulation [10]. Combined statistical and systematic uncertainties are shown.

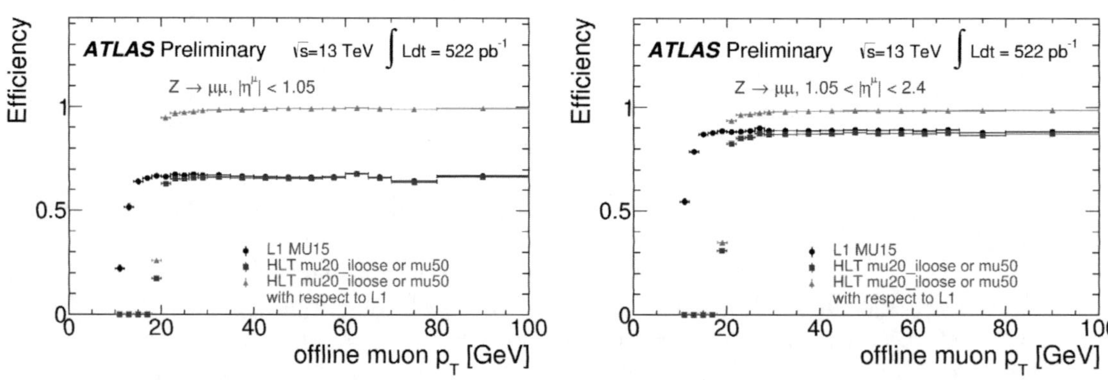

Figure 6. Efficiency for single muon triggers as function of the p_{T} of the offline muon candidates in the barrel (left) and endcap (right) regions [13]. Only statistical uncertainties are shown.

Tau candidates are identified at the HLT using a boosted decision tree (BDT) built from calorimeter and track quantities using different working points (loose, medium and tight) [14]. Figure 7 shows the BDT score and the tau trigger efficiency measured in a data sample enriched in $Z \to \tau\tau \to \mu\tau_{\mathrm{had}}$ events and compared to MC simulation. The efficiency is computed with respect to offline reconstructed tau candidates with $p_{\mathrm{T}} > 25\,\mathrm{GeV}$, one or three tracks and passing the offline medium identification requirement. The corresponding online tau candidate is required to have $p_{\mathrm{T}} > 25\,\mathrm{GeV}$, between one and three tracks and pass the online medium identification requirement.

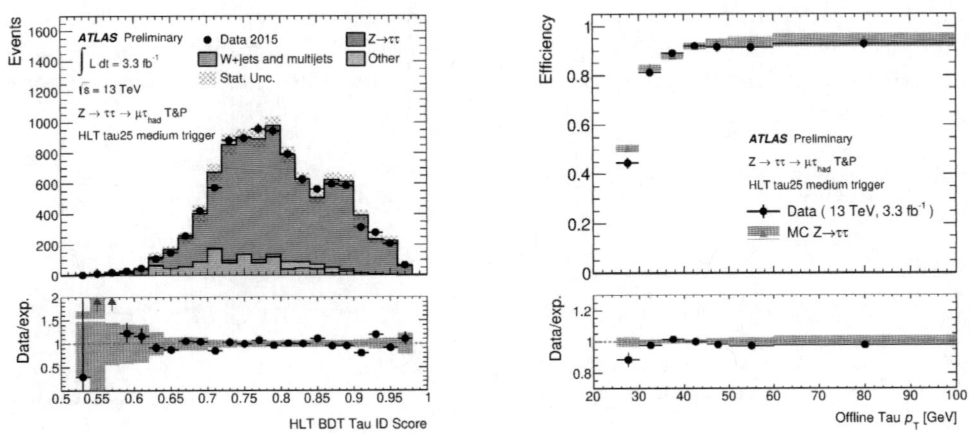

Figure 7. BDT tau identification score for online tau candidates passing the HLT tau trigger with $p_{\mathrm{T}} > 25\,\mathrm{GeV}$ and online medium identification requirement (left) and the corresponding trigger efficiency as function of the p_{T} of the offline tau candidates (right) [15]. Only statistical uncertainties are shown.

HLT jets are formed from topological energy clusters [16] at the electromagnetic energy scale. The HLT jets are then calibrated to the hadronic scale by first applying a jet-by-jet area subtraction procedure followed by a jet energy scale weighting that is dependent on the HLT jet p_{T} and η [17]. The per-event trigger efficiency turn-on curves compared between data and MC simulation for three typical thresholds are shown in Fig. 8. Each efficiency is determined using events retained with a lower threshold trigger that is found to be fully efficient in the phase space of interest.

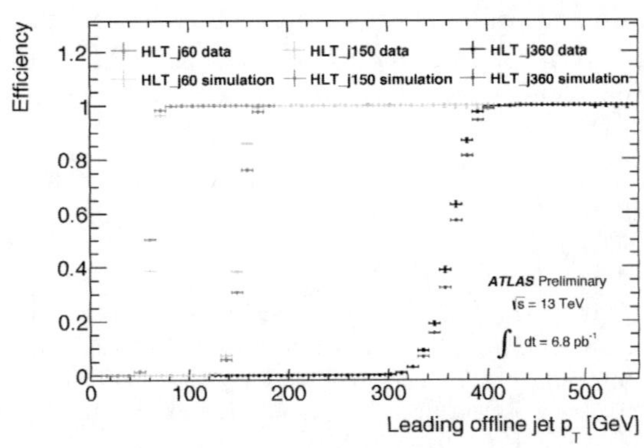

Figure 8. Efficiency for different single jet triggers as function of the offline leading jet p_{T} measured in data and MC simulation [18].

In the $E_{\mathrm{T}}^{\mathrm{miss}}$ trigger, one of the main challenges is the pile-up mitigation. Therefore, different approaches have been studied: cell-based (default), jet-based (mht), and topocluster-based algorithms with (tc_PS) and without (tc) a pile-up subtraction scheme. Figure 9 shows the $E_{\mathrm{T}}^{\mathrm{miss}}$ trigger efficiency turn-on curves with respect to the $E_{\mathrm{T}}^{\mathrm{miss}}$ reconstructed offline without muon corrections. The dataset has been selected using the lowest unprescaled single muon trigger and events are also required to satisfy a $W \to \mu\nu$ selection. The different turn-on curves have been obtained for the lowest unprescaled trigger of various HLT $E_{\mathrm{T}}^{\mathrm{miss}}$ algorithms for a constant HLT threshold of 70 GeV activated during the 25 ns runs and for different HLT thresholds leading to the same trigger rate.

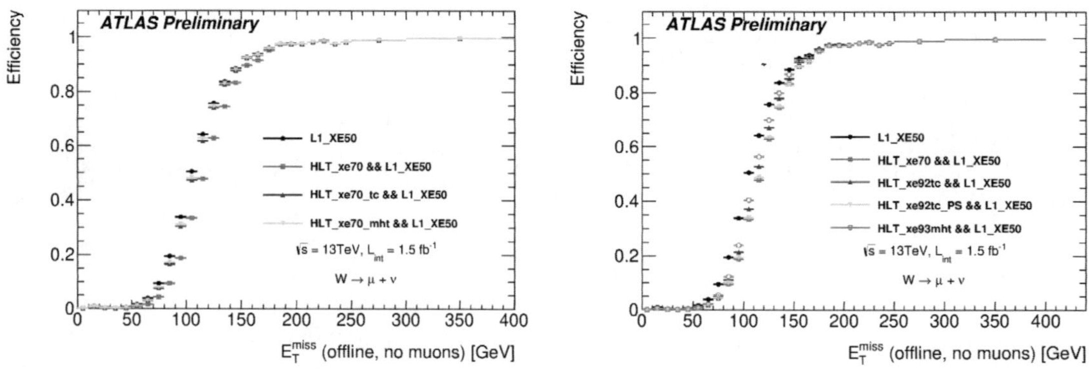

Figure 9. Efficiency for $E_{\mathrm{T}}^{\mathrm{miss}}$ triggers with respect to the $E_{\mathrm{T}}^{\mathrm{miss}}$ reconstructed offline for the different HLT algorithms with a constant HLT threshold (left) and with an equal trigger rate (right) [19]. Only statistical uncertainties are shown.

4. Conclusions

Many improvements have been deployed in the ATLAS trigger system during the LHC shutdown to keep trigger thresholds as low as possible and selections close to the full reconstruction procedure. Some of these improvements include upgrades in the Level-1 Calorimeter and Muon trigger systems, a new Level-1 topological trigger and an upgraded Level-1 central trigger. The HLT architecture has been unified and thus became more performant and flexible.

The trigger system has been successfully commissioned with 2015 data. The first performance studies of the different trigger signatures have been presented using Run-2 $\sqrt{s} = 13$ TeV data.

References

[1] ATLAS Collaboration 2008 *JINST* **3** S08003.
[2] L. Evans and P. Bryant (editors) 2008 *JINST* **3** S08001.
[3] ATLAS Collaboration 2012 *Eur. Phys. J. C* **72** 1849.
[4] ATLAS Collaboration 1998 CERN-LHCC-98-014 http://cdsweb.cern.ch/record/381429.
[5] ATLAS Collaboration 2003 CERN-LHCC-2003-022 http://cdsweb.cern.ch/record/616089.
[6] https://twiki.cern.ch/twiki/bin/view/AtlasPublic/L1CaloTriggerPublicResults.
[7] https://twiki.cern.ch/twiki/bin/view/AtlasPublic/L1MuonTriggerPublicResults.
[8] ATLAS Collaboration 2013 CERN-LHCC-2013-007 http://cdsweb.cern.ch/record/1552953.
[9] ATLAS Collaboration 2012 ATLAS-CONF-2012-123 http://cdsweb.cern.ch/record/1473426.
[10] https://twiki.cern.ch/twiki/bin/view/AtlasPublic/EgammaTriggerPublicResults.
[11] ATLAS Collaboration 2015 ATL-PHYS-PUB-2015-041 http://cdsweb.cern.ch/record/2048202.
[12] ATLAS Collaboration 2015 ATL-PHYS-PUB-2015-037 http://cdsweb.cern.ch/record/2047831.
[13] https://twiki.cern.ch/twiki/bin/view/AtlasPublic/MuonTriggerPublicResults.
[14] ATLAS Collaboration 2015 ATL-PHYS-PUB-2015-045 http://cdsweb.cern.ch/record/2064383.
[15] https://twiki.cern.ch/twiki/bin/view/AtlasPublic/TauTriggerPublicResults.
[16] W. Lampl *et al.* 2008 ATL-LARG-PUB-2008-002 http://cdsweb.cern.ch/record/1099735.
[17] ATLAS Collaboration 2015 ATL-PHYS-PUB-2015-015 http://cdsweb.cern.ch/record/2037613.
[18] https://twiki.cern.ch/twiki/bin/view/AtlasPublic/JetTriggerPublicResults.
[19] https://twiki.cern.ch/twiki/bin/view/AtlasPublic/MissingEtTriggerPublicResults.

ATLAS EventIndex monitoring system using the Kibana analytics and visualization platform

D. Barberis[1], S.E. Cárdenas Zárate[2*], A. Favareto[3], A. Fernandez Casani[4], E.J. Gallas[5], C. Garcia Montoro[4], S. Gonzalez de la Hoz[4], J. Hrivnac[6], D. Malon[7], F. Prokoshin[2], J. Salt[4], J. Sanchez[4], R. Toebbicke[8], R. Yuan[6]**
on behalf of the ATLAS Collaboration

[1] Università di Genova and INFN, Genova, Italy
[2] Universidad Técnica Federico Santa María, Valparaíso, Chile
[3] Università degli Studi and INFN, Genova, Italy
[4] Instituto de Física Corpuscular, Valencia, Spain
[5] Department of Physics, Oxford University, Oxford, United Kingdom
[6] Laboratoire de l'Accélérateur Linéaire, Orsay, France
[7] Argonne National Laboratory, Argonne, USA
[8] CERN, Europe

*Presenter, primary author
**Corresponding author: Fedor.Prokoshin@cern.ch

Abstract. The ATLAS EventIndex is a data catalogue system that stores event-related metadata for all (real and simulated) ATLAS events, on all processing stages. As it consists of different components that depend on other applications (such as distributed storage, and different sources of information) we need to monitor the conditions of many heterogeneous subsystems, to make sure everything is working correctly. This paper describes how we gather information about the EventIndex components and related subsystems: the Producer-Consumer architecture for data collection, health parameters from the servers that run EventIndex components, EventIndex web interface status, and the Hadoop infrastructure that stores EventIndex data. This information is collected, processed, and then displayed using CERN service monitoring software based on the Kibana analytic and visualization package, provided by CERN IT Department. EventIndex monitoring is used both by the EventIndex team and ATLAS Distributed Computing shifts crew.

1. Introduction

The EventIndex [1] is a catalogue of the event-related information for the ATLAS experiment [2]. It is organized in a number of relatively independent modules or subsystems, collecting the event data from the datasets on the CERN central server (Tier-0) and on the Grid, sending this information to CERN, processing and storing it to the Hadoop [3] file system (HDFS) and, finally, exposing an interface for the queries by the users and production system. A smooth operation and interaction of these components is necessary for the effective work of the EventIndex and the success of the analyses depending of its data. We developed a system for automatically monitoring states, critical parameters and performance numbers of the EventIndex components. It provides information to the EventIndex operation team through the Kibana [4] visualization package. In addition, global statuses of the critical components are displayed on the ATLAS central system monitoring page for ATLAS computing shifters. The system is in operation since late 2014.

2. Monitoring of the EventIndex components

The EventIndex consists of number of components of different type distributed between computers at CERN and on the Grid. Each of these components requires a different approach for data collecting and processing. Monitoring information is collected by jobs running on one of the EventIndex computers automatically at specified time managed by *acron* scheduler. The gathered information is organised into *xml* files and then pushed to the ATLAS computing services monitoring server running Kibana.

Kibana is a data visualisation software on top of the ElasticSearch engine. It provides a number of tools to display information, like dashboards, queries, filters and time ranges, so information stored in an ElasticSearch cluster can be visualized as interactive plots such as pie charts, histograms, text, maps and other.

The general scheme of the EventIndex monitoring is presented in Figure 1. Below we shortly describe its components.

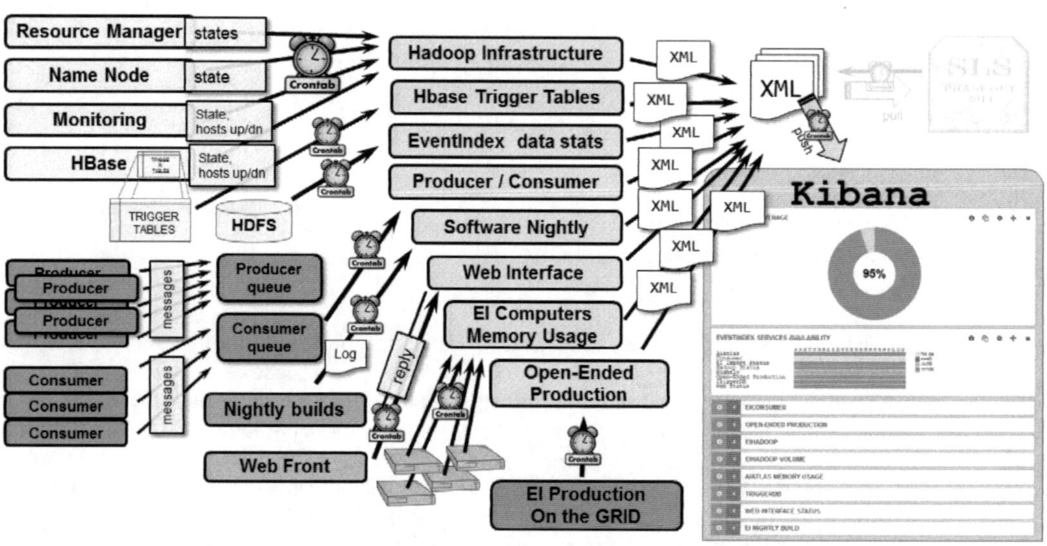

Figure 1: EventIndex monitoring components and data flow

The EventIndex components (consumers, importing and validation, grid production management and monitoring itself, web services) run on a number of servers at both CERN and the Wigner research center. We use some of these machines also for development. Currently we monitor accessibility and memory usage on these servers (see Figure 2) by running system information tools. Low available memory and aggressive swap usage may lead to performance degradation of the services they host, so If swap usage on some of these machines exceeds the limit, an alert for the EventIndex operation team is raised.

The EventIndex gathers information from the “producer” tasks running on the ATLAS central production servers (Tier-0) and on the Grid. The machinery that imports event data on the Grid is called “Open-ended production”. It receives from the Atlas Metadata Interface [5] (AMI) information on datasets that were produced recently and supplies this information to ProdSys2 [6], the second-generation ATLAS distributed production system, which actually runs the tasks extracting event information. In the corresponding part of the monitoring we display information on the number of datasets of different types found and fed to the processing by the EventIndex jobs, and on any errors found in this process. The information is gathered from the log files.

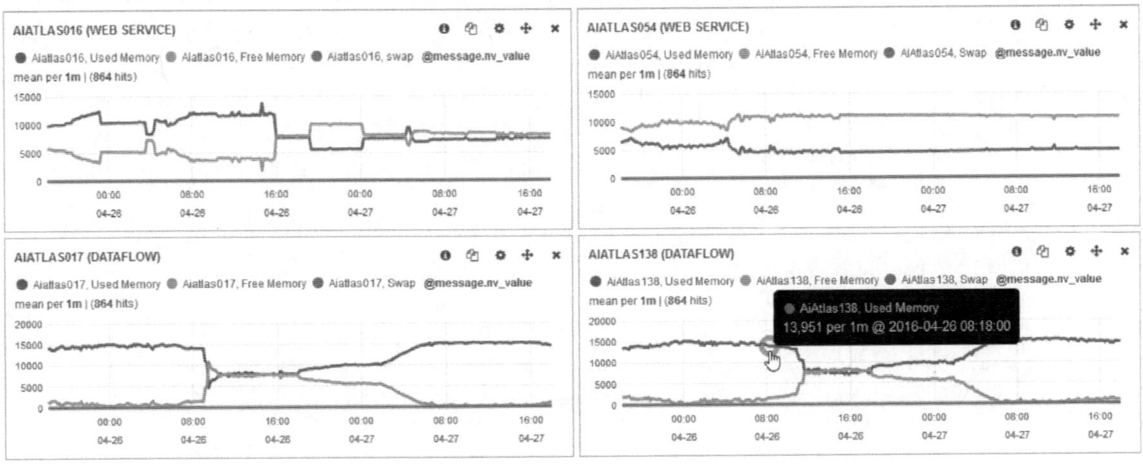

Figure 2: EventIndex servers memory usage monitoring

We track the performance of the producers that gather event information [7], consumers that receive this information from the producers and the messaging system that provides the transport of data between them by processing monitoring messages that are sent to special queues by the consumers. Figure 3 shows typical rates of the messages and amounts of data transfers. If no messages come for a long time or corrupted data appear, an alert is raised for both EventIndex operations and ATLAS computing shifters.

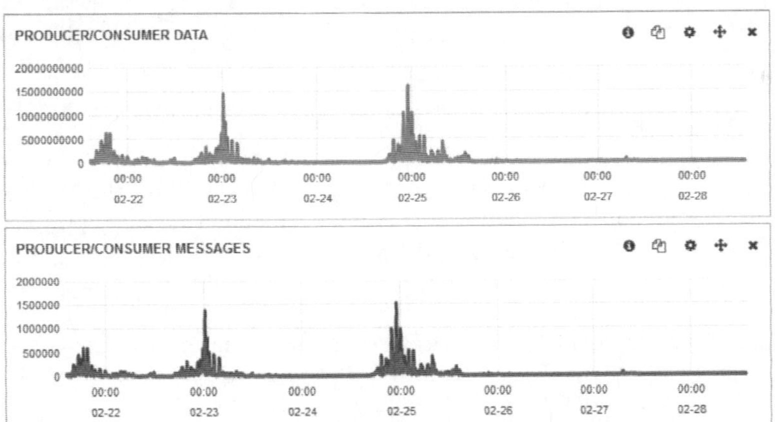

Figure 3: EventIndex Consumers monitoring

The EventIndex data are stored and managed by the CERN Hadoop cluster "lxhadoop". Although Hadoop is designed so that it can automatically fix or bypass most of the problems, still there are critical components that cannot be fixed automatically. Also, experience obtained in the process of the EventIndex commissioning shows that simultaneous failures of few less significant components like one of 24 Hadoop hosts or 22 HBase region servers may still lead to significant degradation of the system performance. So we monitor the Hadoop and HBase master state and the number of inactive hosts and region servers through the corresponding webpages. Significant degradations of this system are reported.

The Hadoop storage system has a storage capacity around 1.5 PB and host most of the EventIndex data. Its size grew to more than 100 TB including transient files created by the consumers, with information on real data from Run 2 reaching 40 TB in half a year. We gather data volume information using HDFS tools. Figure 4 shows the data growth as presented by the monitoring display.

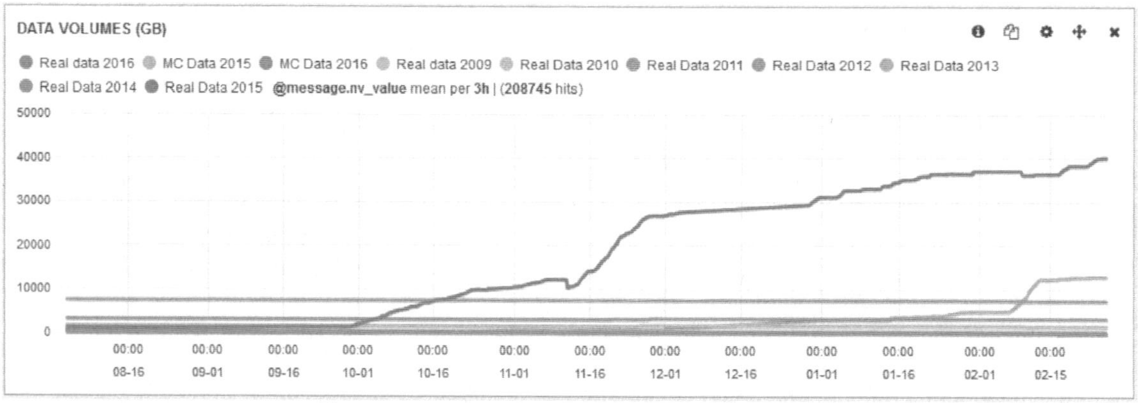

Figure 4: EventIndex data volume monitoring

The trigger decision information of each event appears as a bitmask that is decoded using trigger tables replicated from the ATLAS condition metadata database [8] (COMA). We monitor the status of replication by processing logs and plot the number of trigger tables replicated and the ranges of run numbers and trigger keys for which the trigger information is available.

The EventIndex can be accessed through the command line interface from any Linux computer running CVMFS (CERN network file system [9] optimized to deliver experiment software in a fast, scalable, and reliable way) or using a client package. However, the main way of accessing it is through the web interface and its GUI, or providing plain text commands suitable for the ATLAS distributed analysis system [10][11] (PanDA). We monitor the web interface by periodically sending simple *curl* queries.

3. Conclusions and outlook

System level monitoring in the new Kibana environment at CERN is ready and running, providing information for the EventIndex operation team and for ATLAS shifters. As the EventIndex system evolves and operating experience is accumulated, changes and additions are applied to the monitoring system in order to insure its reliability and ease of use.

References

[1] Barberis D *et al.* 2015 The ATLAS EventIndex: architecture, design choices, deployment and first operation experience, *J. Phys.: Conf. Ser.* **664** 042003 doi:10.1088/1742-6596/664/4/042003.

[2] The ATLAS Collaboration 2008 The ATLAS Experiment at the CERN Large Hadron Collider, *JINST 3 S08003* doi:10.1088/1748-0221/3/08/S08003

[3] Hadoop: http://hadoop.apache.org

[4] Kibana: https://www.elastic.co/products/kibana

[5] Solveig Albrand *et al.* 2010 The ATLAS metadata interface *J. Phys.: Conf. Ser.* **219** 042030 doi:10.1088/1742-6596/219/4/042030

[6] M Borodin *et al.* 2015 Scaling up ATLAS production system for the LHC Run 2 and beyond: project ProdSys2, *J. Phys.: Conf. Ser.* **664** 062005 doi:10.1088/1742-6596/664/6/062005

[7] J Sánchez *et al.* 2015 Distributed Data Collection for the ATLAS EventIndex, *J. Phys.: Conf. Ser.* **664** 042046 doi:10.1088/1742-6596/664/4/042046

[8] E J Gallas *et al.* 2012 Conditions and configuration metadata for the ATLAS experiment, *J. Phys. Conf. Ser.,* **396** 052033 doi:10.1088/1742-6596/396/5/052033

[9] A De Salvo *et al* 2012 Software installation and condition data distribution via CernVM File System in

ATLAS *J. Phys.: Conf. Ser.* **396** 032030 doi:10.1088/1742-6596/396/3/032030
[10] A Klimentov *et al.* 2015 Next Generation Workload Management System For Big Data on Heterogeneous Distributed Computing, *J. Phys.: Conf. Ser.* **608** 012040 doi:10.1088/1742-6596/608/1/012040
[11] De K *et al.* 2015 The future of PanDA in ATLAS distributed computing, *J. Phys.: Conf. Ser.* **664** 062035 doi:10.1088/1742-6596/664/6/062035

ATLAS FTK a – very complex – custom super computer

N Kimura on behalf of the ATLAS Collaboration

Aristotle University of Thessaloniki, Thessloniki, GR

naoki.kimura@cern.ch

Abstract. In the LHC environment for high interaction pile-up, advanced techniques of analysing the data in real time are required in order to maximize the rate of physics processes of interest with respect to background processes. The Fast TracKer (FTK) is a track finding implementation at the hardware level that is designed to deliver full-scan tracks with p_T above 1 GeV to the ATLAS trigger system for events passing the Level-1 accept (at a maximum rate of 100 kHz). In order to achieve this performance, a highly parallel system was designed and currently it is being commissioned within in ATLAS. Starting in 2016 it will provide tracks for the trigger system in a region covering the central part of the ATLAS detector, and will be extended to the full detector coverage. The system relies on matching hits coming from the silicon tracking detectors against one billion patterns stored in custom ASIC chips (Associative memory chip - AM06). In a first stage, coarse resolution hits are matched against the patterns and the accepted hits undergo track fitting implemented in FPGAs. Tracks with p_T > 1GeV are delivered to the High Level Trigger within about 100 μs. Resolution of the tracks coming from FTK is close to the offline tracking and it will allow for reliable detection of primary and secondary vertexes at trigger level and improved trigger performance for b-jets and tau leptons. This contribution will give an overview of the FTK system and present the status of commissioning of the system. Additionally, the expected FTK performance will be briefly described.

1. Introduction

Online event selection for interesting physics in the ATLAS experiment [1] is a very challenging task. The LHC [2] will run with proton-proton collision energy of 13-14 TeV, a bunch-crossing period of 25 ns and luminosities exceeding 10^{34} cm^{-2} s^{-1}. The expected interaction pile-up will reach 80 proton-proton interactions per bunch crossings at Run3 (2020-2022), which will make it more difficult to have efficient online selection of rare events of interesting physics. FTK [3] is an ATLAS trigger upgrade project. It is developed to reconstruct tracks with transverse momentum above 1 GeV for all events accepted by the Level-1 (L1) ATLAS trigger system at a rate of up to 100 kHz. The system performs tracking by a highly parallelized hardware system using custom and commercial electronics. The fast, full event tracking performed by the FTK makes it possible to trigger on some rate processes which require event-wide tracking information like b-jets, and allows in general development of pile-up robust triggers.

2. The ATLAS Detector

ATLAS is one of the general-purpose the particles generated at center of mass energy of 13-14 TeV. In ATLAS, particles are identified and have their momentum and energy measured by the inner tracking detector, electromagnetic and hadronic calorimeters, and the muon system. The inner tracker is inside a 2-T solenoidal magnetic field coaxial withe beam axis. The FTK system is using the ATLAS silicon Inner Detector [4], which consists of Pixel modules and Strip modules, organized as a barrel with co-axial cylindrical layers and two end-caps with discs.

To select interesting physics events, ATLAS has a two-level trigger system. The L1 trigger system is hardware-based and has a maximum output rate of 100 kHz using only coarse information from calorimeters and muon system. In the next stage of event processing, High Level Trigger (HLT), which is implemented as software running in a farm of several thousand CPUs, uses information from all detector systems to select events which are saved to disc at a rate of ~1 kHz.

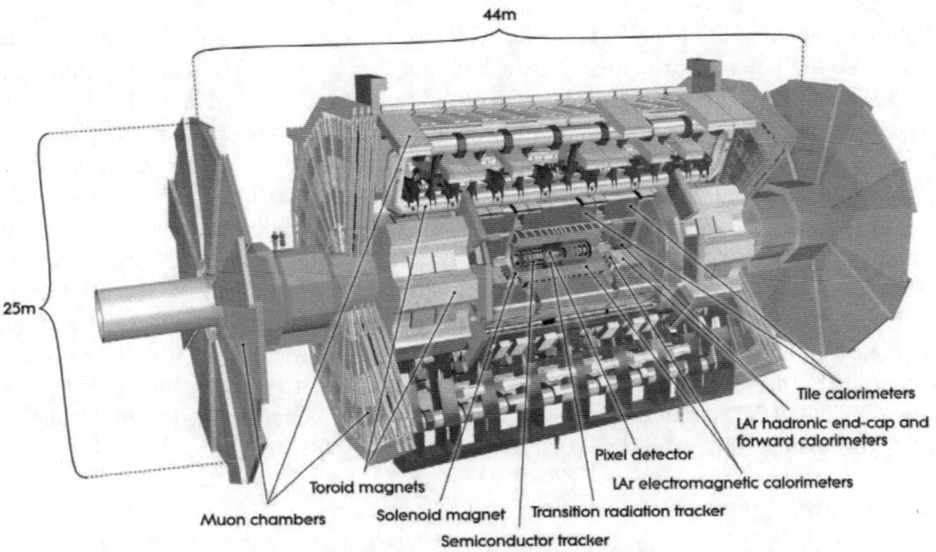

Figure 1. ATLAS Detectors

3. Pattern Recognition and Track Fitting

FTK is a hardware system that reconstructs tracks in real time using information from the inner tracking silicon pixel and strip detectors. For every event which passes the L1 trigger, FTK will provide accurate and complete track information (η, ϕ, p_T, d_0 and z_0) to HLT. This dual requirement of processing speed and accuracy is achieved by FTK system using two-stage processing described below.

3.1. Pattern Recognition

The FTK system finds coarse resolution tracks, called roads, by pattern recognition. A pattern is a list of hits, one for each silicon tracker layer used in the pattern recognition. First, possible patterns are generated by Monte Carlo simulation (MC) using about 50 billion muons, and are stored in a pattern bank. In order to limit the number of necessary pattern the actual readout channels are grouped to form coarser granularity hits. In real physics data, all such hits in the silicon trackers are sent to the pattern bank and are checked against all patterns stored in the bank, similar to a Bingo game. Matching of the patterns is performed concurrently for all patterns once hits are made available in the full-custom ASIC Associative Memory (AM) chips [5].

3.2. Track Fitting

For each pattern with all its elements matched, or all except one, the track parameters (η, ϕ, p_T, d_0 and z_0) are calculated using the full resolution hits using the 8 layer. All hit combinations, made of one hit per layer within the pattern, are fit to determine a goodness of fit. The tracking on the matched pattern is simplified by transforming the helix track parameters and quality estimator calculations in to a set of scalar products of the form

$$p_i = \sum_j C_{i,j} x_j + q_i$$

where X_j are the hit coordinates in the each detector layer, $C_{i,j}$ and q_i are pre-calculated terms, and p_i is either a helix parameter or a term of the χ^2 quality estimator. This approximation can be implemented in modern FPGAs, exploiting their speed and parallelism, to achieve approximately 1 fit/ns.

4. FTK System

Figure 2 shows a schematic of the FTK system. Hit data from the silicon pixel and strip detectors are transmitted from the detector on optical fibers and received by the FTK input mezzanine (IM) where they are clustered. The clustered hits are delivered to each DF board and distributed to the FTK parallelized Data Organizer (DO). The hits are converted to to coarser resolution hits, called Super-Strips (SS) which are appropriate for pattern recognition. The AM board performs the pattern recognition using hits from 8 out of the 12 layers, and sends back matched patterns to the DO. The Track Fitter (TF) uses full resolution hits from the DO to reject fake tracks in each matched pattern. Then, the Second Stage board (SSB) estimates the final track parameters using the full 12 layer information. All reconstructed track information is then merged and sent to the HLT from the FTK-HLT Interface Crate (FLIC). To deal with large amount of input data as well as the large number of track candidates due to the hit combinations at high luminosity, FTK is a highly parallelized system. The information from the ATLAS detector is segmented into 64 regions, and each region is covered by 2 Processor Units (PU) containing 4 units with 16 AM chips each working in parallel. Finally, the pattern recognition is performed with a parallelism of order 5000.

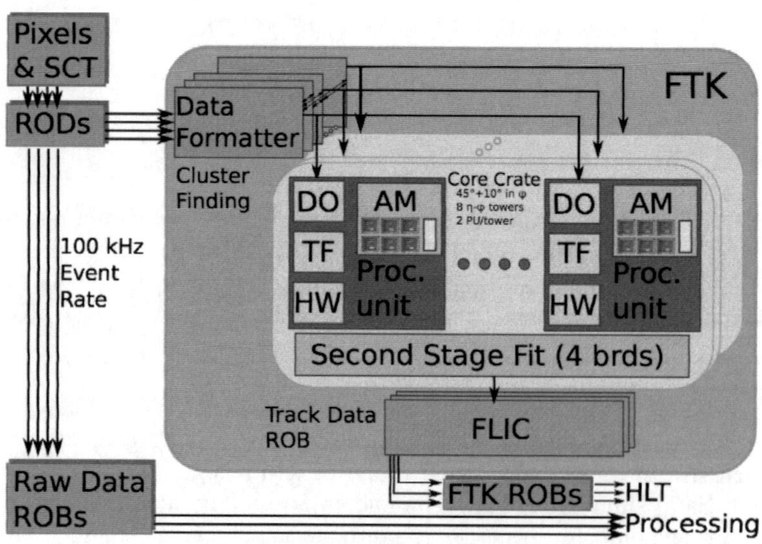

Figure 2. Schematic of FTK System

5. FTK Tracking Performance

5.1. FTK Timing Emulation

The timing of FTK system has been emulated using Z→$\mu\mu$ with pile-up as the luminosity 3×10^{34} $cm^{-2}s^{-1}$. By accounting for the time to process each word of data for each FTK components (DF, DO, AM, TF, SSB) operating at the full L1 trigger rate of 100 kHz, it is determined that FTK will take on average 50 μ sec per event to perform global tracking by highly parallelized system, as shown in Figure 3. This is fast enough for the HLT requirement order of a few milliseconds.

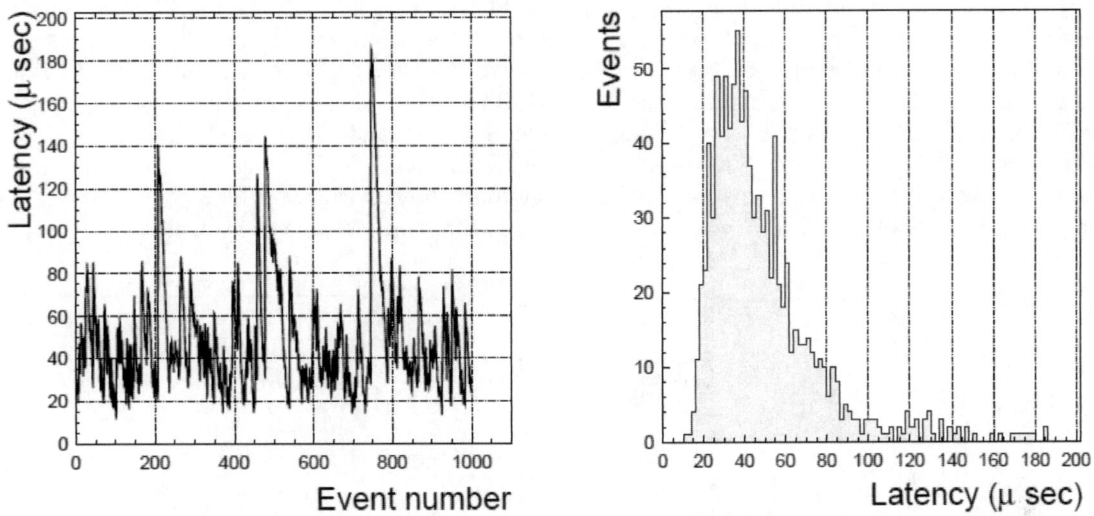

Figure 3. Distribution of FTK latency emulation for $Z \rightarrow \mu\ \mu$ events with 69 pile-up. Event number vs Latency (left), and Latency histogram of 1000 events (right) [3].

5.2. FTK track performance

In order to study the FTK tracking efficiency and resolution single muon and pion MC samples were simulated with a random distribution in the particle helix parameters. Figure 4 shows track finding efficiency with respect to offline tracks and curvature resolution of FTK and offline tracks. There is some inefficiency compared to offline in especially low p_T region, because of differences in the clustering algorithm, the lack of low p_T patterns, and other simplifications. Nevertheless, the efficiency of track finding by FTK is above 90% in all regions. The FTK resolutions for all parameters are also close to with offline, as a shown in Figure 4.

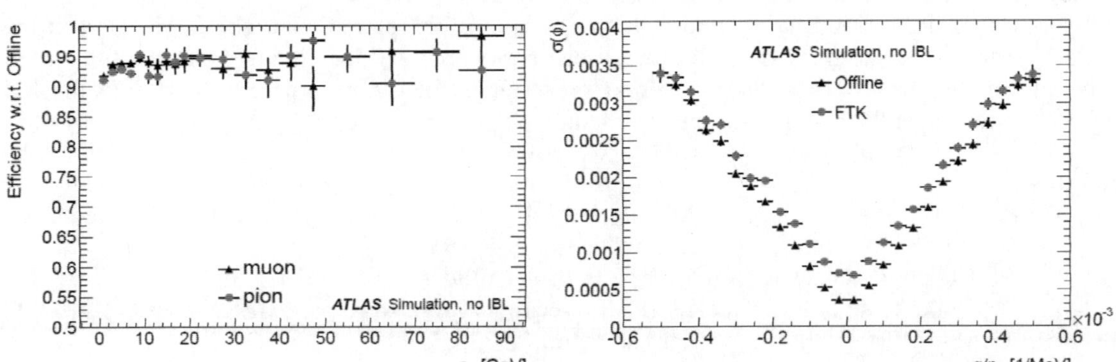

Figure 4. Pion and muon FTK track finding efficiencies with respect to Offline track (left). The ϕ resolution as a function of curvature for FTK and offline tracks (right) [3].

5.3. Example application

New physics that couples to heavy fermions may be rich in final-state *b* quarks, but not necessarily other easily identifiable objects which can be used in the trigger. Given the large QCD production rate of light quarks and gluons, it is important that the ATLAS trigger selects efficiently jets from *b*-quarks while provides large rejection against other jets. The *b* quark jets make characteristic tracks due to the long life time of *b*-quark. The tracks in a *b*-jet are characterised by larger transverse impact parameter, d_0, as compared to tracks in jets from light quarks. Figure 5 shows d_0 significance distribution and trigger efficiency of exotic 4*b* signature (*G*(m=300 GeV)→*hh*→4*b*). The d_0 significance is the significance of the minimum distance of tracks from the collision point in the plane transverse to the beam axis. The *b*-jets exhibit an excess in the positive *d_0* side and this fact can be used for a *b*-jet trigger in the HLT. FTK d_0 significance is compatible with offline as you can see in Figure 5. With FTK active, the CPU power is not spent on the tracking, and thus the HLT can apply a *b*-jet requirement using *d_0* from FTK tracks on the all event that pass L1 trigger before a p_T threshold cut. As a result, due to the lower p_T of the selection a higher trigger efficiency for the exotic 4*b* signature (*G*(m=300 GeV)→*hh*→4*b*) can be achieved with a manageable trigger rate.

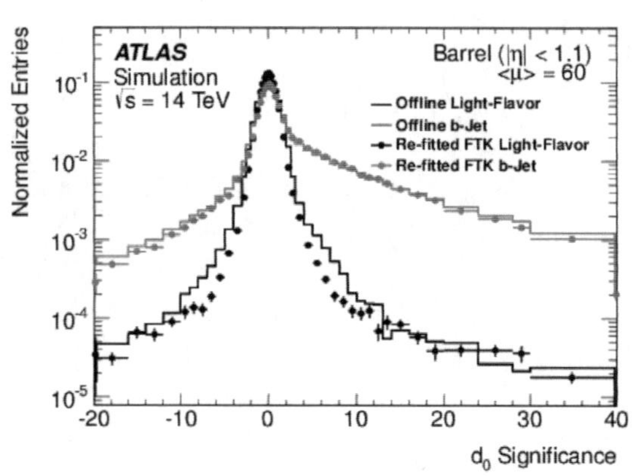

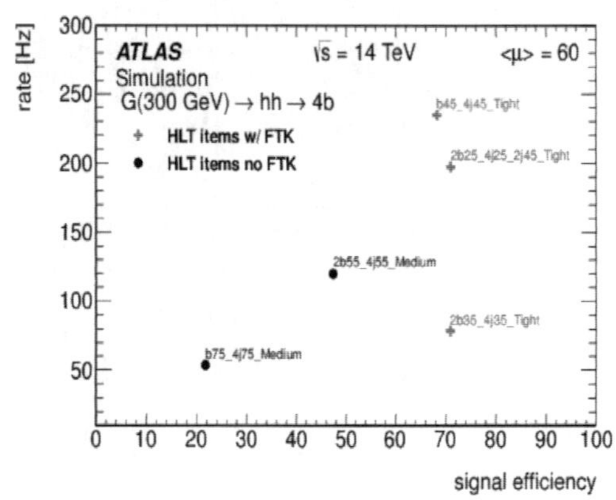

Figre 5. Offline and FTK d_0 significance distribution of light flavour and *b*-jet (left). Signal efficiency and rates for b-jet triggers with and without the FTK (right) [6].

6. Summary and Plan

FTK provides full track information to the HLT for tracks with p_T above 1 GeV. The design of the FTK system was described and its expected performance was presented. All FTK boards are produced or ready for production. Currently, commissioning of the system is ongoing with both final and prototype boards. Installation has started in the beginning of 2016, and data taking will start in the middle of year for the barrel region. After that, processor boards will increase to cover the full region by the end of the year. Finally, the number of processing units will increase (128 PU) to the full specification to handle the high pile-up environment.

References

[1] ATLAS Collaboration, The ATLAS Experiment at the CERN Large Hadron Colooider, 2008 JINST 3 S08003. https://cdsweb.cern.ch/record/1129811
[2] ATLAS Collaboration, The CERN Large Hadron Collider, 2008 JINST 3 S08001.
[3] ATLAS Collaboration, Fast TracKer (FTK) Techinical Design Report, CERN-LHCC-2013-007 ATLAS-TDR-021. https://cds.cern.ch/record/1552953
[4] ATLAS Collaboration, ATLAS Inner Detector Techical Design Report, CERN-LHCC-97-16 and CERN-LHCC-97-17. https://cds.cern.ch/record/331063 and 331064
[5] Morsani F, Amendolia S R, Galeotti S, Passuello D, Ristori D et al., The AM chip: A VSLSI associative memory for track finding. Nucl Instrum Meth A 315 (1992) 446
[6] FTK Public Results https://twiki.cern.ch/twiki/bin/view/AtlasPublic/FTKPublicResults

Automated Finite State Workflow for Distributed Data Production

L Hajdu[1,*], L Didenko[1], J Lauret[1], J Amol[2,3], W Betts[1], H J Jang[2], and S Y Noh[2,3]

[1] Software and Computing Group, RHIC/ STAR experiment, Brookhaven National Lab, PO Box 5000, Upton, NY 11973-5000, USA

[2] Korea Institute of Science and Technology Information, 245 Daehangno, Yuseong, Daejeon 305-806, KOREA

[3] Korea University of Science and Technology, Yuseong, Dajeon 305-350, KOREA

[*] Corresponding Author: Mr. Levente B. Hajdu, LBHajdu@BNL.gov

Abstract. In statistically hungry science domains, data deluges can be both a blessing and a curse. They allow the narrowing of statistical errors from known measurements, and open the door to new scientific opportunities as research programs mature. They are also a testament to the efficiency of experimental operations. However, growing data samples may need to be processed with little or no opportunity for huge increases in computing capacity. A standard strategy has thus been to share resources across multiple experiments at a given facility. Another has been to use middleware that "glues" resources across the world so they are able to locally run the experimental software stack (either natively or virtually). We describe a framework STAR has successfully used to reconstruct a ~400 TB dataset consisting of over 100,000 jobs submitted to a remote site in Korea from STAR's Tier 0 facility at the Brookhaven National Laboratory. The framework automates the full workflow, taking raw data files from tape and writing Physics-ready output back to tape without operator or remote site intervention. Through hardening we have demonstrated 97(±2)% efficiency, over a period of 7 months of operation. The high efficiency is attributed to finite state checking with retries to encourage resilience in the system over capricious and fallible infrastructure.

1. Introduction

The STAR experiment, located in one of the interaction regions of the Relativistic Heavy Ion Collider (RHIC), started taking physics events in year 2000. Since then it took approximately two billion events per year until 2012, when upgrades skyrocketed the number of recorded events past six billion events a year. In 2016 the number of events recorded by the STAR detector is expected to climb again. With the rising number of events, the reconstruction time is also increasing. Hence there is considerable demand to distribute some of the reconstruction load from the RCF (RHIC Computing Facility) to STAR Tier 1 sites in order to increase parallelization and thus decrease the total time between data taking and analysis.

STAR has been offloading simulation workload to Tier 1 sites from the very beginning of its operation, even before the inception of the Open Science Grid (OSG). Simulation was preferentially selected as a target for offloading because relative to real captured datasets, simulated datasets are orders of magnitude smaller. The smaller dataset size also means the whole output of the dataset can be stored under one path on centralized disk. For simulation input there are no input events, only non-unique input parameters and a random seed (initializer), hence there is no need in the workflow to deliverer a unique input file to each job. A dataset of arbitrary size can be produced. Because of simulation jobs' non-unique input, inefficiencies are dealt with by over requesting data to account for the number lost due to inefficiencies. In short, simulation is an easily streamlined, failure tolerant target to offload to satellite grid connected computing sites, and hence it is always considered first.

In contrast, reconstruction of real data has an input file of significant size. Each input file is unique. The number of input files is limited to what was taken during the run and may represent the bare minimum to reach a physics goal. Thus it is important to achieve efficiency close to 100 percent. The trick of over submission as done with simulation is not an option. And in the event of failure of reconstruction jobs, there must be recovery, preferably automated. These requirements make reconstruction of real data a more complicated task.

Local reconstruction has a simpler path routing input to the job and copying output files back, as the whole path is limited to within a single site. Whereas offloaded reconstruction has to use a national or international backbone, which is less reliable because the number of routing points is greater. Besides the added complexity of the network, there is also added complexity in the software stack as a grid framework must be leveraged, in order to handle additional requirements such as user account mapping, authentication, grid file transfer, and job transmission to the remote batch system. The question is can a grid job processing framework be designed to overcome inefficiencies brought by the added complexity to achieve efficiency approaching the reconstruction efficiency of the local batch system? To answer this question STAR developed a framework and offloaded one of its datasets to its Tier 1 site at KISTI (Korea Institute of Science and Technology Information).

2. KISTI

The Korea Institute of Science and Technology Information (KISTI), a STAR Tier 1 site, provided the worker node resources for this production, specifically about 1024 slots. All sites are different in terms of the level of effort they provide, additional (grid) software stack, and user access rights internally and through gateways to the outside world. Though computing power for the production run was plentiful, manpower was limited along with central storage space. Taking this profile into account with our implementation we wanted to avoid deploying new software at KISTI that would require administrative intervention, such as delegated file transfer services such as BeStMan SRM [1]. These services allow caching the output files at the local site so that the job slot can be freed up right after reconstruction without needing to deal with the copy within the run time of the job. However in this instance, without deploying delegated file transfer services, the copy would be handled within the runtime of the job. Optimizing a dedicated network path is always key to the success of large productions, but even more important in this case because it adds to the jobs run time, so STAR reached out to ESnet [2] and KREONET in order to configure a static routing path. The other feature of delegated file transfer services is that the in-job-runtime copy has to replicate the fault tolerance for the transfer. A production that takes many months will see many failures such as crashed file transfer gatekeepers and network outages. To replicate this fault tolerance, a copy script was designed to detect all the failure states for the globus-url-copy [3] command, such as freezes and unexpected return states, and a set of re-tries was implemented. Each time the job re-tries it would try one of three randomly selected gatekeepers and each time the copy failed it adds one minute to the delay before the next retry. The job will retry for up to 24 hours, which is approximately half of the average job's run time. Small network interruptions will not cause the loss of output files, nor will the crash of one gatekeeper because the random selection emulates load balancing with fail-over. This method was found to be practical for production. The time spent copying both input and output was typically less than 1% of the total run time of the job. The total input for this production is 397.3 TB, and the total output is 213.7 TB. For the movement of data in both directions the methods employed with KISTI worked very well with little measurable loss.

3. Stages of production

Figure 1 shows the stages of STAR's grid production framework. Many of the components are used for other functions and have been reused here, because they are well debugged, actively maintained, proven to work over a decade, and cut down on development time. These components include the STAR file catalog [4], STAR data carousel [5] [6], STAR Unified Meta Scheduler [7], HTCondor [8] and, Globus [9].

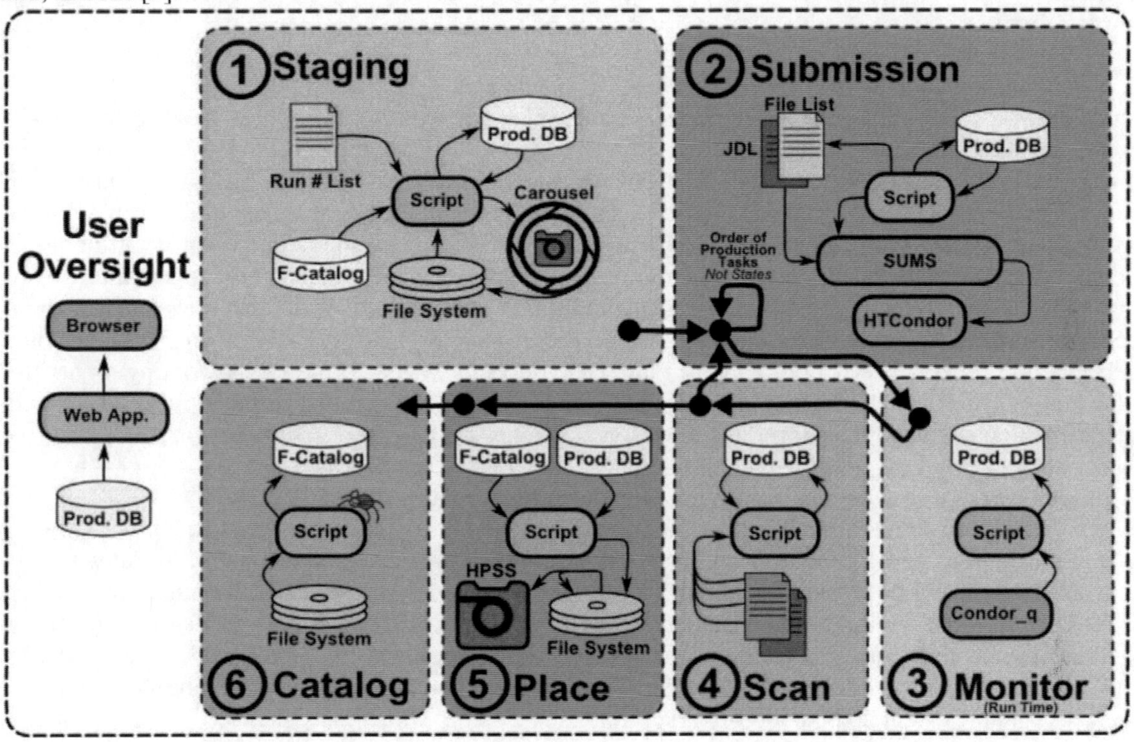

Figure 1. Stages of STAR's production system – each box and stages are explained in details as sub-sections of section 3.

3.1. Input File Staging

The first task illustrated in box one, is staging input files. The input files are initially stored on tape. They have the extension *.daq which stands for data acquisition, meaning it is data read out of STAR's detectors. The STAR file catalog holds the list of files belonging to the dataset that needs to be reconstructed. It holds LFNs (logical file names), PFNs (physical file names) as well as a breadth of MetaData (MD) users may key on. The file list from the STAR file catalog is passed via a script which is part of the grid production framework to the STAR data carousel and an entry stub for each file is made in the production DB. The STAR data carousel is a queueing tool which optimizes requests to restore files from tape (HPSS) by reordering the requests in the queue to minimize the number of mount and dismount operations needed to restore the requests. Once the files are restored from tape to central disk the entry in the production database is marked as restored and input staging is completed.

3.2. Job Submission

Once the input is staged we move on to submission. With the stub in the production database and the input file restored, a script will check the production database for restored files and compose a list of the restored files. The list is passed to SUMS, the STAR Unified Meta Scheduler. SUMS provides a uniform interface to STAR collaborators across different sites, batch systems, and grid interfaces. For the purpose of this framework SUMS transforms the input file into an actual job by generating a script which runs the reconstruction program, passes parameters, initiates transfer of input and output files, handles message passing for the purpose of monitoring, and does other wrapper housekeeping functions such as clean-up of temporary disk space. Each of these generated job scripts are submitted by SUMS to the HTCondor batch system. The HTCondor batch system in turn via an underlying level of Globus grid libraries passes the job to the KISTI site, or another grid site as specified in the job description file, also generated by SUMS at the time of submission. Globus handles mapping of user accounts via grid certificates [10]. The production database is then updated to reflect the state of the job as submitted.

3.3. Job Runtime Monitoring

Once the job is submitted we move on to the next phase in box number three. A script will run a condor_q command and parse the output recovered, it will also iterate over all of the jobs in the production database marked as submitted but not done and try to correlate them and update their states, which can be idle, run, held, and done. The job also sends back messages in the form of small files that are copied back the same way as output files that provide fine detail about what the job is doing internally. Though using file transfer to pass messages may seem inelegant it is guaranteed to be available because it is used for transferring the output files and the same script that retries failed copies can be reused to harden message passing. These messages are useful because if jobs run over their expected run times, there is information in the database as to what step the job is working on and for how long. The internal job states ideally progress from none, to input copy started, input copy stopped, reconstruction started, output copy started, and output copy stopped.

3.4. Job Output Scanning

When the monitor sees the job has exited the queue, the job passes to the next phase which is scanning of the output files and logs. The reconstruction and batch system logs are scanned for known error messages. If the output files or logs do not exist, the job is marked as failed. If the binary output files exist, their sizes are compared with the stated output file sizes from the log. File checksum comparison between source and destination of the binary output files is not necessary as this is done internally by the copy itself – only a check for truncation due to early termination of the copy is needed. If any of the above failures are noted the job is marked for resubmission, which the system will try to do up to a fixed number of times in case of failure. There needs to be a limit on the number of attempts to resubmit the job because some jobs are irrevocably broken regardless of the number of times the job is resubmitted such as a job receiving a corrupted input file. We have noted 6 of these jobs out of 105,632 or 0.0056%. After these seven tries the system gives up.

3.5. Output File Placement

If the job finishes and is marked as successful after one or more submissions it can pass on to the next phase illustrated in box number five, placement. At this point the reconstructed file with extension *.MuDst, which holds the reconstructed events, is sitting in the output buffer. It is moved to another central disk location for general analysis by users, from there a copy is sunk back to HPSS and it can even be moved to distributed disk if the analysis is to be immediate. The production database is once again updated, to indicate a copy has been pushed back to HPSS tape. If this full path is done, reconstruction is considered a success.

3.6. Output File Cataloguing

The STAR data management system spiders will eventually find the output file and register it in the STAR file catalog so users can request it. This is illustrated in box six, cataloguing. This process is fully automated.

4. Dataflow

The diagram in Figure 2 shows STAR's grid production system dataflow. Starting from the left, the input files are originally located on HPSS tapes with the list of files to be restored coming from the STAR file catalog. The carousel will recover as many files as can possibly be fit into the grid production system's DAQ buffer on NFS disk at any one time. This will only be a small fraction of the total dataset. SUMS picks up the list of files and submits one file per job to the remote site's batch system where they eventually execute on the worker nodes. While the job is running it passes back messages. Once the job is done it copies the output MuDst file back to the local Tier 0 site (BNL). The file is placed into the output buffer. Once the scanner passes over the file it is copied to both centralized disk, where it can be used for physics analysis, and to HPSS tape for backup. At multiple stages along the way the scanner is checking the throughput for validity, advancing the dataflow with its movement scripts, and retrying steps where possible and necessary.

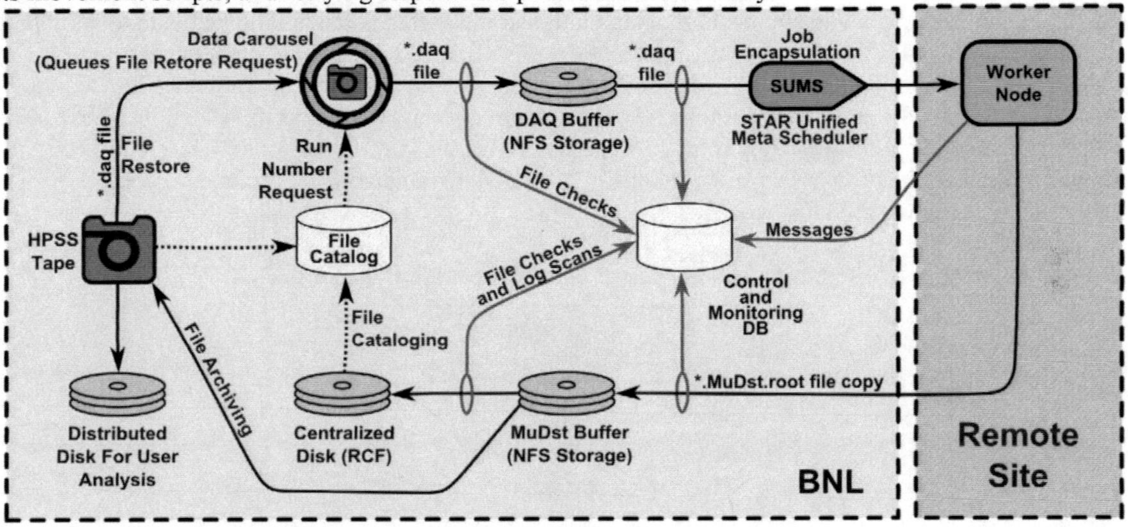

Figure 2. STAR's production framework data flow

5. Finite State Checking

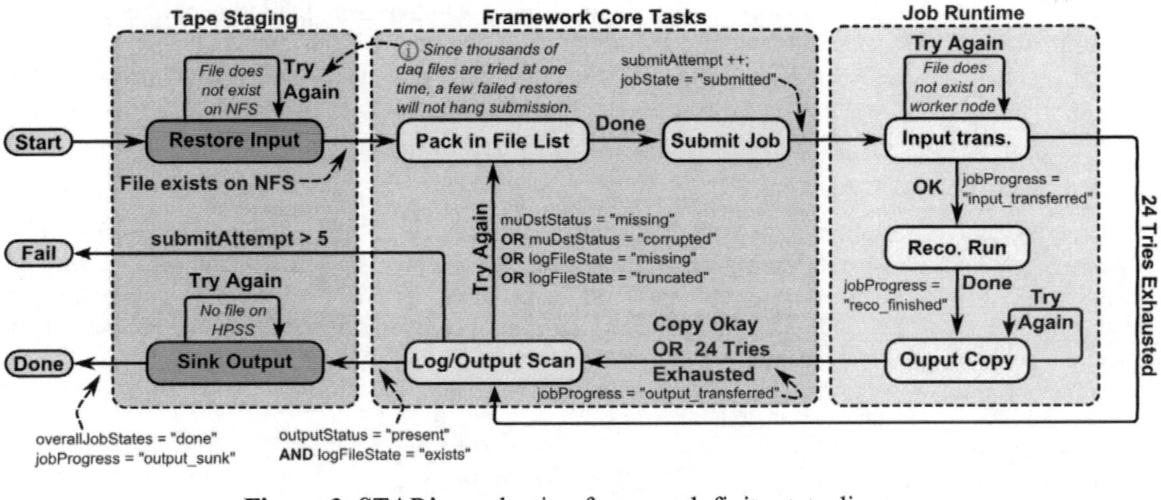

Figure 3. STAR's production framework finite state diagram

Because of the added complexity of grid production, stability can sometimes be poor. A high degree of efficiency could be gained by using multiple passes always resubmitting the failed portion. However,

Journal of Physics: Conference Series **762** (2016) 012006 doi:10.1088/1742-6596/762/1/012006

this is a poor solution as it wastes resources having to stage and resubmit each failed file, whereas in a finite state checking model, the workflow only repeats the smallest number of steps required to achieve successful reconstruction. In order to isolate the finite states, the states must be identified, and there must be checks inside the framework, such as those shown in Figure 3, to determine if the state finished successfully, or whether a state needs to be reattempted. These checks include file system-based checks for file existence and size, queue state polling, and log file parsing.

6. Outcomes

The first production campaign to which the STAR Grid data production framework was assigned consisted of proton on proton collisions at 200 GeV. The dataset consisted of 105,632 *.daq input files totalling 397.3 TB transferred to KISTI for reconstruction. Each of these input files in turn produced a MuDst output file and a log file, both of which had to be transferred back from KISTI to BNL. The sum of the MuDst files which were transferred back to BNL is 213.7 TB. The average run time for one job is about 48 hours, not including time spent idle in the queue. This production was allocated 1,025 slots which were fully saturated with jobs, excluding normal unavoidable batch system overhead. And the total production time was 9 months consuming 6,179,544 wall CPU hours. This length of time allowed for a variety of error states to manifest so that the framework's efficiency at detecting and recovering these errors could be assessed.

Table 1. Automated (re-)submission attempts per job.

Number Of Jobs	Percent (%)	Job Submissions
102,043	96.6023	1
2,895	2.7406	2
493	0.4667	3
103	0.0975	4
83	0.0785	5
6	0.0056	6
3	0.0028	7
6	0.0056	Not Runnable

A job is considered successful if for a given input file on tape both the output files and reconstruction log also exist on tape and are not truncated in any way, over 99% of jobs achieved success. The compounded slot utilization efficiency, taking into account the resubmissions was calculated at 97(±2)%. As this represents a slightly higher level of efficiency than local production, the outcome is very good indeed.

7. Conclusion

STAR has vast computing resource requirements and a history of offloading work to other sites whenever possible. We presented a finite state workflow for processing production jobs leveraging STAR's KISTI Tier 1 site resources. This production framework leveraged past knowledge and features finite state checking and recovery using an automated yet simple set of components. This approach achieved an unprecedented 97(±2)% level of utilization efficiency. The effort shows that job efficiency running on the grid can be as good as local production.

Acknowledgments

We thank the KISTI Center in Korea, the Open Science Grid consortium, ESnet, and KREONET for providing resources and support. This work was supported in part by the Office of Nuclear Physics within the U.S. DOE Office of Science.

References

[1] Hjort E, Lauret J and Hajdu L 2006 Data and Computational Grid decoupling in STAR – An Analysis Scenario using SRM Technology. International Conference on Computing in High Energy and Nuclear Physics (Tata Institute of Fundamental Research, Mumbai, India)

[2] Newman H 2004 Networking for High Energy and Nuclear Physics as Global E-Science (International Conference on Computing in High Energy and Nuclear Physics, Interlaken, Switzerland)

[3] Allcock W 2003 GridFTP: Protocol Extensions to FTP for the grid (Global Grid Forum GFD) page 20

[4] Porter J 2001 STAR Database Implementations with MySQL (International Conference on Computing in High Energy and Nuclear Physics, CERN)

[5] Lauret J and Yu D. 2010 ERADAT and Data Carousel systems at BNL: A tool and UI for efficient access to data on tape with fair share policies capabilities (ACAT 2010 proceedings, PoS ACAT 023)

[6] Yu D and Lauret J 2011 Tape storage optimization at BNL (CHEP 2010 proceedings) *J. Phys. Conf. Ser.* 331 042045 doi:10.1088/1742-6596/331/4/042045

[7] Lauret J and Hajdu L 2006 Meta-configuration for dynamic resource brokering: the SUMS approach (International Conference on Computing in High Energy and Nuclear Physics, Tata Institute of Fundamental Research, Mumbai, India)

[8] Frey J, Tannenbaum T, Foster I, Livny M and Tuecke S 2001 Condor-G: A computation management agent for multi-institutional grids *Proc. of the Tenth IEEE Symposium High Performance Distributed Computing (HPDC)* (San Francisco) pp 7–9

[9] Foster I 2006 Globus Toolkit Version 4: Software for Service-Oriented Systems (IFIP International Conference on Network and Parallel Computing, Springer-Verlag LNCS 3779) pp 2-13

[10] Tuecke S, Welch V, Engert D, Pearlman L and Thompson M 2004 RFC3820 Internet X.509 Public Key Infrastructure (PKI) Proxy Certificate Profile

Belle II Software

T Kuhr[1] and M Ritter[2] for the Belle II Software Group
[1,2]Ludwig-Maximilians University Munich, Faculty of Physics, Excellence Cluster Universe, Boltzmannstr. 2, 85748 Garching, Germany
E-mail: [1]`Thomas.Kuhr@lmu.de`, [2]`Martin.Ritter@lmu.de`

Abstract. Belle II is a next generation B factory experiment that will collect 50 times more data than its predecessor, Belle. The higher luminosity at the SuperKEKB accelerator leads to higher background levels and requires a major upgrade of the detector. As a consequence, the simulation, reconstruction, and analysis software must also be upgraded substantially. Most of the software has been redesigned from scratch, taking into account the experience from Belle and other experiments and utilizing new technologies. The large amount of experimental and simulated data requires a high level of reliability and reproducibility, even in parallel environments. Several technologies, tools, and organizational measures are employed to evaluate and monitor the performance of the software during development.

1. Introduction

The Belle II detector [1] and the SuperKEKB accelerator are currently under construction at the KEK laboratory in Tsukuba, Japan. The aim of this next generation B-factory experiment is to collect 50 times more data than its predecessor Belle [2] and to use this data to search for new physics in a variety of B meson, charm hadron, or τ lepton decays with unprecedented precision. The higher luminosity at the SuperKEKB accelerator leads to an increase in background radiation and higher event rates and requires a major upgrade of the detector and the computing system. As a consequence, also the software for the acquisition, simulation, reconstruction, and analysis of data has to be upgraded substantially. Early in the project it was realized that an evolution of the Belle software would not meet all our goals. Therefore, most software components are new implementations, which take the experience from Belle and other experiments and advances in technology into account.

2. Processing flow

Recorded events are processed one by one using a sequential set of algorithms called “modules.” As all events are independent from each other we can benefit from new architectures with multiple cores by processing different events in parallel. To share the large amounts of static memory and simplify the implementation of the single modules we parallelize using the `fork` system call which enables copy-on-write sharing of memory pages between processes. We employ a single input process, a set of parallel worker processes and a single output process. Data is passed between theses processes using a shared memory buffer. Fig. 1 shows the scaling of performance and memory consumption when running simulation and reconstruction.

To make sure that the random state is well defined in parallel processing we initialize the state of our random generator deterministically for each event using a 1024 bit SHAKE256 [3]

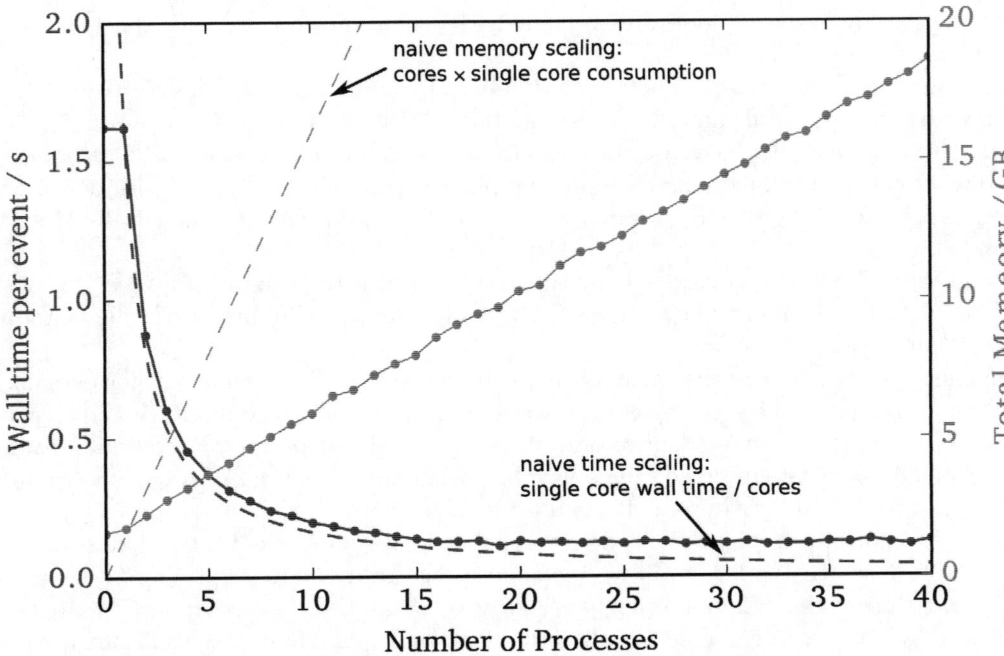

Figure 1. Scaling of performance with multiple worker processes. The blue solid line shows the wall time needed to process one event depending on the number of worker processes and dashed blue line shows ideal scaling when dividing the wall time needed using one worker process by the number of processes. Similarly the solid red line shows the total memory consumption and the dashed red line represents the naive scaling when multiplying the single process consumption by the number of processes. Zero processes denotes disabled multi processing where the whole program runs in one process so there are zero worker processes.

hash of a seed string and the unique event identifier. The hash serves directly as internal state for an Xorshift1024* [4] generator which is passed to the subsequent processes.

3. Tools

Several state-of-the-art tools are used to facilitate the collaborative software development at Belle II. Modern tools and programming practices are also important to attract and educate new students.

We use modern compilers, currently GCC 5.2 [5], clang 3.7 [6], and the Intel compiler version 16.0, Update 2 [7]. They allow us to benefit from C++11 features in our code. The parallel usage of different compilers also helps to maintain the portability of the code. The steering of the framework, like the selection of modules and the setting of their parameters, is done in Python 3 [8], a powerful and frequently used scripting language. This interface provides the possibility to create higher-level meta-frameworks on steering file level, e.g. for physics analyses.

The code is built with SCons [9] which provides a few nice features; it uses Python and therefore does not introduce yet another language in our system, it builds the code in one step and not via an intermediate Makefile like cmake [10] does, and it can work with a central location of the source code plus a partial, updated version in a local directory so that developers only have to check out the part of the code they are working on. Based on SCons, we have developed a build system that is very easy to use because many functionalities are automated. Basically the user only has to put the code in the right folder structure and specify the libraries that should be linked.

We rely on several external products, like ROOT [11], GEANT4 [12], and EvtGen [13] which are built with a custom Makefile. A consistent, tagged set of these external products is called an externals version and the basis for our Belle II software. The externals can be compiled from source, but often the developers install a pre-compiled version that we provide for a selected set of Linux distributions. Scripts for the installation and setup of external and Belle II software are in place.

All code of the Belle II software is maintained in a central Subversion [14] (SVN) repository at KEK and can be displayed via a ViewVC [15] code browser. We are currently switching to Git [16] as our version control system.

Developing software in a large project is very different from writing code just for personal use, like analysis programs. To keep people aware that they are part of a community and to produce maintainable code, common guidelines and rules are established. While coding conventions usually cannot be strictly enforced, rules for the formatting style are checked on commits to the SVN repository. The astyle tool [17] is used for C++ code and pep8 [18] for Python code. One reason for choosing astyle, besides being an open source tool, was that it is not just a style checker, but formats the code. We have included astyle in a small script that developers can use to format their code. Therefore it is very easy to make the code compliant with the style rules. A similar approach was used for Python with the PythonTidy [19] tool, but abandoned recently because of unsatisfactory formatting results. Now we use pep8 for the style check and autopep8 [20] for the formatting, although there are cases where autopep8 is not able to produce a pep8 compliant result.

Doxygen [21] was chosen as a tool for the documentation of the source code. Although there are probably better documentation tools for Python code it allows to use the same tool for C++ and Python code. Documentation that is beyond the scope of Doxygen is maintained in a TWiki [22]. A Redmine issue tracker [23] was set up to make sure that discovered problems are not forgotten. Developers are reminded once a week about issues assigned to them that are due or have no due date set.

Several tools are applied for the software quality control. googletest [24] can be used for unit tests. The framework can automatically run steering files that are located in a test folder and check whether the log output is identical to the expected one. A check for memory management issues is executed with Valgrind [25].

Finally, we have developed a framework for regression testing. It executes, either locally or on a batch system, a set of scripts that can be added to the SVN repository together with the other code. There are production scripts that create simulated data files and plot scripts that read the data files and create output ROOT files with validation plots. One- or two-dimensional plots are supported together with ntuples for monitoring numerical values. XML headers in the scripts provide meta information like the dependencies between the scripts. The plots in all ROOT files are collected and displayed on a web page. The results from older revisions of the code are overlayed so that regressions or improvements can be detected. Additionally, a reference plot, contact information, a description, and instructions for checking the quality can be provided. A (significant) deviation from the reference is indicated by a yellow (red) plot frame.

For an efficient and effective software quality control it has to be highly automated. We use the Buildbot tool [26] for this task. It is very flexible and it fits well into our environment because it uses Python for its configuration. Our setup consists of one server and four slaves with different operating systems: SL5, SL6, Ubuntu 14.04, and a 32bit version of SL5 and Ubuntu 14.04. Several actions are triggered either by commits or at given times:

- On commits an incremental build is started and failures of compilation or tests are reported by email to the committer. This catches, for example, cases where the developer forgot to include a new file in the commit.
- Commits also trigger full builds from scratch on four different systems with three different compilers where all commits in a 10-minute interval are grouped together. A Python class on the Buildbot server keeps track of compilation error and warning messages and informs the developer if any new ones show up. The errors and warnings are also reported on a web page with links to relevant locations in the source code.
- Each night a full build with different compilers is executed. Furthermore, cppcheck [27] is applied to the code, tests are executed, the geometry is checked for overlaps, documentation is generated with Doxygen and messages about missing documentation are recorded, and SCons debug output is analyzed to determine code dependencies and report missing or unnecessary library links. If any issues are detected, an email is sent to the person responsible for the corresponding part of the code base. The result of all checks is displayed on a web page.
- We also use the Buildbot to update the installation of setup scripts, externals, and releases at sites.
- Whenever a new externals version is tagged, it is automatically compiled on the four slaves with different operating systems and tarballs of the binaries are made available for download.
- Monthly builds are automated such that the latest tagged versions of the parts of the code are collected and built. In case of a failure this procedure is repeated the next day. If the build is successful the code is committed to SVN and the new version announced on a mailing list together with a report of changes that are provided by the responsible persons on a TWiki page.

4. Organization

A bunch of tools is not sufficient for a successful collaborative software development. In particular for the collaborative aspect an organization that on the one hand provides sufficient guidance to prevent a divergent development and on the other hand leaves enough freedom to not suppress the creativity of the developers is essential.

The Belle II software group has sub-groups for different tasks: the conditions database, the signal event generators, the detector simulation, the generation and simulation of background, the track reconstruction, and the alignment and calibration. Many further tasks are covered by individuals or small groups so that there is no need for a formal group definition. Developers are often also members of a detector or physics group which facilitates the communication of the software group with the whole Belle II collaboration. In addition we have an official contact person from each sub detector group to make sure that the software group is aware of all the problems faced by the other groups and can provide common solutions if possible.

The source code currently consists of about one million lines of code and is structured in packages. A package, for example, contains the code related to a detector, the core framework, or the event generators interfaces. Each of the about 30 packages has a librarian who is responsible for the code inside it. The librarian can control who has write access to the package and has the right to make a tag of the package.

The tags are used to produce monthly builds or releases. As the name indicates, monthly builds are schedule-driven. They provide a certain pace for the developers. They are also installed on the central computing system at KEK and provide well-defined code versions for users and developers. Besides a successful compilation, no quality requirements are checked for monthly builds. In contrast, releases are feature-driven and a more thorough quality assessment is performed. The set of requested features and checks is defined in a consensus between users and developers.

Virtual and in-person meetings are a key component for the exchange of information and the coordination of activities. Progress and problems in the daily work are discussed informally in a weekly developers meeting. The sub-groups often have dedicated (bi-weekly) video meetings. The tracking group also organizes face-to-face meetings two or three times a year. A good opportunity to discuss software topics in person are the one-week software and computing workshops which are held at KEK in autumn and at one of the other collaborating institute in spring. The software sessions at the collaboration meetings are often joint with other groups to foster the information flow in both directions.

5. Summary

Large software projects are a challenge and the distribution of the Belle II software developers all around the world does not make it easier. We have established a set of tools that lets the developers focus on their main work and provides them with detailed feedback on the code quality in a highly automated way. An organizational structure for collaboration and coordination is in place. Since the start of the project in 2008 about 100 different people have contributed, from undergraduate students to professors. The enthusiasm and devotion of developers will be the key for the success of the Belle II software project.

6. References

[1] Abe T *et al.* 2010 Belle II Technical Design Report *Preprint* `arXiv:1011.0352`
[2] Abashian A *et al.* 2000 *Nucl. Instrum. Meth. A* **479** 117
[3] NIST Computer Security Division 2015 *SHA-3 Standard: Permutation-Based Hash and Extendable-Output Functions.* `doi:10.6028/NIST.FIPS.202`.
[4] Sebastiano V 2014 An experimental exploration of Marsaglia's xorshift generators, scrambled *Preprint* `arXiv:1402.6246`.
[5] https://www.gnu.org/software/gcc/
[6] http://clang.llvm.org/
[7] https://software.intel.com/intel-compilers/
[8] https://www.python.org/
[9] http://www.scons.org/
[10] http://www.cmake.org/
[11] Brun R and Rademakers F 1997 *Nucl. Instrum. Meth. A* **389** 81.
[12] Agostinelli F *et al.* [GEANT4 Collaboration] 2003 *Nucl. Instrum. Meth. A* **506** 250.
[13] Lange D J 2001 *Nucl. Instrum. Meth. A* **462** 152.
[14] https://subversion.apache.org/
[15] http://viewvc.org/
[16] http://git-scm.com/
[17] http://astyle.sourceforge.net/
[18] https://pypi.python.org/pypi/pep8
[19] https://pypi.python.org/pypi/PythonTidy/
[20] https://pypi.python.org/pypi/autopep8
[21] http://www.stack.nl/ dimitri/doxygen/
[22] http://twiki.org/
[23] http://www.redmine.org/
[24] https://code.google.com/p/googletest/
[25] http://valgrind.org/
[26] http://buildbot.net/
[27] http://cppcheck.sourceforge.net/

Bringing Experiment Software to the Web with VISPA

M Erdmann, B Fischer, R Fischer, C Glaser, F Heidemann, G Müller, T Quast, M Rieger, M Urban, D van Asseldonk, R F von Cube and C Welling

III. Physics Institute A, RWTH Aachen University, Germany

E-mail: `rieger@physik.rwth-aachen.de`

Abstract. The Visual Physics Analysis (VISPA) software is a toolbox for accessing analysis software via the web. It is based on latest web technologies and provides a powerful extension mechanism that enables to interface a wide range of applications. It especially meets the demands of sophisticated experiment-specific use cases that focus on physics data analyses and typically require a high degree of interactivity. As an example, we developed a data inspector which is capable of browsing interactively through event content of several data formats, e.g., MiniAOD which is utilized by the CMS collaboration. Visual control of a chain of user analysis modules as well as visualization of user specific workflows support users in rather complex analyses at the level of $t\bar{t}H$ cross section measurements. The VISPA extension mechanism is also used to embed external web-based applications which benefit from dynamic allocation of user-defined computing resources via SSH. For example, by wrapping the JSROOT project, ROOT files located on any remote machine can be inspected directly through a VISPA server instance. We present the techniques of the extension mechanism and corresponding applications.

1. Introduction

Scientific workflows can strongly benefit from the advantages of web applications. Especially location-independent access via web browsers and centralized software installation on dedicated servers are convincing arguments. As an example, *IPython Notebooks* [1] become more and more popular in the physics community.

However, specialized experiment-specific software requires an entirely different level of infrastructure in order to profit from web-based approaches. The VISPA system is capable of coping with the demands of complex experiment environments in the web [2].

The article is structured as follows. Section 2 describes the architecture of the VISPA software as well as its key concepts. Subsequent sections 3 to 5 present developed extensions that demonstrate its suitability as an enviroment for experiment-specific software.

2. The VISPA Software

Following a server-client approach, the VISPA software consists of two parts. The server is written in Python and is implemented using the *CherryPy* framework [3]. The browser-based client is written in JavaScript with graphical components defined in HTML5 [4] and CSS3 [5]. Communication between the two relies on *Hypertext Transfer Protocol* (HTTP) requests and *WebSockets*. User-related information, i.e. login data, permissions and preferences, are stored in a relational database which is accessed via *object-relational mapping* (ORM) using the *SQLAlchemy* toolkit [6]. The internal user management system enables administrators to group users into logical entities, such as *Projects* and *Groups*, and handles allocation of permissions.

Two key concepts are the ability to connect to user-defined resources, called *workspaces*, and a mechnism that allows for extending functionalities. Both of them are described in detail in the following.

2.1. Workspace Architecture
The purpose of the VISPA server is to provide a working environment for multiple users. Thus, in order to preserve server responsiveness, requests for large amounts of data and computing power should be treated separately. Workspaces constitute the means to including dynamic and scalable resources into the server-client architecture. A schematic overview is shown in figure 1.

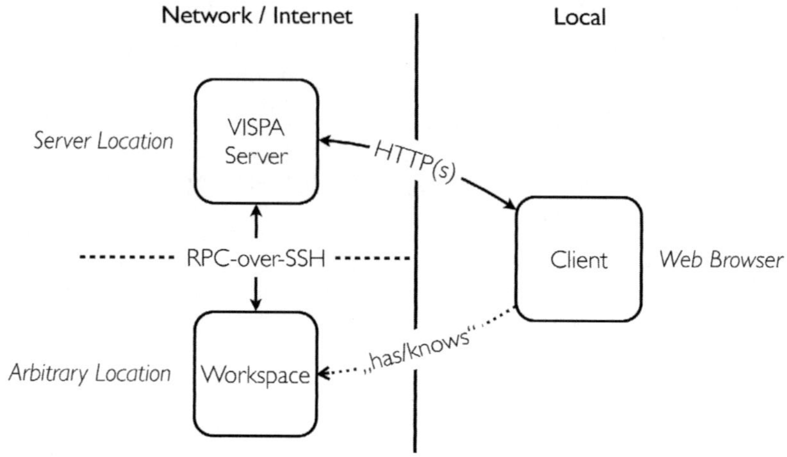

Figure 1: Schematic overview of the workspace architecture [7].

A workspace is a worker node that runs a Python interpreter and is accessible through a network via SSH. These requirements can be met by numerous different machines: desktop computers, computing clusters, and even tablets and smartphones. Resource utilization and file access are granted as the server relies on the user login to interact with workspaces via *remote procedure calls* (RPC). The connection between server and clients as well as between server and workspaces is encrypted via SSL and SSH, respectively.

Using the graphical interface of the client, users can connect to predefined workspaces and, unless explicitly forbidden by an administrator, define and connect to custom workspaces.

2.2. Extension Mechanism
Usually, experiment-specific software is specially tailored for its particular use case. Any system or environment that aims to cover several use cases simultaneously should exhibit appropriate flexibility. Within VISPA, this is achieved by separating core functionality from implementation of specific features using an extension machanism.

Extensions are installed on the machine that runs the VISPA server instance and are typically available for every user that can access the server. They are composed of three parts: code and graphics markup that extend the client (JavaScript/HTML/CSS), code that handles the communication on the server (Python), and, optionally, code that is transferred to and executed on workspaces (Python). The purpose of the latter part is to outsource and encapsulate actual extension logic causing the server to take on the role of a mediator between requests and resources.

Typical basic extensions are file browser, file selector, code editor, and terminal. As they are essential for most common workflows, they are pre-installed on the VISPA server. But due to the vast development of web techniques over the past years as well as the abundant supply of existing web-based software packages, even more complex and specialized extensions become feasible. Possible applications range from visualization of static data to highly interactive interfaces, such as graphical steering of modular analyses which is presented below.

3. Analysis Designer and Data Browser

The extensions presented in this section rely on the *Physics eXtension Library* (PXL), which is a collection of C++ classes addressing analyses in high energy physics (HEP) and astroparticle physics [8]. It features elementary objects such as `pxl::LorentzVector` or `pxl::Event` classes, as well as a module system for structuring complex analysis code into reusable building blocks. Modules can be written in both, C++ and Python. Objects stored in *pxlio* files can be nested and connected to each other via *ownership* and *relations*. They also have the ability to store *user records*, i.e., key-value pairs of arbitrary data types. Therefore, PXL provides a consistent interface to objects and containers which can be organized in various experiment-specific setups.

Chains of analysis modules can be created graphically using the *Analysis Designer* (fig. 2a). Users can add, remove and arrange modules using drag and drop gestures. Options give rise to code flexibility beyond compile time. Connections between modules control the data flow during execution initiated via the `pxlrun` command.

Input and output files of these analysis chains can be inspected by the *Data Browser* (fig. 2b). In the case of HEP data, browsing is performed on a per-event basis. Different views on the event and its contents are presented in dedicated windows. Currently, Feynman-like graphs, plain content listings and detailed property listings are supported. However, the code structure allows for the addition of further views such as, e.g., a three-dimensional event display. Based on the non-restrictive design of PXL objects as explained above, event content stored in other data formats is convertible to *pxlio* and thus, visualizable by the extension.

4. Analysis Workflow Management and Visualization

For mastering a specific research question, management of underlying scientic workflows poses a complex challenge in today's physics working environments. Current HEP analyses utilize a significant amount of resources in terms of CPU time, memory consumption, and disk space. The logical solution to resource requirements is parallelization on large-scale computing centers. Main challenges are, e.g., ten thousands of datasets to be processed, application of multivariate analysis concepts with execution dependencies for training and evaluation data, and the complexity of reconstruction and analysis algorithms. However, physicists are usually required to steer their individual workloads manually, i.e., start (remote) processes, monitor their execution, handle potential failures, and eventually obtain intermediate output data to be passed to further analysis. In addition, all *dependent* workloads have to be initiated once the current one finished successfully. On a large scale, this *management* task is not only time-consuming for the operating physicist, but, in contrast to automated approaches, also represents a risk for errors, e.g., the loss of information on the interplay between particular workloads. In fact, in certain cases the replicability of physics results might not be guaranteed.

We developed the Analysis Workflow Management tool (AWM) which aims at reducing the complexity described above. It provides guidance on structuring arbitrary workloads, so-called *steps*, into directional workflows. Dependencies are resolved automatically by evaluating the dataflow, i.e., data input and output, between particular steps. Scalability is ensured by introducing resources as abstract locations. *Storage locations*, e.g. local hard-drives or remote data centers, define where data can be stored to and accessed from. *Run locations*, e.g. local computers or computing clusters, describe where data is processed. Along with an internal bookkeeping system, AWM provides a high degree of automation of typically overhead tasks like job creation and submission, or file transfers. The information exchanged between steps is defined by the user. An integrated software managing system allows for custom user-installed software including installation of required dependencies on run locations. It is written in Python and allows for embedding any executable in customized steps.

In order to display relations of the steps that constitute a workflow, we integrated a *Workflow Visualizer* as an extension for VISPA. It provides a novel approach of illustrating the structure of individual physics analyses (fig. 2c). Graphs are generated by traversing a workflow, collecting information on inputs and outputs of steps to infer relations, and passing them encoded as *dot* files to algorithms of the *graphviz* package [9]. The resulting *scalable vector graphics* (svg) is then displayed in the web browser. In case of complex workflows, users can interactively change the appearance of a graph by combining multiple steps refering to similar tasks into so-called *clusters*. It is also possible to query detailed information about each step.

5. Embedding External Functionality: JSROOT

The number of portings of successful desktop software projects to the web increased rapidly during the last years. A physics-related example is the ROOT framework [10]. In ROOT, one and two dimensional histograms as well as flat datastructures can be visualized with a high degree of interactivity. This particular feature, called `TBrowser`, was ported to the web using JavaScript and HTML, and released as JSROOT. It is even capable of rendering three dimensional objects using *WebGL*. In order to visualize data, software installation and target files have to be located on the same machine. In certain cases, target files need to be copied to the software location or vice versa.

As explained in section 2, VISPA simplifies remote data access using workspaces while requiring only one software installation on a single VISPA server instance. Technically, JSROOT can be used as a plugin. We developed a VISPA extension consisting of only a thin code layer that embeds this plugin which now benefits from exchangeable workspaces (fig. 2d). It supplies the same features and rich interactions as the standalone JSROOT application.

6. Conclusions

The VISPA project provides an ecosystem for experiment-specific software. Its architecture can be considered a "glue" between multiple web-based applications and combines them into a single, customizable environment to exploit synergies. Custom extensions have been developed to address several use cases for physics analyses.

The event content of HEP data files can be interactively visualized with the *Data Browser*. The *Analysis Designer* provides a graphical interface to modular analysis chains based on PXL. For the purpose of coping with the complexity of modern physics analyses, an *Analysis Workflow Manager* has been developed. This tool allows physicists to define analysis steps, structure them within workflows, and control their execution across arbitrary locations. These workflows can be displayed using a dedicated VISPA extension. With the example of JSROOT, we showed that the extension mechanism is also capable of embedding existing software. Executing JSROOT through VISPA enhances flexibility as it benefits from the concept of workspaces.

Acknowledgments

We wish to thank the conference organizers for their kind support. This VISPA project is supported by the Ministerium für Wissenschaft und Forschung, Nordrhein-Westfalen, the Bundesministerium für Bildung und Forschung (BMBF), the Helmholz Alliance Physics at the Terascale, and the Deutsche Forschungsgemeinschaft.

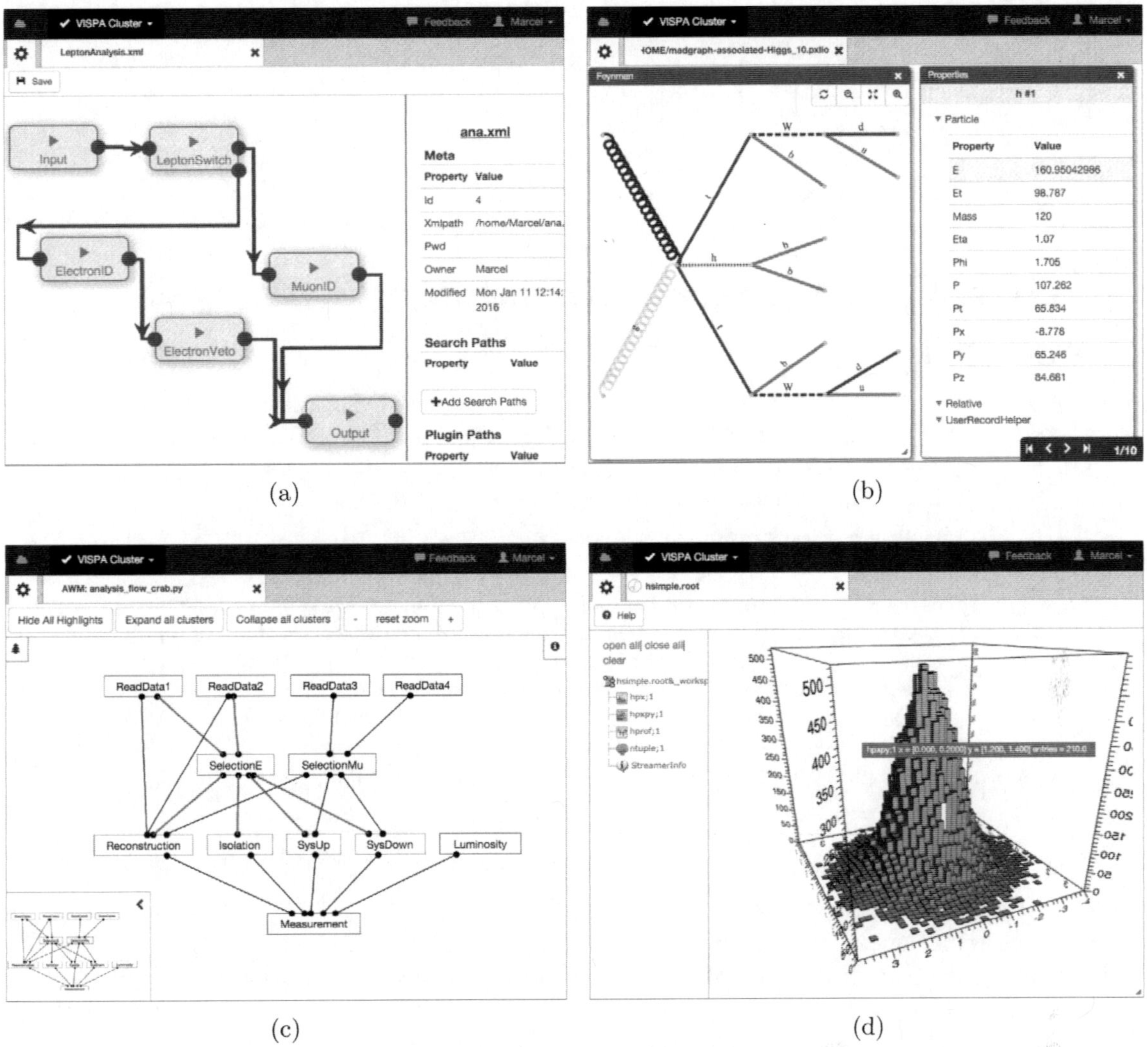

Figure 2: Analysis-specific extensions of the web-based implementation of VISPA. a) Analysis Designer, b) Data Browser, d) Workflow Visualizer, and d) JSROOT.

References

[1] Fernando P and Granger, B E. (2007), *IPython: a System for Interactive Scientific Computing*, IEEE Comput. Sci. Eng., vol. 9, no. 3, pp21-29, doi:10.1109/MCSE.2007.53, `http://ipython.org`

[2] Bretz H-P et al. (2012), *A Development Environment for Visual Physics Analysis*, JINST **7** T08005, doi:10.1088/1748-0221/7/08/T08005, arXiv:1205.4912v2, `http://vispa.physik.rwth-aachen.de`

[3] CherryPy, `http://www.cherrypy.org`

[4] World Wide Web Consortium, *HTML5 Specification*, `http://www.w3.org/TR/html5`

[5] World Wide Web Consortium, *CSS3 Specification*, `https://www.w3.org/Style/CSS`

[6] SQLAlchemy, `http://www.sqlalchemy.org`

[7] Erdmann M et al. (2014), *A Web-Based Development Environment for Collaborative Data Analysis*, J. Phys.: Conf. Ser. **523** 012021, doi:10.1088/1742-6596/523/1/012021

[8] Erdmann M et al. (2012), *Physics eXtension Library (PXL)*, `http://vispa.physik.rwth-aachen.de/pxl`

[9] Gansner E R and North S C (2000), *An open graph visualization system and its applications to software engineering*, Software Pract Exper, vol. 30, no. 11, pp1203-1233, `http://www.graphviz.org`

[10] Brun R and Rademakers F (1996), *ROOT - An Object Oriented Data Analysis Framework*, Nucl. Inst. & Meth. in Phys. Res. A 398 81-86, `http://root.cern.ch`

CLARA: CLAS12 Reconstruction and Analysis Framework

V Gyurgyan[1], S Mancilla[2] and R Oyarzún[2]

1 Thomas Jefferson National Accelerator Facility, Newport News, VA, USA
2 Universidad Técnica Federico Santa María, Valparaíso, Chile

E-mail: gurjyan@jlab.org

Abstract. In this paper we present SOA based CLAS12 event Reconstruction and Analyses (CLARA) framework. CLARA design focus is on two main traits: real-time data stream processing, and service-oriented architecture (SOA) in a flow based programming (FBP) paradigm. Data driven and data centric architecture of CLARA presents an environment for developing agile, elastic, multilingual data processing applications. The CLARA framework presents solutions capable of processing large volumes of data interactively and substantially faster than batch systems.

1. Introduction

Software systems, including physics data processing (PDP) applications have a distinct component based structure. Most data processing applications can be described as a combination of, more or less, independent software units that are cooperating to achieve data processing goals. These components in most cases require different optimization criteria to increase performance of the overall data processing application. It is arguable difficult to optimize functionally different components of an application within the same process. The resulting code becomes bloated, less readable, difficult to test, debug and maintain. We think that by designing a PDP application, based on independent (or loosely coupled), specialized software components as processes much needed agility and efficiency can be achieved. Under "loose coupling" we assume that components do not have interdependences other than a data interface. Thus data processing application components can be considered as service providers that offer their services through a defined data interface.

2. Basic Concepts

The CLARA framework uses SOA[1] to enhance the efficiency, agility, and productivity of data processing applications. Services are the primary means through which data processing logic is implemented. CLARA application services are running in a context that is agnostic to the global data processing application logic. Services are loosely coupled and can participate in multiple algorithmic compositions. Legacy processes or applications can also be presented as services and integrated into a data processing application. Services can be linked together and presented as one, complex, composite service. This framework provides a federation of services, so that service-based data processing applications can be united while maintaining their

Journal of Physics: Conference Series **762** (2016) 012009 doi:10.1088/1742-6596/762/1/012009

individual autonomy and self-governance. CLARA is a multilingual framework that supports services written in Java, Python and C++ languages.

2.1. Design architecture

CLARA architecture consists of three layers, illustrated in Figure 1. The first layer is the service-bus that provides an abstraction of the xMsg publish-subscribe messaging system[2]. xMsg is using the ZeroMQ[3] socket library and implements messaging patterns like topic pub-sub, workload distribution, and request-response. Every service from the service layer or component from the orchestration-layer communicates via the service-bus, which acts as a messaging tunnel between CLARA components. The service-layer houses the inventory of services used to build data processing applications. An administrative registration service stores information about every registered service in the service layer, including address, description and operational details.

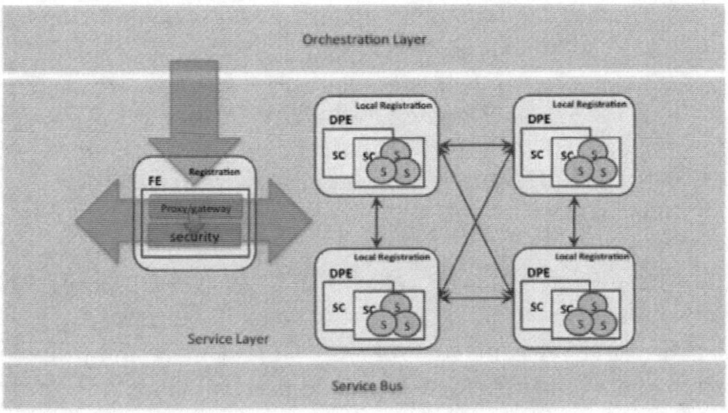

Figure 1. CLARA architecture: Front-End DPE (FE), Data Processing Environment (DPE), Service Container (SC), Service (S).

The orchestration of data analyses applications is accomplished by the application controller (orchestrator), which resides in the orchestration-layer. Front-End (FE) is a specialized data processing environment (DPE) that houses a master registration service on top of the global registration database. The growing need to fuse, correlate and process widely distributed data in the CLARA environment will undoubtedly demand data security measures. CLARA provides normative data-proxy and security services that can be deployed in the FE. CLARA security services implement user authentication, data encryption and transient data envelope integrity.

2.2. Data envelope

The CLARA framework provides developers ability to interact with services using the publish-subscribe message exchanges. In a distributed communication model, success largely depends on the design of the message structure: a communication envelope that describes not only transferred data, but also communication and service operational details. In order for a service communication to be truly useful, every service has to share/use the same vocabulary for expressing the communication details (i.e. common message-interface). CLARA transient data envelope is the main message interface between services. The mutual understanding and acceptance of this object couples the framework services. The CLARA data envelope is illustrated in the Figure 2.

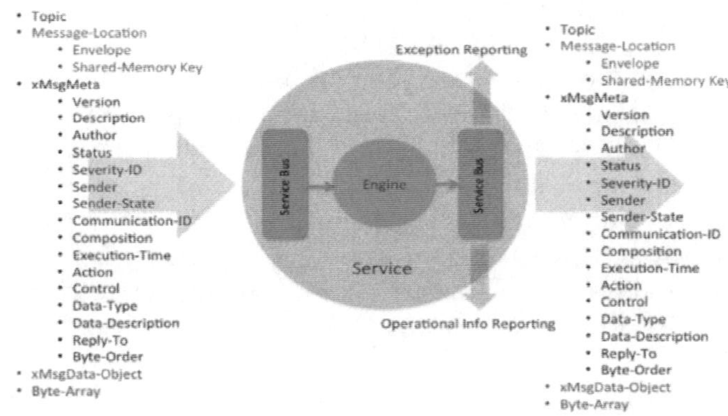

Figure 2. CLARA data envelope: Envelope can have two or 4-part configuration based on the data transferred modes between local or remote services. Local: 1) topic, 2) shared-memory key. Remote: 1) topic, 2) "envelope", 3) xMsg metadata, 4) serialized object.

2.3. Basic components

The CLARA platform is composed of multiple distributed data processing environments (DPE). DPE provides a run-time environment for services and service-containers and allows several services to concurrently execute on the same environment. The use of shared memory prevents unnecessary copying of the data during service communications. The DPE runs one or many service-containers that are designed for logical groupings of services. Service-containers and services may also be distributed across multiple computing nodes for the purposes of scaling horizontally to handle increased data volume. CLARA application orchestrator can start service-containers in a specified DPE, and monitor and track functionality of services by subscribing to specific events from a service-container, reporting the number of requests to a specific service, as well as notifying when a successful execution of a particular service (or its failure) has occurred. Each service-container deploys one or many services, presenting their computing algorithms encapsulated in service-engines. CLARA service implements SOA SaaS as a way of delivering on-demand, ready-made data processing solutions (service-engines) as CLARA services.

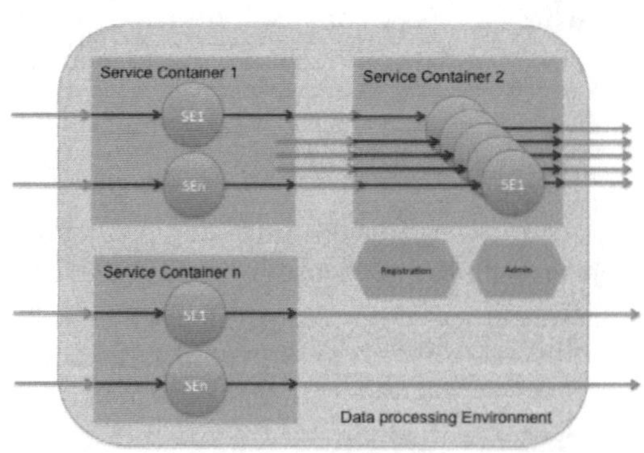

Figure 3. CLARA basic components: services (SE1, SEn) and data streams (arrows).

Journal of Physics: Conference Series **762** (2016) 012009 doi:10.1088/1742-6596/762/1/012009

The CLARA data processing application user uses a *service*, but does not control the operating system, hardware or network infrastructure on which they are running. The quality of the data-processing application (including syntactic, semantic qualities, and performance) depends highly on the quality of constituent service, delegating a request to its engine. It is, therefore, critical to test and validate a *service-engine* before deploying it as a CLARA service. *Service-engines* must be validated with respect to workflow, thread-safety, integrity, reliability, scalability, availability, accuracy, testability and portability. The DPE also contains a proxy for service connections, eliminating direct connections between services. The use of the proxy solves service dynamic discovery problems. The DPE proxy acts like a simple stateless message switch (much like an HTTP proxy). Each *service-container* running in a DPE contains the map of deployed *service* objects. Every *service* object subscribes to service requests and processes the request by getting the *service-engine* object from the object pool, and executing the user provided engine in a separate thread.

2.4. Resource Discovery
The core of the CLARA registration and discovery mechanism is the normative registration service that the services and service containers are registering with. The registration service, which is started within the DPE, functions as a naming and directory facility to uncover services. Services and service-containers in the CLARA registry are described using unique names, types and descriptions of their functionalities. The service is advertised by its service description in the registry. Querying the name and a service functional description defines the service discovery process. Registration database is duplicated in the front-end DPE (FE), controlled by the master registration service. Each registration service periodically pushes its entire database to the FE.

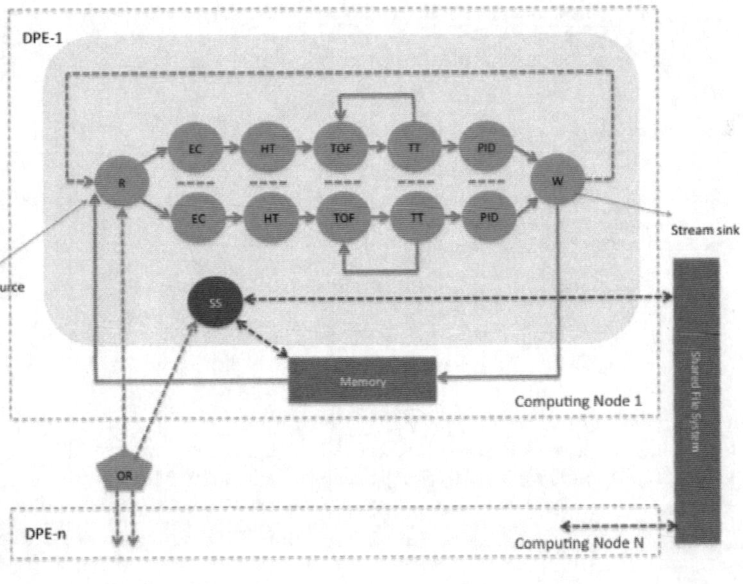

Figure 4. CLAS12 data processing application. Services: stream builder (R), stream writer (W), electromagnetic-calorimeter reconstruction (EC), hit based tracking (HT), time of flight reconstruction (TOF), time based tracking (TT), particle identification (PID), staging (SS), application orchestrator (OR).

3. CLAS12 data processing application

The CLAS12 data processing cover all stages of the physics data processing, including physics and detector simulations, high-level software triggers, reconstruction, analysis and visualization. CLARA based CLAS12 data processing application is a composite application (see Figure 4), that requires assembly and orchestration. Assembling is the process of combining different detector component reconstruction services into a workable unit. The orchestration consists of insuring that all the assembled services work together collaboratively to solve a given problem within a specific scenario. CLAS12 application is designed to scale vertically and horizontally. Vertical scaling assumes multithreading within a single node (desktop operation mode), having stream builder (R) and stream writer (W) services as the only serial components in the system.

3.1. CLAS12 application elasticity

CLAS12 reconstruction application was first deployed in a single 32-core node. The simulated data set was selected to contain events having at least one charged track in the CLAS12 detector. The measurements were conducted reconstructing a single file one event at a time through 32 events at a time in the single CLARA container. The result of this measurement is illustrated in the Figure 5, showing a linear dependence between the average processing time and the number of cores. Figure 5 also shows the performance of the CLAS12 reconstruction on a 16 node Intel Xeon 2x6 2.67GHz CPU cluster, with multiple DPEs deployed to process a set of 64 files using all cores in the nodes. Here we also obtained a linear scaling of the tracking performance.

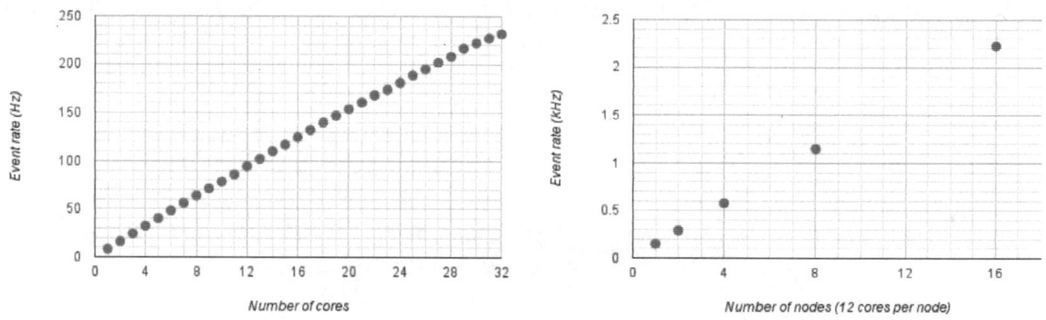

Figure 5. CLAS application scalability.

References

[1] Erl T 2007 *SOA: Principles of Service Design* (Prentice Hall)
[2] Gyurjyan V xMsg: General Purpose Publish-Subscribe Messaging System (unpublished)
[3] Hintjens P 2013 *ZeroMQ: Messaging for Many Applications* (O'Reilly Media)

Cluster Optimisation using Cgroups at a Tier-2

G Qin, G Roy, D Crooks, S C Skipsey, G P Stewart and D Britton
School of Physics and Astronomy, Kelvin Building, University of Glasgow, GLASGOW, G12 8QQ, United Kingdom
E-mail: gang.qin@glasgow.ac.uk, gareth.roy@glasgow.ac.uk

Abstract. The Linux kernel feature Control Groups (cgroups) has been used to gather metrics on the resource usage of single and eight-core ATLAS workloads. It has been used to study the effects on performance of a reduction in the amount of physical memory. The results were used to optimise cluster performance, and consequently increase cluster throughput by up to 10%.

1. Introduction

Beginning in 2015, the ATLAS experiment started to deliver eight-core production workloads to the sites of the WLCG. The purpose of moving to eight-core production was to significantly reduce the overall memory usage of production payloads, and to remain within the '1-CPU to 2 GB of physical RAM' regime. After the introduction of eight-core payloads, UKI-SCOTGRID-GLASGOW saw an overall reduction in site utilisation, caused by CPU cores being unused when the ideal mix of single and eight-core jobs was not available. In an attempt to maximise site throughput, a study was undertaken comparing the memory allocation requested by ATLAS eight-core payloads to the amount of memory actually used by such jobs.

Accurately measuring resource usage required a technology that would not double-count shared memory, which had been previously identified as an issue by ATLAS[1]. Control Groups (cgroups) [2] is a Linux kernel feature that supports resource allocation among user-defined groups or processes. Importantly, cgroups also accurately reports the memory usage of running processes in a variety of different ways, which allows accurate usage numbers to be extracted.

To gain a deeper understanding of cgroups and to better understand the performance of ATLAS eight-core simulation payloads, a study was carried out which capped the available physical memory of jobs. This allowed the performance and behaviour of such jobs in memory-starved environments to be examined.

Subsequently, cgroups were enabled on the entire production HTCondor (hereafter referred to as 'Condor') cluster in late 2014. Using the existing cgroups hierarchy put in place by Condor, a system was created to collect and aggregate metrics extracted from all running batch jobs. From data collected between January and December 2015, memory and CPU usage footprints were extracted and the type of ATLAS[3] workload was identified. This data has been used to optimise the amount of memory allocated to ATLAS workloads (initially ATLAS single-core production workloads) in order to increase cluster utilisation by up to 10%.

Journal of Physics: Conference Series **762** (2016) 012010 doi:10.1088/1742-6596/762/1/012010

2. Performance in a Memory-constrained Environment

A series of tests were carried out to study how ATLAS eight-core simulation workloads could be affected by constraining the amount of available physical memory. Tests were carried out on a dual Intel Xeon X5650 system with 12 physical cores (or 24 Hyper-Threaded cores), 24 GB of DDR3 SDRAM and 24 GB of available swap space on disk. The test platform was configured as a Condor worker node with a cgroup hard limit which was varied between 2600 MB and 1000 MB in steps of 200 MB. Initial results were obtained by running a single ATLAS AthenaMP eight-core simulation workload which would have requested 8 CPU cores and 16 GB of physical memory. For each workload which was run, the resident set size (RSS) memory usage, swap space usage (SWAP), total memory usage (MEMORY, equal to RSS + SWAP) and CPU usage (CPU) was extracted for the lifetime of the job. In this case, the RSS is the actual physical memory usage and similar in concept to Proportional Set Size (PSS)[4] used in batch systems like SLURM.

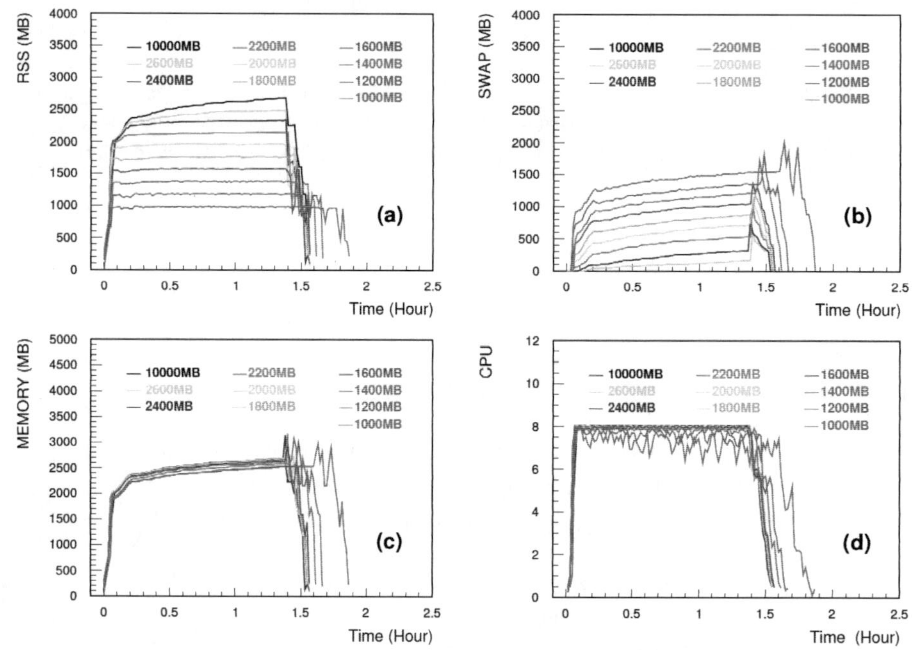

Figure 1. (a) RSS footprints of a single 8-core ATLAS simulation workload; (b) SWAP footprints of these workload; (c) MEMORY footprints of these workload; (d) CPU footprints of these wokload.

Figure 1 illustrates the results obtained from this study. From Figure 1(c) it can be observed that the peak memory usage was approximately 3500 MB even though the requested memory allocation would have been 16 GB, which leads to a 4.5 times under-utilisation of memory. From Figure 1(a) it can be seen that the overall running time of the workload is not greatly impacted until the memory limit is reduced below 1400MB, and Figure 1(d) shows that below this the CPU utilisation is affected for the whole of the job's lifetime due to the need to constantly stage data from the swap partition. These results demonstrate that, for this particular payload, a 16 GB requirement for memory usage under-utilises system resources, and that moderate amount of swapping does not necessarily affect the running time of the simulation.

3. Resource Usage of ATLAS Jobs

Having examined the performance of typical ATLAS workloads on an isolated node, the next step involved investigating how adjustments to the cgroups configuration would affect the performance of our cluster as a whole. This entailed the development of tools to extract various pieces of useful information from both the local cluster and from remote ATLAS servers. Parameters for approximately 2.5 million jobs such as MaximumMemory, MaximumCpuTime, AverageMemory, and Runtime were collect during 2015.

3.1. ATLAS Eight-core Simulation

During 2015, details of about 55,000 successfully-completed ATLAS eight-core simulation jobs were collected. Figure 2(a) shows the distribution of MaximumMemory with a maximum value of 11 GB and a mean value of 5.66 GB, along with the distribution of AverageMemory which has a maximum value of 10 GB and a mean of 4.81 GB. The overall AverageCpu usage is illustrated in Figure 2(b), with the corresponding normalised efficiency being 94%. Figure 2(c) shows that the mean Runtime of these jobs is 4.3 hours. Runtimes vary up to 48 hours, but only 1.65% run for longer than 24 hours.

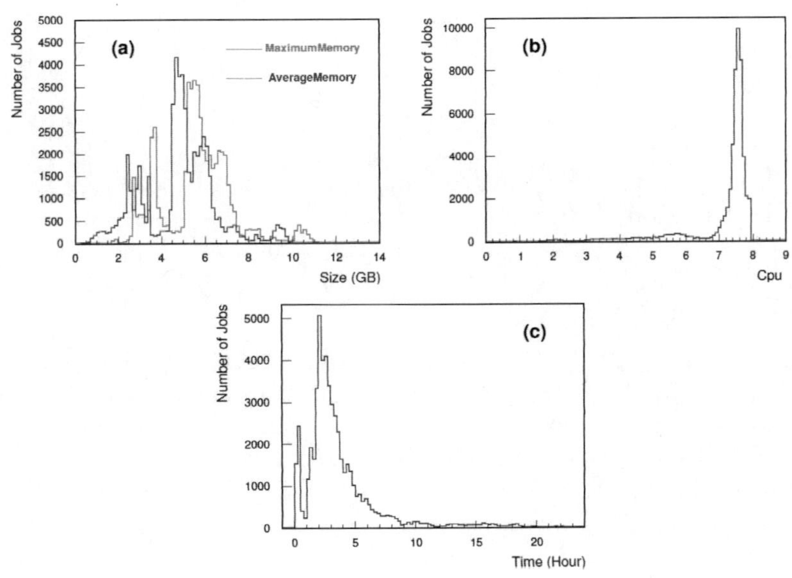

Figure 2. (a) Distribution of MaximumMemory & AverageMemory; (b) Distribution of AverageCpu; (c) Distribution of Runtime.

3.2. ATLAS Eight-core Reconstruction Job

In the same period, details of about 29,000 successfully-completed ATLAS eight-core reconstruction jobs were collected. Figure 3(a) shows that the mean MaximumMemory value is about 11 GB with a maximum of 19 GB, while AverageMemory shows a mean of 6 GB and a maximum of 14 GB. The overall AverageCpu usage is illustrated in Figure 3(b) with a calculated normalised CPU efficiency of 61%. Figure 3(c) shows the Runtime of these jobs, with a mean value of ~2.8 hours. In a very small proportion of jobs, a maximum Runtime of up to 48 hours was observed, but only 0.3% of jobs run for longer than 24 hours, with a peak Runtime centred at ~1.5 hours. At UKI-SCOTGRID-GLASGOW, all production queues have a 48-hour maximum Runtime; given the results from Figure 2 and Figure 3, this could be easily reduced to 24 hours, or even much lower, to allow better scheduling of resources.

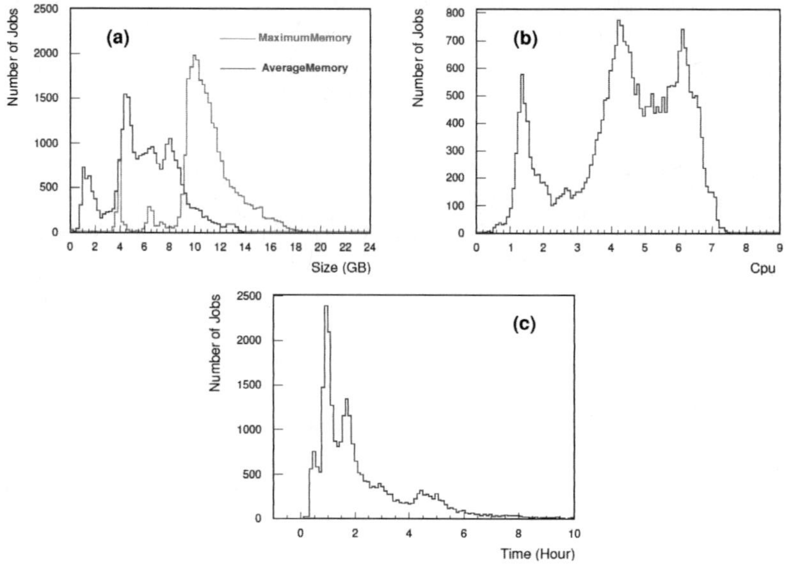

Figure 3. (a) Distribution of MaximumMemory & AverageMemory; (b) Distribution of AverageCpu; (c) Distribution of Runtime.

3.3. ATLAS Single-core Production

During 2015, details of about 294,000 successfully-completed ATLAS single-core production jobs were collected. Figure 4(a) shows that the mean MaximumMemory value is ~0.98 GB with a maximum of ~2.4 GB, while AverageMemory shows a mean of 0.83 GB and a maximum of ~2.3 GB. AverageCpu usage has a mean value of 0.96, as illustrated in Figure 4(b). Figure 4(c) shows the Runtime of these jobs, with a mean value of ~5.63 hours; about half finished within two hours, while 3.5% have a Runtime in excess of one day.

3.4. ATLAS Single-core Analysis

During 2015, details of about 463,000 successfully-completed ATLAS analysis single-core jobs were collected. Figure 5(a) shows that the mean MaximumMemory value is ~0.41 GB with a maximum of ~1.5 GB, while AverageMemory shows a mean of 0.32 GB and a maximum of ~1.5 GB. The overall AverageCpu usage is 0.79, as illustrated in Figure 5(b). Figure 5(c) shows that approximately 90% of jobs finished within five hours, and only 1.7% of jobs ran for longer than one day.

4. Evaluation of Memory Utilisation

Table 1 summarises the results obtained from the above studies. From Table 1 it can be seen that the amount of memory requested for ATLAS 8-core reconstruction workloads correctly estimates their requirement. However, for all other ATLAS workloads, the requested memory over-estimates the requirement significantly, and in the case of single-core analysis does so by a factor of two. In the case of analysis, this over-estimation is a function of the payloads executed at UKI-SCOTGRID-GLASGOW and may not be indicative of all other sites.

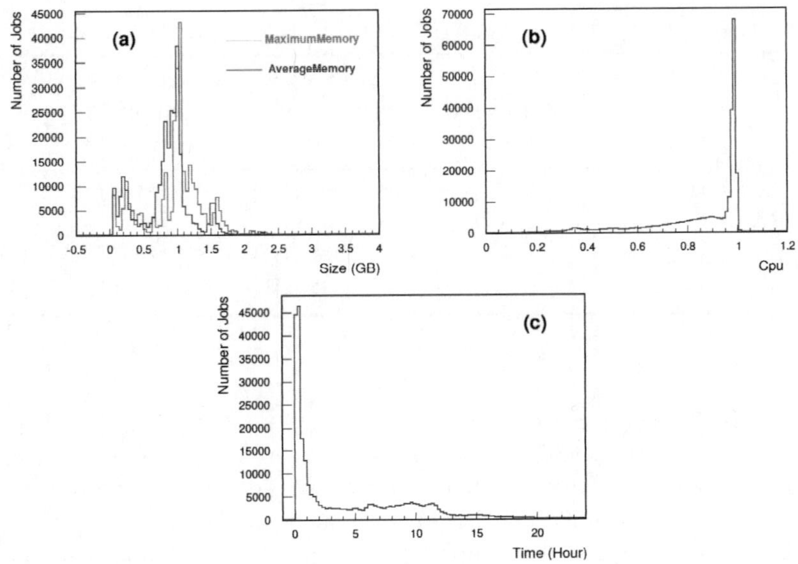

Figure 4. (a) Distribution of MaximumMemory & AverageMemory; (b) Distribution of AverageCpu; (c) Distribution of Runtime.

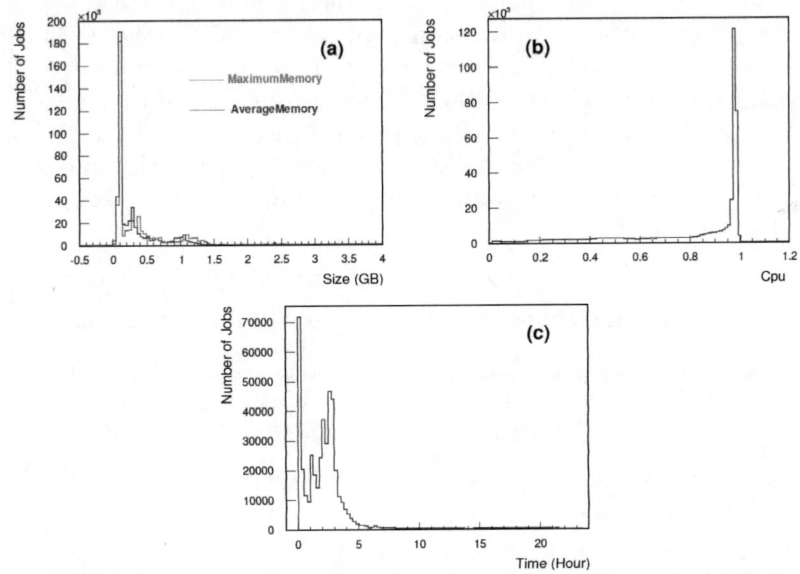

Figure 5. (a) Distribution of MaximumMemory and AverageMemory; (b) Distribution of AverageCpu; (c) Distribution of Runtime.

There are two changes that ATLAS could make to improve utilisation of system resources. Firstly, if they were to distinguish between eight-core simulation and eight-core reconstruction jobs at the queue or pilot level, an improvement in utilisation of up to 31% could be obtained. Secondly, if the memory requested for single-core jobs was reduced to ~2GB, a significant increase in system utilisation could be achieved.

Table 1. Memory requested by various types of ATLAS job compared to memory actually used.

	Job Description	Cores Requested	Memory Requested	Maximum Memory Used	Average Memory Used
1	8-core Simulation	8	16 GB	11.0 GB	5.66 GB
2	8-core Reconstruction	8	16 GB	19.0 GB	11.00 GB
3	1-core Production	1	3 GB	2.4 GB	0.98 GB
4	1-core Analysis	1	4 GB	1.5 GB	0.41 GB

5. Optimisation of Cluster Resources

As discussed in the introduction, with the inclusion of eight-core payloads an overall reduction in site utilisation occurred when the correct mix of single-core to eight-core payloads was unavailable. This was due to older generations of hardware not being easily quantised into units of eight cores. Assuming the data presented on memory utilisation was indicative of all existing ATLAS workloads, it was possible to better utilise local resources by modifying the amount of memory allocated to each job. At first, reducing the memory allocation for single-core jobs to 1 GB of physical memory (based on average usage) allowed all unused cores to be used, resulting in a significant increase in overall cluster utilisation. However, this reduction in memory resulted in jobs being killed due to worker nodes running out of memory and excessively swapping. In order to better utilise resources without causing failed jobs, a value of 1.8 GB per single-core job was selected (Table 2 illustrates suggestions for additional limits for other ATLAS workloads). By doing this, it was observed that the number of used cores increased to ~460 with only ~180 unused cores, representing an improvement in cluster utilisation of ~10%. Cluster operation in this configuration was observed over a period of six months, and it was found that this setting provided the greatest improvement in utilisation without risking instability due to the overloading of machines.

Table 2. Suggested memory allocations for various types of ATLAS job, compared to requested memory.

	Job Description	Cores Requested	Memory Requested	Memory Allocated	Saving
1	8-core Simulation	8	16 GB	12.0 GB	25%
2	8-core Reconstruction	8	16 GB	16.0 GB	0%
3	1-core Production	1	3 GB	1.8 GB	40%
4	1-core Analysis	1	4 GB	1.5 GB	63%

6. Conclusions

In this paper we have introduced a system used to extract memory and CPU footprints from the UKI-SCOTGRID-GLASGOW HTCondor cluster. We have examined the detailed resource usage of single and eight-core ATLAS workloads. Based on the results, the site has adjusted its scheduling policy with the result that an increase in cluster utilisation of approximately 10% has been obtained.

References

[1] Status and Future Evolution of the ATLAS Offline Software, R. Seuster et al., J. Phys.: Conf. Ser., 664(7) 072044, 2015.
[2] Red Hat Enterprise Linux 6 Resource Management Guide, P. Ondrejka et al., 2013.
[3] ATLAS Computing Technical Design Report, 2005.
[4] Formulating Optimized Storage and Memory Space Specifications for Linux Network Embedded Systems, K. Tsiligkos and A. Meliones, HPCC/CSS/ICESS, pp. 580–584., IEEE, 2014.

Data Locality via Coordinated Caching for Distributed Processing

M Fischer, E Kuehn, M Giffels, C Jung
Karlsruhe Institute of Technology, Steinbuch Centre for Computing,
Hermann-von-Helmholtz-Platz 1, 76344 Eggenstein-Leopoldshafen, Germany

E-mail: {max.fischer, eileen.kuehn, manuel.giffels, christopher.jung}@kit.edu

Abstract. To enable data locality, we have developed an approach of adding coordinated caches to existing compute clusters. Since the data stored locally is volatile and selected dynamically, only a fraction of local storage space is required. Our approach allows to freely select the degree at which data locality is provided. It may be used to work in conjunction with large network bandwidths, providing only highly used data to reduce peak loads. Alternatively, local storage may be scaled up to perform data analysis even with low network bandwidth.

To prove the applicability of our approach, we have developed a prototype implementing all required functionality. It integrates seamlessly into batch systems, requiring practically no adjustments by users. We have now been actively using this prototype on a test cluster for HEP analyses. Specifically, it has been integral to our jet energy calibration analyses for CMS during run 2. The system has proven to be easily usable, while providing substantial performance improvements.

Since confirming the applicability for our use case, we have investigated the design in a more general way. Simulations show that many infrastructure setups can benefit from our approach. For example, it may enable us to dynamically provide data locality in opportunistic cloud resources. The experience we have gained from our prototype enables us to realistically assess the feasibility for general production use.

1. Introduction

End user data analysis tasks in HEP are commonly processed by hundreds of jobs on a batch cluster, reading data over network from file servers. As we have shown in earlier work [1, 2], an analysis on a modern institute cluster easily saturates network capacity. With moving simulation jobs to opportunistic resources [3], we expect saturation from analysis jobs to be more frequent. To enable efficient analyses in the future, we therefore investigated data locality as a means to eliminate dependency on network resources.

Data locality approaches reduce overall remote I/O by executing jobs as close as to their input data as possible. Ideally, the machine executing a job and hosting its data are the same. Several frameworks such as Hadoop [4] already provide data locality based processing, and have proven the feasibility of this approach.

However, we have found such frameworks to be inadequate for end user analyses. For example, the extent of software modifications required would effectively eliminate portability to and from other infrastructure. Thus, we have developed an alternate approach to data locality that integrates into regular batch processing.

Our approach uses coordinated caches to provide data locality for a fraction of data. This exploits that at any time, only a few sets of data contribute to overall throughput (see Figure 1). By eliminating remote accesses to them, network capacity remains available for less frequently used data. This is a similar strategy to the current mixture of simulation and analysis tasks.

We have implemented a prototypical middleware [5], targeting the HTCondor batch system [6]. This prototype is deployed on a portion of the local KIT CMS analysis groups' batch system. It has since been successfully used for the CMS Jet Energy Scale calibration analyses performed at KIT.

In this paper, we focus on discussing advantages and disadvantages of our approach and prototype. Section 2 details the features inherent to our approach in general. In section 3, we discuss our current prototype implementation. Finally, Section 4 provides a short conclusion.

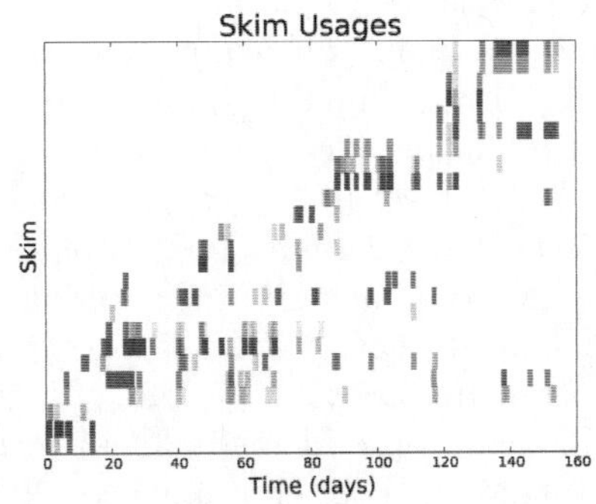

Figure 1: Read accesses from jobs to skim versions: Over time, users create new skims for their analyses. Only some of these are used frequently, however. Often, it takes several intermediate iterations before a skim is replaced.

2. Coordinated Caching in Batch Systems

Individual features of our approach have been subject of past research. Coordinated caching has been shown to be effective in distributed systems, e.g. web services [7]. Applicability of data locality frameworks for HEP has been investigated in abundance [8, 9]. Limited data locality via a middleware has been attempted using cache servers [10]. Other caches for batch processing provide applications used across several jobs [11, 12] Our work is set apart mainly by the scope in terms of size and subjects of caching.

2.1. Scope and Granularity

Our approach is to have a single cache target the batch system as a data consumer. This sets us apart from coordinated caches that target data providers such as web services. The systen itself can be compared to a scaled up operating system page cache. A page cache targets *applications* accessing *blocks* via *read* system calls to process *files*. Our cache targets *workflows* accessing *files* via *jobs* to process *datasets*. Under the hood, the system is composed of several caches, one on each worker node. These are joined together by a coordination service.

The biggest advantage is the scaled up decision layer, selecting files for caching. Even in a small cluster, using a few cores for managing the caches is negligible overhead. Likewise, storing file meta-data in the scale of MB is negligible compared to file sizes at the scale of GB. Since jobs operate on the scale of minutes to hours, the system does not need to respond any faster. This allows for sophisticated caching logic.

The biggest challenge originates from our cache volume being actually several distinct volumes. Coordinating these is not a technical challenge, but a scheduling problem. Since we want to avoid remote accesses, jobs and data must be closely aligned.

2.2. Data Selection and Hit Rates

Using coordinated caches for data locality adds another dimension to data handling. It is not sufficient to have a file anywhere in the cache. Instead, it must be available on the host jobs are trying to access it from. This is the key reason why coordination is required for distributed caching of unique input to be effective. If files and jobs were placed randomly, the chance of a file and corresponding job being on the same worker node is inversely proportional to the number of nodes. When a job requires several files, the expected fraction of files locally available converges to the inverse number of nodes. Even for small clusters, this makes the impact of uncoordinated caching negligible.

In most setups, network throughput still is substantial. To maximize overall throughput, it is best not to read everything from cache, but instead read some data over network. This can easily be achieved by caching only a portion of the data deemed relevant. There are two extremes to this: On one end, caches provide just enough data to not overburden the network. This maximizes cache volume, as only a fraction of each dataset must be provided. On the other end, caches provide as much data as possible, freeing the maximum of network resources. This is optimal if there are many workflows unsuitable for caching, as these can use the network fully.

2.3. User Workflow Integration

Being designed as a cache, our system relies on intercepting access requests. This contrasts with dedicated data locality solutions, which require explicit requests to the middleware.

As we treat the entire job as an access, requests are intercepted at different points. On the one hand, jobs are intercepted as meta-data in the batch system. This provides extensive information, e.g. estimated runtime. On the other hand, the file accesses of jobs executing on worker nodes are intercepted. This implements the actual rerouting to local data.

Intercepting requests has the advantage of being transparent to end users. It does not make a technical difference to jobs whether our system is present. The only notable difference is an increase in performance if files are provided from our cache. This ensures optimal portability.

The downside of a transparent system is that it cannot directly interact with user workflows. For example, data locality frameworks actively set job input to match the distribution of datasets. Our system instead has to optimize data placement to match the splitting already used by jobs.

3. HTDA Prototype

Our approach is prototypically implemented as the High Throughput Data Analysis middleware [5]. At its core is a generic node application, which is deployed on worker nodes and service machines. The HTDA nodes implement all facilities to join together to a single pool.

Each node runs one or several modules which implement the actual services:

- A *Provider* on each worker node, which adds, maintains and removes local copies of files.
- A *Locator* per submission node, which tracks the files available on worker nodes.
- A *Coordinator* per pool, which decides what files to cache and where to do so.

We have deployed our middleware on a portion of the local KIT CMS *HTCondor* cluster. The HTDA section is composed of 4 worker nodes (see Table 1) running *Provider* nodes. A total of 7 file servers mounted via NFS are used.

Table 1: Test Cluster Worker Node

OS		Scientific Linux 6 (Kernel 2.6.32) or CentOS 7 (Kernel 4.4.2)
CPU	2x	Intel Xeon E5-2650v2 @ 2.66 GHz (à 8 cores, 16 threads)
Memory	8x	8GB RAM
SSD	1x	Samsung SSD 840 PRO 512 GB or
	2x	Samsung SSD 840 EVO 256 GB
Network	1x	Intel X540-T1 (10GigE/RJ45)

3.1. Middleware Performance

Being a prototype, we have implemented the middleware in *Python*. This is motivated by the need for rapid development and ease of maintenance. The prototype makes heavy use of abstraction, both between node and module as well as between modules themselves. The implementation is mature enough for stable deployment and operation.

Experiences in terms of system requirements have been unexpectedly good. The only performance critical component, the *Provider* node, has negligible overhead (see Table 2). Its CPU consumption is linear to the frequency of validating files. For our tests, this was once every 5 minutes. Since files are rarely deleted by users before being discarded by our cache, this could be reduced by at least one magnitude.

Table 2: Module Resource Consumption, according to the *ps* utility

module	CPU	RSS
Provider	3.5%	120 MB
Locator	1.0%	60 MB
Coordinator	14.1%	1 GB

3.2. Request Interception

We have implemented the interception of requests via two means: Job meta-data is intercepted via hooks in the *HTCondor job_router* daemon. Application access to data is intercepted on the worker node by a union file system.

Using hooks in the *job_router* has several advantages over the common method of parsing the job queue. Most importantly, we do not have to repeatedly scan the queue for jobs. The selection and tracking is efficiently handled by *HTCondor* itself. The hooks are automatically called on job submission and removal as well as regularly while it is running. This allows for all our services to be event driven.

Since the hooks connect to the pool, communication can easily be optimized. Our hooks are executed often, but skip several updates after updating a job successfully. This naturally leads to spreading out requests if our system cannot service them fast enough. Additionally, hooks can address any end-point of the pool for load balancing. Finally, we can limit how many jobs are connected to our system by *HTCondor* at any time. We can thus handle an arbitrary number of queued jobs.

The downside is that only one type of hook may be active per job. Using hooks, it is not possible to track one job by multiple systems, e.g. our cache and an opportunistic resource provider. This would require creating an intermediate hook calling the services' hooks.

Intercepting read requests via a union file system has proven ideal for performance. We have used Another Union File System in all our setups. It performs the redirection to cache or storage inside the VFS layer of the kernel. Any overhead from this is too small to measure. It is worth noting that this technique is not production ready on Scientific Linux 6. The combination of its kernel and the available *AUFS* 2 may deadlock. However, we have since switched to CentOS 7 and *AUFS* 4, which works flawlessly.

3.3. Compatibility with Volatile Resources

The HTDA middleware is robust against nodes unexpectedly entering and leaving the pool. Any node may keep on functioning on its own. *Provider* nodes maintain existing files, allowing *Locator* and *Coordinator* nodes to work with their last known state for some time. This makes the system intrinsically suitable for deployment on opportunistic resources.

For better adaptability to such resources, the handling of file meta-data and ownership may be improved in the future. At the moment, we assume reading from remote caches has no benefit over reading from the original source. Thus files and their meta-data are owned by the

Provider maintaining them. To allow shared caches accessed from multiple hosts, it would be better to have both owned by the cache device. An attached *Provider* would take ownership only temporarily. If the opportunistic worker node hosting it shuts down, the data may persist on the shared device and a new *Provider* may take ownership.

4. Conclusion

Data locality is an important approach for scalable data analysis. To integrate data locality into HEP workflows, we have created a new approach to transparently enhance batch systems. This is based on a pool of coordinated caches, providing files used by jobs locally on worker nodes.

There are several advantages intrinsic to our approach, which make it suitable for use in HEP. Since we target the batch system as a consumer, our system must only provide frequently used data. An arbitrary number and volume of data servers may provide infrequently used data. The system is by design transparent to users and existing workflows. It can thus be added seamlessly to existing infrastructure without negative side effects.

We have implemented our system as a prototype and successfully used it for CMS run 2 analyses. Being the first of its kind, there are several features that may be improved or expanded in the future. These mainly concern the applicability to other setups, such as opportunistic resources or shared cache volumes. The system itself is mature enough for active use in dedicated batch systems.

Acknowledgments

The authors would like to thank all people and institutions involved in the project Large Scale Data Management and Analysis (LSDMA), as well as the German Helmholtz Association, and the Karlsruhe School of Elementary Particle and Astroparticle Physics (KSETA) for supporting and funding this work.

References

[1] Fischer M, Metzlaff C, Kühn E, Giffels M, Quast G, Jung C and Hauth T 2015 *J. Phys.: Conf. Ser.* **664** 092008
[2] Fischer M, Giffels M, Jung C, Kühn E and Quast G 2015 *J. Phys.: Conf. Ser.* **608** 012018
[3] Giffels M, Hauth T, Polgart F and Quast G 2015 *J. Phys.: Conf. Ser.* **664** 022022
[4] The Apache Software Foundation 2015 Apache hadoop URL `https://hadoop.apache.org`
[5] Fischer M, Metzlaff C and Giffels M 2015 HPDA middleware repository URL `https://bitbucket.org/kitcmscomputing/hpda`
[6] Thain D, Tannenbaum T and Livny M 2005 *Concurr. Comput.: Pract. Exper.* **17** 2–4
[7] Paul S and Fei Z 2001 *Comput. Commun.* **24** 256 – 268
[8] Russo S A, Pinamonti M and Cobal M 2014 *J. Phys.: Conf. Ser.* **513** 032080
[9] Lehrack S, Duckeck G and Ebke J 2014 *J. Phys.: Conf. Ser.* **513** 032054
[10] Yang W, Hanushevsky A B, Mount R P and the Atlas Collaboration 2014 *J Phys.: Conf. Ser.* **513** 042035
[11] Blomer J and Fuhrmann T 2010 A fully decentralized file system cache for the cernvm-fs *2010 Proc. of 19th Int. Conf. Comput. Commun. and Netw. (ICCCN)* pp 1–6
[12] Weitzel D, Bockelman B and Swanson D 2015 Distributed caching using the htcondor cached *Proc. for Conf. Parallel and Distrib. Process. Techn. and Appl.*

Dynamic provisioning of a HEP computing infrastructure on a shared hybrid HPC system

Konrad Meier[2], Georg Fleig[1], Thomas Hauth[1], Michael Janczyk[2], Günter Quast[1], Dirk von Suchodoletz[2], Bernd Wiebelt[2]

[1]Karlsruhe Institute of Technology, Institut für Experimentelle Kernphysik, Wolfgang-Gaede-Str. 1, 76131 Karlsruhe, Germany

[2]Albert-Ludwigs-Universität Freiburg, Professur für Kommunikationssysteme, Hermann-Herder-Str. 10, 79104 Freiburg im Breisgau, Germany

E-mail: `konrad.meier@rz.uni-freiburg.de`

Abstract

Experiments in high-energy physics (HEP) rely on elaborate hardware, software and computing systems to sustain the high data rates necessary to study rare physics processes. The Institut fr Experimentelle Kernphysik (EKP) at KIT is a member of the CMS and Belle II experiments, located at the LHC and the Super-KEKB accelerators, respectively. These detectors share the requirement, that enormous amounts of measurement data must be processed and analyzed and a comparable amount of simulated events is required to compare experimental results with theoretical predictions.

Classical HEP computing centers are dedicated sites which support multiple experiments and have the required software pre-installed. Nowadays, funding agencies encourage research groups to participate in shared HPC cluster models, where scientist from different domains use the same hardware to increase synergies. This shared usage proves to be challenging for HEP groups, due to their specialized software setup which includes a custom OS (often Scientific Linux), libraries and applications.

To overcome this hurdle, the EKP and data center team of the University of Freiburg have developed a system to enable the HEP use case on a shared HPC cluster. To achieve this, an OpenStack-based virtualization layer is installed on top of a bare-metal cluster. While other user groups can run their batch jobs via the Moab workload manager directly on bare-metal, HEP users can request virtual machines with a specialized machine image which contains a dedicated operating system and software stack. In contrast to similar installations, in this hybrid setup, no static partitioning of the cluster into a physical and virtualized segment is required.

As a unique feature, the placement of the virtual machine on the cluster nodes is scheduled by Moab and the job lifetime is coupled to the lifetime of the virtual machine. This allows for a seamless integration with the jobs sent by other user groups and honors the fairshare policies of the cluster. The developed thin integration layer between OpenStack and Moab can be adapted to other batch servers and virtualization systems, making the concept also applicable for other cluster operators.

This contribution will report on the concept and implementation of an OpenStack-virtualized cluster used for HEP workflows. While the full cluster will be installed in spring 2016, a test-bed setup with 800 cores has been used to study the overall system performance and dedicated HEP jobs were run in a virtualized environment over many weeks. Furthermore, the dynamic integration of the virtualized worker nodes, depending on the workload at the institute's computing system, will be described.

1. Introduction

Current and upcoming experiments in High Energy Physics (HEP) require large amounts of processing and storage capacity to handle the recorded data and to provide sufficient simulation and analysis capabilities to their scientists. The Institute of Experimental Particle Physics (EKP) is actively involved in the Compact Muon Solenoid (CMS) experiment at the LHC collider in Geneva (Switzerland) and the Belle II experiment at the Super-KEKB collider in Tsukuba (Japan). The CMS experiment resumed operation, after improvements to the LHC accelerator in 2014 with an increased center-of-mass-energy of $\sqrt{s} = 13\,\mathrm{TeV}$, more concurrent collisions and an increase in factor three of the stored measurement data. These factors make the event reconstruction more challenging, and more computing resources are required to process and analyze the recorded data in a timely fashion. The Belle II particle detector is being built at the moment (02/2016) and is expected to start taking data in 2018. The factor 40 increase in luminosity, compared to the predecessor experiment Belle, also poses a challenge to the available computing resources. In conjunction with the recorded collision data, both experiments require an equal amount of Monte-Carlo simulation events to compare the observations with theoretical models.

The classical model, used with great success during the last decades, is to provide these computing resources via data centers dedicated to HEP computing and running a HEP-specific software stack. Due the independent nature of HEP computing workloads, they can be easily partitioned in many small jobs and workflows can be evenly distributed across different data centers and there is no need for a concurrent execution as no inter-process communication (IPC) is required. Therefore, HEP jobs can be considered to belong to the High-throughput Computing (HTC) class, which describes a loosely coupled computing workflow. Logically, dedicated HEP data centers are tailored to suit the HTC workloads and have batch systems which can manage small jobs efficiently.

In contrast, many other user groups in academia require computing clusters that efficiently serve the High-performance Computing (HPC) class of workloads, which can be characterized as tightly-coupled processing across as significant part of the cluster, which needs low latency interconnection between nodes.

Furthermore, the software environment typically used on HPC clusters is very different from the typical HEP experiment configuration. HPC-focused data-centers use an operating system best suitable to support the installed hardware, for example the IPC components. The cluster operator also provides compilers, software libraries and even whole applications suites for its users. In contrast, most HEP experiments require a specific operating system and centrally provide the experiment software, including compilers and libraries, which have to be used on all sites in order to guarantee the software portability and consistency of the results across all data centers. Furthermore, HEP experiment software is updated in regular intervals and new software releases must be available in a matter of hours on every site. This release mode cannot be properly supported if software must be installed by local experts.

These conditions make it challenging to employ classical HPC-based centers to fulfill the computation needs of HEP experiments. Yet, HPC data centers can provide a significant addition to the computing budget of HEP experiments and their usage is therefore highly desired.

In the following, a new approach to integrating the HEP software setup and workflows, based on virtualization technology, will be outlined and its implementation in a shared cluster system will be described.

2. Hybrid HPC Setup with OpenStack

To allow classic HPC computing (bare metal) and virtual machines on the same cluster the HPC operating modell is extended by a virtualization layer. In order to run virtual machines on a compute node, a virtualization hypervisor is installed on every compute node. Users that require a HPC system for computation can use the middle layer with direct access to hardware to run computation jobs. Users that require a special software environment and do not need direct access to hardware can use the top layer that provides Virtual Research Environments.

To manage virtual machines (network, images, life-cycle) OpenStack [1] is used as a virtualization management framework. OpenStack is chosen for its modular structure. Components required for the cluster virtualization, such as network, image handling or a web interface, can be used while others can be ignored without interference with the overall system. This allows to build a virtualization environment for the cluster that meets the specific requirements.

The challenge in this setup is to interconnect the virtualization environment and the classic HPC environment to allow running classic bare metal compute jobs and VMs on the same cluster at the same time without interference. In order to achieve this, it is necessary to integrate the scheduling of virtual machines into the already existing HPC scheduler. A single resource management is handling resources used by bare metal computation and VMs. The advantage is that, it is not necessary to assign a compute nodes to run bare metal jobs or VMs. Thus a static partitioning of the cluster between virtualization and classic HPC nodes is not required.

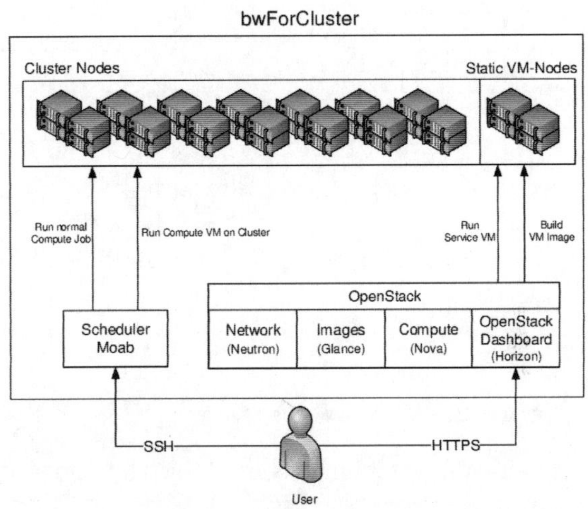

Figure 1. Cluster Structure

The workflow from the user's perspective is given in figure 1. If the user has a prebuild VM image he can directly upload the image to the cluster using the OpenStack Dashboard. If an image is not present, the user is able to use the cluster environment to run a VM on a static VM-Node (management server) in order to install the required software. The image of that VM can then be used as a new image for new VMs running on the cluster.

To run a compute VM on the cluster the user connects to the login server of the cluster and submits a compute job to the scheduler that includes a request for a virtual machine.

2.1. Integration of OpenStack

In a traditional HPC cluster the resource management is done via a scheduler. A user submits a job to the scheduler and the scheduler starts compute jobs on dedicated resources if they are available. Since users in the hybrid cluster are able to submit jobs to the scheduler for bare metal computation, it is required that the scheduler is also responsible for the resources allocated by virtual machines. If virtual machines are started on the cluster without the knowledge of the scheduler, resources could get assigned twice. For CPU resources this would slow down the computation. For memory resources it is even worse if more memory is allocated than physically available and the running jobs are canceled.

To solve the problem it is necessary that the scheduler is aware of the resources that virtual machines are using. A script was developed to submit a request for a virtual machine as a normal cluster job. In this way, the scheduler handles a request for a new virtual machine like any other cluster job and is not required to have knowledge of the virtualization environment. This makes the overall hybrid HPC cluster concept flexible and independent of the actual scheduler used.

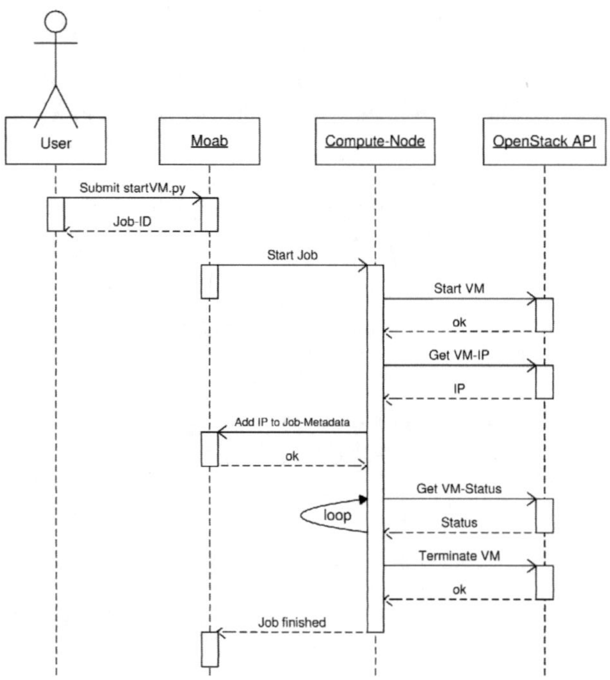

Figure 2. Workflow for starting VMs on the Cluster

The workflow of the script is given in figure 2. The script is submitted by the user as a normal cluster job to the scheduler. By doing so, the user is able to specify a wall time, number of CPU cores and amount of memory. These values are read by the job script and mapped to a matching OpenStack flavor. This way it is possible for the user to define the size of the virtual machine by the time he submits the job to the scheduler and the scheduler is aware of the resources used by the virtual machine.
After submission the scheduler assigns a free compute node to the job and starts the script on this node. The script reads the job metadata (requested wall time, CPU cores, memory) and requests OpenStack to start a virtual machine on the node the script is executed on.

If the job is then executed by the scheduler on the compute node, the script reads the environment variables HOSTNAME, PBS_NP, PBS_JOBID. The host name is required by OpenStack to start the VM on the correct compute node. The job ID is used to generate the VM name. Thus, it is possible to find a compute job for every VM name in the resource manager. PBS_NP corresponds to the requested number of CPUs and is mapped to an OpenStack-Flavor[1]. To start the virtual machine, the following information is passed to the OpenStack API: flavor, generated VM name, compute node hostname and the ID of the image.

Once the VM has started successfully, information of the VM network configuration is polled. If an IP address is successfully assigned to the VM the address is added to the metadata of the compute job. This allows the user to get the IP of a VM by requesting the job status from the scheduler. The IP can be used to login to the VM for monitoring and debugging purposes.

The status of the VM is monitored by the startVM-script. If the status is ACTIVE the script waits 30 seconds before checking the status again. In case of the status ERROR the VM is deleted. The status SHUTDOWN indicates that the VM was shutdown and all computations within the VM are finished. In this case the VM is deleted and the script exits, to signal the scheduler that the resources are free again. The shutdown of the VM can thus be used to signal the system that the VM is no longer needed. If a VM is not shutdown and reaches the maximum wall time, the job is canceled by the scheduler. For this a SIGTERM signal is sent to the job. This single is captured by the startVM script and the VM is deleted. The same procedure is triggered if the job is canceled by the user.

2.2. Testbed Setup

The described hybrid HPC cluster is developed and tested as testbed cluster with 1248 Cores at the computing center at the University of Freiburg. The final cluster, named bwForCluster, with 15000 cores will be installed in spring 2016.

[1] Describes a predefined set of resources of CPU cores, RAM and Storage.

To test the system concept, tests were performed over many weeks. The following section describes the setup and discusses first results.

3. Integration of an HPC Cloud into HEP Workflows

In this section we switch from the provider perspective to the user perspective. The access to the new cloud-based resources should be seamless and hassle-free for HEP users so that no adaptations of the existing workflows are required. Additionally, it should be possible to request resources on-demand in a dynamic manner. To achieve this, a flexible batch system as well as a central cloud manager are necessary.

3.1. A Flexible Batch System: HTCondor

HTCondor is a modular open-source workload management system using a client-server architecture [2]. Two important features make it the batch system of choice for the dynamic integration of cloud resources: It easily handles the addition and removal of worker nodes in the computing pool and it enables the access to resources in a private network or behind a firewall. Both are typical scenarios when using cloud resources for virtualized worker nodes.

3.2. A Cloud Meta-Scheduler: ROCED

A central component is required which dynamically manages virtual machines depending on the demand for computing resources. For this purpose, the cloud meta-scheduler ROCED (**R**eponsive **O**n-demand **C**loud **E**nabled **D**eployment) has been developed at the EKP since 2010 [3]. ROCED is written in a modular structure, the interfaces to batch systems and cloud sites are implemented as modules. This makes ROCED independent of a specific user group or workflow. For the described use case, the HTCondor and a special HPC-cloud module were implemented. The ROCED source code is available on GitHub [4].

3.3. Virtual Machine Management Challenges

Encapsulating VMs in user jobs increases the complexity of managing these machines. The underlying resources are shared with other communities, which means that a VM request might not get processed immediately when the cluster is under heavy load. In addition, the run time of VMs is limited to the maximum wall time allowed for user jobs on the cluster, which is 4 days. As a result, static booking of VMs for several weeks is not possible. These conditions needed to be taken into account when provisioning resources.

3.4. Integration of a Remote HPC Cloud

The following list describes the interaction of all components, illustrated in Figure 3:

(i) HEP user submits computing job to HTCondor, specifying a requirement to send the job to a cloud resource.

(ii) ROCED continuously monitors the demand for computing resources.

(iii) If required, a batch job containing a VM request (presently 4 vCPUs, 8 GB memory, but flexible) is sent to the job scheduler of the bwForCluster.

(iv) The startVM script is scheduled on the cluster and requests an OpenStack-based VM on the allocated worker node.

(v) After booting, the VM integrates in the physics institute's HTCondor batch system and acts as an additional worker node.

(vi) HEP jobs get scheduled on this virtualized HEP worker node. The VM can be reused for multiple user jobs, it shuts down when it is idle for more than 5 minutes. In addition, the job slots are set to drain mode when the underlying batch job is close to the wall time limit.

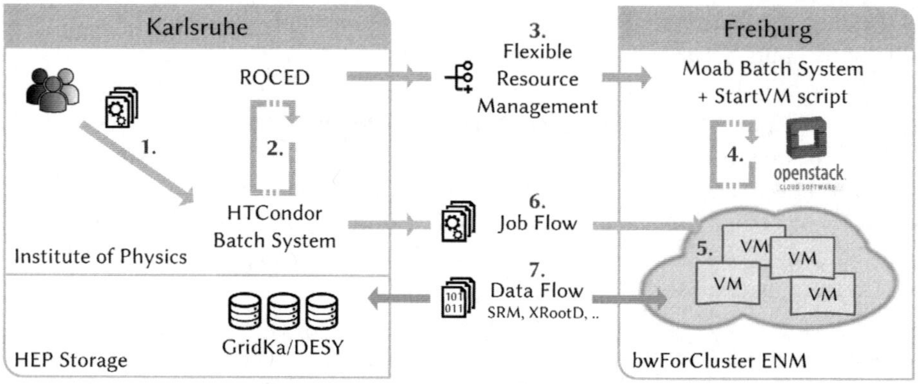

Figure 3. Interplay of tools and technologies for accessing cloud resources at the bwForCluster ENM. Accessing virtualized HEP worker nodes is completely transparent to the end user. No adaptations to the existing workflows of job submission are required.

(vii) Data can be read from and written to any HEP storage via common HEP protocols such as XRootD [5] and SRM [6].

3.5. Building the Virtual Machine Image

The virtual machine disk image containing the HEP experiment software stack is created in an automated manner. This is done by using the OZ toolkit [7] to install the SLC 6.7 operating system [8], as well as CernVM-FS [9] to access Grid UI and CMS software and the HTCondor client to integrate the VM at boot-time. Additional scripts for monitoring, auto-shutdown and auto-drain were implemented. The templates for the images can easily be shared between various cloud sites to ensure the exact same software environment on each site.

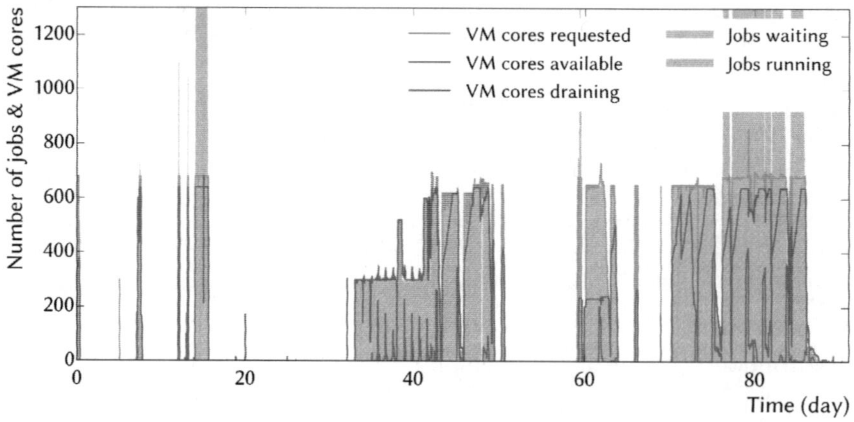

Figure 4. Allocation of virtualized HEP worker nodes at the bwForCluster. Up to 640 CPU cores were used. The discontinuous behavior originates from the many different HEP workflows and submission scenarios. For example testing the workflow with a few jobs and then submitting the whole analysis later or increasing the size of a Monte-Carlo simulated sample by adding more computing jobs after a first batch is finished.

Journal of Physics: Conference Series **762** (2016) 012012 doi:10.1088/1742-6596/762/1/012012

4. Stability and Long-Term Evaluation

The described setup has been successfully in production use for over 5 months: in total 412 000 h of CPU time have been used. Various kinds of HEP workflows such as Monte-Carlo event generation and detector simulation and skimming of datasets were processed. The complete usage overview is shown in Figure 4. The discontinuous usage of resources is typical for HEP users and is one of the reasons why sharing computing power between multiple communities to buffer peak loads is necessary.

5. Conclusion

The hybrid approach presented in this paper combines the advantages of classical HPC data centers, namely complete software environment and batch system, with the flexibility of virtualization to allow groups of scientists, such as HEP users, to deploy a software setup tailored to their needs. The dynamic nature of integration of such a hybrid center into HEP workflows guarantees that only the required resources are booked and freed for other user groups in times of low demand. The booking of virtual machines on a shared cluster is not limited to the used Moab or OpenStack system and can also be applied to other clusters using different software setups.

[1] OpenStack website `https://www.openstack.org` (25/02/2016)
[2] HTCondor website `http://research.cs.wisc.edu/htcondor` (17/02/2016)
[3] Hauth T, Quast G, Kunze M, Bge V, Scheurer A and Baun C 2011 *Journal of Physics: Conference Series* **331** 062034 URL `http://stacks.iop.org/1742-6596/331/i=6/a=062034`
[4] Erli G, Fleig G, Hauth T and Riedel S Roced cloud meta-scheduler project website `https://github.com/roced-scheduler/ROCED` (17/02/2016)
[5] XRootD website `http://xrootd.org` (28/02/2016)
[6] SRM website `https://sdm.lbl.gov/srm-wg/index.html` (28/02/2016)
[7] OZ website `https://github.com/clalancette/oz` (17/02/2016)
[8] ScientificLinux Cern website `https://linux.web.cern.ch/linux/` (25/02/2016)
[9] CVMFS website `http://cernvm.cern.ch/portal/filesystem` (25/02/2016)

Elastic Extension of a CMS Computing Centre Resources on External Clouds

G. Codispoti[1,2], R. Di Maria[1,2,a], C. Aiftimiei[3,4], D. Bonacorsi[1,2], P. Calligola[1], V. Ciaschini[3], A. Costantini[3], S. Dal Pra[3], D. DeGirolamo[3], C. Grandi[1], D. Michelotto[3], M. Panella[3], G. Peco[1], V. Sapunenko[3], M. Sgaravatto[5], S. Taneja[3], and G. Zizzi[3]

[1] INFN Bologna, Bologna, I-40127, Italy
[2] Universit di Bologna, Bologna, I-40127, Italy
[3] CNAF, Bologna, I-40127, Italy
[3] IFIN-HH, Magurele, Romania
[5] INFN Padova, I-35131, Italy
[a] now at Imperial College, London, UK

E-mail: Giuseppe.Codispoti@bo.infn.it

Abstract. After the successful LHC data taking in Run-I and in view of the future runs, the LHC experiments are facing new challenges in the design and operation of the computing facilities. The computing infrastructure for Run-II is dimensioned to cope at most with the average amount of data recorded. The usage peaks, as already observed in Run-I, may however originate large backlogs, thus delaying the completion of the data reconstruction and ultimately the data availability for physics analysis. In order to cope with the production peaks, CMS - along the lines followed by other LHC experiments - is exploring the opportunity to access Cloud resources provided by external partners or commercial providers. Specific use cases have already been explored and successfully exploited during Long Shutdown 1 (LS1) and the first part of Run 2.

In this work we present the proof of concept of the elastic extension of a CMS site, specifically the Bologna Tier-3, on an external OpenStack infrastructure. We focus on the "Cloud Bursting" of a CMS Grid site using a newly designed LSF configuration that allows the dynamic registration of new worker nodes to LSF. In this approach, the dynamically added worker nodes instantiated on the OpenStack infrastructure are transparently accessed by the LHC Grid tools and at the same time they serve as an extension of the farm for the local usage. The amount of resources allocated thus can be elastically modeled to cope up with the needs of CMS experiment and local users. Moreover, a direct access/integration of OpenStack resources to the CMS workload management system is explored.

In this paper we present this approach, we report on the performances of the on-demand allocated resources, and we discuss the lessons learned and the next steps.

1. Introduction

Scientific experiments nowadays require ever-increasing computing resources, and several of them work in a "burst" modality in which there are periods of peak usage where resource usage greatly increases with respect to periods of "normal" usage. Traditional scientific (non-commercial) computing centres (CC) may find it difficult to size themselves so that they may absorb the peak usage of the experiments without generating excessively long queues and therefore compromising usage of the computing centre for the other users. In order to cope with the production peaks, CMS [1] - along the lines followed by other LHC experiments as well - is exploring the opportunity to access Cloud resources provided by external partners or commercial providers. Specific use cases have already been explored and successfully exploited during Long Shutdown 1 (LS1) and the first part of Run 2. In this work we present the proof of concept of the elastic extension of a CMS site, specifically the Bologna Tier-3, on an external OpenStack [2] infrastructure. This extension is performed through three logical steps: virtualization of the local resources and dynamic allocation; dynamic extension of a CMS farm in external Openstack resources; direct access to external Openstack resources.

2. Elastic estension of the Bologna Tier-3

In order to cope with the request peaks, INFN - along the lines of other WLCG [3] communities - is exploring the opportunity to access Cloud resources by external partners or commercial providers. In CMS we realised a prototype to extendi an existing site's batch system to external resources. Specifically we focus on the "Cloud Bursting" of a CMS Grid site using a newly designed LSF configuration that allows the dynamic registration of new worker nodes (WNs) to LSF. In this approach, the dynamically added WNs instantiated on Cloud infrastructure are transparently accessed by the LHC Grid tools and at the same time they serve as an extension of the farm for the local usage. The amount of resources allocated thus can be elastically modelled to cope up with the needs of CMS experiment and local users.

The first step performed is the virtualization of the CMS resources of a Tier-3 [4] centre serving both as a CMS Grid site and local Farm. As a general rule, we are keeping the images as lightweight as possible by using remote services instead of local installation. For this reason we are relying on LDAP [5] and Grid Pool Account services for the users mapping; experiment software access through CVMFS [6]; central authentication systems (ARGUS); and LSF access through remote mounting and central configuration. In order to enable the Grid site usage, we provide a custom installation of the GLEXEC [7] package. In order to enable the local farm use case and to be able to perform effective tests, we access the local storage also via nfs export from a dedicated machine. Following these guidelines, we created images for the user interface (UI) and the Worker Nodes (WNs) , and we tested the usability of the system through a workflow belonging to the Top Quark mass measurement analysis.

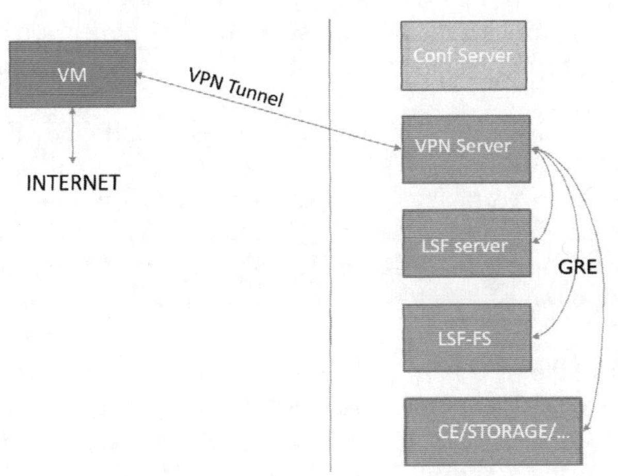

Figure 1. The VPN used for the Tier-3 Bursting: a configuration server provides the VM at boot with all needed info (configs, addresses,...); a VPN server is then contacted by the VM in order to be registered in the private network; a VPN tunnel connects VMs and a VPN server; a GRE tunnel enable the connection among the VMs and necessary services (LSF server, CE, ...). As soon as the configuration is completed, LSF Master sees the new node and start sending jobs.

We used then a dedicated LSF configuration which allows to resize a standard site by bursting out to third parties (Figure 1). This is achieved by instantiating virtual machines (VM) on a remote site and making them part of the local farm. Each VM at startup will contact a Configuration Service in the bursting site: after passing an authorization phase, it receives a set of configuration files and commands allowing the VM to establish a connection with a VPN server inside the CC. The VPN server has established a GRE (generic routing encapsulation) tunnel with the machines that must be visible to the VM to allow it to work and adjust the proper routes. The VPN connection will make the VM and the OpenVPN [8] server visible to each other, but it will have no further effect on the network connectivity of the VM. This will ensure that all and only the necessary traffic will reach the CC, therefore limiting the problems that may be caused by an increased network latency due to the geographical distance from the CC. This mechanism allows a CC to accept workloads greatly superior to those it has been created to accept.

The basic functionalities have been tested by executing the aforementioned analysis workflow using one VM as UI and one as WN, and a dedicated LSF master and queue.

The WN was initially instantiated in the local Tier-3 cluster. Then the test was repeated with the VMs instantiated in the CNAF Cloud infrastructure, an Infrastructure as a Service (IaaS) based on OpenStack Havana.

The first problem hit is related to the access to the local storage. The reduced performances outside of the CC perimeter of the GPFS (IBM general parallel file system) storage accessed through the VPN server affect the whole storage system and represents a real bottleneck that is a show stopper for the extension of the local farm.

The next step was the actual Cloud Bursting of the Grid site, as the natural evolution of this proof of concept and the solution proposed for the production peaks absorption. The images need a very small tuning (adding cloud-aware packages, removing local storage access) to properly run in Openstack. The Bologna site LSF Master has been reconfigured in order to allow the dynamic extension of the nodes. The VM instantiated were included first in a test queue of the master LSF and tested with direct Grid submission. Right after we added the newly instantiated nodes to the production LSF queue used for Grid submission.

We submitted a Top Quark skimming workflow, accessing remote data through xrootd [9] and copying the results with a standard Grid command to the destination storage. The test setup included five virtual nodes statically allocated with QuadCore CPU with 4 cores and 8 GB RAM.

Since the virtual nodes were inserted in a production queue, the jobs were distributed between standard WN and virtual nodes, with only about 5% to the latter. Over more than 3000 jobs submitted, 172 jobs ran in the virtual nodes. Since the jobs were submitted in bunches, they hit all the variability of a production systems, encountering the concurrency of other users jobs, variation of the load.

We measured the job efficiency as a ratio between the cpu time over the cpu wall clock time for both the jobs running on the Openstack nodes and the jobs running on the normal WNs. The distribution of the efficiencies does not show a peak, as it can be observed in Figure 2, given all the fluctuation affecting a productions system. Nevertheless, a relevant part of the jobs reached more than 95% efficiency while almost their totality has an efficiency greated than 80%. No job failures on the virtual nodes were observed.

3. The Bologna Tier-3 as a pure Cloud site

Finally we tested the possibility to allocate an on-demand, brand new, CMS site as a service in Openstack, decoupled from the existing Grid site. The goal was to access this new site, created in the Openstack infrastructure using the standard CMS Workload Management tools (full description can be found here [10]) as described in Figure 3.

The main difference with the already described prototype is the complete absence of a dedicated batch system and of the Grid infrastructure logical parts such as the Computing

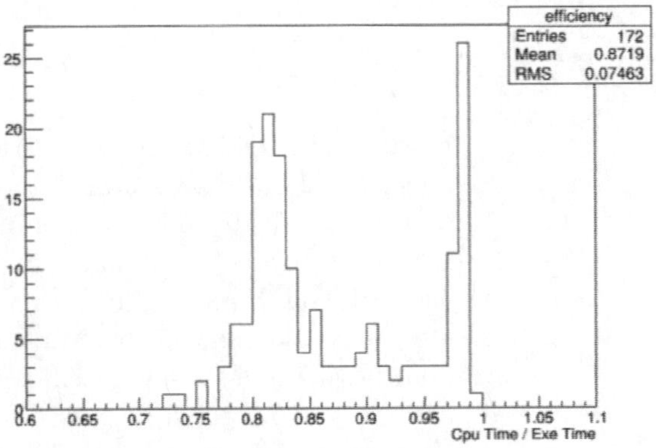

Figure 2. Efficiency of the jobs run over the virtual nodes instantiated as extension of the Tier-3 in Openstack. The efficiency is high even if not clearly concentrated in a peak, but more subject to the behaviour of a production system.

Elements (CE).

Instead of using the standard Grid CE to access the resources, we used directly OpenStack through the EC2 (Amazon Elastic Compute Cloud [11]) interface of the OpenStack-Havana infrastructure. EC2 allows scalable deployment of applications by providing a Web service through which GlideIn-WMS can boot an AMI-type (Amazon Machine Image) image to create a virtual machine. The GlideIn-WMS can thus elastically instantiate virtual machines running the HTCondor [12] startd processes that are able to fetch users' jobs. Moreover, the GlideIn-WMS can create, launch, and terminate these VMs as needed. The virtual nodes image can be thus even shrunk in terms of required packages.

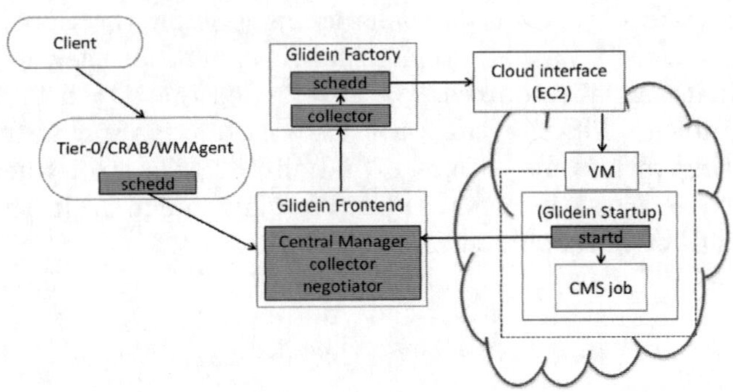

Figure 3. Direct access of Cloud resources: the standard Grid Computing Element is replaced by the OpenStack EC2 interface, the nodes instantiation is delegated to the infrastructure (e.g. Openstack).

A first attempt was performed using the old CMS analysis submission tool, Crab2 [13], a private test instance of the GlideInWMS, and relying on an Openstack Havana infrastructure. Once the prototype was tested, we required access to the CERN CMS Integration Test Bed (ITB) GlideInWMS and for the first time used Crab3 [14], the newly redesigned tool for CMS jobs submission. Given the reduced availability of the ITB GlideInWMS, shared with all the other CMS-wide projects, we limited the number of jobs submitted to 800, executing the very same Top Quark skimming workflow used for the bursting case. We did not observe any failure. Being the jobs running in a controlled and protected environment, they all run with an average efficiency of 98% (Figure 4). As a further exercise, we proved that the infrastructure can also accept more memory demanding jobs, using reconstruction workflow in the context of the Super-LHC upgrade simulation, just increasing from 8 GB to 12 GB the ram allocated for the virtual machines.

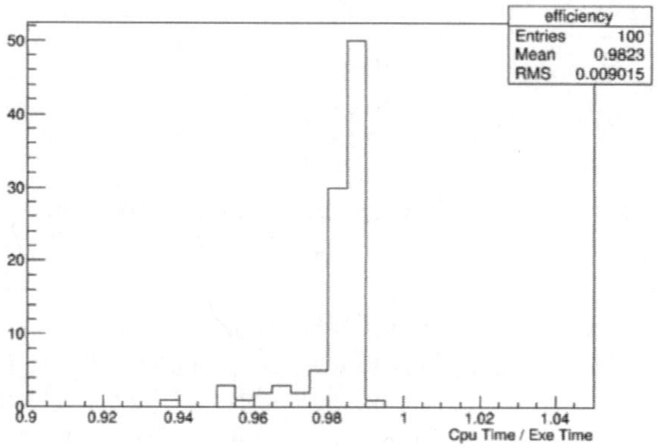

Figure 4. Efficiency of the jobs run over the Tier-3 instantiated as a pure Cloud site. The environment is more protected than in he case of the bursting and the job efficiency is peaked around 98%, demontrating a very high efficiency of jobs and no sensible loss due to the virtualization.

4. Conclusions

The possibility to dynamically extend the computing resources is already crucial for the high energy physics experiments and is becoming more and more appealing for many eScience fields. While big experiments may require to acquire external resources to absorb production peaks, other fields may totally rely on opportunistic resources, specifically provided through Cloud systems, to cope with occasional needs instead of maintaining proprietary farms. In this paper we discussed several possible approaches as we implemented and tested in the Bologna Computing Centre. In conclusion we proved that a CMS Computing Centre using LSF as a local batch system can be efficiently and dynamically expanded on external Cloud-aware resources without a serious loss of performance. The resources dynamically allocated can thus be used for the normal operation of the site. The Bologna Tier-3 has been a realistic use case for the CNAF OpenStack infrastructure. The LSF extension served as development environment for the tools and setup used for Tier-1 extension over commercial resources. We also proved that Openstack can host an on-demand, brand new, decoupled CMS site "as a service". After this successful experience, the Bologna Tier-3 is evaluating to become a pure Cloud site in order to reduce maintenance costs and profit from the CNAF Tier-1 infrastructure. We plan to extend in the next future the CMS specific case to a multi-VO environment.

References

[1] The CMS Collaboration, *The CMS experiment at the CERN LHC*, JINST 3 (2008) S08004.
[2] OpenStack, http://www.openstack.org/
[3] J. D. Shiers, *The Worldwide LHC Computing Grid*, Comp. Phys. Comm. 177 (2007) 219-223.
[4] D. Bonacorsi, *WLCG Service Challenges and Tiered architecture in the LHC era*, IFAE, Pavia, April 2006.
[5] C. Donley, *LDAP Programming, Management, and Integration*, Manning Pubs (2002), ISBN 1-930110-40-5.
[6] P. Buncic, C. Aguado Sanchez, J. Blomer, L. Franco, A. Harutyunian, P. Mato, Y. Yao, *CernVM - a virtual software appliance for LHC applications*, J. Phys.: Conf. Ser. 219 (2010) 042003.
[7] D. Groep, O. Koeroo, G. Venekamp, *gLExec: gluing grid computing to the Unix world*, J. Phys.: Conf. Ser. 119 (2008) 062032.
[8] OpenVPN software, http://openvpn.net/
[9] J. Andreeva et al, *Monitoring of large-scale federated data storage: XRootD and beyond*, J. Phys.: Conf. Ser. 513 (2014) 032004.
[10] M. Cinquilli, D. Evans, S. Foulkes, D. Hufnagel, M. Mascheroni, M. Norman, Z. Maxa, A. Melo, S. Metson, H. Riahi, S. Ryu, D. Spiga, E. Vaandering, S. Wakefield, R. Wilkinson, *The CMS workload management system*, J. Phys.: Conf. Ser. 396 (2012) 032113.
[11] Amazon Elastic Compute Cloud, http://aws.amazon.com/ec2/
[12] HTCondor, http://research.cs.wisc.edu/htcondor/
[13] Giuseppe Codispoti et al., *CRAB: A CMS Application for Distributed Analysis*, IEEE Trans. Nucl. Sci. 56 (2009) 2850-2858. doi:10.1109/TNS.2009.2028076.
[14] M. Mascheroni et al, *CMS distributed data analysis with CRAB3*, J. Phys.: Conf. Ser. 664 (2015) 062038.

Electromagnetic Physics Models for Parallel Computing Architectures

G Amadio[1], A Ananya[2], J Apostolakis[2], A Aurora[2], M Bandieramonte[2], A Bhattacharyya[3] C Bianchini[1 6], R Brun[2], P Canal[4], F Carminati[2], L Duhem[5], D Elvira[4], A Gheata[2], M Gheata[2 7], I Goulas[2], R Iope[1], S Y Jun[4], G Lima[4], A Mohanty[3], T Nikitina[2], M Novak[2], W Pokorski[2], A Ribon[2], R Seghal[3] O Shadura[2], S Vallecorsa[2], S Wenzel[2], and Y Zhang[2]

[1]Parallel Computing Center at São Paulo State University (UNESP), São Paulo, Brazil
[2]CERN, EP Department, Geneva, Switzerland
[3]Bhabha Atomic Research Centre (BARC), Mumbai, India
[4]Fermilab, MS234, P.O. Box 500, Batavia, IL, 60510, USA
[5]Intel Corporation, Santa Clara, CA, 95052, USA
[6]Mackenzie Presbyterian University, São Paulo, Brazil
[7]Institute of Space Sciences, Bucharest-Magurele, Romania

E-mail: syjun@fnal.gov

Abstract. The recent emergence of hardware architectures characterized by many-core or accelerated processors has opened new opportunities for concurrent programming models taking advantage of both SIMD and SIMT architectures. GeantV, a next generation detector simulation, has been designed to exploit both the vector capability of mainstream CPUs and multi-threading capabilities of coprocessors including NVidia GPUs and Intel Xeon Phi. The characteristics of these architectures are very different in terms of the vectorization depth and type of parallelization needed to achieve optimal performance. In this paper we describe implementation of electromagnetic physics models developed for parallel computing architectures as a part of the GeantV project. Results of preliminary performance evaluation and physics validation are presented as well.

1. Introduction

The latest evolution of parallel computing architectures and software technologies makes a variety of concurrent programming models applicable not only to high performance computing in advanced scientific computing research, but to the conventional computing intensive application such as the high energy physics (HEP) event simulation. Faced with ever-increasing demands of computing power required for present and future experimental (HEP) programs, there have been coherent efforts to promote efficient use of common resources and solutions to tackle next generation hardware [1] and to advance HEP community software [2]. Within HEP software ecosystem, event simulation is one of the most time consuming parts of the work flow, but a large portion of code used for HEP detector simulation is independent from the details of individual experiments and can be shared substantially. As one of domain specific research and development activities, the GeantV project [3] (GeantV) launched in 2013 studies performance gains from propagating multiple tracks from multiple events in parallel, improving instruction throughput and data locality in the particle transportation process. Using code specialized to take advantage of the hardware specific features, it aims to leverage both the vector pipelines in modern processors and the massively parallel capability of coprocessors, including the Xeon Phi and general purpose GPUs. GeantV has three major components that need to be adapted for concurrent programing models: concurrent framework, vectorized geometry, and physics modeling. In this paper we focus on the techniques explored to enhance electromagnetic (EM) physics models used in HEP utilizing emerging parallel hardware architectures.

2. Parallelization of Electromagnetic Physics Models

HEP detector simulation models the passage of particles (tracks) through matter. The typical HEP event consists of a set of particles produced by a primary collision and subsequent secondary interactions or decays. Geant4 [4, 5], the most widely used simulation toolkit in contemporary HEP experiments, processes all tracks of an event sequentially even though multiple events can be processed simultaneously (event-level parallelism) using multi-processors or multi-threading. On the other hand, GeantV explores particle-level parallelism by grouping similar tracks and processing them concurrently with fine-grained parallelism to maximize locality of both data and instructions. To take full advantage of SIMD (Single Instruction Multiple Data) or SIMT (Single Instruction Multiple Threads) architectures, identical instructions without conditional branches should be executed on multiple data simultaneously. Therefore, vectorization of physics models requires algorithms adapted to utilize instruction level parallelism as much as possible.

An essential component of particle interactions with matter is the generation of the final state described by a physics model associated with the selected physics process for a given particle. As electrons and photons are primary components of the particles produced in typical collider detectors, simulation of EM physics processes is one of leading consumers of computing resources during HEP event simulation. In most EM physics models, the atomic differential cross section of the underlying physics process plays a central role in modifying the kinematic state of the primary particle or producing secondary particles. In Geant4, combined composition and rejection methods [6, 7, 8] are often used to sample variables following probability distribution functions used in EM physics models, as inverse functions of their cumulative distributions are not analytically calculable in general. However, composition and rejection methods are not suitable for SIMD computing architectures due to repetitive conditional trials until a random draw is selected. Alternative sampling techniques that can be effectively parallelizable for SIMD and SIMT architectures are the alias method [9, 10] and the shuffling technique similar to the pack pattern used in parallel programing models [11].

The alias sampling method is similar to the acceptance-rejection method, but it uses an alias outcome for the rejected case which is thrown away in the traditional acceptance-rejection method. It recasts the original probability density function with N equally probable events, each with likelihood $c = 1/N$, but keeps information of the original distribution in the alias table that consists of the alias index and the non-alias probability.

Even though the alias method with a finite bin size reproduces an input probability distribution function within a desired precision, it is subject to have biased outcomes if p.d.f. is significantly nonlinear within a bin. One alternative sampling method that still incorporates the combined composition and rejection technique is the shuffling technique. It eliminates accepted trials from a collection and reorganize failed elements in contiguous memory for the next vectorization loop and repeat the process until all retained elements are accepted. The shuffling process will reproduce the original p.d.f. at the cost of the overhead of gathering data in each interaction and can be used when the alias method has a bias.

To simulate kinematical distributions of secondary particles produced by EM physics models, both the alias method and the shuffling method are tested and their results are compared in different energy regions. A list of EM physics processes and models for high energy electrons and photons that are currently implemented is summarized in Table 1.

Table 1. A list of electromagnetic physics processes and models of electron and photon processes that are implemented and tested for SIMD and SIMT architectures.

Primary	Process	Model	Secondaries	Survivor
γ	Compton Scattering	Klein-Nishina	e^-	γ
	Pair-Production	Bethe-Heitler	e^-e^+	–
	Photo-Electric Effect	Sauter-Gavrila	e^-	–
e^-	Ionization	Moller-Bhabha	e^-	e^-
	Bremsstrahlung	Seltzer-Berger	γ	e^-

3. Implementation

Besides performance and scalability, another important objective of GeantV design is portability. To enable the use of multiple modern hardware architectures, GeantV uses backends [14, 15], which are software layers between the platform-independent generic simulation code and the hardware-specific details and their software-related constructs like SIMD intrinsics, MIC pragmas or CUDA `C++` extensions. The main purpose of the backends is to isolate all the complexity of low-level, high performance details behind simplified abstractions which are then available for use by carefully designed, generic kernels. The Vc library [12] is used for the choice of the vector backend for this paper, but UME::SIMD [13] is also being integrated to promote explicit SIMD vectorization.

There are three categories of methods implemented in vector physics package: initialization and management, interfaces to physics processes, and kernels. Kernels are high-performance versions of performance-critical algorithms and based on the data structures offered by the backends. In order to take full advantage of the performance capabilities of the underlying hardware, some important choices were made:

- Inlined functions are used extensively, to avoid the overhead due to function calls.
- Static polymorphism is used instead of virtual inheritance. Virtual function calls inside the kernels are explicitly avoided since kernels themselves are coded in terms of `C++` templates, with a specific backend as the template parameter.
- Branching of execution flow is strictly minimized and mask operations are used whenever necessary.
- The structure of arrays (SOA) is used for track data
- Gather and scatter operations are used to rearrange queried data into a contiguous memory segment and to store the vector of results back into the original track data, respectively.

4. Physics Validation

One of essential requirements of vectorized EM physics models is to verify the accuracy and precision of simulated physics results. Since implementations of EM physics models are designed to be architecture-independent, they can be executed in the same way for different backends, allowing direct validation of simulation results within statistical uncertainties due to the use of different random number generators - the vector and CUDA backend use random numbers generated by `Vc` and `CURAND` [16], respectively. To test the correctness of the algorithms' implementation, we extended the validation by executing the same operations using the original Geant4 sampling method. Figure 1 shows kinematic distributions (energy and angle) of scattered photons of the Klein-Nishina model obtained by Geant4 algorithm and the vector backend using the alias sampling method. Based on Pearson's chi-squared test [17], kinematical distributions of simulated events using different sampling techniques are statistically validated for each model at selected monoenergetic inputs.

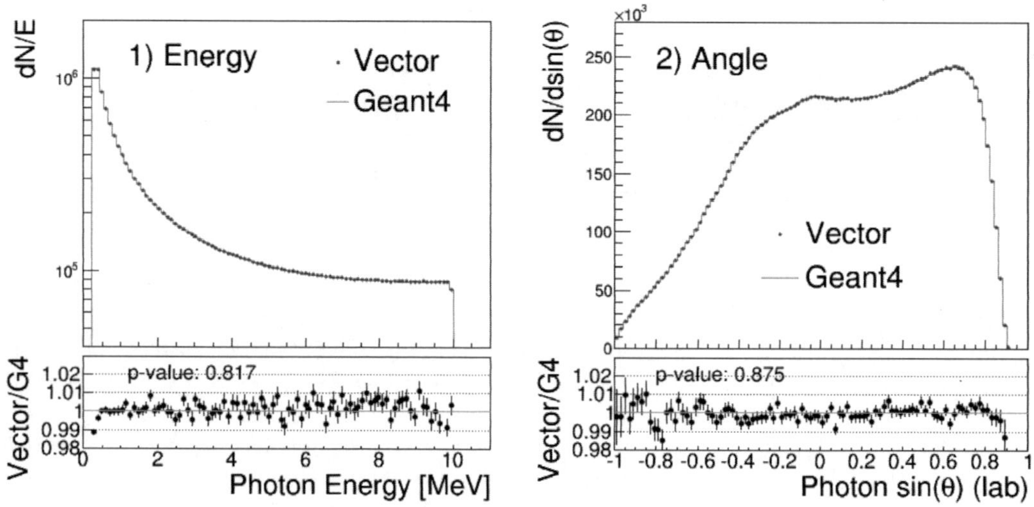

Figure 1. Examples of physics validation with kinematical distributions of simulated events: 1) Energy and 2) Angle of scatter photons obtained by the Klein-Nishina model (Geant4 vs. Vector) and their ratios and p-values of Pearson's chi-squared test with the monoenergetic input photon energy at 10 MeV.

5. Performance

To have efficient parallel code, performance analysis is a critical part of the development cycle. As the primary measure of the performance, we define the relative speedup as the ratio between the time taken by a model with a specific backend (Scalar, Vector, CUDA and etc) and by the Geant4 algorithm to execute the same task. For the purpose of performance measurements, particles are generated with a set of exponentially falling spectra within valid energy ranges for each model. Even though the relative speedup is not an absolute measure of the speedup, because the efficiency of sampling varies as a function of the energy, it can be used as a general guideline for performance comparisons, to identify potential problems, and to tune models optimized for a specific architecture. Figure 2 shows preliminary performance results of vectorized code tested on Intel® Xeon Phi (5110P) and CUDA code on NVidia GPU (Kepler K20M) for simulating interactions and sampling secondary particles using the alias method - the host used for performance evaluate is Intel® Xeon E5-2620 for both Xeon Phi and GPU.

6. Summary

As the GeantV project assembles the various pieces of the infrastructure: framework, geometry and physics, we demonstrated feasibility of implementing high-performance electromagnetic physics models for SIMD/SIMT architectures with common source code. Preliminary performance evaluation shows that parallelized physics code improves simulation speed using

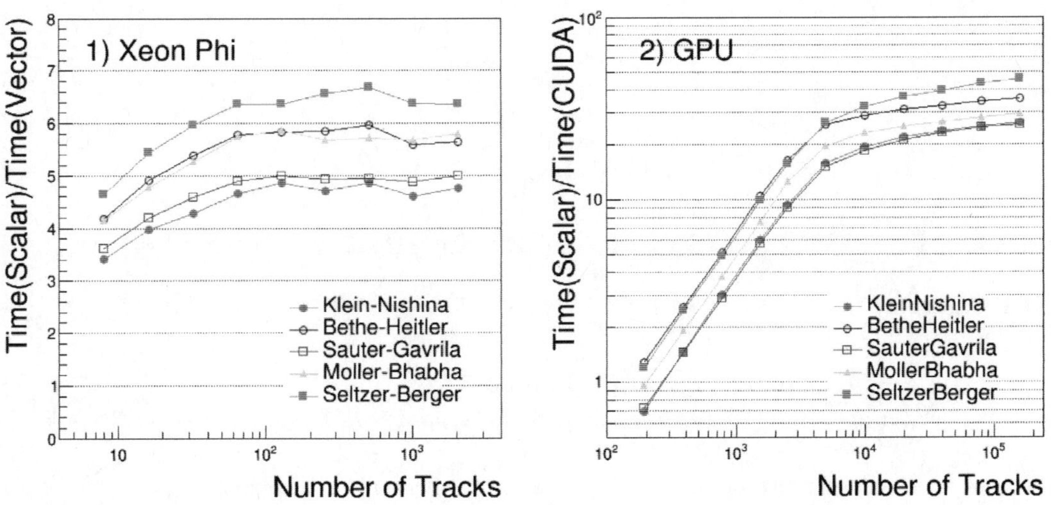

Figure 2. Relative speedup for simulating particles that undergo EM processes using alias sampling method: 1) the ratio between the CPU time taken by the vector code and by the scalar code on Intel Xeon Phi (5110P, one-thread), 2) the ratio between the time taken by the CUDA code on NVidia GPU (K20, 2496-cores) and by the scalar code on Intel Xeon (E5-2620, one-core). The energy range of input particles is [2,20] MeV where all models are valid.

the vector pipeline of SIMD and massively-parallel threads of SIMT. For example, as shown in this paper, vectorizing physics code improves simulation speed by a factor around 4-6 on Xeon Phi and around 20-50 on GPU depending on the size of tracks for different models. As integration of EM vector physics models into the GeantV framework is being carried out, computing performance will be re-evaluated to understand the impact of parallelization and to have insights for optimization.

References

[1] Habib S *et al* 2015 High energy physics forum for computing excellence *arXiv:1510.08545*
[2] The HEP Software Foundation `http://hepsoftwarefoundation.org`
[3] Apostolakis J *et al* 2014 A concurrent vector-based steering framework for particle transport *J. Phys.: Conf. Series* **523** 012004
[4] Agostinelli S *et al* 2003 Geant4 - A Simulation Toolkit *Nucl. Instrum. Methods Phys. Res.* A **506** 250-303
[5] Ahn S *et al* 2014 Geant4-MT: bringing multi-threading into Geant4 production *Joint International Conference on Supercomputing in Nuclear Applications and Monte Carlo 2013 (SNA + MC 2013)* 04213
[6] Butcher J C and Messel H 1960 *Nucl. Phys.* **20** 15
[7] Messel H and Crawford D 1970 Electron-Photon shower distribution, Pergamon Press
[8] Ford R and Nelson W 1985 SLAC-265, UC-32
[9] Walker A J 1977 *ACM Trans. Math. Software.* **3** 253-256
[10] Brown F J, Martin W R, and Calahan D A 1981 *Trans. Am. Nucl. Soc.* **38** 354-355
[11] McCool M, Robinson A D, Reinders J, Structured parallel programing, ISBN-9780124159938, p. 187
[12] Kretz M and Lindenstruth V 2012 *Software: Practice and Experience* **42** 1409-1430
[13] UME::SIMD `https://bitbucket.org/edanor/umesimd`
[14] Wenzel S 2014 Towards a high performance geometry library for particle-detector simulation *16th International workshop on Advanced Computing and Analysis Techniques in physics research (ACAT)*
[15] de Fine Licht J 2014 First experience with portable high-performance geometry code on GPU *GPU Computing in High Energy Physics 2014*
[16] NVIDIA CUDA Toolkit 5.0 CURAND Guide `http://docs.nvidia.com/cuda/pdf/CURAND_Library.pdf`
[17] Pearson K 1900 *Philosophical Magazine Series 5* **50** 157175

eSciMart: Web Platform for Scientific Software Marketplace

A P Kryukov[1] and A P Demichev[2]

Skobeltsyn Institute of Nuclear Physics, Lomonosov Moscow State University, Leninskie gory, 1, Moscow, 119991, Russia

E-mail: [1] kryukov@theory.sinp.msu.ru, [2] demichev@theory.sinp.msu.ru

Abstract. In this paper we suggest a design of a web marketplace where users of scientific application software and databases, presented in the form of web services, as well as their providers will have presence simultaneously. The model, which will be the basis for the web marketplace is close to the customer-to-customer (C2C) model, which has been successfully used, for example, on the auction sites such as eBay (ebay.com). Unlike the classical model of C2C the suggested marketplace focuses on application software in the form of web services, and standardization of API through which application software will be integrated into the web marketplace. A prototype of such a platform, entitled eSciMart, is currently being developed at SINP MSU.

1. Introduction

In many cases researchers need to run a large number of similar computing tasks to solve some problem in a particular application area. In this case, pre-installation and configuration of commonly used application packages on a computing resource and the availability of specialized web services as well as appropriate user web interfaces allow owners of computing resources to increase efficiency of their use (Software as a Service model; SaaS). The set of specialized interconnected web services and web-application interfaces is called a web platform for remote access to computing resources. Development of web platforms for scientific research in different areas are widely underway around the world (see, e.g., [1] and refs therein). Examples of such developments are nanoHUB (nanohub.org), Yabi (ccg.murdoch.edu.au/yabi) and e-Science Central (www.esciencecentral.co.uk). In particular, such a platform could be very useful for Monte Carlo simulation problems with parameters sweep in the field of high energy physics.

There exist many software applications that were created by scientists and engineers. However, most of them are not available for public access for various reasons. A number of web portals, including those mentioned in the preceding paragraph, are intended for providing free software. It is clear that a distribution on such a basis of a software suitable for a full-fledged research, especially for commercial use, is impossible. Among other things, this is the reason that NanoHUB has focused on educational purposes and, accordingly, the relatively simple software that is unsuitable for commercial use. On the other hand, many developers of application software or databases in various branches of knowledge have the intention to provide its product for use by other researchers, but under certain conditions, in particular, of receiving a revenue for the use of the provided product. The problem is that the software developers (individual programmers or small teams) as a matter of rule have no time as well as financial and administrative resources to develop and support market infrastructure in full scope.

Therefore, the basic idea of the work, presented in this paper, is to develop a web platform where users of resources (application software and databases), presented in the form of web services, and their providers be able to directly interact, while the web platform have to perform a range of tasks typical for the free open market. A prototype of such a platform, entitled eSciMart, is currently being developed at SINP MSU.

2. Web-Platform Basic Principles, Architecture and Algorithms

The Internet witnesses the unprecedented boom of e-commerce in the framework of customer-to-customer (C2C) model. This type of e-commerce involves electronically-facilitated transactions between individuals, often through a third party. One common example is online auctions, such as Ebay, where an individual can list an item for sale and other individuals can bid to purchase it. Another example is the site italki.com which provides a free marketplace for teachers and learners of a foreign language (distance learning via Internet, e.g. via Skype; more than 2000000 learners of foreign languages have used this site). C2C is expected to increase in the future because it minimises the costs of using third parties.

Unlike the classical model of C2C the marketplace, suggested in our work, focuses on application software in the form of web services. We shall use the following nomenclature: consumer is a platform user requesting software as a service via the platform, provider is a user providing a software to consumers by means of the platform. There are two ways for offering an application software by means of the web platform. In the first case, the software is deployed on provider's own resources. Thus the provider offers not only a software, but also the computing resources. In this case the provider can use different types of resources, e.g., supercomputers, clusters, standalone servers, cloud systems. Of course, these computing facilities are not necessarily in property of the provider but can be leased resources (e.g., in a public cloud). This way is especially suits those providers who want to exclude completely access both users and platform administrators to the software (both to source code and binary executables). It is worth noting that in the case of provider's own resources a problem of reliability of delivery of services may appear. In the second case, the software is deployed on platform resources. In this case provider offers only a software as a service; a fee for the use of computing resources goes to the owners of the platform resources. The web platform have to provide a service for remote installation of the application software by users (providers) themselves. In this case the type of the resources is practically limited to the cloud systems. Indeed, an independent remote installation of application software on supercomputers by users is highly problematic both in terms of security and from a technical point of view: for a deployment of a software the user must deeply know the features of the architecture and software of the specific supercomputer. A natural solution to these problems is the use of virtual machines [2] that provide both safety (isolation of installed software) and feasibility of using the operating system required for the application software to be installed. Thus, since cloud infrastructure provides tools for managing virtual machines, the web platforms allowing software installation must provide access to the cloud systems in the framework "Platform as a Service" (PaaS) model.

General architecture of the web platform for the marketpalce have the three-layer architecture (see fig. 1).

The first layer is the frontend that provides the user web interface; the second layer consists of the platform engine responsible for job submitting, obtaining results, execution of workfows, etc., the system of remote software deployment, the administration system (responsible for user management, tool configuration, audit trails, etc.) and the system of providers/consumers interactions; the third layer is a resource manager that exposes data and compute resources to

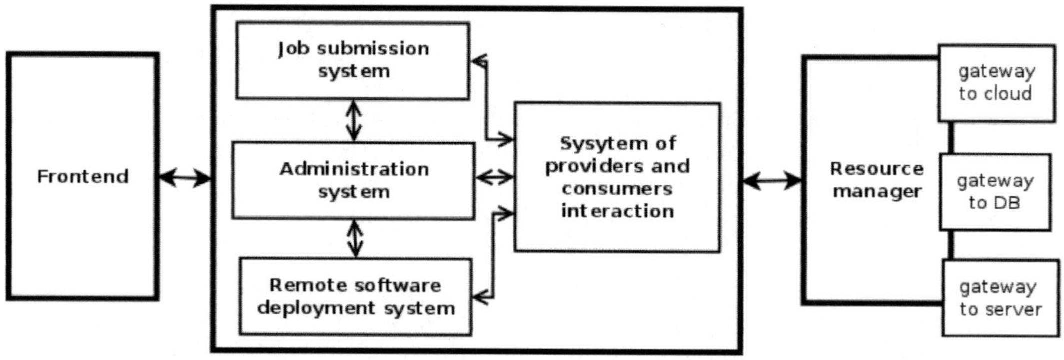

Figure 1. A general architecture of the web marketplace

the preceding layer. Also the web platform has security infrastructure (see below).

Each of the systems of the second layer, in turn, consists of a set of the functional modules. For example, the system of job submission consists of modules of task launching, workflow processing, access to scientific databases, etc. (see, e.g., [1] and refs therein). The most important system from the point of view of the purposes and tasks of the web marketplace is the system of providers/consumers interactions.

The list of modules of this system includes: module of software registration; service catalog with search engine; logging (event tracking) module; billing module; integrator of conditions for workflows; module of verifying compliance with the terms of service; software deployment module; module of a tool setup; module of e-wallets; repository of the software codes; module for review and blog posting and for assigning rating marks.

A standardization of API through which application software is integrated into the web marketplace is realized by means of the process of software-into-tool transformation. An individual prepared job (launch of an application software package, access to a storage, etc.) is called a "tool". After a deployment of a software on a resource, the creation of a associate tool starts from its description in a machine readable form via a special web interface (web forms) including the following information:

- unique name of a tool;
- name of an application package installed;
- parameters of the tool (and their acceptable values);
- formats of input and output files for the tool;
- description of particular requirements for input files;
- control of a user (or user group) access to the tool;

as well as presetting all technical details, in particular,

- web addresses of resources where the software is installed,
- storage to save output files,
- particular formats of database queries, etc.

In fact, the creation of a tool together with the platform functionality convert the associated application software or database into SaaS.

Having access to tools provided by a web platform, the user only specifies particular values of input parameters or files containing input data; the rest of the job description is generated automatically. Thus, the preparation of a job, its submission, execution control, and obtainment of the result are done using a common web browser. In addition to launching application packages, a tool can include other operations, for example, database query, visualization of results, and obtaining of on-line data from a scientific installation. All that is left for users to do is to formulate in the natural language the essence of a particular request. It is especially important that web services and corresponding interfaces allow users to avoid dipping into the intricacies of installing and launching each of these packages, so they can concentrate only on research problems.

Algorithm of the platform operation when a provider is offering his SaaS is depicted in figure 2. This figure corresponds to the case when the provider uses the cloud resources of the platform for software deployment. In the case of provider's own resources, the provider deploys the software on this resources and provide an access to the software via the SSH protocol. If the resource is not a standalone server but an HPC system with a batch manager, the latter should be in the list of managers supported by the platform. It is worth noting that the platform supports workflows, which consist of sequences of tasks (for example, access to databases with subsequent use of results as input data for computing tasks). Thus a provider may not only a single tool but also debugged and ready to execution workflows, constructed of provider's own application software or/and software from third-party providers in the spirit of the BPaaS model [3].

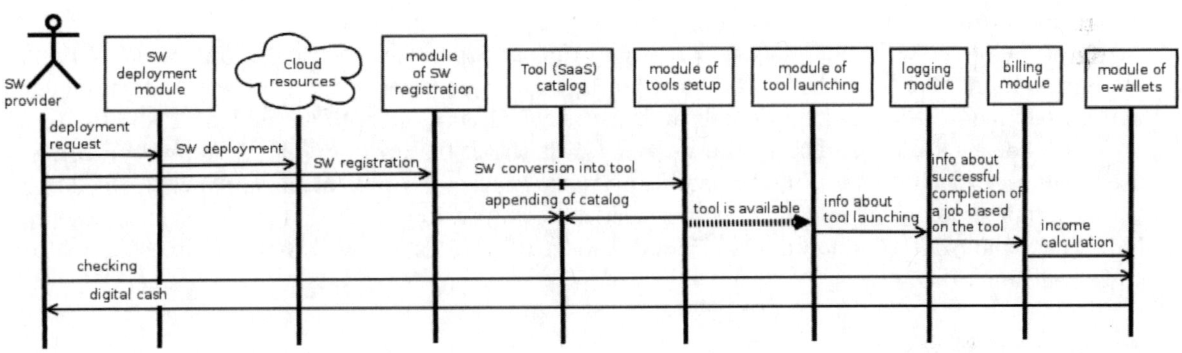

Figure 2. The algorithm of a provider and market modules operation.

Each user of the platform (both a customer and a provider) has the electronic (virtual) wallet (e-wallet) containing the conditional means (digital cash) within a platform. Users can transfer their funds to the e-wallets of other users. In reality users also shall have opportunity to transfer funds from the e-wallets to real payment systems (like PayPal) or to bank accounts, but in this work we do not consider this process.

To use the platform tools, a consumer chooses from the catalog appropriate tools; designs out of these tools a workflow or select predefine business process (BPaaS); gets terms of services and price for their use; accepts or deny the terms of conditions; if accepts, transfer digital money from his e-wallet to the provider's e-wallet (the module of verifying compliance with the terms of services checks that the conditions are accepted and ample means have been transferred to the providers' e-wallets); starts the workflow and obtains a result.

An administrator of the web platform provides user registration, moderation of software installation, user support and Quality of Service (QoS) obligations from the side of providers offering their own computing resources.

The web marketplace together with computer resources where the application software and databases are deployed is a special case of the distributed computing systems (DCS). To provide secure access to the resources of DCS with the account of the rights of a given user as well as of the politics of a given service or resource a DCS security infrastructure is needed. This infrastructure should be sufficiently reliable and, at the same time, user friendly. Within creation of the DCS, the new technique of authentication of users [4] which is based on login/password pair together with a sessional key is developed. Key difference of this method from the most popular approach based on public key infrastructures (PKI) [5] is the non-utilization of proxy certificates for execution of user requests. Another new principle which is suggested to be used for security arrangement is the use of hashes for signing requests in the framework of a DCS. For confirmation of legality of each request in DCS signatures in the form of specially created unique hashes with unlimited validity period are used. The use of the unique hashes allows to solve, on the one hand, a problem of protection of request contents while processing within DCS, and, on the other hand, solves the problems caused by limited period of validity of usual proxy certificates. Registration of the unique hashes allows to provide authentication and authorization of requests.

In general the suggested approach leads to essential simplification both the registration of new users in the system, and their operations in the distributed system. Some lowering of the security caused by the sessional key without password is compensated by restriction of its validity time. After validity expiration of the key, the user requests a new key either via the special web interface, or through API of the appropriate service.

3. Conclusion

A prototype of the web marketplace, presented in this paper and entitled eSciMart, is currently being developed at SINP MSU. The basis of the prototype platform is web service technologies and, in particular, the service-oriented architecture (SOA) [6], as well as the REST architectural style [7]. For the data interchange between services the JSON format (http://www.json.com) is used. The approach to creation of this web-platform market of application software is based both on the original solutions and on the synthesis and adaptation of the solutions used in research hubs (e.g., nanoHUB; nanohub.org), cloud and grid systems, as well as in on-line application stores and auctions.

Acknowledgments

This work is supported by the RFBR grant No.15-07-09309.

References

[1] Polyakov S P, Demichev A P, and Kryukov A P 2015 *Procedia Computer Science* **66**429
[2] James E S and Ravi N 2005 *Virtual Machines: Versatile Platforms For Systems And Processes* (San Francisco: Morgan Kaufmann)
[3] Barton T and Seel C 2014 *Enterprise Modelling and Information Systems Architectures (EMISA 2014). Lecture Notes in Informatics Proceedings Series of the German Informatics society* vol P-234 (Bonn: Kollen Druck & Verlag GmbH) p145
[4] Dubenskaya J Demichev A Kryukov A and Prikhodko N 2015 *Procedia Computer Science***66**525
[5] Buchmann J A Karatsiolis E and Wiesmaier A 2013 *Introduction to Public Key Infrastructures* (Heidelberg: Springer)
[6] Erl T 2009 *SOA Design Patterns* (New York: Prentice Hall)
[7] Richardson L and Ruby S 2007 *RESTful Web Services* (Cambridge: O'Reilly Media)

Journal of Physics: Conference Series **762** (2016) 012016 doi:10.1088/1742-6596/762/1/012016

Federated data storage and management infrastructure

A Zarochentsev[1], A Kiryanov[2,3], A Klimentov[3,4], D Krasnopevtsev[3,5], P Hristov[6]

[1] Saint-Petersburg State University
[2] Petersburg Nuclear Physics Institute
[3] National Research Centre "Kurchatov Institute"
[4] Brookhaven National Laboratory, BNL
[5] National Research Nuclear University MEPhI
[6] European Centre for Nuclear Research, CERN

Abstract. The Large Hadron Collider (LHC), operating at the international CERN Laboratory in Geneva, Switzerland, is leading Big Data driven scientific explorations. Experiments at the LHC explore the fundamental nature of matter and the basic forces that shape our universe. Computing models for the High Luminosity LHC era anticipate a growth of storage needs of at least orders of magnitude; it will require new approaches in data storage organization and data handling. In our project we address the fundamental problem of designing of architecture to integrate a distributed heterogeneous disk resources for LHC experiments and other data-intensive science applications and to provide access to data from heterogeneous computing facilities. We have prototyped a federated storage for Russian T1 and T2 centers located in Moscow, St.-Petersburg and Gatchina, as well as Russian / CERN federation. We have conducted extensive tests of underlying network infrastructure and storage endpoints with synthetic performance measurement tools as well as with HENP-specific workloads, including the ones running on supercomputing platform, cloud computing and Grid for ALICE and ATLAS experiments. We will present our current accomplishments with running LHC data analysis remotely and locally to demonstrate our ability to efficiently use federated data storage experiment wide within National Academic facilities for High Energy and Nuclear Physics as well as for other data-intensive science applications, such as bio-informatics.

1. Overview and description of the problem

The processing, management and analysis of data in the current Mega-Science-scale projects require integration of computing centers of different sizes, power and architecture into a single computing environment (cyberinfrastructure). When designing such an environment one must consider not only the disk and computing resources, but also the bandwidth of the global computing network and the throughput and data access time between the computing centers. Aristotle asserted that "the whole is greater than the sum of its parts", so the integration of heterogeneous computing centers into a single federated distributed cyberinfrastructure (FDCI) will allow more efficient utilization of computing and disk resources for a wide range of scientific applications.
Existing problems when creating FDCI are as follows:

1. lack of "building blocks" from which you can build such a federation;
2. localized solutions, not going beyond immediate use and a specific task (and/or the center), in other words, it is usually an applied solutions with a low abstraction level of interfaces and software modules;

3. the integration in FDCI considered only after the creation of computing infrastructure, and not at the stage of development of the overall architecture of the system;
4. using the characteristics of the storage and computing resource as the parameters determining the power of the whole computing system, without taking into account the characteristics of global computer networks, data transmission speeds and the possibility of remote access to them.

Thus we need to solve four abovementioned problems and provide access to information about FDCI parameters to dynamic data and computing resource management applications.
During the FDCI architecture plot we have to follow the following fundamental principles:

- uniform method and abstraction level of resource management;
- common workload and data management system in heterogeneous computing environment;
- integrable and expandable tools to manage FDCI software.

Altogether, this comprises the problem of a fundamental nature. The lack of adequate solutions so far leads to economic and functional losses. Requirements for such a system are typical for scientific applications in the areas of science that require storage, analysis, processing and data management in petabyte or exabyte range.
In 2015 in the framework of the Laboratory "Big Data Technologies for mega-science class projects" in NRC "Kurchatov Institute" a work has begun on the creation of a united disk resource federation for geographically distributed data centers, located in Moscow, St. Petersburg, Dubna, Gatchina (all above centers are part of the Russian Data Intensive Grid of WLCG) and Geneva, its integration with existing computing resources and provision of access to this resources for applications running on both supercomputers and high throughput distributed computing systems (Grid). At first stage a federated disk infrastructure for data storage was plotted out and deployed.

2. Objective

The objective of these studies was to create a federated storage system with a single access endpoint and an integrated internal management system.
With such an architecture, the system looks like a single entity for an end user, while in fact being put together form geographically distributed resources. The system should possess the following properties:

- fault tolerance through redundancy of key components;
- scalability, with the ability to change the topology without stopping the entire system;
- security with mutual authentication and authorization for data and metadata access;
- optimal data transfer routing, providing the user direct access to the closest (optimal) data location;
- universality, which implies validity for a wide range of research projects of various sizes, including, but not limited to the LHC experiments [1].

3. Underlying technology choice

Today, there are particular solutions for abovementioned problems. For example, some of the LHC experiments use storage systems on top of which they build their own data catalogs. ALICE experiment [3] for instance uses a storage system based on xrootd, which keeps track of only physical file names (PFN), and utilizes a separate metadata catalog integrated in ALIEN infrastructure. This approach does not satisfy the requirements of flexibility and scalability. Solutions that satisfy all of the stated requirements are:

- storage based on HTTP Dynamic Federations (DynaFed) [4];
- EOS storage system [5];
- dCache storage system [10].

So far as EOS storage system is a complete integrated project, tested and used by all four major LHC experiments, ALICE, ATLAS, CMS and LHCb, the authors decided to start with it as a prototype software platform.

4. Federated data storage prototype

For federated storage prototype we have chosen geographically distributed resource centers located at the vertices of a triangle (figure 1): PNPI and SPbSU in St.-Petersburg district, NRC KI and JINR in Moscow district and CERN in Geneva. CentOS 6 and EOS storage system have been used as a software platform. During the deployment of the federation it was necessary to define the topology of storages and a common authentication scheme. Taking into account that the resources provided by federation participants are roughly equivalent, it was decided to use a simple layout: in each organization one management server (MGM) and one storage server (FST) are deployed. As EOS does not support simultaneous operation of multiple peer MGMs within one segment, one of the MGMs operates in master mode (primary) and others operate in slave mode (secondary) with automatic metadata synchronization. This solution allows to have a single access endpoint through the primary MGM, and improves system fault tolerance with the ability to use one of the secondary MGMs as primary in the case of failure.

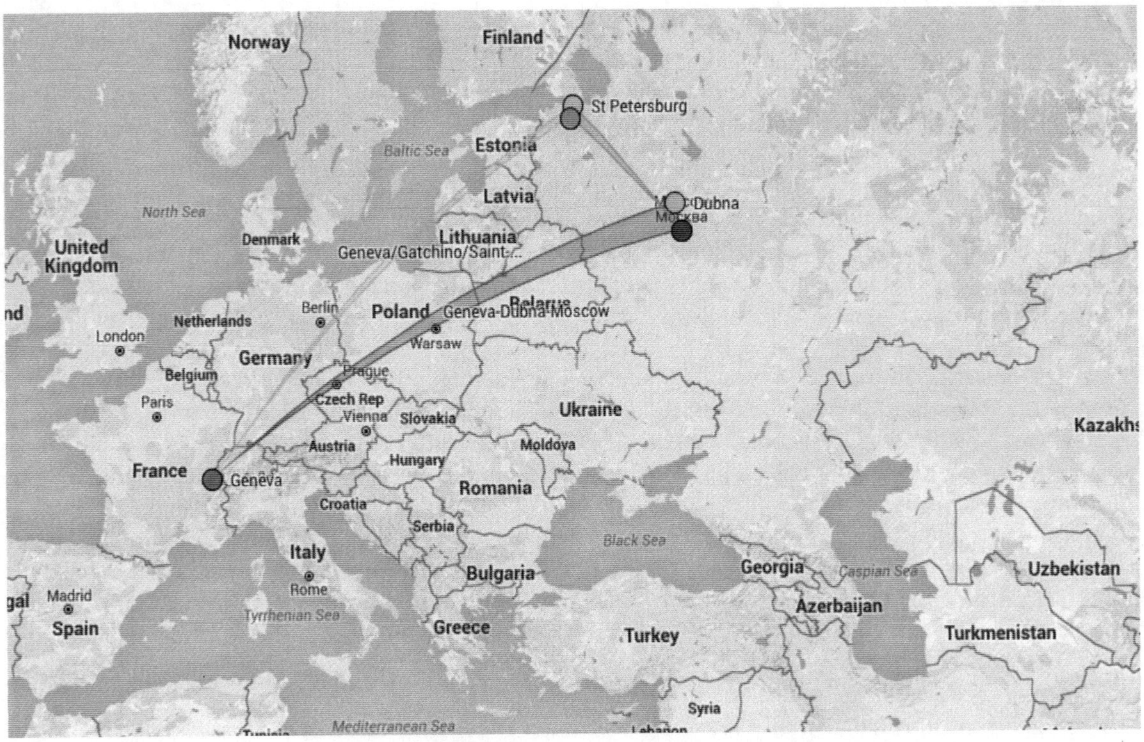

Figure 1. Participants of the federation prototype.

Normally, the client request first goes to the top-level MGM server on which the authentication and authorization are performed, enabling access to the metadata. Top-level MGM then redirects the client to the appropriate (optimal) storage server (FST), thus ensuring optimal data transfer routing.

EOS system includes support for four authentication methods: SSS, Unix, Kerberos and GSI. Unix authentication scheme is not secure enough for WAN use and can only be used within a single data center. SSS scheme is commonly used for authentication between the servers using the shared secret key, while GSI, the most secure one, can be used to authenticate both servers and users via standard X.509 digital certificates. Kerberos scheme is only useful in cases where integration with Kerberos is required.

For the federation prototype the GSI scheme was chosen as the most secure and widely used in the WLCG [6]. GSI authentication implies the need for signed X.509 certificate on each client and server. Our prototype uses certificates signed by Certification Authorities from a standard EUGridPMA list, which is also used in WLCG.

5. Testing procedure

The aim of the test is to verify reliability and obtain quantitative characteristics of performance efficiency of the data analysis jobs for the ATLAS and ALICE experiments. At first testing is carried out with the help of synthetic benchmarks, including remote access over the WAN, followed by efficiency measurements by real-life experiment applications.

In order to test file I/O performance we used a synthetic Bonnie++ [7] test, which is capable of measuring file and metadata access rate on a filesystem. EOS supports mounting as a virtual filesystem via Linux FUSE [8] mechanism, which makes it possible for Bonnie++ to test both file read-write speeds (FST access) and metadata transaction rates (MGM access) independently.

PerfSONAR [9] suite was used for an independent network performance measurements. Two significant metrics that were measured between each pair of resource centers are bandwidth and latency. These metrics allow to understand the impact of network parameters and topology on performance test results.

Reliability test was performed by simulating a failure of management servers with metadata migration and role switch between master and slave MGMs.

The applicability of such federation for real-life physics experiments was tested with application suites from ATLAS and ALICE experiments.

5.1. Processing and analysis of the ATLAS experiment data

ATLAS test application performs a proton-proton event reconstruction using so-called “raw” data as an input. Specific of this task is the need for reconstruction of all particle jets in the Transition Radiation Tracker (TRT) detector. Reconstruction of a signal from each of the proportional drift tubes in TRT is one of the most challenging and highly CPU-bound task, especially in high-luminosity conditions.

Reconstruction if performed in several stages, each of which requires intensive read and write of data stored in the federation. In addition, test application also analyzes the kinematic distribution in TRT, which allows to compare results obtained with federated and traditional storages.

Input and output data, be it a single file or a dataset, may be accessed both locally, on the filesystem, and remotely via xroot protocol.

Upon completion test application produces a comprehensive log file containing information about consumed computing resources on three key stages: environment initialization, event reconstruction and finalization. In addition, during event processing, application records in the log file information about resources consumed during processing of each individual event.

5.2. Processing and analysis of the ALICE experiment data

ALICE test application sequentially reads events from a file, analyzes them and produces information about the most "interesting" event according to the specified selection parameters. This application was specifically invented to evaluate the performance of the storage system. Just like before, input dataset can be accessed both locally and remotely via xroot protocol.

In order to run ATLAS and ALICE test applications we have deployed a standard user interfaces nodes containing all the necessary software stack: user and CA certificates, EOS client, Bonnie++ for synthetic tests, and LHC experiment software via CVMFS, which fully corresponds to the classical data processing environments used in both experiments. Following the developers’ recommendations perfSONAR nodes were deployed on dedicated physical machines.

6. Test results

So far we have tested FST and MGM servers, located at CERN, SPbSU and PNPI. We have evaluated the following resource combinations:

- SPbSU (FST and MGM on the same server);
- PNPI (FST and MGM on the same server);
- PNPI (MGM master and FST on different servers) + SPbSU (MGM slave and FST on different servers);
- CERN (MGM master) + PNPI (MGM slave and FST on different servers) + SPbSU (FST).

In addition, user interface nodes with at least 4GB of RAM and a gigabit network connection were deployed on all three resource centers composing “the small triangle": CERN, PINP and SPbSU. Tests were performed both before and after consolidation of resources under a single management server. Also, a "hot" role change was tried between MGM without loss of data of accessibility.

6.1. *Configuration and testing of the WAN Federation (perfSONAR measurements)*

Since the first test results are only available for PNPI, SPbSU and CERN network data channel performance measurements are only of interest between these centers. Network channel conditions were measured with perfSONAR installed on all participating centers of the federation.

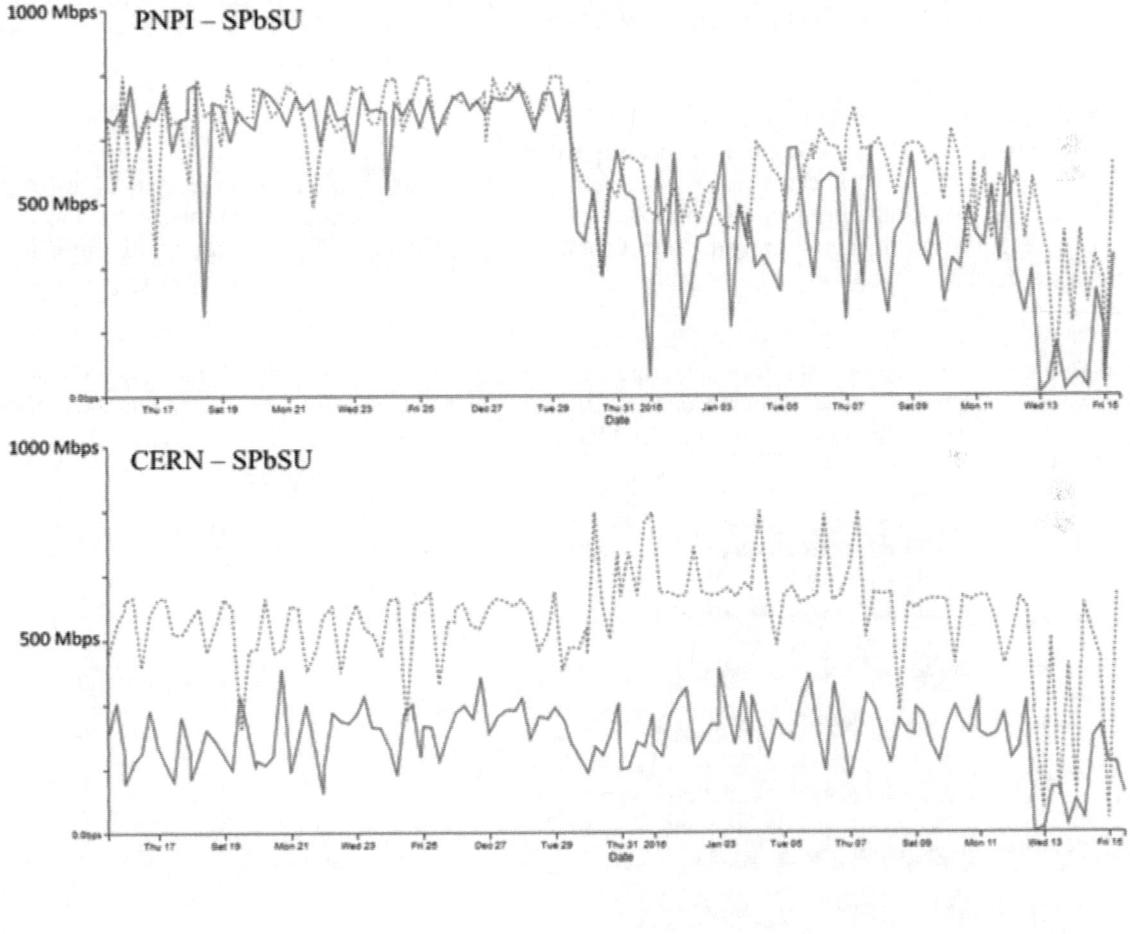

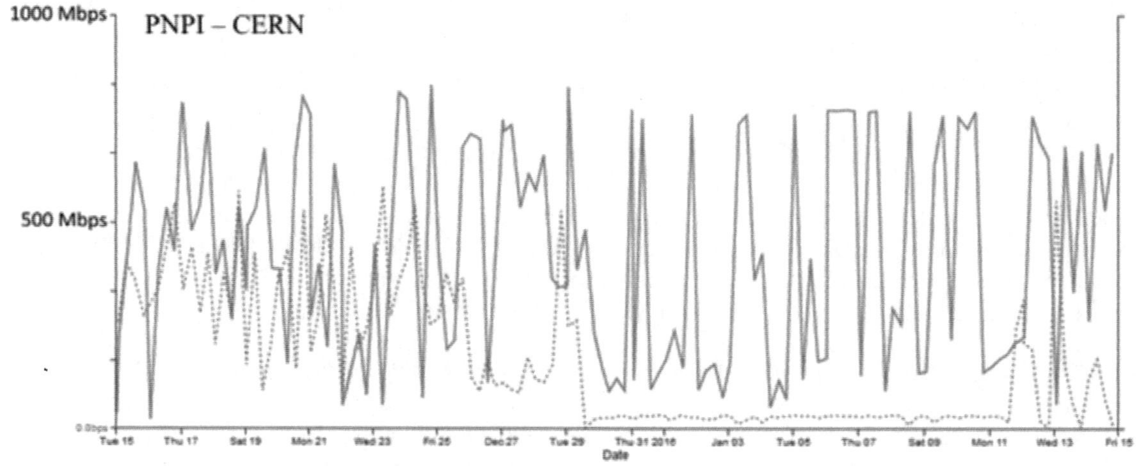

Figure 2. Network link bandwidth measured between 15th of December 2015 and 15th of January 2016. Solid and dotted lines represent forward and backward direction respectively.

In figure 2 we can see that:

- data transfer rates are not symmetric;
- communication speed changes significantly with time.

This behaviour of network channels is due to the fact that participating resource centers are not dedicated and also handle production workload. It makes our test results harder to interpret but at the same time gives us the opportunity to test federation in the real-world conditions with saturated network channels.

6.2. Bonnie++

For Bonnie++ test the most informative are file (figure 3) and metadata (figure 4) read-write performance diagrams, since in our scenario files and metadata are stored at different locations. During these tests CERN was used as a master MGM.

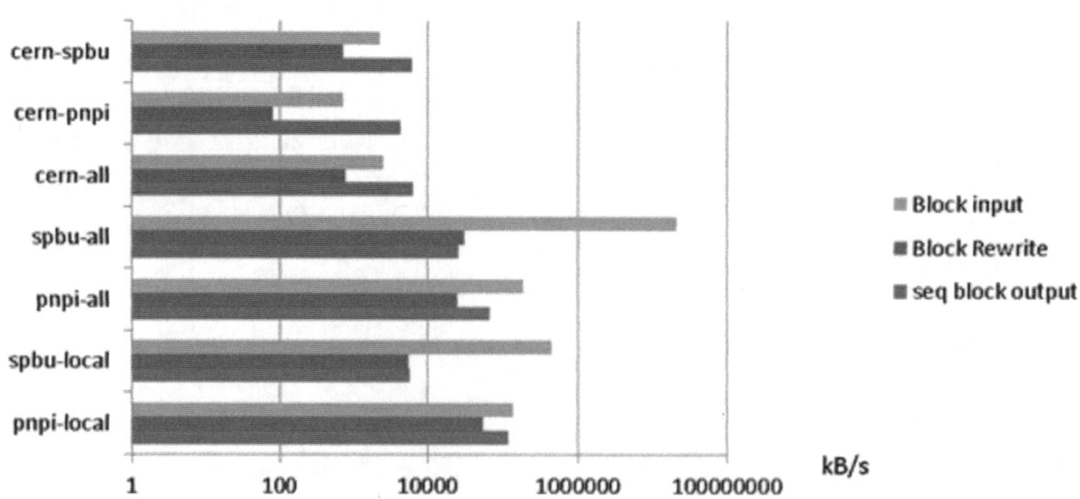

Figure 3. Bonnie++ test results for block file I/O. On the vertical axis we put a site combination (local means standalone) and test type. On horizontal axis we put a data transfer rate in KBps (logarithmic scale).

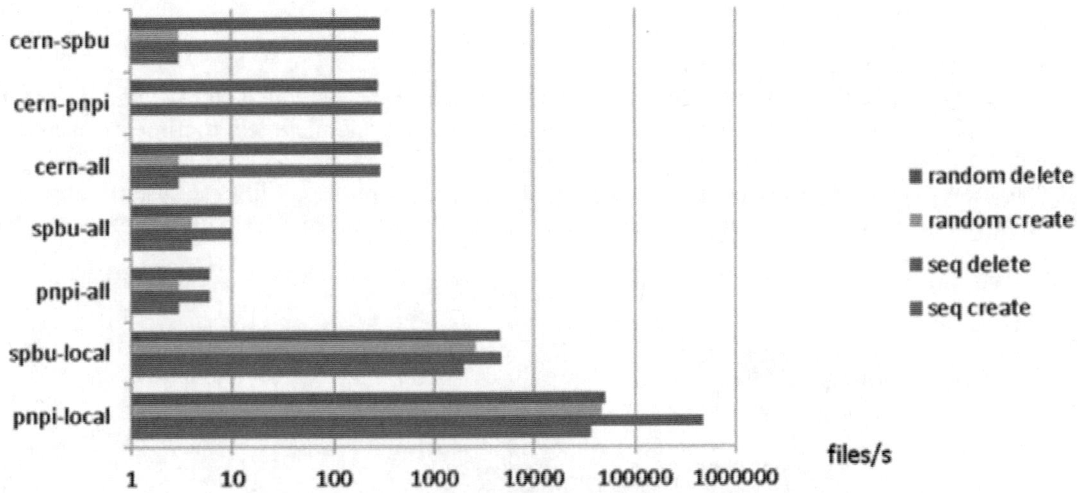

Figure 4. Bonnie++ test results for metadata transaction rate. On the vertical axis we put a site combination (local means standalone) and test type. On the horizontal axis we put a transaction rate in operations per seconds (logarithmic scale).

6.3. ATLAS event reconstruction test

During ATLAS event reconstruction test we had to reasonably choose application output parameters that appeared the most suitable for our needs. With a single input file the most reasonable performance estimate was an application initialization time, i.e. the duration of the stage where input data are preloaded from the storage. Another particularity in comparison with Bonnie++ test is that the input files do not necessarily have to be available in the locally mounted filesystem and can be read remotely via xroot protocol. Therefore, in the results of this test we have one additional parameter: access type, which is either FUSE or xrootd. During these tests managers were installed alongside the storage servers, as we were measuring performance of individual storages rather than federation.
The dependence of the test application initialization time on various client-server combinations and data access protocol is shown in figure 5.

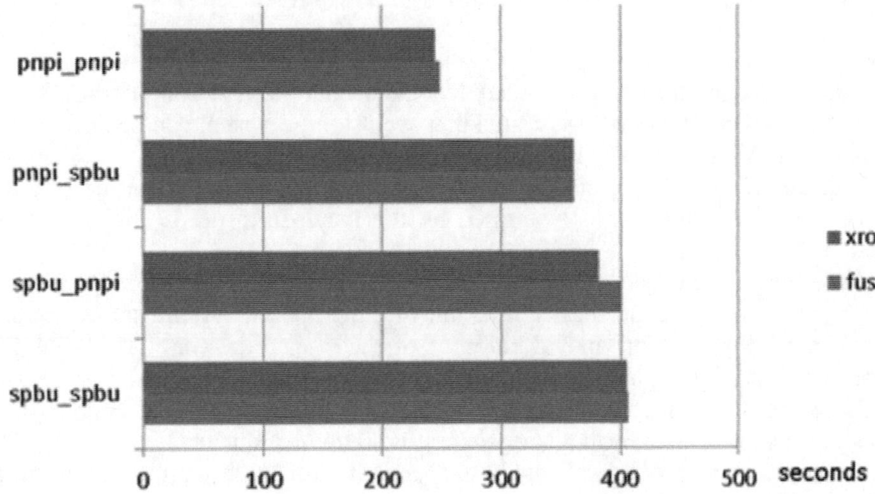

Figure 5. ATLAS test application initialization time. On the vertical axis we put a client-server combination and data access protocol. On the horizontal axis we put time in seconds.

6.4. ALICE event analysis and retrieval test

These tests were specifically written to test the performance of the federated storage and do not have such a complex output structure as in ATLAS tests. Therefore a complete test runtime measured with a system 'time' utility was taken as a primary parameter.

Figure 6 shows the results of reading from a single FST at PNPI but from different user interfaces and through different MGMs. Tests were performed using the dataset of 25 files with a gross size of ~34 GB.

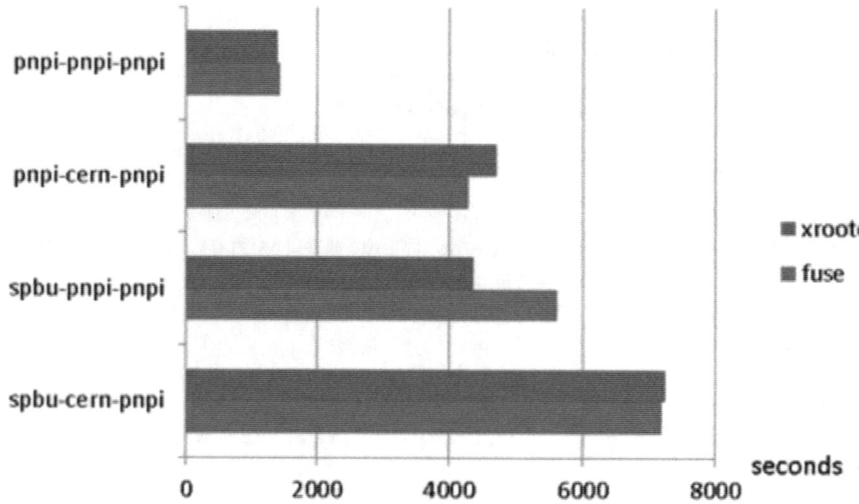

Figure 6. Runtime of the ALICE test. On the vertical axis we put a UI-MGM-FST combination and data access protocol. On the horizontal axis we put time in seconds.

Noteworthy is the fact that with other things being equal relocation of the MGM server entails a significant (more than twice) difference in runtime. Thus, there is a clear correlation between the ALICE test performance and the geographical location of MGM server.

7. Conclusions

The first prototype of geographically distributed federated data storage comprising of CERN and RDIG centers has been set up. It was demonstrated that such storage system can be efficiently used for data processing and analysis by the LHC scientific applications.

Performed synthetic and real-life application tests from ATLAS and ALICE have shown a reasonably high performance of the federation mostly limited by the local disk speeds and the bandwidth of network connections.

Bonnie++ tests have shown the expected dependency of metadata access speed on the speed of access to the management server that already speaks in favour of the proposed system, since metadata access times are not bound by the location and availability of the storage servers. Also, file I/O speeds correlate reasonably with the throughput of the network channels. On the other hand, is not yet entirely clear why there's so big difference in the performance of the remote and local clients. In this case, additional testing is considered with further study of configuration parameters.

Taking into account all already obtained test results authors conclude on the applicability of the federated storage for the considered usage scenario.

Journal of Physics: Conference Series **762** (2016) 012016 doi:10.1088/1742-6596/762/1/012016

Acknowledgments
This work was funded in part in part by the Russian Fund of Fundamental Research under contract "15-29-07942 офи_м" and U. S. DOE, Office of Science, High Energy Physics and ASCR under Contract No. DE-AC02-98CH10886 and SPbSU under research grant 11.38.242.2015.
The authors express appreciation to SPbSU Computing Center and PNPI ITAD Computing Center for provided resources.

References

[1] LHC – The Large Hadron Collider, http://lhc.web.cern.ch/lhc/
[2] The ATLAS Collaboration (G. Aad et al.). *J. Inst.* **3** S08003 2008 The ATLAS Experiment at the CERN Large Hadron Collider.
[3] Kenneth Aamodt, A Abrahantes Quintana, R Achenbach, S Acounis, D Adamová, C Adler, M Aggar- wal, F Agnese, G Aglieri Rinella, Z Ahammed, et al. *J. Inst.* **3** S08002 2008 The ALICE experiment at the CERN LHC.
[4] DynaFed, https://svnweb.cern.ch/trac/lcgdm/wiki/Dynafeds
[5] EOS, https://eos.web.cern.ch
[6] Jamie Shiers *C. Phys. Comm.* 177:219–223 2007 The worldwide LHC computing grid (worldwide LCG).
[7] Bonnie++, http://www.coker.com.au/bonnie++/
[8] FUSE, http://fuse.sourceforge.net/
[9] perfSONAR, http://www.perfsonar.net/
[10] dCache, https://www.dcache.org/

Evaluating non-relational storage technology for HEP meta-data and meta-data catalog

M A Grigorieva[1], M V Golosova[1], M Y Gubin[2], A A Klimentov[1,3], V V Osipova[2], E A Ryabinkin[1]

[1]National Research Centre "Kurchatov Institute", Moscow, Russia
[2]National Research Tomsk Polytechnic University, Tomsk, Russia
[3]Brookhaven National Laboratory, Upton, NY, USA

E-mail: maria.grigorieva@cern.ch, marina.golosova@cern.ch, maksim.gubin@cern.ch, alexei.klimentov@cern.ch, vikosi@tpu.ru, rea@grid.kiae.ru

Abstract. Large-scale scientific experiments produce vast volumes of data. These data are stored, processed and analyzed in a distributed computing environment. The life cycle of experiment is managed by specialized software like Distributed Data Management and Workload Management Systems. In order to be interpreted and mined, experimental data must be accompanied by auxiliary metadata, which are recorded at each data processing step. Metadata describes scientific data and represent scientific objects or results of scientific experiments, allowing them to be shared by various applications, to be recorded in databases or published via Web. Processing and analysis of constantly growing volume of auxiliary metadata is a challenging task, not simpler than the management and processing of experimental data itself. Furthermore, metadata sources are often loosely coupled and potentially may lead to an end-user inconsistency in combined information queries. To aggregate and synthesize a range of primary metadata sources, and enhance them with flexible schema-less addition of aggregated data, we are developing the Data Knowledge Base architecture serving as the intelligence behind GUIs and APIs.

1. Introduction

The volume of metadata generated at all stages of the mega-science experiment's life cycle in High Energy and Nuclear Physics (HENP) and other compute-intensive scientific applications such as bioinformatics and astro-particle physics, is comparable to the scientific data volume of the XX century experiments. Metadata have become an integral part of the overall architecture of storage and preservation of scientific information. Efficient metadata access and handling is the vital challenge for data-intensive science. Processing and analysis of huge volume of auxiliary metadata, which is constantly grows, is no less crucial challenging task than the management and processing of experimental data itself.

Scientific experiments are using software packages for distributed processing and data management, for instance, PanDA Workload Management System (WMS) in the ATLAS collaboration [1], AliEn[1] Grid framework in ALICE [2], CRAB (CMS Remote Analysis Builder)[2] in CMS. These packages orchestrate workload and data management in a heterogeneous computing environment, including payloads execution on the Grid, academic and commercial clouds, and supercomputers; and store transient and persistent meta-information in various databases needed by the international collaborations.

For the majority of scientific communities the actual issue is the lack of connectivity between metadata, describing data processing cycle, and metadata, representing the life cycle of scientific research in general, including annotation, indexing and publication of the results.

In this paper, we will explore the possible scenarios to solve the above problem by developing methods of integration of heterogeneous and distributed metadata sources, and infrastructure storage of meta-information, relevant to the specific research and experimental activities of scientific collaborations.

Though our initial motivation was related to the ATLAS needs, the proposed solution and architecture is more general and it will be applicable to a wide range of cutting edge large scientific experiments.

2. Metadata Sources

LHC Experiments have a set of metadata sources [3,4], with data, recorded at each phase of the experiment lifecycle: from the proposal of hypothesis and detector prototype, the selection of engineering solutions, modeling of experimental installation, to launch processing and data analysis tasks, writing of publications, conference papers and preprints. ATLAS metadata sources are listed below as an example.

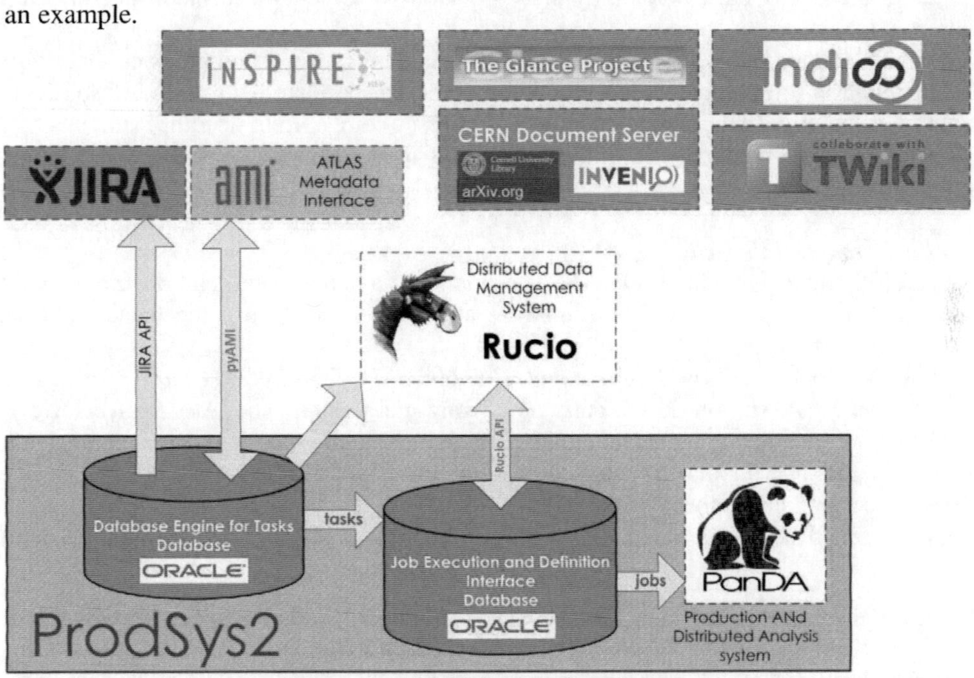

Figure 1 – ATLAS Metadata Sources

Data processing metadata sources:

- AMI[3] - Metadata Interface and database
- Rucio[4] - Distributed Data Management system

[1] http://alien.web.cern.ch/
[2] http://home.fnal.gov/~gutsche/computing/crab/crab_old.html
[3] http://ami.in2p3.fr/index.php/en/
[4] http://rucio.cern.ch/

- ProdSys2: DEFT, JEDI, PanDA [5,6,7] - Production system and its components, it is a workflow driven system and it's used to deal with the whole variety of ATLAS data processing and analysis activities: Monte-Carlo simulation, data (re)processing, High Level Trigger[5] Grid processing, physics groups derivation[6] and train production, user's analysis.

Information resources (publications, conference proceedings, meetings notes and actions items, etc.) metadata sources:

- GLANCE[7] - search engine for the ATLAS Collaboration publications.
- Indico[8] tools (allows to manage complex conferences, workshops and meetings)
- Inspire[9], containing 500,000 HEP documents
- CDS (CERN Document Server)[10] containing 200,000 CERN documents
    - arXiv[11] – openly accessible electronic archive and distribution server for research articles.
    - Invenio[12] - is an open source software package that provides the tools for management of digital assets in an institutional repository.
- CERN Twikis (a tool for web page collaborative writing)
- JIRA ITS[13] (Issue Tracking Service)

3. ATLAS Metadata sources integration challenge

As it is shown on Figure 1, these metadata sources are loosely coupled and information may not be timely synchronized, so they potentially may provide to an end-user inconsistency in requested information.

To aggregate and synthesize a range of primary metadata sources, and enhance them with flexible schema-less addition of aggregated data, we are developing the Data Knowledge Base (DKB) serving as the intelligence behind GUIs and APIs. We will present our current accomplishments to aggregate and to store LHC experiments meta-information and as a backend for future knowledgeable data storage.

3.1 Scientific Research Lifecycle

We have carefully investigated the ATLAS experiment metadata and the relationship between primary and aggregated metadata sources. These analyze allowed us to identify the basic components of a scientific research lifecycle.

Activities are conducted by *Physics and Combined Performance Groups* (for instance: Top, ETmiss, BPhysics). A set of initial data processing parameters such as *Projects* (for example, "mc14_8TeV"), information, associated with Software Release versions, and conditions data are recorded as auxiliary metadata. To start the production cycle, which includes a pipeline of data transformations and data analysis tasks, *Group Manager* defines *a Request* to be executed by the *Production System*. Request is translated to the set of tasks and jobs to be brokered by the PanDA

---

[5] The ATLAS High Level Trigger (HLT) is a distributed real-time software system that performs the final online selection of events produced during proton-proton collisions at the LHC

[6] There are more than 12 physics groups in the ATLAS Collaboration, bulk data processing and simulation is done centrally, but then each group runs group specific code to produce 'events collection' it is called derived data in ATLAS, the process is called derivation production

[7] https://atglance.web.cern.ch/atglance/

[8] https://indico.cern.ch/

[9] http://hep-inspire.net

[10] https://cds.cern.ch

[11] http://arxiv.org/

[12] http://invenio-software.org/

[13] http://information-technology.web.cern.ch/services/JIRA-service

WMS on Grid and opportunistic resources. Each task is accompanied by input data - organized in *Containers* - collection of files, produced or processed under the same conditions (software release, transformation step, conditions data, etc.).

The processes of hypothesis formulation, determination data transformation and data analysis parameters, adoption of the data samples, and other activity related to the organization and conducting the research, are registered in Twiki Pages and Indico. Interim results are published as Conference Notes. The final step of the experimental research lifecycle is a *Publication* in scientific magazines.

3.2 *Analysis/paper registry supporting the recording of paper-associated containers/datasets*

Specificity of scientific data is necessity of the eternal storing the RAW data, obtained directly from the detector, and have not been processed and transformed. Any RAW data changes leads to a scientific data, which scientists use in experiments. With the development of science and technology, the gap between the RAW and scientific data is constantly increases, and the procedures of RAW data processing have become more sophisticated. A complete data handling cycle includes RAW data preservation, data processing and analysis and information exchange. The requirement for eternal storage of the RAW data (as well as software version) and the access to it determines the basic principles of scientific research - reproducible scientific results.

The result of a chain of scientific data transforms are publications. Moreover, each publication can be based on the analysis of different versions of the data or use the data, obtained through various transformations. All of this meta-information must be fixed to ensure repeatability and reproducibility.

Currently meta-information, describing the lifecycle of the data processing and catalogs publications, operates separately. It would be useful to have some method of associating, in a searchable way, analysis teams and publications with datasets on the Grid.

4. Data Knowledge Base Architecture

DKB is designed to integrate information from many sources in a meaningful way and to perform meta-data analysis and meta-data mining. The structure and data flow of DKB is shown on Figure 2.

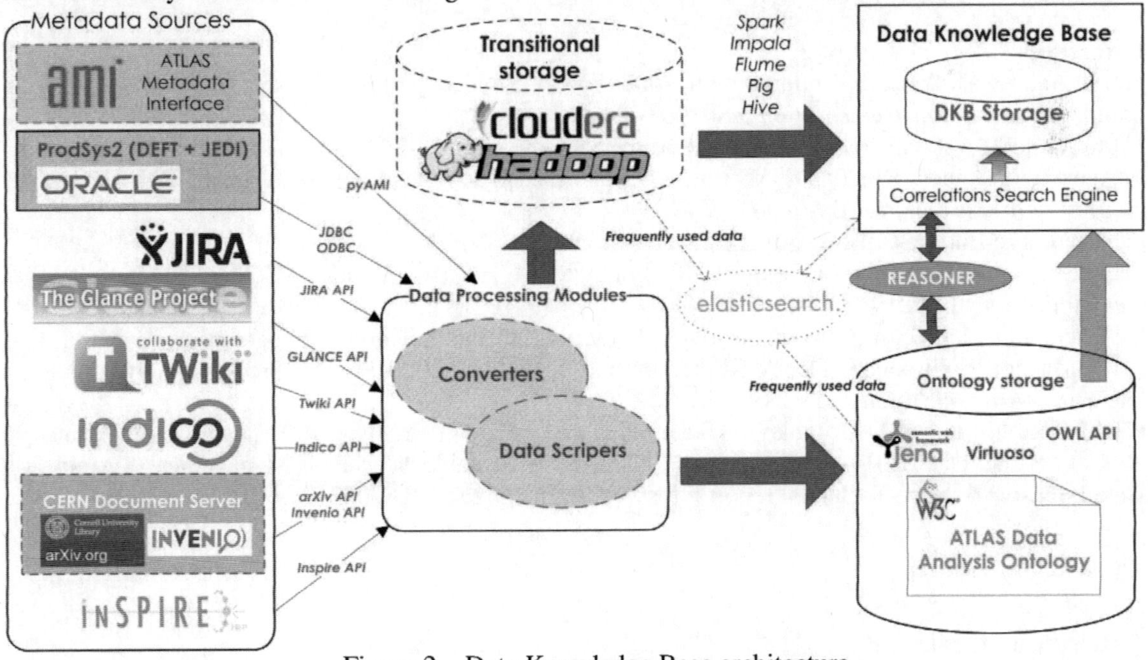

Figure 2 – Data Knowledge Base architecture

Ontology Storage, particularly OWL (Web Ontology Language) DB, is used to store information from multiple sources and to transform it into interconnected dataset suitable for further analysis. At this moment the exact platform for implementing OWL DB is under consideration. Possible candidates include Apache Jena[14], Virtuoso[15], and OWL API[16]. The data that is supplied to the OWL DB from various sources (physics papers metadata from GLANCE, meeting agendas from Indico, documents from CDS and arXiv) is usually initially available in the form of human readable texts. Before being stored into OWL DB, all this data is preprocessed by scrapers and converters that produce OWL metadata from natural language documents. After the preprocessed data is stored in the OWL DB, it is transformed into single interconnected dataset with reasoner employing domain ontologies and inference rules to create connections between heterogeneous data.

Each data source that is used in conjunction with OWL DB has an ontology associated with it, describing objects that can be obtained from the data source and their relationships with objects of other domains, most importantly rules for determining identity of objects and determining when separate objects in database refer to a single concept.

The bulk of ProdSys2 and AMI, after minor transformation, is uploaded to transitional storage - Hadoop cluster. There, it is processed to reveal, which data would be accessed by physics groups at various dates, and the ways to process the data.

Due to long response times of Hadoop and OWL DB, they are unsuitable for interfacing with users directly. The most frequently requested data is uploaded to ElasticSearch[17] cluster, providing a distributed, multitenant-capable full-text search engine, based on Lucene. It will allow short response times and generally improve user experience when interacting with DKB.

DKB serves as central part of the database structure, accepting user requests and transforming them into queries aimed at OWL DB, ElasticSearch or Hadoop, respectively. It is also responsible for combining query results and presenting them to user.

Acknowledgements
We wish to thank all our colleagues from ATLAS experiment, PanDA team and CERN IT. This work was funded in part by the Russian Ministry of Science and Education under contract № 14.Z50.31.0024.

References
[1] G Aad et al. ATLAS Collaboration 2008 The ATLAS experiment at the CERN large hadron collider *Journal of Instrumentation* **3** S08003
[2] K Aamodt, A Abrahantes Quintana, R Achenbach, S Acounis, D Adamová, C Adler, M Aggarwal, F Agnese, G Aglieri Rinella, Z Ahammed, et al. The ALICE experiment at the CERN LHC 2008 *Journal of Instrumentation* **3** S08002
[3] The Large Hadron Collider http://lhc.web.cern.ch/lhc/
[4] J Shiers 2007 The worldwide LHC computing grid (worldwide LCG) *Computer physics communications* **177** 219–223
[5] The PanDA Project http://www.usatlas.bnl.gov/twiki/bin/view/PanDA/WebHome
[6] T Maeno, P Nilsson, K De, A Klimentov and T Wenaus 2009 PanDA Production and Analysis backend *Journal of Physics* **219** 210
[7] M Borodin, K De, D Golubkov, T Maeno, D Oleynik, A Petrosyan, A Klimentov, T Wenaus, P Nilsson, S Jha, et al. 2014 PanDA Beyond ATLAS: A Scalable Workload Management System For Data Intensive Science Technical report https://cds.cern.ch/record/1670021?ln=ru

[14] https://jena.apache.org/
[15] http://semanticweb.org/wiki/Virtuoso.html
[16] http://owlapi.sourceforge.net/
[17] https://www.elastic.co/products/elasticsearch

Exploiting multicore compute resources in the CMS experiment

J E Ramírez[1], A Pérez-Calero Yzquierdo[2,3], J M Hernández[3] on behalf of the CMS Collaboration.

[1] University of Puerto Rico Mayagüez, Mayagüez, Puerto Rico, USA
[2] Port d'Informatió Científica (PIC), Universitat Autónoma de Barcelona, Bellaterra (Barcelona), Spain
[3] Centro de Investigaciones Energéticas, Medioambientales y Tecnológicas, CIEMAT, Madrid, Spain

E-mail: juaneduardo.ramirez@upr.edu

Abstract. CMS has developed a strategy to efficiently exploit the multicore architecture of the compute resources accessible to the experiment. A coherent use of the multiple cores available in a compute node yields substantial gains in terms of resource utilization. The implemented approach makes use of the multithreading support of the event processing framework and the multicore scheduling capabilities of the resource provisioning system. Multicore slots are acquired and provisioned by means of multicore pilot agents which internally schedule and execute single and multicore payloads. Multicore scheduling and multithreaded processing are currently used in production for online event selection and prompt data reconstruction. More workflows are being adapted to run in multicore mode. This paper presents a review of the experience gained in the deployment and operation of the multicore scheduling and processing system, the current status and future plans.

1. Introduction

The experimental collaborations taking data at the Large Hadron Collider (LHC) have begun transitioning to multi-core aware application frameworks and scheduling systems capable of executing multi-threaded applications across multiple cores[1, 2, 3]. This transition has been driven by an ever increasing volume of complex data resulting from the higher energy and luminosity delivered by the LHC compared to Run 1. The increase of intensity results in a higher number of concurrent collisions per event (pile-up). Figure1(a) shows a high pile-up event from CMS data with 78 reconstructed vertices. These busy events lead to a higher per-event processing time and memory usage. On the hardware side, the evolution of multicore technology has been driven by the power limitations on increasing the density of transistors on single core processors to enhance its performance, and as a result, the need to look for alternative solutions to increase CPU performance.

Utilizing multicore CPU capability requires software changes to transition from sequential programming to parallel processing. The multicore applications developed to adapt the code to this new architecture design offer several advantages. In a threaded process, memory is shared between threads and this reduces the memory consumption per core[4], thereby avoiding running into memory limitations as depicted in Fig.1(b). Another advantage concerns the experiment's

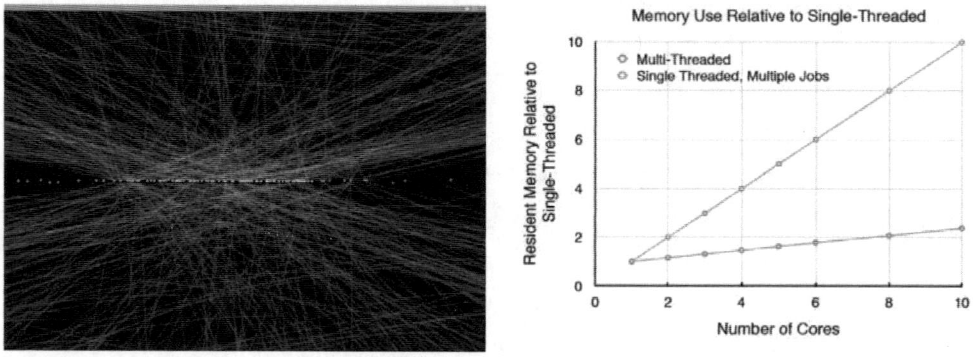

(a) High pileup CMS event with 78 reconstructed vertices

(b) Memory use of a multi-threaded application compared to n simultaneous single-threaded tasks

Figure 1. Drivers motivating use of multicore

Workload Management System (WMS). The number of jobs to be handled by the system is reduced and the output files produced are of larger sizes, which then requires less managing and merging operations. The software evolution to utilize multicore CPUs scattered across the Grid has required the modification of the applications themselves as well as the development of new resource allocation, scheduling tools and procedures.

For the LHC Run 2 data taking period that began in early 2015, the initial CMS priority in the deployment of multicore processing and scheduling was the provision of sufficient computing power to perform multithreaded prompt data reconstruction tasks. In 2015 it was expected to use 100% of the Tier-0 and 50% of the Tier-1's available CPUs. Thus, for the first phase of multicore deployment, the focus has been on T0 and T1s sites (see [5, 6] for a description of CMS tiered computing model and mesh configuration). Multicore resources are also starting to be deployed at CMS Tier-2 sites, with the objective of switching simulation and digitization tasks to multithreaded algorithms as well during 2016. Single core and multicore jobs will coexist during Run 2 and therefore, a strategy for scheduling both types of job is mandatory. Eventually, the use of multicore pilots by CMS will expand to manage majority of the resources at the supporting computing sites.

2. CMS model for multicore resource provisioning

The CMS workload management system (WMS) employs software agents (WMAgents[7]) to manage centralized workflows populating job queues, assigning job priorities, handling errors, job retries, merging output files and log collection. The allocation of execution nodes and the scheduling of jobs is done through GlideinWMS[8, 9, 10]. Resources are allocated at the grid sites by means of pilot submission as a result of job pressure. Jobs are matched to resources by managing a transient pool of computing resources controlled by *pilot jobs*, whose function is to schedule user jobs.

The model for the integrated scheduling of CMS jobs with different core count requests is the use of multicore pilots capable of internal partitioning of the resources into dynamic slots. A single pilot will run multiple payloads, as shown in Fig. 2 for the example of a 4-core pilot. Two multicore jobs are initially run by creating two slots of two cores each, followed by two additional single core jobs by further fragmenting one of the slots after the first 2-core job is finished, This continues until the pilots approaches the end of its defined lifetime. As a result, running multicore partitionable pilots on CPU resources allows managing all CMS workflows with a simplified model.

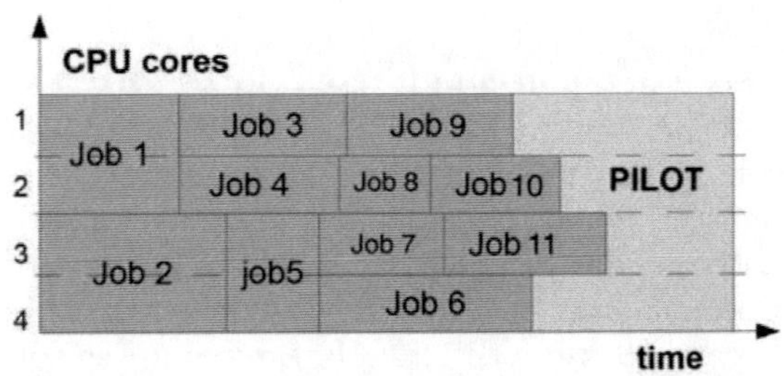

Figure 2. Schematic of a multicore partitionable pilot

Additional advantages of this model are: a total control of job scheduling priorities by CMS, as opposed to the scheduling at site level; avoiding competition of single and multicore pilots for resources at sites and for matching jobs once running, which simplifies the scheduling optimization; a reduction in the number of pilots needed to run the global pool of resources, which contributes to the scalability of the system.

The main disadvantages are the inherent inefficiency coming from the draining of retiring pilots and the potentially slow ramp up of allocated resources (in the case of irregular pilot submission patterns) in multi-VO sites with policies to protect their compute farms from excessive resource draining. A previous study[11] observed however negligible inefficiencies when pilot lifetimes are about ten times or longer than the average job length.

A potential limitation in the model is the inability to pull new multicore jobs once the pilot has internally fragmented the resource. However, the pool of pilots is continuously renewed (since pilots have a finite lifetime). Therefore providing fresh non-fragmented pilots (see Fig. 3). This avoids the costly need of induced defragmentation inside running pilots, which however could still be applied in case of urgency in executing a particular multi-threaded workflow.

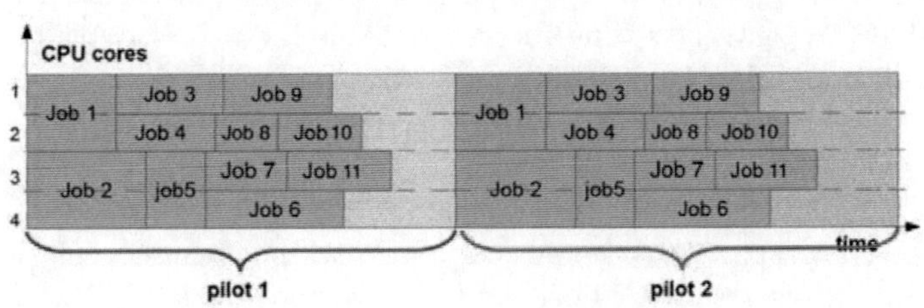

Figure 3. Schema of CMS model for multicore resource provisioning

The dynamic behaviour of the system with a dependence on a number of parameters (e.g. pilot and average job lifetimes, pilot and payload core counts, etc.) implies that its tuning is essential in order to achieve optimal performance.

3. Results for 2015

CMS has successfully deployed and tested its multicore architecture at the T0 and T1 computing sites during 2015. We present here four main results:

- **Deployment at T1 sites.** The multicore resource provisioning model has been incrementally deployed at T1s as they joined the pool of multicore-capable sites and tuned their batch systems to optimal performance when handling multicore requests. Figure 4(a) shows the monthly average number of cores allocated to running multicore pilots at all CMS T1s in the second half of 2015.

- **Mixed scheduling of multicore and single core jobs within multicore pilots**. In order to assess the multicore pilot model readiness to schedule a mixed workload consisting of multicore and single core jobs, prompt reconstruction 4-core jobs were tested at large scale at all CMS Tier-1 sites prior to the start of 2015 data taking, as shown in Fig.4(b). The target, set at 50% of the pledged T1 CPU cores, was essentially achieved during the moments of highest multicore job pressure. However, the small amount of data collected during 2015 made the urgency to use of T1 CPUs for prompt reconstruction lower than expected, so finally prompt data reconstruction was executed mainly at the T0.
- **Multicore Pilot job scheduling efficiency**. To measure the efficiency of multicore pilot internal payload scheduling, single core jobs were submitted with enough pressure to saturate the allocated resources of about 1000 CPU cores. Multicore pilots were typically 40 hours long, while single core jobs run with a variety of job lengths between 1 to 2 hours. The test, run over the lifetime of the pilots, yielded negligible scheduling inefficiencies, as shown in Fig.4(c).
- **Execution of multicore prompt data reconstruction jobs at the T0 during 2015**. As shown in Fig.4(d), a variety of core counts were employed for jobs running the T0 CPUs. Prompt data reconstruction for proton-proton collisions was typically executed as 4-core jobs, while other multicore job sizes were also employed, in particular for the end-of-year heavy ion run, with 8-core jobs.

4. Future outlook

In 2016 it is expected that CPU resources at the T1 sites will be regularly used to run multithreaded tasks, such as data reconstruction jobs and also simulated data processing, as the Monte-Carlo simulation and digitization steps have been ported to run efficiently in multithreaded mode. The commissioning work done during 2015 with T1 sites ensures that both the sites and the multicore scheduling system are ready to reach this goal. In 2016 the pool of available multicore resources will be increased to include the major T2 sites. Finally, while CMS tools for multicore job management are ready, a complete deployment of monitoring tools is needed in order to continue optimizing the scheduling algorithms.

Acknowledgments

This work was supported in part by the U. S. National Science Foundation Award No. ACI-1440552, the Spain Ministery of Science and Innovation grant FPA2010-21816-C02-00, Spain Ministery of Economy and Competitiveness grant FPA2013-48082-C2-1/2-R and EU 7th Framework Programme INFRA-2010-1.2.1: Distributed computing infrastructure Contract Number RI-261323 (EGI-InSPIRE). The Port d'Informació Científica (PIC) is maintained through a collaboration between the Generalitat de Catalunya, CIEMAT, IFAE and the Universitat Autonoma de Barcelona.

References

[1] Pérez-Calero Yzquierdo A, Hernández J, Holzman B, Majewski K, McCrea A and the CMS Collaboration 2014 *Journal of Physics: Conference Series* **513** 032074 URL `http://stacks.iop.org/1742-6596/513/i=3/a=032074`

[2] Hernández J M, Evans D and Foulkes S 2012 *Journal of Physics: Conference Series* **396** 032055 URL `http://stacks.iop.org/1742-6596/396/i=3/a=032055`

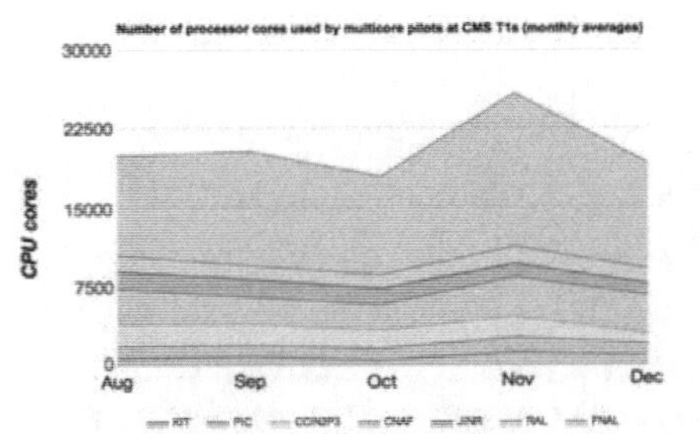

(a) Monthly average number of processor cores used by multicore pilots at T1s

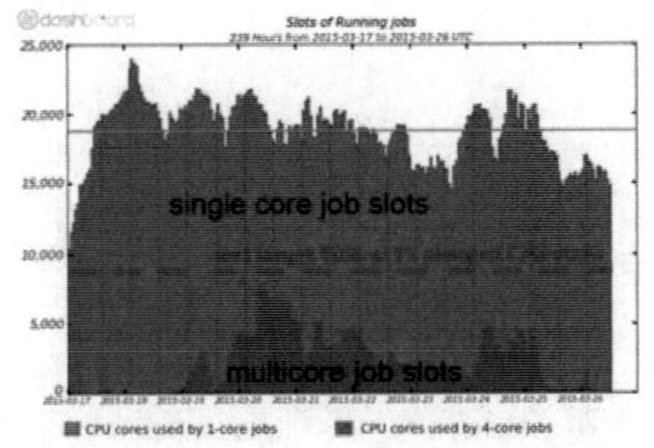

(b) CPU cores in use by single and 4-core jobs running at T1s

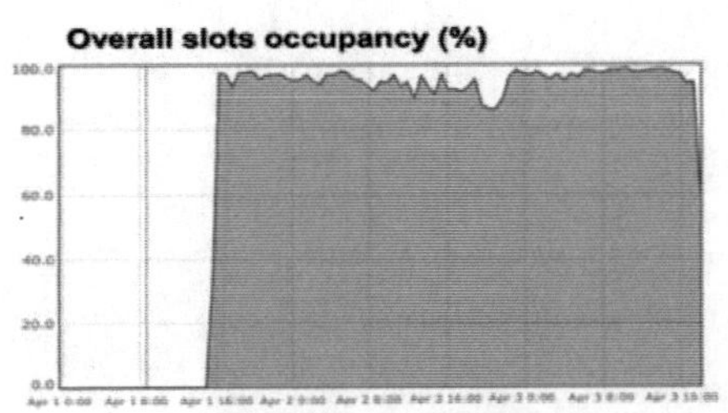

(c) Percentage of overall slots occupancy as a measure of pilot internal scheduling performance

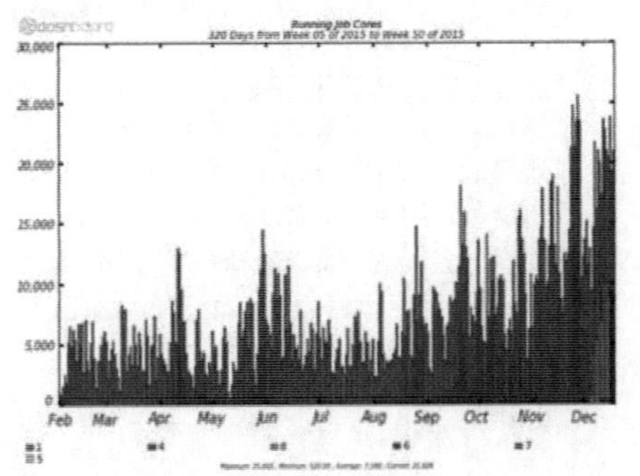

(d) Running jobs at the T0 as a function of time during 2015, colored by number of cores per running job

Figure 4. Results for 2015

[3] Forti A, Yzquierdo A P C, Hartmann T, Alef M, Lahiff A, Templon J, Pra S D, Gila M, Skipsey S, Acosta-Silva C, Filipcic A, Walker R, Walker C J, Traynor D and Gadrat S 2015 *Journal of Physics: Conference Series* **664** 062016 URL `http://stacks.iop.org/1742-6596/664/i=6/a=062016`

[4] Jones C D, Contreras L, Gartung P, Hufnagel D and Sexton-Kennedy L 2015 *Journal of Physics: Conference Series* **664** 072026 URL `http://stacks.iop.org/1742-6596/664/i=7/a=072026`

[5] Bayatyan G L, Della Negra M, Fo, Herv A and Petrilli A (CMS Collaboration) 2005 *CMS computing: Technical Design Report* Technical Design Report CMS (Geneva: CERN) submitted on 31 May 2005 URL `http://cds.cern.ch/record/838359`

[6] Ayllon A A, Salichos M, Simon M K and Keeble O 2014 *Journal of Physics: Conference Series* **513** 032081 URL `http://stacks.iop.org/1742-6596/513/i=3/a=032081`

[7] Fajardo E, Gutsche O, Foulkes S, Linacre J, Spinoso V, Lahiff A, Gomez-Ceballos G, Klute M and Mohapatra A 2012 *Journal of Physics: Conference Series* **396** 042018 URL `http://stacks.iop.org/1742-6596/396/i=4/a=042018`

[8] Sfiligoi I, Bradley D, Holzman B, Mhashilkar P, Padhi S and Wurthwein F 2009 *Computer Science and Information Engineering, 2009 WRI World Congress on* vol 2 pp 428–432

[9] GlideinWMS Homepage URL `http://www.uscms.org/SoftwareComputing/Grid/WMS/glideinWMS/`

[10] HTCondor Homepage URL `http://research.cs.wisc.edu/htcondor/`

[11] Pérez-Calero Yzquierdo A, Hernández J M, Khan F A, Letts J, Majewski K, Rodrigues A M, McCrea A and Vaandering E 2015 *Journal of Physics: Conference Series* **664** 062046 URL `http://stacks.iop.org/1742-6596/664/i=6/a=062046`

GeantV: from CPU to accelerators

G Amadio[1], A Ananya[2], J Apostolakis[3], A Arora[2], M Bandieramonte[3], A Bhattacharyya[4], C Bianchini[1,5], R Brun[3], P Canal[6], F Carminati[3], L Duhem[7], D Elvira[6], A Gheata[3], M Gheata[3,8], I Goulas[3], R Iope[1], S Jun[6], G Lima[6], A Mohanty[4], T Nikitina[3], M Novak[3], W Pokorski[3], A Ribon[3], R Sehgal[4], O Shadura[3], S Vallecorsa[3], S Wenzel[3], Y Zhang[3]

[1] UNESP, Universidade Estadual Paulista (BR)
[2] IIT- Indian Institute of Technology (IN)
[3] CERN - European Organization for Nuclear Research - Geneva, Switzerland
[4] Bhabha Atomic Research Center (IN)
[5] Mackenzie Presbyterian University (BR)
[6] Fermi National Accelerator Laboratory (US)
[7] Intel Corporation
[8] Institute of Space Sciences (RO)

Andrei.Gheata@cern.ch

Abstract. The *GeantV* project aims to research and develop the next-generation simulation software describing the passage of particles through matter. While the modern CPU architectures are being targeted first, resources such as GPGPU, Intel© Xeon Phi, Atom or ARM cannot be ignored anymore by HEP CPU-bound applications. The proof of concept *GeantV* prototype has been mainly engineered for CPU's having vector units but we have foreseen from early stages a bridge to arbitrary accelerators. A software layer consisting of architecture/technology specific backends supports currently this concept. This approach allows to abstract out the basic types such as scalar/vector but also to formalize generic computation kernels using transparently library or device specific constructs based on Vc, CUDA, Cilk+ or Intel intrinsics. While the main goal of this approach is portable performance, as a bonus, it comes with the insulation of the core application and algorithms from the technology layer. This allows our application to be long term maintainable and versatile to changes at the backend side. The paper presents the first results of basket-based *GeantV* geometry navigation on the Intel© Xeon Phi KNC architecture. We present the scalability and vectorization study, conducted using Intel performance tools, as well as our preliminary conclusions on the use of accelerators for *GeantV* transport. We also describe the current work and preliminary results for using the *GeantV* transport kernel on GPUs.

1. Vertical scaling for particle transport simulation

LHC has intensively used horizontal scaling for the most important HEP computing tasks, namely simulation, reconstruction and analysis. The LHC computing GRID [1] allowed running hundreds of thousands of concurrent jobs taking advantage of distributed computing resources and relying on the fact that events are uncorrelated. Measurements [2] have revealed however a rather poor usage of the silicon per socket, interpreted to be due to instruction fetch and data locality issues. Correlating this with the lack of awareness of HEP software to features common to modern processors such as instruction level parallelism and SIMD, we concluded that there was a very large potential for improvement.

The *GeantV* simulation prototype introduced the concept of vectorized processing in particle transport [3], enabled by a so-called *basketizing* procedure that feeds vectors of particles to floating-point intensive geometry and physics algorithms. The goal is to make use of the SIMD vector units available on modern processors and to optimize the use of caches by profiting from the geometry and physics locality as much as possible. The design of the *GeantV* vector prototype has been described in detail in other papers [4,5]. The idea is to leverage the vertical dimension of scalability by making maximum use of the silicon available on a single socket, and to study the feasibility of using coprocessors and accelerators such as Xeon Phi and GPU's. The project aims to eventually provide the path to a long-term solution for particle transport simulation working in both general purpose and HPC environments.

The different pieces composing *GeantV* - such as concurrent scheduling, geometry and physics - are currently being built and in different stages of development. This is an iterative process, as we learn that some approaches are better than others. In terms of porting the vectorized CPU prototype on accelerators, running full *GeantV* in native mode is the preferred but not always possible solution. For the GPU we are offloading modules of *GeantV* as kernels, coming with an additional overhead of having to handle data transfers between the host and device, as shown in the next section.

2. GeantV in native and offload mode

The design of *GeantV* has foreseen creating *baskets* (vectors of tracks located in the same geometry volume of having similar physics properties) of filtered tracks and dispatching them to processing units. These can be threads of the same process, different processes or arbitrary co-processors acting as accelerators. The track data is passed to a broker that can collect as many tracks as deemed efficient for the co-processor, taking care of shipping the data to and from the device. This run mode can only be efficient if the data transfer back and forth is asynchronous and concurrent in order to mitigate the data transfer latency. The GPU broker, for example, collects baskets until the number of contained tracks is equal to the number of GPU cores, and then serializes them to GPU. Each track is given to a different thread, which propagates it until the next geometry boundary or physics process. The broker receives the tracks back and gets them "*basketized*" again by the CPU scheduler, as shown in Figure 1.

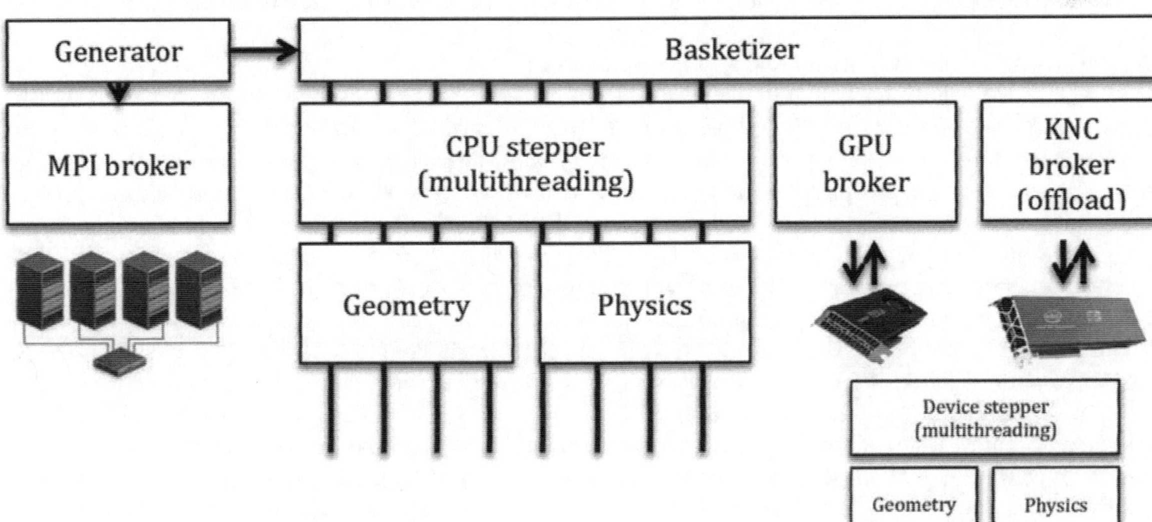

Figure 1 ***GeantV*** **scheduling in native and offload modes. The coprocessor brokers can work fine grain processing a set of baskets and sending transported tracks back to host, or MPI mode fetching data from an event of file server.**

The workflow described above has been tested for the GPU *Kepler* cards using *CUDA7*, and currently is being optimized. The same approach was initially foreseen also for the Intel® KNC device broker but it was eventually dropped due to difficulties in managing asynchronous data transfers from the host, using the KNC offload mechanism. We decided to rather use the Xeon Phi co-processor in native/offload modes to test the behavior of specific *GeantV* geometry and physics workloads, without asynchronous communication. This helped us understanding performance features on this architecture for some of our main software components, targeting a full *GeantV* deployment on the future Intel®KNL. For the offload mode the device library was cross-compiled on the host and the deployment of the work was done using *OpenMP's #pragma offload* directive within a dedicated *MicVec* backend.

Another use case foreseen for *GeantV* is running in native mode on some of the accelerators. This will be the preferred run mode on *KNL* considering the overhead introduced by offloading. The pre-requisites are the native compilation of the full *GeantV* library on the device and a custom tuning of the scheduling parameters to optimize the workflow depending on the device architecture.

An important feature of the *GeantV* library allowing compiling the same code on both host and device modes is the concept of backend interfaces [6, 7]. The idea behind this is that the main differences across platforms related to data types, library-dependent implementations, or scalar versus vector handlers, can be wrapped into backend-dependent structures. Template specializing on the backend type makes then a common source code available across libraries and different technologies in both scalar and SIMD vector mode when available. A backend approach insulates the code base from the heterogeneous technology layer and makes maintenance of the code much easier. It is clear that a considerable effort shifts into the development and maintenance of the backend interfaces – currently the supported ones being the *scalar*, *CUDA*, *Vc*, *MicVec* and *UME::SIMD* (using Intel native intrinsics).

We are currently re-shaping the *GeantV* scheduler to allow running in native mode on the KNL Xeon Phi, implying a deep redesign to provide better scalability and NUMA awareness. On the GPU side, the CUDA backend allows already running a full transport step (propagation of a vector of particles to the current geometry volume limits or the simulation of the next physics process) but currently the rescheduling has to be done still on the host. We do not exclude the possibility to eliminate this limitation in future. Dispatching work to accelerators running in native mode can be done either by MPI or their native communication interface, considering that the input data required by particle transport is lightweight compared to the amount of processing performed per track.

3. Physics and geometry benchmarks on accelerators

We have investigated in a first step the performance of *GeantV* geometry and physics benchmarks on *Kepler K20* and *Intel®Xeon Phi C0PRQ-7120* (KNC) cards. We have tried to understand the performance limits on such devices for elementary algorithms such as geometry intersection for different solids or simple physics models. The two architectures are complementary as many-core topologies, exposing different parallelism use cases. While the *Kepler* provides 2496 lightweight cores which are suitable for track-level parallelism, the *Intel®KNC* provides 60 cores running up to 4 threads per core with access to 512 bits vector registers, allowing to profit from the *GeantV* basket model, enabling multi-track vectorization.

For the geometry case, we have tested the speedup compared to *Geant4* [8] and *ROOT* [9] implementations across different platforms and for different vector lengths. Table 1 presents such a comparison, combining both scalar optimizations (with respect to the equivalent ROOT algorithms) and vector gains for a simplified detector setup using a simple navigation, which examines all daughter volumes (i.e. without geometry voxel optimizations). Standard intersection benchmarks for solids such as boxes, trapezoids or tubes are showing very good acceleration of often more than a factor of 4 on KNC. We note that for the most complex algorithms such as the polycone or polyhedron intersection, the vector gains are reduced (sometimes less than x2) due to their multiple branches, which gets penalized when mapped to mask operations. In such complex cases we are seeking alternatives, e.g to profit from vectorizing inner loops on surfaces or to provide better algorithms. Figure 2 presents the performance for the different box shape algorithms on the K20, showing how the die can be saturated with large track containers running in a single kernel.

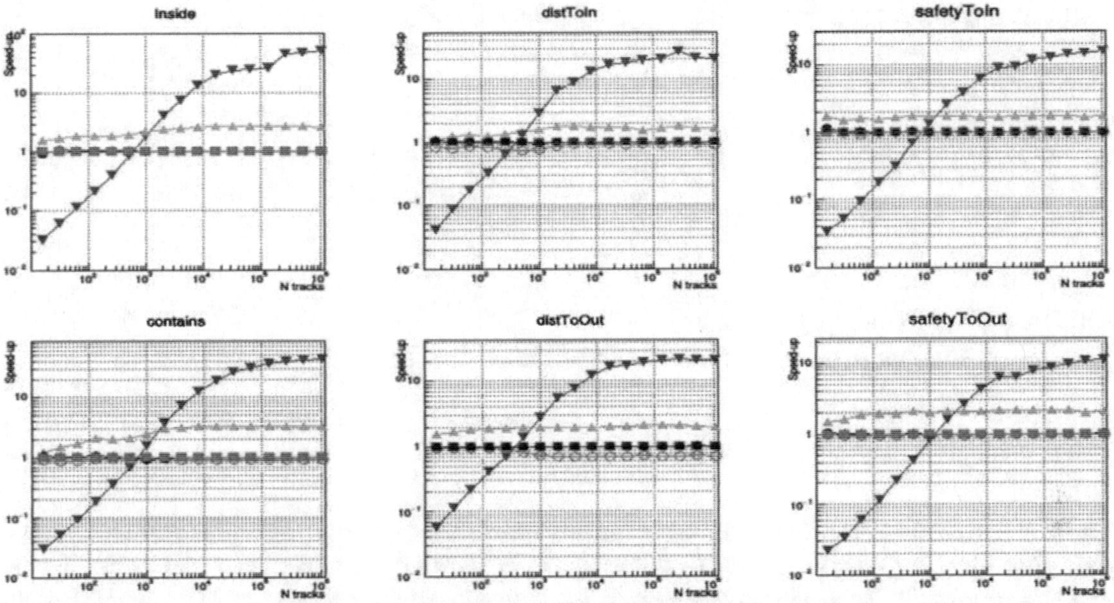

Figure 2 Performance of box navigation algorithms on K20 GPU compared to scalar/vector modes on CPU and compared to equivalent ROOT algorithms. The speedup is computed as $T_{scalar}/T_{vector/GPU}$

The latest versions of CUDA allow running multiple kernels on the same die, for workloads processing smaller data units such as the *GeantV* baskets. Note that, due to its lower performance per core the GPU needs at minimum about 10^3 tracks to match the scalar CPU performance.

Architecture	16 particles	1024 particles	SIMD max
Intel © Ivy-Bridge (AVX)	2.8	4.0	4.0
Intel © Haswell (AVX2)	3.0	5.0	4.0
Intel © Xeon Phi (Intel®IMCI)	4.1	4.8	8.0

Table 1 Overall performance for particles traversing a simplified detector, compared to ROOT v5.34.17. We compare the speedup of the vector version compared to the scalar version of ROOT running on the same architecture.

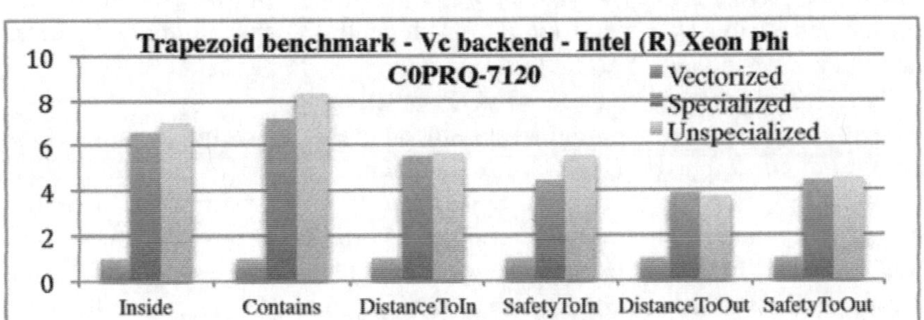

Figure 3 Relative time normalized to the vectorized case for different trapezoid shape intersection algorithms (Xeon®Phi C0PRQ-7120 P

The development of vector-aware physics algorithms has recently started in the *GeantV* project. We are trying to optimize at first common algorithms for distribution sampling, random number generators and an initial set of physics model implementations for gamma processes.

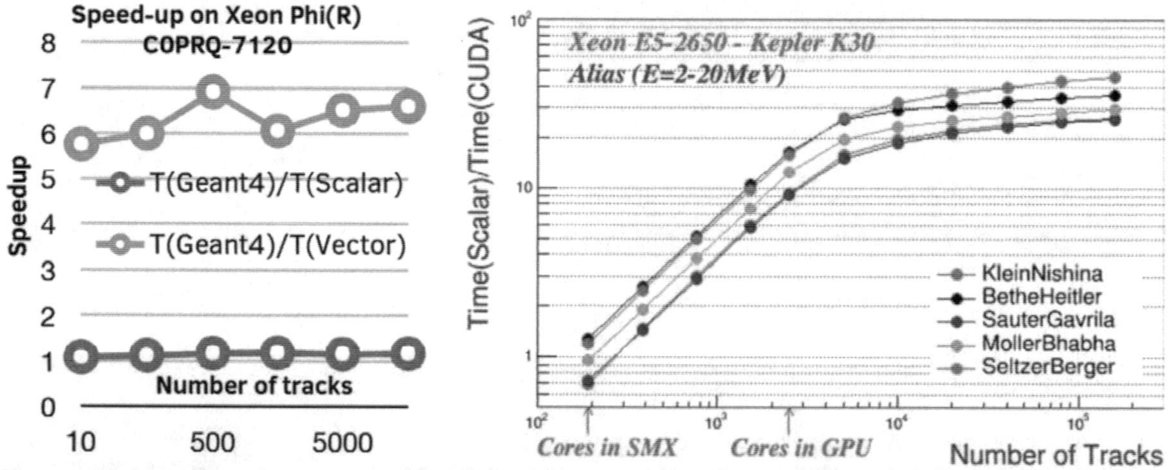

Figure 4 Vectorized Compton process on KNC (left). Performance of the alias sampling method for different process cross-sections (right).

Figure 4 (left) presents the performance of the vectorized Compton physics model algorithm on KNC card, which uses the vector unit of the accelerator close to its ideal performance, while the scalar version has basically the same performance as the corresponding *Geant4* one. We have tested alias sampling on different electromagnetic models' distributions (having quite different profiles) on the *Kepler* card [10]. The performance profile keeps the same shape but can vary with as much as 50% for different distributions.

4. Performance of combined algorithms on accelerators: the X-ray benchmark.

An important question is the relevance of vector optimizations of elementary algorithms for the complete simulation workflow: is it possible to maintain the performance improvements shown above in the combined algorithm case? It is clear that some of the processing steps connecting different vectorized algorithms may represent scalar bottlenecks that will have direct impact on performance. It is very important to first identify these scalar bottlenecks, then try to minimize them.

We have tried to understand the vector optimization procedure by running a composite benchmark on the Intel® Xeon Phi *COPRQ-7120* card. The standard Xeon-based benchmark uses a part of the full detector geometry and runs multi-particle propagation traversing through it as shown in Figure 5. This operation is performed during the *GeantV* stepping but there it competes with the active physics models. This benchmark performs just the geometry navigation part of the task and creates an "X-ray" virtual image on the other side. A pixel's gray level depends on the number of different volumes traversed by each track. We looked at some basic quantities such as scalability and throughput, and moved on to in-depth *VTune* profiling to understand the bottlenecks.

The detailed results of this procedure are not reproduced in this paper but we have clearly identified the particle relocation algorithm as one of the important scalar bottlenecks. A possibility to optimize relocation is to understand geometry navigation from a heuristic perspective. As a simple example, a calorimeter detector is composed of possibly millions of elements, out of which often just a few logically different volumes. In addition, these volumes are disposed symmetrically and in a lattice manner, so that each cell has a very limited number of neighbors. We are currently investigating and prototyping methods to optimize navigation for this kind of topologies, giving very promising results so far and very likely be used in future by the framework.

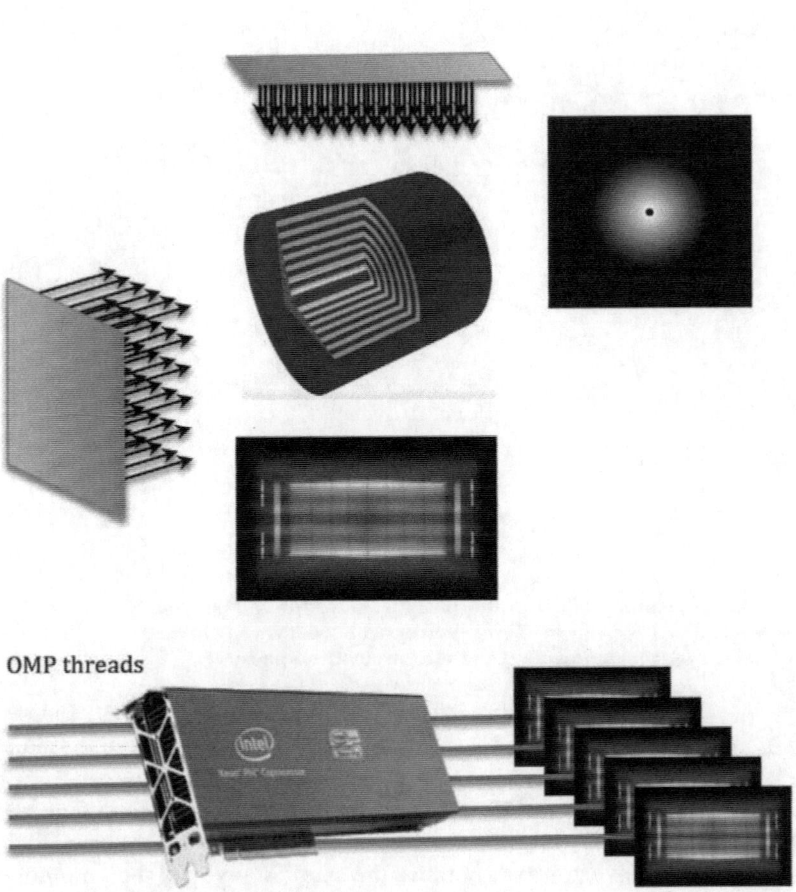

Figure 5 The X-Ray benchmark tests geometry navigation in real detector geometry, which is one of the main components of GeantV. The name comes the fact that one takes a detector geometry module and shoots virtual rays (with starting points disposed in a grid) along a given direction and with a given resolution (input parameters). Each ray is propagated from boundary to boundary using the VecGeom navigator, and the number of crossings is counted until the volume is exited on the other side. The model for parallelism uses *OpenMP* to dispatch different scans to different threads.

For our optimization purpose, we have used a simplified geometry setup emulating concentric cylindrical tracking detectors, where crossing volume boundaries are predictive. We attached a basketizer to each of these detectors, aiming to form vectors of tracks of fixed size before invoking geometry navigation algorithms. This reproduces to some extent the actual re-basketizing schema used in *GeantV*, including the rundown phase when the partially filled baskets have to be processed at the end of the simulation.

The setup is presented in Figure 5. The exercise's objective was to saturate the processing pipelines of a Xeon Phi card. To achieve this goal we steered multithreading using OpenMP, parallelizing on the loop of X-ray scans to be performed: each thread had to complete full scans over a fixed-size grid of 1024x1024 pixels. We have tested both the compact and balanced OpenMP thread scheduling, using an increasing number of threads, up to 4 threads per physical core of the Xeon Phi card. We have measured both scalability and the image throughput, comparing them with the same benchmark done on the *IvyBridge* dual processor machine hosting the KNC card.

5. Results and conclusions

The benchmark was run with both compact and balanced OMP modes. In balanced mode it showed an excellent scalability approaching the ideal case up to the physical core count. We observed expected limited performance increase as more threads are allocated. The balanced model converges towards the compact model as all thread slots are filled. The conclusion was that running Xeon Phi fully saturated brought the maximum benefit for this application. The throughput performance for a saturated KNC is equivalent (for this setup) to the dual Xeon E5-2650L@1.8GHz server which hosts the card.

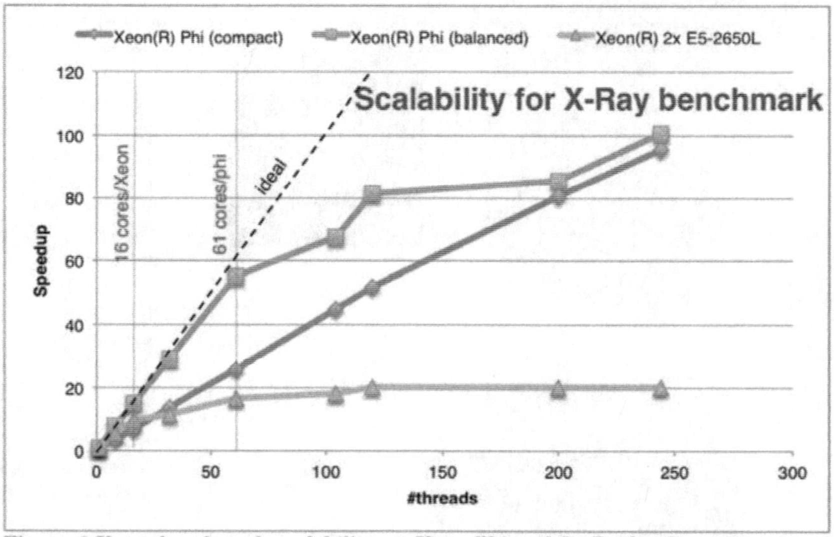

Figure 6 X-ray benchmark scalability on Xeon Phi and IvyBridge for compact and balanced thread allocation. The Xeon server shows unexpected benefits even running beyond its hyper-threading regime, hinting to vectorization as main responsible.

To understand the vectorization features of the model, we have run the benchmark and studied three different cases:

- *Scalar case*: Simple loop over pixels, generating a ray for each pixel and transporting it in scalar mode to the exit of the detector.
- *Ideal vectorization case*: Filling each vector with N times the same X-ray, and then providing this as input to our geometry propagator vector API. This is of course not realistic as the work for each track is redone the vector size times but this was useful to be used as a reference for the maximum achievable vectorization.
- *Filtering/regrouping case*: Filling vectors of tracks grouped per geometry volume and regrouping tracks to the appropriate baskets as particles are entering the next volume (similar to *GeantV approach*). Transport is triggered when vectors reach a given size.

In Figure 7 we plotted the speedup of the ideal and realistic cases with respect to the scalar approach with the same number of threads. We also plotted the vector performance on the IvyBridge host for comparison. While some of the vector efficiency is lost compared to the ideal case, we observed gains up to 4.5 from vectorization in basketized mode, with vector starvation showing up beyond the core count. A more detailed analysis using Intel *VTune* is ongoing in order to understand better the source of contention.

An important conclusion to our benchmark was that navigation specialization could bring benefits to the vectorization performance of *GeantV*. While studying a non-trivial yet idealized case, we have confirmed that a new level of optimization for geometry navigation is possible and is needed for obtaining additional performance on both standard Xeon processors and Xeon Phi cards. In addition, the results on KNC are promising for future migration of *GeantV* code to many core architectures such as the Intel®Knights Landing.

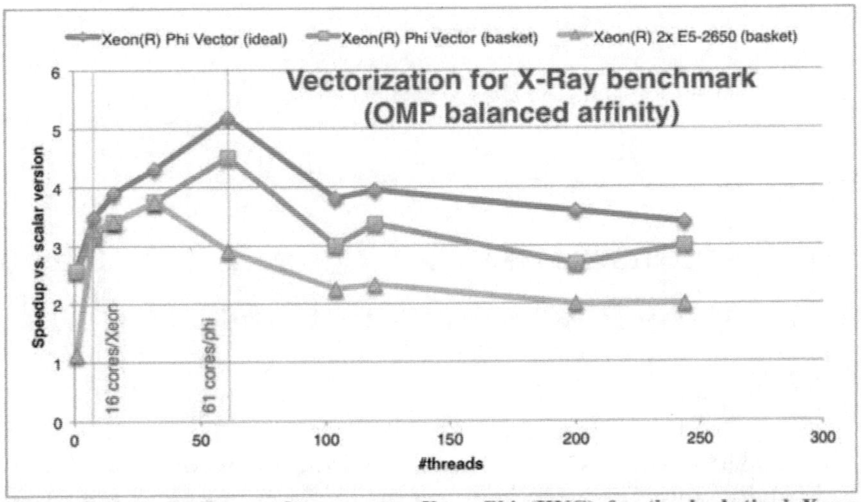

Figure 7 Vectorization performance on Xeon Phi (KNC) for the basketized X-ray benchmark, compared to idealized performance of the same setup and with IvyBridge. The speedup is defined as T_{scalar}/T_{vector} for the same number of threads. The decreasing slope at large number of threads is not an absolute loss in performance but a relative one (i.e. the scalar version scales better than the vector)

In parallel with the *GeantV* steering framework, the development of vectorized algorithms based on our approach using backends is well advanced in case of geometry and in an initial phase in case of physics. Both geometry and physics benchmarks demonstrate good performance so far on co-processors. Development plans for the next phase aim to provide full built-in accelerator awareness for *GeantV*-based simulation on both GPGPU and KNL processors.

Acknowledgements
The authors wish to acknowledge the contribution of Intel to the *GeantV* project through the Intel Performance Computing Centre (IPCC) program. We also want to acknowledge the very useful technical contributions of the CERN openlab, which allowed performing these benchmarks.

References
[1] M Lamanna 2004 *N.I.M. in Phys. Res.* **A 534**(1):1-6.
[2] S Jarp, A Lazzaro, J Leduc and A Nowak 2012 *Comp. High Energy and Nucl. Phys. (CHEP)* https://indico.cern.ch/event/149557/session/6/contribution/486
[3] Apostolakis J, Brun R, Carminati F and Gheata A 2012 *J.Phys: Conf. Ser.* **396** 022014 http://iopscience.iop.org/1742-6596/396/2/022014
[4] Apostolakis J, Brun R, Carminati F, Gheata A and Wenzel S 2014 *J. Physics: Conf. Ser.* **513** 052006
[5] J Apostolakis et al 2015 *J. Phys.: Conf. Ser.* **608** 012003
[6] Wenzel S 2014 Towards a high performance geometry library for particle-detector simulation *16th Intl. workshop on Adv. Comp. and Analysis Techniques in phys. Res. (ACAT)*
[7] de Fine Licht J 2014 First experience with portable high-performance geometry code on GPU *GPU Computing in High Energy Physics*
[8] S Agostinelli et al 2003 *Nuclear Instruments and Methods* **A 506** (53pp)
[9] http://root.cern.ch
[10] G Amadio et al 2016 Electromagnetic Physics Models for Parallel Computing Architectures *17th Intl. workshop on Adv. Comp. and Analysis Techniques in phys. Res. (ACAT)*

Hardware Demonstrator of a Level-1 Track Finding Algorithm with FPGAs for the Phase II CMS Experiment

D. Cieri*[1,2] on behalf of the CMS Collaboration and of the Time Multiplexed Track Trigger group

[1] University of Bristol, Bristol, United Kingdom; [2] Rutherford Appleton Laboratory, Didcot, United Kingdom

*Supported by the EU FP7-PEOPLE-2012-ITN project nr. 317446, INFIERI, "Intelligent Fast Interconnected and Efficient Devices for Frontier Exploitation in Research and Industry"

E-mail: davide.cieri@stfc.ac.uk

Abstract. At the HL-LHC, proton bunches collide every 25 ns, producing an average of 140 pp interactions per bunch crossing. To operate in such an environment, the CMS experiment will need a Level-1 (L1) hardware trigger, able to identify interesting events within a latency of 12.5 µs. This novel L1 trigger will make use of data coming from the silicon tracker to constrain the trigger rate. Goal of this new *track trigger* will be to build L1 tracks from the tracker information.

The architecture that will be implemented in future to process tracker data is still under discussion. One possibility is to adopt a system entirely based on FPGA electronic.

The proposed track finding algorithm is based on the Hough transform method. The algorithm has been tested using simulated pp collision data and it is currently being demonstrated in hardware, using the "MP7", which is a µTCA board with a powerful FPGA capable of handling data rates approaching 1 Tb/s.

Two different implementations of the Hough transform technique are currently under investigation: one utilizes a systolic array to represent the Hough space, while the other exploits a pipelined approach.

1. Introduction

The High Luminosity (HL–LHC) upgrade [1] to the Large Hadron Collider will operate at an increased instantaneous luminosity, up to 7 times the nominal value, in order to collect an integrated luminosity of 3000 fb^{-1} in the decade following 2025. Proton bunches at the HL–LHC will cross every 25 ns, producing an average of 140 pile-up pp collisions per bunch crossing (BX).

In order to meet these experimental challenges, the CMS collaboration [2] will need to address the aging of the present detector and to improve the ability of the apparatus to isolate and precisely measure the products of the most interesting collisions (CMS Phase II Upgrade [3]).

In fact, by 2025 the detector will have suffered significant radiation damage, therefore many components will need to be completely replaced. In particular, CMS will have a new tracker [4]

with an increased granularity in both pixel and outer tracker system, which will guarantee the required reconstruction performance.

A major upgrade will involve also the online L1 trigger system. Presently only the muon chambers and the calorimeters are part of the L1 trigger. Indeed, the combinatorial background at the HL–LHC will not allow to keep the current trigger system, since it will be impossible to maintain the trigger rate within the nominal range without degrading physics performance by rising substantially the trigger thresholds. Therefore, for the first time, data coming from the Tracker will be included in the L1 trigger of a high luminosity hadron experiment.

Before transmitting tracking information to the trigger processors, the amount of data needs to be reduced. This will be achieved through an on-detector selection on transverse momentum of particles p_T. Therefore the Outer Tracker system (Fig. 1a) will be formed by double sensor modules (p_T modules [3]), capable of rejecting signals from particles with a p_T lower than 2 GeV/c. A correlation between hits in sensors of one module, which are consistent with a high-p_T track, is called a *stub* (see Fig. 1b).

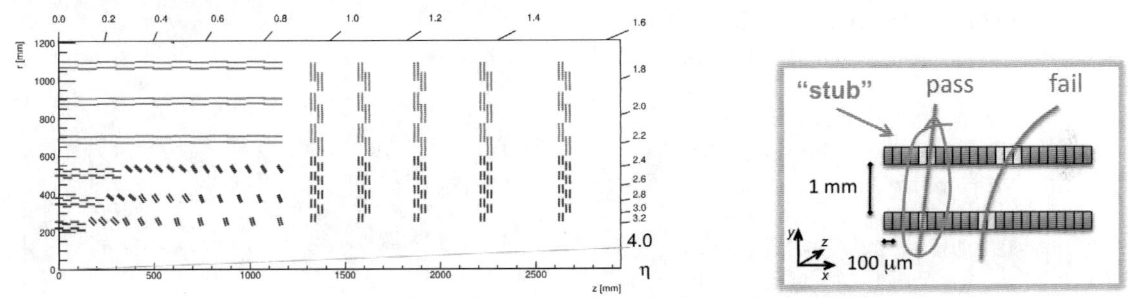

(a) (b)

Figure 1: *(a) Cross section of the upgraded CMS Outer Tracker in* $r-z$*, showing the two different* p_T *modules developed: pixel–strip modules in blue and strip–strip modules in red. (b) Sketch illustrating the correlation of signals in two sensors and the "stub" selection in a single* p_T *module.*

The architecture that will be used to process the tracker data is still subject to discussion. An intriguing proposal is to use a time multiplexed system, similar to the one already adopted by the CMS Phase I calorimeter trigger [5], and composed entirely by FPGA boards.

2. The Time Multiplexed Track Trigger

The fundamental idea of a Time Multiplexed Trigger (TMT) [6] is that all data from a single event flow to a single destination (node) for processing. This requires two processing layers with a passive switching network between them, the Pre-Processor (PP) and the Main-Processor (MP). The PPs take data directly from the front-end modules, organise and format them, before distributing to the MPs, where the track finding algorithm will be implemented. Each MP receives data from many PPs belonging to a single event. Data from the next event flow to a second MP, out of phase by one LHC clock cycle. Each MP will then have a period of time (*Time Multiplexed period*) proportional to the total number of MP boards to run the algorithm.

This particular system is well matched to FPGA processing. Therefore, a conceptual architecture can be built by means of the best currently readily available. Our choice ended up being the µTCA board MP7 [7], originally designed for the CMS L1 calorimeter trigger upgrade, which contains a powerful Virtex 7 FPGA with 72 I/O optical links (total bandwidth ~ 1 Tb/s).

3. L1 Track Finding with Hough Transform

The goal of the track trigger is to build L1 tracks from stub data. The Hough Transform (HT) [8] is the method chosen to build track candidates. The simplest version of this technique permits detection of straight lines. In the real space a line is defined a set of points (x, y), which the HT represents in terms of the slope–intercept parameters (m, q). In this way each point in the real space represents a line in the parameters space, so that a line in the real space will be identified by the crossing point in the parameter space.

Dividing the tracker into 5 trigger regions in pseudo–rapidity, each MP will receive around 3000 stubs per event. Each MP will bin then the stubs in 36 ϕ_0 segments of around 0.17 rad each, where ϕ_0 is the estimated azimuthal angle of the track at the production point. The Hough Transform is then performed in each of these ϕ_0 segments. Thanks to this ϕ_0 binning the algorithm will have to operate just with an average $\sim$ 80 stubs, with a maximum of 200 at a time. In the first stage of the algorithm the Hough Transform is applied to the stub coordinates (r, ϕ), which have the best resolution in CMS, independently for each η–ϕ_0 sector.

Charged particles in CMS are bent by the strong magnetic field. It is possible in any case to assume for high–p_T particles ($p_T \geq 3$ GeV/c) a linear relation between ϕ and r.

$$\phi(r) = \frac{cBQ}{2p_T} r + \phi_0, \tag{1}$$

where c is the speed of light, B the CMS magnetic field, Q the electric charge of the particle, p_T its transverse momentum and ϕ_0 is the estimated azimuthal angle at production. Here the slope parameter $m = cBQ/2p_T$ is bound by the lower p_T limit of 3 GeV/c, while the intercept q depends on the size of the ϕ_0 segment under consideration.

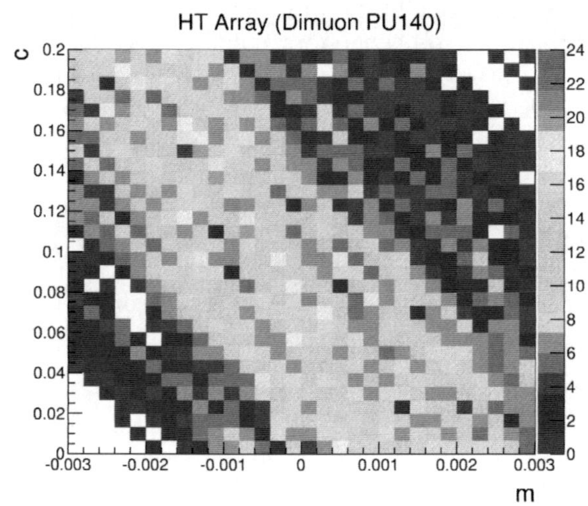

Figure 2: *A typical filled Hough Transform for one single segment in ϕ_0 for one trigger region. On the z–axis number of stubs is represented. On average of ∼80 stubs fill this histogram, occupying 90% of cells. Filtering stages are required to identify L1 track candidates.*

The algorithm histograms the stub data in (m, q) bins in a 2-D array of 32 × 32 cells. For each column in m, a value of q is calculated,

$$q = -mr + \phi - \phi_0^{min}, \tag{2}$$

where ϕ_0^{min} is the lower bound of the current ϕ_0 segment. A track is found when stubs cluster in the 2-D histogram (see Fig. 2). In most cases the stubs present in a HT cell are not consistent with real tracks. Indeed they are usually associated to random combinations of hits belonging to pile–up tracks. Therefore several filtering stages are required to accept only cells compatible with true high–p_T particles.

Bend Filter

The first filter to be applied is the *Bend Filter*. The local bend of a stub can be easily calculated knowing the hits and sensors separation. The condition for a stub to be kept in a HT cell is that the stub bend δ_s would be consistent with the predicted bend δ_c based on the q/p_T of the cells and the (r, z) coordinates of the stub.

$$|\delta_s - \delta_c| < \sigma, \quad (3)$$

where σ is the assumed bend resolution.

Z Filter
Another filter exploits the use of a third coordinate (η or z), which has not been utilised so far. A possible solution would be to calculate the estimated value of z of the track at a chosen radius R per each stub, assuming that the tracks originate from the beam spot.

$$z_R = \frac{R}{r} z. \quad (4)$$

Later stubs with a value of z_R far away from the cell average are then discarded.

R Filter
High–p_T tracks should be able to leave signals in all tracker layers they pass through. Therefore it is handy to require a HT cell to contain stubs belonging to 5 or more tracker layers, in order to be marked for readout.

Algorithm Performance
The algorithm has been tested using simulated data samples of different signal and pile-up content. Figure 3a shows the distribution of the algorithmic efficiency as a function of the pile-up for three different kinds of events (single muon, single electron and $t\bar{t}$). Results obtained show a generally good algorithmic efficiency, larger than 97% for muons and $t\bar{t}$. In contrast, electron candidates have a lower efficiency (~85%), mainly because of the Bremsstrahlung effect that deviates the electron's trajectory.

The distribution of the number of candidates per event as function of the pile–up content is displayed in Fig. 3b. The track filtering stages significantly reduce the number of track candidates, even if more improvements are still required to further diminish the fake rate.

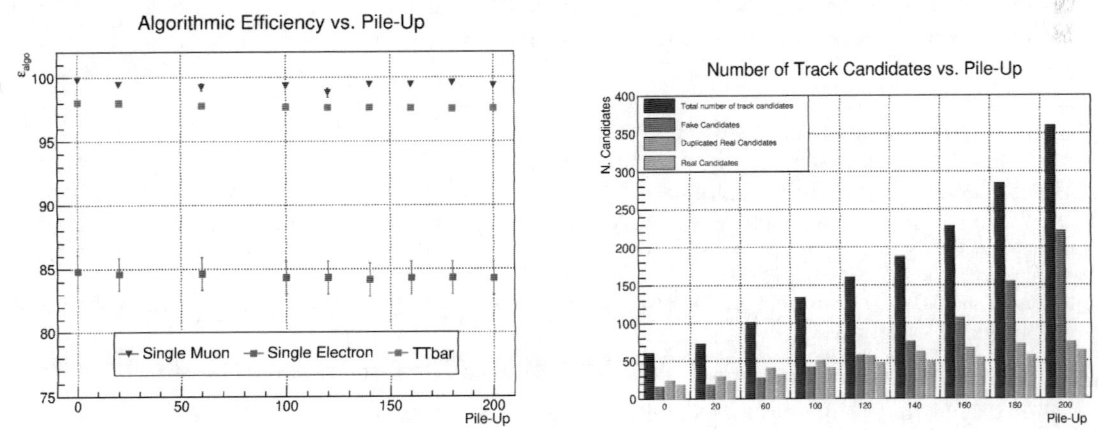

(a) (b)

Figure 3: *(a) Track Finding Algorithmic Efficiency, shown as a function of the Pile-Up content, in different samples. (b) Distribution of number of track candidates per event found by the HT algorithm as a function of the pile–up content. Also the distributions of the number of real, duplicate and fake candidate tracks per event are shown.*

4. Hardware Demonstration

Two different firmware implementations of the Hough Transform algorithm have been developed so far: the systolic array and the pipelined Hough Transform. The former reproduces in hardware the Hough Transform array in a bi-dimensional structure (q cells $\times$ m cells). In this case the same histogram is used for all ϕ sector, having then just one array per MP7. Nevertheless preliminary results have been obtained on a hardware demonstrator using the pipelined firmware. In this implementation each pair of I/O links of the MP7 is assigned to a single ϕ_0 segment. The Hough Transform algorithm is then run in parallel independently for each segment. Due to the resource limitations of the MP7, it is possible to implement a maximum of 9 segments per chip, therefore every board can process data belonging to one region in pseudo–rapidity and 9 segments in ϕ_0.

Preliminary hardware results show a good agreement with the floating point algorithm simulation. In fact the pipelined firmware has been able to find $\sim$92% of the candidate tracks identified by the simulation. Further work is therefore needed to assure an exact matching between the firmware and the simulation.

5. Conclusions

A Level–1 Track Finding algorithm for the proposed Time Multiplexed Track Trigger for the phase II CMS experiment has been presented herein. The proposed algorithm, based on the Hough Transform technique, is capable to detect tracks with transverse momenta above 3 GeV/c with an excellent efficiency, keeping the total number of track candidates within an acceptable level.

A hardware demonstrator has been set up, by means of MP7 boards. Two different implementations of the HT algorithm have been developed. Preliminary hardware results have shown a encouraging agreement with the software simulation. Further work is still required to obtain a perfect matching between the two.

Acknowledgements

The research leading to these results has received funding from the People Programme (Marie Curie Actions) of the European Union's Seventh Framework Programme FP7/2007-2013/ under REA grant agreement nr. 317446 INFIERI "INtelligent Fast Interconnected and Efficient Devices for Frontier Exploitation in Research and Industry".

References

[1] Bruning O 2015 The High luminosity LHC project description *Proceedings, 6th International Particle Accelerator Conference (IPAC 2015)* IPAC-2015-FRXC2
[2] CMS Collaboration 2008 *J. Inst.:* **3** S08004
[3] Butler J and Contardo D and Klute M and Mans J and Silvestris L 2015 *Technical proposal for the phase-II upgrade of the CMS detector* (Geneva, CERN) CERN-LHCC-2015-010
[4] Tricomi A 2014 *J. Inst:* **9** C03041
[5] Paramesvaran S 2016 The next step in real time data processing for large scale physics experiments *Proceedings of this conference*
[6] Hall G, Newbold D, Pesaresi M and Rose A 2014 *J. Inst.:* **9** C10034
[7] Compton K *et al.* 2012 *J. Inst:* **7** C12024
[8] Hough P 1962 Method and means for recognizing complex patterns *US Patent:* 3,069,654

Journal of Physics: Conference Series **762** (2016) 012021 doi:10.1088/1742-6596/762/1/012021

Integration Of PanDA Workload Management System With Supercomputers for ATLAS and Data Intensive Science

A Klimentov[1], K De[2], S Jha[3], T Maeno[1], P Nilsson[1], D Oleynik[2,5], S Panitkin[1], J Wells [4] and T Wenaus[1]

[1] Physics Department, Brookhaven National Laboratory, Upton, NY, US

[2] Physics Department, University of Texas at Arlington, Arlington, TX 76019, USA

[3] Department of Electrical and Computer Engineering Rutgers University, NJ 08854-8058, USA

[4] Oak Ridge Leadership Computing Facility, Oak Ridge National Laboratory, Oak Ridge, TN, 37831-6161

[5] Laboratory of information Technologies, Joint Institute for Nuclear Research, Dubna 141980, Russia

E-mail : alexei.klimentov@cern.ch

Abstract. *The LHC, operating at CERN, is leading Big Data driven scientific explorations. Experiments at the LHC explore the fundamental nature of matter and the basic forces that shape our universe.*

ATLAS, one of the largest collaborations ever assembled in the sciences, is at the forefront of research at the LHC. To address an unprecedented multi-petabyte data processing challenge, the ATLAS experiment is relying on a heterogeneous distributed computational infrastructure. The ATLAS experiment uses PanDA (Production and Data Analysis) Workload Management System for managing the workflow for all data processing on over 150 data centers. Through PanDA, ATLAS physicists see a single computing facility that enables rapid scientific breakthroughs for the experiment, even though the data centers are physically scattered all over the world. While PanDA currently uses more than 250,000 cores with a peak performance of 0.3 petaFLOPS, LHC data taking runs require more resources than grid can possibly provide. To alleviate these challenges, LHC experiments are engaged in an ambitious program to expand the current computing model to include additional resources such as the opportunistic use of supercomputers.

We will describe a project aimed at integration of PanDA WMS with supercomputers in United States, in particular with Titan supercomputer at Oak Ridge Leadership Computing Facility. Current approach utilizes modified PanDA pilot framework for job submission to the supercomputers batch queues and local data management, with light-weight MPI wrappers to run single threaded workloads in parallel on LCFs multi-core worker nodes. This implementation was tested with a variety of Monte-Carlo workloads on several supercomputing platforms for ALICE and ATLAS experiments and it is in full production for the ATLAS since September 2015.

We will present our current accomplishments with running PanDA at supercomputers and demonstrate our ability to use PanDA as a portal independent of the computing facilities infrastructure for High Energy and Nuclear Physics as well as other data-intensive science applications, such as bioinformatics and astro-particle physics.

1.Introduction

The largest scientific instrument in the world – the Large Hadron Collider [1] – operates at the CERN Laboratory in Geneva, Switzerland. The ATLAS [2] experiment at the LHC explore the fundamental nature of matter and the basic forces that shape our universe. To address an unprecedented multi-petabyte data processing challenge, experiments are relying on the computational grids infrastructure deployed within the framework of the Worldwide LHC Computing Grid (WLCG) [3].

The WLCG infrastructure will be sufficient for the planned analysis and data processing, but it will be insufficient for Monte Carlo (MC) production and any extra activities. Additional computing and storage resources are therefore required. To alleviate these challenges, ATLAS is engaged in an ambitious program to expand the current computing model to include additional resources including the opportunistic use of supercomputers and high-performance computing clusters (HPC's).

2.PanDA Workload Management System

PanDA (Production and Distributed Analysis) is the workload management system developed in 2005 for the ATLAS experiment. Its move to production represented a paradigm shift in ATLAS computing by federating the O(100) heterogeneous, independent computing centers of the WLCG into a unique job submission system. Since then, PanDA manages both the analysis jobs from around 3000 users and the production jobs from a few groups of power-users, and brokers them to the best available resource. Furthermore, PanDA is more than a job submission system, and is capable of managing complex tasks, defining different steps and dependencies between the jobs inside a task.

PanDA is built as a central, well-tuned service that knows the status of the distributed resources and manages these intelligently by matching jobs and sites to ensure quick turnaround and full resource usage (figure 1). In addition PanDA implements global fair-shares, job priorities and policies. PanDA communicates with so called pilot jobs [4,5], which are running on the worker nodes and request the real payload once a worker node health-check has been completed.

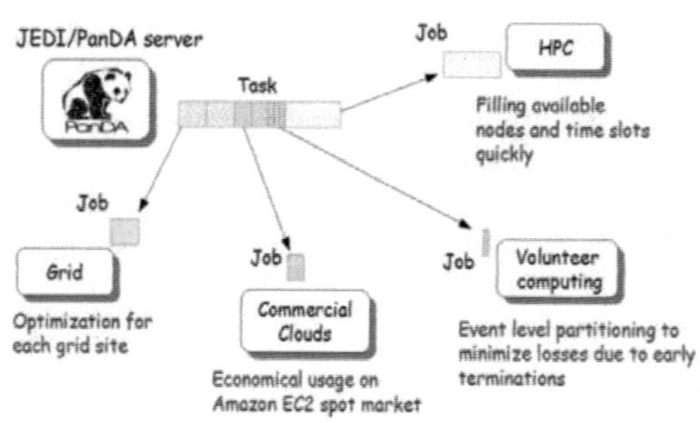

Figure 1: Dynamic job definition adapted to the computing resource

Using of pilot-based workflows helps to improve job reliability, optimize resource utilization, allows for opportunistic resources usage, and mitigates many of the problems associated with the inhomogeneities found on the Grid. Extending PanDA beyond the Grid will further expand the potential user community and the resources available to them.

3.Extending PandDA to Titan supercomputer

Titan is the most powerful supercomputer for open science (number 2 in Top 500 list) hosted at the Oak Ridge Leadership Computing Facility in Oak Ridge National Laboratory, USA [5].

Titan is a hybrid-architecture Cray XK7 system with a theoretical peak performance exceeding 27 petaflops. Titan features 18,688 compute nodes, (each with one 16-core AMD Opteron CPU and 1 NVIDIA Kepler K20X GPU), 299,008 x86 cores, a total system memory of 710 terabytes, and a high-performance proprietary network [6]. Worker nodes use Cray's Gemini interconnect for inter-node MPI messaging but have no network connection to the outside world. Titan is served by the shared Lustre filesystem that has 32 PB of disk storage and by the HPSS tape storage that has 29 PB capacity. Titan's worker nodes run Compute Node Linux which is a run time environment based on the Linux kernel derived from SUSE Linux Enterprise Server.

Taking advantage of modular and extensible design of PanDA, the PanDA pilot code and logic has been enhanced with tools and methods relevant for HPC. The pilot runs on Titan's front-end nodes which allows it to communicate with the PanDA server, since front end nodes have connectivity to the Internet. The interactive front-end machines and the worker nodes use a shared file system which makes it possible for the pilot to stage-in input data that is required by the payload and stage-out the produced output at the end of the job. The ATLAS Tier 1 computing center at Brookhaven National Lab is currently used for data transfer to and from Titan, but in principle that can be any Grid site. The pilot submits ATLAS payloads to the worker nodes using the local batch system (PBS) via the SAGA (Simple API for Grid Applications) interface [7][8]. Figure 2 shows the schematic diagram of PanDA components on Titan.

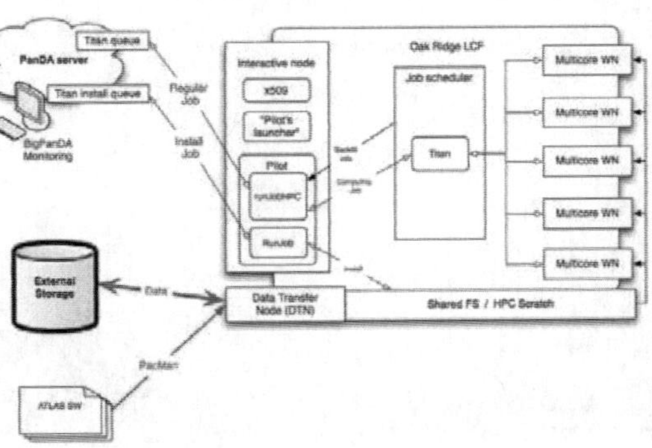

Figure 2 Titan integration with PanDA

The majority of experimental high energy physics workloads do not use Message Passing Interface (MPI). They are designed around event level parallelism and thus are executed on the Grid independently. Typically, detector simulation workloads can run on a single compute node using multiprocessing.

For running such workloads on Titan we developed an MPI wrapper that allows us to launch multiple instances of single node workloads simultaneously. MPI wrappers are typically workload specific since they are responsible for setup of workload specific environment, organization of per-rank worker directories, rank specific data management, input parameters modification when necessary and cleanup on exit. The wrapper scripts is what pilot actually submits to a batch queue to run on Titan. The pilot reserves the necessary number of worker nodes at submission time and at run time a corresponding number of copies of the wrapper script will be activated on Titan. Each copy will know its MPI rank (an index that runs from zero to a maximum number of nodes or script copies) as well as the total number of ranks in the current submission . When activated on worker nodes each copy of the wrapper script, after completing the necessary preparations, will start the actual payload as a subprocess and will wait until its completion. In other words the MPI wrapper serves as a "container" for non-MPI workloads and allows us to efficiently run unmodified Grid-centric workloads on a parallel computational platforms, like Titan.

Leadership Computing Facilities (LCF), like Titan, are geared towards large scale jobs by design. Time allocation on an LCF machines is very competitive and large scale projects are often preferred. This is especially true for Titan at OLCF which was designed to be the most powerful machine in the world, capable of running extreme scale computational projects. As a consequence, on average, about 10% of capacity on Titan is unused due to mismatches between job sizes and available resources. The worker nodes sit idle because there are not enough of them to handle a large scale computing job. On Titan, these 10% correspond to estimated 300M core hours per year. Hence, a system that can occupy those temporarily free nodes would be very valuable. It would allow delivery of more CPU cycles for scientific research, while simultaneously improving resource utilization efficiency on Titan. This offers a great possibility for PanDA to harvest these opportunistic resources on Titan.

During the early steps of this project aimed to integration of Titan with PanDA, it was demonstrated that PanDA is capable of improving the overall utilization of Titan by running in backfill mode. In this mode, PanDA submits and executes jobs that fit within unused cycles between large leadership class jobs. New functionality was added to the PanDA Pilot to collect information in real time about available unused worker nodes on Titan. This allows PanDA to predict precisely the size and duration of jobs submitted to Titan based on available free resources. Dynamic occupation of backfill resources of Titan shown on Figure 3.

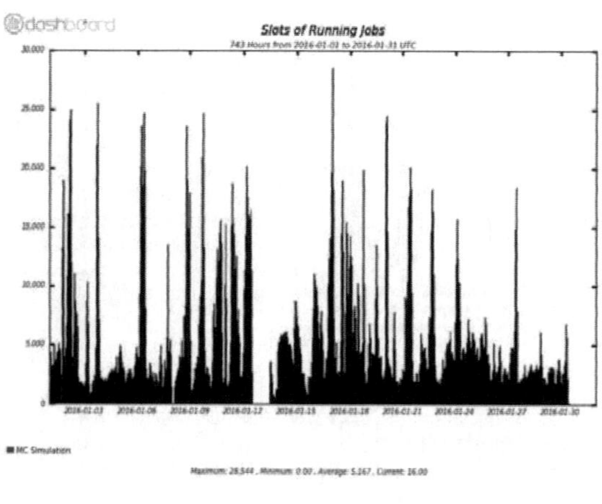

Figure 3. Job slots (CPU cores) handled by PanDA on Titan during January 2016

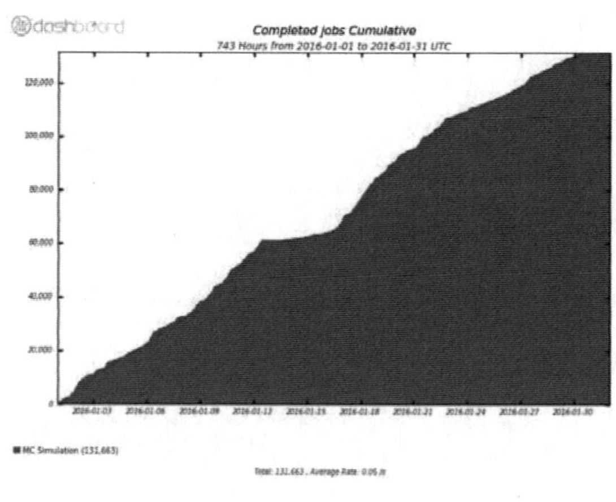

Figure 4. Number of completed jobs on Titan in January 2016

Titan was fully integrated with the ATLAS PanDA based Production and Analysis system and now the ATLAS experiment routinely runs Monte-Carlo simulation tasks there. All operations, including data transfers to and from Titan, are transparent to the ATLAS Computing Operations team and physicists.

4.Conclusion

PanDA WMS was successfully integrated with one of most powerful supercomputer in the world, and this enables Titan to process data from the ATLAS experiment as a valuable computing resource.

Just in one month in January 2016, more than 130,000 PanDA jobs were completed on Titan (Figure 4)

One PAnDA pilot was able to manage up to 300 simultaneously launched PanDA jobs filling up to 4800 cores. With seven simultaneously launched pilots this solution allows to consume >2M CPU/hours per month of backfill resources on Titan.

Initial results approve that advised architecture works, but also shows which parts should be improved for increasing of efficiency:

- Improving of warming-up procedures, environment setup and initial checks before launching of payload may take significant time for big chunks of jobs;
- Data Management procedures, especially Stage-Out.
- IO operations as in Pilot, so in payloads
- Communication with PanDA server, to update information about jobs in bulk.

References

[1] Maeno T et al 2007 PanDA: Distributed Production and Distributed Analysis System for ATLAS *International Conference on Computing in High Energy and Nuclear Physics* (Victoria, Canada)

[2] ATLAS Collaboration 2008 *J. Inst. 3* S08003.

[3] 2005 LHC Computing Grid: Technical Design Report. document LCG-TDR-001, *CERN-LHCC-2005-024 (The LCG TDR Editorial Board)* ISBN 92-9083-253-3

[4] Nilsson P et al 2007 Experience from a pilot based system for ATLAS *International Conference on Computing in High Energy and Nuclear Physics* (Victoria, Canada)

[5] Top500 List *http://www.top500.org/*

[6] Titan at OLCF web page *https://www.olcf.ornl.gov/titan/*

[7] The SAGA Framework web site *http://saga-project.github.io*

[8] Merzky A Weidner O Jha S 2015 SAGA: A Standardized Access Layer to Heterogeneous Distributed Computing Infrastructure *Software-X* DOI: 10.1016/j.softx.2015.03.001

Acknowledgements

This project was funded in part by U. S. DOE, Office of Science, High Energy Physics and Advanced Scientific Computing Research (ASCR) under Contract No. DE-AC02-98CH10886. This research used resources of the Oak Ridge Leadership Computing Facility at the Oak Ridge National Laboratory, which is supported by the Office of Science of the U.S. Department of Energy under Contract No. DE-AC05-00OR22725.

The impact of Moore's Law and loss of Dennard scaling: Are DSP SoCs an energy efficient alternative to x86 SoCs?

L Johnsson[1], **G Netzer**[2]

[1] Department of Computer Science, University of Houston, Houston, TX 77204, USA
[2] PDC, Royal Institute of Technology, SE 100 40 Stockholm, Sweden

E-Mail: johnsson@cs.uh.edu

Abstract Moore's law, the doubling of transistors per unit area for each CMOS technology generation, is expected to continue throughout the decade, while Dennard voltage scaling resulting in constant power per unit area stopped about a decade ago. The semiconductor industry's response to the loss of Dennard scaling and the consequent challenges in managing power distribution and dissipation has been leveled off clock rates, a die performance gain reduced from about a factor of 2.8 to 1.4 per technology generation, and multi-core processor dies with increased cache sizes. Increased caches sizes offers performance benefits for many applications as well as energy savings. Accessing data in cache is considerably more energy efficient than main memory accesses. Further, caches consume less power than a corresponding amount of functional logic. As feature sizes continue to be scaled down an increasing fraction of the die must be "underutilized" or "dark" due to power constraints. With power being a prime design constraint there is a concerted effort to find significantly more energy efficient chip architectures than dominant in servers today, with chips potentially incorporating several types of cores to cover a range of applications, or different functions in an application, as is already common for the mobile processor market. Digital Signal Processors (DSPs), largely targeting the embedded and mobile processor markets, typically have been designed for a power consumption of 10% or less of a typical x86 CPU, yet with much more than 10% of the floating-point capability of the same technology generation x86 CPUs. Thus, DSPs could potentially offer an energy efficient alternative to x86 CPUs. Here we report an assessment of the Texas Instruments TMS320C6678 DSP in regards to its energy efficiency for two common HPC benchmarks: STREAM (memory system benchmark) and HPL (CPU benchmark)

1. Introduction.

Energy efficiency has become a primary concern in all stages of computer systems design and operation, from solid-state device design, see, e.g. [1,2,3,4] to data-center operations, see e.g. [5,6,7,8,9,10,11,12], in regards to cost, environmental impact, and sustainability of energy supply. With an annual growth rate estimated at 6.6% for ICT (Information and Computing Technologies) [13] and an estimated 4.7% share of the total electric energy consumption in 2012, by 2040 ICT would consume 30% of todays' total energy consumption. Even assuming a continued growth rate of total electric energy consumption at the current level, about 3%/year [14], a substantial portion of the total electric energy consumption would be consumed by ICT. These concerns have engaged the semiconductor industry, the platform industry as well as large consumers of computing, such as the internet companies, in finding more energy efficient solutions to computing. Facebook [15,16,17], Google [18], and Microsoft [9,10,11,19] as well as others have opted to design their own platforms, data centers [6,7,20,21,22], and software systems [23,24]. The CMOS (Complementary Metal-Oxide Semiconductor) processor design challenges in light of continued Moore's law scaling [25] without Dennard scaling [26] are illustrated well in [27] that also describes how the exponential growth in transistors per die have been used historically and the resulting decrease in energy efficiency. The impact of the energy challenges on computer architecture and System-on-Chip (SoC) design has also been highlighted in, e.g. [28,29,30], whereas [31] highlights the impact of the energy efficiency challenges on software.

Dennard scaling [26] through which the electric field of a transistor remains constant while reducing the physical dimension, allowed for a doubling of the number of transistors per unit area for each CMOS technology generation while keeping the power dissipation per unit area constant. Further, the combination of the scaling of physical dimensions and voltage the switching speed was also reduced so the performance per unit area increased by about a factor of about 2.8 for each technology generation. However, by about 2004 the feature sizes for CMOS had scaled down to a size where leakage power reached a level where further reduction in voltage was no longer feasible, Figures 1 and 2, and heat density reached unsustainable levels, Figure 3. The industry's response was multi-core processors, with the IBM Power4 in 2001 [34] being the first followed by dual-core x86 based CPUs by AMD in 2004 [35] and Intel in 2005 [36]. The change in CMOS processor characteristics is well captured in Figure 4.

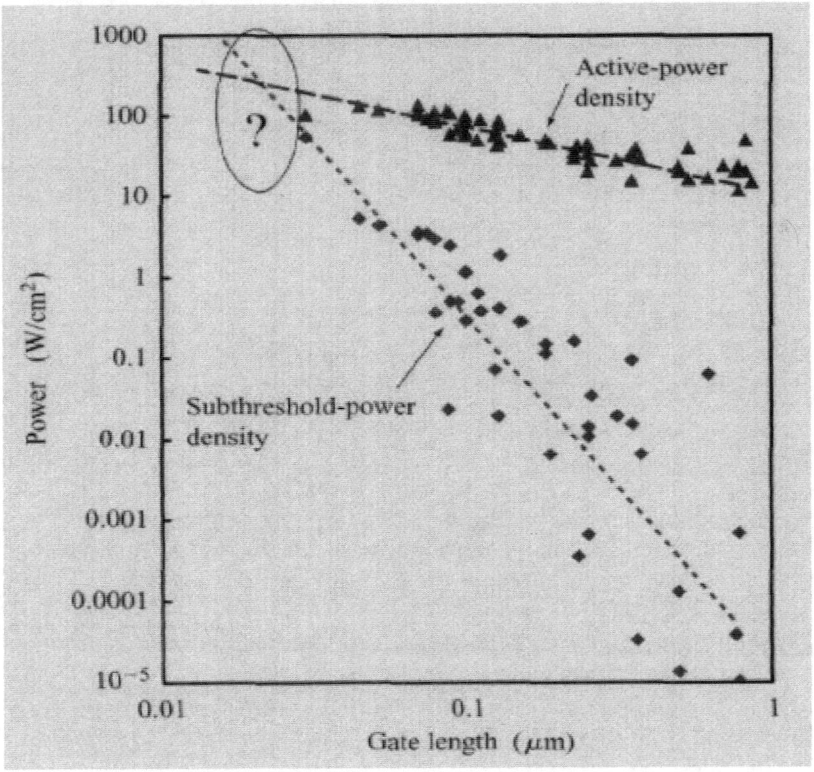

Figure 1. Dennard scaling of active power and leakage [32].

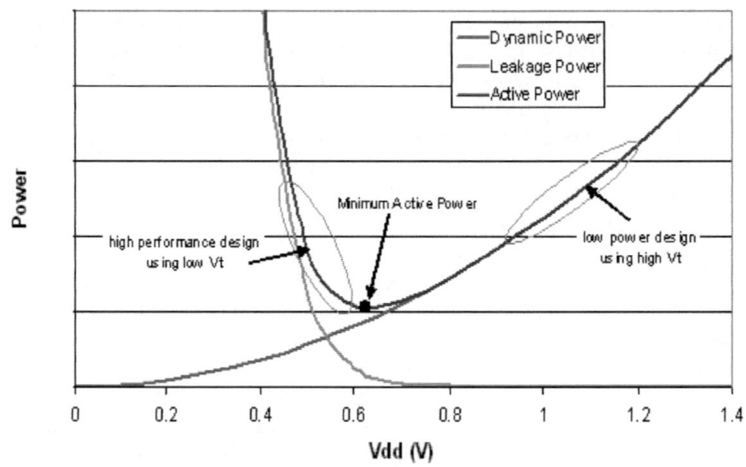

Figure 2. Power consumption as a function of threshold and supply voltage [33].

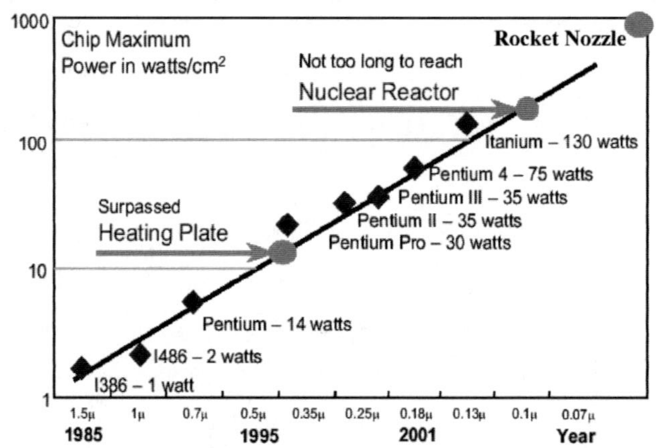

Figure 3. Power densities for Intel CPUs. Source: Shekhar Borkar, Intel.

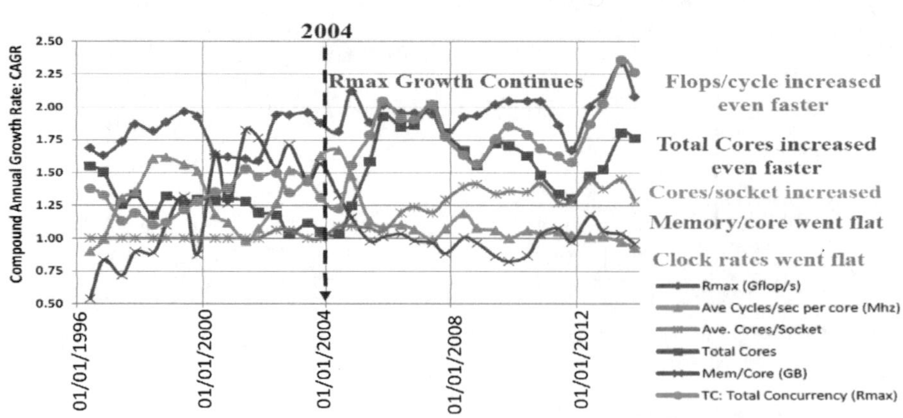

Figure 4. The evolution of CPU characteristics 1996 – 2012, P. Kogge [37].

The power constraints led to the concept of "dark silicon" [38], i.e., silicon that is underutilized. In todays' CPUs (22 nm technology) about 20% is dark [39], a percentage that is rapidly increasing as feature sizes keep being scaled down [40] such that by the end of the decade predictions range from 50% [39,41] to 90+% dark [38,40] silicon for processor dies. This fact poses interesting challenges and opportunities in processor design. The energy requirements for computations carried out on "big", "small" and specialized cores can be quite different. Further, on die memory (cache) tend to consume less energy than corresponding amount of functional logic since only small fractions of memory are actively used at any given time. Caches can not only lead to increased performance but also reduced need for very energy consuming main memory accesses. Though L1 and L2 cache sizes per core have remained relatively unchanged over time, L3 caches have increased significantly with some current processor dies having over 100MB of cache [42]. The difference in average power density of functional logic and caches is illustrated in Figure 5.

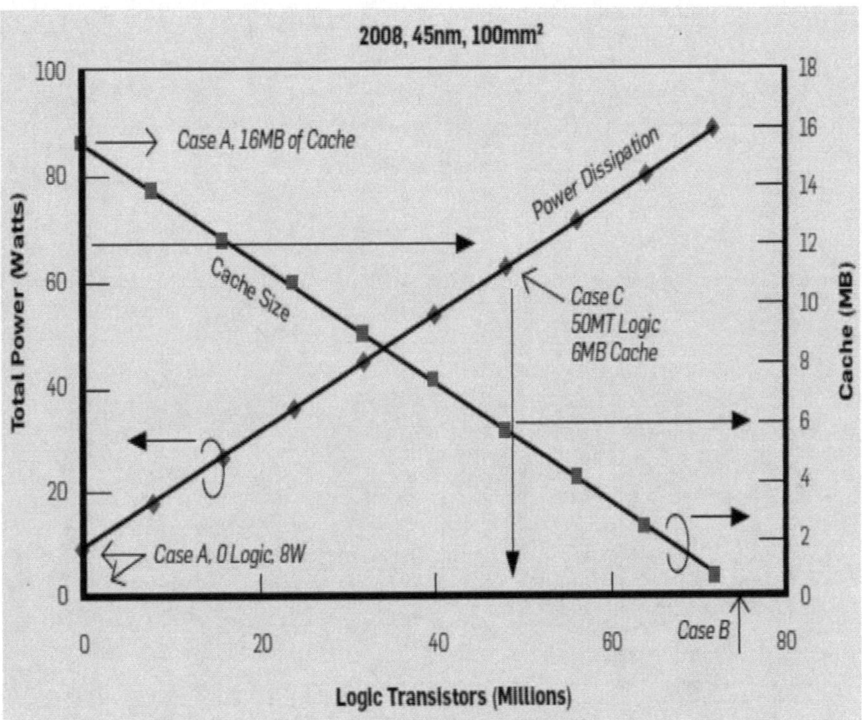

Figure 5. Functional logic and cache trade-offs for constant power dissipation [27].

The energy efficiency benefits of specialization can yield several orders of magnitude improvement in performance, power consumption, or both as seen in Figure 6. For the embedded and mobile markets the use of specialization to achieve high energy efficiency is common as exemplified by the schematic drawing of a Texas Instruments (TI) System-on-Chip (SoC), Figure 7, and the Qualcomm Snapdragon 820 SoC, Figure 8. A comparison of the nominal 64-bit floating-point energy efficiency and computational density for processors considered for evaluation in the PRACE project is shown in Table 1.

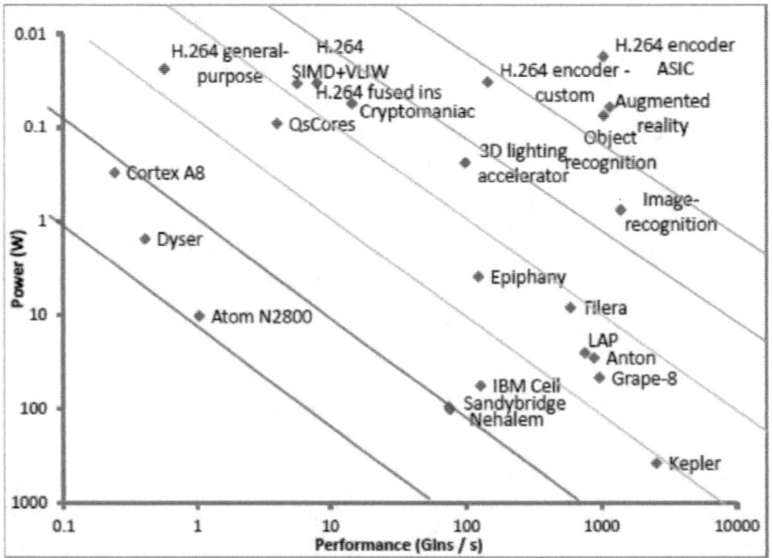

Figure 6. Power and performance of a wide range of accelerators (scaled to 45 nm) [43].

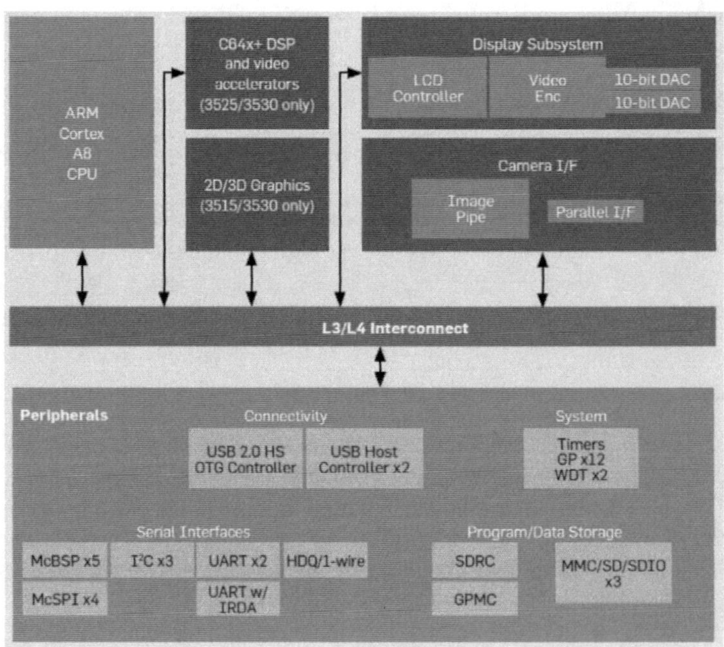

Figure 7. Schematic of a Texas Instruments System-on-Chip [27].

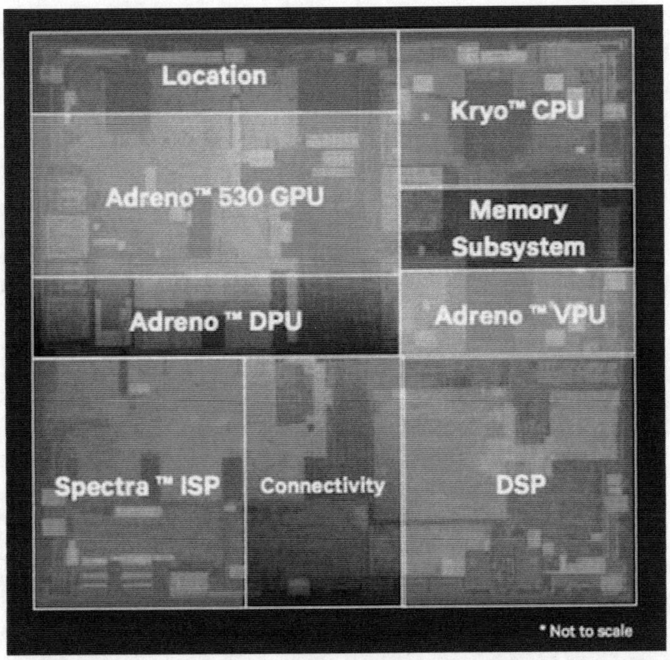

Figure 8. Qualcomm Snapdragon 820 mobile power and thermal optimized SoC [44].

Table 1. Nominal double precision floating-point performance of processors evaluated or considered for evaluation in the PRACE 1st Implementation Phase project [45].

CPU	Current [2012]			Next Generation [2012]		
	Lith. [nm]	E. Eff. [GF/J]	Comp. Dens. [GF/mm^2]	Lith. [nm]	E. Eff. [GF/J]	Comp. Dens. [GF/mm^2]
AMD Interlagos (16C, 2.7 GHz)	32	1.3	0.55			
AMD FirePro 7900	40	3.0	1.19			
AMD S9000				28	3.6	2.21
AMD Trinity	32	1.0	0.40			
IBM Blue Gene/Q	45	3.7	0.57			
IBM Power7	45	1.4	0.48			
Intel Sandy Bridge (8C, 3.1 GHz)	32	1.3	0.46			
Intel Ivy Bridge (4C, 3.5 GHz)				22	1.45	0.70
NVIDIA Fermi	40	2.66	1.26			
NVIDIA Kepler				28	5.2	2.36
NVIDIA Tegra 2	40	≈1	0.04			
TI TMS320C6678	40	6	≈3			
Xilinx Vertex-6	40	5-10				
Xilinx Vertex-7				28	≈13	0.236

Below, we report briefly on an assessment of the energy efficiency of a DSP based SoC in the context of High-Performance Computing (HPC) applications. The assessment used the TI TMS320C6678 DSP, a 10W SoC manufactured in 40 nm CMOS technology with a potential peak 64-bit floating-point performance of 60 GF/s and hence a nominal peak energy efficiency of 6GF/J. As a comparison the contemporary x86 CPU by Intel, the 32nm Sandy Bridge, has a nominal peak energy efficiency of 1.3 GF/J, and the 45nm Blue Gene/Q a nominal energy efficiency of 3.7 GF/J. In section 2 we describe the features of the TI SoC essential for our benchmark implementations, and in Section 3 the instrumentation used to measure power consumption. Section 4 describes the outcome and techniques used to achieve high efficiency and energy efficiency for the STREAM memory system benchmark [46] and Section 5 the result of and techniques used in the High-Performance Linpack (HPL) compute intensive benchmark [47]. Section 6 gives the conclusions.

2. The Texas Instruments DSP TMS320C6678

The TMS320C6678 multi-core SoC, referred to as the 6678 below, contains eight TMS320C66x DSP cores. These 8-wide Very Large Instruction Word (VLIW) processors can execute four IEEE 754 64-bit floating-point additions and two multiplications, together with two 64-bit data accesses per clock cycle in a single hardware thread [48]. The 16-stage instruction pipeline is almost free of interlocks and makes extensive use of *delay slots* to hide instruction latencies. At a maximum common core clock frequency of 1.25 GHz, the SoC delivers up to 60 GF/s, double precision, within the 10W TDP (Thermal Design Power) envelope, and a nominal energy efficiency of 6 GF/J. Figure 9 shows a functional block diagram of the 6678.

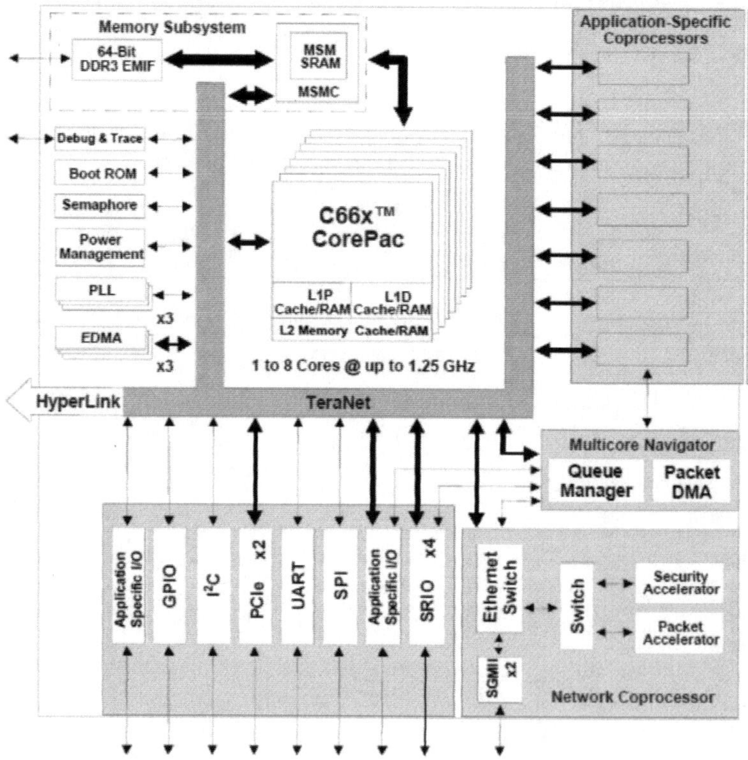

Figure 9. A functional block diagram of the TI TMS320C6678 SoC [49].

A single 64-bit DDR3 interface supports up to 8GiB at up to 1600 MT/s. Of the SoC communication features the proprietary 50 Gb/s Hyperlink [50] is the most interesting for clustering of DSPs for HPC and offers a 0.1 B/Flop off-chip communication to double-precision floating-point capability. This communication-to-compute ratio is almost identical to that of the Blue Gene/Q [51]. The 6678 core private L1 and L2 memories, Table 2, can be configured as cache or fast scratch-pad memory, referred to as SRAM below, or a combination thereof [52]. The memory latencies in Table 2 are from load to use, with L1 latencies fully overlapped by five delay slots for branches and four for loads. Stores do not require a delay slot.

Table 2. TI 6678 core private memory characteristics [53].

	Unit	L1I	L1D	L2
Total Size	KiB	32	32	512
Cache Block Size	B	32	64	128
Associativity		Direct	2-way	4-way
Replacement Policy			LRU	LRU
Allocation Policy		Read	Read	Read/Write
Clock		f_{ck}	f_{ck}	$f_{ck}/2$
Latency SRAM	Cycles	6	5	8-15.5
Latency Cache	Cycles	6	5	12-17.5
Bandwidth	B/Cycle		16	16

Core-initiated memory operations pass via its eXtended Memory Controller (XMC), which also contains a simple hardware prefetcher, to the 6678 Multi-core Shared Memory Controller (MSMC) that manages accesses to the 4MiB on-chip Multi-core Shared Memory (MSM), as well as to the external DDR3 memory. Figure 10 illustrates the data paths of a core (one shown) as well as the 6678 interconnect fabric, the 2 Tbps non-blocking TeraNet (which in fact consists of three networks). The digit after the name in Figure 10 denotes the fraction of the core clock frequency, f_{CK}, of a particular TeraNet (TeraNet 2 runs at $f_{CK}/2$, TeraNet 3 at $f_{CK}/3$). Communication between a core, other cores and on-chip peripheral units use the TeraNets via a core's External Memory Controller (EMC). The TeraNet also connects to the MSM and external memory and hence provides an alternate path to the one via the XMC to MSM and external memory. However, while the XMC port provides a bandwidth of $32*f_{CK}/2$ B/s, the EMC port is connected to the TeraNet 3 and hence is limited to a bandwidth of $16*f_{CK}/3$ B/s. Despite the limitation of the path via the EMC it can offer higher realized bandwidth than the XMC path as will be shown in sections four and five. Using the EMC path a single core cannot fully utilize the DDR3-1333 memory bandwidth of 10.67 GB/s, but two cores could potentially do so.

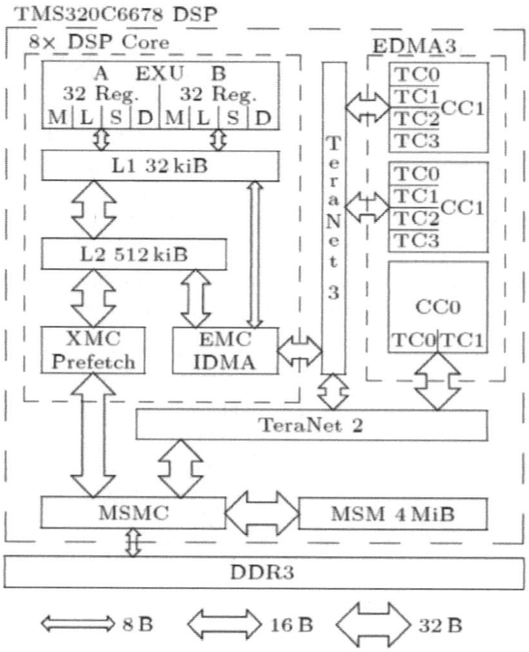

Figure 10. A schematic of the data paths of the 6678 SoC [54].

The 6678 has two DMA engines: 1) an Internal DMA (IDMA) engine [52] that can transfer data between the core private L1 and L2 memories, and between those and the EMC and 2) an Enhanced DMA engine, EDMA3 [55]. The EDMA3 controller can move data between all memories including the core private memories. The EDMA3 engine is very versatile and supports strided block transfers as shown in Figure 11. It has three independent Channel Controllers (CC) that each can serve multiple Transfer Controllers (TC) that execute the data transfers. The CCs support *chaining* by which the completion of a transfer can trigger another transfer, possibly on another channel. *Linking* enables the automatic reload of new transfer specifications from the controller's Parameter RAM (PaRAM) making implementations of linked lists of transfers possible. Strided block transfers, chaining and linking are used to achieve high efficiency and competitive energy efficiency in the benchmarks reported below.

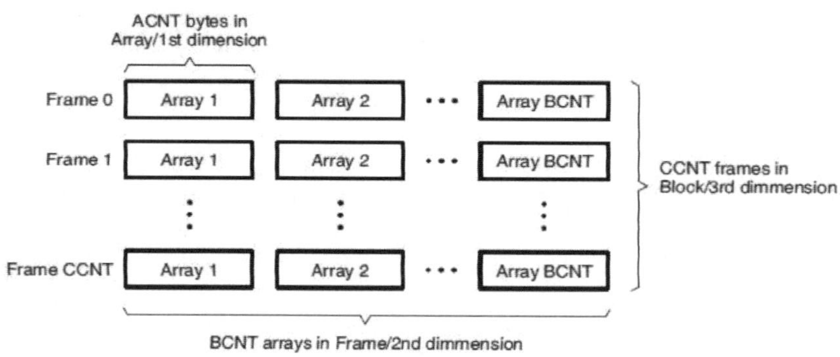

Figure 11. The EDMA3 strided block transfer capabilities [55].

Our experimental setup consists of an Advantech Evaluation Module (EVM) containing the 6678 and 1GiB DDR3 memory [56]. The four 2Gb Samsung K4B2G1646C-HCH9 DDR3 memories run at 666MHz (1333MT/s) giving a peak bandwidth of 10.67 GB/s. The eight bank DDR3 memory has a page size of 8KiB. Knowledge of the page size is essential to understand some aspects of the performance behavior of the benchmarks.

3. The Power Measurement setup

The 6678 has 16 power domains: one for each of the cores, one for the MSM, one for the Hyperlink, one for the SRIO (Serial Rapid I/O), one for the PCIe, one for Peripheral Logic, one for the Packet Engine, and one for the Trace Logic and one for remaining logic. There is no voltage control. Each core can be powered off or on independently, a feature we used in the STREAM benchmark in assessing the energy efficiency of the 6678. We also powered off logic not used in the benchmarks, such as the SRIO, Peripheral Logic, and the Packet Engine.

The different power domains are supplied by three power rails that were instrumented as shown in Figure 12. The logic associated with each of the three power rails is shown in Table 3 and depicted in Figure 13. With our instrumentation we estimate the error to less than 1% for the measured parts. For details about the instrumentation and the error estimation see [57].

Table 3. Logic associated with the three power rails supplying the SoC (Variable, Fixed and Memory)

Rail	Name	Voltage [V]	Usage
Variable	CVDD	0.9 – 1.1	All SoC logic
Fixed	CVDD1	1.0	On-chip memories
	VDDT1		HyperLink termination
	VDDT2		Other termination
Memory	DVDD15	1.5	DDR3 Memory and I/O
	VDDR1		HyperLink SerDes
	VDDR2		PCIe SerDes
	VDDR3		SGMII SerDes
	VDDR4		SRIO SerDes
Input		12	DC input to EVM
Other			Uninstrumented parts
			Calculated from above
			DC-DC converter losses

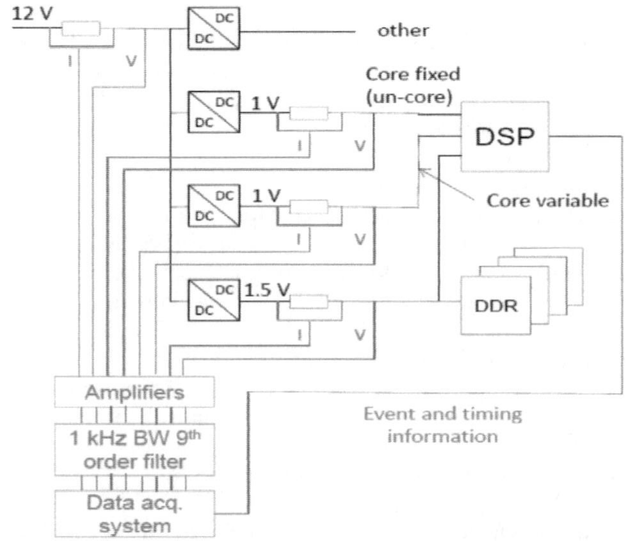

Figure 12. Instrumentation of the EVM for power measurements.

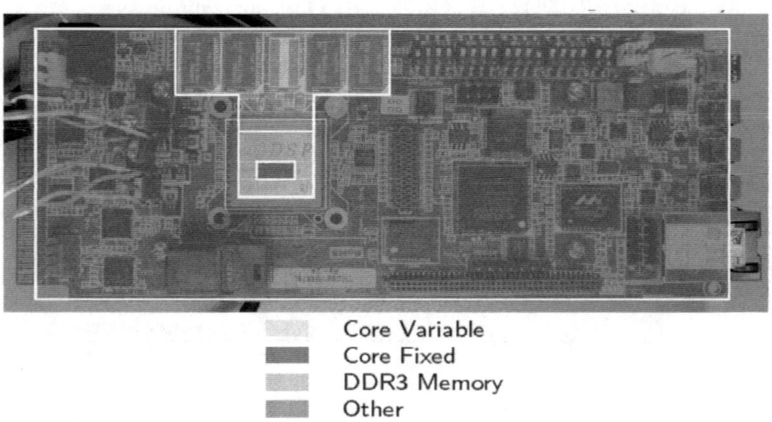

Figure 13. The EVM parts associated with the three power rails.

4. The STREAM Benchmark

The Stream memory system benchmark [46] commonly used to assess memory system performance for stride one access with an emphasis on main memory access consists of the operations shown in Table 4. Despite their low stride that is ideal for the way data is assigned to memory, their regularity, and simplicity, the operations present serious challenges for modern cache based computer architectures. Since there is no data reuse the main objective of caches is not met. Further, caches typically use a read-modify-write policy in writing to main memory causing an extra load not accounted for in the STREAM benchmark hence limiting the peak achievable bandwidth for COPY and SCALE to 2/3rd of the peak main memory bandwidth and to 3/4th of the peak for SUM and TRIAD. The lack of data reuse makes STREAM operations very latency sensitive. To address the memory performance challenges for applications with limited or no data reuse most modern processors employ prefetching, some very sophisticated prefetching techniques that tracks memory access patterns and adapt the prefetching accordingly to avoid stalls caused by waiting for main memory loads. Some also employ cache bypass mechanism for store operations to avoid the cache problems on writes to main memory. Despite these mechanisms, high STREAM performance is difficult to achieve as is evident from Table 5 based on reported performance data and estimates of energy efficiencies when we did not find any reported power consumption data. We estimated the power consumption for STREAM to 50% of the processor TDP. STREAM is not CPU intensive.

Table 4. The STREAM Benchmark operations.

Name	Operation	Data Size [B]	Flops
COPY	$a_i = b_i$	16	0
SCALE	$a_i = qb_i$	16	1
SUM	$a_i = b_i+c_i$	24	1
TRIAD	$a_i = b_i+qc_i$	24	2

Table 5. STREAM data from literature with estimates of energy consumption.

	Bandw. (BW) [GB/s] Peak	Meas.	Eff. [%]	E. Eff. [GB/J]	Lith. [nm]		Ref.
IBM Power 7	409.6	122.8	30	≈0.10	45	Power est.	[58][59],[60]
Intel Xeon Phi	352.0	174.8	50	≈1.20	22	at 50% of	[58],[61]
Intel E5-2697v2	119.4	101.5	85	≈0.30	22	TDP	[62]
NVIDIA Tegra 3	6.0	1.6	27	0.05	40	Power and BW meas.	[63]
TI TMS320C6678 Cache	10.7	3.0	28	0.40	40	Power and	[45]
TI TMS320C6678 DMA	10.7	10.2	96	1.26	40	BW meas. at 1 GHz	[53]

As seen from Table 5, the efficiency of the "6678 cache" STREAM implementation yields an efficiency at the low end of the range among processors in the table, though the energy efficiency of the 40 nm 6678 is about the same as the 22 nm Intel E5-2697v2 (Ivy Bridge). Thus, for the 6678 to represent an interesting alternative from an energy efficiency point of view the memory bandwidth must be utilized close to 100% without a comparable increase in power consumption. If that could be achieved, then the 6678 would compete with the 22 nm Intel Xeon Phi in regards to energy efficiency. As seen in Table 5 the 6678 DMA alternative indeed realizes this potential.

As shown in Table 6 much of the poor performance of the "6678 cache" can be attributed to latency [53]. This fact also implies that the XMC prefetching is not very effective. The only poor performance prediction from measured latencies is that for the DDR3 bandwidth. The reason for the poor prediction is that the DDR3 measured stall time did not incur a significant number of DDR3 bank conflicts. Table 7 shows the impact on DDR3 memory latency as a function of memory stride and resulting bank conflicts.

Table 6. Stall times: estimated and measured. Bandwidth: per core bandwidth estimated based on stall times, bandwidth measured for 8 cores and average measured bandwidth/core [53].

	Stall Time [ns]		Bandwidth [GB/s]		
	Spec. [64]	Meas.	Est./core	Meas./8	Meas.
L1D	0	0	16.0	15.5	124.0
L2 SRAM	3-10.5	7	5.8	6.0	47.6
L2 Cache	7-12.5	9	4.9	5.8	46.5
MSM	-	10	4.6	4.8	38.7
DDR3	-	23	1.7	0.38	3.0

Table 7. Measured latencies as a function of stride for the EVM.

Memory	Stride in Bytes								
	64	128	256	512	1ki	2ki	4ki	8ki	16ki
L1D SRAM	5	5	5	5	5	5	5	5	5
L2 SRAM	12	12	12	12	12	12	12	12	12
MSM	15	28	28	28	28	28	28	28	28
DDR3-1333	28	49	92	93	95	99	106	121	122

By using the EDMA3 and configuring the on-die L1 and L2 memories mostly as scratchpad memory, not only can prefetching be managed such that stalls are avoided, but the DDR access by the eight cores managed such that bank conflicts are minimized and not affecting the performance. In our EDMA3 implementation it is in fact the master of the benchmark execution that signals the cores when data is available in their caches. Conversely, they signal the EDMA3 when data is ready to be copied back to main memory. No communication between cores is necessary. Communication between a core and the EDMA3 is made via flags local to the core. To further decouple core and EDMA3 operations multi-buffering is used so that for each core the core and the EDMA3 use different buffers at any given time. To overlap overhead and latencies in setting up and executing data transfers two transfer controllers are used. The chaining feature of EDMA3 is used to reduce latencies by having the completion of one transfer event directly start another. The linking feature is used to avoid waiting time for loading transfer descriptors. Details of the implementation can be found in [53].

Despite the fact that a single core bandwidth to DDR3 memory is limited to $16*f_{CK}/3$ B/s due to the TeraNet 3, the EDMA3 bandwidth for a single core in our implementation is still higher than what was achieved using the XMC $32*f_{CK}/2$ B/s path. The limited success of the XMC pre-fetcher, the read-modify-write cache policy, and the DDR3 bank conflicts caused by the accesses from the eight cores all contributes to this less than stellar performance (though the efficiency is comparable to some very sophisticated designs). Using the EDMA3 theoretically two cores operating at 1 GHz suffices to saturate the DDR3-1333 bandwidth. Table 8 and Figure 14 give some of the results of our EDMA3 implementation. Our implementation achieves an efficiency of 96.8% for a single core for which the bandwidth is limited by the TeraNet 3. For two cores for which the TeraNet 3 and DDR3 bandwidths are the same the implementation achieves an efficiency of 83.0%. For four active cores the efficiency is 89.0%, for six active cores 92.1% and for eight active cores 95.6%. The "dips" in performance for an odd number of cores is due to load imbalance between the two transfer-controllers used in our EDMA3 STREAM implementation. Increasing the core clock frequency improves the performance somewhat, largely due to the fact that the TeraNet bandwidth scales with the core clock frequency.

Table 8. Performance and energy efficiency for STREAM on 128 MiB vectors and different active core counts with non-active cores powered off for f_{CK} = 1 GHz [53]. DDR3 bandwidth 10.67 GB/s.

Cores	Measured BW [GB/s]				Energy Efficiency [GB/J]			
	COPY	SCALE	SUM	TRIAD	COPY	SCALE	SUM	TRIAD
1	5.16	5.16	5.12	5.12	1.01	1.01	1.01	1.00
2	8.86	8.86	8.41	8.41	1.47	1.47	1.43	1.42
3	6.99	6.99	6.89	6.89	1.15	1.15	1.15	1.14
4	9.50	9.50	9.11	9.11	1.41	1.41	1.38	1.37
5	7.99	7.99	7.85	7.85	1.17	1.17	1.16	1.16
6	9.83	9.83	9.58	9.58	1.33	1.33	1.31	1.30
7	8.67	8.67	8.66	8.66	1.15	1.15	1.15	1.14
8	10.20	10.20	10.04	10.04	1.26	1.26	1.25	1.24

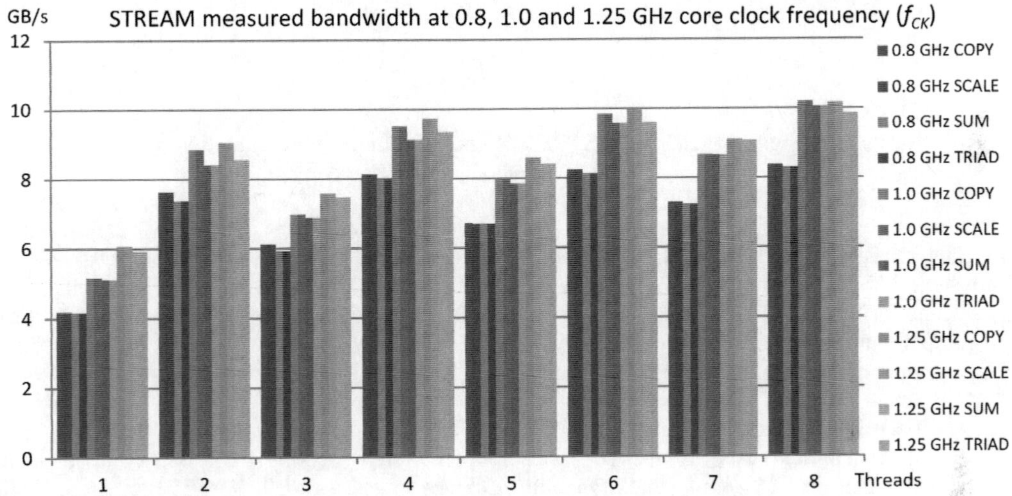

Figure 14. EDMA3 STREAM performance data as a function of core clock frequency and active cores for DDR3-1333 (10.67 GB/s).

Table 8 and Figure 15 shows that the energy efficiency peaks at 1.47 GB/J for COPY and SCALE, 1.43 GB/J for SUM and 1.42 GB/J for TRIAD with two active cores. As additional cores are activated and used the energy consumption increases more than what is gained in DDR3 memory bandwidth utilization. It is also clear that 1 GHz core clock frequency for two or more active cores is optimum. For a single core the increased bandwidth offered by the increased TeraNet bandwidth at the higher clock frequency yields a higher energy efficiency, i.e., the realized bandwidth increases more than the power consumption.

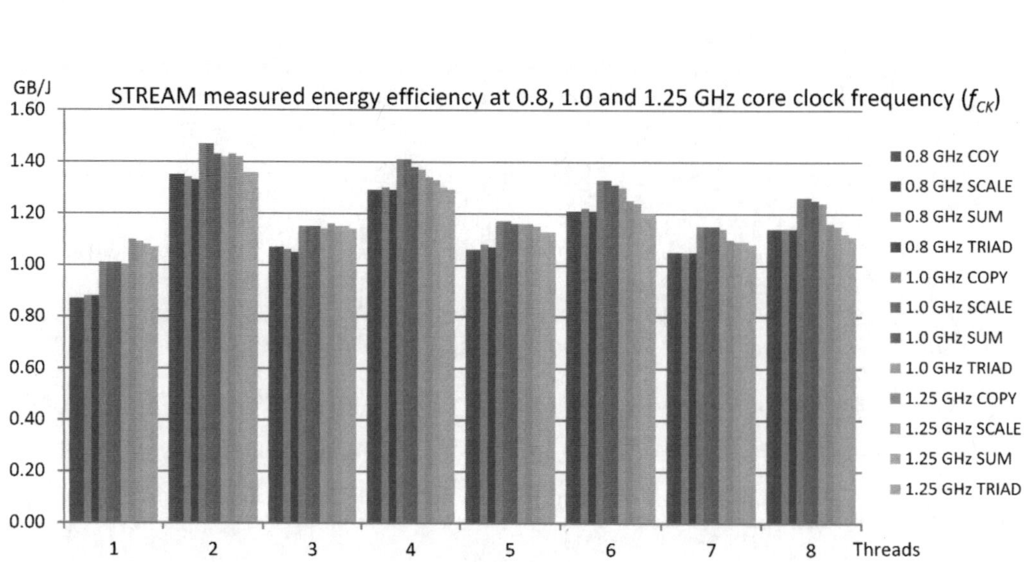

Figure 15. EDMA3 STREAM energy efficiency as a function of core clock frequency and active cores.

Our measurements show that an idle 6678 core consumes about 110 mW. For COPY the power consumption going from idle to fully active increases by about 200 mW at f_{CK} = 1 GHz.

Thus, for STREAM using the EDMA3 feature the 40nm 6678 is competitive with the 22 nm Intel Xeon Phi in regards to energy efficiency, and considerably more efficient in utilizing the main memory bandwidth (about 95% of peak compared to about 50% of peak).

5. The HPL Benchmark

The HPL benchmark (used for the Top500 computer system ranking [65]) is a compute intensive benchmark that on x86 architectures typically achieves in excess of 90% efficiency [66]. For most current architectures high efficiency is required for high energy efficiency (as shown above for STREAM). Like for STREAM a straightforward compilation of the HPL benchmark for the 6678 resulted in poor efficiency, less than 25% at f_{CK} = 1 GHz. To get an efficiency that could make the 6678 a viable competitor in regards to energy efficiency, an efficiency comparable to that of x86 platforms is required. To achieve this goal efficient use of the EDMA3 engine, as for STREAM, was necessary, but not only for DDR3 memory access, but also for overlapping memory accesses with computation. In addition, EDMA3 was used to carry out data transposition in moving data from main memory to L2 SRAM using the MSM as a staging memory and the strided block transfer capability of the EDMA3 [54]. Unlike STREAM, HPL has the potential for significant data reuse since it requires about $\approx 2/3N^3$ operations for N^2 data. For our HPL implementation the 6678 L1 and L2 core private memories are largely configured as SRAM for storing matrix blocks that are reused. The MSM also serves a critical role, mostly as a staging memory in allocating matrix blocks retrieved from main memory to cores in a cyclic fashion, and when needed used for rearranging data from block row-major to block column major ordering. Out HPL implementation also make use of the IDMA engine. It is used to pre-fetch matrix blocks from L2 SRAM to L1 SRAM such that all data loads are from L1 SRAM without stalls. Stores are directed directly to L2 SRAM via a write-buffer bypassing L1 SRAM.

As for most efficient HPL implementation ours also depend on an efficient matrix-multiplication (DGEMM [67]) implementation. Our matrix-multiplication routine achieves an efficiency of about

95% at f_{CK} =1 GHz and hence is comparable to that observed for Intel CPUs and higher than the efficiency on some other x86 CPUs as well as other architectures [66]. With un-optimized forward and backsolve our HPL implementation achieves a peak efficiency of 77%, and a peak energy efficiency of 2.92 GF/J [54] which compares favorably with Intel Sandy Bridge systems, 0.97 GF/J [69] and Blue Gene/Q systems 2.10 GF/J [68].

The 6678 computational rate as a function of matrix size for three core clock rates are shown in Figure 16, the corresponding efficiencies in Figure 17 and energy efficiencies in Figure 18

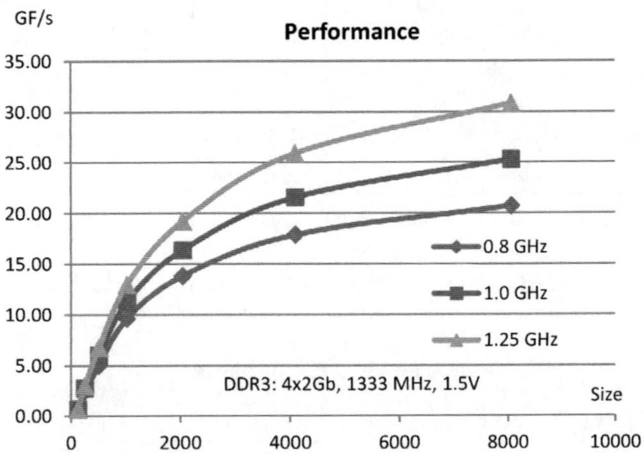

Figure 16. The 6678 HPL performance as a function of matrix size and core clock rate.

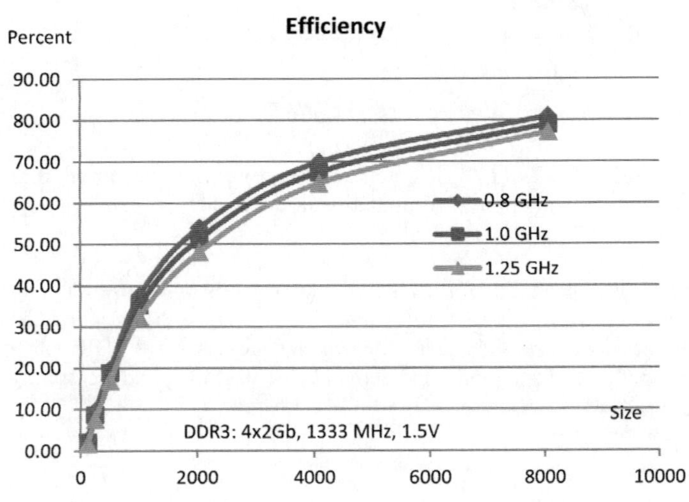

Figure 17. The 6678 HPL efficiency as a function of matrix size and core clock rate.

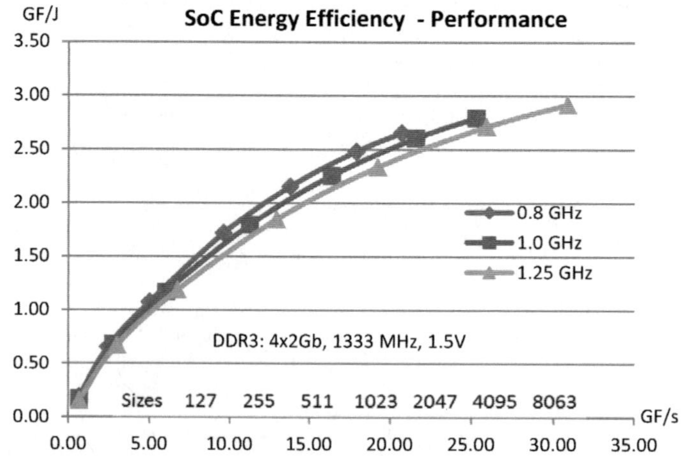

Figure 18. The 6678 HPL energy efficiencies as a function of matrix size and core clock rate.

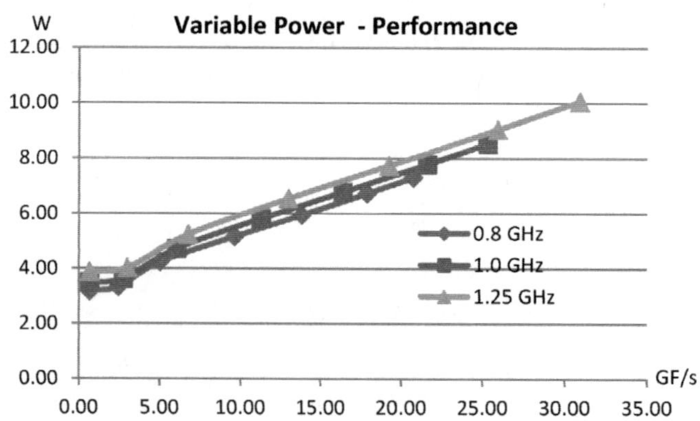

Figure 19. The 6678 HPL Variable (core) power consumption as a function of computational rate and clock rate.

Figure 19 shows, as expected, that for HPL the power consumption, except at low rates and small matrices increases linearly with the computational rate. Figure 20 supports the claim that HPL is limited by the computational rate since as the matrix size increases the DDR3 power dissipation in fact decreases once the matrix and computational demands increases beyond a certain size.

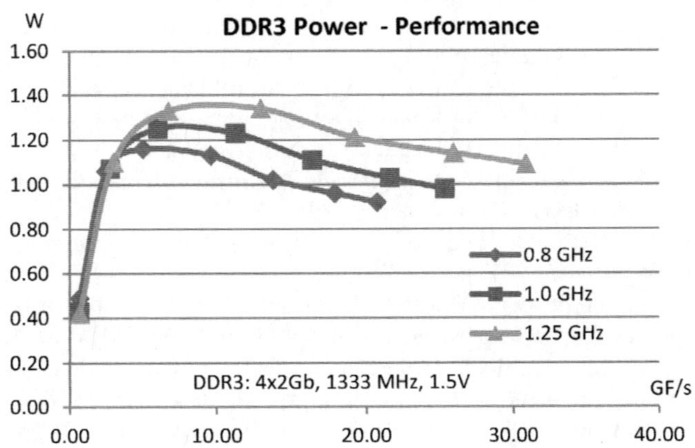

Figure 20. The EVM DDR3 memory power consumption as a function of computational and clock rates.

6. Conclusion

Compared to the highly successful and feature rich x86 architecture with a TDP of up to about 150W the 10W TMC320C6678 DSP from Texas Instruments does offer competitive energy efficiencies for the two common HPC benchmarks reported here; STREAM and HPL. In fact, the 40 nm 6678 SoC offered comparable or better energy efficiency than 22 nm x86 based designs and even better energy efficiency than the proprietary Blue Gene/Q. Compared to the latter the 6678 is a high volume product with a cost even significantly below high end x86 CPUs from Intel. But, in regards to HPC applications the software eco-system is quite immature compared to that for x86 architecture based systems. Further, some features in the x86 architecture that simplifies achieving high efficiency are not present in the 6678 SoC requiring a greater effort in achieving comparable efficiencies, and a more detailed understanding of the architecture and how to use it effectively. However, TI provided tools for the 6678 are sufficient to achieve efficiencies comparable to that on x86 architectures for STREAM and HPL and we conjecture also for many HPC applications.

References

[1] W.M. Holt. Moore's Law: A path going forward. *2016 IEEE International Solid-State Circuits Conference (ISSCC)*, pp 8-13 January 2016. http://ieeexplore.ieee.org/xpl/login.jsp?tp=&arnumber=7417888&url=http%3A%2F%2Fieeexplore.ieee.org%2Fstamp%2Fstamp.jsp%3Ftp%3D%26arnumber%3D7417888

[2] K. Bourzac. Intel: Chips will have to sacrifice speed gains for energy savings. *MIT Review*, February 5 2016. https://www.technologyreview.com/s/600716/intel-chips-will-have-to-sacrifice-speed-gains-for-energy-savings

[3] R. Merritt. Moore's Law goes Post-CMOS. *EE-Times*, February 1 2016. http://www.eetimes.com/document.asp?doc_id=1328835

[4] F. von Trapp. From the Ashes of More's Law: More than Moore has Arrived. 3DInCites, February 12 2016. http://www.3dincites.com/2016/02/from-the-ashes-of-moores-law-more-than-more-has-arrived/

[5] J. Gao. Machine Learning Applications for Data Center Optimization. May 28 2014. http://static.googleusercontent.com/media/www.google.com/en//about/datacenters/efficiency/internal/assets/machine-learning-applicationsfor-datacenter-optimization-finalv2.pdf

[6] S. Levy. Google Throws Open Doors to Its Top-Secret Data Center. *Wired,* October 17 2012. http://www.wired.com/2012/10/ff-inside-google-data-center/

[7] L.A. Barroso and U. Hölzle. The Datacenter as a Computer: An Introduction to the Design of Warehouse-Scale Machines, *In Synthesis Lectures on Computer Architecture* ed Mark D Hill (Morgan & Claypool Publishers) 2009.

[8] L.A. Barroso and U. Hölzle. The case for energy proportional computing. *IEEE Computer* **vol 40** No. 12 pp 33 – 37 2007. http://ieeexplore.ieee.org/xpl/login.jsp?tp=&arnumber=4404806&url=http%3A%2F%2Fieeexplore.ieee.org%2Fxpls%2Fabs_all.jsp%3Farnumber%3D4404806

[9] A.M. Putnam, A.M. Caulfield, E.S. Chung, D. Chiou, K. Constantinides, J. Demme, H. Esmaeilzadeh, J. Fowers, G.P. Gopal, J. Gray, M. Haselman, S. Hauck, S. Heil, A. Hormati, J.-Y. Kim, S. Lanka, J. Larus, E. Peterson, S. Pope, A. Smith, J. Thong, P.Y. Xiao, and D. Burger. A Reconfigurable Fabric for Accelerating Large-Scale Datacenter Services. *41st Annual International Symposium on Computer Architecture (ISCA)*, pp. 13-24 June 2014. http://research.microsoft.com/apps/pubs/default.aspx?id=212001

[10] K. Ovtcharov, R. Olatunji, J.-Y. Kim, J. Fowers, K. Strauss, and E.S. Chung. Accelerating Deep Convolutional Neural Networks Using Specialized Hardware. February 23, 2015. http://research.microsoft.com/apps/pubs/default.aspx?id=240715

[11] A. Putnam. Large-Scale Reconfigurable Computing in a Microsoft Datacenter. *HotChips* August 2014. http://www.hotchips.org/wp-content/uploads/hc_archives/hc26/HC26-12-day2-epub/HC26.12-5-FPGAs-epub/HC26.12.520-Recon-Fabric-Pulnam-Microsoft-Catapult.pdf

[12] The Open Compute Project. http://www.opencompute.org/

[13] B. Lannoo, S. Lambert, W. Van Heddeghem, M. Pickavet, F. Kuipers, G. Koutitas, H. Niavis, A. Satsiou, M.T. Beck, A. Fischer, H. de Meer, P. Alcock, T. Papaioannou, N.H. Viet, T. Plagemann, and J. Aracil. Information and Communication Technologies, Network of Excellence in Internet Science, D8.1. Overview of ICT energy consumption. February 5 2013. http://www.internet-science.eu/sites/eins/files/biblio/EINS_D8%201_final.pdf

[14] B. Lannoo. Energy Consumption of ICT Networks. TREND Final Workshop, Brussels, October 24 2013.

[15] H. Li. Server Board Design. *HotChips,* August 2011. http://www.hotchips.org/wp-content/uploads/hc_archives/hc23/HC23.17.2-tutorial2/HC23.17.230.Open%20Compute%20Motherboard.pdf

[16] E. Tal. Facebook Storage Tiers. *HotChips*, August 2011. http://www.hotchips.org/wp-content/uploads/hc_archives/hc23/HC23.17.2-tutorial2/HC23.17.220.Open%20Compute%20Storage.pdf

[17] J. Brodkin. Who needs HP and Dell? Facebook now designs all its own servers. *Arstechnica*, February 14 2013. http://arstechnica.com/information-technology/2013/02/who-needs-hp-and-dell-facebook-now-designs-all-its-own-servers/

[18] S. Shankland. Google uncloaks once-secret server. *CNET*, December 11 2009. http://www.cnet.com/news/google-uncloaks-once-secret-server-10209580/

[19] T.P. Morgan. How Microsoft Is Using FPGAs To Speed Up Bing Search. *EnterpriseTech*, September 3 2014. http://www.enterprisetech.com/2014/09/03/microsoft-using-fpgas-speed-bing-search/

[20] A. Michael. The Open Computer Project: Project Motivation and high level system design *HotChips*, August 2011.
http://www.hotchips.org/wp-content/uploads/hc_archives/hc23/HC23.17.2-tutorial2/HC23.17.210.Open%20Compute%20Overview.pdf

[21] A. Singh, J. Ong, A. Agarwal, G. Anderson, A. Armistead, R. Bannon, S. Boving, G. Desai, B. Felderman, P. Germano, A. Kanagala, J. Provost, J. Simmons, E. Tanda, J. Wanderer, U. Hölzle, S. Stuart, and A. Vahdat. Jupiter Rising: A Decade of Clos Topologies and Centralized Control in Google's Datacenter Network, *SIGCOMM*, August 2015.
http://conferences.sigcomm.org/sigcomm/2015/pdf/papers/p183.pdf

[22] R. Miller. Inside Microsoft's Chicago Data Center. *Data Center Knowledge*, 2009.
http://www.datacenterknowledge.com/inside-microsofts-chicago-data-center

[23] J. Dean and S. Ghemawat. MapReduce: Simplified Data Processing on Large Clusters. *CACM*, **vol 51** No.1 pp 107-113 2008.
http://dl.acm.org/citation.cfm?id=1327492

[24] F. Chang, J. Dean, S. Ghemawat, W.C. Hsieh, D.A. Wallach, M. Burrows, T. Chandra, A. Fikes, and R.E. Gruber. Bigtable: A Distributed Storage System for Structured Data. ACM Transactions on Computer Systems, **vol 26** No.2 article No. 4 June 2008.
http://dl.acm.org/citation.cfm?id=1365816

[25] G.E. Moore. Cramming more components onto integrated circuits. *Electronics*, **vol 38** No 8 April 1965.

[26] R.H. Dennard, F.H. Gaensslen, H.-N. Yu, V.L. Rideout, E. Bassous, and A.R. Leblanc. Design of ion-implanted MOSFETs with very small physical dimensions *IEEE J. Solid- State Circ* **vol 9** p.256 1974.
http://ieeexplore.ieee.org/xpl/abstractAuthors.jsp?tp=&arnumber=752522&url=http%3A%2F%2Fieeexplore.ieee.org%2Fiel4%2F5%2F16251%2F00752522.pdf%3Farnumber%3D752522

[27] S. Borkar and A. Chien. The Future of Microprocessors *CACM* **vol 54** No 5 pp 67-77 May 2011.

[28] R Hameed, W. Qadeer, M. Wachs, O. Azizi, A. Solomatnikov, B.C. Lee, S. Richardson, C. Kozyrakis, and M. Horowitz. Understanding Sources of Inefficiency in General-Purpose Chips. *ISCA '10 Proceedings of the 37th annual international symposium on Computer architecture* pp 37-47 2010.
http://dl.acm.org/citation.cfm?id=1815968

[29] W.J. Dally, J. Balfour, D. Black-Shaffer, J. Chen, R.C. Harting, V. Parikh, J. Park, and D. Sheffield. Efficient Embedded Computing. *Computer* **vol 41** no 7 pp 27-32 July 2008.

[30] W.J. Dally. Challenges for Future Computing Systems. *HiPEAC* 2015 Keynote. https://www.hipeac.net/2015/amsterdam/keynotes/

[31] J. Larus. It's the End of the World as we know it. *HiPEAC* 2015 Keynote.
https://www.hipeac.net/2015/amsterdam/keynotes/

[32] E.J. Nowak. Maintaining the benefits of CMOS scaling when scaling bogs down *IBM Journal of Research and Development* **vol 46** no 2/3 pp 169-180 2002.

[33] D. Hillman. Integrated Power Management, Leakage Control and Process Compensation Technology for Advanced Processes. *Design & Reuse*, March 16 2009.
http://www.design-reuse.com/articles/20296/power-management-leakage-control- process-compensation.html

[34] Power4, The first Multi-core, 1 GHz processor
http://www-03.ibm.com/ibm/history/ibm100/us/en/icons/power4/

[35] J. Kovacs. AMD Demonstrates World's First x86 Dual-Core Processor. August 2004.
http://ir.amd.com/phoenix.zhtml?c=74093&p=irol-newsArticle&ID=608198

[36] Dual Core Era Begins, PC Makers Start Selling Intel-Based PCs April, 2005.
http://www.intel.com/pressroom/archive/releases/2005/20050418comp.htm

[37] P. Kogge. Reading the Entrails: How Architecture Has Evolved at the High End IPDPS Keynote 2014.
http://www.ipdps.org/ipdps2014/IPDPS2014keynote-Kogge.pdf
[38] R. Merritt. ARM CTO: power surge could create 'dark silicon'. *EE-Times* October 22 2009. http://www.eetimes.com/document.asp?doc_id=1172049
[39] H. Esmaeilzadehy, E. Blemz, R. St. Amantx, K. Sankaralingamz, and D. Burger. Dark Silicon and the End of Multicore Scaling *ISCA '11 Proceedings of the 38th annual International Symposium on Computer Architecture* pp 365-376 2011 also
IEEE Micro **vol 32** no 3 pp 122-134 2012.
http://dl.acm.org/citation.cfm?id=2000108
[40] M.B. Taylor. A Landscape of the New Dark Silicon Design Regime. *IEEE Micro* **vol 33** no 5 pp 8-19 2013.
http://ieeexplore.ieee.org/xpl/articleDetails.jsp?tp=&arnumber=6583151
[41] International Technology Roadmap for Semiconductors
www.itrs2.net
[42] J. Stuecheli. Power8. *HotChips*, August 2013.
http://www.hotchips.org/wp-content/uploads/hc_archives/hc25/HC25.20-Processors1-epub/HC25.26.210-POWER-Studecheli-IBM.pdf
[43] A. Guha, Y. Zhang,cR.Yao ur Rasool, and A. Chien. Systematic Evaluation of Workload Clustering for Extremely Energy-Efficient Architectures. *ACM SIGARCH Computer Architecture News*, **vol 41** Issue 2 pp 22-29 May 2013.
[44] **A.** Cunningham. Snapdragon 820 is official: A look at its GPU (and how much the chip matters). *Arstechnica* August 12 2015
http://arstechnica.com/gadgets/2015/08/snapdragon-820-is-official-a-look-at-its-gpu-and-how-much-the-chip-matters
[45] L. Johnsson and G. Netzer. Report on Prototypes Evaluation D9.3.3 PRACE April 2013.
http://www.prace-ri.eu/prace-1ip-deliverables/
[46] J. McCalpin. Memory Bandwidth and Machine Balance in Current High Performance Computers. *IEEE Computer Society Technical Committee on Computer Architecture (TCCA) Newsletter* pp 19-25 December 1995.
[47] A. Petitet, R.C. Whaley, J. Dongarra, and A.Cleary. HPL - A Portable Implementation of the High-Performance Linpack Benchmark for Distributed Memory Computers version 2.2, February 2016.
http://www.netlib.org/benchmark/hpl/
[48] Texas Instruments. TMS320C66x DSP CPU and Instruction Set Reference Guide Literature Number: SPRUGH7. November 2010.
[49] Texas Instruments. Multicore Fixed and Floating-Point Digital Signal Processor. March 2014. http://www.ti.com.cn/cn/lit/ds/symlink/tms320c6678.pdf
[50] Texas Instruments. KeyStone Architecture, Hyperlink *User Guide* Literature Number: SPRUGW8C. June 2013.
http://www.ti.com/lit/ug/sprugw8c/sprugw8c.pdf
[51] R. Haring. The Blue Gene/Q Compute Chip, *HotChips*, August 2011.
http://www.hotchips.org/wp-content/uploads/hc_archives/hc23/HC23.18.1-manycore/HC23.18.121.BlueGene-IBM_BQC_HC23_20110818.pdf
[52] Texas Instruments. TMS320C66x DSP CorePac *User Guide*, Literature Number: SPRUGW0C. July 2013.
http://www.ti.com/lit/ug/sprugw0c/sprugw0c.pdf
[53] G. Netzer, L. Johnsson, D. Ahlin, E. Stotzer, P. Varis, and E. Laure. Exploiting DMA for Performance and Energy Optimized STREAM on a DSPIEEE *2014 International Parallel Distributed Processing Symposium Workshop* pp 805-814 May 2014.

http://ieeexplore.ieee.org/xpl/login.jsp?tp=&arnumber=6969464&url=http%3A%2F%2Fieeexplore.ieee.org%2Fxpls%2Fabs_all.jsp%3Farnumber%3D6969464

[54] G. Netzer. Efficient LU-Factorization for Texas Instruments Keystone Architecture Digital Signal Processors, M.Sc. Thesis, KTH June 2015. http://www.divaportal.org/smash/get/diva2:837145/FULLTEXT01.pdf

[55] Texas Instruments. KeyStone Architecture Enhanced Direct Memory Access (EDMA3) Controller *User Guide* Literature Number:SPRUGS5B. May 2015. http://www.ti.com.cn/cn/lit/ug/sprugs5b/sprugs5b.pdf

[56] TMDXEVM6678LE. TMX320C6678 Evaluation Module [Online] Available: http://www2.advantech.com/Support/TI-EVM/6678le_of.aspx

[57] G. Netzer, L. Johnsson, D. Ahlin, and E. Laure. Instrumentation for accurate energy-to-solution measurements of a Texas Instruments TMS320C6678 digital signal processor and its DDR3 memory. *E2SC '14 Proceedings of the 2nd International Workshop on Energy Efficient Supercomputing* (IEEE Press) pp 89-98 November 2014. http://dl.acm.org/citation.cfm?id=2689720

[58] J.D. McCalpin. STREAM: Sustainable memory bandwidth in high performance computers. https://www.cs.virginia.edu/stream/

[59] IBM. IBM Power7 data from IBM Power 750 and 760 Technical Overview and Introduction Literature Number 4985. May 2013.

[60] W.J. Starke. Power7: IBMs next generation, balanced POWER server chip. *HotChips*, August 2009. http://www.hotchips.org/wp-content/uploads/hc_archives/hc21/3_tues/HC21.25.800.ServerSystemsII-Epub/HC21.25.835.Starke-IBM-POWER7SystemBalancev13_display.pdf

[61] R. Krishnaiyer, E. Kultursay, P. Chawla, S. Preis, A. Zvezdin, and H. Saito. Compiler-based Data Prefetching and Streaming Non-temporal Store Generation for the Intel R Xeon Phi Coprocessor. *IEEE 27th International Parallel and Distributed Processing Symposium Workshops & PhD Forum* (NewYork, NY) IEEE pp 1575–1586 May 2013. http://ieeexplore.ieee.org/xpl/login.jsp?tp=&arnumber=6651054&url=http%3A%2F%2Fieeexplore.ieee.org%2Fxpls%2Fabs_all.jsp%3Farnumber%3D6651054

[62] Dell Performance and Power Efficiency of Dell Power Edge servers with E5-2600v2. October 2013.

[63] PRACE Final Report on Prototype Evaluation D9.3.4. December 2013.

[64] Texas Instruments. TMS320C66x DSP Cache *User Guide* Literature Number: SPRUGY8. November 2010. http://www.ti.com/lit/ug/sprugy8/sprugy8.pdf

[65] Top500 www.top500.org

[66] The HPC Challenge Benchmark http://icl.cs.utk.edu/hpcc/

[67] DGEMM http://www.netlib.org/lapack/explore-html/d7/d2b/dgemm_8f_source.html

[68] The Green 500 www.green500.org

Simulations and study of a new scheduling approach for distributed data production

Dzmitry Makatun[1,3], Jérôme Lauret[2], Hana Rudová[4] and Michal Šumbera[3]

[1]Faculty of Nuclear Physics and Physical Engineering, Czech Technical University in Prague
[2]STAR, Brookhaven National Laboratory, USA
[3]Nuclear Physics Institute, Academy of Sciences, Czech Republic
[4]Masaryk University, Czech Republic

E-mail: `dzmitry.makatun@fjfi.cvut.cz`

Abstract. Distributed data processing has found its application in many fields of science (High Energy and Nuclear Physics (HENP), astronomy, biology to name only those). We have focused our research on distributed data production, an essential part of computations in HENP. Using our previous experience, we have recently proposed a new scheduling approach for distributed data production which is based on the network flow maximization model. It has a polynomial complexity providing required scalability with respect to the size of computations. Our approach improves the overall data production throughput due to three factors: transfer input files in advance before their processing (allows to decrease I/O latency); balancing of the network traffic (includes splitting the load between several alternative transfer paths); and transfer files sequentially in a coordinated manner (allows to reduce the influence of possible network bottlenecks). In this contribution, we present the results of our new simulations based on the GridSim framework which is one of the commonly used tools in the field of distributed computations. In these simulations we study the behavior of standard scheduling approaches compared to our recently proposed approach in a realistic environment relying on the data from the STAR and ATLAS experiments and considering the influence of the background traffic. The final goal of the research is to integrate the proposed scheduling approach into the real data production framework. In order to achieve this we are constantly moving our simulations towards real use cases, study scalability of the model and the influence of the scheduling parameters on the quality of the solution.

1. Introduction

Modern experiments in HENP heavily rely on distributed model of computations [1]. Data production (often called event reconstruction) is an essential part of such computations, while similar computational tasks can be found in many other fields. Despite the data production differs in many aspects from other types of computations (user analysis, simulations) it is often approached by general scheduling techniques, and that may lead to sub-optimal performance and increase the requirements for computational infrastructure (CPU, network, storage). For this reason, some experiments use custom setups in order to increase computational efficiency of data production for a given distributed infrastructure [2]. Such an approach raises a scalability issue, since reconfiguration of the setup in case of changes to the infrastructure (addition of new sites, resource fluctuations) requires an increased effort.

In our research we use typical properties of distributed data production in order to develop a scheduling approach that can optimize resource utilization and automatically adapt to the current state of the computational infrastructure. In this approach we use the following assumptions:

- A job duration and the size of its output can be estimated knowing the size of its input data, using statistics from previously finished jobs.
- Each input file has to be processed exactly once.
- All the input data are placed at a centrally located source site and the output has to be transferred back to that location. The underlying model also allows multiple data sources (data replication) and multiple output destinations, such cases will be studied in future research.

The task of the scheduler is to distribute input data between available computational sites considering the real network topology, bandwidth, storage space and computational throughput in order to complete processing as fast a possible. The detailed description of the developed approach can be found in [3] and [4], while here we provide a brief summary of the model (Section 2) and of the setup for simulations (Section 3). In Section 4 we present results of simulations of distributed data production studying an influence of background traffic. Section 5 provides results of simulations with Grid of a large scale (repeating the Tier-1 network of one of the largest HENP experiments) in order to verify scalability of the developed approach.

2. Model summary

Since the order of particular jobs and their assignment to a particular computational site is not important for data production, in our approach we plan the resource (CPU, storage, network) load only and then distribute input files accordingly. The planning is done cyclically, i.e. a plan for a limited planning time interval ΔT is produced at the beginning of this interval each time. The plan relies on the system state at the moment when it is created, but not on previously issued plans.

The computational Grid can be represented as a graph, where weights of links correspond to the maximal amount of data that can be transferred within ΔT, calculated as $bandwidth \times \Delta T$. For planning of input transfers (Figure 1) we connect all the computational sites to a dummy sink and a dummy source to input storage(s) via dummy links. The capacities of dummy links represent constraints on storage, CPU throughput and data availability. For planning of output transfers (Figure 2) we swap the sink and the source and inverse directions of the dummy edges. In this case, their capacities impose constraints on the storage and amount of available/expected output data. We apply a network flow maximization approach [5] to both input/output problems and the solution is the amount of input/output data (F_i) to be transferred over each network link during ΔT in order to maximize the overall computational throughput.

The network flow maximization problem is known to have a polynomial complexity on the network size. It is an advantage since, in the general case, scheduling problems are NP-hard.

In order to execute the plan, there is a dedicated service running at each computational site, called "handler", also responsible for sending statistics and status data to the planner. As the handler receives a plan, it knows the amount of input/output data F_i to be transferred over each outgoing link to the neighboring nodes. Each time a new input file arrives over one of the incoming links, the planner either submits it for processing (if there are free CPUs, see Figure 3) or, otherwise, forwards it over one of the outgoing links with a remaining flow $F_i > 0$ and then decreases this flow by the size of the file (see Figure 4). If there are no outgoing links with $F_i > 0$ the file is kept at the local queue until a CPU becomes free or a new plan is issued. In this manner the handler maintains a local queue of input files, so that a CPU can start the next job as soon as a previous one is finished, therefore, a transfer overhead is eliminated.

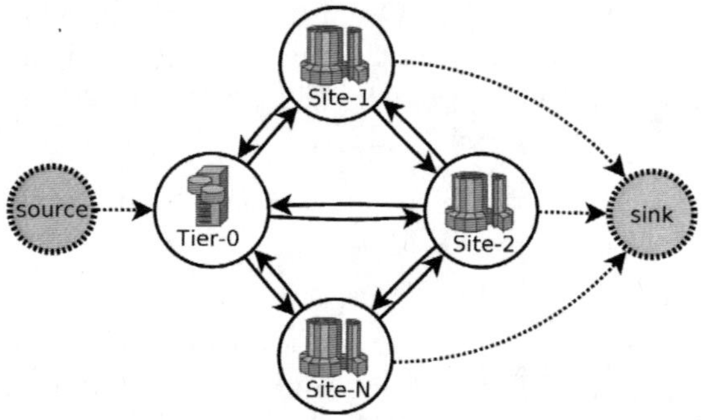

Figure 1. Network for the *input* flow maximization problem.

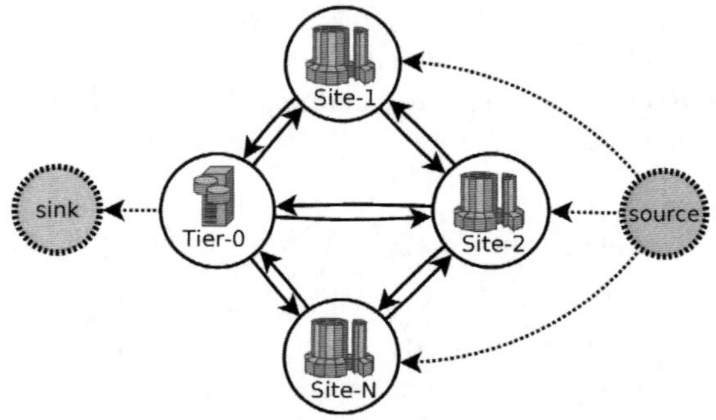

Figure 2. Network for the *output* flow maximization problem.

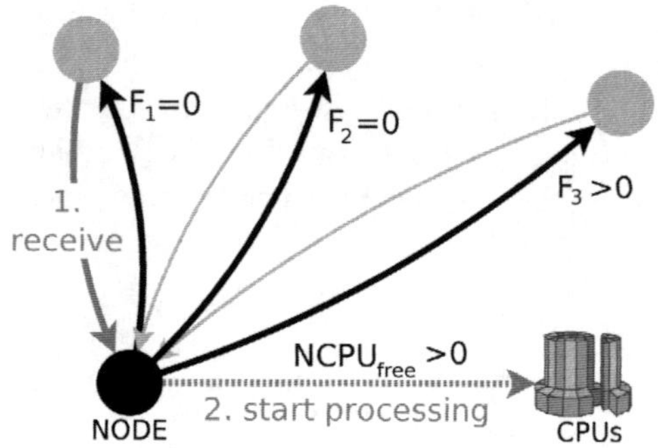

Figure 3. Handler submits a received input file for processing.

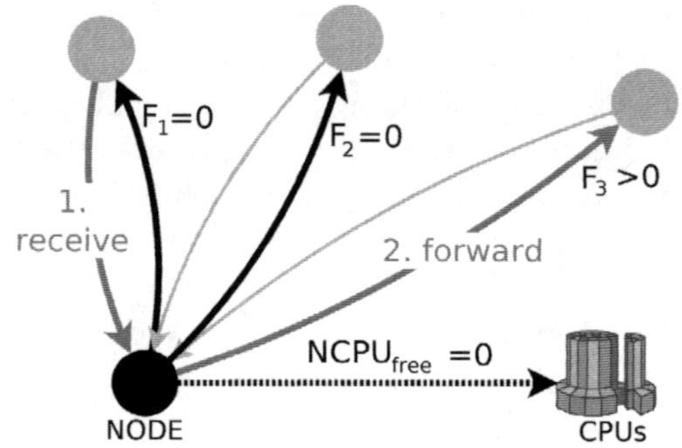

Figure 4. Handler forwards a received input file to another site.

3. Simulation setup

For testing the proposed scheduling approach, we did perform simulations of distributed data production in GridSim [6], a commonly used tool for simulation of computational Grids. As input parameters, we have also used data collected during and extracted from data production for the STAR experiment at the KISTI computational facility [7]. The data production was performed during three months in 2014 and records of 77,000 jobs were collected. For larger simulations a larger set of jobs can be generated using Monte-Carlo selection from the original one.

In addition to the proposed approach (called PLANNER), two other standard job scheduling policies were simulated for comparison. Both of them use the same approach for allocation of CPUs to jobs – "when a CPU is free, send an input file from the Tier-0 for processing", but data transfers are modeled differently:

SEQ: A network link is modeled as a *space-shared* resource, i.e. if multiple files are submitted for a transfer over the same link simultaneously, they are dispatched one by one in the First-In-First-Out order.

PAR: A network link is modeled as a *time-shared* resource, i.e. multiple transfers can be performed simultaneously, sharing the bandwidth of the link.

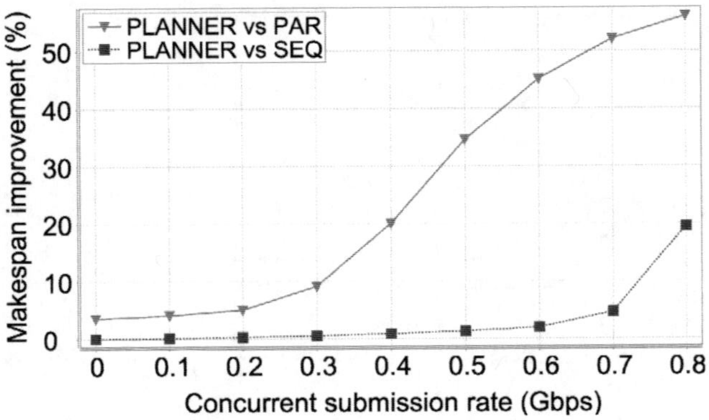

Figure 5. Results of simulations with background traffic. The plot shows how much makespan was gained due to usage of our PLANNER instead of the other scheduling approaches depending on the amount of background traffic. The PLANNER systematically shows better performance especially in the simulations with high background.

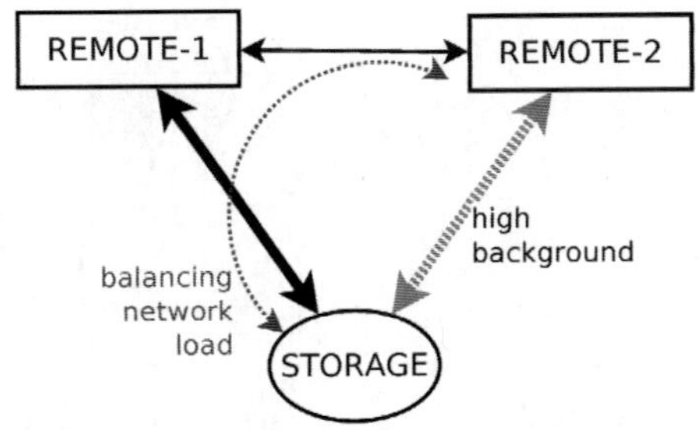

Figure 6. A usecase for a future planner improvement in order to provide load balancing in networks with high background traffic.

4. Simulations with background traffic

Network infrastructure is often shared between several experiments, as a result transfer latency can increase when several network activities are ongoing simultaneously. We have studied the influence of the background traffic on the efficiency of data production comparing different scheduling approaches with the help of simulations. The simulated Grid consisted of a single remote computational site with 1,000 CPUs and 14 TB disc cache connected via 1 Gbps network link to the central storage (Tier-0). Both sites were sending unrelated files of 12 GB size every 1,000 seconds to each other, the number of those files to be sent at once was changing in different simulations. The performance of the simulated scheduling approaches is compared in Figure 5. As one can see in the plot, our PLANNER shows better performance in all the simulations and the gain in performance increases as the level of background traffic grows. The PLANNER's makespan increases just by 0.17 % when changing from 0 to 0.8 Gbps of the concurrent transfer submission rate while the makespan for the SEQ and PAR is much higher corresponding to 24 % and 120 %, respectively. This is achieved due to file transfers in advance before computation, so that the jobs are not delayed by the network latency.

It is important to consider the maximal possible network latency when setting up the planning time interval ΔT. If ΔT is too small, the transfers started at one planner cycle may be completed at the next one and that can compromise the preciseness of the plan execution. The simulations have shown that ΔT set to $6 - 12$ hours is large enough to avoid such issues.

In future we plan to perform simulations with background traffic in more complex Grid topologies (many remote sites) and implement an additional reasoning into the planner such that it can identify network links with high background traffic and decrease the load on those links by utilizing alternative transfer paths as illustrated in Figure 6.

5. Scalability test

In order to test the scalability of the planner we have performed simulations using data on Tier-1 sites of one of the currently largest experiments in HENP. We have used statistics of 10 computational sites obtained from online monitoring tools. However, the designed parameters of such a network are well balanced and a transfer latency does not influence the

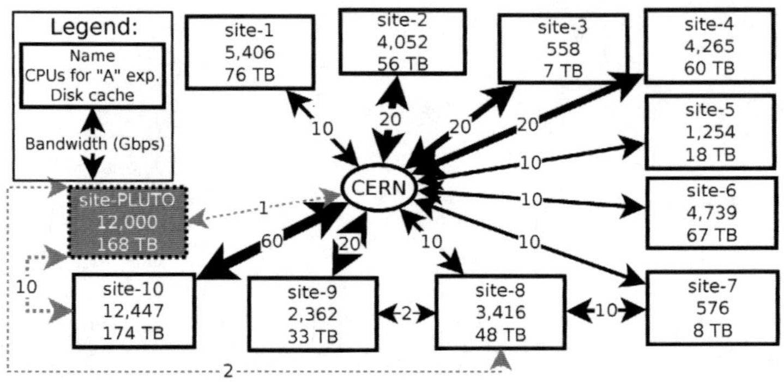

Figure 7. The simulated Grid of Tier-1 sites of one of the largest HENP experiments. The data were obtained from online monitoring tools and the dummy "PLUTO" site was added to simulate usage of a facility outside of the primary network infrastructure. The main purpose of the simulations is to test the scalability and advantage of load balancing provided by our planner.

computational performance significantly. Under these conditions all the simulated approaches provide similar performance. For a better illustration of the planner abilities, we have added a dummy site called "PLUTO" to the initial Grid. The site has a significant amount of CPUs but a limited connection (1 Gbps) to the Tier-0 while it has a reasonable connection to two other Tier-1 sites. Such situation can often emerge in real life, when there are computational resources provided at sites outside of the CERN network infrastructure. The final Grid setup used in the simulations is provided in Figure 7. The amount of disk cache at sites was set proportional to the number of CPUs since we assume that only a part of the overall storage can be dedicated to data production.

The plot in Figure 8 shows how the total CPU utilization in the simulated Grid changes with time for SEQ and PLANNER scheduling approaches. Unfortunately, the results for PAR approach are not presented here as the underlying GridSim model for the parallel network mode appeared to be too computationally demanding. However, this approach has shown the worst performance in all previous simulations [3]. As one can observe in Figure 8, the PLANNER has reached 100 % of the overall CPU performance since it was able to utilize all the CPUs at the "PLUTO" site using alternative (indirect) transfer paths. Such approach allows to use more computational resources outside of the primary network infrastructure. The makespan improvement in the considered case is 21 %, an increase in throughput which would not have been easily possible without reasoning and a planner.

During the simulations with the realistic size of Grid an average runtime of the planner was measured. The result shows that planning for 12 hours of data production is done within 7 milliseconds. This allows to apply this approach for an online planning of data production in the real environment.

6. Conclusion

In previous work [3] we have proposed a new approach to scheduling of data production on distributed resources. GridSim simulations and data from the real infrastructure were used to test the approach and its implementation. The simulations have shown that the proposed approach systematically provides better makespan and can optimize over a wide range of infrastructure configurations in {CPU, disk, network bandwidth} phase-space compared to classic job scheduling approaches.

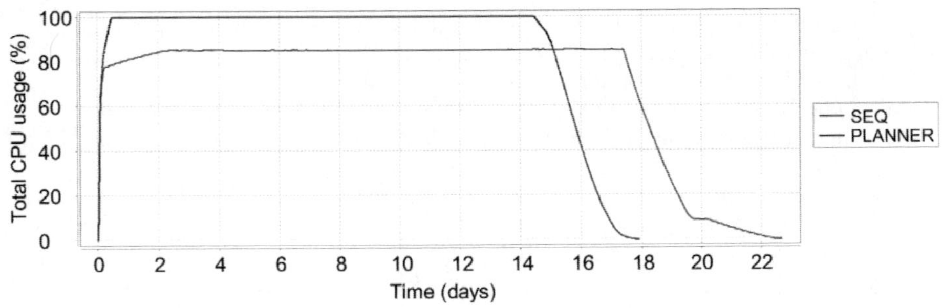

Figure 8. Results of simulations of the large-scale Grid. The planner achieves a higher CPU utilization due to network load balancing.

The recent research has shown that the approach can provide a significant performance improvement when a shared network infrastructure is used. It can better adapt to background traffic, since planning is done in advance and data are transferred before computations. It can also provide a network load balancing, including usage of network paths which are not always intuitive (optimization on secondary network paths), that allows to utilize sites with poor connectivity to the Tier-0 site more efficiently. Such optimization is automated based on current Grid topology and state. Simulations of data production in a Tier-1 Grid of one of the largest HENP experiments has shown that the proposed approach is scalable and can be implemented for online planning in real infrastructures.

In future we plan to continue the work on the proposed approach, conduct simulations with multiple data sources/destinations (Tier-0s), simulate data production of more experiments, implement advanced reasoning about background traffic and study an influence of internal planner parameters (such as ΔT). The final goal is to integrate the planner into existing Resource Management Systems and deploy it into the data production framework of the STAR experiment.

Acknowledgments

This work has been supported by the Czech Science Foundation (13-20841S, P202/12/0306), the MEYS grant LG15001 and the Office of Nuclear Physics within the U.S. Department of Energy. Furthermore, the authors would like to acknowledge the Institute of Physics of the Czech Academy of Sciences for kindly providing access to the GOLIAS computer farm for running simulations.

References

[1] I Bird, P Buncic, F Carminati, M Cattaneo, P Clarke, I Fisk, M Girone, J Harvey, B Kersevan, P Mato, R Mount, and B Panzer-Steindel. Update of the Computing Models of the WLCG and the LHC Experiments. Technical Report CERN-LHCC-2014-014. LCG-TDR-002, CERN, Geneva, Apr 2014.

[2] Jan Balewski, Jerome Lauret, Doug Olson, Iwona Sakrejda, Dmitry Arkhipkin, et al. Offloading peak processing to virtual farm by STAR experiment at RHIC. *Journal of Physics: Conference Series*, 368(012011), 2012.

[3] Dzmitry Makatun, Jérôme Lauret, Hana Rudová, and Michal Šumbera. Distributed data production planning for High Energy and Nuclear Physics. *Journal of Scheduling*, 2015. (Submitted for publication).

[4] Dzmitry Makatun, Jérôme Lauret, Hana Rudová, and Michal Šumbera. Model for planning of distributed data production. In *Proceedings of the 7th Multidisciplinary International Scheduling Conference (MISTA)*, pages 699–703, 2015.

[5] R K Ahuja, T L Magnati, and J B Orlin. *Network flows: theory, algorithms, and applications.* Prentice Hall, 1993.

[6] Rajkumar Buyya and Manzur Murshed. GridSim: A toolkit for the modeling and simulation of distributed resource management and scheduling for grid computing. *The Journal of Concurrency and Computation: Practice and Experience (CCPE)*, 14, 2002.

[7] Levente Hajdu, Lidia Didenko, Jerome Lauret, J Amol, W Betts, H J Jang, and S Y Noh. Automated finite state workflow for distributed data production. In *ACAT*, 2016. (Submitted for publication).

Journal of Physics: Conference Series **762** (2016) 012024 doi:10.1088/1742-6596/762/1/012024

Multi-threaded software framework development for the ATLAS experiment

G A Stewart[1], J Baines[2], T Bold[3], P Calafiura[4], A Dotti[5], S A Farrell[4], C Leggett[4], D Malon[6], E Ritsch[7], S Snyder[8], V Tsulaia[4], P Van Gemmeren[6], B M Wynne[9] for the ATLAS Experiment

[1]University of Glasgow, University Avenue, Glasgow G12 8QQ, Scotland
[2]Rutherford Appleton Laboratory, Harwell, Didcot OX11 0QX, United Kingdom
[3]AGH University of Science and Technology, 30-059 Kraków, Poland
[4]Lawrence Berkeley National Laboratory, 1 Cyclotron Rd, Berkeley, CA 94720, United States
[5]SLAC National Accelerator Laboratory, 2575 Sand Hill Rd, CA 94025, United States
[6]Argonne National Laboratory, 9700 Cass Ave, Lemont, IL 60439, United States
[7]European Organization for Nuclear Research (CERN), CH-1211 Geneva 23, Switzerland
[8]Brookhaven National Laboratory, Upton, New York, United States
[9]University of Edinburgh, Edinburgh EH9 3FD, Scotland

E-mail: graeme.andrew.stewart@cern.ch

Abstract.
ATLAS's current software framework, Gaudi/Athena, has been very successful for the experiment in LHC Runs 1 and 2. However, its single-threaded design has been recognised for some time to be increasingly problematic as CPUs have increased core counts and decreased available memory per core. Even the multi-process version of Athena, AthenaMP, will not scale to the range of architectures we expect to use beyond Run2.

ATLAS examined the requirements on an updated multi-threaded framework and laid out plans for a new framework, including better support for High Level Trigger use cases, in 2014. In this paper we report on our progress in developing the new multi-threaded task parallel extension of Athena, AthenaMT.

Implementing AthenaMT has required many significant code changes. Progress has been made in updating key concepts of the framework, allowing different levels of thread safety in algorithmic code. Substantial advances have also been made in implementing a data flow centric design, as well as on the development of the new 'event views' infrastructure. These event views support partial event processing and are an essential component to support the High Level Trigger's processing of certain regions of interest. A major effort has also been invested to have an early version of AthenaMT that can run simulation on many core architectures, which has augmented the understanding gained from work on earlier ATLAS demonstrators.

1. Introduction

Recent developments in microprocessor technology have underlined the trends that have been identified since the mid-2000s: that while the density of trasnsistors on CPUs has continued to rise more or less exponentially (Moore's Law[1]), clock speeds have stalled (Figure 1).

The fundamental limitations of clock speed, which are thermal in origin, underline that the next challenges for computing are almost entirely based around power efficiency. This drives new generations of CPU designs, which are both low power and have low memory per CPU core.

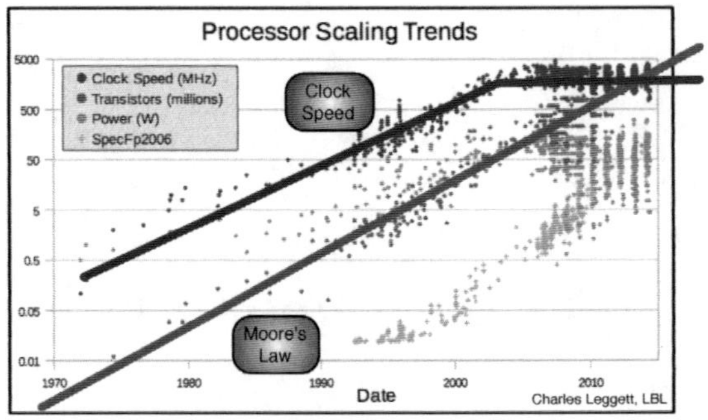

Figure 1. Scaling of key characteristcs of microprocessors since 1970 [2].

In addition to the challenges posed by these constraints, increasing application performance today means taking advantage of advanced CPU features, such as multi-cores and wide vector registers.

In this paper we describe how ATLAS experiment [3] at CERN is adapting to the software challenges from this technological evolution.

2. The Athena Framework

The ATLAS software framework, Athena[4], which is based on the Gaudi framework[5], was designed in the early 2000s. Thus it predates the stall of single core clock speed and is firmly based on a serial processing model. In spite of that, it has successfully processed billions simulated and collider events and played a key role in the discovery of the Higgs Boson[6] and hundreds of other ATLAS observations made during LHC Run1, from 2009 to 2012.

During LHC Run2, 2015 to 2018, increasingly difficult processing conditions have been overcome with the deployment of a multi-process version of Athena, AthenaMP[7]. This uses a multi-process model, where after initialisation multiple AthenaMP worker processes are forked to run the event loop. This allows the sharing of the memory pages allocated for large static structures, such as detector geometry and magnetic field, between worker processes, taking advantage of the Linux kernel's *copy on write* feature. However, there are limitations to how much memory can be saved in this fashion – the change of a single bit will result in an entire memory page being unshared, and C++'s memory model does not allow for fine control over which data objects are assigned to which physical pages. In addition, the further evolution of the LHC machine will increase the complexity of the events that ATLAS will observe and the physics goals of ATLAS necessitate in increase in the data taking rate of the experiment. This increase in event complexity and event rate will be a huge challenge for the experiment, in regards to the memory and CPU requirements of reconstructing events.

3. Future Framework Requirements

Aware of these challenges ATLAS launched a study group, the *Future Framework Requirements Task Force* in 2014. This group's report summarised the main requirements and recommendations of the software processing framework that ATLAS would deploy for Run3, 2021 to 2023:

- The increasingly difficult memory per core environments in the future would require ATLAS to develop a *multi-threaded* framework, in order to maximise memory savings.
- That such a framework should endeavour to provide advanced capabilities, such as event processing in partial regions of the detector, so that the ATLAS online and offline

frameworks could share code and development effort even more than was possible in Runs 1 and 2.

- That a continued development of multi-threading in the Athena framework offered the best chance of success, as, for a running experiment, continuity of operations had to be ensured and disruptive breaks in code continuity would be prohibitively expensive in development and validation effort.
- That in order to be able to exploit the largest range of practical memory and processor options the future framework should foresee exploiting parallelism both at the event level (multiple events processed at once) and the sub-event level. The sub-event parallelism is further divided into inter-algorithm parallelism, directly managed by the framework when algorithms can run independently; and intra-algorithm parallelism, where an algorithm itself exploits opportunities for concurrency.

4. AthenaMT

In the wake of the future frameworks report, the collaboration launched a development effort towards its future framework, which has been christened AthenaMT (*Athena Multi-Threaded*). Here we report on some of the main design discussions and conclusions that have been reached on the framework, as well as on early performance tests.

4.1. Underlying Components

As AthenaMT envisages exploiting parallelism at many levels, including triggered by an algorithm itself, it is important that threading is therefore coordinated across the execution of a program. If each component started its own thread pool the machine's resources would easily become over-contended, affecting memory footprint and throughput. Such a problem is far better suited to a task-based library model than one where thread control is delegated to components. It is also more suited to task queues than loop based parallelism, as much of the available parallelism is from unrelated tasks executing on different events or parts of an event. For the moment the choice has been made to base our threading on the Intel Threaded Building Block library[8].

4.2. Framework Components

Gaudi's original component model and state engine included a number of features that were used heavily in ATLAS software, but prove to be problematic in a multi-threaded environment. One of these was the concept of a *public tool*, which was a single Tool instance shared by multiple algorithms. Such tools in ATLAS were frequently used to store data locally, which would be passed between algorithms, which worked fine when the algorithm execution order was fixed and events were processed sequentially. However, in the multi-threaded case, when algorithms may be processing different events, such a design pattern breaks down. In addition a data dependency between these algorithms has been hidden from the scheduler, which may execute algorithms in the wrong order.

In order to overcome this problem, public tools will be removed from the framework. All tools will be *private* to algorithms and communication between algorithms must go via the event store *service*. It is worth pointing out that services in multi-threaded Gaudi will be event context aware, where as algorithms and tools will be only aware of the event that they are currently processing. Then, developers will be required to migrate public tools to private tools or to services, which ever is more appropriate.

A model of how AthenaMT components process within a single event is shown in Figure 2.

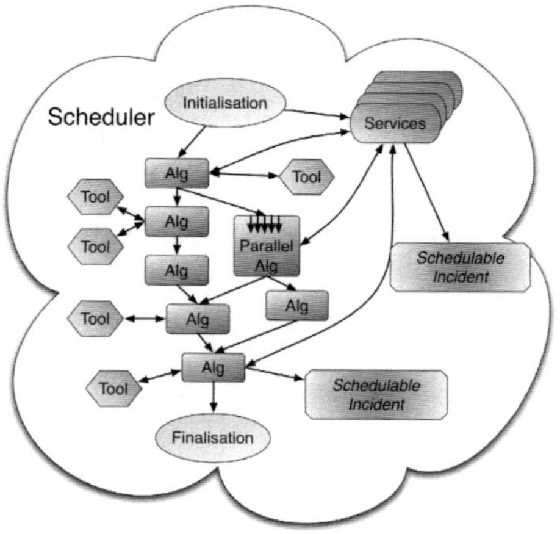

Figure 2. Framework elements involved in the processing of a single event in AthenaMT.

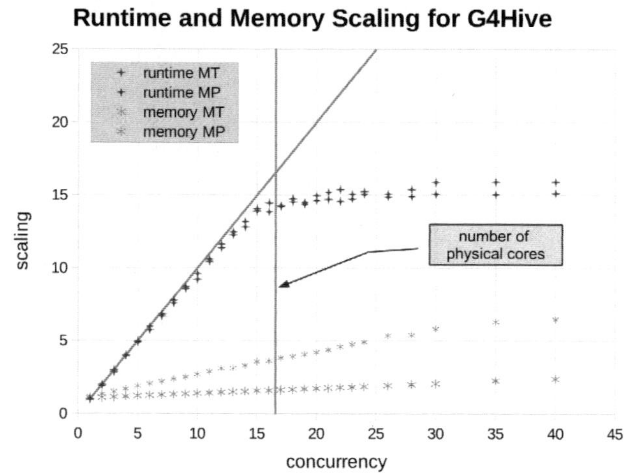

Figure 3. Event throughput and memory scaling comparing AthenaMP (multi-process) and AthenaMT (multi-threaded) approaches for ATLAS simulation.

4.3. Cloneable and Re-entrant Algorithms

For algorithms themselves we now envisage three classes, expressing different levels of thread safety:

Legacy A legacy algorithm can only have one instance and can only process one event at a time; if more than one event wishes to use a legacy algorithm the scheduler is forced to pause them.

Cloneable A cloneable algorithm can have multiple instances, each one of which can be used to process a different event; however, as each clone is a separate instance, memory usage is increased for cloneable algorithms.

Re-entrant A re-entrant algorithm has a single instance, but can be used to process multiple events simultaneously; such algorithms are ideal for both lowering memory consumption and for increasing throughput, but are the most difficult to program as their execution has to be thread safe.

Re-entrant algorithms also put design requirements onto the data handle implementation (data handles declare data dependencies and provide access to event store data), as the algorithm's execute method has to be `const`.

4.4. Event Views

In order to support the requirement of the ATLAS High Level Trigger, which selects events in the online system, it is necessary to process only certain *regions of interest*, which have been flagged by the ATLAS level 1 trigger. To do this, AthenaMT will support *Event Views*, which encapsulate only part of the event data. Development of these views is ongoing, but early prototypes supported the idea of the views themselves being objects within the main event store. Further, the prototype views interface has been made the same as any other ATLAS data proxy, thus algorithms can be made to use a view without any special coding required. Thus the goal, to share algorithmic code between online and offline systems is well on the way to being accomplished. Proper integration with the scheduler is underway.

4.5. Time Varying Data
Time varying data, such as detector conditions, currently relies heavily on Gaudi *incidents*, which trigger the retrieval of data when the algorithm or tool that requires some conditions data finds it is processing an event which is out of the current *interval of validity* (IOV). This blocks a thread while such data is retrieved and does not work well when events being processed may themselves cross IOV boundaries. In AthenaMT, conditions data will be considered as a normal data input to an algorithm, albeit using a different store, supporting multiple IOVs at once. Then conditions data will always be present before an algorithm that needs it is allowed to execute.

5. AtlasG4 Multi-Threaded
Although many pieces of AthenaMT are still under design and development, many simpler test cases and prototypes have been evaluated in order to validate the approach that has been adopted. This started with GaudiHive[9] prototypes, which demonstrated good scaling with limited LHCb reconstruction. An ongoing ATLAS prototype aims at developing a multi-threaded version of ATLAS simulation, suitable for running on Intel's Xeon Phi architecture, where memory is severely constrained. Results from these tests, shown in Figure 3, demonstrate both excellent throughput and memory scaling.

6. Summary and Conclusions
To face the challenges of future computing, with severe power and memory limitations, ATLAS has started to develop a new multi-threaded processing framework, AthenaMT. As well as supporting multi-threading, partial event processing in regions of interest will support HLT use cases better. Prototypes of this framework have already demonstrated excellent throughput and memory scaling. Many other aspects of the framework are being prototyped and implemented now. However, a huge migration effort will be required to port ATLAS's 3 million lines of C++ code to the new framework in time for Run3, and for this a substantial training effort and design review will also be required.

References
[1] Moore G E 1965 *Electronics* **38**
[2] Calafiura P, Lampl W, Leggett C, Malon D, Stewart G and Wynne B 2015 *Journal of Physics: Conference Series* **664** 072031 URL `http://stacks.iop.org/1742-6596/664/i=7/a=072031`
[3] ATLAS Collaboration 2008 *Journal of Instrumentation* **3** S08003 URL `http://stacks.iop.org/1748-0221/3/i=08/a=S08003`
[4] Calafiura P, Lavrijsen W, Leggett C, Marino M and Quarrie D 2005 The athena control framework in production, new developments and lessons learned *Computing in high energy physics and nuclear physics. Proceedings, Conference, CHEP'04, Interlaken, Switzerland, September 27-October 1, 2004* pp 456–458 URL `http://doc.cern.ch/yellowrep/2005/2005-002/p456.pdf`
[5] Barrand G *et al.* 2000 GAUDI - The software architecture and framework for building LHCb data processing applications *Proceedings, 11th International Conference on Computing in High-Energy and Nuclear Physics (CHEP 2000)* pp 92–95 URL `http://lhcb-comp.web.cern.ch/lhcb-comp/General/Publications/longpap-a152.pdf`
[6] Aad G *et al.* (ATLAS) 2012 *Phys. Lett.* **B716** 1–29 (*Preprint* 1207.7214)
[7] Binet S, Calafiura P, Jha M K, Lavrijsen W, Leggett C, Lesny D, Severini H, Smith D, Snyder S, Tatarkhanov M, Tsulaia V, VanGemmeren P and Washbrook A 2012 *Journal of Physics: Conference Series* **368** 012018 URL `http://stacks.iop.org/1742-6596/368/i=1/a=012018`
[8] Intel Threaded Building Blocks URL `https://www.threadingbuildingblocks.org/`
[9] Clemencic M, Hegner B, Mato P and Piparo D 2014 *Journal of Physics: Conference Series* **513** 052028 URL `http://stacks.iop.org/1742-6596/513/i=5/a=052028`

Journal of Physics: Conference Series **762** (2016) 012025 doi:10.1088/1742-6596/762/1/012025

Performance and Advanced Data Placement Techniques with Ceph's Distributed Storage System

M D Poat[1], J Lauret [1]
[1] Brookhaven National Laboratory, P.O. Box 5000, Upton, New York 11973-5000, USA

E-mail: mpoat@bnl.gov

Abstract. The STAR online computing environment is a demanding concentrated multi-purpose compute system with the objective to obtain maximum throughput with process concurrency. Motivation for extending the STAR online compute farm from a simple job processing tool for data taking, into a multipurpose resource equipped with a large storage system would lead any dedicated resources to become an extremely efficient and an attractive multi-purpose facility. To achieve this goal, our compute farm is using the Ceph distributed storage system which has proven to be an agile solution due to its effective POSIX interface and excelling its object storage with IO concurrency. With this we have taken our cluster one step further in terms of IO performance by investigating and leveraging new technologies and key features of Ceph. With the acquisition of a 10Gb backbone network we have ensured to eliminate the network as a limitation. With further acquisition of large fast drives (1TB SSDs) we will show how one can customize the data placement options Ceph has to offer such as primary affinity, mounting OSD journals on SSDs, and Cache Tiering along with non-Ceph related local disk caching techniques. We will discuss the latest performance results along with the expected results by using each technique. We hope this paper will serve the community's interest for the Ceph distributed storage solution.

1. Introduction

The Solenoidal Tracker at RHIC (STAR) experiment at the Relativistic Heavy Ion Collider (RHIC) located at Brookhaven National Laboratory is an international collaboration that has been collecting vast amounts of data for the past ~16 years. The online infrastructure within STAR is made up of many heterogeneous compute systems and sub-systems. In previous work which was presented at the CHEP 2015 conference [1], we have shown how we took our online compute infrastructure and transformed it from a simple job processing farm into a multipurpose resource including a CephFS POSIX storage system offering users ~80TB of redundant storage. This storage system can be used for online QA, real-time calibration or a simple local all-purpose storage system. Since its inception, STAR has strived to further extend and leverage its local resources by pushing the limitations of the Ceph storage system to a new level in terms of IO performance. We have acquired a few different models of 1TB SSD drives intended for use within the Ceph cluster. Ceph has multiple data placement techniques such as primary affinity which ensures which OSD or drive receives the first write of a replicated object, Ceph also has a Cache Tiering method which allows you to create a Ceph pool of

[1] To whom any correspondence should be addressed.

fast drives (SSD) and overlay the pool over a pool of slow drives (HDD). Lastly, since Ceph OSDs use journals when storing data and the journals placed on standard HDDs could result in slow writes, it is possible to put the journals on SSDs with the expectation of write a performance impact.

In this study we aim to compare previous IO results with each of the Ceph data placement techniques along with the comparison of non-Ceph related techniques all with the goal to increase the IO performance of our Ceph cluster while retaining redundancy and safe data.

2. Architecture

The Ceph distributed storage system was built with a specific architecture in mind to ensure the robustness, scalability, and reliability it promises. Every component of Ceph is infinitely scalable and by design contains no single point of failure. The three main components required for Ceph to run is the Monitor service, OSD service and the Metadata Service (needed for CephFS). The storage servers run the OSD service (Object Storage Daemon) which is used to store the actual data on disks. Typically, one OSD service is run per disk or per storage unit (RAID, etc.). The monitors are used to maintain the master copy of the cluster maps, the monitors distribute the cluster maps to all of the storage nodes and clients in order to keep track of where the data is being stored. Lastly, when adding the CephFS POSIX layer to Ceph, a metadata server is required. A Ceph metadata server (MDS) is used to store the POSIX semantics of the files stored in CephFS.

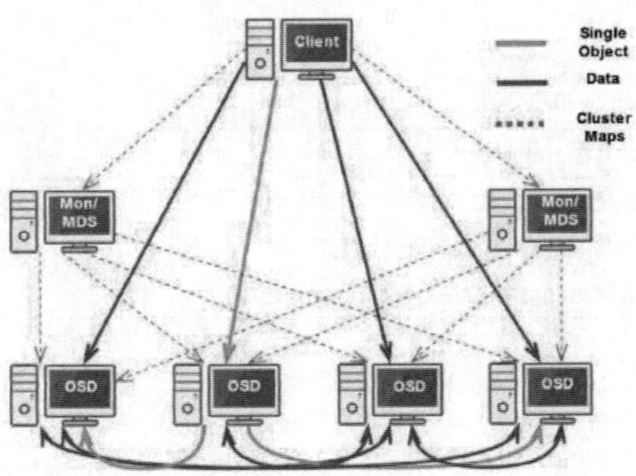

Figure 1: Architecture of the Ceph distributed storage system.

As mentioned previously, Ceph has no single point of failure. As seen in Figure 1, when clients read/write data into Ceph they contact the storage nodes directly opposed to going through a proxy or a middleman. If a storage node goes down, the data on the backend will be reshuffled to ensure consistent replication and the clients will just continue writing to the other storage nodes. Both the monitor servers which distribute the cluster maps, and the metadata servers which store the POSIX semantics, distribute their data to all storage nodes and clients. The monitor and metadata servers will work in parallel with redundancy, if one server goes down the other(s) will continue to work seamlessly.

3. Use of SSDs

SSDs have been known for their increase in single and multi-threaded IO performance. However, what we have learned is that not all SSDs perform at the same levels. We have obtained a few different models of SSD drives to give us an idea range of how a mixed set of SSDs perform compared to our 2TB Seagate ST2000NM0001 SAS HDDs in our Ceph cluster. The SSD drives obtained are all consumer grade drives; we have 4 - 1TB Crucial M550 models [2], 20 -1TB Mushkin

Journal of Physics: Conference Series **762** (2016) 012025 doi:10.1088/1742-6596/762/1/012025

MKNSSDRE1TB models [3], and 4 - 480 GB Mushkin MKNSSDCR480GB-DX [4] models. While during its release, the 1TB Crucial drives cost ~$500 while the 1 TB Mushkin drives cost ~$250, both claim similar IO performance specifications. We tested the performance of each drive using multiple tests and the results were quite interesting. The performance impact of single thread writes to each drive (including the 2 TB HDD) was minimal. However, as the number of threads increase at a particular record size the performance of each drive is quite different. For example, at a 10 thread count with each thread composed of 1024k blocks the 1TB Crucial drives perform at 180 MB/s while the 1TB Mushkin drives only reach speed of 65 MB/s which is just slightly faster than the regular HDD at 55 MB/s. Lastly the 480 GB Mushkin drives hit 120 MB/s in this test. We summarize our results in Table 1. With these results in mind, it is clear that we need to be careful when trying to disentangle the IO performance results of Ceph and need to be sure we are using the correct hardware for the correct IO range. We are not expecting the IO performance within Ceph to be the same as the bare disks since the way Ceph writes to the disks creates wait time ensuring the writes have actually been written and synced properly.

Table 1. Write performance of 4 separate storage drives using Iozone. Performance per drive running 10 Iozone write threads at 1024 KB block sizes shown in MB/s.

Storage Drives	MB/s
1TB Crucial M550	180
480GB Mushkin MKNSSDCR480GB-DX	120
1TB Mushkin MKNSSDRE1TB	65
2TB Seagate ST2000NM0001 HDD	55

4. Data Placement Techniques

4.1. General

The Ceph community has an abundance of different data placement techniques and methods for storing data across specific drives, pools or nodes. In a cookie cutter stock Ceph cluster, all Ceph data pools are created with replication 3 and are spread across the storage nodes and OSDs thus distributing the data evenly across the cluster [5]. In this setup, if an OSD falls out of Ceph (due to a failing hard drive for example), Ceph will immediately rebalance the cluster to ensure consistent replication.

4.2. Primary Affinity

When a Ceph Client reads or writes data, the client always contacts the primary OSD in the acting set [6]. When writing, the client will write data to the first OSD where that OSD will then replicate the data out to two more OSDs fulfilling the replication 3 requirement. When reading, the client will only read from the primary OSDs. The primary affinity configuration has a range which can be set between 0 – 1. An OSD with the primary affinity setting of 0 means that the OSD will not be used as a primary while a setting of 1 means that it will be used as a primary. However, if you set half of your OSDs to 1 and the other half to 0.5, the first set of OSDs will be the primary 75% of the time while the others will be the primary 25% of the time.

Using the Primary Affinity setting we replaced 20 OSDS in an 80 OSD cluster composed of the 2TB Seagate SAS HDD with 20 of the 1TB Mushkin MKNSSDRE1TB SSD drives, we set the primary affinity of the SSD OSDs to 1 while all other OSDs were set to 0. Once the configuration was set we ran a series of single and multiple clients writes including single and concurrent writes per client using both RADOS object storage and the CephFS POSIX storage.

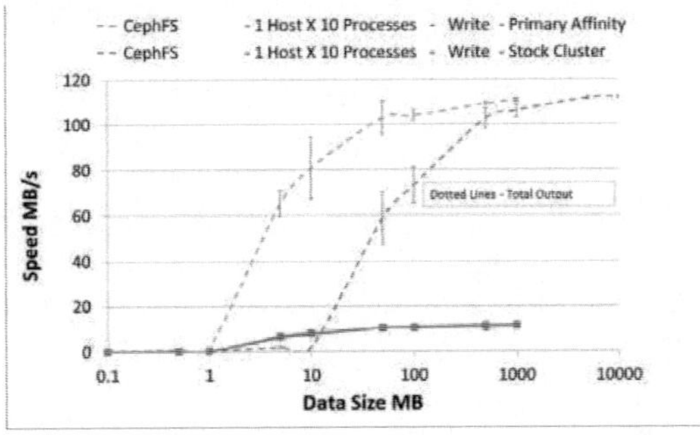

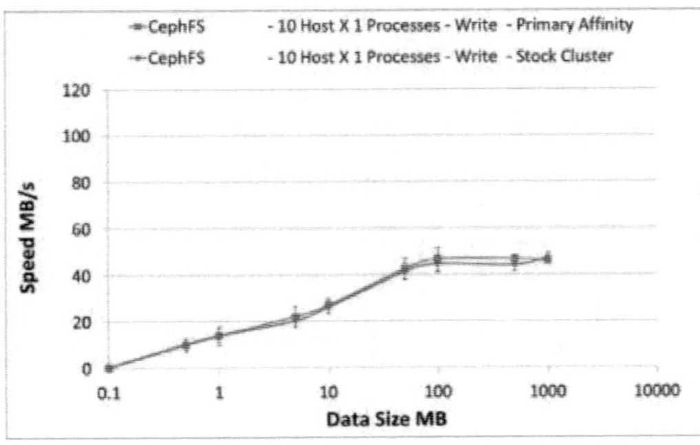

Figure 2: The left plot represents 1 client running 10 concurrent write processes into a CephFS cluster with 20 SSD OSDs with primary affinity set 1 while the other 60 HDD OSDs primary affinity is set to 0 vs. the same performance tests into a stock CephFS cluster composed of 80 HDD OSDs. The right plot is showing the performance of each client's performance when running 10 clients at 1 write process each into CephFS with primary affinity (20 SSDs & 60 HDDs) vs. a stock CephFS cluster of 80 HDDs. Both plots are in MB/s as a function of data file size. The dash curve on the left hand-side plot indicates aggregate IO per client while the solid line curves are normalized values per process.

Performance measurements were taken from tests run against both the RADOS object storage and CephFS storage using multiple tools, however all tests pertain to the CephFS approach. In Figure 2, it can be seen that the performance of CephFS using primary affinity settings with SSDs has been unchanged whether it is a single client writing or multiple clients writing. With further investigation, we found that Ceph uses "weights" for each individual OSD which determines the fill ratio. We currently are using 2 TB HDDs and 1 TB SSDs in our cluster. In our case, Ceph assigns the weight of 2 to the HDD OSDs and a weight of 1 to SSD OSDs (if, for example, we had 512 GB drives the weight would be 0.5). CephFS ensures that the cluster is filled evenly and this may actually cause Ceph to ignore the primary affinity settings by setting a priority on OSD weights and cluster fill ratio (this was seen when closely watching RADOS object store writes, although this was not a predictable behavior). Secondly, when writing into Ceph the SSD OSDs may be finished writing but still waiting for the HDD OSDs to sync for replication 3 virtually creating no performance gain. It is difficult to extrapolate the CephFS writing process as CephFS clients open many connections to many or nearly all storage nodes when writing into the cluster causing difficulty when trying to track down where the data is going. If the primary affinity setting is being respected, then the unchanged write performance would be due to the time to sync the HDDs for replication.

4.3. SSD Journals

In CephFS, when a client writes into the cluster the file is broken down into objects (chunks) and is written to the OSDs [7]. When Ceph writes to an OSD, there are two sequential IO operations, first writing to the journal and second writing to the filesystem itself. The journals will not release the data to the OSDs filesystem until it has acknowledged the write to itself; this can cause a slowdown in overall write performance. If there is heavy IO into the Ceph cluster, many journal operations would occur (Ceph writes small random IO into the journals). With this in mind, we created 3 OSDs per node each composed of the 2TB Seagate SAS HDD, and set each of the journals of the OSDs to run on 1

Mushkin MKNSSDRE1TB SDD. A 3OSD:1SSD per node configuration in hope for a write performance gain.

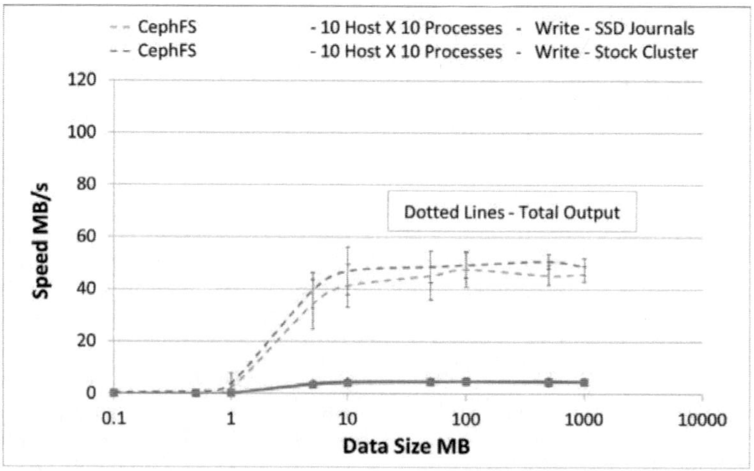

Figure 3: The plot represents 10 clients running 10 concurrent write processes into CephFS with the journal operations on SSDs vs. a stock CephFS cluster. The plot is showing the performance in MB/s as a function of data file size. The dash curves are representing the aggregate write speed per client of all 10 processes, the solid line curves are the speed per client of each write process.

When accounting for the error bar it can be seen that there is no write performance gain with this configuration. With further investigating, we found that when Ceph performs journal operations it uses the write options D_SYNC and O_DIRECT (D_SYNC does not allow any other writes until the initial write is finished and writing with O_DIRECT skips using the kernel page cache). In the test shown in Figure 3, 10 clients running 10 write processes into Ceph (100 concurrent writes) may not actually be enough concurrent writes to see a performance impact. Due to the small size of our cluster, when writing an excess of 100 concurrent writes, the Ceph monitors report blocked requests and heavy IO wait.

4.4. Cache Tiering

Cache Tiering involves creating a Ceph storage pool composed of fast drives (SSDs) that is overlaid on top of a backing storage pool composed of slower drives (cheap HDD) [8]. With this configuration the administrator can set a specific write back policy on how and when to flush data to the backing pool. The cache flush can be based on fill ratio, data popularity, cache age, etc. but the data can be primarily stored on the SSD pool and only flush to the HDD when needed. Unfortunately, at this time it does not appear that Cache Tiering is compatible (not implemented) within CephFS as Ceph is restricting the storage pools to be part of both a Cache Tier and CephFS simultaneously. With this restriction taken into account, we still wanted to simulate a Cache Tier by creating a CephFS storage system with a small SSD only pool and compare it to a small HDD only pool. The assumption being that in the deal Cache Tiering scenario, the IO would be directed to the SSD first in this case.

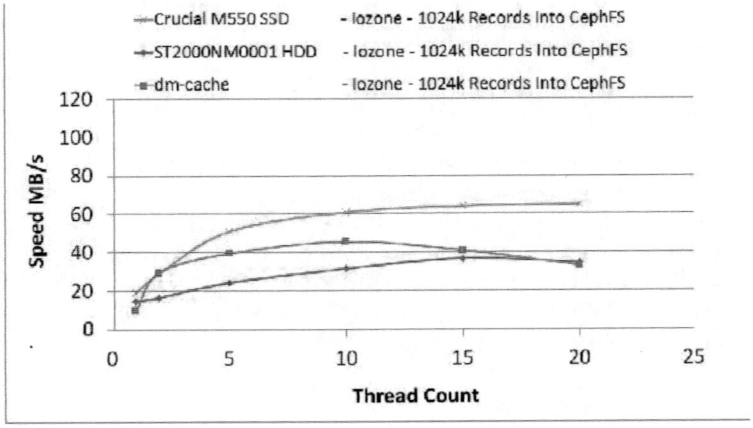

Figure 4: The plot represents one client using Iozone to test the write performance of CephFS composed of a 4 Crucial M550 SSD pool vs. CephFS composed of a 4 Seagate ST2000NM0001 HDD pool. We have further compared the performance with a CephFS composed of a dm-cache pool. The plot is shown in MB/s as a function of increased thread containing 1024k block sizes.

As it can be seen in Figure 4, there is indeed an expectation of a write performance gain for an SSD only CephFS (blue curve) as the thread count increases compared to the HDD only CephFS (red curve). This is partly due to using the Crucial M550 drives which have a better IO performance over the Mushkin MKNSSDRE1TB which have been used in many of the tests. The performance impact is present and this data is useful as we now know Ceph performs better when all the writes are written to faster performing drives.

5. dm-cache

We now know the use of SSDs can impact Ceph IO performance along with the understanding of the expected performance range with different SSD drives. The remaining question is how could we emulate Cache Tiering within a CephFS deployment without this feature being natively supported? One could achieve the same effect by using an external caching mechanism such as dm-cache [10], acting at a lower hardware level. While similar to a Ceph cache tier, dm-cache is local to each node and uses SSDs to act as a cache in front of HDDs, presenting to the higher level IO component (CephFS in this case) a logical device. In other words, this would circumvent the primary affinity and OSD weight mechanisms as Ceph would have no idea of the presence of SSDs. By using dm-cache, we could ensure the writes go to the SSDs first but note one downside: if we overlay one SSD over 3 HDD and create 3 OSDs on one machine and the SSD fails, we would lose all OSDs attached to that single machine from the Ceph cluster. In Figure 4, we are showing the performance of a 'dm-cache only CephFS' (green curve) compared to the 'SSD only CephFS' (blue curve) or 'HDD only CephFS' (red curve). The dm-cache policy in this case was set to a high write-promote setting which will allow streaming IO to be written to the SSD and immediately flush back to the HDD. While the performance of the 'dm-cache only CephFS' does show a performance gain over the 'HDD only CephFS" setup, the performance is not at the same level as the 'SSD only CephFS', this is due to the write-back from the SSD to the HDDs. Note that at a high thread count (15 - 20 threads) the performance of the 'dm-cache CephFS' begins to degrade and is similar to the 'HDD only CephFS'.

6. Conclusion

In this paper, we have tested multiple Ceph data placement techniques with the objective to increase the IO write performance of our current Ceph cluster with the use of SSDs. While the data placement techniques appear intuitive and our expectations were high, we found that not all of these techniques align with the expected results. The primary affinity configuration did seem promising at first, however the OSD weights for balancing and the syncing with the HDD may be the cause of no performance impact. The OSD journals on SSD drives may only create a performance impact with better performing SSDs or only with an extremely large number of threads where our small Ceph cluster can't even keep up with. Cache Tiering was found to show a performance impact, but does not work with CephFS. Fortunately, we were able to revert to and prove that a local caching technique, such as using dm-cache, increases the IO performance of our system within caveats. In our preliminary testing, dm-cache does show positive performance impact, but may be risky if the SSDs were to fail. A less risky solution may be a 1:1 ratio of SSD:HDD but this would create a cost impact to STAR we were are trying to avoid.

Acknowledgements

This work was supported by the Office of Nuclear Physics within the U.S. Department of Energy's Office of Science.

References

[1] Poat M D, Lauret J and Betts W 2015 *POSIX and Object Distributed Storage systems – Performance Comparison Studies With Real-Life Scenarios in an Experimental Data Taking Context Leveraging OpenStack Swift & Ceph* J. Phys.: Conf. Ser 664 042031
[2] Crucial M550 SSD Storage http://www.crucial.com/usa/en/storage-ssd-m550
[3] Mushkin MKNSSDRE1TB SSD Storage http://www.poweredbymushkin.com/index.php/catalog/item/44-reactor/1076-reactor-1tb-7mm.html
[4] Mushkin MKNSSDCR480GB-DX SSD Storage http://www.poweredbymushkin.com/index.php/component/djcatalog2/item?id=665:7mm-chronos-deluxe-480gb&cid=9:chronos-deluxe
[5] Singh K 2015 *Learning Ceph* (Birmingham, UK)
[6] Ceph Primary Affinity http://docs.ceph.com/docs/master/rados/operations/crush-map/#primary-affinity
[7] Ceph OSD Journals on SSD http://docs.ceph.com/docs/master/start/hardware-recommendations/#solid-state-drives
[8] Ceph Cache Tiering http://docs.ceph.com/docs/master/rados/operations/cache-tiering/
[9] Iozone http://www.iozone.org/
[10] dm-cache https://en.wikipedia.org/wiki/Dm-cache

Journal of Physics: Conference Series **762** (2016) 012026 doi:10.1088/1742-6596/762/1/012026

Reducing power usage on demand

G Corbett, A Dewhurst

Science and Technology Facilities Council, Rutherford Appleton Laboratory, Harwell Oxford, Didcot OX11 0QX, UK

E-mail: greg.corbett@stfc.ac.uk, alastair.dewhurst@stfc.ac.uk

Abstract. The Science and Technology Facilities Council (STFC) datacentre provides large-scale High Performance Computing facilities for the scientific community. It currently consumes approximately 1.5MW and this has risen by 25% in the past two years. STFC has been investigating leveraging preemption in the Tier 1 batch farm to save power.

HEP experiments are increasing using jobs that can be killed to take advantage of opportunistic CPU resources or novel cost models such as Amazon's spot pricing. Additionally, schemes from energy providers are available that offer financial incentives to reduce power consumption at peak times.

Under normal operating conditions, 3% of the batch farm capacity is wasted due to draining machines. By using preempt-able jobs, nodes can be rapidly made available to run multicore jobs without this wasted resource. The use of preempt-able jobs has been extended so that at peak times machines can be hibernated quickly to save energy.

This paper describes the implementation of the above and demonstrates that STFC could in future take advantage of such energy saving schemes.

1. Introduction

The STFC datacentre provides large-scale High Performance Computing facilities for the scientific community. In June 2015, the power consumption of the compute resources was 913 kW, with an additional 500 kW for cooling. This means the datacentre has a Power Usage Effectiveness (PUE) of 1.55. Over the last two years, the power consumption of the centre has increased by approximately 25% despite the physical footprint remaining the same. This power increase corresponds to a $£247,000$ increase in running costs over the two years.

This paper documents a pilot scheme to reduce the energy consumption of the STFC datacentre to both save money and reduce environmental impact. The pilot scheme focused on a subset of the datacentre, specifically the worker nodes of the UK Tier 1 batch farm, which forms part of the Worldwide LHC Computing Grid[1]. The batch farm uses HTCondor[2] for workload management. The batch farm consists of approximately 600 physical worker nodes that provide 15500 job slots. When full the power consumption is $\approx$ 200kW.

The motivation behind the pilot scheme was to decide if STFC could take advantage of a UK National Grid scheme[5] where large commercial users are paid to reduce their power usage when necessary during peak winter months. The National Grid scheme is called "Demand Side Balancing Reserve" (DSBR) and was developed in 2013 to protect consumers from experiencing problem with their electricity supply. This was due to the margin between energy generation and demand tightening. The increase in the amount of renewable energy generated also means that the supply is less reliable. Peak power demand is between 4pm and 8pm during winter months and if activated DSBR will pay users if they can provide a 1MW drop in the power they use within 15 minutes. There are different price bands, ranging from 0.25 per kWh through to 2.50 per kWh, depending on the severity of the problems. Users can choose which price band depending on the cost to them but the higher they pick the less likely they are to be selected. The scheme was first used in November 2015 [6].

There were two main parts of the project. To develop a means of powering down idle machines and powering them back up when required later. To develop a system of identifying preempt-able jobs on the batch farm so that they can be killed if required. The work undertaken for this project could be applied to other parts of the STFC datacentre as well as other datacentres.

2. Preempt-able jobs

HTCondor has features which allow the running of jobs that can be paused if a higher priority job comes along and restarted when resources become available again. However, it requires the software to be written in a specific way, which can be difficult for large collaborations to implement. Due to the security model employed by the LHC experiments, jobs cannot be paused for an extended period of time as the job's authentication credentials will expire. Jobs that are paused at one site cannot be transferred to another site to keep running if capacity is available there. This means that this form of preemption is not widely adopted. It should also be noted that in the past, if an experiments jobs were considered preempt-able at a site, it normally meant they were not the primary user of the site and therefore received worse support.

The arrival of competitively priced large scale cloud computing has significantly changed things. One particular example that has been looked at by ATLAS is Amazon's Elastic Cloud spot pricing[4]. It allows users to bid on the price of CPU resources. If the price rises above what you are prepared to pay your jobs will be killed although you will only be charged for complete CPU hours used. This has lead to the development of the ATLAS Event Service[7] which uploads the results of every event it processes and very little work would be lost in the event the job is killed. The implementation of preempt-able jobs at STFC, is therefore designed to mimic this use case.

3. Power benchmarking and monitoring

The first step was to measure how the power consumption of a worker node varies depending on the load it is put under. A multi-meter was connected to one of the chassis in the datacentre. Each chassis contains four worker nodes. While measuring the power usage each node in the chassis ran identical tests.

HEP-SPEC06 [8] benchmarking tool was used to generate load on the worker nodes. The power usage of the machines was measured when they were idle and with the HEP-SPEC06 benchmark using 8, 16, 24, and 32 threads. The results are shown in Figure 1. An idle worker node uses just over 90W. The CPUs have hyper-threading turned on which is why the power usage increase slows down after 16 cores are being used. HEP-SPEC06 is linearly proportional to the power used. If the worker nodes are made to hibernate, they used less than 5% of their idle power.

An idle machine uses 35% the power of a fully loaded machine. To maximise the energy efficiency of the batch farm, packing jobs on to as few a nodes as possible and hibernating the rest is optimal. It should be noted that in general Grid jobs are not as CPU efficient as the HEP-SPEC06 benchmark. While it is difficult to quantify this, it does mean that an experiment will get more useful work out of a fully loaded machine than the HEP-SPEC06 measurement might suggest. This does not change the conclusion to pack jobs on to as few nodes as possible.

To measure the power usage of the batch farm overall, information was collected from the PDUs in each rack via snmp and aggregated via Cacti[9]. Each rack contains two PDUs and several chassis. Each chassis is connected to both PDUs. Measuring the power usage of the sum

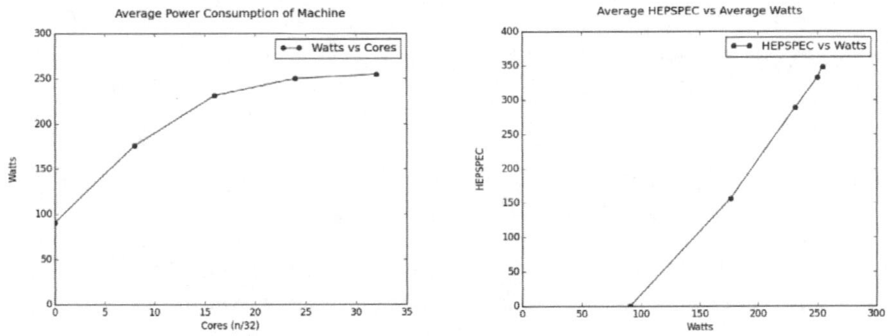

Figure 1. (left) Plot showing how the power usage increases as the number of threads (and hence cores) of the HEP-SPEC06 benchmarking are increased. (right) Plot showing how the measured HEP-SPEC06 changes as a function of energy used.

of the two PDU will give the power usage of each rack, but it is not possible to collect finer granularity information.

4. Hibernating idle machines

As described in the previous section, hibernating idle machines will provide a significant saving in power. Once every 5 minutes, the startd daemon checks the number of jobs a machine is running to determine if it is idle and if so for how long it has been in this state. It also checks how long it has been since the last hibernation. This is done via a shell script which parses the HTCondor logs looking for the last hibernation time. Finally, startd also checks that the machine is in a valid state to run jobs by checking the health status. For the machine to be hibernated it needs to be idle for at least 5 minutes and to have been awake for at least thirty minutes. This latter check prevents a machine from cycling between hibernating and up too quickly, potentially causing damage to the machine. It also must be able to run jobs, this prevents machines that may have been disabled for other reasons being hibernated.

Hibernating machines can be woken up via the HTCondor Rooster daemon[3]. To wake up hibernating machines, there needs to be jobs waiting in the queue for more than 15 minutes that could be run on the hibernating machines (i.e. HTCondor would have matched a job to that worker node if it had been awake). One machine is woken up per Rooster cycle which is configured to run every 5 minutes. The machine chosen is based on number of largest number of CPUs. Machines with more CPUs are newer so should be more energy efficient. The machine being woken up needs to have been hibernating for at least 30 minutes.

The automatic hibernation of idle machines was enabled at the start of December 2015. Over the Christmas period there was a dip in the number of jobs being sent to batch farm and some machines became idle. Figure 2 shows the used and idle capacity of the batch farm over this period as well as the energy usage of a generation of worker nodes. As the farm started to drain these worker nodes became idle and were powered down. A mistake in the configuration of the HTCondor Rooster (the configuration file was only put on one of the two head nodes) caused a delay in the worker nodes being woken up when the farm subsequently filled again.

One concern with repeatedly hibernating idle machines is that this may increase the chance of hardware failure. As a result only the older worker nodes - those that are over 3 years old have the ability to be hibernated enabled. This is around 40% of the worker nodes. The intention is to monitor the hardware failure rates over the next few years to see if hibernation has any impact.

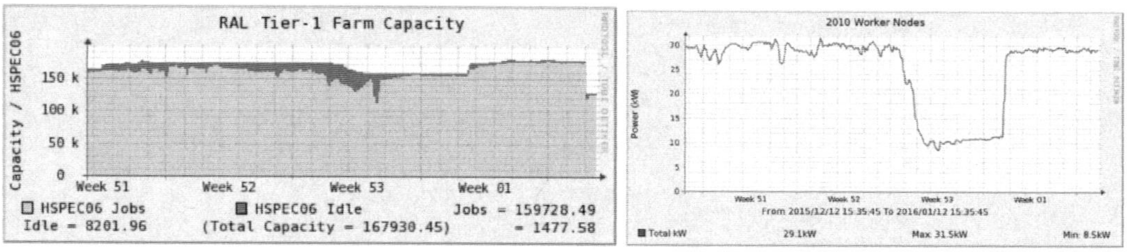

Figure 2. (left) Plot showing the total used and idle capacity of the Tier 1 batch farm (right) Plot showing how the measured HEP-SPEC06 changes as a function of energy used.

From the observed power saving shown in Figure 2, and from previous years data on the amount of time the batch farm is not completely full, we estimate that there is an energy saving of around 15000kWh a year from hibernating idle nodes.

5. Preempt-able jobs on draining worker nodes

The batch farm runs a mixture of single and multi-core (normally 8 core) jobs. HTCondor is configured to allow dynamic provisioning of multi-core jobs. If the batch farm is running at or near capacity, it is highly unlikely that several CPU slots will become available on the same node at the same time to allow a multi-core job to start. In this situation HTCondor needs to select some nodes to drain (i.e. stop starting new jobs) while it allows single core jobs to start on other nodes. Previously a node that was being drained would have wasted resources. With the introduction of preempt-able jobs these can be used to fill worker nodes while they are being drained. Worker nodes also need to be drained periodically so that they can be rebooted for Kernel and other updates.

Worker nodes are drained via a dedicated python script which replaces the HTCondor defrag daemon. When a worker node needs to be drained an attribute is added to the machine ClassAd so that only preempt-able jobs can be started. Once the required number of slots on the machine are either empty of running preempt-able jobs then the preempt-able jobs are killed.

When selecting a node to drain the aim is to identify the node that will drain the fastest. This minimises the wasted resources as well as allowing the multi-core job (which should have the highest priority) to run. The python script has an algorithm which it uses to decide which node should be drained. From the experience we have at STFC, the LHC experiments do not accurately specify the length of time they think a job will take. All jobs are normally submitted with the absolute maximum wall-time specified. Using this to calculate when jobs are likely to finish is not useful. Instead, a simpler algorithm which assumes that all jobs are equally likely to finish at any time is used. Therefore, the more running jobs there are relative to the number of slots required, the faster that machine is likely to be drain. This algorithm was modified to take into account the preempt-able jobs that are now being run:

$$\text{Rank} = \frac{[\text{Total Slots}] - [\text{Preemptable Jobs}] - [\text{Free Slots}]}{[\text{CPUs requested}] - [\text{Preemptable Jobs}] - [\text{Free Slots}]}.$$

Machines with the highest Rank will be drained next. If $[\text{CPUs requested}] - [\text{Preemptable Jobs}] - [\text{Free Slots}] \leq 0$ there is no need to drain a node (as their are enough slots already available) so the Rank will not be calculated.

Under normal operating conditions there are around 15 worker nodes ($\approx 0.5\%$ of batch farm capacity) being drained so that multi-core jobs can be started. Worker nodes are also rebooted around 6 times a year for kernel and security patching. To completely drain a worker node takes around 36 hours. This has previously meant that up to 3% of the batch farm capacity is wasted due to worker nodes being drained. These resources can now be used by preempt-able jobs.

6. Killing jobs to reduce power usage

The final aim of the project was to see if we could reduce the energy usage on the batch farm rapidly (within 15 minutes) assuming we had received a signal as part of an energy saving scheme. A large number of test preempt-able jobs were submitted to the farm and allowed to start running. A script was then run which would kill all preempt-able jobs and prevent any more from starting. From the time the script was run it took just under 11 minutes for all the preempt-able jobs to be killed. It took under 20 minutes for the energy usage to drop to its lowest point. While some further optimisation may be required this does show that it should be possible to take part in schemes like this. While it was not monitored for this test, it can be assumed that any energy saving from a reduction in the amount of work being done by the air-conditioning would take significantly longer to manifest.

The average energy usage drop observed per job killed was 0.1W. The test was done when the batch farm was busy so jobs were being run on worker nodes which were effectively full. None of the nodes were full with preempt-able jobs, so none of them went into hibernation and the energy reduction was simply due to the fact that the worker nodes were less loaded. Figure 1 shows that this is the least effective way to save energy. Preempt-able jobs are preferentially matched to older hardware, which is the least energy efficient. In order to make significant savings, we estimate that at least 40% of the running jobs should be preempt-able. With this volume of preempt-able jobs the batch system can easily fill entire nodes with them, which allows for the maximum energy saving. Currently there are a limited number of ways experiments can take advantage of preempt-able slots but this is growing.

7. Summary and future work

This paper described the successful implementation of both preempt-able jobs and the hibernation of idle worker nodes on the STFC Tier 1 batch farm. The hibernation of idle worker nodes should save 15000kWh per year while the implementation of preempt-able jobs allows 3% of the batch farm's capacity that was previously being wasted to be utilised. This paper also described a test to reduce power usage on demand. While very small scale it did demonstrate that we could reduce the power usage within the 15 minute time frame requested. We will be encouraging experiments to submit more preempt-able jobs in future.

We will be monitoring hardware failure rates on the farm to see if the hibernation of machines causes it to increase. We also aim to run preempt-able jobs on the idle cloud resources available at STFC.

References

[1] WLCG: http://wlcg.web.cern.ch/
[2] HTCondor: https://research.cs.wisc.edu/htcondor/
[3] http://research.cs.wisc.edu/htcondor/manual/v8.4/3_15Power_Management.html
[4] Amazon Spot Pricing: https://aws.amazon.com/ec2/spot/
[5] Demand Side Balancing Reserve: http://www2.nationalgrid.com/UK/Services/Balancing-services/System-security/Contingency-balancing-reserve/
[6] http://www.theguardian.com/business/2015/nov/04/national-grid-issues-urgent-call-for-extra-power
[7] P Calafiura and [ATLAS Collaboration] 2015 The ATLAS Event Service: A new approach to event processing *J. Phys.: Conf. Series* **664** 062065
[8] HEP-SPEC06: https://w3.hepix.org/benchmarks/doku.php
[9] Cacti: http://www.cacti.net/

Scaling up ATLAS Event Service to production levels on opportunistic computing platforms

D Benjamin[1], J Caballero[2], M Ernst[2], W Guan[3], J Hover[2], D Lesny[4], T Maeno[2], P Nilsson[2], V Tsulaia[5], P van Gemmeren[6], A Vaniachine[7], F Wang[3] and T Wenaus[2] on behalf of the ATLAS Collaboration

[1] Duke University, Durham, NC 27708, United States of America
[2] Brookhaven National Laboratory, Upton, NY 11973, United States of America
[3] University of Wisconsin, Madison, WI 53706, United States of America
[4] University of Illinois at Urbana–Champaign, Champaign, IL 61801, United States of America
[5] Lawrence Berkeley National Laboratory, Berkeley, CA 94720, United States of America
[6] Argonne National Laboratory, 9700 South Cass Avenue, Argonne, IL 60439, United States of America
[7] Tomsk State University, Lenina Avenue 36, 634050, Tomsk, Russian Federation

E-mail: sacha.vaniachine@cern.ch

Abstract. Continued growth in public cloud and HPC resources is on track to exceed the dedicated resources available for ATLAS on the WLCG. Examples of such platforms are Amazon AWS EC2 Spot Instances, Edison Cray XC30 supercomputer, backfill at Tier 2 and Tier 3 sites, opportunistic resources at the Open Science Grid (OSG), and ATLAS High Level Trigger farm between the data taking periods. Because of specific aspects of opportunistic resources such as preemptive job scheduling and data I/O, their efficient usage requires workflow innovations provided by the ATLAS Event Service. Thanks to the finer granularity of the Event Service data processing workflow, the opportunistic resources are used more efficiently. We report on our progress in scaling opportunistic resource usage to double-digit levels in ATLAS production.

1. Introduction

Using opportunistic resources efficiently and fully for ATLAS processing is an important means for maximizing ATLAS computing throughput within budget constraints. Examples of such opportunistic resources are:

- High Level Trigger (HLT) farm, available opportunistically during LHC Run 2;
- High Performance Computing (HPC), also known as supercomputers;
- Opportunistic grid resources, e.g. non-ATLAS OSG sites;
- Cloud resources, particularly, the Amazon EC2 spot market;
- Volunteer computing, also known as ATLAS@Home via BOINC.

Fine-grained event-based processing enables agile, efficient utilization of opportunistic resources that appear and disappear with unpredictable timing and quantity. Event consumers can be injected into such a resource when it becomes available, and participate in ongoing task processing, concurrently delivering outputs to an aggregation point until the resource disappears, with negligible losses if the resource disappears too suddenly.

2. Event Service

The new ATLAS production system prodsys2 [1] includes a PanDA extension JEDI [2] that adds to the PanDA server the capability to intelligently and dynamically break down tasks (as defined by the prodsys2 task definition component DEFT [3]) based on optimal use of the processing resources currently available. JEDI can break down tasks not only at the job level but also at the event level. The initial application of this capability is for event-level job splitting: jobs are defined in terms of the event ranges within files that they should process, and their dispatch and execution otherwise proceeds as normal. However this capability also presents the opportunity to manage processing and dispatch work at the event or event cluster level.

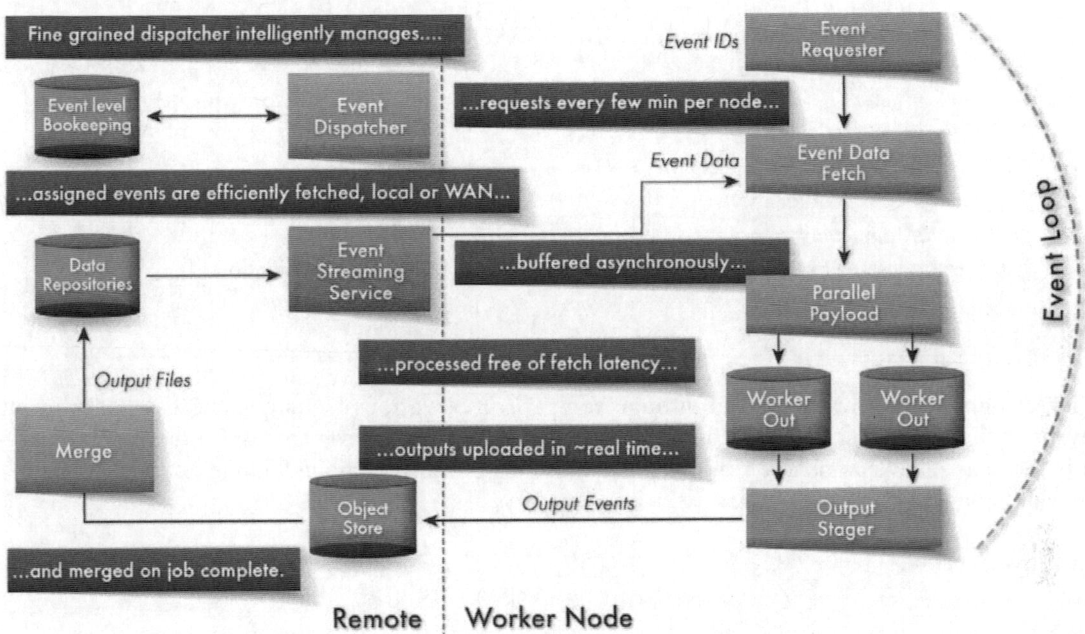

Figure 1. Event Service architecture including the Event Streaming Service (work-in-progress).

2.1. Requirements

We have to be agile in how we use these opportunistic resources:

- Rapid setup of software, data and workloads (quick start) when they become available;
- Quick exit when resources are about to disappear;
- Robust against resources disappearing with no notice: minimal losses;
- Use resources until they disappear – harvest unused cycles, filling resources with fine-grained workloads.

2.2. Implementation

The Event Service (ES) [4] leverages excellent networks for efficient remote data access, distributed federated data access (using xrootd), and highly scalable object store technologies for data storage that architecturally match the fine-grained data flows, to support dynamic, flexible, distributed workflows that adapt in real time to resource availability and leverage remote data repositories with no data locality or pre-staging requirements (Figure 1).

2.3. Commissioning
During Event Service commissioning more that 150 test and/or validation tasks has been submitted and/or processed. Commissioning optimized the set of configuration parameters for production tasks for Event Service workflow execution. Table 1 shows the representative set of the validated configuration parameters for HPC and AWS sites.

Table 1. Event Service task configuration parameters.

Parameter	HPC	AWS
cmtconfig	x86_64-slc6-gcc48-opt	x86_64-slc6-gcc48-opt
spacetoken	ATLASDATADISK	ATLASDATADISK
cloud	US	US
site	NERSC_Edison	BNL_EC2E1_MCORE
skipScout	yes	yes
coreCount	8	8
nEventsPerWorker	1	1
nEsConsumers	1	1
esProcessingType	validation	validation
disableReassign	yes	yes
maxAttemptES	10	

Further optimization was achieved through comparison of various techniques necessary for event-by-event navigation. After detailed comparison of remote access to EventIndex [5], use of in-situ tag files [6], or use of events position in the input files, we chose the last technique as most robust for large scale production.

3. Lessons Learned

3.1. Results
Event Service production commissioning and operation scales up steadily. On the Amazon AWS platform the production level achieved up to 50k concurrent cores (Figure 2). Thanks to the attractive pricing of the AWS EC2 spot instances, the total costs were lower than those associated with certain ATLAS dedicated resources. To scale up the AWS productions level further, preparations are ongoing for the 100k core production run.

At the HPC platform at NERSC we successfully commissioned scaling tests of up to 50k concurrent cores, with production level achieving up to 25k concurrent cores. Following improvements in I/O throughput, the slope of simulation production CPU consumption sharply increased with two equal contributions: Yoda [7] and Generator [8] (Figure 3).

3.2. Work in progress
On the opportunistic HLT farm we are working with simulation production and scaling up in tests now ongoing towards 35k core scale. (Off) Grid platforms at various stages of commissioning and operation: backfill on conventional grid sites, US Tier 3, BNL shared Tier 3, campus resources and non-ATLAS Open Science Grid sites. The ARC/Event Service/Yoda integration is working, with ARC Control Tower [9] integration is underway. Also the ATLAS@Home [10] implementation is underway.

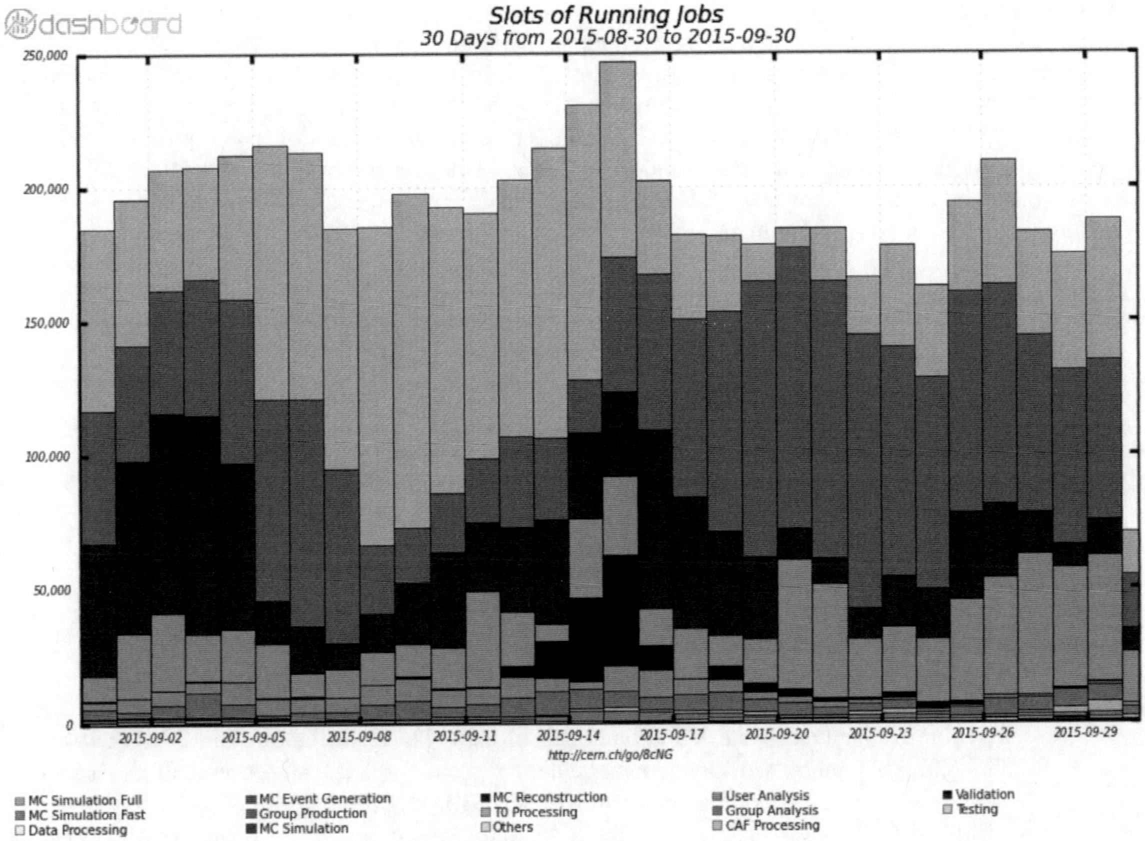

Figure 2. Event Service production level on AWS of up to 50k CPU-cores (black) contributed to the traditional of ATLAS resources of about 200k CPU-cores (colors).

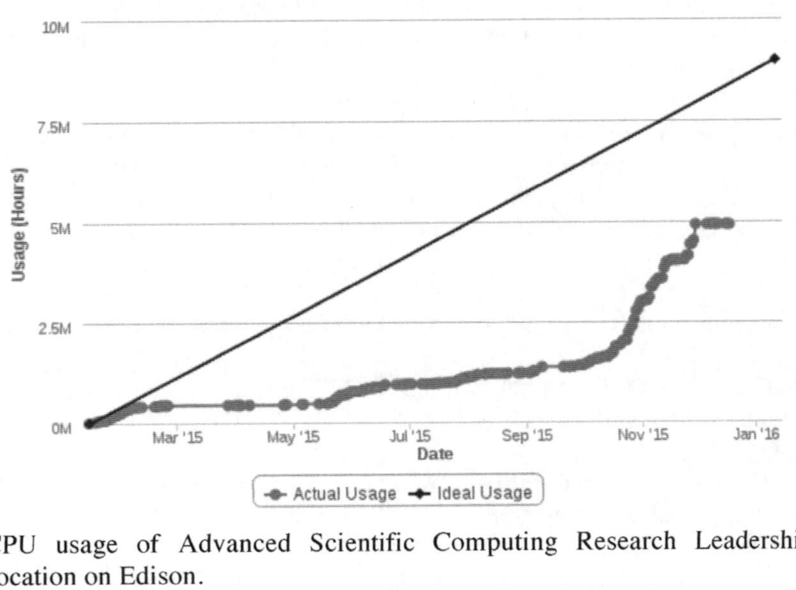

Figure 3. CPU usage of Advanced Scientific Computing Research Leadership Computing Challenge allocation on Edison.

4. Conclusions

The ATLAS Event Service enables cost-effective exploitation of opportunistic computing on heterogeneous, high capacity and high value platforms such as HPCs ('hole filling'), commercial clouds (dynamic market based spot pricing) and volunteer computing (no resource availability guarantees). In the next months we will continue efficient utilization of preemptible resources, particularly opportunistic such as HPC and AWS spot instances as well as backfilling dedicated ATLAS resources to keep them truly full. In general, the Event Service will serve as the basis for a uniform resource pool consuming in an automated and flexible way a predefined task pool.

Acknowledgments

We thank all our colleagues who contributed to the ATLAS Event Service development and operations. This work was funded in part by the U. S. Department of Energy, Office of Science, High Energy Physics under Contract No. DE-AC02-06CH11357.

References

[1] Borodin M, De K, Garcia Navarro J, Golubkov D, Klimentov A, Maeno T, South D and Vaniachine A on behalf of the ATLAS Collaboration (2015) Unified System for Processing Real and Simulated Data in the ATLAS Experiment. Proc. of the DAMDID/RCDL'2015 Conf. *CEUR-WS.org* **1536**:157;

Borodin M, De K, Garcia Navarro J, Golubkov D, Klimentov A, Maeno T and Vaniachine A on behalf of the ATLAS Collaboration (2015) Multilevel Workflow System in the ATLAS Experiment. Proc. of the ACAT2014 Workshop. *J. Phys.: Conf. Ser.* **608**:012015

[2] Maeno T, De K, Klimentov A, Nilsson P, Oleynik D, Panitkin S, Petrosyan A, Schovancova J, Vaniachine A, Wenaus T and Yu D on behalf of the ATLAS Collaboration (2014) Evolution of the ATLAS PanDA workload management system for exascale computational science. Proc. of the CHEP2014 Conf. *J. Phys.: Conf. Ser.* **513**:032062

[3] De K, Golubkov D, Klimentov A, Potekhin M and Vaniachine A on behalf of the ATLAS Collaboration (2014) Task Management in the New ATLAS Production System. Proc. of the CHEP2014 Conf. *J. Phys.: Conf. Ser.* **513**:032078

[4] Calafiura P, De K, Guan W, Maeno T, Nilsson P, Oleynik D, Panitkin S, Tsulaia V, van Gemmeren P and Wenaus T on behalf of the ATLAS Collaboration (2015) The ATLAS Event Service: A new approach to event processing. Proc. of the CHEP2015 Conf. *J. Phys.: Conf. Ser.* **664**:062065

[5] Barberis D *et al.* (2015) The ATLAS EventIndex: architecture, design choices, deployment and first operation experience. Proc. of the CHEP2015 Conf. *J. Phys.: Conf. Ser.* **664**:042003

[6] Cranshaw J, Doyle A T, Kenyon M J, Malon D, McGlone H and Nicholson C (2008) Integration of the ATLAS tag database with data management and analysis components. Proc. of the CHEP2008 Conf. *J. Phys.: Conf. Ser.* **119**:042008

[7] Calafiura P, De K, Guan W, Maeno T, Nilsson P, Oleynik D, Panitkin S, Tsulaia V, van Gemmeren P and Wenaus T on behalf of the ATLAS Collaboration (2015) Fine-grained event processing on HPCs with the ATLAS Yoda system. Proc. of the CHEP2015 Conf. *J. Phys.: Conf. Ser.* **664**:092025

[8] Childers J T, Uram T D, LeCompte T J, Papka M E and Benjamin D P (2015) Simulation of LHC events on a millions threads. Proc. of the CHEP2015 Conf. *J. Phys.: Conf. Ser.* **664**:092006

[9] Nilsen J K, Cameron D and Filipčič A (2015) ARC Control Tower: A flexible generic distributed job management framework. Proc. of the CHEP2015 Conf. *J. Phys.: Conf. Ser.* **664**:062042

[10] Adam-Bourdarios C, Cameron D, Filipčič A, Lancon E and Wu W on behalf of the ATLAS Collaboration (2015) ATLAS@Home: Harnessing Volunteer Computing for HEP. Proc. of the CHEP2015 Conf. *J. Phys.: Conf. Ser.* **664**:022009

Journal of Physics: Conference Series **762** (2016) 012028 doi:10.1088/1742-6596/762/1/012028

The ATLAS EventIndex: data flow and inclusion of other metadata

D Barberis[1], S E Cárdenas Zárate[2], A Favareto[3], A Fernandez Casani[4], E J Gallas[5], C Garcia Montoro[4], S Gonzalez de la Hoz[4], J Hrivnac[6], D Malon[7], F Prokoshin[2*], J Salt[4] , J Sanchez[4], R Toebbicke[8] , R Yuan[6]
on behalf of the ATLAS Collaboration

[1] Università di Genova and INFN, Genova, Italy
[2] Universidad Técnica Federico Santa María, Valparaíso, Chile
[3] Università degli Studi and INFN, Genova, Italy
[4] Instituto de Física Corpuscular, Valencia, Spain
[5] Department of Physics, Oxford University, Oxford, United Kingdom
[6] Laboratoire de l'Accélérateur Linéaire, Orsay, France
[7] Argonne National Laboratory, Argonne, USA
[8] CERN, Europe

*Corresponding author: Fedor.Prokoshin@cern.ch

Abstract. The ATLAS EventIndex is the catalogue of the event-related metadata for the information collected from the ATLAS detector. The basic unit of this information is the event record, containing the event identification parameters, pointers to the files containing this event as well as trigger decision information. The main use case for the EventIndex is event picking, as well as data consistency checks for large production campaigns. The EventIndex employs the Hadoop platform for data storage and handling, as well as a messaging system for the collection of information. The information for the EventIndex is collected both at Tier-0, when the data are first produced, and from the Grid, when various types of derived data are produced. The EventIndex uses various types of auxiliary information from other ATLAS sources for data collection and processing: trigger tables from the condition metadata database (COMA), dataset information from the data catalogue AMI and the Rucio data management system and information on production jobs from the ATLAS production system. The ATLAS production system is also used for the collection of event information from the Grid jobs. EventIndex developments started in 2012 and in the middle of 2015 the system was commissioned and started collecting event metadata, as a part of ATLAS Distributed Computing operations.

1. Introduction

The ATLAS experiment [1] runs a complex system that employs detectors with millions of read-out channels, and gathers billions of events in data taking periods that may extend to several years. It produces several petabytes of data per year, with different physics conditions and data formats. Different types of metadata (data about data) make possible the effective usage of this tremendous amount of information. Dataset metadata deal with data storage organization. Run-wise metadata deal with slowly changing parameters and links to the data-taking conditions. It is useful to have a catalogue containing information about the basic properties of event data and references to their storage locations, and also providing a means of rapid search through the event metadata.

The EventIndex [2] is a system designed to be a complete catalogue of ATLAS events, real and simulated data at all processing stages. However, experience obtained in the process of commissioning and first period of use, input from the physics groups and performance constraints resulted in the exclusion of some of the short-living data (like derived data for workgroups - *DAODs*) from being processed by the EventIndex. We also try to avoid repeated processing of the same information (like trigger decisions) from different production stages and software tags.

2. The EventIndex Data

The event record contains event identifiers: (run number, event number, trigger stream, luminosity block), trigger decisions for different levels and references (GUID + internal pointer) to the events at different processing stages in permanent files, generated by central production. For the Level-1 trigger data we store the final trigger decision and also additional information on events rejected by veto or prescaling. For the High-Level Trigger data in addition to "physics" triggers, "pass-through" and "resurrected" triggers are also stored.

The event record has size of approximately 350 bytes per event. This gives about 350 gigabytes of uncompressed information per billion (10^9) events, multiplied by number of processings and formats. The real amount of storage space needed depends on the chosen technology, replication multiplicity and internal data format and indexing.

There are a number of use cases for the EventIndex:

- Event picking: a request for reference to the specific event in given format and processing type.
- Event counting: counting the number of times that a specific trigger decisions were passed. The information is collected by the EventIndex using trigger tables from the Condition Metadata Store (COMA) [3]. Results are provided to users also by COMA.
- Duplicate finding: detecting duplicated data within the scope of a dataset or physics collection. The information is supplied to the users through the ATLAS Metadata Interface (AMI) [4].

Different data types [5] are processed by the EventIndex and may be treated differently. Top priority for the EventIndex are Analysis Oriented Data (AOD), both for real and simulated (MC) data. The indexing of AODs provides also indexing of RAW data through the provenance information that is picked up. MC datasets containing generated events (EVNT) are indexed for duplicate event detection. Routine indexing of derived data (DAODs) is currently not performed, because of a short useful lifetime and large amount of data (10x AODs). Still the capability to index DAOD's is preserved; a very small subsample is indexed for that purpose. We avoid repeated processing of the trigger data for the same events, so we have trigger processing enabled only for real data AODs from Tier-0 (including spill-over) and for MC AODs. The trigger processing is disabled for real data AOD from reprocessing and for DAODs.

3. EventIndex structure and dataflow

The EventIndex infrastructure consists of a few autonomous components, providing the collection and processing of EventIndex data. The modular approach allowed sharing the developments between different institutions. The main components and dataflow are presented in Figure 1.

The information is collected by event processing tasks ("producers") which can run at Tier-0 (initial reconstruction at CERN) or on Grid sites (downstream processing). Event metadata collected by the producers is sent via the ActiveMQ [6] message broker to the Hadoop [7] store at CERN. At CERN messages are read by Consumer program and stored as Hadoop Mapfile (a file-based map from keys to values) objects. Data are validated for completeness and duplicated events, and decoding of the trigger information performed using trigger tables from COMA. Users can access EventIndex information through the query services: command line interface and web services with graphical and text/plain output suitable for the ATLAS PanDA [8] (Production and Distributed Analysis) system [9].

Figure 1: EventIndex components and dataflow

Datasets produced at the Tier-0 computing center at CERN are indexed by tasks running EventIndex Producer transformations.

On the Grid the EventIndex uses ProdSys2 [10], a second generation ATLAS production system, to index datasets as soon as they appear in the AMI catalogue and are marked as valid and completed. Information on data produced/modified daily on the Grid is gathered by automatic scripts by querying the AMI metadata catalog for VALID datasets from specific datasets from mc_15 and data_15 projects, and soon mc_16 and data_16 too. Then a catalogue in the ATLAS Distributed Data Management system [11] (Rucio) is queried for the identifiers of the datasets to feed the EventIndex processing jobs. Collected datasets are attached to the special "technical containers" by cron script, to be picked up by the EventIndex production tasks. There is one technical container for each type of data to be indexed and trigger information processing setting. A new set of technical containers is being created when a new version of the EventIndex producer appears. For each dataset that appears in the technical container, ProdSys2 creates a Grid task that extracts event data and sends information to the "consumers" running on servers at CERN [12].

This operation is done in the following steps (see Figure 2):

- The producer process extracts EventIndex information from each event in a file and fills an intermediate file on the worker node. Event data are then read from the intermediate file, trigger masks are compressed, duplicate information is dropped and records are encoded into JSON to build the text message.
- EventIndex uses the Streaming Text Oriented Messaging Protocol [13] (STOMP) to transport information between producer and consumers. The message size is set small (1-10 kB).
- Messages are received by the ActiveMQ message broker and queued for the consumers.
- Messages are finally read by EventIndex consumers, decoded and organized into Hadoop MapFile objects.
- Consumers also run a validation step, verifying that the EventIndex for a collection (dataset) contains all the files (GUIDs) and all the events, and detecting multiple processing of the same file and job failures.

The consumers send back statistical information, containing the counts of messages, files and events to the broker to report the status. This is used, in particular, by the ATLAS computing services monitoring system.

Measuring the system performance for sending real events shows that it is able to process 200k events per second or 60 MB of data per second for the typical configuration: 1 broker, 6 producers, 4 consumers and 50k events/job.

The Hadoop data management platform was chosen as the storage technology for the EventIndex. This decision was driven by the following reasons:

- This platform is provided and supported by CERN-IT
- The DDM (Distributed Data Management) project also uses Hadoop
- There are plenty of tools to organize the data, index them internally and search them
- It showed a satisfactory performance in a prototype populated with a year of ATLAS data

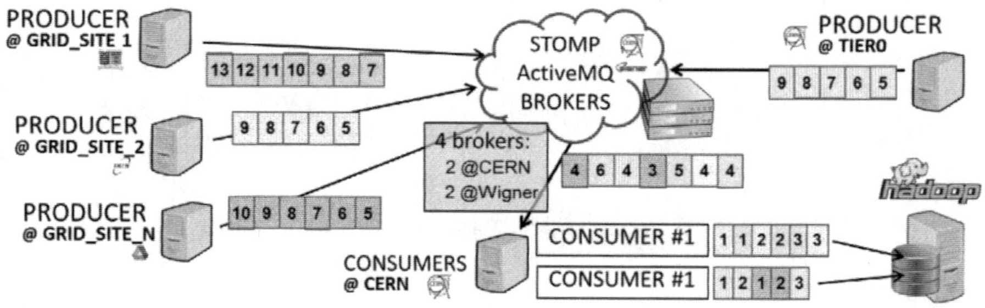

Figure 2: EventIndex messaging system and consumers

Data are stored as Hadoop mapfiles in HDFS (Hadoop File System). The trigger information receives additional processing: trigger decisions are decoded using tables from COMA and the resulting lists of trigger chain names are added to the mapfiles.

Data are internally catalogued by HBase [14], creating metadata about HDFS files as keyed indexes, thus enhancing the performance: searches based on keys give almost immediate results (seconds). Complex searches instead use MapReduce [15] and require 1-2 minutes for typical event collections.

Since the beginning of 2015 we store approximately 50 TB of EventIndex data: 46 TB corresponding to the real data and 4 TB to the MC data. We currently also keep the consumer data, that takes approximately the same amount of space and will eventually be removed.

Users can access EventIndex information through the query services:

- Command line interface, available on the Hadoop cluster and remotely from AFS, CVMFS and a downloadable tar file;
- Web services with graphical and text/plain output suitable for the Panda system.

Remote access goes via a proxy server to a web server at the Hadoop cluster to assure the Hadoop cluster isolation from the outside environment and fallback & load-balancing.

Searches are executed in two steps. First, the HBase catalogue finds the mapfiles to search for. The execution time depends on the search criteria and may take from less than 1 s (exact key-based search) to 20 s (full scan search). Then Hadoop executes the search in the selected files. It can be a key-search (almost immediate) if searching directly on RunNumber-EventNumber or MapReduce jobs if searching with other attributes, like trigger decisions.

In the **Table 1** the event lookup results for the search for events from a list are presented. Events are specified by run-event pairs, and in this case only AODs were searched. Requests were done from a remote machine for general ATLAS use, not from the restricted EventIndex cluster.

Number of events	Found GUIDs	Searched runs	Searched tag Files	Time (seconds)
1	3	1	6	2
10	39	1	6	2
50	185	1	6	3
100	381	2	12	4
500	1803	9	54	11
1000	3808	22	146	31
5000	9957	50	319	120
8879	24179	66	426	193

Table 1 Event lookup benchmark results

It should be noted that HBase, Hadoop and Tomcat keeps some results in cache/memory so the second (overlapping) search is usually faster than the first one.

4. Data exchange with other systems

The EventIndex uses information from other systems in order to find and correctly process data. In particular, trigger information is decoded using information from COMA [16]. The trigger decoding data flow is presented in Figure 3.

Trigger decisions are stored in data files as trigger masks. If a bit of the mask is set, this means that one of the sets of trigger conditions (trigger chains) has been passed in this event. The position of this bit in mask (chain counter) indicates which chain has been passed. Depending on the trigger configuration, the same chain counter may correspond to a different chain name. The relation is uniquely defined by the trigger table corresponding to the trigger key (SMK). Trigger masks from event record are decoded, and chain counters are converted to chain names. The list of trigger chain names, obtained after decoding are then stored in updated event records.

The EventIndex uses trigger tables replicated from COMA to HBase for the trigger decision decoding. The trigger tables use SMK as key and ChainName:ChainCounter pairs as columns. Additionally there is a Run:SMK table for verification. As of the beginning of 2016 the trigger tables contain information on 616 SMKs and 10526 runs. Trigger table replication is done by a *cron* job running daily. Automatic replication is in progress since March 2015.

The EventIndex is a source of data on trigger event counts for COMA. The EventIndex program does the counting on a server within the CERN network providing information through a RESTful service that can be accessed from outside CERN.

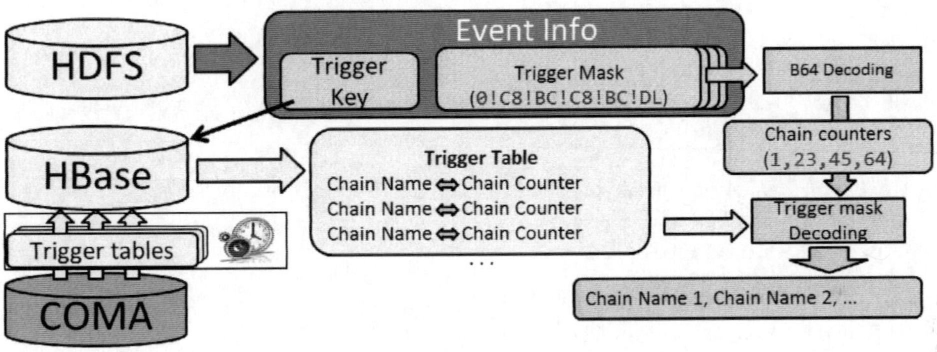

Figure 3: Trigger information decoding

We use AMI information for identifying new and changed datasets that have to be processed by the EventIndex. We query AMI through a python interface for a list of complete dataset containers of certain types with last modification time on the previous day and flagged as "valid". As large datasets in ATLAS can be produced by a number of tasks, containers are made of one or few task ID (TID) datasets that have to be specified as an input for the EventIndex production job. Information on such datasets is obtained ATLAS data management system (Rucio).

The current procedure based on dataset modification times has a number of drawbacks. It is difficult to detect datasets that became corrupted or obsolete. Dataset modification times may change for reasons not related to their contents, requiring unnecessary re-importing of data. There is no feedback to AMI that may be useful to tag corrupted data based on the result of the EventIndex verification.

It was proposed to introduce a special table in the AMI for information exchange between the two systems. Information on the new, changed and invalid datasets will be placed by AMI to a special field, AMI_DS_STATE, with a timestamp. The EventIndex will put dataset import status into another field, EI_DS_STATE, providing to AMI information on data consistency problems detected by the EventIndex. This machinery is currently in the process of implementation by the AMI Team.

5. Conclusions and outlook

The development of the EventIndex started in 2012 and at the beginning of 2015 all its components were operative. The import of data from the previous years of LHC operation started in February 2015 and now is complete. Run 2 data import started when it became available in summer 2015 and now it flows continuously from the Tier-0 and Grid and is available to users almost in real time.

All major components exist and work:

- Data Collection: Producer transforms run at Tier-0 and on the Grid. Consumers reads data from the ActiveMQ servers, validate them and store them to HDFS
- Storage System: Data organization in Hadoop and indexing in the catalogue is running smoothly, with recent significant stability improvement. Trigger decoding services have been developed, tested and are currently used in data import.
- Query System: Command line and web interface are fully operative and used. We constantly improve its functionality and adding new features based on users' feedback.
- Monitoring: System level monitoring in the new Kibana [17] environment at CERN is ready and running, providing information for the EventIndex operation team and for ATLAS shifters.

The EventIndex development team continues its work on further improving the system. This includes optimization of the internal data representation, further automation of the data flow, system interconnections and monitoring, and automatic checks of production completeness.

References

[1] The ATLAS Collaboration 2008 The ATLAS Experiment at the CERN Large Hadron Collider, *JINST 3 S08003*

[2] Barberis D *et al.* 2015 The ATLAS EventIndex: architecture, design choices, deployment and first operation experience, *J. Phys.: Conf. Ser.* **664** 042003

[3] Gallas E J *et al.* 2012 Conditions and configuration metadata for the ATLAS experiment, *J. Phys. Conf. Ser.,* **396** 052033

[4] Albrand S *et al.* 2010 The ATLAS metadata interface *J. Phys.: Conf. Ser.* **219** 042030

[5] Catmore S *et al.* 2015 A new petabyte-scale data derivation framework for ATLAS, *J. Phys.: Conf. Ser.* **664** 072007

[6] ActiveMQ: http://activemq.apache.org

[7] Hadoop: http://hadoop.apache.org

[8] Klimentov A *et al.* 2015 Next Generation Workload Management System For Big Data on Heterogeneous Distributed Computing, *J. Phys.: Conf. Ser.* **608** 012040

[9] De K *et al.* 2015 The future of PanDA in ATLAS distributed computing, *J. Phys.: Conf. Ser.* **664** 062035

[10] Borodin M *et al.* 2015 Scaling up ATLAS production system for the LHC Run 2 and beyond: project ProdSys2, *J. Phys.: Conf. Ser.* **664** 062005

[11] Garonne V *et al.* 2014 Rucio – The next generation of large scale distributed system for ATLAS Data Management, *J. Phys. Conf. Ser.* **513** 042021

[12] Sánchez J *et al.* 2015 Distributed Data Collection for the ATLAS EventIndex, *J. Phys.: Conf. Ser.* **664** 042046

[13] STOMP: http://stomp.github.io

[14] HBase: http://hbase.apache.org

[15] Dean J and Ghemawat S 2004 MapReduce: Simplified Data Processing on Large Clusters *Communications of the ACM Volume 51 Issue 1, January 2008 Pages 107-113*

[16] Barberis D *et al.* 2015 Integration of the EventIndex with other ATLAS systems, *J. Phys.: Conf. Ser.* **664** 042045

[17] Kibana: https://www.elastic.co/products/kibana

ACAT2016 IOP Publishing
Journal of Physics: Conference Series **762** (2016) 012029 doi:10.1088/1742-6596/762/1/012029

The design and performance of the ATLAS Inner Detector trigger for Run 2 LHC Collisions at $\sqrt{s} = 13$ TeV

C Kilby on behalf of the ATLAS Collaboration

Royal Holloway University of London, Department of Physics, Egham Hill, Egham, Surrey TW20 0EX, UK

E-mail: callum.kilby@cern.ch

Abstract. The design and performance of the ATLAS Inner Detector trigger algorithms running online on the High Level Trigger processor farm with the LHC Run 2 data with collisions at both 50 ns and 25 ns are discussed. The HLT Inner Detector tracking algorithms are essential for the identification of nearly all physics signatures in the ATLAS trigger system. In order to deal with the expected higher rates for LHC Run 2, the Inner Detector trigger was redesigned during the 2013-15 long shutdown to satisfy the demands of the higher energy LHC operation. The detailed performance of the tracking algorithms with the Run 2 data taken in 2015 for the different trigger signatures in terms of both efficiency, and resolution is presented. The online processing times for running trigger tracking for the different trigger signatures are discussed in detail. Where appropriate, comparisons are made between the new strategies for Run 2, and the strategies adopted previously in Run 1. These comparisons demonstrate successful application and superior performance of the strategies adopted for Run 2.

1. Introduction

The LHC [1] is placed at the forefront of collision energy and intensity, allowing its associated experiments to push the boundaries of knowledge in many areas of High Energy Physics. However, these conditions provide highly demanding environments for the data-taking performed by the experiments. In particular, the trigger systems which determine whether to keep or reject collision events must cope with the high rate of collisions, while achieving accurate and efficient reconstruction of the events.

After Run 1 of the LHC between 2010 and 2013, the accelerator has been upgraded and prepared for running with higher collision energy and intensity in Run 2, which commenced data taking in June 2015. The Run 2 conditions greatly increase the demands on detector trigger systems, requiring upgrades and new approaches to be employed. A comparison of Run 1 and Run 2 conditions is shown in Table 1. In Run 1, the ATLAS Inner Detector (ID) trigger systems coped very well with the LHC running conditions, and provided excellent performance in physics reconstruction. The same performance must be kept after the upgrades and new approaches employed for Run 2.

In this document, the changes to the ATLAS ID trigger for Run 2 are discussed. The ID Trigger upgrades for Run 2 have been discussed previously [2, 3], and additional plots and information can be found there.

Table 1. Comparisons of LHC run parameters for Run 1 and Run 2.

LHC parameter	Unit	Run 1	Run 2
E_{CM}	[TeV]	7	13
Bunch separation	[ns]	50	25
Peak luminosity	[cm^{-2} s^{-1}]	7×10^{33}	2×10^{34}
Interactions per crossing (at peak luminosity)	[interactions]	≈ 21	≈ 55
ATLAS detector input rate	[MHz]	20	40

2. The ATLAS Detector, Inner Detector, and Inner Detector Trigger

The layout of the ATLAS ID is shown in Figure 1, showing the layers of silicon based pixel and SCT detectors, and straw tube based TRT detectors. The overall system is fully hermetic in the azimuthal range, and extends to a coverage of $|\eta| < 2.5$.[1] The signals registered in the ID are used to reconstruct tracks, collision vertices, and particle decay vertices through application of tracking algorithms. For Run 2 an innermost pixel system called the Insertable B Layer (IBL) [4] has been added, adding a fourth pixel barrel layer. This layer starts approximately 2 mm from the beam pipe, with the additional hits provided by the layers allowing for more robust track finding, with better impact parameter[2] resolution and therefore more precise vertex reconstruction.

The ID trigger performs track and vertex reconstruction as a part of the total ATLAS trigger system, where the reconstructed tracks and vertices will be used in further processing for the decision to accept or reject an event. As the trigger is run online[3], the ID trigger must perform the track and vertex reconstruction within tight timing constraints, whilst keeping good performance to ensure the overall trigger system accepts events of interest.

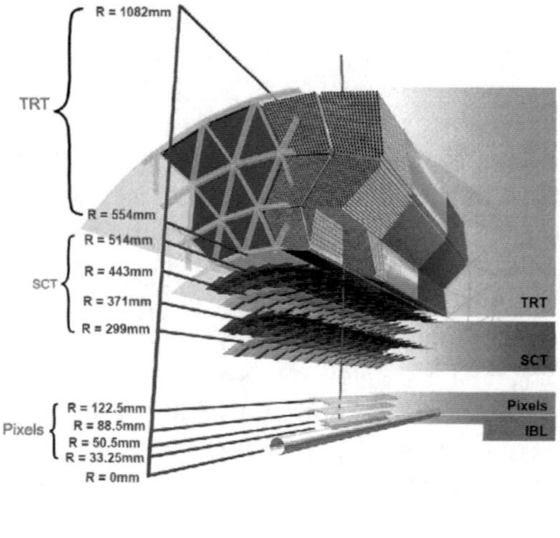

Figure 1. A sketch of a segment of the ATLAS ID barrel modules, showing the radial layout of the detection sub-systems [5]. The central grey cylinder is the LHC beam-pipe. Visible is the IBL pixel layer, which is newly added for Run 2. Not shown are the pixel, SCT, and TRT end-cap modules placed at each end of the barrel, aligned perpendicular to the beam-pipe.

[1] ATLAS uses a right-handed coordinate system with its origin at the nominal interaction point (IP) in the centre of the detector and the z-axis along the beam pipe. Cylindrical coordinates (r, ϕ) are used in the transverse plane, ϕ being the azimuthal angle around the z-axis. The pseudorapidity is defined in terms of the polar angle θ as $\eta = -\ln\tan(\theta/2)$.

[2] The distance of closest approach of a track to some reference point.

[3] Online refers to processing done before an event has been read-out and recorded by ATLAS, while offline refers to any processing done after an event has been read-out and recorded.

3. Upgrades and New Approaches for Run 2

The most significant upgrade of the ATLAS trigger systems has been the redesign of the computational architecture. In Run 1, the trigger system consisted of a hardware-based Level 1 (L1) trigger which was followed by fast software-based Level 2 (L2) processing which used, on demand, only fragments of detector information. Finally, a software-based Event Filter (EF) stage where entire event information was available was used. The L1 trigger decision time was less than 2.5 μs; the L2 stage decision time was approximately 75 ms; and the EF stage decision time was approximately 1 s.

For Run 2, it was decided to merge the two software-based stages into a single High Level Trigger (HLT). The HLT is run from a dedicated server cluster, where each node is dedicated to processing a single event provided by L1. The merged system approach avoids repeated data preparation, access, and readout that occurred when L2 and EF were separated in Run 1. The L2 and EF track reconstruction approaches are kept by way of L2-style Fast Track Finder and EF-style precision tracking algorithms. A new hardware-based HLT pre-processor, the Fast Tracker (FTK), will be added during Run 2 [6]. The FTK will provide the HLT with tracks that are rapidly found within the full volume of the ID using template-based pattern recognition in massively parallel operation, followed by futher track fitting. The HLT tracking algorithms can then be seeded from this rough track information. Figure 2 shows a detailed overview of the ID trigger architecture in Run 2.

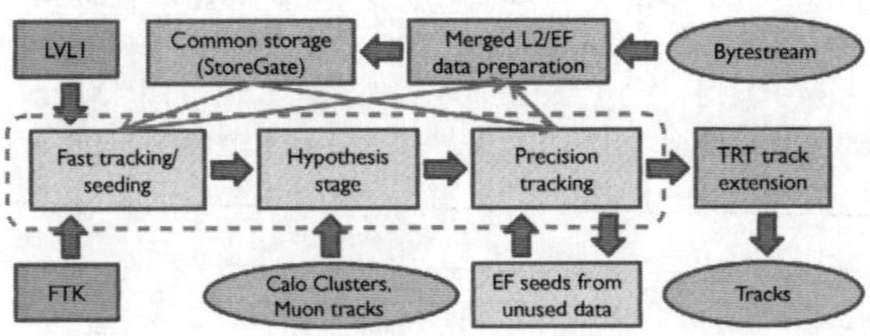

Figure 2. Detailed overview of the ID trigger architecture for Run 2 [7]. The LVL1 and FTK boxes show inputs from the respective trigger components. Bytestream refers to the binary readout of the ATLAS detector.

The new trigger architecture allows for more advanced approaches in the ID trigger track reconstruction. In particular, the merging of the L2 and EF storage allows for a multi-step approach for the reduced detector-volume Regions of Interest (RoIs) used for the tracking. In Run 1, all trigger chains used a single RoI, with the spatial extent determined by the target physics signature (for example muons). In Run 2 some trigger chains, in particular for tau triggers, have been upgraded to use two RoIs in a two-step method. In Run 1, the tracking for tau chains was run in a single RoI which was long in z and wide in η and ϕ. For Run 2, firstly an RoI which is long in z, but narrow in η and ϕ is used. The Fast Track Finder algorithm is run in this RoI, and a track of interest is chosen. A second RoI which is short in z, but wide in η and ϕ is defined around this track, and the Fast Track Finder algorithm is run again, followed by precision tracking. As such, the tracking algorithms are run in a significantly reduced volume compared to Run 1. A comparison of the Run 1 and Run 2 RoIs is shown in Figure 3, showing the positioning of the second step RoI, and the differences in volume.

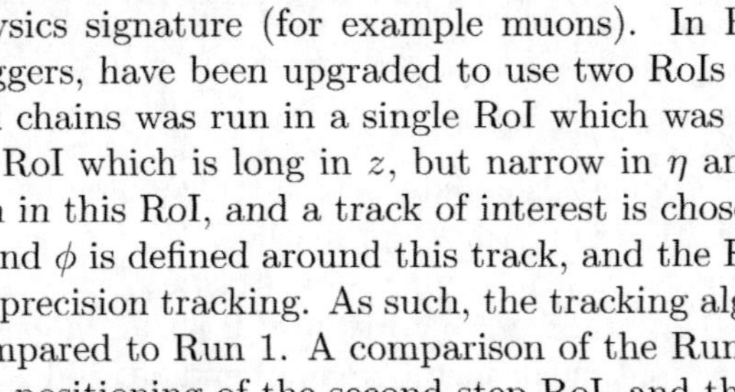

Figure 3. Sketches of the Regions of Interest (RoIs) used in tau trigger chains for Run 1 and Run 2 [8]. The single-stage Run 1 approach is compared to the two-stage Run 2 approach.

Significant improvements in algorithm timing were achieved through optimisation of the code by a number of methods. General optimisation of code 'hot spots' was achieved through use of the profiling tool `Valgrind` [9]. A much faster linear algebra library, `Eigen` [10], was used. The `Eigen` library also included automated code re-factoring. The compiler used was upgraded from `GCC 4.3` to `GCC 4.8` [11]. The code compilation was also carried out on 64-bit CPU hardware.

After all above improvements, the decision time for the entire HLT is approximately 200 ms.

4. Results From Simulation and Run 2 Data

Figure 4 shows a number of plots that show the improvements in processing timings due to the architectural redesign, two-step RoI approach, and code optimisations mentioned in Section 3.

Figure 4(a) shows the total processing time per event for an electron trigger, comparing Run 1 and Run 2 approaches, using $\sqrt{s} = 14$ TeV Monte Carlo simulated data. Both approaches were run with the code optimisations. As such, the timing difference seen is due to the architechtural upgrades. It is seen that the mean processing time improves by approximately 65%. Total timing per event performance and improvement is similar for other physics signatures.

Figure 4(b) shows the processing time per RoI for the precision tracking in a tau trigger, comparing single-step and two-step approaches, using Run 2 $\sqrt{s} = 13$ TeV data. It is seen that the mean processing time improves by approximately 60%. The improvement for the Fast Track Finder algorithm has also been plotted, and a mean improvement of approximately 33% was seen [8]. When combining these results, the overall mean improvement is approximately 37%.

Figure 4(c) shows the processing time per algorithm-call for the pattern recognition stage of the track reconstruction, which has a notably high algorithm run time, using $\sqrt{s} = 14$ TeV Monte Carlo simulated data. The Run 1 approaches built with and without the code optimisations are compared, a mean improvement of approximately 68% is seen.

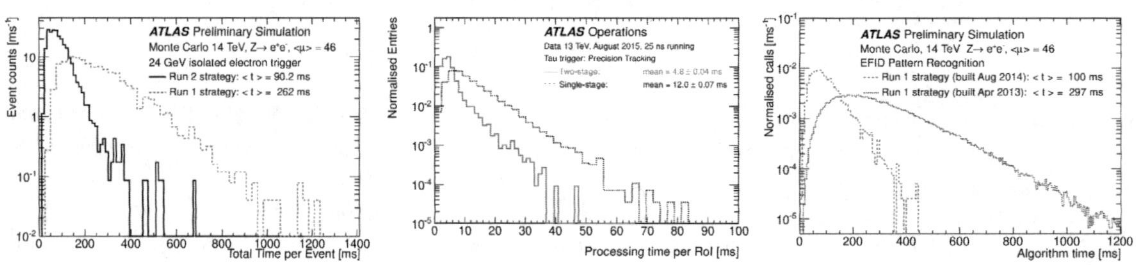

(a) (b) (c)

Figure 4. (a) Total processing time per event for a 24 GeV isolated electron trigger, comparing Run 1 (dashed red line) and Run 2 (solid black line) architectural approaches. (b) Processing time per RoI for precision tracking in a tau trigger, comparing single-step (dashed black line) and two-step (solid red line) approaches. (c) Processing time per algorithm-call for an intensive tracking algorithm, comparing Run 1 approaches before (fine dashed blue line) and after (dashed red line) code optimisations. (a) and (c) use 14 TeV $Z \rightarrow e^+e^-$ Monte Carlo simulated data, (b) uses $\sqrt{s} = 13$ TeV 25 ns bunch spacing data recorded by ATLAS in August 2015 [8].

Figure 5 shows performance for the ID trigger track reconstruction, using Run 2 $\sqrt{s} = 13$ TeV data. The performance is shown for a muon trigger, which uses a single RoI, and runs the Fast Track Finder algorithm followed by precision tracking. The track reconstruction efficiency as a function of η is shown, and it is seen that efficiency is >99% over the whole η range. Efficiency performance is similar for other physics signatures. The resolution of track transverse impact parameter, which is defined in the x-y plane with respect to the LHC beamline, is shown as a function of η, and it is seen that resolution is on the order of tens of μm and does not significantly degrade at higher η values. The efficiency and impact parameter resolution are defined with reference to muon candidate tracks found by the offline tracking algorithms, with the resolution defined as the difference between the value of the impact parameter for online and offline tracks.

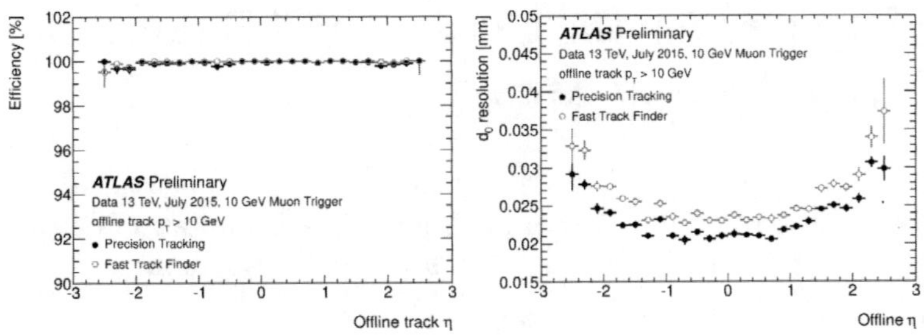

(*a*) (*b*)

Figure 5. (*a*) The efficiency of tracking algorithms and (*b*) the resolution of the transverse impact parameter with respect to beamline of tracking algorithms for a 10 GeV muon trigger, as a function of offline muon candidate track η [8]. Red unfilled circles show results for the Fast Track Finder, black filled circles show results for the precision tracking. The results were produced using $\sqrt{s} = 13$ TeV 25 ns bunch spacing data recorded by ATLAS in July 2015.

5. Conclusions

LHC Run 2 has placed high demands on the ID trigger computing hardware and software used for data-taking within the ATLAS detector. Effective redesigns and new approaches have allowed it to meet these demands, and to show an excellent increase in timing performance when compared to the approaches used during Run 1. The ID trigger also continues to show excellent performance, maintaining high reconstruction efficiency and resolution. It is therefore ready to meet further expectations as conditions continue to grow more stringent as the LHC moves towards peak Run 2 running conditions. It is also well prepared for the introduction of further upgrades, particularly the introduction of the FTK in 2016.

References

[1] Evans L and Bryant P (editors) 2008 LHC Machine *JINST* **3** S08001
[2] Sowden B 2015 The Performance and Development of the Inner Detector Trigger Algorithms at ATLAS for LHC Run 2. Tech. Rep. ATL-DAQ-PROC-2015-055. arXiv:1511.01136. CERN. URL `https://cds.cern.ch/record/2064436`
[3] Penc O 2015 The design and performance of the ATLAS Inner Detector trigger for Run 2. Tech. Rep. ATL-DAQ-PROC-2015-058. CERN. URL `https://cds.cern.ch/record/2104217`
[4] Capeans M et al 2010 ATLAS Insertable B-Layer Technical Design Report. Tech. Rep. CERN-LHCC-2010-013. ATLAS-TDR-19. CERN. URL `https://cds.cern.ch/record/1291633`
[5] ATLAS Collaboration 2015 Track Reconstruction Performance of the ATLAS Inner Detector at $\sqrt{s} = 13$ TeV. Tech. Rep. ATL-PHYS-PUB-2015-018. CERN. URL `https://cds.cern.ch/record/2037683`
[6] Shochet M et al 2013 Fast TracKer (FTK) Technical Design Report. Tech. Rep. CERN-LHCC-2013-007. ATLAS-TDR-021. CERN. URL `https://cds.cern.ch/record/1552953`
[7] ATLAS Collaboration 2013 Studies for the development of the Inner Detector trigger algorithms at ATLAS. Tech. Rep. ATL-DAQ-PUB-2013-002 CERN URL `https://cds.cern.ch/record/1602918`
[8] ATLAS Collaboration 2016 HLT Tracking Public Results. *Online* URL `https://twiki.cern.ch/twiki/bin/view/AtlasPublic/HLTTrackingPublicResults`
[9] Seward J et al 2004 Valgrind: A Program Supervision Framework. *Online* URL `http://www.valgrind.org`
[10] Guennebaud G 2013 Eigen: a C++ linear algebra library. *Online* URL `http://downloads.tuxfamily.org/eigen/eigen_CGLibs_Giugno_Pisa_2013.pdf`
[11] GCC Steering Committee 1987 Gnu Compiler Collection (GCC). *Online* URL `https://gcc.gnu.org/`

The software system for the Control and Data Acquisition for the Cherenkov Telescope Array

P Wegner[1], M Füßling[1], I Oya[1], L Hagge[2] U Schwanke[3], J Schwarz[4], G Tosti[5], V Conforti[6], E Lyard[7], R Walter[7] for the CTA Consortium;

P Oliveira Antonino[8], A Morgenstern[8]

[1] DESY Zeuthen Germany
[2] DESY Hamburg Germany
[3] Humboldt Universität zu Berlin, Germany
[4] INAF - Osservatorio Astronomico di Brera, Italy
[5] University of Perugia, Italy
[6] INAF - IASF Bologna, Italy
[7] ISDC, Univ. Geneva
[8] Fraunhofer IESE, Kaiserslautern, Germany

E-mail: Peter.Wegner@desy.de

Abstract. The Cherenkov Telescope Array (CTA), as the next generation ground-based very high-energy gamma-ray observatory, is defining new areas beyond those related to physics. It is also creating new demands on the control and data acquisition system. CTA will consist of two installations, one in each hemisphere, containing tens of telescopes of different sizes. The ACTL (array control and data acquisition) system will consist of the hardware and software that is necessary to control and monitor the CTA array, as well as to time-stamp, read-out, filter and store the scientific data at aggregated rates of a few GB/s. The ACTL system must implement a flexible software architecture to permit the simultaneous automatic operation of multiple sub-arrays of telescopes with a minimum personnel effort on site. In addition ACTL must be able to modify the observation schedule on timescales of a few tens of seconds, to account for changing environmental conditions or to prioritize incoming scientific alerts from time-critical transient phenomena such as gamma-ray bursts. This contribution summarizes the status of the development of the software architecture and the main design choices and plans.

1. Introduction

This paper gives a short overview on the Cherenkov Telescope Array (CTA) project [1] and then describes the software system for the CTA Array Control and Data Acquisition work package (ACTL).

The CTA project is an initiative to build the next generation ground-based very high-energy gamma-ray instrument. CTA opens a new era in gamma-ray astronomy. It will serve as an open observatory to a wide astrophysics community and will provide deep insight into the non-thermal high-energy universe. The present generation of imaging atmospheric Cherenkov telescopes (H.E.S.S., MAGIC and VERITAS) has in recent years opened the realm of ground-based gamma-ray astronomy in the energy range above a few tens of GeV. The CTA design foresees a factor of 5-10 improvement in sensitivity compared to the current very high-energy gamma-ray domain. It needs to be optimized for a wide energy band, which leads to the design of telescopes of three different sizes.

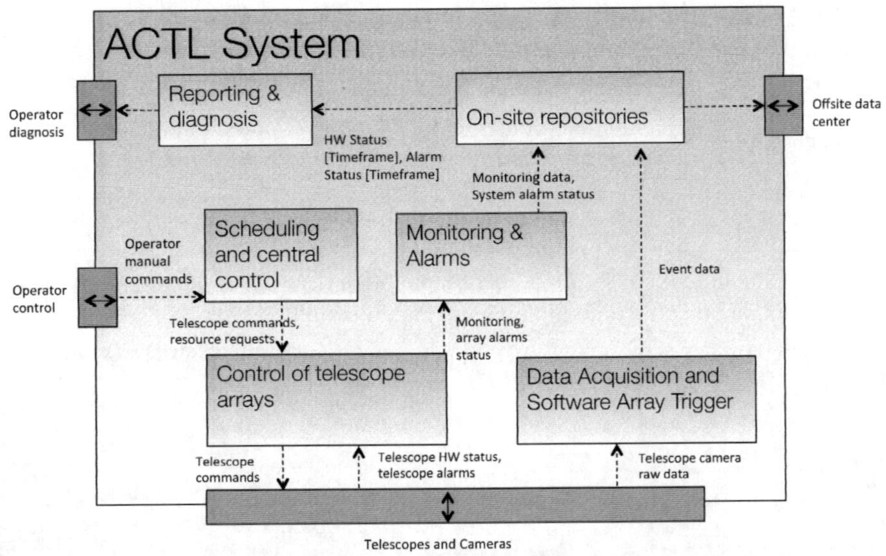

Figure 1. ACTL building blocks

CTA will consist of two arrays for complete sky coverage: a southern array consisting of about 100 telescopes and a northern array with about 20 telescopes. Detailed contract negotiations have been started with the southern site candidate, which is located at less than 10 km southeast of ESO's existing Paranal Observatory in the Atacama Desert, and the northern site candidate, which is located on the existing site of the Instituto Astrofisica de Canarias Observatorio del Roque de los Muchachos on the island of La Palma. Sites in Namibia (south) and Mexico (north) will be kept as viable alternatives.
The CTA Consortium consists of over 1200 members from more than 200 institutes in 32 countries.

2. Array Control and Data Acquisition – ACTL

The system to control, monitor and readout the CTA array (Array Control and Data Acquisition system, ACTL) will provide means for (see Fig. 1 for a schematic view):

- Data Acquisition (DAQ)
- Central control and execution of observations
- Control and monitoring of all Cherenkov telescopes and auxiliary devices
- Automatic scheduling of observations
- Central array triggering and nanosecond time synchronization
- On-site repositories for data storage and an environment to execute a real-time analysis

The much larger size and complexity of CTA compared to current installations lead to numerous challenges for ACTL:

- Boundary conditions involve much larger latencies, more difficult synchronization, higher data rates, larger data volume and a raw data bandwidth comparable to that of the LHC ATLAS (before the recent upgrade)
- CTA supports more telescopes of different types than present installations, encompassing heterogeneous control and readout architectures

- CTA is intended to act as an open observatory, which requires the provision of high-standard services and very high reliability beyond the levels of current gamma-ray experiments

An advanced and well-conceived software design based on adequate software architecture together with the use of software frameworks and standards are essential ingredients for reaching the required high level of reliability and availability of ACTL products. For all underlying hardware for processing, data storage, networking and triggering, commodity systems are being used.

3. ACTL software model

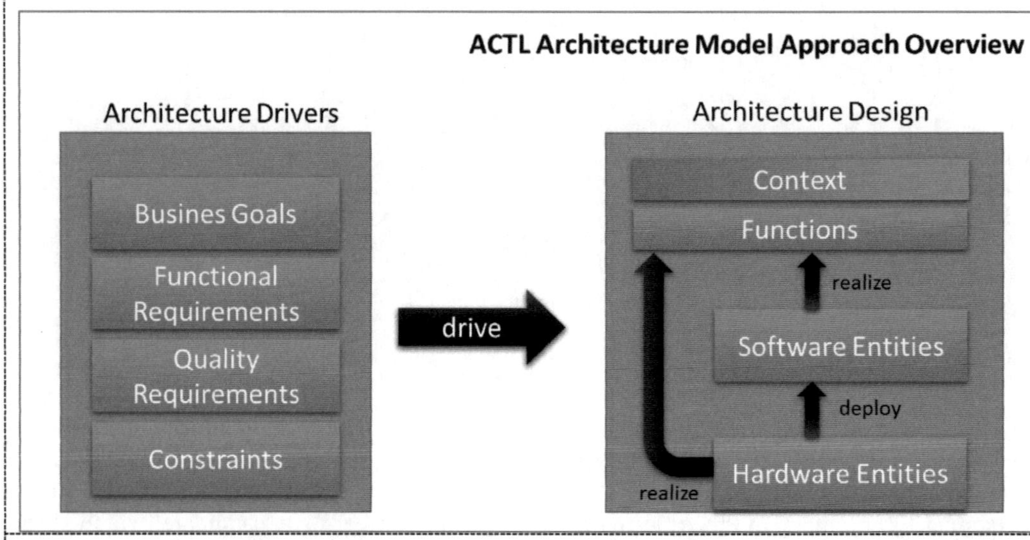

Figure 2. Schematic view of the ACTL system model viewpoints and the associated design steps.

Software architecture can be defined as the fundamental organization of a system embodied in its components, their relationships to each other, and to the environment, and the principles guiding its design and evolution.

A preliminary version of the ACTL architecture model has been introduced and established as a common means of communication within ACTL. The model is being used for establishing and driving the system development, and as foundation for project management. The architecture model provides intuitive schematic visualization of the ACTL system, and allows vision sharing within the ACTL collaboration, as well as with the ACTL stakeholders in the CTA project team and beyond.

The model represents the entire ACTL system. It comprises the drivers for the ACTL architecture (e.g. requirements, specification and use cases) together with a description of the ACTL functionality, and of the software and hardware components that have to be provided (see Fig. 2). The model follows the SPES2020 approach for describing embedded (real-time) systems [2], using a tailored [3] SysML notation (Systems Engineering Modeling Language) [4] and the modeling software "Enterprise Architect" [5].

Figure 3 illustrates one of the diagrams contained in the views of the ACTL architecture model. In the context view, the ACTL system is depicted as a black box (big box in the center), together with the external elements, which can be either human stakeholders, depicted as stick figures in the diagram, or external systems, which are depicted as the smaller boxes outside the main box. The information exchange between the ACTL system and the external entities are indicated by the Ports (blue boxes) at the boundaries of the ACTL system, along with the directed connections (arrows) and the data elements transported by each connection.

The idea of using a model-based approach aimed at supporting two goals. The first is to define a systematic architecture specification for the entire ACTL project to support:

- Identification of all architecture drivers, such as requirements, use cases and stakeholders, and tracing them to functionalities, and then to software and hardware components.
- Providing a framework for all ACTL product owners to further develop the detailed design of their products
- Providing guidelines to be followed by all ACTL software developers in the development and implementation of their software components and to document each design choice and its implications

On the other hand the architecture model helps us to reach project-related goals, which include:

- Develop an implementation plan and enable progress control
- Identify unplanned work and assign responsibilities
- Evaluate ACTL project priorities, main risks and their mitigation
- Provide means for product acceptance (verification and validation)
- Allow deriving cost and effort estimates for realizing the ACTL system

4. ACTL frameworks

The high-level ACTL software is developed on top of the ALMA Common Software (ACS [6]), a software framework for the implementation of distributed data acquisition and control systems. ACS has been developed by ESO and has been successfully applied in projects of similar scale (e.g. ALMA [7]). ACS distributions (provided by ALMA computing) are executed on a variant of the Linux operating system (RedHat Enterprise Linux re-distribution), which will therefore be used on all full-scale ACTL computers.

ACS is based on a container-component model and supports the programming languages C++, Java and Python (the latter in particular for scripting). The high-level ACTL software will be executed predominantly on the central computer cluster and will access most hardware devices via OPC Unified Architecture (OPC UA, see [8]). OPC UA is an industrial standard, so OPC UA servers can either be provided by device vendors or be developed by the device teams. In most cases, the functionality of the OPC UA servers (variables, methods etc.) defines the interface between ACTL and a hardware device (e.g. a weather station or CCD camera) that must be controlled and read-out.

5. Conclusions

The design of the ACTL software accounts for the need to develop, maintain, and operate a complex but very reliable software system with an efficient use of limited resources. To allow achieving its goals, the ACTL project has adopted a model-driven development dynamic based on the SPES2020 methodology using the SysML notation.

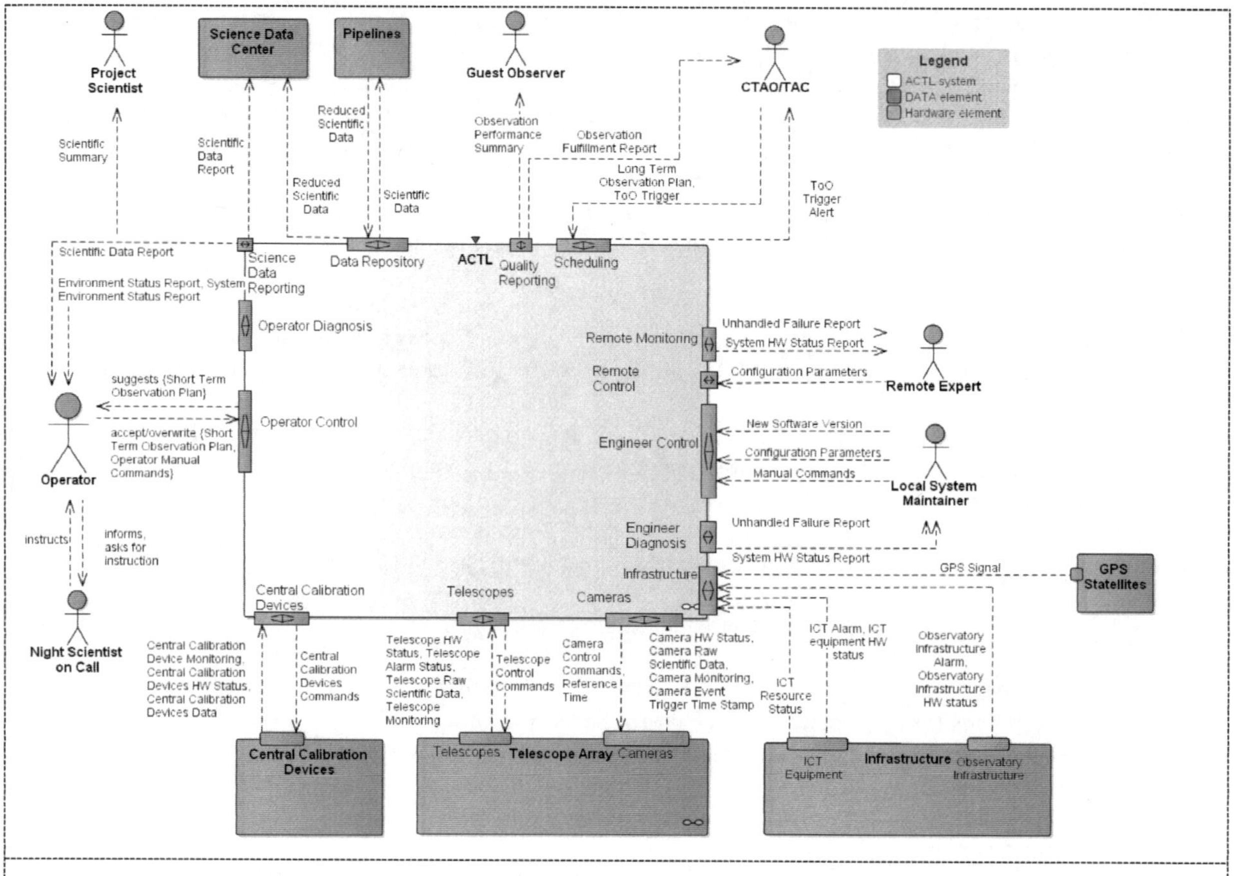

Figure 3. Example element of the ACTL architecture model for CTA: the context diagram. See text for details.

Acknowledgments

We gratefully acknowledge support from the agencies and organizations listed under Funding Agencies at this website: http://www.cta-observatory.org/

References

[1] B S Acharya et al., Introducing the CTA concept, APh 2013 (43) 3.

[2] Klaus Pohl, Harald Honninger, Reinhold Achatz, and Manfred Broy. Model-Based Engineering of Embedded Systems: The SPES 2020 Methodology. Springer, Berlin Heidelberg, 2012

[3] SysML – "OMG Systems Modeling Language 1.4", OMG 2015

[4] Thomas Kuhn, Pablo Oliveira Antonino. Model-Driven Development of Embedded Systems. Embedded Software Engineering Congress 2014. Sindelfingen, Germany, December 2014.

[5] Enterprise Architect – http://www.sparxsystems.com

[6] Chiozzi G, Jeram B, Sommer H *et al.* (2004). *The ALMA common software: a developer-friendly CORBA-based framework*. In H. Lewis & G. Raffi (editors), *Advanced Software, Control, and Communication Systems for Astronomy*, volume 5496 of *Society of Photo-Optical Instrumentation Engineers (SPIE) Conference Series*, pp. 205–218

[7] Wootten A & Thompson A R (2009). *The Atacama Large Millimeter/Submillimeter Array*. IEEE Proceedings, **97**, 1463

[8] OPC UA foundation: http://www.opcfoundation.org/

Using NERSC High-Performance Computing (HPC) systems for high-energy nuclear physics applications with ALICE

Markus Fasel for the ALICE Collaboration
Lawrence Berkeley National Laboratory
E-mail: mfasel@lbl.gov

Abstract. High-Performance Computing Systems are powerful tools tailored to support large-scale applications that rely on low-latency inter-process communications to run efficiently. By design, these systems often impose constraints on application workflows, such as limited external network connectivity and whole node scheduling, that make more general-purpose computing tasks, such as those commonly found in high-energy nuclear physics applications, more difficult to carry out. In this work, we present a tool designed to simplify access to such complicated environments by handling the common tasks of job submission, software management, and local data management, in a framework that is easily adaptable to the specific requirements of various computing systems. The tool, initially constructed to process stand-alone ALICE simulations for detector and software development, was successfully deployed on the NERSC computing systems, Carver, Hopper and Edison, and is being configured to provide access to the next generation NERSC system, Cori. In this report, we describe the tool and discuss our experience running ALICE applications on NERSC HPC systems. The discussion will include our initial benchmarks of Cori compared to other systems and our attempts to leverage the new capabilities offered with Cori to support data-intensive applications, with a future goal of full integration of such systems into ALICE grid operations.

1. Introduction

The processing and analysis of event-based data from High Energy Nuclear physics (HENP) experiments is well suited for commodity computing hardware, configured to match the memory and I/O bandwidth requirements per CPU of the workload. The event structure of the data and processing tasks are embarrassingly parallel, whereby event sets can be processed in independent jobs run on separate CPUs, on different nodes, or even at different facilities. As a result, the experiments have come to rely on conventional compute clusters that range in capacity from a few hundred to many thousand CPUs, which can be distributed over a number of facilities without a general loss of capability.

Supercomputers or High-Performance Computing (HPC) systems are large-scale clusters that have been optimized for massively parallel processing tasks that require concurrent use of large numbers of compute cores to operate efficiently. As such, HPC systems are specialized hardware for targeted use cases that rely on low-latency communication between their processors. However, the scheduling of many large-scale jobs on a single HPC system can leave resource gaps that can be available for use by small-scale tasks, such as those in HENP applications. This condition provides an opportunity for HENP experiments to make use of these resources, while, at the same time, increasing the overall utilization of those systems.

While the potential for opportunistic use of HPC resources can be significant, these systems are generally configured with restrictions on access and utilization not present on more conventional, commodity based clusters. These restrictions can include vendor-specific environments and workflow limitations that make these systems difficult to integrate into the computing infrastructures developed by the HENP experiments to leverage their distributed resources. We report here the results of some initial work done by the ALICE collaboration to make use of HPC systems operated at the US Department of Energy's National Energy Research Scientific Computing Center, NERSC. The goal of this work is to evaluate the efficacy of these HPC systems for our HENP applications, carried out within a framework developed to simplify an eventual integration into the larger distributed workflow of the experiment.

NERSC is the primary scientific computing center for the DOE Office of Science and the principal provider of high performance computing services to their science programs. NERSC currently operates two HPC systems, Edison and Cori, at its facility at Lawrence Berkeley National Laboratory (LBNL). Edison[1], commissioned in 2013, consists of over 130k CPU cores with peak performance of over 2.5 PF. Cori, named after the US biochemist Gerty Cori, is the newest NERSC HPC system and is being deployed in two phases. Cori Phase 1 [2], available since November 2015 and referred to as the "Data Phase", consists of over 1600 Intel Xeon Haswell™ nodes and has been designed with data intensive features such as a large memory per core configuration, a 28 PB Lustre scratch file system, and an 800 TB SSD-based Burst Buffer file system for workflows that demand high I/O capacities. The phase II [3] Cori system is expected to arrive in late 2016 and will include about 9000 nodes, each with a large number (60) of Intel Phi Knights Landing™ cores. Thus, Cori Phase II will be optimized for massive parallel computing and is therefore an excellent testbed for AliceO^2 [4], the next generation ALICE Software framework.

Along side the HPC systems, NERSC operates a large mid-range system of more conventional configuration, in which the HENP compute cluster, PDSF, exists. PDSF is an evergreen system that has been operated for over 15 years for the High Energy and Nuclear Physics communities at LBNL and consists of about 3000 compute cores with more that 3.0 PB of disk storage. PDSF is shared by several experiments and is part of the ALICE Grid as a WLCG ALICE Tier-2 center. The proximity of ALICE storage and grid resources on PDSF to the NERSC HPC systems makes NERSC an ideal site for evaluating use of HPC by ALICE computing operations.

2. The ANALISA tool running nuclear physics applications on HPC systems

HPC compute clusters often have features that limit their usability by distributed workflows developed for serial applications. Such features include whole-node scheduling, limited external network connection, time-based scheduling and historically small amounts of memory per CPU core. Several of these limitations require adaptions in the workflow, particularly during the integration of HPC systems into the ALICE computing grid.

In order to make job submission transparent to the user, we developed a tool called ANALISA, which takes into account the various features of all supercomputers currently supported. The tool consists of two parts, a submission layer and a worker layer. The submission layer runs on interactive (submit) hosts to manage job submission, preparation of the user software and the sandbox to contain the sub-job contents. The worker layer is executed on the compute nodes (or intermediate "mom" nodes depending on the cluster) and provides the MPI interface. It also handles the transfer of input files to the sandbox and output files to their final storage location. Fig. 2 shows the general workflow of our tool.

Steering of the job submission is done using a configuration file. The content can be expressed either in a key-value format, as an xml-tree or as a json string. The minimal content must provide

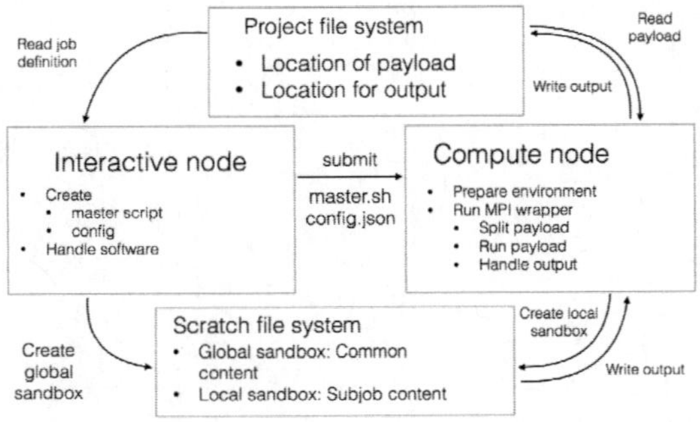

Figure 1. Workflow of ANALISA: The submitter prepares the job sandbox on the scratch file system, splits the job into n-master jobs and submits the master jobs to the batch queue. The worker provides the actual MPI interface. It copies the payload from the file system to the sandbox, runs the payload assigned to the different cores within the master job and copies the output back the output registered via the job definiton.

the set of information necessary to schedule the job, such as the amount of CPU cores, the time requested for each job, the name of the job and pointer to the executable. Other parameters like a set of input files and an output location are optional.

3. Experience with ALICE simulation jobs on NERSC supercomputers

ALICE simulation jobs consist of a variety of components: event generators (Pythia6/8 [5, 6], HIJING [7], DPMJet [8], others), collision systems (pp, p-Pb, Pb-Pb) center-of-mass energies, transport engines (GEANT3/4 [9, 10]), and software versions of ROOT and AliRoot. Therefore, the simulation tool must support a large number of configurations to be executed on the NERSC computing clusters. For our tests, a cocktail of different configurations was created as a test suite to cover a range of typical use cases. All collision systems are included in this test suite, while for pp collisions two different center-of-mass energies were used. The tests were limited to minimum bias collisions, acknowledging that biased simulations might skew to a higher multiplicity and consequently longer job execution times, not relevant to our current investigation. The settings of the test suite are listed in Tab. 1.

Tests were performed on the supercomputers Cori and Edison, as well as the local batch farm PDSF for direct comparison with a more conventional cluster. In all cases, the same versions of ROOT, AliRoot and GEANT3 were used. In the initial test, the software was obtained from a local build system component of the simulation tool on Cori and Edison but directly from CVMFS on PDSF. The ALICE conditions database, OCDB, was mirrored onto Edison and Cori scratch file systems, while direct access via CVMFS on PDSF was possible.

Collision system	Center-of-mass energy	Generator	Number of events / job
pp	7 TeV	Pythia6 Perugia2011	100
pp	8 TeV	Pythia8 Monash2013	100
p-Pb	5.02 TeV	DPMJET	100
PbPb	2.76 TeV	HIJNG	5

Table 1. Test setup used for our initial benchmarks: Jobs differ in the collision system, the center-of-mass energy and the event generator including tune.

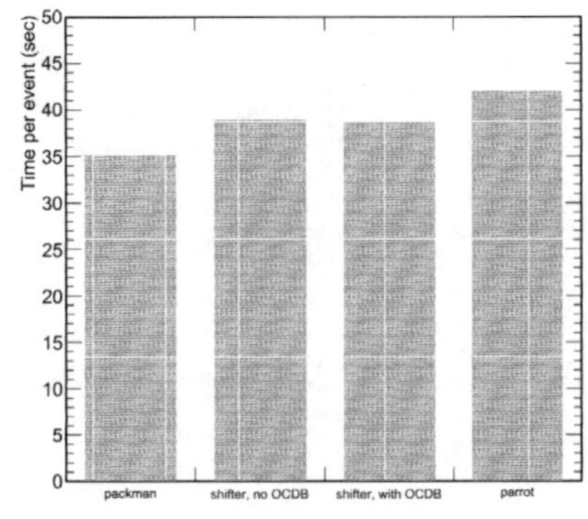

Figure 2. Left: Time per event spent for simulation jobs of the four different scenarios on the three platforms Cori, Edison and PDSF. The time takes into account event generation, particle transport, detector simulation and reconstruction. Right: Time per event for the test case pp, $\sqrt{s} = 7$ TeV using different methods for the distribution of the software stack and the OCDB.

The left panel in Fig. 2 shows the mean time spent per event from running our test cocktail on the different platforms. In all test cases, Cori performed the fastest of the three clusters. As mentioned above, PDSF is an evergreen cluster composed of several generations of CPUs. When we restrict the tests to a subset of machines on PDSF with the same CPU type as used in Cori, the same performance can be reached.

4. Software distribution on HPC systems

One critical task for distributing work, both onto a cluster and across many resources, is the consistent distribution of experiment software to the actual compute nodes. For the test already described, locally pre-compiled packages on each cluster's scratch system were used for software distribution. This option is optimal for development work in which strict consistency with officially released software is not required. In production, however, use of software that is centrally managed by the experiment is required.

The LHC experiments have adopted CVMFS [11] as their software distribution tool, on which software is built centrally at CERN and is efficiently migrated out to intermediate caches that the compute systems dynamically access. Normal use of CVMFS requires a kernel modification and a disk cache local to each node, neither of which generally exists on HPC systems. Our additional tests focus on making software distributed by CVMFS directly available on Cori.

NERSC has worked with Cray to develop a tool called Shifter [12] based on linux container technology [13] in which the experiments can customize some aspects of the operating environment. While use of Shifter does not allow CVMFS to run natively, it does give the experiments some additional options, two of which were investigated. In one option, the full ALICE repository is unpacked directly into a Shifter image and accessed during runtime as a normal file system. The second option used a preload procedure to deploy the ALICE repository onto a filesystem, mounted as a repository with a tool, parrot [14], installed into the Shifter image.

The tests were run on Cori using the pp-scenario at 7 TeV listed in Tab. 1. In the case where CVMFS is included in the image, the test is split into two parts: one with just the software stack taken from the image and the other with both the ALICE software and OCDB included in image. The results of the tests are shown in the right panel of Fig. 2 and compared with the locally built test (labeled packman) as a measure of the optimal performance which can be reached on Cori. All options for CVMFS distribution were successfully commissioned; however, the performance found with the locally-built software stack has not yet been achieved, suggesting further optimizations of the software distribution via CVMFS onto NERSC supercomputers are needed.

5. Integration of the NERSC supercomputer Cori into the ALICE Computing Grid

A future goal of this work is the integration of Cori into the ALICE Grid facility [15]]. Accessing the distributed software through the CVMFS system is just a first step. For example, the test jobs presented here take their payload from the local file system while, for grid jobs, it is assigned dynamically from the remote AliEn task queue. Thus, additional integration will need to cope with scheduling differences between normal Grid and HPC systems. To simplify that work, Cori is initially intended only for simulations.

6. Conclusions

To summarize, we have successfully commissioned a tool for running serial ALICE jobs on HPC systems designed for parallel processing. We have tested the tool with a suite of ALICE simulation jobs in which the payload is the same as would be run centrally from the ALICE Grid and found that the single job efficiency was similar to that obtained on the more conventional PDSF cluster. We have tested and verified two methods to deliver the centrally managed ALICE software stack onto NERSC HPC system, which represents significant initial steps towards the integration of NERSC supercomputers into the ALICE computing grid.

7. Acknowledgements

This work is supported by the Office of Nuclear Physics within the U.S. DOE Office of Science under contact number DE-AC03-76SF00098. We wish to thank the NERSC Center at LBNL and specifically L. Gerhardt of NERSC for her help with the CVMFS installation into Shifter.

References

[1] Online, accessed Feb 21, 2016 URL `http://www.nersc.gov/users/computational-systems/edison/configuration/`
[2] Online, accessed Feb 21, 2016 URL `http://www.nersc.gov/users/computational-systems/cori/cori-phase-i/`
[3] Online, accessed Feb 21, 2016 URL `http://www.nersc.gov/users/computational-systems/cori/cori-phase-ii/`
[4] Buncic P, Krzewicki M and Vande Vyvre P 2015 Technical Design Report for the Upgrade of the Online-Offline Computing System Tech. Rep. CERN-LHCC-2015-006. ALICE-TDR-019 CERN Geneva URL `http://cds.cern.ch/record/2011297`
[5] Sjöstrand T, Mrenna S and Skands P 2006 *JHEP* **05** 026 (*Preprint* `hep-ph/0603175`)
[6] Sjöstrand T, Mrenna S and Skands P 2007 *Comput. Phys. Commun.* **178** 852–867. 27 p URL `https://cds.cern.ch/record/1064095`
[7] Wang X N and Gyulassy M 1991 *Phys. Rev.* **D44** 3501–3516
[8] Roesler S, Engel R and Ranft J 2000 *Advanced Monte Carlo for radiation physics, particle transport simulation and applications. Proceedings, Conference, MC2000, Lisbon, Portugal, October 23-26, 2000* pp 1033–1038 (*Preprint* `hep-ph/0012252`) URL `http://www-public.slac.stanford.edu/sciDoc/docMeta.aspx?slacPubNumber=SLAC-PUB-8740`
[9] Brun R, Hagelberg R, Hansroul M and Lassalle J C 1978 *Simulation program for particle physics experiments, GEANT: user guide and reference manual* (Geneva: CERN) URL `https://cds.cern.ch/record/118715`
[10] Agostinelli S *et al.* (GEANT4) 2003 *Nucl. Instrum. Meth.* **A506** 250–303
[11] Blomer J, Aguado Sanchez C, Buncic P and Harutyunyan A 2011 *J. Phys. Conf. Ser.* **331** 042003
[12] Botts J, Jacobsen D and Gerhardt L talk at the Workshop HEPIX Fall 2015 URL `https://indico.cern.ch/event/384358/session/12/contribution/59/`
[13] Online, accessed Feb 21, 2016 URL `http://www.docker.io`
[14] Online, accessed Feb 21, 2016 URL `http://ccl.cse.nd.edu/software/parrot/`
[15] Saiz P, Aphecetche L, Buncic P, Piskac R, Revsbech J E and Sego V (ALICE) 2003 *Nucl. Instrum. Meth.* **A502** 437–440

A new Riemann fit for circular tracks

R Frühwirth[1] **and A Strandlie**[2]

[1] Institute of High Energy Physics of the Austrian Academy of Scienes, Nikolsdorfer Gasse 18, A-1050 Wien, Austria

[2] NTNU Norwegian University of Science and Technology, Department of Technology, Economy and Management, P. O. Box 191, N-2802 Gjøvik, Norway

E-mail: rudolf.fruehwirth@oeaw.ac.at, are.strandlie@ntnu.no

Abstract. We present in this contribution a new Riemann track fit which operates on centered and scaled measurements. With these transformations, the fit becomes invariant under translations and similarity transforms of the measurements. We show in a simulation study in a generic, cylindrical detector that the modified Riemann fit is more precise than the standard Riemann fit, in particular if the hit resolution is large.

1. Introduction

The fitting of circular arcs to a set of measurements is of high importance in high-energy physics experiments due to the fact that tracking detectors often are embedded in homogeneous magnetic fields, resulting in a circular track in the bending plane of the particle trajectories. Even if the homogeneity of the magnetic field is imperfect, fast fits of circular tracks are important, e. g. for quality checks in track finding, in a track trigger, in defining a fast reference track for a Kalman filter [1], or for creating a preliminary track for a broken line fit [2]. Other types of detectors such as RICH detectors create circular patterns without the presence of any magnetic field, but with the same need of applying circle fit algorithms for the analysis of the data.

The problem of fitting a set of measurements to a circular arc is non-linear. Non-linear approaches to solving this problem are in general iterative and quite time-consuming. If the radius of curvature of the circle is not too small, the track model can be linearized around a reference track, making possible the application of linear methods such as a global least-squares fit or the Kalman filter. Alternatively, direct approaches such as the conformal mapping method [3], the Karimäki method [4] or the Riemann track fit [5] can be applied.

We present in this contribution a new version of the Riemann track fit method, which is based on centering and scaling of the data in order to achieve invariance under translations and similarity transforms. In a simulation experiment of a generic tracking detector system we show that this new method in some cases is significantly more precise than the standard Riemann track fit method [5].

2. Track fitting on the Riemann sphere

The Riemann sphere sits on the (x, y)-plane. Its south pole is the origin, and its north pole is $(0, 0, 1)$. Points in the plane are mapped to the sphere by a stereographic projection. If the track measurements in the (x, y)-plane are given in polar coordinates,

$$(R_i, \Phi_i), \quad i = 1, \ldots, N$$

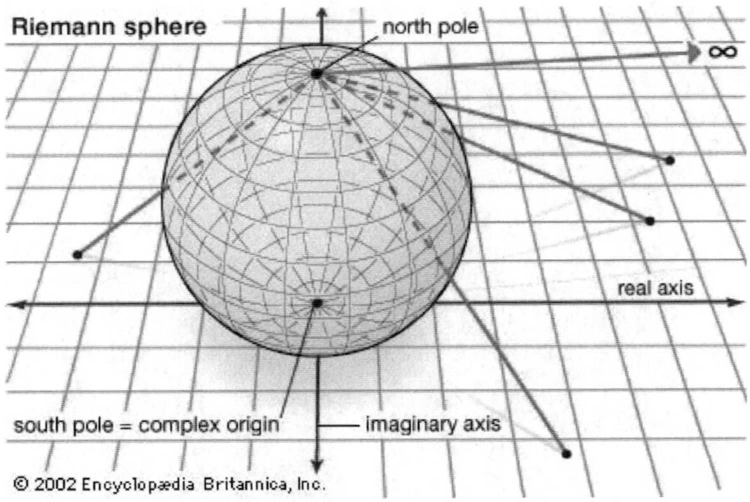

Figure 1. An illustration of the Riemann sphere and the lines defining the transformed points. The imaginary axis corresponds to the x-axis, the real axis corresponds to the y-axis of the bending plane of the track. By courtesy of Encyclopaedia Britannica, Inc., copyright 2002; used with permission.

the mapping to the Riemann sphere is defined as:

$$x_i = R_i \cos \Phi_i/(1 + {R_i}^2)$$
$$y_i = R_i \sin \Phi_i/(1 + {R_i}^2)$$
$$z_i = {R_i}^2/(1 + {R_i}^2)$$

The transformed point (x_i, y_i, z_i) is the point of intersection of the straight line from the point (R_i, Φ_i) in the plane to the north pole of the sphere with the sphere itself, as indicated in Figure 1. The mapping is conformal and maps circles in the plane to circles on the sphere, illustrated in Figure 2. Since a circle on the sphere is the intersection of a unique plane with the sphere itself, the mapped points lie in a plane, or close to a plane if the measurement errors are taken into account. Due to this correspondence of a circle in the plane and a plane in space, the circle fit is transformed into fitting a plane to the mapped points.

All points with a position vector $\boldsymbol{r}$ satisfying $\boldsymbol{n}^{\mathrm{T}}\boldsymbol{r} + c = 0$ are lying in the same plane, where $\boldsymbol{n}$ is a unit vector normal to the plane and c is the signed distance from the plane to the origin. The plane is fitted to the transformed set of measurements by minimizing the following objective function:

$$S = \sum_{i=1}^{N} (1 + {R_i}^2)^2 {d_i}^2 = \sum_{i=1}^{N} p_i {d_i}^2$$

where d_i is the distance from the point $\boldsymbol{r}_i = (x_i, y_i, z_i)^{\mathrm{T}}$ to the plane. The solution to this minimization problem is that the normal vector $\boldsymbol{n}$ is the unit eigenvector corresponding to the smallest eigenvalue of the sample covariance matrix:

$$\boldsymbol{A} = \frac{1}{N} \sum_{i=1}^{N} p_i (\boldsymbol{r}_i - \boldsymbol{r}_0)(\boldsymbol{r}_i - \boldsymbol{r}_0)^{\mathrm{T}}$$

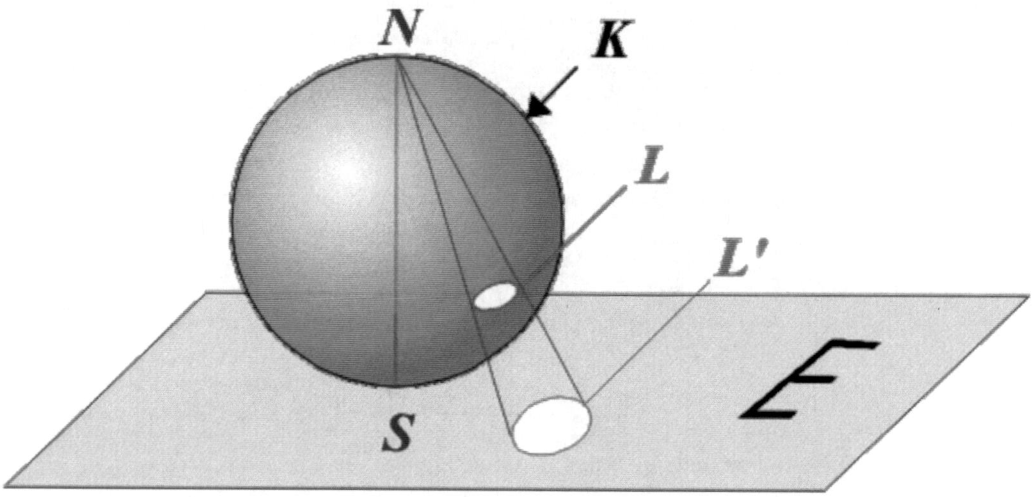

Figure 2. An illustration of the Riemann sphere K (with north pole N and south pole S) and the important feature that a circle L' in the plane E maps onto a circle L on the sphere.

with $\boldsymbol{r}_0 = (x_0, y_0, z_0)^{\mathrm{T}}$ and

$$x_0 = \sum_{i=1}^{N} p_i x_i / \sum_{i=1}^{N} p_i, \quad y_0 = \sum_{i=1}^{N} p_i y_i / \sum_{i=1}^{N} p_i, \quad z_0 = \sum_{i=1}^{N} p_i z_i / \sum_{i=1}^{N} p_i$$

Given $\boldsymbol{n}$, c is computed by:

$$c = -\boldsymbol{n}^{\mathrm{T}} \boldsymbol{r}_0$$

The parameters $\boldsymbol{n}$ and c of the plane can then be mapped to a set of parameters of the corresponding circle in the (x, y)-plane [5].

3. The modified Riemann fit

Chernov [6] recommends centering and scaling of the measurements before mapping to the Riemann sphere in order to achieve invariance of the fit under translations and similarity transformations. For this purpose, it is convenient to transform the measured points to Cartesian coordinates:

$$X_i = R_i \cos \Phi_i, \; Y_i = R_i \sin \Phi_i, \quad i = 1, \ldots, N$$

Centering is then achieved by:

$$X_{c,i} = X_i - \overline{X}, \; Y_{c,i} = Y_i - \overline{Y}, \quad i = 1, \ldots, N$$

with

$$\overline{X} = \frac{1}{N} \sum_{i=1}^{N} X_i, \quad \overline{Y} = \frac{1}{N} \sum_{i=1}^{N} Y_i$$

After collecting all centered measurements $X_{c,i}$ and $Y_{c,i}$ in vectors $\boldsymbol{X}_c$ and $\boldsymbol{Y}_c$, the scaling is done by defining a scaling factor s:

$$s = b / \sqrt{(\boldsymbol{X}_c{}^{\mathrm{T}} \boldsymbol{X}_c + \boldsymbol{Y}_c{}^{\mathrm{T}} \boldsymbol{Y}_c)/N}$$

$$\boldsymbol{X}_{cs} = s \cdot \boldsymbol{X}_c, \; \boldsymbol{Y}_{cs} = s \cdot \boldsymbol{Y}_c$$

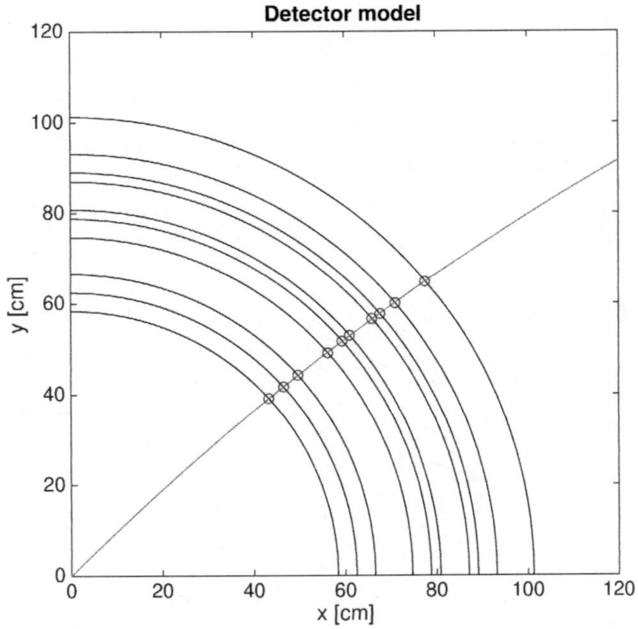

Figure 3. A simulated track in a generic type of cylindrical detector. The positions of the measurements are at the intersections between the track and the detector elements, indicated by open rings in the figure.

with b being an arbitrary, preselected constant. We have chosen the value $b = 0.5$, which has turned out to work well in our simulation studies. After centering and scaling, the measurements are transformed back to polar coordinates before performing the Riemann fit.

4. Simulation study in a generic cylindrical detector

In order to focus on the basic performance of the circle fit, we have simulated a generic type of a cylindrical detector system embedded in a perfectly homogeneous magnetic field. In the bending plane, the track model is in this case a circle. In order to test out the performance of the algorithms under a variety of conditions, we have generated a sample of 10000 tracks coming from the origin with radii of curvature in a range from about 1.5 m to about 750 m. This corresponds to arcs between less than 0.1 degrees and about 20 degrees, following a reasonably flat distribution in this range. There are about 10–12 hits per track, and the hit resolution, which is the standard deviation of the position error, varies between 0 mm, i.e. infinite precision, and 1.6 mm. We assume no background and thereby implicitly a perfect pattern recognition. An example of a simulated track is shown in Figure 3.

We have compared the modified fit with the standard Riemann fit by considering the generalized variances of the residuals, which are the difference of the estimated and the true track parameters in both the standard and the modified fit. The generalized variance is the determinant of the sample covariance matrix of the residuals. Figure 4 shows the ratio of the generalized variance of the modified fit with respect to that of the standard fit as a function of the hit resolution. The figure shows that the modified Riemann fit performs better at all values of the measurement uncertainties, and the improvement grows larger with increasing resolution.

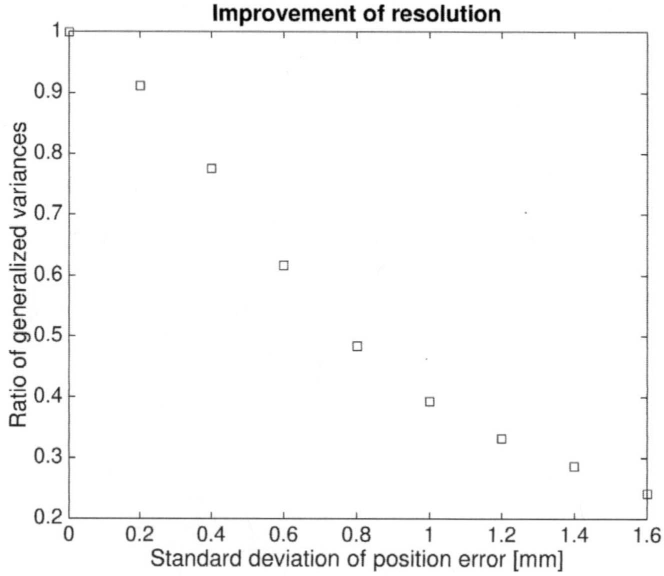

Figure 4. The generalized variance of the modified Riemann fit divided by the generalized variance of the standard Riemann fit as a function of the hit resolution, i.e. the standard deviation of the position error.

5. Conclusions and outlook

We have in this contribution presented a new Riemann track fit which operates on centered and scaled measurements [6], making the fit invariant under translations and similarity transforms of the measurements. With these transformations, the fit becomes more precise than the traditional Riemann fit, in particular for large measurement uncertainties.

For the sake of completeness, a track fit also should be able to provide the covariance matrix of the track parameters. For the standard Riemann fit, the necessary Jacobians for this purpose have been derived earlier [7]. We intend to derive similar Jacobians for the new Riemann track fit in a future study.

References

[1] Frühwirth R 1987 *Nucl. Instrum. Methods Phys. Res.* A **262** 444
[2] Kleinwort C 2012 *Nucl. Instrum. Methods Phys. Res.* A **673** 107
[3] Hansroul M, Jeremie H, and Savard D 1988 *Nucl. Instrum. Methods Phys. Res.* A **270** 498
[4] Karimäki V 1991 *Nucl. Instrum. Methods Phys. Res.* A **305** 187
[5] Strandlie A, Wroldsen J, Frühwirth R and Lillekjendlie B 2000 *Comp. Phys. Commun.* **131** 95
[6] Chernov N 2010 *Circular and linear regression: Fitting circles and lines by least squares* (Boca Raton: CRC Press)
[7] Strandlie A and Frühwirth R 2002 *Nucl. Instrum. Methods Phys. Res.* A **480** 734

A novel method for event reconstruction in Liquid Argon Time Projection Chamber

M Diwan, M Potekhin, B Viren, X Qian and C Zhang
Brookhaven National Laboratory, Upton, NY11973, USA
E-mail: xqian@bnl.gov

Abstract. Future experiments such as the Deep Underground Neutrino Experiment (DUNE) will use very large Liquid Argon Projection Chambers (LArTPC) containing tens of kilotons of cryogenic medium. To be able to utilize sensitive volume of that size, current design employs arrays of wire electrodes grouped in readout planes, arranged with a stereo angle. This leads to certain challenges for object reconstruction due to ambiguities inherent in such a scheme. We present a novel reconstruction method (named "Wirecell') inspired by principles used in tomography, which brings the LArTPC technology closer to its full potential.

1. Introduction

The new method of event reconstruction in Liquid Argon Time Projection Chamber (LArTPC) presented here is motivated by the stringent requirements of the DUNE experiment [1]. The DUNE primary science program [2] includes precision measurements of the parameters that govern neutrino oscillations, search for proton decay, and detection and measurement of the ν_e flux from a core-collapse supernova in our galaxy. One of the central components of DUNE is a massive Far Detector (40kt fiducial volume) based on Liquid Argon Time Projection Chamber (LArTPC) located deep underground at a distance of 1,300 km from the neutrino source [3]. This device has the potential to reconstruct tracks and showers with high level of detail and efficiency, as well as to provide a precise measurement of ionization charge necessary for good particle identification based on ionization energy loss (dE/dx). The single-phase LArTPC design which is our focus here is an ionization chamber instrumented with planar arrays of wire electrodes [3].

2. Operation of the LArTPC with wire readout

2.1. Principle of Operation

The LArTPC principle of operation is schematically illustrated in Fig.1. Ionization electrons are produced in the medium by ionizing particles associated with an event. Due to an electrostatic field maintained in the chamber, they start drifting to the right, while positive ions travel towards the cathode on the left hand side. The right-hand side of the diagram depicts planar arrays of sensor wires. Spacing between ajacent wires in a plane is of the scale of a few mm.

Just like any charge moving in the vicinity of a conductor, drifting electrons create signals on the wires belonging to the induction wire planes "U" and "V". When they finally reach the collection plane "Y" (where they are removed from the detector volume after completing their drift in the liquid argon medium), a signal is also produced on its wires. The ionization pattern can then be reconstructed in the plane perpendicular to the drift direction by analyzing signals on (U,V,Y), which form a stereo angle configuration illustrated in Fig.2 that is a schematic view of the three planes along the drift direction. The dot in the middle represents a localized ionization charge.

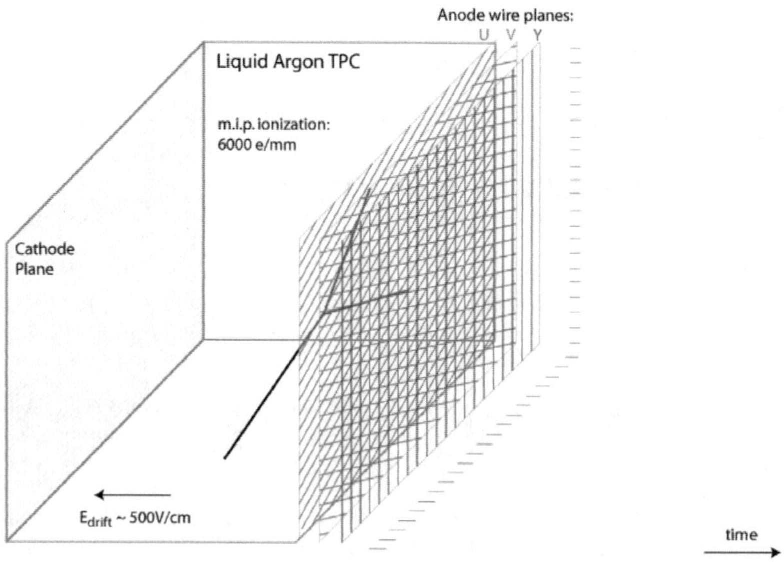

Figure 1. The LArTPC Principle of Operation

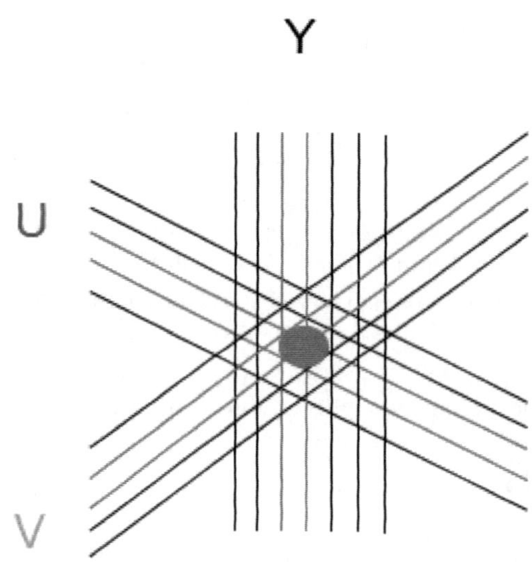

Figure 2. Collection and Induction Wire Planes as viewed along the drift direction

Every induction and collection wire is read out as an individual channel, with waveforms recorded at ~2 MHz digitization frequency. There are also planar Cathode Plane Assemblies which are necessary to create a near-uniform electrostatic field in the chamber, however they are not instrumented.

Reconstruction along the axis of the drift direction can be done using the information contained in the time evolution of the signals on the (U,V,Y) wire planes, since each time bin corresponding to the ADC digitization cycle can be mapped to the coordinate in that direction, through the known (calibrated) value of drift velocity. In general, signals on individual wires produce complex waveforms, and are subject to electronics noise, shaping parameters in the amplifier chain and other factors. Evaluating the charge in a particular time bin involves complex signal processing, and is beyond the scope of this paper (see [4] for a few details).

2.2. The Challenge
It is not difficult to see that while the wire readout design makes construction of very large LArTPCs possible by keeping the channel count within practical limits, it does so at the cost of information loss, since now instead of pads or pixels one has to reconstruct the position of the particle "hits" in 2D using three projections. If the number of sensor wires that registered a signal is ~N, the number of potential hit locations scales as $\sim N^2$, which naturally leads to ambiguities.

3. The "Wirecell" method

3.1. Reconstruction in LArTPC
Event reconstruction in LArTPCs with wire readout is primarily geared towards neutrino interactions with multiple final state particles emerging from a single vertex. Reconstruction methods which existed prior to introduction of the method described in this paper were inspired by tracking algorithms utilized in High Energy Physics, and indeed reused existing software libraries in some cases. A typical approach starts with identifying track candidates in each of the three projections which were described above. This is done by identifying groups of wires in each projection which had "hits" is adjacent time bins, effectively relying on the continuity of the object being reconstructed at this stage in the reconstruction process. Hypotheses are applied regarding whether the object corresponds to a track or to a shower. Then, an attempt is made to match individual 2D (time-coordinate) projections in 3D [5]. As mentioned in 2.2, there are ambiguities in this process which may result in reconstruction artifacts.

3.2. Parallels to Tomography
The method of collecting signals from LArTPC has parallels with Computerized Axial Tomography (CAT) applications. Each time slice in LArTPC is read out separately and effectively represents a separate unit of data. This can be compared to one of the original (and simplest) methods of CAT called Multiplanar Reconstruction – MPR – in which the volume under study is treated as a stack of 2D slices. Revisiting Fig.2 one can see that the same charge pattern in an individual 2D slice is sampled in three different projections. Each wire is measuring a line integral of the ionization charge located in its vicinity. This has significant similarity to absorption tomography where attenuation of radiation intensity also corresponds to a line integral along the direction of the beam. Clearly, the number of projections is extremely limited (to three) which makes application of mainstream CAT reconstruction techniques such as Radon transform [6] not practical. It must be noted however that sparse and limited-angle tomography does have its applications and domain-specific mathematical methods have been developed to deal with such extremely ill-posed problems.

CAT applications typically feature voxelization of the object under study due to natural granularity of the apparatus and also to make numerical calculations feasible. In multiplanar reconstruction, this effectively leads to treating each individual 2D slice as a set of pixels, defined according to the type of measurement. Application of this technique (introduction of cells) is the foundation of the Wirecell method for LArTPC described here [7].

3.3. The "Wirecell" algorithm

The term "time slice" is used here to describe data (signal values) belonging to a single time bin – a single digitization cycle of the ADCs processing the signals from wire planes. These data contain information about what is essentially a 2D image (a slice of the 3D object under study). In Wirecell, a tessellation algorithm is used to describe each time slice as a set of polygon pixels in 2D. Each wire is associated with a subset of pixels. Reconstruction is done in two steps:

- Solving the inverse problem in each time slice, i.e. performing 2D imaging (reconstructing pixel values) using the three projections of the charge distribution in the time slice.
- Combining the time slices to obtain a 3D model of the ionization pattern in the Liquid Argon Volume, which can then be used to extract physics information about the event.

The inverse problem of determining pixel charge values based on limited number of projections (three sets of wires) is ill-posed. Therefore, the following algorithm is applied which helps to reduce the number of degrees of freedom: groups of adjacent candidate cells with signal above a certain threshold are identified. They are grouped into "merged cells" (i.e. the charge values are combined to effectively form a *larger pixel*). Note that many of these cells will still be ghosts due to ambiguities as described above (see 2.2). Wires mapped to these merged cells are also merged.

Note that the values of charge in "merged cells" are not known as a result of a direct measurement and need to be computed, based on the values of signals on "merged wires" (groups of adjacent wires that registered signals). The relationship between the wires and cells can be represented as $W_m = G\,C_m$, where W_m is the vector of values attributed to charges on merged wires (measured directly), G is the geometry matrix mapping wires to cells, and the C_m is the vector of values for merged cells, for which a solution must be found. In vast majority of cases this cannot be solved by matrix inversion. Instead, the Wirecell method employs the Markov Chain Monte Carlo (MCMC) approach, which aims to minimize the following value:

$$\chi^2 = (W_m - GC_m)^T V^{-1} (W_m - GC_m)$$

V stands for the covariance matrix for signals on merged wires. Each step in the MCMC process involves removing a random cell and recalculating G to evaluate the metric as described above. If the χ^2 value improves, such cell is permanently eliminated under the hypothesis that it was a "ghost cell". Once a sufficient number of cells have been eliminated, matrix inversion becomes possible which completes the solution. A reconstructed 2D image is formed, and a stack of these images forms the 3D picture of the physics event in LArTPC. It is important to note that in this method, no physics assumptions are made either about the type of particle or the number of tracks in the event before the event is assembled in 3D.

The optimization technique as described above can be modified to take advantage of certain apriori information about the object, which is another common approach in the field of tomography. For example the hypothesis of object continuity in 3D can be included in the solution by adding a penalty value to χ^2 in cases there there are no active cells in neighboring 2D slices, adjacent to the one being considered for elimination during the MCMC step.

The Wirecell algorithm contains a number of elements which are computationally intensive, e.g. each step in the Markov Chain involves recalculating matrices and other operations. For that reason, work is currently under way to adapt Wirecell for use on modern parallel computing platforms, including GPUs.

3.4. Status and future directions

The Wirecell method has been applied to reconstruction of a variety of event classes using Monte Carlo data. In the first round of analysis, efficiency of track reconstruction in this method is at or above the levels stipulated by the requirements of the DUNE experiment. It is currently being adapted for event reconstruction in the μBooNE experiment [8], and results are expected to become available soon.

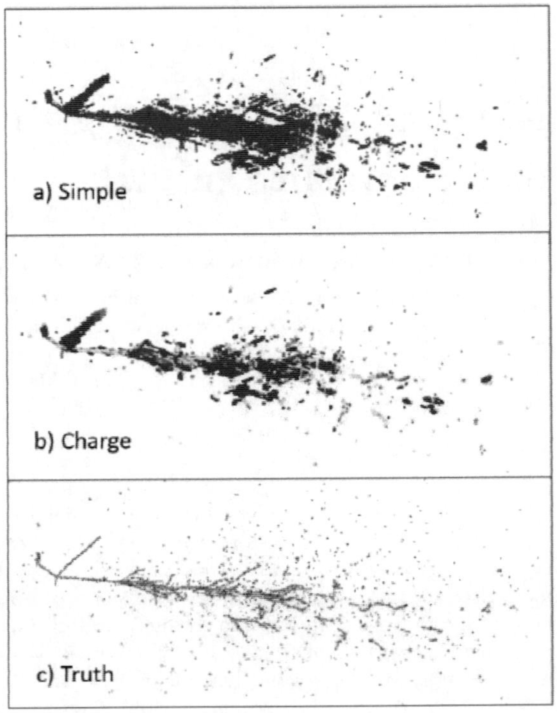

Figure 3. Comparison of hit-based reconstruction ("simple"), Wirecell ("charge") and MC Truth ("truth"). Blue color in (a) and (b) represents the reconstructed charge density distribution. Note the improvement from (a) to (b).

4. Summary

A novel reconstruction method has been developed which applies tomography principles to neutrino event reconstruction in Liquid Argon Time Projection Chambers with wire readout. Markov Chain Monte Carlo technique is used for the inverse problem solution in 2D, with a full 3D image formed as a stack of 2D slices. First results are promising and indicate that this method can bring the LArTPC technology closer to its full potential for physics research.

References

[1] DUNE CDR Vol 1 – The LBNF and DUNE Projects: `http://arxiv.org/abs/1601.05471`
[2] DUNE CDR Vol 2 – The Physics Program for DUNE at LBNF: `http://arxiv.org/abs/1512.06148`
[3] DUNE CDR Vol 4 – The DUNE Detectors at LBNF: `http://arxiv.org/abs/1601.02984`
[4] μBooNE public notes 1016 and 1017
[5] F. Arneodo et al. (The ICARUS-Milano Collaboration) "Performance Of A Liquid Argon Time Projection Chamber Exposed To The WANF Neutrino Beam", Phys.Rev.D74:112001,2006
[6] William H. Press, " Discrete Radon transform has an exact, fast inverse and generalizes to operations other than sums along lines", Proceedings of the National Academy of Sciences of the United States of America, 1924919254, doi: 10.1073/pnas.0609228103
[7] Wirecell: `http://www.phy.bnl.gov/wire-cell/`
[8] μBooNE experiment: `http://www-microboone.fnal.gov`

Experiments using machine learning to approximate likelihood ratios for mixture models

K Cranmer[1], J Pavez[2], G Louppe[1] and W K Brooks[3]

[1] Physics Department, New York University, New York, NY 10003, U.S.A.
[2] Informatics Department, Universidad Técnica Federico Santa María, 1680 Av. España, Valparaíso, Chile
[3] Physics Department, Universidad Técnica Federico Santa María and Center for Science and Technology of Valparaíso, 1680 Av. España, Valparaíso, Chile

E-mail: juan.pavezs@alumnos.usm.cl

Abstract. Likelihood ratio tests are a key tool in many fields of science. In order to evaluate the likelihood ratio the likelihood function is needed. However, it is common in fields such as High Energy Physics to have complex simulations that describe the distribution while not having a description of the likelihood that can be directly evaluated. In this setting it is impossible or computationally expensive to evaluate the likelihood. It is, however, possible to construct an equivalent version of the likelihood ratio that can be evaluated by using discriminative classifiers. We show how this can be used to approximate the likelihood ratio when the underlying distribution is a weighted sum of probability distributions (e.g. signal plus background model). We demonstrate how the results can be considerably improved by decomposing the ratio and use a set of classifiers in a pairwise manner on the components of the mixture model and how this can be used to estimate the unknown coefficients of the model, such as the signal contribution.

1. Introduction

In High Energy Physics (HEP) and many other fields, hypothesis testing is a key tool when reporting results from an experiment. Likelihood ratio tests are the main technique for hypothesis testing and they are the most powerful statistic for simple hypothesis testing. For composite hypothesis testing the profile or generalized likelihood ratio test is commonly used. When computing the likelihood ratio the data distribution $p(x|\theta)$ must be evaluated, where θ are parameters of the probability distribution. However, it is common in HEP to have physics simulations that allow to sample high dimensional vectors from the distribution $p(x|\theta)$ while not having a description that can be directly evaluated. Commonly it is impossible or computationally expensive to compute the likelihood ratio in this setting.

A common use of likelihood ratios in HEP is signal process identification. In this task the hypothesis testing procedure is used to evaluate the signal process significance by contrasting the background-only (null) hypothesis versus the signal plus background (alternative) hypothesis. In this setting, the underlying distribution can be seen as a signal and background mixture model defined as

$$p(x \mid \mu, \nu) = \mu p_s(x \mid \nu) + (1 - \mu)\, p_b(x \mid \nu), \tag{1}$$

where $p_x(x \mid \nu)$ correspond to the signal distribution and $p_b(x \mid \nu)$ is the background distribution, both parametrized by nuisance parameters ν which describe uncertainties in the underlying physics predictions or response of measurement devices. The parameter μ is the mixture coefficient corresponding to the signal component of the distribution. In this case the generalized likelihood ratio test takes the form of

$$\Lambda(D) = \prod_{e=1}^{n} \frac{p(x_e|\mu = 0, \hat{\hat{\nu}})}{p(x_e|\hat{\mu}, \hat{\nu})}, \tag{2}$$

where D is a data set of i.i.d observations x_e, $\hat{\hat{\nu}}$ is the conditional maximum likelihood estimator for ν under the null hypothesis θ_0 ($\mu = 0$) and $\hat{\nu}, \hat{\mu}$ are the maximum likelihood estimators for ν and μ. This approach has been used extensively to assert the discovery of new particles in HEP [1], such as in the discovery of the Higgs boson [2, 3].

As previously mentioned, the original distributions for signal and background can only be approximated by simulations. Most of the likelihood ratio tests at the LHC are made on the distribution of a single feature that discriminates between signal and background observations. For this, the simulated data is used together with interpolations algorithms in order to approximate the parametrized model and then use it in the hypothesis testing procedure [4].

Recently, it has been shown that a discriminative classifier trained to classify between signal and background can be used to obtain an equivalent likelihood ratio test (see eq. 2.9 of [5]). Given a classifier trained to learn a monotonic function of the per event ratio $p(x_e|\theta_0)/p(x_e|\theta_1)$, it can be proved that the likelihood ratio test on the conditional distributions of the classifier score is equivalent to the original likelihood ratio. Moreover, many of the commonly used classifiers learn to approximate some monotonic function of the per-event ratio.

In this work we show how these results can be used to approximate the likelihood ratio when the underlying distribution is a weighted sum of probability distributions (mixture model). We also show that by training a set of classifiers in a pairwise manner on the components of the mixture model it is possible to improve the results of the approximation.

2. Decomposed likelihood ratio test for mixture models

A generalized version of the signal and background mixture model of eq. (1) for several components is

$$p(x|\theta) = \sum_{i=1}^{k} w_i(\theta) p_i(x|\theta), \tag{3}$$

where $w_i(\theta)$ are the mixture coefficients for each one of the components parametrized by θ. In [5] it is shown that the likelihood ratio between two mixture models

$$\frac{p(x|\theta_0)}{p(x|\theta_1)} = \frac{\sum_{i=1}^{k} w_i(\theta_0) p_i(x|\theta_0)}{\sum_{j=1}^{k'} w_j(\theta_1) p_j(x|\theta_1)}, \tag{4}$$

is equal to the composition of pairwise ratios for each one of the components which, in turn, is equivalent to the composition of ratios on the score distribution of pairwise trained classifiers

$$\frac{p(x|\theta_0)}{p(x|\theta_1)} = \sum_{i=1}^{k} \left[\sum_{j=1}^{k'} \frac{w_j(\theta_1)}{w_i(\theta_0)} \frac{p_j(x|\theta_1)}{p_i(x|\theta_0)} \right]^{-1} = \sum_{i=1}^{k} \left[\sum_{j=1}^{k'} \frac{w_j(\theta_1)}{w_i(\theta_0)} \frac{p_j(s_{i,j}(x;\theta_0,\theta_1)|\theta_1)}{p_i(s_{i,j}(x;\theta_0,\theta_1)|\theta_0)} \right]^{-1}. \tag{5}$$

In the case that the only free parameters of the mixture model are the coefficients $w_i(\theta)$, then each distribution $p_i(s_{i,j}(x;\theta_0,\theta_1)|\theta)$ is independent of θ and can be pre-computed and used afterwards in the evaluation of the likelihood ratio. Moreover, when numerator and denominator correspond to the same distribution, the values can be directed replaced by unity, avoiding unnecessary computations. Also, for a two-class classifier the values of $s_{j,i}(x;\theta_0,\theta_1)$ for one of the classes can be replaced by the values of $s_{i,j}(x;\theta_0,\theta_1)$ for the opposing class, in which case it is only necessary to train the classifiers for $i < j$. This saves a lot of computation time and reduces the variance that is introduced by differences between $s_{i,j}(x;\theta_0,\theta_1)$ and $s_{j,i}(x;\theta_0,\theta_1)$ due to imperfect training. In the case of background-only versus signal plus background hypotheses it is common that the signal coefficient $w_j(\theta_1)$ is a very small number compared to the background coefficients. In these conditions a classifier trained on data from the full mixture model will have difficulty identifying the signal since most of the useful discriminative data will be located in a small region of the feature space while the decomposed model will not face these issues.

It is possible to estimate the signal and background coefficients by using the maximum likelihood method on the ratios, as follows

$$\hat{\theta} = \arg\max_{\theta} \prod_{e=1}^{N} \frac{p(x_e|\theta)}{p(x_e|\theta_1)} = \arg\max_{\theta} \prod_{e=1}^{N} \frac{p(s(x_e;\theta,\theta_1)|\theta)}{p(s(x_e;\theta,\theta_1)|\theta_1)}, \quad (6)$$

for some fixed value of θ_1.

The complete algorithm to approximate the likelihood ratio using pairwise trained classifiers can be separated into three independent stages: classifier training, score distribution estimation, and computation of the composition formula. Since these steps are independent, each one can be solved as a different problem. In the first step any classifier satisfying the monotonic requirement can be used. In the second stage the probability distribution of the score on data from θ_0 or θ_1 can be approximated using any univariate density estimation technique such as histograms or kernel density estimation [6, 7].

3. Experiments

In this section two examples of how the method works on data generated from known distributions will be presented. In both cases we study the results of using the decomposition formula to compute the ratios. Then, we compare the results to the exact (in this case known) ratios and to the density ratios obtained by training a classifier on data from the full mixture model. All studies were conducted using a simple multilayer perceptron model. That classifier shows a good tradeoff between quality of the ratios and simplicity of the model (good results were also obtained using boosted decision trees, logistic regression and support vector machines). The probability models were implemented with the `RooFit` [6] probabilistic programming language and the classifiers were implemented using `Theano` [9] (a framework to build neural network models) and `scikit-learn` [10] (a general framework for machine learning in python). The code is available for replication of the results at [11].

First, we present a simple case in which each component is a univariate distribution. We consider a mixture model consisting of three distributions, where $p_0(x)$ and p_1(x) are univariate Gaussian distributions while $p_2(x)$ is a decaying exponential. The mixture models are composed by the weighted sum of those distributions where $p_0(x)$ correspond to the signal component. The mixture models with coefficients $W(\theta_0) = [0., 0.3, 0.7]$ for the background-only hypothesis and $W(\theta_1) = [0.1, 0.27, 0.63]$ for the background plus signal hypothesis are shown in Figure 1a.

Three neural networks were trained on 200000 examples sampled from the pairs of single distributions and one on examples sampled from the full models (it is important to use the same amount of training data in order to allow a fair comparison). Each neural network consists of 1 hidden layer of 4 units and stochastic gradient descent with a learning rate of 0.01 and no regularization was used in training (the networks were implemented with `Theano`). The distribution of the score is estimated using histograms on a different dataset with 100000

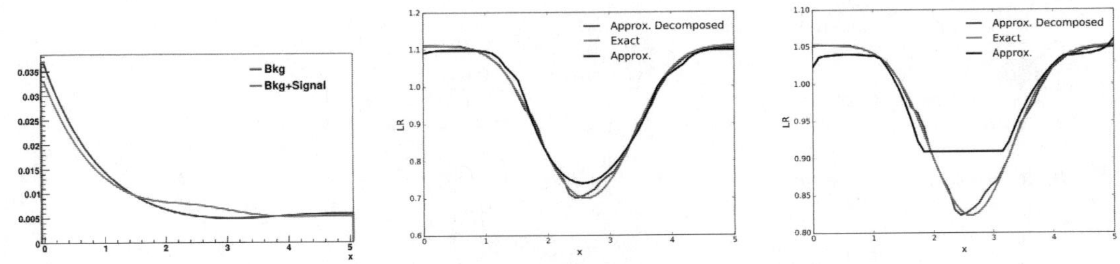

(a) (b) Signal weight of 0.1. (c) Signal weight of 0.05.

Figure 1: (1a) shows the mixture model distributions. (1b) and (1c) show a density ratio comparison for different values of the signal weight.

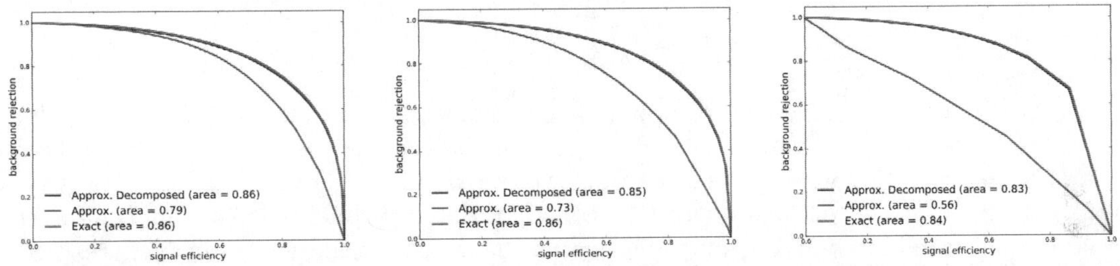

(a) Signal weight of 0.1. (b) Signal weight of 0.05. (c) Signal weight of 0.01.

Figure 2: Background rejection versus signal efficiency curve comparison for different values of the signal weight.

samples (the number of bins were carefully chosen in order to allow a good approximation while minimizing Poisson fluctuations). The composed ratio using eq. (5), the ratio estimated using a classifier trained on data from the full model, and the exact density ratio are shown in Figure 1b and Figure 1c for different values of the signal contribution (0.1 and 0.05) while keeping the ratio between the backgrounds contributions fixed.

It can be seen that the ratios obtained by the composition method are better (closer to the exact ratios) than those obtained using a model trained on data from the full mixture models and this become clearer when the signal contribution is smaller.

Next, the same experiments are repeated but now considering a much harder mixture model consisting of three distributions, each one composed of the sum of three 10-dimensional multivariate Gaussian distributions. Again, we used neural networks with 1 hidden layer with 40 units, a learning rate of 0.01 and a ℓ_2 regularization of 0.0001 in training. To evaluate the approximated ratios the background rejection versus signal efficiency curves are used, employing the density ratio as discriminative variable. The values for each one of the three cases (decomposed, full and exact) are shown in Figure 2, for a signal contribution of 0.1, 0.05 and 0.01.

The values of each one of the coefficients of the mixture model can be estimated by using the method of maximum likelihood as explained in Section 2. In Figure 3a the contour plot for the log-likelihood ratio obtained using the composed density ratios and the exact density ratios are shown. Histograms of the maximum likelihood estimators (MLEs) for approximated and exact ratios and for the signal coefficient and one of the background coefficients on 200 different pseudo-datasets of size 1000 are shown in Figure 3b and Figure 3c. It is seen from these histograms that the estimations are unbiased.

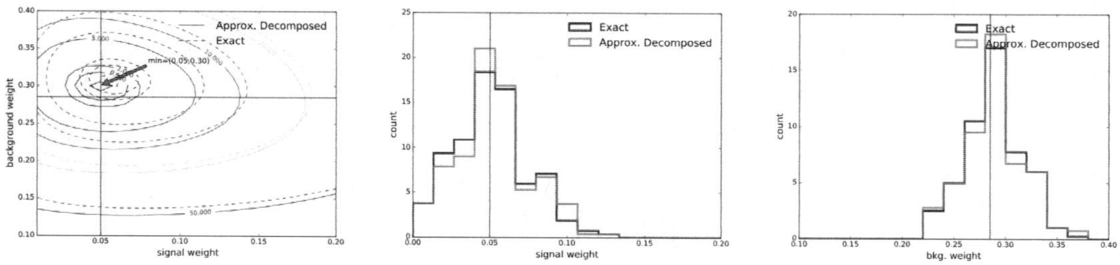

(a) $-2 \log \Lambda(D)$ (b) Signal weight MLEs. (c) Bkg. weight MLEs.

Figure 3: (3a) shows the values of $-2 \log \Lambda(D)$ given the signal and one of the bkg. weights. (3b) and (3c) show histograms for maximum likelihood estimated values for the signal and background weights respectively.

4. Conclusions

We have shown the power of using discriminative classifiers in order to approximate the likelihood ratio. In the case of mixture models we have proved that the decomposed version of the ratio can greatly improve the quality of the results. Using the same method we demonstrated how to estimate the unknown parameters of the model. Initial experiments have been conducted on simulated data from different Higgs production mechanisms, obtaining encouraging results. An open source `Python` package was created to facilitate the usage of this method [12].

5. Acknowledgments

KC and GL are supported by the DIANA grant ACI-14503. JP and WB were partially supported by the Center for Science and Technology of Valparaíso (CCTVal) under Fondecyt grant BASAL FB0821. JP would like to thank Hector Allende and Carlos Valle for helpful discussions.

References

[1] Cowan G, Cranmer K, Gross E and Vitells O 2010 *Eur.Phys.J.* **C71** 1554 (*Preprint* 1007.1727) URL http://arxiv.org/abs/1007.1727

[2] The ATLAS Collaboration 2012 *Phys.Lett.* **B716** 1–29 (*Preprint* 1207.7214)

[3] The CMS Collaboration 2012 *Phys.Lett.* **B716** 30–61 (*Preprint* 1207.7235)

[4] Cranmer K, Lewis G, Moneta L, Shibata A and Verkerke W 2012 CERN-OPEN-2012-016, http://inspirehep.net/record/1236448

[5] Cranmer K, Pavez J and Louppe G 2016 *ArXiv e-prints* http://arxiv.org/abs/1506.02169 (*Preprint* 1506.02169)

[6] Verkerke W and Kirkby D P 2003 *eConf* **C0303241** MOLT007 (*Preprint* physics/0306116)

[7] Cranmer K S 2001 *Comput. Phys. Commun.* **136** 198–207 (*Preprint* hep-ex/0011057)

[8] Verkerke W and Kirkby D 2003 *Statistical Problems in Particle Physics, Astrophysics and Cosmology, edited by L. Lyons and MK Ünel* 186–190

[9] Bergstra J, Breuleux O, Bastien F, Lamblin P, Pascanu R, Desjardins G, Turian J, Warde-Farley D and Bengio Y 2010 Theano: a CPU and GPU math expression compiler *Proceedings of the Python for Scientific Computing Conference (SciPy)* oral Presentation

[10] Pedregosa F, Varoquaux G, Gramfort A, Michel V, Thirion B, Grisel O, Blondel M, Prettenhofer P, Weiss R, Dubourg V, Vanderplas J, Passos A, Cournapeau D, Brucher M, Perrot M and Duchesnay E 2011 *Journal of Machine Learning Research* **12** 2825–2830

[11] Pavez J 2016 DecomposingTests: Reproducible results for ACAT 2016 http://dx.doi.org/10.5281/zenodo.47945, https://github.com/jgpavez/DecomposingTests/tree/ACAT2016

[12] Louppe G, Cranmer K and Pavez J 2016 carl: a likelihood-free inference toolbox http://dx.doi.org/10.5281/zenodo.47798, https://github.com/diana-hep/carl

Image Processing, Computer Vision, and Deep Learning: new approaches to the analysis and physics interpretation of LHC events

A. Schwartzman[1], M. Kagan[1], L, Mackey[2], B. Nachman[1] and L. De Oliveira[3]

[1] SLAC National Accelerator Laboratory, Stanford University, 2575 Sand Hill Road, Menlo Park, CA 94025, USA

[2] Department of Statistics, Stanford University, Stanford, CA 94305, USA

[3] Institute for Computational and Mathematical Engineering, Stanford University, Stanford, CA 94305, USA

E-mail: `sch@slac.stanford.edu`

Abstract. This review introduces recent developments in the application of image processing, computer vision, and deep neural networks to the analysis and interpretation of particle collision events at the Large Hadron Collider (LHC). The link between LHC data analysis and computer vision techniques relies on the concept of *jet-images*, building on the notion of a particle physics detector as a digital camera and the particles it measures as images. We show that state-of-the-art image classification techniques based on deep neural network architectures significantly improve the identification of highly boosted electroweak particles with respect to existing methods. Furthermore, we introduce new methods to visualize and interpret the high level features learned by deep neural networks that provide discrimination beyond physics-derived variables, adding a new capability to understand physics and to design more powerful classification methods at the LHC.

1. Introduction

The Large Hadron Collider is exploring physics at the energy frontier, probing some of the most fundamental questions about the nature of our universe. The datasets of the LHC experiments are among the largest in all science. Each particle collision event at the LHC is rich in information, particularly in the detail and complexity of each event picture, making it ideal for the application of machine learning techniques to extract the maximum amount of physics information. Machine learning has been widely used for the analysis of particle collision data. Examples include unsupervised learning algorithms, such as jet clustering [1], and supervised algorithms for different particle identification and classification tasks, including the identification of different kind of quark flavors, such as b- and c-jets [2], tau leptons [9], and photons [4], among many others. Particle classification algorithms in particle colliders have traditionally been designed using physics intuition to extract a small set of observables or features that were then input into a multivariate classifier. For example, in the case of b flavor tagging in the ATLAS experiment, a Boosted Decision Tree consisting of 24 input variables is used [5]. The 24 input variables provide information about the physics of b-quark decays that is distinct from the light-quark and gluon jet backgrounds. Much of the work in developing such techniques relies on finding the physics variables that best capture the differences between different classes of particles.

During the last several years, spectacular advances in the fields of artificial intelligence, computer vision, and deep learning have resulted in remarkable performance in image classification and vision tasks [6], in particular through the use of deep convolutional neural networks (CNN) [7]. One of the key innovations in these methods is that they operate directly from raw images instead of requiring the extraction of features from the images. Motivated by the success of these techniques, this paper investigates the applicability of deep neural networks for image classification to the analysis of LHC data. LHC events are in fact ideal for the use of computer vision methods, since each LHC event picture can be interpreted as an image from a digital (100M pixel) camera.

2. Jets at the LHC

One of the most important signatures for the analysis of data from the Large Hadron Collider (LHC) are collimated streams of particles known as jets. Individual jets identify quarks and gluons emitted in high-energy interactions. Combinations of jets are used to identify unstable massive particles such as the top quark, the W, Z, and Higgs bosons. For example, W bosons are reconstructed by the two jets from the decay process $W \rightarrow qq$, where q represents a quark. These heavy particles, appear in many models of new physics that predict the existence of new exotic sub-atomic particles. Almost every analysis of LHC data rests on the efficient and accurate reconstruction of jets and their properties. An example of an event display from the ATLAS experiment[1] containing two high p_T[2] jets displayed in different colors is shown in Figure 1.

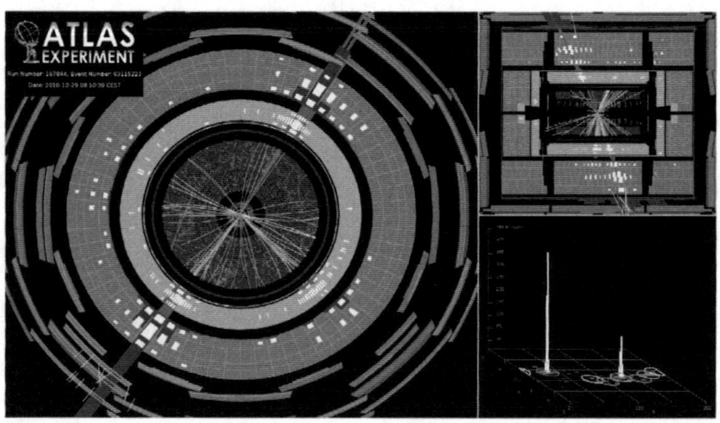

Figure 1. ATLAS event display containing two high p_T jets.

The event display shows three different views: a transverse (relative to the beam axis) $x - y$ view on the left panel, a parallel $x - z$ view on the top right panel, and a *lego* view on the bottom right panel. The lego view shows the projection of the energy deposits on the calorimeter into a two-dimensional (η, ϕ) plane that is obtained by unfolding the detector along the ϕ direction. This view is particularly relevant for image processing and computer vision approaches as it can be related to a fixed size *detector image*. The variable η is called *pseudo-rapidity* and is related to the polar angle θ.

[1] https://atlas.web.cern.ch/Atlas/GROUPS/PHYSICS/CONFNOTES/ATLAS-CONF-2011-047/fig_28.png

[2] Analyses of hadron collider data use the jet transverse momentum, p_T, rather than total energy. The jet p_T is defined as the the magnitude of jet momentum transverse to the beam axis: $p_T = \sqrt{p_x^2 + P_y^2}$.

Jets are reconstructed using dedicated algorithms that group calorimeter energy deposits into clusters of fixed two-dimensional spatial (η, ϕ) size, corresponding to individual jets. Circles in the bottom right panel of Figure 1 represent the area of jets found with the *anti-k_t* clustering algorithm with radius parameter $R = 0.4$, whereas in yellow is shown the magnitude of the transverse energy of the calorimeter deposits within jets. For a review of jet algorithms see [8].

At the LHC, electroweak heavy particles (W, Z, and Higgs bosons, and top quarks) can be produced with large Lorentz boosts. The boost of the electroweak particle determines the distance in the (η, ϕ) plane between their decay products. Highly boosted heavy electroweak particles can hence lead to signatures of *jet substructure*, consisting of single (large radius) jets containing multiple nearby *sub-jets*. For example, experimental signatures boosted W bosons will contain two close-by sub-jets. Figure 2 shows an event display containing two boosted top quarks[3]. The lego view in the top right panel shows top-jet candidates as green circles with red circles indicating individual sub-jets. The size of the yellow squares is proportional to the transverse component of the energy deposits. A three-prong structure is clearly visible for the jet labeled as *hadronic top candidate*. In this case, the boosted jet corresponds to the process $t \rightarrow Wb \rightarrow qqb$.

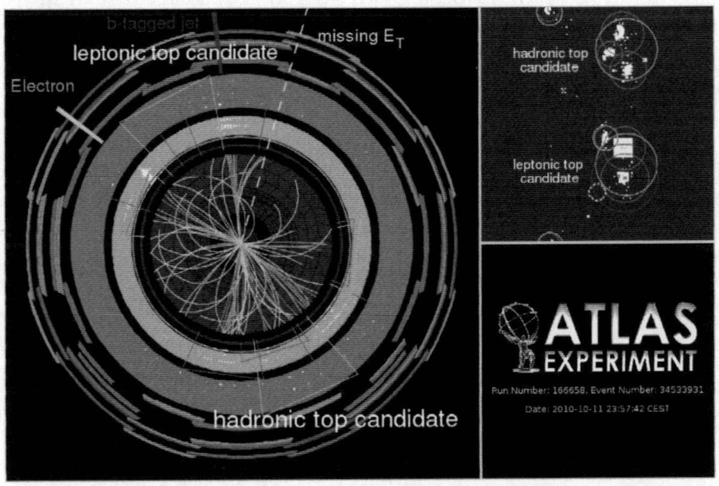

Figure 2. ATLAS event display containing two boosted top-quark jets.

The most difficult challenge to the identification of boosted jets from heavy electroweak particles at the LHC is the rejection of the very large background from multi-jet production in QCD. The key idea to classify boosted top or boson jets is to identify the internal structure of individual jets. The main observables used in jet substructure analyses that best separate signal (top and boson jets) from background (quark and gluon jets, generically referred as *QCD jets*) are the jet mass, the n-prong structure of the jets, for example measured with the variable *n-subjettiness* [9], and the radiation pattern within the jet. For a review of jet substructure techniques see [10][11][12].

3. Data samples

This review considers simulated jet data produced from Monte Carlo simulations of LHC collisions. Boosted W bosons were generated as signal using using PYTHIA 8.170 [13] at $\sqrt{s} = 14$ TeV via the decay of a hypothetical heavy W' boson forced to decay to a hadronically decaying W boson ($W \rightarrow qq$) and a Z boson which decays invisibly to neutrinos. Multijet production

[3] https://atlas.web.cern.ch/Atlas/GROUPS/PHYSICS/CONFNOTES/ATLAS-CONF-2011-073/fig_20.png

of quarks and gluons were generated as background using the same PYTHIA generator and center-of-mass energy as for W bosons.

4. Jet images

In order to use image processing and computer vision approaches for jet tagging at the LHC, a new data representation is introduced: the *jet-image* [14]. Jet images build on the notion of a detector as pixels on a digital camera, and jets as images, enabling the connection between the fields of computer vision and jet substructure and jet physics. Jet images are defined by a 25×25 grid of size (0.1×0.1) in $(\eta \times \phi)$ space centered around the axis of R=1.0 anti-k_t jets. The intensity of the pixels given by the transverse momentum p_T of the pixel cell. Prior to the application of computer vision classification techniques, a series of pre-processing steps to account for the space-time symmetries of jets images are applied. Pre-processing for the specific case of the identification of 2-prong jets such as those from the decay of W bosons, is defined by a *translation* that places the leading p_T subjet at the center of jet-image, a *rotation* such that the second p_T leading subjet is placed down below, and a *flip* operation such that the right side of the image has higher total p_T than the left. The goal of these pre-processing steps is to make it easier for image classification algorithms to focus on the important differences in physics between images. Jet-image pre-processing is illustrated in Figure 3 which shows the average jet-image of W (signal) and QCD (background) jets in a narrow p_T and mass bin before and after the the pre-processing steps.

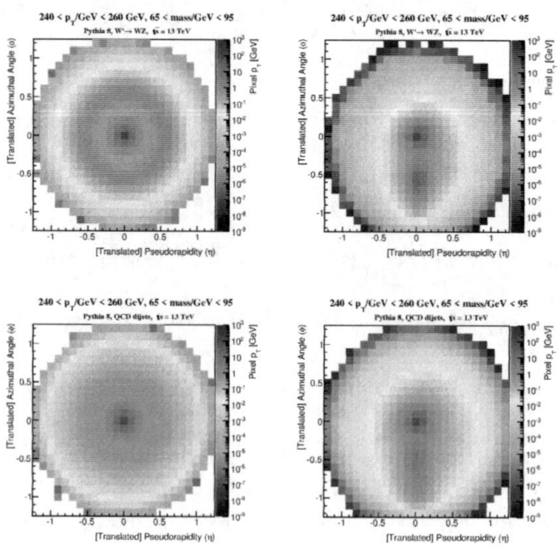

Figure 3. Average jet-image for W jets (top panel) and QCD jets (bottom panel) with $240 < p_T < 260$ GeV and $65 <$mass< 95 GeV before (left) and after (right) pre-processing.

5. Jet tagging using deep neutral networks

The concept of jet-images has enabled the use image classification methods for the identification (*tagging*) of boosted W boson and top quarks at the LHC. In the first case, through the use of *Fisher jets* [14]: a linear classifier inspired by facial recognition algorithms. In the second case, by the use of neural networks [15]. An earlier use of image-based event reconstruction from the OPAL Collaboration at the LEP collider is described in [16]. This paper focuses on the application of modern computer vision algorithms based on deep neural networks for jet-image classification.

Using the pre-processed jet-image representation of jets, two different architectures of deep neural networks for image classification were trained to separate W from QCD jets: Convolutional Neural Networks (CNN), and Fully Connected (FC) MaxOut networks. This represents a significant departure from state-of-the-art jet tagging at LHC in that the full data image is used as input to the networks rather than a set of physics-defined jet substructure variables. In analogy to image recognition problems in computer vision, deep neural networks trained on the full jet-image are expected to learn high level representation of the data and capture new, more discriminant information and features between signal and background.

For a detailed description of the deep network architectures see [17]. The MaxOut architecture consists of two FC layers with MaxOut activations units followed by two FC layers with rectified linear units (ReLU), followed by a FC sigmoid layer for classification. The number of units per layer are 256, 128, 64, and 25. The CNN network consists of three convolution layers followed by two fully connected layers. Each convolution layer utilize 64 filters of sizes 11×11, 3×3, and 3×3 respectively. The large size of the filters in the first layer is motivated by the typical scale of the most prominent features in W jets, and the fact that images are very sparse, with an average occupancy of about 5%. Each convolutional layer consists of three sequential units: convolution, max-pool, and dropout, with ReLU activation functions. The max-pool down sampling in each convolution layer is (2,2), (3,3), and (3,3). The FC hidden layer consists of 64 units. Sigmoid units are used in the final classification layer. Figure 4 shows a diagram illustrating the CNN architecture.

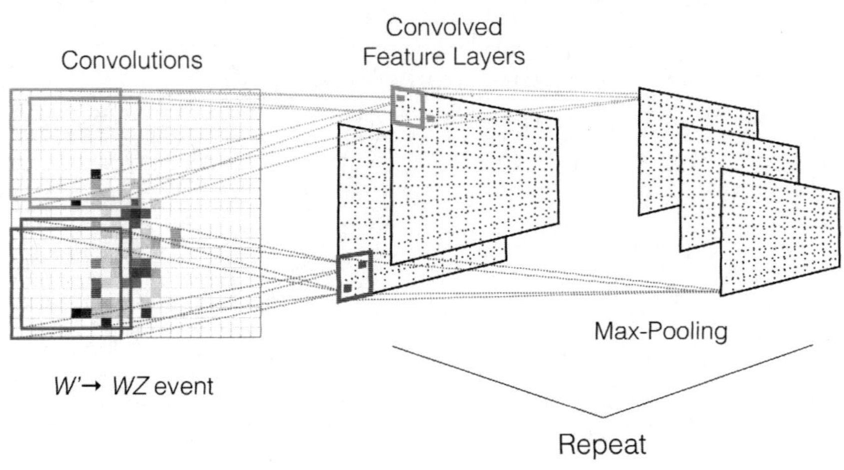

Figure 4. Convolutional neural network architecture applied to a jet-image.

The convolution layer can be intuitively understood as follows. In the first convolution layer, filters of 11×11 pixels are scanned across the jet-image in a sliding window approach. At each point in the jet-image, the convolution between the filter and the patch of the image behind it is computed. This process results in a new image for each filter. The filter values are learned during the CNN training. The following max pool layer then considers patches of size 2×2 in a sliding window approach and only the maximum value within the 2×2 patch is used to form the max-pool resulting image. In the next convolutional layer, new filters are run across all previous convolution images and a similar max-polling is performed. An unique aspect of CNN architectures is that the first convolution layer only have access to local information of size given by the filter size. In the subsequent layers, when the previous convolutions are combined, units have access to more global information, until reaching the final FC layer where units combine information of the whole image. This architecture enables a hierarchical representation of the data where the first layers focus on local features for discrimination, for example individual subjets or local radiation patterns, whereas deeper layers learn higher level features such as distances between subjets or more global structure features within jets.

6. W boson jet tagging perfomance

The performance of the deep neural networks to identify W boson and reject QCD jets was evaluated for jets with 250 GeV $< p_T <$ 300 GeV and within a 65 GeV $< m <$ GeV mass window containing the peak of the W boson mass. The primary figure of merit to compare the performance of different jet tagging techniques is the ROC curve, which measures the rejection of background QCD jets as a function of the efficiency to correctly classify signal W jets for different thresholds applied to the output of the classifier. Background rejection is defined as the inverse of the efficiency to incorrectly classify background jets as signal.

Figure 5 shows ROC curves comparing the performance of several single jet substructure variables (mass, n-subjettiness τ_{21}, and the distance ΔR between the two leading p_T subjets) and the two neural networks architectures. An additional *normalized* convolutional deep neural network (Convnet-norm) was also considered. In this setting, the pixel intensities were normalized by the total image intensity before training. Figure 5 shows that the performance of the three DNN classifiers significantly outperform all individual single jet substructure variables. We also observe that the MaxOut architecture performs slightly better than the convolutional networks. The curved denoted *Fisher* corresponds to the Fisher jet-image discriminant described in reference [14].

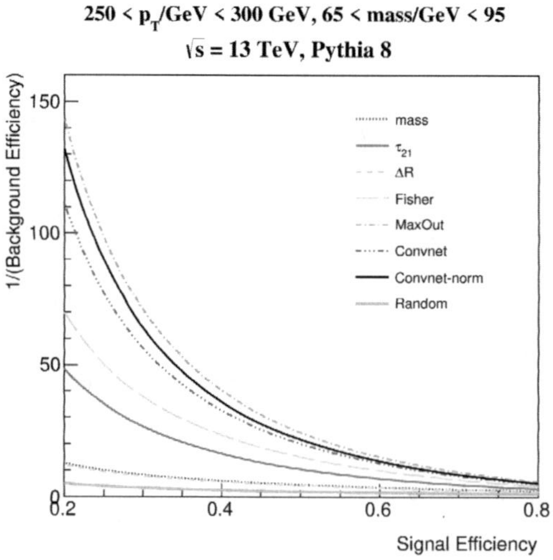

Figure 5. ROC curves for state-of-the-art jet substructure variables and computer vision deep neural network discriminants.

Figure 6(a) compares the performance of the DNN classifiers with two-variable combinations of jet substructure variables. While two-variable combinations achieve better discrimination power than individual variables, DNN classifiers still perform significantly better, indicating that the deep networks are learning new physics information not captured in the jet substructure variables considered.

In order to understand if the DNN have learned the jet substructure variables, Figure 6(b) shows ROC curves for combinations of the CNN classifier and each of the individual jet substructure variables. It can be seen than adding τ_{21} and ΔR to the CNN classifier does not improve the ROC curve, indicating that the network has fully learned these two variables. However, there is clear improvement in classification performance when the CNN is combined with the jet mass. This suggests that, while the CNN is able to learn new physics features to significantly outperform the combination of mass and τ_{21}, it does not fully learn the jet mass.

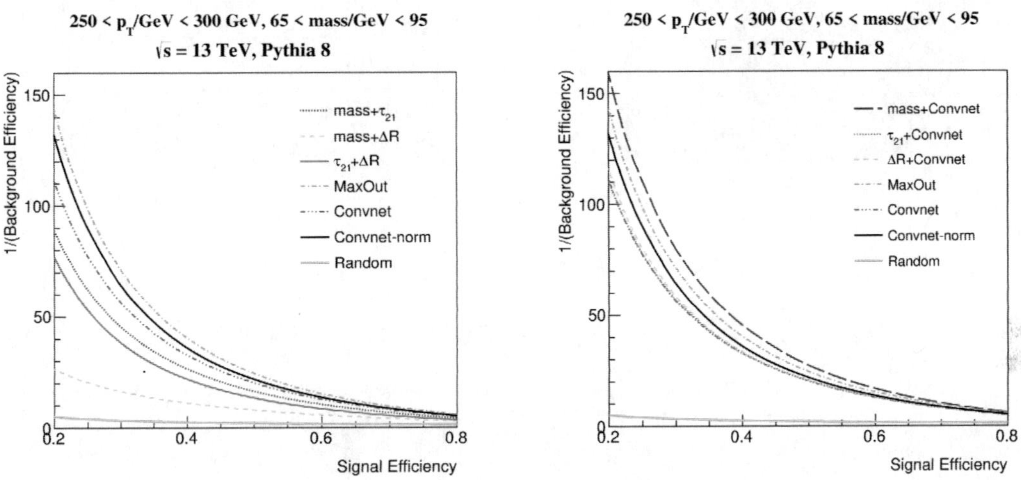

(a) (b)

Figure 6. (a) ROC curves for two-variable combination of jet substructure features and computer vision deep neural network discriminants. (b) ROC curves for combinations of CNN output with individual jet substructure features.

7. Visualization of learned features

Image-based approaches to LHC data analysis can provide unique new ways to visualize what physics information is learned by the different classifiers. Here only a few examples are given. For a more detailed discussion see [17]

Figure 7 shows the conditional distribution of n-subjettiness τ_{21} and jet mass given the CNN output. The strong non-linear relationship between CNN and τ_{21}, indicates that the CNN has learned the information contained in n-subjettiness. On the other hand, the correlation between jet mass and CNN is significantly weaker, in agreement with the previous observation that jet mass is not fully learned by the deep neural networks architectures considered.

A very powerful way to visualize the physics information learned in deep representations is the *deep correlation jet-image*, constructed as the Pearson correlation coefficient of the pixel intensities with the deep neural network output, as shown in Figure 8. This image shows how discriminating information is contained within the network. In particular, one finds that the the localization and size of the second leading p_T subjet (located at the bottom of the image) is highly correlated with high CNN activations corresponding to signal W jets. This is consistent with our expectation for soft QCD gluon emission compared with hard emissions for the case of W jets. Broad blue regions around and outside the leading two subjets are indicative of gluon radiation expected in QCD jets.

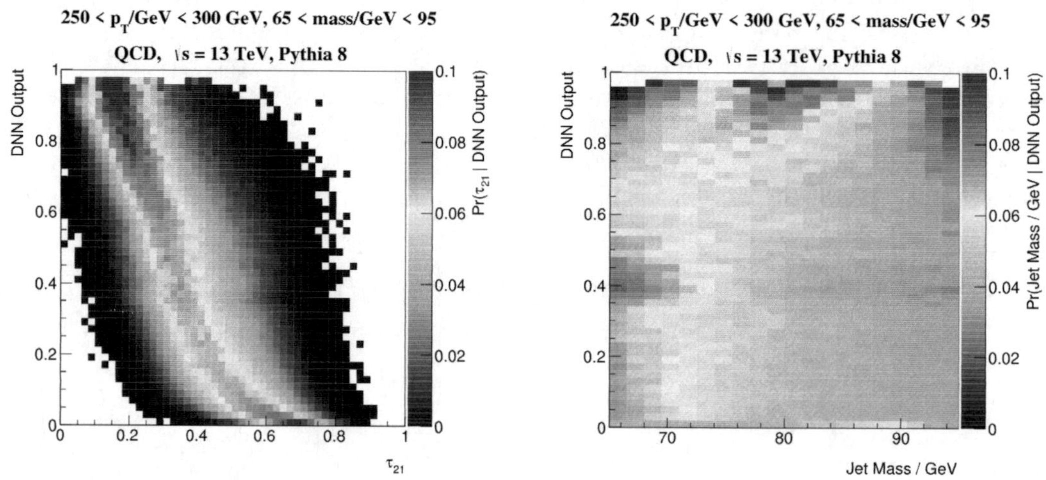

(a) (b)

Figure 7. CNN output as a function of τ_{21} (a) and mass (b).

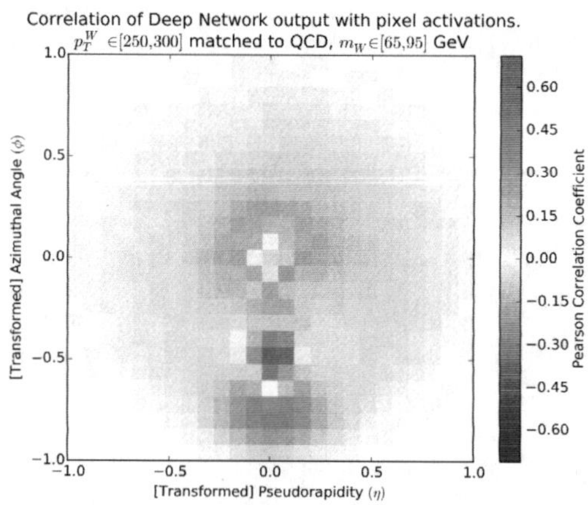

Figure 8. Deep correlation jet-image computed as the linear correlation between the CNN output and the pixel intensity. Red regions are signal-like, whereas blue regions are background-like

In order to understand what new physics information DNN are learning beyond jet mass and τ_{21}, we considered a highly *restricted phase space* so that τ_{21} and mass does not vary significantly over this reduced phase space. Figure 9 shows the ROC curves for the deep neural networks and the individual jet substructure variables for a subset of jets with very narrow values of p_T, mass, and τ_{21}. The small range of τ_{21} between 0.19 and 0.21 selects both signal and background jets that have a very similar (and significant) 2-prong structure. In other words, this restrictive selection ensures that signal and background jets look almost identical in terms of jet mass and n-subjettiness. As expected, jet mass and τ_{21} provide almost no discrimination power in Figure 9, whereas both CNN and MaxOut networks are still able to distinguish between signal and background within this restrictive phase space, confirming that they are learning new information beyond mass and τ_{21}.

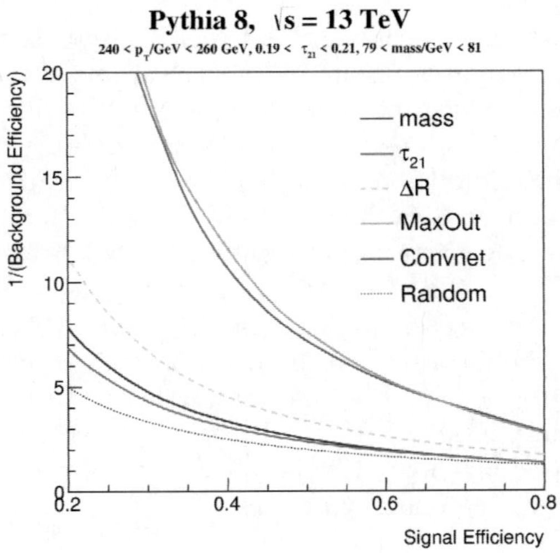

Figure 9. ROC curves in a highly restrictive phase space of 79 GeV $< m <$ 81 GeV and 0.19 $< \tau_{21} <$ 0.21.

Figure 10 shows the deep correlation jet-images obtained in three small windows of τ_{21}. These images provide spacial information indicative of where in the image the networks are finding discriminating features beyond mass and n-subjettiness. A prominent feature in these images is the red region in-between the two leading p_T subjets in the cases of small and moderate τ_{21} values, related to the color flow pattern differences between W and QCD jets. Large values of τ_{21}, corresponding to jets that exhibit a strong one-prong structure, also show a particular radiation pattern not exploited in the two state-of-the-art jet substructure variables considered. These images can suggest further development of novel jet substructure variables that are able to capture these differences between signal and background.

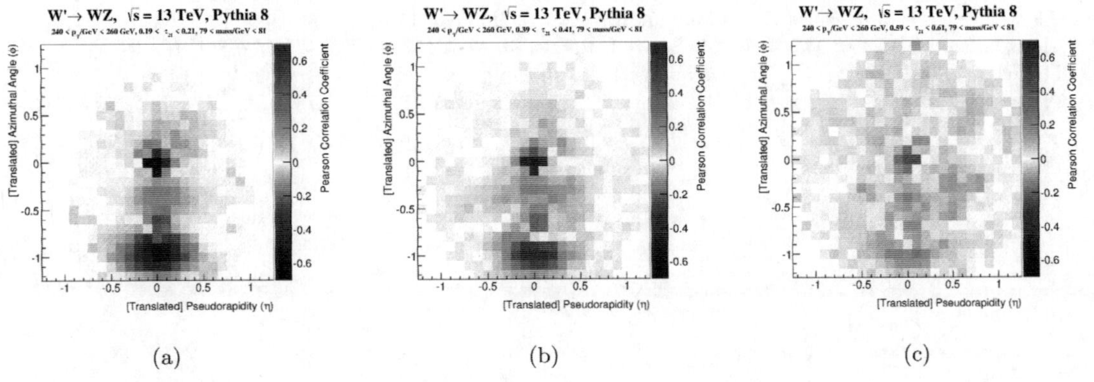

Figure 10. Deep correlation jet-image in a highly restrictive phase space of 79 GeV $< m <$ 81 GeV and 0.19 $< \tau_{21} <$ 0.21 (a), 0.39 $< \tau_{21} <$ 0.41 (b), and 0.59 $< \tau_{21} <$ 0.61 (c).

8. Conclusions

This paper presents a new paradigm to study jets at the LHC. Representing jets as images enables the application of advanced computer vision and deep learning methods for jet classification, adding a new capability to understand LHC events and to design new, more powerful, classification methods. Jet-image deep neural networks classifiers significantly outperform state-of-the-art W boson jet classification methods as they are able to learn new discriminant features not captured in the physics motivated variables. Studies suggest that this additional information is related to the different color-flow patterns between W and QCD jets. While the DNN classifiers studied in this paper are very powerful, it was shown that they do not exploit all the information available for discrimination. Jet mass, in particular, is not fully learned by any of the DNN architectures considered. In addition to improved classification performance, DNN jet-image analysis provides new ways to visualize the information learned by the DNNs and to connect these visualizations with the physics of jets. The ideas presented in this paper have broad applicability and could be used for many different analyses where events can be represented as images both within and beyond high energy physics.

9. Acknowledgements

This work was supported by the Stanford Data Science Initiative and by the US Department of Energy (DOE) grant DE-AC02-76SF00515.

10. References

[1] Cacciari M, Salam G P, and Soyez G 2008, *JHEP* **0804** 063
[2] ATLAS Collaboration 2016, *JINST* **11** P04008
[3] CMS Collaboration 2016, *J. Instrum.* **11** P01019
[4] CMS Collaboration 2015, *J. Instrum.* **10** P08010
[5] ATLAS Collaboration 2015, *ATL-PHYS-PUB-2015-022*
[6] Russakovsky O, et. al. 2015, *arXiv:1409.0575* [cs.CV]
[7] LeCun Y, Boser B, Denker J S, Henderson D, Howard R E, Hubbard W, et al. 1989, *Neural Comput.* **1** 541-551
[8] Salam G P 2010, *Eur. Phys. J. C.* **67** 637-686
[9] Thaler J, Tilburg K V 2011, *JHEP* **1103** 015
[10] Abdesselam A, et. al. 2011, *Eur. Phys. J. C.* **71** 1661
[11] Altheimer A, et. al. 2012, *J. Phys. G.* **39** 063001
[12] Altheimer A, et. al. 2014, *Eur. Phys. J. C.* **74** 2792
[13] Sjostrand T, Mrenna S, and Skands P Z 2008 *Comput. Phys. Commun.* **178** 852-867
[14] Cogan J, Kagan M, Strauss E, Schwartzman A, 2015 *JHEP* **02** 118
[15] Almeida L G, Backovic M, Cliche M, Lee S J, Perelstein M 2015, *arXiv:1501.05968* [hep-ph]
[16] Thomson M A 1996, *Nuclear Instruments and Methods in Physics Research A* **382** 553-560
[17] Oliveira L , Kagan M, Mackey L, Nachman B, Schwartzman A, 2015, *arXiv1511.05190* [hep-ph]

Journal of Physics: Conference Series **762** (2016) 012036 doi:10.1088/1742-6596/762/1/012036

Reweighting with Boosted Decision Trees

Alex Rogozhnikov[1,2]

[1] National Research University Higher School of Economics (HSE), RU
[2] Yandex School of Data Analysis (YSDA), RU

E-mail: `axelr@yandex-team.ru`

Abstract. Machine learning tools are commonly used in modern high energy physics (HEP) experiments. Different models, such as boosted decision trees (BDT) and artificial neural networks (ANN), are widely used in analyses and even in the software triggers [1].

In most cases, these are classification models used to select the "signal" events from data. Monte Carlo simulated events typically take part in training of these models. While the results of the simulation are expected to be close to real data, in practical cases there is notable disagreement between simulated and observed data. In order to use available simulation in training, corrections must be introduced to generated data. One common approach is reweighting — assigning weights to the simulated events.

We present a novel method of event reweighting based on boosted decision trees. The problem of checking the quality of reweighting step in analyses is also discussed.

1. Introduction

Reweighting distributions (also known as event reweighting or importance weighting) is a general procedure, but its major use-case for particle physics is to modify the output of the Monte Carlo (MC) simulation to reduce disagreement with real data (RD) collected at a collider.

There are many applications in HEP including searching for rare decays (decays with an extremely low probability in the standard model of elementary particles), when a classifier is trained on MC data to discriminate signal decays from background. However, the simulation is often imperfect (see [2] for more details) and corrections should be introduced. To calibrate the reweighting a similar physics process is considered, for which both real data and simulation can be obtained. For instance, in rare decays a normalization channel is selected — a decay with the similar kinematic characteristics (see [3] for an example)[1].

Reweighting techniques have applications outside HEP: i.e. in sociology a survey reweighting is used to reduce a non-response bias [5]. In what follows HEP terminology is used, but approaches discussed in the paper are applicable to any reweighting.

Mathematically, the problem is equivalent to estimating the density ratio $f_{\mathrm{RD}}(x)/f_{\mathrm{MC}}(x)$ as a function of the variables participating in reweighting. A density ratio estimation is a general problem in machine learning (ML) with numerous applications (see [6]).

[1] Data-MC inconsistencies can also be taken into account by the calibration of subdetector response, Monte Carlo generator tunes [4] or discarding regions, where disagreement is the worst.

2. Basic approach to event reweighting

An approach widely used in High Energy Physics is reweighting with bins. The space of variables is split into bins, in each bin the weights of the simulated events are multiplied by

$$\text{multiplier}_{\text{bin}} = \frac{w_{\text{bin, RD}}}{w_{\text{bin, MC}}}$$

to compensate the difference ($w_{\text{bin, RD}}$ and $w_{\text{bin, MC}}$ — total weight of events in a bin for RD and MC distributions). In other words, both densities $f_{\text{RD}}(x)$ and $f_{\text{MC}}(x)$ are estimated using histograms and then divided (this gives another name of this approach — "histogram division").

Reweighting using bins is intuitive and easy-to-use, however, has very strong limitations:

- very few variables can be reweighted in practice, typically one or two;
- choosing which variable(s) to use in reweighting is complex: reweighting one variable often brings disagreement in others;
- the amount of data needed to reliably estimate a density function with a histogram grows exponentially with the number of variables, which is commonly referred to as the "curse of dimensionality".

To fight the last problem, one can reduce the number of bins along each variable, but this drives to a rough reweighting rule, insufficient to cover discrepancies.

3. Reusing classification ML techniques to reweight distribution

Density estimation is a complex problem and it should be avoided in cases when only the ratio is of interest. A general and natural method of density ratio estimation is based on reusing general-purpose ML techniques. In [2] this method was proposed and was successfully applied to particle physics problems.

Some general-purpose classification techniques (i.e. BDT and ANN) trained to discriminate MC and RD can provide probabilities $p_{\text{MC}}(x)$, $p_{\text{RD}}(x)$ that a given event x belongs to MC or RD. The probabilities can be used to estimate required density ratio:

$$\frac{f_{\text{RD}}(x)}{f_{\text{MC}}(x)} \sim \frac{p_{\text{RD}}(x)}{p_{\text{MC}}(x)}.$$

This approach successfully overcomes the curse of dimensionality, but provides inaccurate predictions when density ratio is high. One possible explanation is as follows: while regions with the high ratio are significant for reweighting, those are not of high importance for classification task: guessing correct class within regions with high / low ratio is easier, since most of the events belong to one class, and classification algorithms focus on the other regions. For example, when training ANN or GBDT, these regions provide smaller contribution to the loss function, thus are given less attention.

4. BDT reweighter

In this section a machine learning algorithm is proposed to solve the specific problem of reweighting. To address the problems of the histogram reweighting approach, the space of variables is split into a few large regions. These regions are not obtained by a simple splitting of each variable into several bins, but in correspondence with the problem.

A decision tree is used to split the regions. Recall that decision trees split the space of variables into the regions by checking simple conditions. Each region is associated with some leaf of the tree.

To find the regions that are suitable for reweighting, the symmetrized χ^2 is greedily optimized[2]:

$$\chi^2 = \sum_{\text{leaf}} \frac{(w_{\text{leaf, MC}} - w_{\text{leaf, RD}})^2}{w_{\text{leaf, MC}} + w_{\text{leaf, RD}}}. \tag{1}$$

This metric is maximized to find the regions important for reweighting. If in some leaf (region) the amount of MC events $w_{\text{leaf, MC}}$ is much higher than the amount $w_{\text{leaf, RD}}$ of RD events, the MC weights in this region must be decreased. The corresponding summand in the χ^2 will be high reflecting the importance of this region for reweighting.

The BDT reweighter makes use of many such trees which are trained one-by-one by repeating the following steps many times:

(i) build a shallow tree to maximize the symmetrized χ^2

(ii) compute predictions in the leaves: $\text{leaf\_pred} = \log \frac{w_{\text{leaf, MC}}}{w_{\text{leaf, RD}}}$

(iii) reweight the MC distribution (compare this step with AdaBoost [9]):

$$w = \begin{cases} w, & \text{if event from RD distribution} \\ w \times e^{\text{pred}}, & \text{if event from MC distribution} \end{cases} \tag{2}$$

For each event pred is equal to the prediction of a leaf containing this event.

The last two steps work in the same way as reweighting with bins, the distinction being that the bins are selected differently. Also, since logarithm is taken, the predictions of the different trees are summed up, as is usually done in the boosting.

In the BDT reweighter, each tree in the sequence is trying to cover the discrepancies that were not resolved on the previous iterations. The complexity of a decision tree can be adjusted by varying the depth and the minimal number of samples in the leaf, making this approach highly tunable.

5. Comparison of multidimensional distributions

The goal of reweighting is to have the MC distribution coincide with the RD distribution.

Comparing one-dimensional distributions is simple and can be done either by looking at the distributions or by computing one of well-known two-sample tests like Kolmogorov-Smirnov, Anderson-Darling or Cramer–von Mises. However, in the applications all the distributions are multidimensional. Comparing only projections is obviously not enough to be sure that the distributions are identical.

At the same time, there are no useful multidimensional two-sample tests. Given the whole pipeline of our analysis, two-samples tests are not necessary, because the question of interest is not whether the MC and RD distributions are different (those are different). The question is whether an ML technique used later in the analysis (i.e. to detect signal decay, typically it is BDT or ANN) is able to use the discrepancy between RD and MC. Thus it is only needed to check that after reweighting a classifier used in the analysis is not able to find the difference between the distributions. For this purpose a classifier is trained to discriminate RD and MC.[3] Its quality is checked by inspecting the ROC curves on a holdout sample.[4]

[2] For comparison: in the gradient boosting algorithm decision trees are built by greedily minimizing mean squared error.

[3] This also gives another perspective of reweighting as an adversarial process, involving discriminator and reweighter. Training of adversarial neural networks is actively studied now, see [8].

[4] See also [7] for an alternative ML-based approach to compare multidimensional distributions.

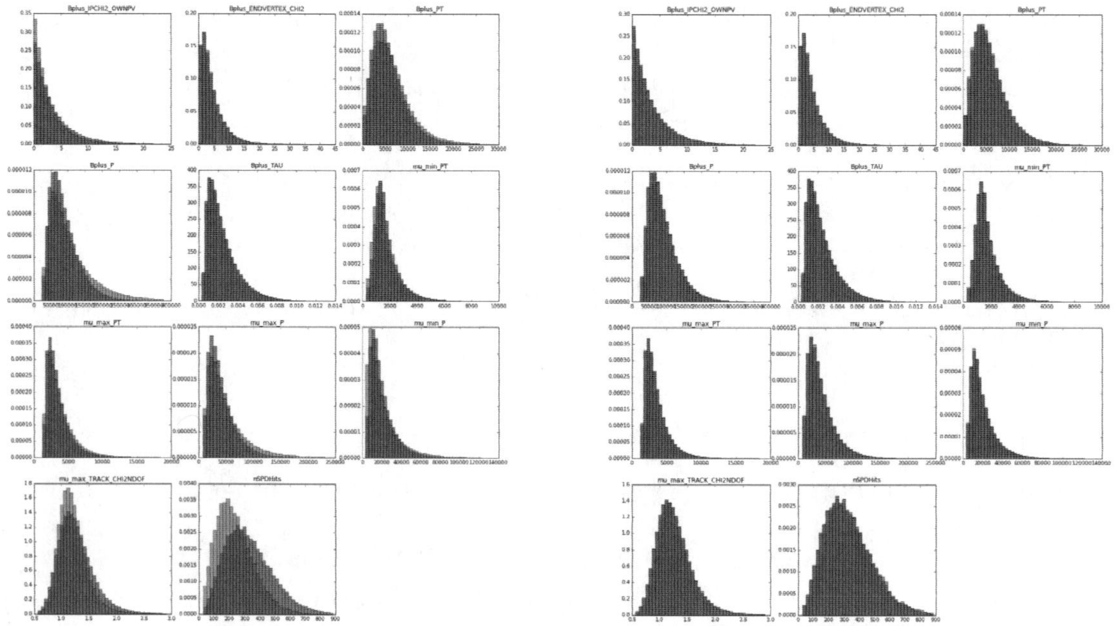

Figure 1. Comparison of real data (blue) and simulated (green) distributions before and after using the BDT reweighter.

Overfitting is an issue that becomes obvious when using advanced methods of reweighting. One should measure the quality of the classification on a holdout (data sample that was not participating in training) to get unbiased estimations. The same approach works for reweighting: the quality of reweighting should be checked on the data that did not participate in training of the reweighting rule. The different cross-validation techniques like folding are also applicable.

6. Case study

As an example the 11-dimensional distribution is taken (simulated and real data). Figure 1 demonstrates how the distribution of different features (variables) has changed after reweighting with the new method.

In table 2 the Kolmogorov-Smirnov distances are provided. Reweighting with bins is done for last two variables, while two other approaches use all 11 variables. Finally, the quality of reweighting is checked as it was proposed earlier and the ROC curves are built on a holdout (figure 3).

7. Conclusion

Two problems are discussed in the paper:

(i) event reweighting for multidimensional distributions

(ii) the comparison of multidimensional distributions

It is demonstrated that both problems are effectively addressed by means of machine learning, while typically these steps in the analysis are considered outside of the scope of ML.

Also, the novel method of reweighting is proposed: a modification of BDT algorithm, which alters the procedures of boosting and decision tree building. This method outperforms known reweighting approaches and makes it possible to reweight dozen of variables. When compared on the same problems, it requires less data to achieve the same quality.

Ready-to-use implementation of introduced algorithm is available in the `hep_ml` package [10].

Feature name	original	bins reweight	reuse ML	BDT reweighter
Bplus_IPCHI2_OWNPV	0.0796	0.0642	0.0463	0.0028
Bplus_ENDVERTEX_CHI2	0.0094	0.0175	0.0490	0.0021
Bplus_PT	0.0586	0.0679	0.0126	0.0053
Bplus_P	0.1093	0.1126	0.0044	0.0047
Bplus_TAU	0.0037	0.0060	0.0324	0.0044
mu_min_PT	0.0623	0.0604	0.0017	0.0036
mu_max_PT	0.0483	0.0561	0.0053	0.0035
mu_max_P	0.0906	0.0941	0.0084	0.0036
mu_min_P	0.0845	0.0858	0.0058	0.0043
mu_max_TRACK_CHI2NDOF	0.0956	0.0042	0.0128	0.0043
nSPDHits	0.2478	0.0098	0.0180	0.0075

Figure 2. Kolmogorov-Smirnov distances for each variable before reweighting and after applying different reweighting techniques. Only the last two variables are used during reweighting with bins.

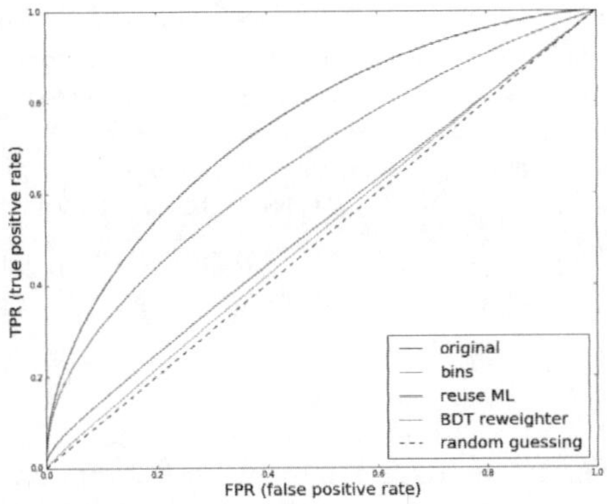

Figure 3. Checking the quality of reweighting with ML. The ROC curves for the classifier trained to discriminate RD and MC are computed on a holdout. Reweighting with bins significantly reduces initially high discrepancy, but the classifier still can easily find the difference. ML-based solutions provide significantly better results, though reusing ML approach has minor issues in the left bottom corner.

References

[1] Likhomanenko, T., Ilten, P., Khairullin, E., Rogozhnikov, A., Ustyuzhanin, A., Williams, M. (2015). LHCb Topological Trigger Reoptimization. In Journal of Physics: Conference Series (Vol. 664, No. 8, p. 082025). IOP Publishing.

[2] Martschei, D., Feindt, M., Honc, S., Wagner-Kuhr, J. (2012). Advanced event reweighting using multivariate analysis. In Journal of Physics: Conference Series (Vol. 368, No. 1, p. 012028). IOP Publishing.

[3] Blake, T., Bettler, M.-O., Chrzaszcz, M., Dettori, F., Ustyuzhanin, A., and Likhomanenko, T. (2015). Flavours of physics: the machine learning challenge or the search of $\tau \to 3\mu$ decays at LHCb

[4] ATLAS collaboration. (2012). Summary of ATLAS Pythia 8 tunes (Vol. 14). ATL-PHYS-PUB-2012-003.

[5] Kizilcec, R. Reducing non-response bias with survey reweighting: Applications for online learning researchers. Proceedings of the first ACM conference on Learning @ scale conference. ACM, 2014.

[6] Sugiyama, M., Suzuki, T., Kanamori, T. Density Ratio Estimation in Machine Learning, Cambridge University Press, Cambridge, UK, 2012.

[7] Friedman, J. (2003). On Multivariate Goodness-of-Fit and Two-Sample Testing. In Statistical Problems in Particle Physics, Astrophysics, and Cosmology (Vol. 1, p. 311).

[8] Goodfellow, I., Pouget-Abadie, J., Mirza, M., Xu, B., Warde-Farley, D., Ozair, S., Courville, A., Bengio, Y. (2014). Generative adversarial nets. Advances in Neural Information Processing Systems (pp. 2672-2680).

[9] Freund, Y., Schapire, R. E. (1996). Experiments with a new boosting algorithm. In ICML (Vol. 96, pp. 148-156).

[10] `https://github.com/arogozhnikov/hep_ml`

Constrained fits with non-Gaussian distributions

R. Frühwirth[1] **and O. Cencic**[2]
[1] Institute of High Energy Physics, Austrian Academy of Sciences, Vienna, Austria
[2] Institute of Water Quality, Resources and Waste Management, TU Wien, Vienna, Austria
E-mail: `rudolf.fruehwirth@oeaw.ac.at`

Abstract. Non-normally distributed data are ubiquitous in many areas of science, including high-energy physics. We present a general formalism for constrained fits, also called data reconciliation, with data that are not normally distributed. It is based on Bayesian reasoning and implemented via MCMC sampling. We show how systems of both linear and non-linear constraints can be efficiently treated. We also show how the fit can be made robust against outlying observations. The method is demonstrated on a couple of examples ranging from material flow analysis to the combination of non-normal measurements. Finally, we discuss possible applications in the field of event reconstruction, such as vertex fitting and kinematic fitting with non-normal track errors.

1. Introduction

Observed data are often in conflict with known conservation laws such as mass, momentum or energy balance. Also, observed data are often not distributed according to a Gaussian or normal distribution. In data reconciliation (DR) contradictions are resolved by statistically adjusting the data, assuming their uncertainty is described by a probability density function, not necessarily a normal one. The method outlined here is based on the following assumptions:

- There are N measured or unmeasured variables that take values in a subset $D \subseteq \mathbb{R}^N$.
- The $I \leq N$ measured variables form an I-dimensional random variable with known joint density. The latter is called the prior density.
- The prior density can be either objective, i.e. the model of a measurement process, or subjective, i.e. the formalization of an expert opinion.
- The variables are subject to linear or nonlinear constraints that define a manifold $S \subset \mathbb{R}^N$ of dimension $P < N$.
- The posterior density is obtained by restricting the prior to S and normalizing it to 1.

We illustrate the principle by an example with a simple linear constraint. Assume there are three non-normal observations x_1, x_2 and x_3 with the prior density $f(x_1, x_2, x_3)$. The constraint equation $x_3 = x_1 + x_2$ defines a plane in $\mathbb{R}^3$. The prior density is restricted to this plane and normalized to 1. This gives the posterior density $g(x_1, x_2, x_3)$ and its marginals $g_i(x_i)$, $i = 1, 2, 3$, shown in Fig. 1.

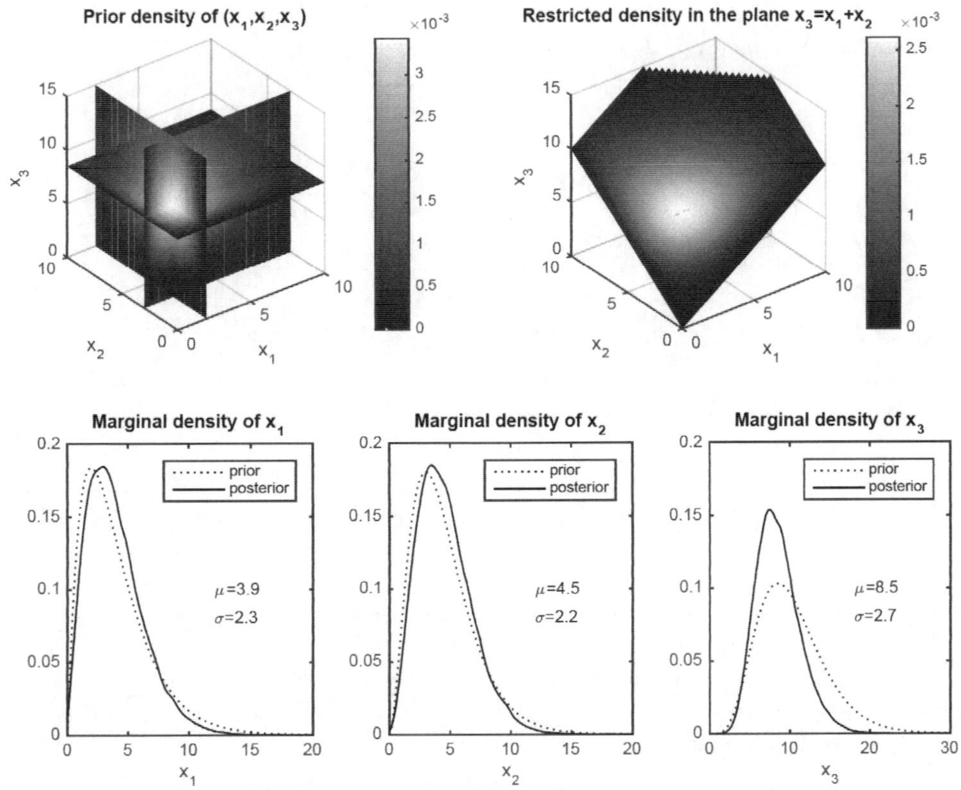

Figure 1. Top left: prior density $f(x_1, x_2, x_3)$. Top right: prior density $f(x_1, x_2, x_3)$ restricted to the plane $x_3 = x_1 + x_2$. Bottom: measured (prior, dotted) and reconciled (posterior, continuous) densities of x_1, x_2, x_3.

2. General linear constraints

2.1. The independence sampler

In the general case the linear constraints are solved explicitely for dependent variables $\boldsymbol{u}$ as a function of free variables $\boldsymbol{w}$:

$$\boldsymbol{u} = \boldsymbol{D}\boldsymbol{w} + \boldsymbol{d}$$

For a joint prior $f(\boldsymbol{u}; \boldsymbol{w})$ the joint posterior of $\boldsymbol{w}$ is given by:

$$\pi(\boldsymbol{w}) \propto f(\boldsymbol{D}\boldsymbol{w} + \boldsymbol{d}; \boldsymbol{w})$$

The explicit calculation of the normalizing constant can be avoided by drawing a random sample from the posterior density $\pi(\boldsymbol{w})$ by Markov Chain Monte Carlo (MCMC). The sample is not independent, as each draw depends on the previous one. The corresponding sample of the dependent variables $\boldsymbol{u}$ is generated via the functional dependence $\boldsymbol{u} = \boldsymbol{D}\boldsymbol{w} + \boldsymbol{d}$. From this sample the joint posterior density of all variables can be estimated, along with the correlations, marginal densities, moments and quantiles. See [1] for an introduction into MCMC, including a comprehensive list of references.

The sampler best suited to the problem of DR is the Independence Sampler (IS). There is a natural proposal density, namely the prior density $p(\boldsymbol{w})$ of $\boldsymbol{w}$. The sampler can be summarized as follows.

Independence sampler

1. Set $i = 1$, choose the sample size L and the starting value $\boldsymbol{w}_1$.
2. Draw a proposal value $\hat{\boldsymbol{w}}$ from the proposal density $p(\boldsymbol{w})$.
3. Compute the acceptance probability α:
$$\alpha(\boldsymbol{w}_i, \hat{\boldsymbol{w}}) = \min\left(1, \frac{\pi(\hat{\boldsymbol{w}})\, p(\boldsymbol{w}_i)}{\pi(\boldsymbol{w}_i)\, p(\hat{\boldsymbol{w}})}\right)$$
4. Draw a uniform random number $u \in [0, 1]$.
5. If $u \leq \alpha$, accept the proposal and set $\boldsymbol{w}_{i+1} = \hat{\boldsymbol{w}}$, otherwise set $\boldsymbol{w}_{i+1} = \boldsymbol{w}_i$.
6. Increase i by 1. If $i < L$, go to 2., otherwise go to 7.
7. Stop sampling.

For more details about the general linear case and additional references, see [2].

2.2. Combination of non-normal observations

The following example shows how to combine non-normal observations by MCMC. Assume that there are three independent measurements x_1, x_2, x_3 of a small nonnegative quantity, for instance a small cross section x. We want to combine the measurements by imposing the constraints $x_1 = x_2 = x_3$. The measurements are described by three experimental densities $f_1(x_1), f_2(x_2), f_3(x_3)$ the support of which is restricted to the positive x-axis. The densities need not be given in closed form, but it must be possible to compute the densities at arbitrary points and to draw random numbers from at least one of them. The procedure is illustrated with the following example priors:

$x_1 \sim$ Ex(1.1) .... Exponential distribution with mean 1.1
$x_2 \sim$ Ga(2,0.5) .... Gamma distribution with mean 1
$x_3 \sim$ TrNorm(0,1.2,0,∞) ... Half normal distribution with $\mu = 0$, $\sigma = 1.2$ and mean 0.96

The posterior marginals under the assumption of independent measurements are shown in Fig. 2. Computing a χ^2-statistic does not make sense with non-normal obervations, as its distribution is difficult to determine. Instead, a measure of goodness of fit can be obtained by computing the discrepancy between the prior densities and the posterior marginals. Possible measures are, among others, the Kolmogorov-Smirnov distance d_{KS} and the Hellinger distance $d_{\mathrm{H}} = \sqrt{1 - \mathrm{BC}}$, where BC is the Bhattacharya coefficient. A small acceptance rate of the sampler also indicates a poor fit. Table 1 shows the discrepancies and some key figures of the posterior marginals.

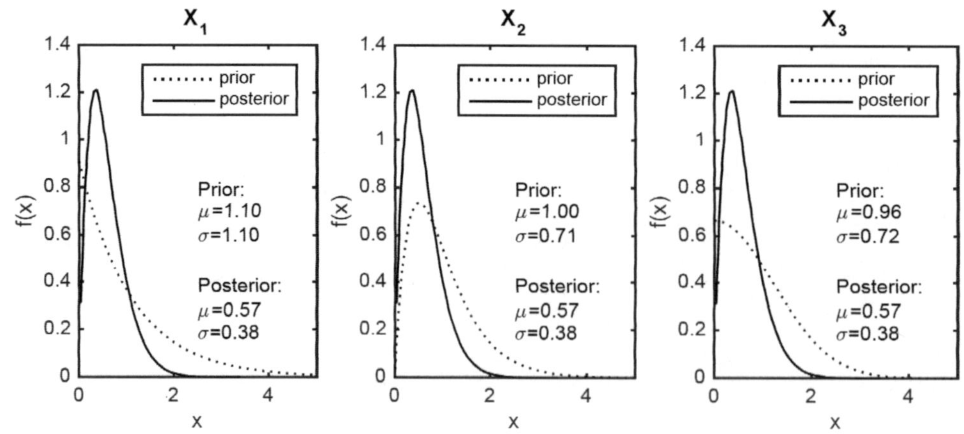

Figure 2. Priors and posterior marginals under the assumption of independent measurements.

Table 1. Posterior discrepancies (left) and key figures of the posterior marginals (right) under the assumption of independent measurements.

	X_1	X_2	X_3
d_{KS}	0.272	0.295	0.280
d_{H}	0.316	0.285	0.287

Mean	Stdev	Median	95%-Quantile
0.572	0.379	0.491	1.308

2.3. Correlated observations
Correlations can be introduced via a copula, making it possible to study the influence of correlations between the measurements on the resulting posterior. The posterior marginals with a Gaussian copula and a global correlation $\rho = 0.25$ are shown in Fig. 3. Table 2 shows the discrepancies and some key figures of the posterior marginals. By varying ρ the influence of possible correlations between the measurements can be systematically studied.

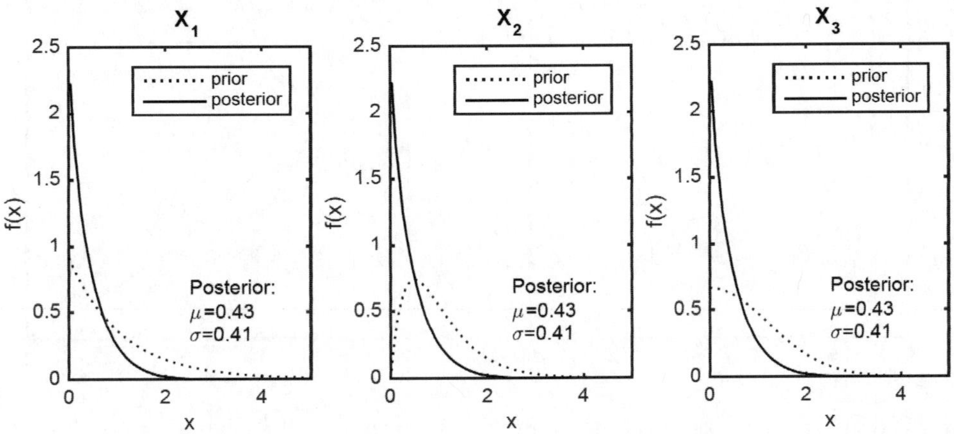

Figure 3. Priors and posterior marginals under the assumption of correlated measurements with all correlations equal to $\rho = 0.25$.

Table 2. Posterior discrepancies (left) and key figures of the posterior marginals (right) under the assumption of correlated measurements with all correlations equal to $\rho = 0.25$.

	X_1	X_2	X_3
d_{KS}	0.272	0.295	0.280
d_{H}	0.325	0.413	0.362

Mean	Stdev	Median	95%-Quantile
0.432	0.415	0.310	1.276

It can be seen that in this example the correlations have a significant influence on the shape and the key characteristics of the posterior.

3. Robust fitting
If there are outlying observations, the acceptance rate of the sampler gets smaller and can even be zero. In this case the fit has to be robustified. The well-known M-estimator [3] can be generalized to non-normal observations. To this end, the residuals

$$r_i = (E_{pr}[x_i] - E_{po}[x_i])/\sigma_{pr}[x_i]$$

are computed for all variables x_i, where $E_{pr}[x_i]$ is the prior expectation, $\sigma_{pr}[x_i]$ the prior standard deviation, and $E_{po}[x_i]$ the posterior expectation of x_i. From the residuals weights are determined according to:

$$w(r_i) = \frac{\varphi(r_i)}{\varphi(r_i) + \varphi(c)} \cdot \frac{\varphi(0) + \varphi(c)}{\varphi(0)}$$

where φ is a symmetric pdf and c a cut value, resulting in a redescending M-estimator [4]. The density φ should not decay too fast in the tails, in order to avoid the weights becoming numerically equal to 0. The weight as a function of the residual is shown in Fig. 4. In this particular case the cut value is $c = 2.5$, and φ is the density of the Student distribution with four degrees of freedom. Once the weights are available, the priors are stretched around their mode by the factor $s_i = 1/\sqrt{w_i}$ and the estimation of the posterior is repeated. This is iterated until convergence. The iteration can be speeded up if the initial weights are computed in the normal approximation [5].

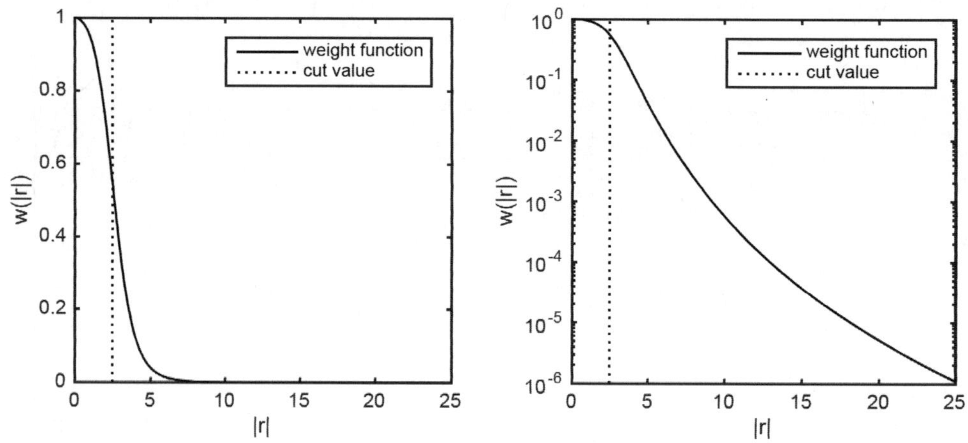

Figure 4. The weight as a function of the absolute residual on a linear (left) and a log (right) scale.

4. Applications in event reconstruction

Track errors are sometimes non-normal, for instance if tracks are fitted with the Gaussian-sum filter. In this case, vertex fitting and kinematic fitting can be extended to non-normal track errors as outlined above. Whereas momentum conservation gives linear constraints on the momentum vectors, energy conservation and vertex constraints are nonlinear. As usual, nonlinear constraints are linearized at a suitable expansion point. The dependent variables can then be computed via a Newton iteration. The sampler is basically the same as in the linear case. The nonlinear case will be discussed in detail in a forthcoming paper [5].

5. References

[1] L. Tierney, *Markov chains for exploring posterior distributions*, The Annals of Statistics 22 (1994) 1701–1762.
[2] O. Cencic and R. Frühwirth, *A general framework for data reconciliation—Part I: linear constraints*, Comp. Chem. Eng. 75 (2015) 196–208.
[3] P. Huber and E.M. Ronchetti, *Robust Statistics*, 2nd ed., John Wiley & Sons, Hoboken NJ, 2009.
[4] F.R. Hampel *et al.*, *Robust Statistics*, John Wiley & Sons, New York, 1986.
[5] O. Cencic and R. Frühwirth, *Data reconciliation of nonnormal observations—Part II: nonlinear constraints, correlated observations and gross error detection by robust estimation*, in preparation.

The CptnHook Profiler - A tool to investigate usage patterns of mathematical functions.

D. Piparo*, V. Innocente*

*** CERN, CH-1211 Geneva 23, Switzerland**

Abstract. Transcendental mathematical functions are one of the main hot-spots of scientific applications. The usage of highly optimised, general purpose mathematical libraries can mitigate this issue. On the other hand, a more comprehensive solution is represented by the replacement of the generic mathematical functions by specific implementations targeting particular sub-domains only.

CptnHook is a tool that helps achieving this goal allowing to monitor the input values of mathematical functions used in a given application, categorised according to the stacktraces leading to their invocations. In this contribution we describe the design of CptnHook, the data format of its profile and how it is possible to perform measurements without instrumenting the users code and imposing the need of recompilation.

We demonstrate that this approach scales on production workflows of LHC experiments and characterise a set of real life measurements, showing where opportunities for improvement lie and how the tool can be used for advanced debugging. We also illustrate how elegant summaries of the measurements can be produced and how ROOT based analysis of the profiles can be performed.

1. The ANR Metalibm Project

CptnHook has been developed within the framework of Metalibm [1]. This project is financed by the French National Research Agency (ANR) and is supported by a consortium of four partners: INRIA, UPMC, LIRMM and CERN.

Metalibm addresses the implementation of elementary mathematical functions such as exponential or trigonometric, but also other functions which are specific to certain applications or domains and cannot be found in existing present and future libraries. The standard mathematical library (libm [2]) is traditionally a rock-solid reference in scientific computing but offers a small set of elementary functions in a small set of precisions. In addition the performance of its implementations is not always ideal.

The runtime performance of mathematical functions is of critical importance, in particular in scientific and financial computing. In many contexts, these functions have always received special attention. For instance, processor manufacturers have teams of engineers who implement an optimized mathematical library version for each new processor released in order to take maximum advantage of the microarchitecture at disposal: e.g. registers, cache hierarchies, ports. This is a time-consuming and error-prone task.

A first objective of this project is therefore to automate the development of mathematical libraries, up to the point where libms are generated in a fraction of the time required by the manual development procedure mentioned above. The performance and precision must also match the hand-coded ones not only when running on general purpose processors, e.g. for the server market segment, but for a wide range of targets including embedded processors, graphical processing units and even FPGA as well as digital circuits.

It is part of the program of work of Metalibm to provide the means for creating special versions of elementary functions, notably optimised for subdomains smaller than the full ones. A posterchild example is represented by the trigonometric functions for a restricted set of angles.

2. The Cost of Floating Point Operations

Floating point operations represent a considerable fraction of the computing time of scientific applications. High Energy Physics is not an exception in this respect: workflows are not exceptional for which 10% to 20% [3] of the time is spent in floating point operations, e.g. the applications performing maximum likelihood fits in presence of complex models and seizable datasets, detailed simulations of passage of particle through matter or simulation of many body systems such as particles in accelerators [4].

In general, the main portion of this time is accountable to mathematical functions: their cost in terms of runtime must be kept under control and minimised whenever possible if performance needs to be achieved.

3. A possible Solution

One possible way to speed up scientific application consists in the adoption of a highly optimised mathematical library such as VDT [3], AMD Libm [5] or Intel MKL [6]. On the other hand, this approach is less than optimal. A more comprehensive solution would be to identify on a case by case basis, e.g. for all sequences of function calls separately, a mathematical function implemented in terms of polyomials [7] tailored to provide just the right result's accuracy needed within the domain required for the calculation at hand.

Operations on the input values such as range reductions, e.g. in presence of trigonometrical functions, and validity checks are expensive calculations. In addition, guaranteeing accurate results over large intervals provides generality but penalizes performance of mathematical functions given the branching this kind of implementations often foresee.

The knowledge of the possible input values of mathematical functions can be immediately be translated in a performance improvement thanks to the optimisations relative to the aforementioned effects.

Unfortunately it is not trivial to determine the range of the input values of functions while implementing an algorithm. It is even harder to figure this out when dealing with large software systems like the software stacks of the LHC experiments.

A flexible, lightweight and easy to use profiler is needed to measure mathematical functions domains.

3.1. The CptnHook Profiler

Without the need of instrumenting nor recompiling the original code, the CptnHook profiler [1] allows to intercept and record for every call to any single argument mathematical function:

- the input value
- the stacktrace leading to the function call. This is necessary to isolate the contexts in which a particular function is invoked. It is indeed in general not enough in presence of a real life scientific application to record in a flat profile the input values of the mathematical functions.

[1] `https://github.com/dpiparo/cptnHook`

The hooking mechanism is based on the preload of the CptnHook shared library and not on code instrumentation. This preload happens transparently for the user (see 1). This library provides functions whose symbols have an identical signature to the one of the mathematical functions. Internally though, before redirecting to the actual mathematical function for the output value calculation, they store the input value as well as the back-trace which lead to the function call.

These features can be offered thanks to the preloading of a CptnHook component which "filters" all the calls to mathematical functions. Such a procedure implies a cost in terms of runtime overhead. It depends on the application and is quantified by a 2x-5x slowdown. The memory overhead is negligible. The data of the profile is stored in a columnwise compressed binary format for optimal usage of disk storage. The tool can be easily generalised: it is indeed straightforward to extend the hooking mechanism to apply to user defined functions with a `double(double)` or `float(float)` signature.

The invocation of CptnHook looks like the following command:

```
> cptnhook o myProfile  myProgram [arguments]
```
(1)

3.2. Analysis of the Results
Once the binary profile has been written, the `cptnhook-analyze` tool can be used to convert the binary profile into a web report or a ROOT [8] file.

The web report, see figure 1, provides high level information about the math function usage, such as:

- Minimum and maximum values of the arguments.
- Overall number of calls.
- Individual frames of the stacktraces.
- Grouping by stacktrace.

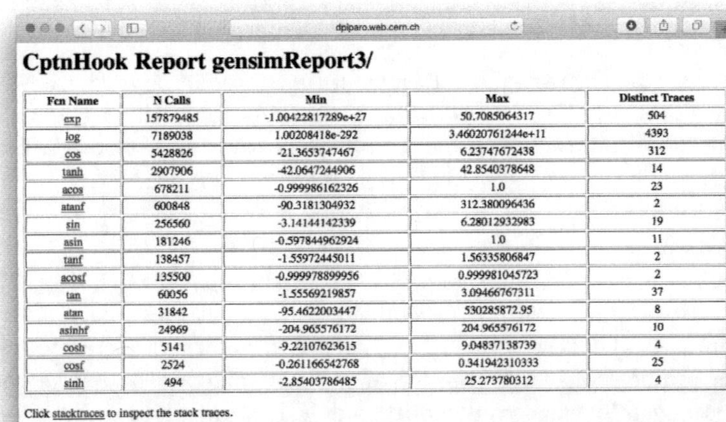
dpiparo.web.cern.ch

CptnHook Report gensimReport3/

Fcn Name	N Calls	Min	Max	Distinct Traces
exp	157879485	-1.00422817289e+27	50.7085064317	504
log	7189038	1.00208418e-292	3.46020761244e+11	4393
cos	5428826	-21.3653747467	6.23747672438	312
tanh	2907906	-42.0647244906	42.8540378648	14
acos	678211	-0.999986162326	1.0	23
atanf	600848	-90.3181304932	312.380096436	2
sin	256560	-3.14144142339	6.28012932983	19
asin	181246	-0.597844962924	1.0	11
tanf	138457	-1.55972445011	1.56335806847	2
acosf	135500	-0.999978899956	0.999981045723	2
tan	60056	-1.55569219857	3.09466767311	37
atan	31842	-95.4622003447	530285872.95	8
asinhf	24969	-204.965576172	204.965576172	10
cosh	5141	-9.22107623615	9.04837138739	4
cosf	2524	-0.261166542768	0.341942310333	25
sinh	494	-2.85403786485	25.273780312	4

Click stacktraces to inspect the stack traces.

Figure 1. Simulation of proton collisions in the CMS detector. Part of the full CptnHook webreport. A table summarising the name of the mathematical function, the number of times it was called, the minimum and maximum input value and the number of distinct stacktraces which led to the invocation of it.

CptnHook allows to obtain information about the single stack traces, figure 2, and the details of the single stack frames too, figure 3.

CptnHook Report log

N Calls	Min	Max	Trace ID
1641406	2.07315779511	73410.8041792	4134
1123015	0.000219112293296	520.133257527	4115
763666	7.15255737305e-07	0.999999701977	4090
475896	0.000800461976442	861.412920114	4098

Figure 2. The list of stack traces that led to the calculation of a logarithm. The row relative to stack 4090 signals a stack trace which was identified 763666 times and for which the input value of the logarithm was between 7e-07 and about one. Cases like this one could benefit from a polynomial approximation.

CptnHook Report Stack 4090

```
cmsRun() [0x40b789]
G4VProcess::ResetNumberOfInteractionLengthLeft()
G4VDiscreteProcess::PostStepGetPhysicalInteractionLength(G4Track const&, double, G4ForceCondition*)
G4SteppingManager::DefinePhysicalStepLength()
G4SteppingManager::Stepping()
G4TrackingManager::ProcessOneTrack(G4Track*)
G4EventManager::DoProcessing(G4Event*)
RunManager::produce(edm::Event&, edm::EventSetup const&)
OscarProducer::produce(edm::Event&, edm::EventSetup const&)
```

Figure 3. The names of the individual stack frames which led to the call of the logarithm. With this information it's possible to understand in which context the replacement with a polynomial expansion could be put in place.

The profile can also be converted to ROOT format, respecting the original column-wise storage of the data adopting ROOT tree data structures. Selections can be then applied on individual argument values or stacktraces and ROOT data visualisation routines leveraged. See for example figure 4.

3.3. A Real Life Example

A production workflow of the CMS [9] experiment was analysed with CptnHook: generation and Geant4 [10] based simulation of proton collisions at a centre of mass of 13 TeV originating top quark pairs. On a regular server machine, the profiler worked correctly demonstrating the ability to deal with applications of the order of millions of code lines like the CMS software stack. A detailed analysis of the results obtained is beyond the scope of this paper, but a preliminary assessment could point out to potential portion of the simulation code where the usage of polynomial expansions might be profitable, notably in the initialisation phase.

4. Conclusions and Future Work

This work is the first step toward a free and open source tool which is able to generate the optimal function for each context allowing to study the input values of mathematical functions resolving the different stacktraces which led to their invocation.

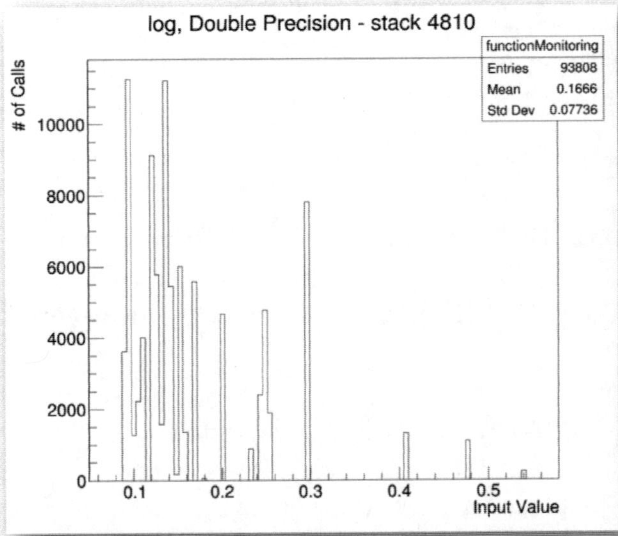

Figure 4. Simulation of proton collisions in the CMS detector. A ROOT histogram showing the input values of the log function for a particular stacktrace, hashed by the number 4810.

The hooking mechanism of CptnHook imposes a reasonable overhead on the profiled application and does not require recompilations nor code instrumentations. The profile written on disk during a measurement is in a compressed binary format and can be transformed either in a browsable web report summarising the principal informations about the calls carried out to mathematical functions or in a ROOT file for in-depth analysis based on non trivial cuts

Some of the main milestones of the future development program are a plug-in mechanism to add functions to be tracked, a richer look and feel of the default web reports e.g. visualising more statistics and control plots and the possibility to reduce the precision of the saved input values to further reduce the size of the reports on disk.

References

[1] Metalibm `http://www.metalibm.org`
[2] The GNU C library `https://www.gnu.org/software/libc`
[3] D Piparo et al.; 2014 J. Phys.: Conf. Ser. 513 052027, "Speeding up HEP experiment software with a library of fast and auto-vectorisable mathematical functions"
[4] A. Lasheen et al.; 2015 Proceedings of HB2014 East-Lansing, MI, USA, "Synchrotron Frequency Shift as a Probe of the CERN SPS Reactive Impedance"
[5] AMD LibM `http://developer.amd.com/tools-and-sdks/cpu-development/libm`
[6] Intel's MKL `http://software.intel.com/en-us/intel-mkl`
[7] S Chevillard et al.; 2010 Mathematical Software - ICMS "Sollya: an environment for the development of numerical codes."
[8] R. Brun et al. 1997 ROOT - An Object Oriented Data Analysis Framework *Nucl. Inst. Meth. In Phys.* **A 389** pp 81-86
[9] The CMS Experiment `http://cms.web.cern.ch`
[10] S Agostinelli et al.; 2003 Nucl. Instrum. Meth. A506 "GEANT4: A Simulation toolkit"

Journal of Physics: Conference Series **762** (2016) 012039 doi:10.1088/1742-6596/762/1/012039

Data Mining as a Service (DMaaS)

E. Tejedor*, D. Piparo*, L. Mascetti*, J. Moscicki*, M. Lamanna*, P. Mato*

*** CERN, CH-1211 Geneva 23, Switzerland**

Abstract. Data Mining as a Service (DMaaS) is a software and computing infrastructure that allows interactive mining of scientific data in the cloud. It allows users to run advanced data analyses by leveraging the widely adopted Jupyter notebook interface. Furthermore, the system makes it easier to share results and scientific code, access scientific software, produce tutorials and demonstrations as well as preserve the analyses of scientists.

This paper describes how a first pilot of the DMaaS service is being deployed at CERN, starting from the notebook interface that has been fully integrated with the ROOT analysis framework, in order to provide all the tools for scientists to run their analyses. Additionally, we characterise the service backend, which combines a set of IT services such as user authentication, virtual computing infrastructure, mass storage, file synchronisation, development portals or batch systems. The added value acquired by the combination of the aforementioned categories of services is discussed, focusing on the opportunities offered by the CERNBox synchronisation service and its massive storage backend, EOS.

1. Introduction

Not only High-Energy Physics is confronted with challenging data processing and analysis needs, often referred to as "big data". Numerous communities are innovating in this field by contributing ideas for alternative analysis workflows and new tools. Among the directions explored, there is a noticeable trend towards web-based interactive analysis backed up by data and computing resources accessed via "the cloud" [1, 2, 3]. Such trends are coupled with the provision of software as a service, which allows users to focus on the solution for the problem to be solved rather than on installation, configuration and operational matters. These circumstances have suggested to us a rethink of our data analysis model, extending the capabilities of our tools to complement the existing techniques in order to allow even more effective approaches in a shared environment leveraging the available IT infrastructures in an optimal way.

In that sense, this paper presents the Data Mining as a Service (DMaaS) project, which aims to provide a software (see section 2) and distributed computing infrastructure (see section 3) for interactive analysis of scientific data in the cloud. DMaaS eases the sharing of results and scientific code, the access to scientific software, the creation of tutorials and demonstrations as well as the preservation the analyses of scientists.

2. The Notebooks: One of the Interfaces of DMaaS

Nowadays, a common approach for interactive data analysis is to combine code, text, plots and rich media in the same document, known as "notebook". Notebooks are usually divided

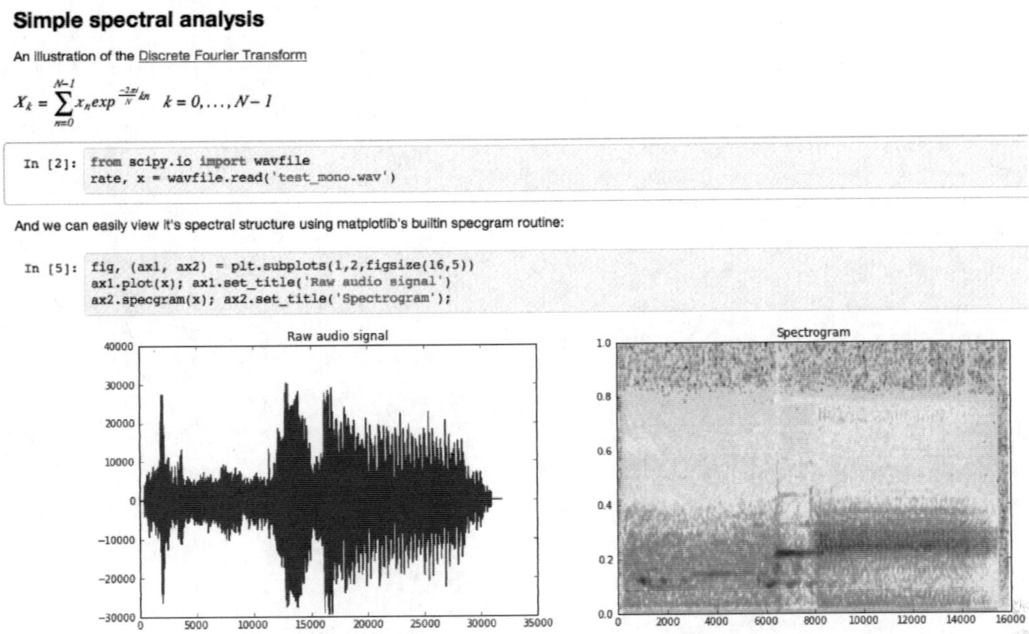

Figure 1. Example of a Jupyter notebook.

in cells where the user can type code, execute it and see the results inline. A particularly successful notebook environment is the one provided by Jupyter [4], which can be thought of as an interactive programming shell running in a web browser. It is an agile tool for both exploratory computation and data mining, and provides a platform to support reproducible research, since all inputs and outputs may be stored in a one-to-one way in notebooks. Figure 1 depicts an example of a Jupyter notebook document where markdown text, formulae, code and plots are combined.

The aforementioned attractive features of the Jupyter notebooks motivated their choice as the main interface of the DMaaS service.

2.1. Integration of ROOT and Notebooks

Jupyter is not restricted to a particular language, but instead it allows to plug in language extensions known as *kernels*. It can also accommodate various ecosystems of tools for data analysis, e.g. R [5] or pandas [6].

In order to make ROOT [7] a first-class citizen of Jupyter too, we developed a software layer to integrate both technologies. As a result, users can write *ROOTbooks* (i.e. ROOT notebooks), characterised by the following features:

(i) *Two languages of choice*: a new Jupyter kernel for ROOT was developed and, as a result, ROOT-based analyses can be written in C++. Additionally, the ROOT Python interface, PyROOT, was enhanced to fully exploit its usage in notebooks.

(ii) *Language interoperability*: it is possible to mix C++ and Python in the same ROOTbook, as well as to use C++ variables/functions/classes from Python and vice versa.

(iii) *Static and interactive data visualisation*: graphics can be displayed in the notebook. ROOTbooks support two modes for graphics: static and JavaScript based; while the former

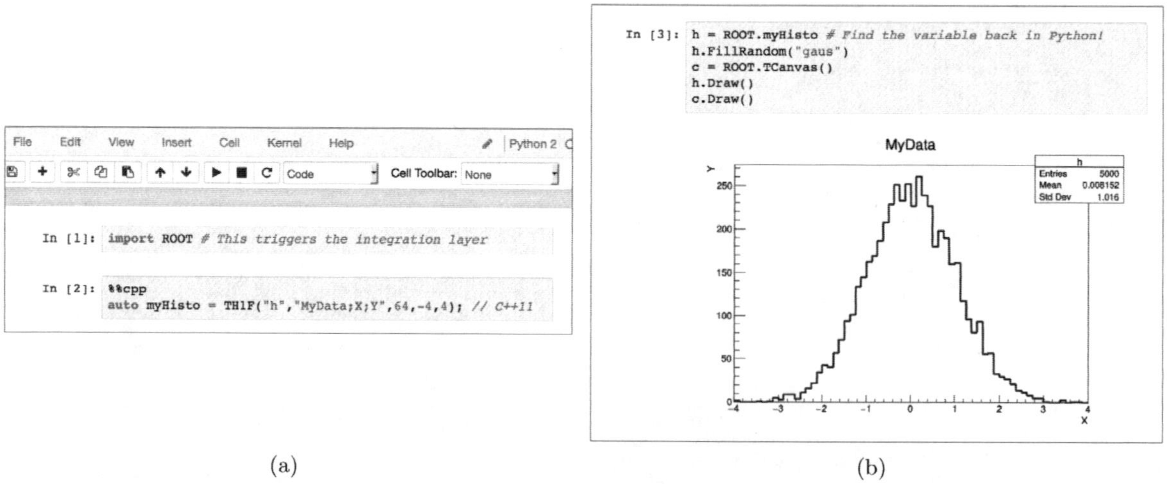

Figure 2. A Python ROOTbook.

embeds an image in the document, the latter shows interactive plots that can be modified (e.g. by zooming) and saved as an image.

(iv) *Tab completion*: equivalent to that of the ROOT prompt.

2.2. Programming Model Example

In order to illustrate the features enumerated in section 2.1, this section shows an example of how to write a ROOTbook in Python and C++.

In Python, the user starts by importing the ROOT Python module so that all the ROOTbook features are activated, as can be seen in the first cell of Figure 2(a). After that, she can choose to either continue writing in Python or to switch to C++ in the next cell; in the latter scenario, the **%%cpp** magic is used: in Jupyter, magics can modify the behaviour of a particular cell, in this case transforming it into a C++ cell that creates a histogram (still in Figure 2(a)). It is worth noting how this C++ cell is highlighted in C++, differently from the first (Python) cell.

The third cell (Figure 2(b)) is an example of Python-C++ interoperability, since the histogram created in C++ is accessed from Python, filled and drawn. As a result of the drawing, the histogram is displayed inline in the notebook.

Special mention is deserved by the interactive graphics of the ROOTbooks, implemented with JSROOT [8]. Continuing with the example, the code in Figure 3(a) defines a function in C++ and uses that function in Python to fit a histogram. After that, if the JavaScript visualisation mode is enabled and the canvas is drawn, the embedded display that appears corresponds to a JavaScript object, with which the user can interact by e.g. inspecting the content of the histogram bins (Figure 3(b)).

Finally, regarding C++ ROOTbooks, they also benefit from all the functionalities described so far, the only difference being the language used.

The integration of ROOT and notebooks reached production quality. These new features were already used to document ROOT features in a modern way, e.g. on the framework's website (see [9]) and to provide an online ROOT demo on Binder [10] (see [11]).

Journal of Physics: Conference Series **762** (2016) 012039 doi:10.1088/1742-6596/762/1/012039

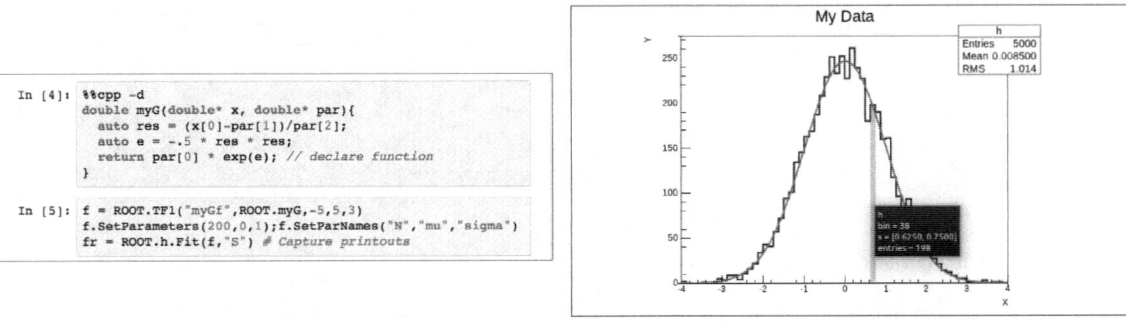

(a) (b)

Figure 3. JavaScript visualisation in a ROOTbook.

3. DMaaS and the Portfolio of CERN Services

Section 2 described the Jupyter notebook interface and its integration with ROOT. The main objective of the DMaaS project is to leverage that interface to provide a service of notebooks (and ROOTbooks) on demand.

In order to develop the DMaaS service, CERN will rely on existing, production-grade technologies such as:

- CVMFS [12] for the distribution of the software in the cloud, including ROOT and LHC experiments' software stacks.
- OpenStack [13] for the provision of the CPU resources necessary to run the users' notebooks and their calculations.
- EOS [14] as mass storage, hosting for example experimental and simulation data.
- CERNBox [15], the CERN file synchronisation and share service, for storing user files.

Figure 4 shows the DMaaS deployment on top of the CERN services' portfolio. The login happens with CERN credentials and the orchestration of users is delegated to JupyterHub [16]. The execution of notebooks is encapsulated in Docker containers [17], scheduled on the OpenStack infrastructure. The individual containers are lightweight: the needed software (e.g. ROOT) is available on CVMFS. Access to the CERN mass storage EOS is available, and the user's home directory in each container is the EOS portion private to each user synchronised via CERNBox. The synchronisation of local directories is one of the gateways to access the cloud.

With such a service and by means of only a web browser, users will be able to write and execute data analyses, see their results inlined (text, graphics) and combine those with explanations about what they are doing, everything in the same document. When finished, notebooks could be saved and shared with other colleagues who could review, modify and re-run those notebooks.

4. Conclusions and Future Work

In the context of the DMaaS project, the ROOT and Jupyter technologies have been successfully integrated: C++ and Python interfaces are supported in production. Documentation, tutorials and online demos showing advanced features like interactive JavaScript graphics have been created.

Within the framework of a coherent view at CERN, a cloud-based data analysis model is being made available to scientists. Its implementation consists in the JupyterHub deployment of DMaaS, which relies entirely on production-grade services offered by the CERN IT department.

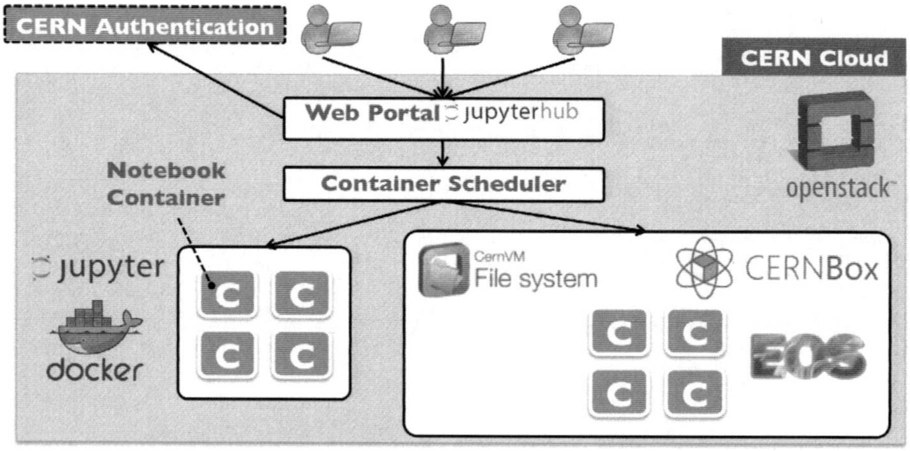

Figure 4. The Jupyter deployment model of DMaaS.

In the future, the current design of DMaaS will be generalised so that it can be integrated with other ecosystems of technologies (different from CERN's). Additionally, a tighter integration of ROOT and other well-established data analysis ecosystems such as pandas will be investigated in the context of the Jupyter notebooks. Finally, the possibility of on-demand exploitation of clusters for distributing calculations spawned from within notebooks will be assessed.

References

[1] Wakari: Web-based Python Data Analysis. `http://wakari.io/`
[2] SageMathCloud: Collaborative Computational Mathematics. `http://cloud.sagemath.com/`
[3] Plotly: Make charts and dashboards online. `http://plot.ly/`
[4] Jupyter: Open source, interactive data science and scientific computing. `http://jupyter.org/`
[5] The R Project for Statistical Computing. `http://www.r-project.org/`
[6] Python Data Analysis Library. `http://pandas.pydata.org/`
[7] R. Brun et al. 1997 ROOT - An Object Oriented Data Analysis Framework *Nucl. Inst. Meth. In Phys.* **A 389** pp 81-86
[8] B Bellenot et al.; 2015 J. Phys.: Conf. Ser. 664 062033,"JavaScript ROOT"
[9] ROOT website, code examples. `http://root.cern.ch/code-examples`
[10] Binder. `http://mybinder.org`
[11] ROOT Binder repository. `http://root.cern.ch/rootbinder`
[12] J Blomer et al.; 2011 J. Phys.: Conf. Ser. 331 042003, "Distributing LHC application software and conditions databases using the CernVM file system"
[13] OpenStack: Open Source Cloud Computing Software. `http://www.openstack.org`)
[14] A Peters et al.; 2015 J. Phys.: Conf. Ser. 664 042042, "EOS as the present and future solution for data storage at CERN"
[15] L Mascetti et al.; 2015 J. Phys.: Conf. Ser. 664 062037, "CERNBox + EOS: end-user storage for science"
[16] JupyterHub: A multi-user server for Jupyter notebooks. `https://github.com/jupyter/jupyterhub`
[17] Docker: build, ship and run any app, anywhere. `https://www.docker.com`

New data reduction protocol for Bragg reflections observed by TOF single-crystal neutron diffractometry for protein crystals with large unit cells

Katsuaki Tomoyori and Taro Tamada
Quantum Beam Science Directorate, National Institutes for Quantum and Radiological Science and Technology
2-4 Shirakata, Tokai, Naka, Ibaraki 319-1106, Japan

E-mail: tomoyori.katsuaki@qst.go.jp

Abstract. In protein crystallography, high backgrounds are caused by incoherent scattering from the hydrogen atoms of protein molecules and hydration water. In addition, the scattering intensity from large unit-cell crystals is very small, which makes it difficult to improve the signal-to-noise ratio. In the case of time-of-flight (TOF) single-crystal neutron diffractometry, the measured spectra cover four-dimensional space including X, Y, and TOF in addition to intensity. When estimating the integrated intensity, 3D background domains in the vicinity of peaks should be clearly classified. In conventional 1D or 2D background evaluation, the evaluation is applied for individual peaks assigned using peak searches; however, it is quite difficult to classify the 3D background domain in TOF protein single-crystal neutron diffraction experiments. We undertook the development of a data reduction protocol for measurements involving large biomacromolecules. At the initial stage of the reduction protocol, appropriate 3D background estimation and eliminations were applied over the entire range of X, Y, and TOF bins. The histograms were then searched for peaks and indexed, and the individually assigned peaks were finally integrated with an effective profile function in the TOF direction. Three-dimensional deconvolution procedures for overlapping peaks associated with large unit cells were implemented as necessary. This data reduction protocol may lead to the improvement of signal-to-noise ratios to enable TOF spectral analysis of large unit-cell protein crystals.

1. Introduction

Neutron protein crystallography has the unique ability to locate the precise positions of hydrogen atoms and hydration water molecules in biomacromolecules. We will construct a high-resolution time-of-flight (TOF) protein single-crystal neutron diffractometer at the Materials and Life Science Experimental Facility (MLF) at the Japan Proton Accelerator Research Complex (J-PARC). Using a spallation neutron source, the diffractometer can efficiently sweep a large reciprocal space in a single measurement via a white thermal neutron beam, which is suitable for diffraction experiments. This new diffractometer will be designed to be the best-in-class for efficient measurement of Bragg data sets at a resolution of 2.0 Å for protein crystals with unit cell constants of 250 Å. These include over 95% of the protein structures, including large proteins such as membrane proteins and protein complexes, deposited in the Protein Data Bank [1]. We will achieve high throughput, minimal peak overlap, and high signal-to-noise ratio by using a large wavelength bandwidth of TOF-sorted neutrons and an array of high-spatial-resolution position-sensitive area detectors covering a large solid angle. Large unit-cell protein crystals lead to weak average peak intensities. In addition, high backgrounds are caused by incoherent scattering from the hydrogen atoms of protein and hydration water molecules. To precisely calculate weak signals in the presence of such high backgrounds, a data reduction protocol is needed to improve the signal-to-noise ratio and thus enable the neutron structural analysis of large proteins.

Two types of procedures exist for the integration of Bragg reflections: the box summation method, which simply designates each peak region within a box, and the profile fitting method. Box summation is frequently used to integrate a designated domain while distinguishing between signals and background. Having determined the background domain, a simple estimate of the integrated intensity is obtained by summing the pixel values of all the pixels in the peak domain of the mask and then subtracting the sum of the background values calculated from the background domain for the same pixels. If the background level is very low compared with the intensity of a spot and the spots are well resolved, this will give as accurate an estimate of the intensity as possible.

In TOF neutron protein crystallography, the intensities of the diffraction spots can be integrated using STARGazer [2], which employs box summation strategies, at MLF/J-PARC, using subroutines from the Integrated Spectral Analysis Workbench (ISAW) software [3]. To carry out the integration in TOF single-crystal diffractometry involving four-dimensional (4D) spaces with XY + TOF plus intensity, the 3D classified domains enclosed by XY + TOF around each spot are box-summed and computed after background elimination. It is important that pixels are not misclassified, as misclassification can lead to systematic errors in the integrated intensity. There are weak Bragg signals with asymmetric shapes over a large, structured, incoherent backgrounds derived from the incident profile in the TOF direction. In addition, specific backgrounds associated with the X-Y accidental coincidence signal are non-negligible. The large variety of continuous background shapes in the vicinity of the peaks should be faithfully reproduced over the entire range of space and time directions for every peak. However, conventional background estimation is inadequate for improving the signal-to-noise ratios of peak integration, especially for weak reflections.

To achieve high signal-to-noise ratios for TOF protein single-crystal neutron diffractometry with large unit-cell proteins, we investigated the methodology of background evaluation and ways of implementing the background algorithm in the data reduction protocol. The optimization of the data reduction protocol, combined with the use of the background estimation and the profile-fitting and deconvolution methods, may lead to the improvement of signal-to-noise ratios to enable TOF spectral analysis of large unit-cell protein crystals.

2. Data reduction protocol targeted for protein crystals with large unit cells

In protein crystallography, there are two distinct procedures for integrating spot intensities, known as 2D (XY) and 3D (XY + "Frame") integration, available in different software packages; e.g., MOSFLM [4] and HKL2000 [5] use the 2D (XY) method, while d*TREK [6] and XDS [7] use the 3D method. The 3D (XY + "Frame") indicates that the data set consists of multiple 2D (XY) data sets in which Bragg spots are recorded. The spectra themselves are 2D (XY) images including only 2D spatial coordinates.

Fig. 1 shows the conventional reduction protocol for protein crystallography. Using the orientation matrix, UB, we can predict where single crystal peaks should be found in detector/TOF space. The UB describes the sample orientation with respect to the diffractometer angles. Given UB, it is possible to calculate the diffractometer angles necessary to rotate a particular scattering vector **Q**, indexed by (*h, k, l*), into the diffraction position. The matrix **B** transforms the given (*h, k, l*) into an orthonormal coordinate system fixed in the crystal. The matrix **U** is the rotation matrix that rotates the crystal's reference frame into the spectrometer's [8]. Providing the background and peak regions are correctly defined, summation integration provides a method for evaluating integrated intensities which is both robust and free from systematic error. For weak reflections, however, many of the pixels in the peak region will contain very little signal (Bragg intensity), but will contributes significantly to the noise because of the Poissonian variation in the background. It is possible to obtain a more accurate estimate of the integrated intensity by using a procedure known as profile fitting. In this method, it is assumed that the shape or profile (in 2D/3D) of the spots is known. The background plane is determined in the same way as for box summation, but the intensity is derived by determining the scale factor, which gives the best fit to the observed spot profile when applied to the known spot profile. This scale factor is then proportional to the profile-fitted intensity for the reflection. However, the 4D (XY + TOF plus intensity) precise summation integration or profile-fitting approach has not yet been used to carry out data reduction of TOF protein single-crystal neutron diffraction data. Herein, the notation "4D" indicates four-dimensional spaces including X, Y, and TOF in addition to intensity.

To apply the 4D Laue neutron diffraction data to the conventional data reduction protocol, as shown Fig. 1, a suitable choice of 3D background domain or 3D signal shape functions around spots is required to carry out precise peak integration of the Bragg reflections. In spot-based background estimation, the background around the peak spots is locally averaged or calculated by least-square method (LSTSQ) inside a designated 3D domain. In protein crystallography, incoherent neutron scattering events due to the hydrogen atoms of protein and surrounding water molecules are continuously distributed over the entire range of the measured TOF region. On the incoherent backgrounds, there are coherent weak Bragg signals, and the peaks exhibit asymmetric tails in the TOF direction due to the neutron moderation process used to extract the available thermal neutrons for diffraction experiments. The tails also prevented us from accurately discriminating between signal and background events. Furthermore, we will employ the 2D neutron detectors in X-Y coincidence techniques. Therefore, the accidental coincidence events associated with signal processing are continuously present as the backgrounds of the 3D Bragg peaks in the spatial directions. Therefore, the conventional method for designating the 3D background domain in the vicinity of the peaks cannot accurately describe the large variety of continuous background shapes underlying a single peak in both spatial and time directions, indicating that conventional background estimation is quite inadequate for improving the signal-to-noise ratio of peak integration for weak reflections in particular. It is clearly important that pixels are not misclassified, as this can lead to systematic errors in the integrated intensity. It is essential to employ a highly precise background evaluation method that can cope with the backgrounds of weak reflections. One or two dimensions can be reduced from the original dimensions to evaluate backgrounds. However, projecting the histogram data onto the lower dimension leads to degradation of the information between signal and background, resulting in systematic errors in the integrated intensity.

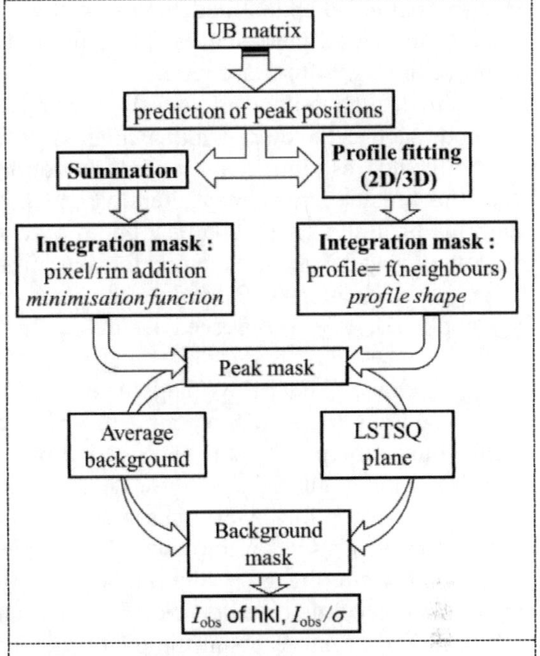

Fig 1. Conventional reduction protocol for protein crystallography

We will intend to decrease the misclassification by appropriately implementing the background elimination procedures with the intact 4D spectra, which apart from peak integration procedure with the profile functions to reproduce the asymmetric tail of Bragg reflections. The optimization of the data reduction protocol tailored to cope with 4D spectra observed by Laue single crystal neutron diffractometry could be inevitable to improve the signal-to-noise ratio. Profile fitting provides a means of improving the signal-to-noise ratio especially for large unit-cell protein crystals. The integration of weak reflections by using profile fitting has been demonstrated. In TOF protein single-crystal neutron diffratometry, we speculate an improvement of totally about 10% signal-to-noise ratio can be expected: a few percent for the discrimination on asymmetric tail, and a few percent for the discrimination on spatial and time direction, respectively. In addition, the approaches based on the scaling of learned peak shapes found from strong neighboring peaks as shown Fig. 1 further could improve the signal-to-noise ratio.

When dealing with continuously varying backgrounds, non-iterative background approximation methods such as the three-window method [9] or polynomial fitting [10] are used. In addition, iterative methods such as simple multi-point smoothing [11] or more sophisticated methods such as the statistics-sensitive non-linear iterative peak-clipping (SNIP) algorithm [12] are used. The SNIP algorithm can iteratively compute optimum continuously varying backgrounds in accordance with the peak widths and is a good method for multidimensional spectra [13]. This algorithm can compute 3D background domains without decreasing dimensions, that is, losing background information. Recently, we applied it to the observed Bragg reflections measured by TOF protein single-crystal neutron diffractometry and found it to be an appropriate method for realizing the criteria mentioned above [14].

Fig. 2 shows a data-flow diagram of the data reduction protocol used in the TOF protein single-crystal neutron diffraction experiment. The left figure shows a typical reduction protocol and the right figure shows the proposed new reduction protocol, which is suitable for the integration of diffraction sets from large-unit-cell proteins. In a typical reduction protocol, the event data from a sample, collected by detectors, were presented as a histogram. Vanadium spectra measured under identical conditions were used to normalize the sample spectra. Vanadium is a purely incoherent scatter and incoherent scattering is by definition isotropic. The scattering intensity seen in each detector is a measurement of detector efficiency, solid angle and analyzer transmission. Vanadium scattering ordinarily takes into account the relative efficiencies of different pixels on the detector, the incident neutron flux profile, $I_0(\lambda)$, and the detector response at different neutron wavelengths ($I_0(\lambda)$ and detector sensitivity correction in Fig. 2). Each 3D data histogram was searched for peaks and these were indexed using a UB matrix. A refined UB matrix was used to determine which (X, Y, TOF) points should be included in the (q_h, q_k, q_l) points in the reciprocal lattice plane of interest. With many reflections determined within each frame, this allowed straightforward determination and refinement of the UB matrix. The cell dimensions were determined using a refinement of the UB matrix with respect to a set of reflections taken from a wide range of data frames, thus ensuring good averaging. For peak integration, on the other hand, a local UB refined for the particular frame under consideration was used. After refinement of the local UB, peak integration was undertaken in one of two ways, using either the peak position predicted from the UB matrix or those found during peak searching. After the peak position was predicted, the peak domain was determined and the background domain was also designated to evaluate the local background in the vicinity of a peak. The backgrounds were subtracted and the peak was box-summed or projected and integrated in a 1D TOF direction with appropriate asymmetric functions [15]. The integrated intensity was finally estimated for each Bragg reflection. The resulting intensities were reduced to structure factors, giving data sets of reflections after various basic corrections associated with the measurements and the principle.

In the proposed new reduction protocol, after correcting the incident neutron flux profile and the pixel and wavelength efficiency of the detectors, as mentioned in the typical reduction protocol, the entire 3D background estimation and subtraction were applied at the initial stage of the data reduction processes (the green colored component in Fig. 2, right), although background estimation was carried out for individual spot areas predicted by the refined UB matrix in a typical protocol. After the background elimination and peak identification processes were complete, the peaks found were then indexed and evaluated either as legitimate peaks or as overlapping peaks of higher order reflections. The potential overlapping peaks were routed through a deconvolution process, which is an efficient algorithm for multidimensional deconvolution (i.e. the Gold algorithm contained in TSpectrum [16]), and re-evaluated (the blue colored component in Fig. 2, right). Each of the single peak regions convolved in a doublet was resolved within an appropriate accuracy after iteration of the Gold deconvolution. The indices were then assigned for the resolved single peaks (Fig. 1 (1)). The fitting parameters including analytical functions thus employed were reflected on the integration processes (Fig. 1 (2)). No background mask was used when integrating the individual peaks (the yellow component in Fig. 2, right); instead, background estimation and elimination were applied at the initial stage. The neutron peak counts for each spot could be summed in the appropriate peak mask domain. Otherwise, the spot could be projected onto a 1D/2D histogram and integrated using a profile fitting approach based on the known analytical shape of the reflections in the TOF direction, which were well understood from the characteristics of the neutron source. We found that the Landau–Vavilov distributions, which are used to describe the energy loss of charged particles traversing a thin absorber, were in excellent agreement with the observed TOF profile measured using an IBARAKI Biological Crystal Diffractometer (iBIX) installed at a coupled moderator source at MLF/J-PARC [17].

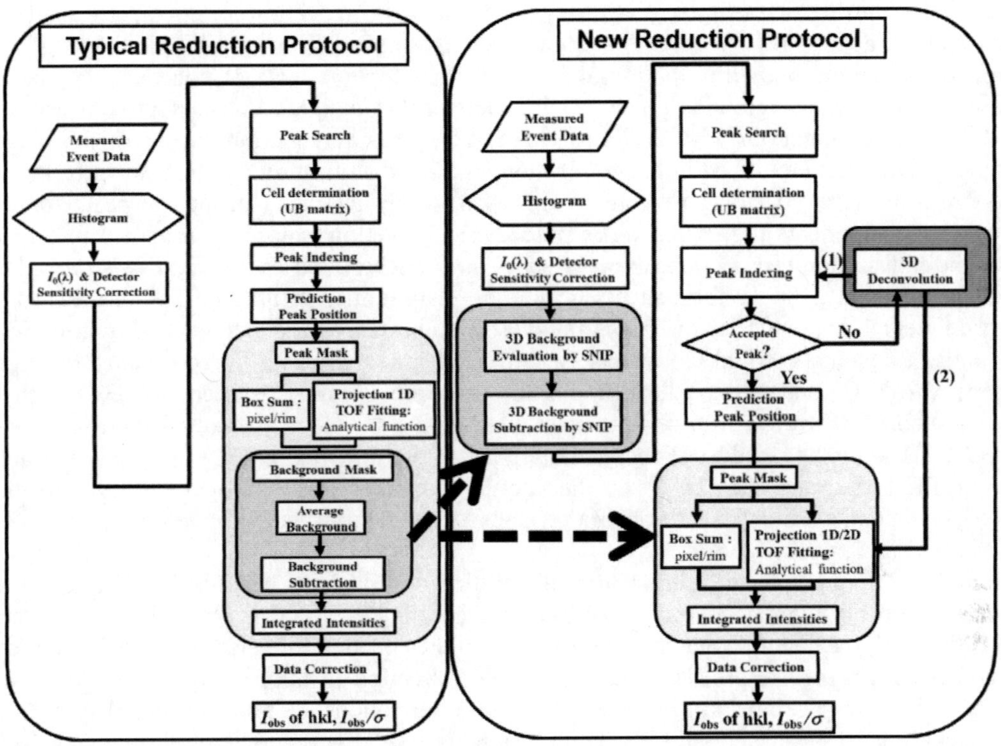

Fig 2. Data-flow of the data reduction protocol in TOF protein single-crystal neutron diffraction

The left panel shows a typical data reduction protocol for a TOF protein single-crystal neutron diffraction experiment. The right panel shows a proposed new reduction strategy for spectra obtained with a TOF protein single-crystal neutron diffractometer for crystals with large unit cells. The background evaluation routines in typical data reduction can be converted into a 3D background evaluation method by SNIP, as shown by the green colored components in the two protocols linked with a dashed arrow. The component for dealing with individual Bragg reflections is converted to the yellow component in the new reduction protocol. The blue colored component shows the routines for deconvoluting the overlapped Bragg reflections. The resolved peaks are re-indexed (1) and the fitting parameters including analytical functions thus obtained are also utilized in integration processes (2).

At the data correction stage in either a typical or the new reduction protocol, semi-empirical absorption corrections and multiple scattering must also be applied, using vanadium and sample scattering and inelasticity. A physics-related correction, such as the Lorentz factor, which is associated with the velocity to sweep the reciprocal space derived from the Laue neutron diffraction principle, must also be taken into account. These data are then further used in a least-squares refinement program to apply a variable wavelength extinction correction based on the Becker–Coppens formalism [18], normally using a Gaussian model with one variable parameter, the mosaic spread.

3. Summary

To integrate weak Bragg reflections from protein crystals with large unit cells using TOF protein single-crystal neutron diffractometry, we revisited the data reduction protocol, focusing on the background estimation to improve the signal-to-noise ratio. We introduced the high precise background estimation method at the initial stage of a new data reduction procedure to establish the precise evaluation of backgrounds under weak reflections from large unit-cell protein crystals (the green colored component in Fig. 2, right). In a typical data reduction protocol, the spot-based background estimation employed in the conventional data reduction protocol, as shown in Fig. 1, was applied for each observed Bragg reflection. This was locally averaged inside the peak domain (the green colored components Fig. 2, left). The procedure (the yellow colored component in Fig. 2, left) for each individual Bragg reflection involves spot-based background estimation. This procedure was divided into two independent components in the new data reduction protocol. With the background elimination method available for the background estimation of weak reflections and multidimensional space, authentic signals could be precisely extracted by analyzing the correlated backgrounds associated with X, Y, and TOF, indicating that this may prevent misclassification between these and the backgrounds. To overcome a serious peak-overlapping problem associated with large unit-cell protein crystals, we introduced a 3D deconvolution algorithm allowing precise integration of the peaks using a profile-fitting approach based on the known analytical shape of the reflections in the TOF directions, well understood from the characteristics of spallation neutron source (the blue colored component in Fig. 2, right).

A precise 4D summation integration or profile-fitting approach, as described in Fig. 1, remains to be realized to enable the data reduction of TOF protein single-crystal neutron diffraction data. The profile-fitting approach employed thus far in our studies is not intended for precise background estimation but for a precise profile shape in the TOF direction to improve the signal-to-noise ratio. Before realizing the profile-fitting approach for weak reflections, it will be necessary to establish precise background estimation and elimination methods. We will try to improve the signal-to-noise ratio more than 10% by employing software algorithm tailored to analyze the 4D spectra observed by Laue neutron diffraction experiment. Further modifications in the data reduction protocol for TOF protein single-crystal neutron diffractometry, which is designed for lattice constants greater than 250 Å, will be investigated.

4. Acknowledgement

This work was partly supported by JSPS KAKENHI Grant Number 15K04737.

References

[1] Tomoyori, K., Kurihara, K., Tamada, T. and Kuroki, R. (2015). *Journal of Physical society of Japan* DOI 036004 10.7566/JPSCP.8.036004
[2] Ohhara, T., Kusaka, K., Hosoya, T., Kurihara, K., Tomoyori, K., Niimura, N., Tanaka, I., Suzuki, J., Nakatani, T., Otomo, T., Matsuoka, S., Tomita, K., Nishimaki, Y., Ajima, T. and Ryufuku, S. (2009). *Nucl. Instr. Meth. Phys. Res., Sect.* A **600**, 195.

[3] Chatterjee, A., Mikkelson, R., Hammonds, J. and Worlton, T. (2002). *Appl. Phys.* A **S194**.
[4] Leslie, A. G. W. (1992). *Jnt CCP4/ESF–EACBM Newsl. Protein Crystallogr.* **26**.
[5] Otwinowski, Z. and Minor, W. (1997). *Methods Enzymol.* **276**, 307–326.
[6] Pflugrath, J. W. (1999). *Acta Cryst.* D **55**, 1718–1725.
[7] Kabsch, W. (1988). *J. Appl. Cryst.* **21**, 916–924.
[8] Busing, W. R. and Levy, H. A. (1967) *Acta Cryst* **22**, 457
[9] Jenkins, R. (1995). *Quantitative X-ray Spectrometry, CRC Press*, Boca Raton
[10] Van Grieken, R. and Markowicz, A. (2001). *Handbook of X-ray Spectrometry, CRC Press,* Boca Raton
[11] Clayton, E., Duerden, P. and Cohen, D. D. (1987). *Nucl Instr Meth Phys Res Sect* B **22** (1), 64-67
[12] Ryan, C. G., Clayton, E., Griffin, W. L., Sie, S. H. and Cousens, D. R. (1988). *Nucl Instr Meth Phys Res Sect* B **34**, 396-402
[13] Morhac, M. and Matousek, V. (2008). *Appl. Spectrosc* **62**, 91-106
[14] Tomoyori, K., Hirano, Y., Kurihara, K. and Tamada, T. (2015). *Journal of Physics* conference series **664** DOI:10.1088/1742-6596/664/7/072049
[15] Gutmann, M. J. (2005) SXD2001, ISIS Facility, Rutherford Appleton Laboratory, Oxford-shire, England
[16] Morhac, M., Matousek, V. and Kliman, J. (2003). *Digital Signal Processing*, **13**, No. 1, 144-171
[17] Tomoyori, K., Kusaka, K., Yamada, T., Hosoya, T., Ohhara, T., Kurihara, K., Tanaka, I., Katagiri, M. and Niimura, N. (2013). *Nucl Instr Meth Phys Res Sect* A **723**, 128-135
[18] Becker, P. and Coppens, P. (1974). *Acta Cryst* A **30**, 129 and 148.

ACAT2016 IOP Publishing
Journal of Physics: Conference Series **762** (2016) 012041 doi:10.1088/1742-6596/762/1/012041

Deconvolving the detector in Fourier space

Joseph Boudreau, Carlos Escobar, James Mueller and Jun Su

Department of Physics and Astronomy, University of Pittsburgh, Pittsburgh Pa 15260, USA

E-mail: boudreau@pitt.edu, cae53@pitt.edu, mueller@pitt.edu, jus48@pitt.edu

Abstract. We discuss algorithms for the analysis of hadronic final states, with application to single top t-channel production, heavy Higgs decaying as $H \to W^+W^-$ in the lepton plus jets mode, and others. In our examples, nature has arranged for the triple differential decay rates in angles we call θ_1, θ_2, ϕ to be a short finite series in orthogonal functions, $\sqrt{2\pi}a^m_{k,l}Y^m_k(\theta_1,\phi)Y^m_l(\theta_2,0)$ (summation implied), where the $a^m_{k,l}$ are complex constants. This observation can be exploited in two ways; first, a technique called orthogonal series density estimation may be employed to extract coefficients $a^m_{k,l}$ of the decay and physics parameters related to these coefficients; second, an angular analog of the convolution theorem may be employed to analytically deconvolve detector resolution effects from an observed signal.

The angular distributions of the decay products of massive elementary particles are a window on fundamental physics. Nowadays, the main goal of precision measurements in Higgs physics and top quark physics is to determine the strength and structure of the couplings amongst fermionic and bosonic fields. One example is the coupling of the Higgs boson to the charged W bosons, for which the effective interaction Lagrangian[1] is:

$$\mathcal{L}_{HWW} = m_W^2\left(\sqrt{2}G_F\right)^{1/2}(1-\frac{v^2}{2\Lambda^2}f_{\Phi,2})HW^+_\mu W^{-\mu} + \frac{g^2v}{2\Lambda^2}\frac{f_W}{2}(W^+_{\mu\nu}W^{-\mu}\partial^\nu H) - \frac{g^2v}{2\Lambda^2}f_{WW}W^+_{\mu\nu}W^{-\mu\nu}$$

(h.c. implied) where the complex constants $f_{\Phi,2}$, f_W, and f_{WW} are *anomalous* couplings. Likewise at the Wtb vertex one has the effective interaction Lagrangian[2]

$$\mathcal{L}_{Wtb} = -\frac{g}{\sqrt{2}}\bar{b}\gamma^\mu(V_LP_L+V_RP_R)tW^-_\mu - \frac{g}{\sqrt{2}}\bar{b}\frac{i\sigma^{\mu\nu}q_\nu}{M_W}(g_LP_L+g_RP_R)tW^-_\mu + h.c;$$

here, the coupling constant $V_L = 1$ in the standard model while the anomalous couplings $V_R = g_L = g_R = 0$. These couplings are affected by new physics and they affect the angular distributions of top quark and Higgs decay products.

$H \to W^+W^-$ and $t \to Wb \to l^+\nu b$ are the two benchmark examples we choose to illustrate the proposed technique for analyzing the angular distribution of Higgs and top quark decay products. In the former example, we assume that the decay occurs in a lepton plus jets mode. In the latter, we assume single top quark production in the t-channel, which produces polarized top quarks at hadron colliders, followed by their semileptonic decay. Each of these modes is characterized by three decay angles: In the case of the Higgs decay, we define θ_1 (θ_2) to be the angle between the W^- (W^+) boson direction in the Higgs frame and the outgoing lepton or quark in the W frame. ϕ is the angle between the decay planes of the two W bosons. In the case of the top quark decay, we define θ_1 to be the angle between a light "spectator" quark from t-channel production, and the W boson in the top quark reference frame; θ_2 is the angle between the W boson direction in the top quark rest frame and the lepton direction in the W boson rest frame. ϕ is the angle between the plane defined by the spectator quark and the W boson, and that defined by the W boson and the lepton.

We use the following shorthand notation. The product of spherical harmonics, $M^m_{k,l}(\theta_1, \theta_2, \phi) \equiv \sqrt{2\pi}Y^m_k(\theta_1, \phi)Y^m_l(\theta_2, 0)$ is referred to as an M-function. Properties of the M-functions derive from those of those of spherical harmonics. The M-functions are orthonormal,

$$\int M^m_{k,l}(\theta_1, \theta_2, \phi)M^{m'*}_{k',l'}(\theta_1, \theta_2, \phi)d\Omega^M = \delta_{k,k'}\delta_{l,l'}\delta_{m,m'} \quad (1)$$

where $d\Omega^M = \sin\theta_1 \sin\theta_2 d\theta_1 d\theta_2 d\phi$; their complex conjugates obey

$$M^{m*}_{k,l}(\theta_1, \theta_2, \phi) = M^{-m}_{k,l}(\theta_1, \theta_2, \phi),$$

and they obey a version of Gaunt's theorem,

$$M^m_{k,l}(\theta_1, \theta_2, \phi)M^{m'}_{k',l'}(\theta_1, \theta_2, \phi) = W^{m,m',M}_{k,l,k',l',L,K}M^M_{K,L}(\theta_1, \theta_2, \phi) \quad (2)$$

where

$$W^{m,m',M}_{k,l,k',l',L,K} = \sqrt{2\pi}G^{m,m',M}_{k,k',K}G^{m,m',M}_{l,l',L},$$

and

$$G^{m,m',M}_{l,l',L} = \sqrt{\frac{(2l+1)(2l'+1)}{4\pi(2L+1)}}C^{m,m',M}_{l,l',L}C^{0,0,0}_{l,l',L}$$

and the constants $C^{m,m',M}_{l,l',L}$ are ordinary Clebsch-Gordan coefficients. In this paper summation over repeated indices is always implied unless otherwise indicated.

The two benchmark decays, as well as plenty more in B-physics such as $B^0 \to J/\psi K^*$, are characterized by three decay angles and nature arranges for the triple differential decay rate at tree level to have a simple form when expressed in M-functions:

$$\rho(\theta_1, \theta_2, \phi) \equiv \frac{1}{\Gamma}\frac{d\Gamma}{d\Omega^M} = a^m_{k,l}M^m_{k,l}(\theta_1, \theta_2, \phi) \quad (3)$$

In $H \to W^+W^-$, the complex coefficients $a^m_{k,l}$ can be expressed in terms of the amplitudes A_{00}, A_{LL}, and A_{RR} for the Higgs to decay into a pair of longitudinal, left handed, or right handed W bosons. All other transitions are forbidden by the conservation of angular momentum. From the helicity formalism one obtains:

$$a^0_{0,0} = \frac{1}{\sqrt{8\pi}}\left(|A_R|^2 + |A_L|^2 + |A_0|^2\right) \qquad a^0_{2,2} = \frac{1}{40\sqrt{2\pi}}\left(|A_R|^2 + |A_L|^2 + 4|A_0|^2\right)$$

$$a^0_{0,1} = a^0_{1,0} = \sqrt{\frac{3}{32\pi}}\left(|A_L|^2 - |A_R|^2\right) \qquad a^1_{1,1} = \frac{3}{\sqrt{128\pi}}\left(A_0A^*_R + A_LA^*_0\right)$$

$$a^0_{0,2} = a^0_{2,0} = \frac{1}{\sqrt{160\pi}}\left(|A_R|^2 + |A_L|^2 - 2|A_0|^2\right) \qquad a^1_{1,2} = a^1_{2,1} = \frac{3}{\sqrt{640\pi}}\left(A_LA^*_0 - A_0A^*_R\right)$$

$$a^0_{1,1} = \frac{3}{\sqrt{128\pi}}\left(|A_R|^2 + |A_L|^2\right) \qquad a^1_{2,2} = \frac{3}{40\sqrt{2\pi}}\left(A_0A^*_R + A_LA^*_0\right)$$

$$a^0_{2,1} = a^0_{1,2} = \sqrt{\frac{3}{640\pi}}\left(|A_L|^2 - |A_R|^2\right) \qquad a^2_{2,2} = \frac{3}{20\sqrt{2\pi}}A_LA^*_R.$$

For single top quark t-channel production the transition amplitudes are $A_{-1,-\frac{1}{2}}$, $A_{1,\frac{1}{2}}$, $A_{0,-\frac{1}{2}}$, and $A_{0,\frac{1}{2}}$ to a left handed W boson and a left handed b quark, a right handed W boson and a right handed b quark, a longitudinal W boson and a left handed b quark, and a longitudinal W boson and a right handed b quark. The angular coefficients are:

$$a^0_{0,0} = \frac{1}{\sqrt{8\pi}}\left(|A_{1,\frac{1}{2}}|^2 + |A_{0,\frac{1}{2}}|^2 + |A_{0,-\frac{1}{2}}|^2 + |A_{-1,-\frac{1}{2}}|^2\right)$$

$$a^0_{0,1} = +\frac{\sqrt{3}}{2}\left(|A_{1,\frac{1}{2}}|^2 - |A_{-1,-\frac{1}{2}}|^2\right)$$

$$a^0_{0,2} = +\frac{1}{2\sqrt{5}}\left(|A_{1,\frac{1}{2}}|^2 - 2|A_{0,\frac{1}{2}}|^2 - 2|A_{0,-\frac{1}{2}}|^2 + |A_{-1,-\frac{1}{2}}|^2\right)$$

$$a^0_{1,0} = +P\frac{1}{\sqrt{3}}\left(|A_{1,\frac{1}{2}}|^2 - |A_{0,\frac{1}{2}}|^2 + |A_{0,-\frac{1}{2}}|^2 - |A_{-1,-\frac{1}{2}}|^2\right)$$

$$a^0_{1,1} = +P\frac{1}{2}\left(|A_{1,\frac{1}{2}}|^2 + |A_{-1,-\frac{1}{2}}|^2\right)$$

$$a^0_{1,2} = +P\frac{1}{2\sqrt{15}}\left(|A_{1,\frac{1}{2}}|^2 + 2|A_{0,\frac{1}{2}}|^2 - 2|A_{0,-\frac{1}{2}}|^2 - |A_{-1,-\frac{1}{2}}|^2\right)$$

$$a^1_{1,1} = -P\frac{1}{\sqrt{2}}\left(A_{1,\frac{1}{2}}A^*_{0,\frac{1}{2}} + A^*_{-1,-\frac{1}{2}}A_{0,-\frac{1}{2}}\right)$$

$$a^{-1}_{1,1} = -P\frac{1}{\sqrt{2}}\left(A^*_{1,\frac{1}{2}}A_{0,\frac{1}{2}} + A_{-1,-\frac{1}{2}}A^*_{0,-\frac{1}{2}}\right)$$

$$a^1_{1,2} = -P\frac{1}{\sqrt{10}}\left(A_{1,\frac{1}{2}}A^*_{0,\frac{1}{2}} - A^*_{-1,-\frac{1}{2}}A_{0,-\frac{1}{2}}\right)$$

$$a^{-1}_{1,2} = -P\frac{1}{\sqrt{10}}\left(A^*_{1,\frac{1}{2}}A_{0,\frac{1}{2}} - A_{-1,-\frac{1}{2}}A^*_{0,-\frac{1}{2}}\right),$$

P being the top quark polarization. In both cases, a real PDF requires that $a^m_{k,l} = a^{-m*}_{k,l}$.

To estimate unknown angular coefficients $a^m_{k,l}$ from a dataset $\mathcal{D} = \{(\theta_{1i}, \theta_{2i}, \phi_i)\}$, $i = 1, N$, the usual recourse is some variant of a maximum likelihood fit. Another approach relies upon the orthonormality of the M-functions. By projecting the density function $\rho(\theta_1, \theta_2, \phi)$ onto a basis of M-functions we can express the coefficients as

$$a^m_{k,l} = \int \rho(\theta_1, \theta_2, \phi) M^{m*}_{k,l}(\theta_1, \theta_2, \phi) d\Omega^M \tag{4}$$

The Monte Carlo estimate of the integral is simply the average of the function $M^{m*}_{k,l}(\theta_1, \theta_2, \phi)$, over a realization of the PDF $\rho(\theta_1, \theta_2, \phi)$. The dataset $\mathcal{D}$ is, precisely, the desired realization. So, the estimate of the coefficients $a^m_{k,l}$ comes down simply to, first, computing the value of $M^{m*}_{k,l}(\theta_1, \theta_2, \phi)$ for each event in the dataset and, second, taking the average. A complete covariance matrix between all of the $a^m_{k,l}$ can be computed at the same time. The technique is referred to as Orthogonal Series Density Estimation[3] (OSDE).

To illustrate this we show, in Fig. 1 on the left, an image of the American film actor Humphrey Bogart (1899-1957). The image defines a two-dimensional PDF. In the middle of Fig. 1 we show a realization of the distribution, generated using Von Neumann rejection. 100K points were generated. On the right we show an OSDE estimate of the distribution, obtained using harmonic functions as a basis.

Returning now to physics, we have used the Pythia Monte Carlo[4] to generate a sample of heavy Higgs bosons with a mass of 200 GeV decaying to W boson pairs. An OSDE analysis of the data sample was performed to determine the coefficients. Results are shown in Fig. 2. In Fig. 3 we implement simple acceptance cuts and smearing (mostly due to missing energy effects). These spoil the simple picture, but we can recover using a signal processing trick akin to the convolution theorem. The convolution theorem says that the Fourier transform of a convolution of two functions, $\widetilde{f \circ g}$, is the product of the Fourier transforms of each function:

$$\widetilde{f \circ g} = \tilde{f} \cdot \tilde{g}.$$

An analogous theorem, the Funk-Hecke theorem[5, 6] treats isotropic smearing on a sphere. If

$$\rho(\theta, \phi) = \sum_{l=0}^{\infty} \sum_{m=-l}^{l} c^m_l Y^m_l(\theta, \phi) \qquad \text{and} \qquad \mathcal{R}(\Theta) = \sum_{l=0}^{M} r_l P_l(\cos\Theta)$$

Figure 1. Demonstration of OSDE. Left: original image. Middle: data generated according to the original image. Right: estimate of the light density obtained with OSDE.

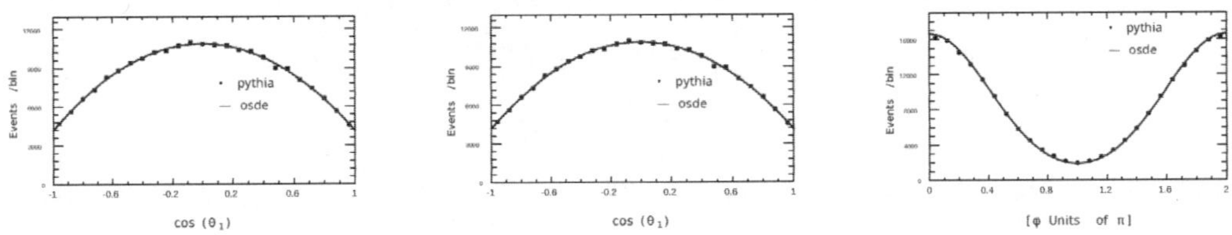

Figure 2. Projections of the joint PDF $\rho(\theta_1, \theta_2, \phi)$ determined using OSDE onto $\cos\theta_1$ and $\cos\theta_2$ (left, middle) and ϕ (right), for a 200 GeV Higgs decaying to W^+W^-.

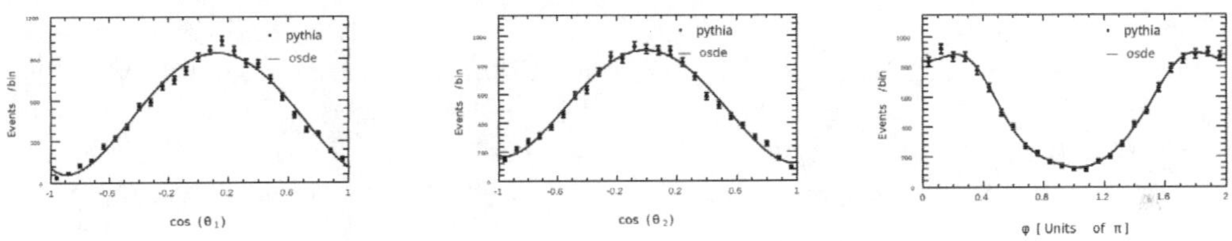

Figure 3. Projection of the joint PDF $\rho(\theta_1, \theta_2, \phi)$ with the inclusion of detector effects determined using OSDE. Obvious distortions occur.

then the convolution is

$$(\rho \circ \mathcal{R})(\theta, \phi) = \sum_{l=0}^{M} \sum_{m=-l}^{l} d_l^m Y_l^m(\theta, \phi) \qquad \text{where} \qquad d_l^m \equiv \frac{2}{2l+1} r_l c_l^m \qquad \text{(no summation)}$$

To handle the case of anisotropic smearing of the function $\rho(\theta_1, \theta_2, \phi)$ we develop our own convolution theorem as follows. We describe the joint PDF for production at true angles θ_{1T}, θ_{2T}, and ϕ_T reconstructed at θ_{1R}, θ_{2R}, and ϕ_R, as a series in *product* M-functions:

$$\mathcal{R}(\theta_{1T}, \theta_{2T}, \phi_T, \theta_{1R}, \theta_{2R}, \phi_R) = r_{k,l,m,k',l',m'} M_{k,l}^{m}(\theta_{1T}\theta_{2T}, \phi_T) M_{k',l'}^{m'}(\theta_{1R}, \theta_{2R}, \phi_R). \qquad (5)$$

and determine the coefficients from Monte Carlo, using OSDE. Using orthonormality (Eq. 1) and Gaunt's theorem (Eq. 2), one can relate the coefficients of a joint PDF to those, denoted by $g_{\kappa',\lambda',\mu',K',L',M'}$, of a conditional PDF, as follows:

$$a_{\kappa,\lambda,\mu} W_{\kappa',\lambda',\kappa,\lambda,K,L}^{\mu',\mu,M} \cdot g_{\kappa',\lambda',\mu',K',L',M'} = r_{1,K,L,M,K',L',M'} \qquad (6)$$

Here, the $a_{\kappa,\lambda,\mu}$ are the known coefficients of $\rho(\theta_{1T}, \theta_{2T}, \phi_T)$ taken from Monte Carlo.

For each value of K, L, M, K', L', M', Eq. 6 is a matrix equation which can be inverted to determine the coefficients $g_{K',\lambda',\mu',K',L',M'}$. To obtain the distribution of reconstructed angles, one must convolve the distribution of true angles with the conditional probability to migrate from the true angles the reconstructed angles. The coefficients of this distribution are:

$$\mathcal{A}_{k,l,m} = g_{K,L,-M,k,l,m} a_{K,L,M}. \tag{7}$$

This is a matrix equation that can be expressed as

$$\vec{\mathcal{A}} = \mathbf{G} \cdot \vec{a} \tag{8}$$

This equation has to be inverted, but the matrix **G** is rectangular, and has an infinite number of rows! So, we restrict to a finite subset of reconstructed coefficients and "invert" by minimizing

$$\chi^2(\vec{a}) = (\vec{\mathcal{A}} - \mathbf{G} \cdot \vec{a})^T \cdot \mathbf{W} \cdot (\vec{\mathcal{A}} - \mathbf{G} \cdot \vec{a}) \tag{9}$$

which has the analytic solution

$$\vec{a} = \mathbf{V}\mathbf{G}^T\mathbf{W}\vec{\mathcal{A}} \quad \text{where} \quad \mathbf{V} = \mathrm{Cov}(\vec{a}) = (\mathbf{G}^T\mathbf{W}\mathbf{G})^{-1} \quad \text{and} \quad \mathbf{C} = \mathbf{W}^{-1} = \mathrm{Cov}(\vec{\mathcal{A}}) \tag{10}$$

Increasing the number of coefficients increases the information used in the fit, but also increases the noise. Some experimentation is required to optimize the cutoff.

1. Conclusion

We have outlined here a procedure for using Fourier tricks, particularly a kind of convolution theorem, to deconvolve the effects of a particle detector from a reconstructed signal. We show this procedure in action in Fig. 4. On the left we show reconstructed coefficients obtained by applying the OSDE procedure on simulated data at reconstruction level. On the right we compare the true coefficients with those obtained by the proposed deconvolution procedure.

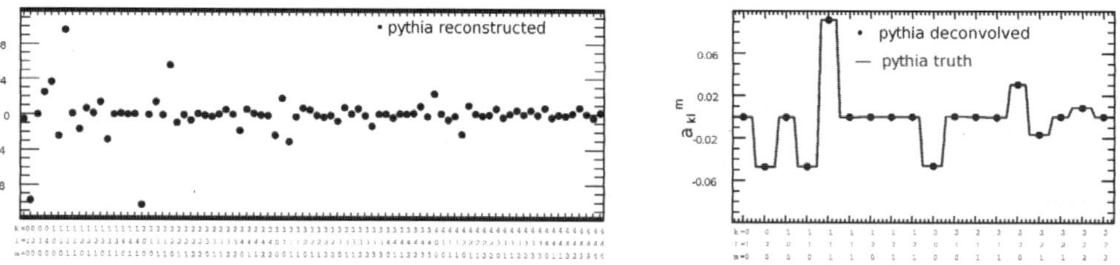

Figure 4. Left: angular coefficients $a^m_{k,l}$ of a simulated $H \to W^+W^-$ signal in a lepton+jets final state, at reconstruction level. Right: deconvolved angular coefficients (blue points) compared to a truth-level estimate (red line). This demonstrates the viability of the method outlined herein.

References

[1] Tyler, C, Eboli O J P, Goncalves D, Gonzales-Fraile J, Plehn T and Rauch M 2015 *Preprint* arXiv:1505.05516
[2] Aguilar-Saavedra J A and Bernabeu J 2010 *Nucl. Phys. B* **840** 349-378
[3] Schwartz S C 1967 *Ann. Math. Statist.* **38** 1261-1265
[4] Sjöstrand T et al 2015 *Comput. Phys. Commun.* **191** 159-177
[5] Hecke E 1917 *Mathematische Annalen* **78** (1) 3 98-404
[6] Basri R and Jacobs D 2003 *IEEE transactions on pattern analysis and machine intelligence* **25** (2) 218-233

Journal of Physics: Conference Series **762** (2016) 012042 doi:10.1088/1742-6596/762/1/012042

Density Estimation Trees as fast non-parametric modelling tools

Lucio Anderlini

Istituto Nazionale di Fisica Nucleare, Sezione di Firenze – via Sansone 1, Sesto Fiorentino, 50019 Italy

E-mail: Lucio.Anderlini@cern.ch

Abstract. A Density Estimation Tree (DET) is a decision trees trained on a multivariate dataset to estimate the underlying probability density function. While not competitive with kernel techniques in terms of accuracy, DETs are incredibly fast, embarrassingly parallel and relatively small when stored to disk. These properties make DETs appealing in the resource-expensive horizon of the LHC data analysis. Possible applications may include selection optimization, fast simulation and fast detector calibration. In this contribution I describe the algorithm and its implementation made available to the HEP community as a RooFit object. A set of applications under discussion within the LHCb Collaboration are also briefly illustrated.

1. Introduction

The fast increase of computing resources needed to analyse the data collected in modern hadron-collider experiments, and the lower cost of processing units with respect to storage, pushes High-Energy Physics (HEP) experiments to explore new techniques and technologies to move as much as possible of the data analysis at the time of the data acquisition (*online*) in order to select candidates to be stored on disk, with maximal, reasonably achievable, background rejection. Besides, research on multivariate algorithms, active both within and outside of the HEP community, is approaching the challenge of operating in distributed computing environments, which represents a further motivation for studies of new classes of algorithms.

Statistical inference of probability density functions underlying experimental datasets is common in *High Energy Physics*. *Fitting* is an example of *parametric* density estimation. When possible, defining a parametric form of the underlying distribution and choose the values for the parameters maximizing the likelihood is usually the best approach. In multivariate problems with a large number of variables and important correlation, however, fitting may become unpractical, and *non-parametric density estimation* becomes a valid, largely employed, solution.

In HEP, the most common non-parametric density estimation, beyond the histogram, is probably *kernel density estimation* [1], based on the sum of normalized kernel functions centered on each data-entry.

In this write-up, I discuss *Density Estimation Trees*, algorithms based on a *multivariate, binary tree* structure, oriented to *non-parametric density estimation*. Density Estimation Trees are less accurate than kernel density estimation, but much faster. Integrating Density Estimation Trees is also trivial and fast, making iterative search algorithms convenient. Finally, storing a

Density Estimation Trees and propagate it through the computing nodes of a distributed system is relatively cheap, offering a reasonable solutions for compressing the statistical information of large datasets.

An implementation of the algorithm in ROOT/RooFit is available through CERN GitLab[1].

2. The algorithm

The idea of iteratively splitting a data sample, making the density estimation to coincide with the average density in each portion of the data space is not new. However, the technique had little room for applications in analyses of datasets up to a few thousands of entries described by small sets of correlated variables.

Recently, *kd*-trees [2] have been used to split large samples into sub-sets consisting of equal fractions of the data entries. The idea underlying *kd*-trees is the iterative splitting of the data-sample using as threshold the median of a given projection. While powerful to solve a vast range of problems, including notably nearest-neighbour searches, the lack of appropriateness of *kd*-trees to estimate probability densities is evident considering samples including multiple data-entries.

Ideally, the density estimation $\hat{f}(\mathbf{x})$ approximating the underlying density function $f(\mathbf{x})$ should minimize the quantity $\mathcal{R} = \int (\hat{f}(\mathbf{x}) - f(\mathbf{x}))^2 \mathrm{d}\mathbf{x}$. It can be shown [3] that, exploiting the Monte Carlo approximation

$$\lim_{N\to+\infty} \frac{1}{N}\sum_{i=1}^{N} g(x_i) = \int g(x)h(x)\mathrm{d}x \qquad \text{where } x_1, x_2, ..., x_N \text{ distribute as } h(x), \tag{1}$$

the minimization of $\mathcal{R}$ is equivalent to growing a Density Estimation Tree iteratively splitting the node ℓ, into the two sub nodes ℓ_R and ℓ_L minimizing the *Gini* index, $G(\ell) = R(\ell) - R(\ell_R) - R(\ell_L)$. Here, $R(\ell)$ represents the *replacement error*, defined by

$$R(\ell) \equiv -\frac{N_\ell^2}{N_{\mathrm{tot}}^2 V_\ell}, \tag{2}$$

where V_ℓ is the hyper-volume of the portion of the data-space associated to the node ℓ, and N_ℓ the number of data entries it includes; N_{tot} is the number of data entries in the whole dataset.

Figure 1 shows an example of how the training is performed.

2.1. Overtraining

As in the case of Classification algorithms, overtraining is the misinterpretation of statistical fluctuations of the dataset as relevant features to be reproduced by the model.

An example of overtraining of Density Estimation Trees is presented in Figure 2. In the presented dataset, the alignment of data-entries in one of the input variables is interpreted as narrow spikes. To compensate spikes, in terms of absolute normalization, the density is underestimated in all of the other points of the parameter space.

Overtraining in decision trees is controlled through an iterative approach consisting in *pruning* and *cross-validation*: finding and removing the branches increasing the complexity of the tree without enhancing the statistical agreement with a set of test samples. Cross-validation is very expensive in terms of computing power and often fails to identify problems of over-training arising close to the root of the decision tree.

Overtraining in density estimation is fought, instead, by defining *a priori* an expected resolution width, neglecting fluctuations under that resolution while building the statistical

[1] gitlab.cern.ch/landerli/density-estimation-trees

 doi:10.1088/1742-6596/762/1/012042

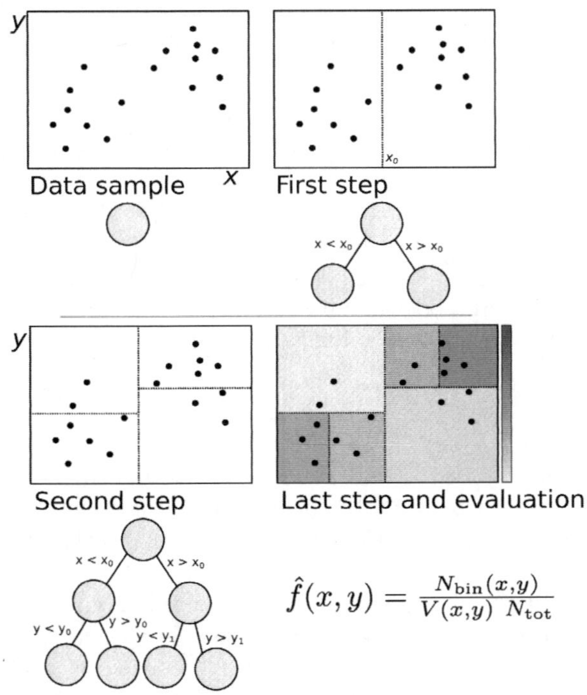

Figure 1. Schematic representation of the training and the evaluation procedures of a Density Estimation Tree.

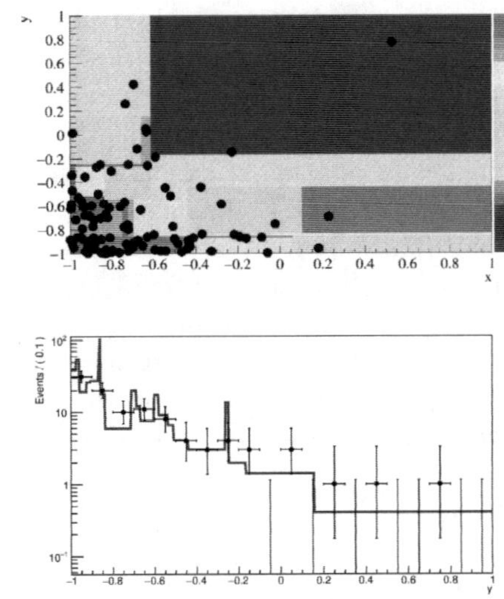

Figure 2. Top, an example of an overtrained Density Estimation Tree. The random alignment of data-entries with respect to one of the variables describing the problem is misinterpreted as a spike in the density estimation, as evident in the projection onto the vertical axis shown on the bottom.

model. For example, kernel density estimation algorithms require a parameter, named *bandwidth* as an input. The bandwidth is related to the width of the kernel function. Abundant literature exists on techniques to optimise the bandwidth for a certain dataset, most of them represent a preliminary step of the density estimation algorithm. Growing a Density Estimation Tree with a minimal leaf width is fast, doesn't require post-processing and it is found to result in better-quality estimations with respect to cross-validation procedures. The same techniques used to compute the optimal bandwidth parameter for kernel density estimation algorithms can be used to define the optimal minimal leaf width of the Density Estimation Tree.

2.2. Integration

As mentioned in the introduction, fast integration of the statistical model built using Density Estimation Trees is one of the strengths of the algorithm. Integration usually responds to two different needs: *normalization* and *slicing* (or *projecting*, or *marginalizing*).

Integrals to compute the overall normalization of the density estimation, or the contribution in a large fraction of the data-space, gain little from exploiting the tree structure of the density estimator. A sum over the contributions of each leaf represents the best strategy.

Instead, integrals over a narrow subset of the data-space should profit of the tree structure of the density estimation to exclude from the integration domain as many leaves as possible, as early as possible. Exploiting the tree structure when performing integrals of slices can drastically reduce the computing time in large density estimation trees.

2.3. Operations with Density Estimation Trees

Combining weighted Density Estimation Trees can be useful to model data samples composed of two or more components. Combination is achieved implementing both scalar and binary operations. There is not much to discuss about scalar operations, where the scalar operation is applied to each leaf independently. Instead, binary operations require the combination of two different Density Estimation Trees, which is not trivial because the boundaries are *a priori* different. The algorithm to combine two Density Estimation Trees consists of the iterative splitting of the terminal nodes of the first tree, following the boundaries of the terminal nodes of the second one. Once the combination is done, the first tree is compatible with the second one and the binary operation can be performed node per node. The resulting tree may have several additional layers with respect to the originating trees, therefore a final step removing division between negligibly different nodes is advisable.

3. Discussion of possible applications

Density Estimation Trees are useful to approach problems defined by many variables and for which huge statistical samples are available. To give a context to the following examples of applications, I consider the calibration samples for the Particle IDentification (PID) algorithms at the LHCb experiment.

PID calibration samples are sets of decay candidates reconstructed and selected relying on kinematic variables only, to distinguish between different types of long-lived particles: electrons, muons, pions, kaons, and protons.

The PID strategy of the LHCb detector relies on the combined response of several detectors: two ring Cherenkov detectors, an electromagnetic calorimeter, a hadronic calorimeter and a muon system [4]. The response of the single detectors are combined into likelihoods used at analysis level to define the tightness of the PID requirements.

Calibration samples count millions of background-subtracted candidates, each candidate is defined by a set of kinematic variables, for example momentum and pseudorapidity, and a set of PID likelihoods, one per particle type. The correlation between all variables is important and not always linear.

3.1. Efficiency tables

The first application considered is the construction of tables defining the probability that the PID likelihood of a candidate, defined by a set of kinematic variables, satisfies a particular requirement.

Building two Density Estimation Trees with the kinematic variables defining the data-space, one with the full data sample (tree $t_{\rm all}$), and one with the portion of data sample passing the PID criteria (tree $t_{\rm pass}$), allows to compute the efficiency for each combination of the kinematic variables by evaluating the Density Estimation Tree obtained taking the ratio $t_{\rm pass}/t_{\rm all}$.

For frequently-changing criteria a dynamic determination of the efficiency can be envisaged. For simplicity, consider the generic univariate PID criterion $y > 0$. In this case a single Density Estimation Tree $d(x_1, x_2, y)$ defined by the kinematic variables (x_1, x_2), and one PID variable y, has to be trained on the calibration sample. The dynamic representation of the efficiency for a candidate having kinematic variables $(\hat{x}_1, \hat{x}_2)$ is the ratio

$$\epsilon(x_1, x_2; y > 0) = \int_{y>0} d(\hat{x}_1, \hat{x}_2, y)\mathrm{d}y \Bigg/ \int_{\mathrm{any}\ y} d(\hat{x}_1, \hat{x}_2, y)\mathrm{d}y. \qquad (3)$$

Thanks to the fast slice-integration algorithm, the computation of this ratio can be included in an iterative optimization procedure aiming at an optimization of the threshold on y.

3.2. Sampling as fast simulation technique

Another important application is related to fast simulation of HEP events. Full simulation, including interaction of the particles with matter, is becoming so expensive to be expected exceeding the experiments' budgets in the next few years. Parametric simulation is seen as a viable solution, as proved by the great interest raised by the DELPHES project [5]. However, parametrizing a simulation presents the same pitfalls as parametrizing a density estimation: when correlation among different variables becomes relevant, the mathematical form of the parametrization increases in complexity up to the point it becomes unmanageable.

Density Estimation Trees are an interesting candidate for non-parametric fast simulation. Let $d(x_1, ..., x_N, y_1, ..., y_n)$ be a Density Estimation Tree trained on a set of candidates defined by *generator* variables $\mathbf{x} \equiv (x_1, ..., x_N)$, and by variables $\mathbf{y} \equiv (y_1, ..., y_n)$ obtained through full simulation. For example, $\mathbf{x}$ could represent the kinematic variables of a track and $\mathbf{y}$ the PID likelihoods.

The aim of fast non-parametric simulation is, given a new set of values $(\hat{x}_1, ..., \hat{x}_N)$ for $\mathbf{x}$, to compute a set of values for $\mathbf{y}$ distributed according to the conditional probability density function $d(y_1, ..., y_n|\hat{x}_1, ..., \hat{x}_N)$.

Once the DET is trained, the tree structure of the density estimator is used to compute for each leaf ℓ the hyper-volume $V_{\ell \cap \hat{\mathbf{x}}}$ of the intersection between ℓ and hyper-plane defined by $\mathbf{x} = (\hat{x}_1, ..., \hat{x}_N)$.

A random leaf $L \in \{\ell\}$ is then chosen with probability proportional to $V_{L \cap \hat{\mathbf{x}}}$, and variables $\mathbf{y}$ are generated following a flat distribution bounded within L.

A set of *generator* variables $\mathbf{x}$ can then be completed by the corresponding $\mathbf{y}$ variables without full simulation, but relying on the joint multivariate distribution learnt by the Density Estimation Tree.

4. Summary and outlook

I discussed Density Estimation Tree algorithms as fast modelling tools for high statistics problems characterized by a large number of correlated variables and for which an approximated model is acceptable. The fast training and integration capabilities make these algorithms of interest for the high-demanding future of the High-Energy Physics experiments. The examples discussed, which benefited from an active discussion within the Particle Identification Group of the LHCb collaboration, explore cases where the statistical features of huge samples have to be assessed in a time shorter than what standard estimators would require. In future, Density Estimation Trees could be sampled to train Regression Multivariate Algorithms, such as Neural Networks, in order to smooth the response and further speed up the query time, at the cost of loosing its fast-integration properties.

Acknowledgements

I thank Alberto Cassese, Anton Poluektov, and Marco Cattaneo for the encouragements in developing this work and for the useful discussions we had.

References

[1] K. S. Cranmer, Comput. Phys. Commun. **136** (2001) 198 doi:10.1016/S0010-4655(00)00243-5 [hep-ex/0011057].
[2] J. L. Bentley, Communications of the ACM **18** (1975) 9 doi:10.1016/S0010-4655(00)00243-5
[3] P. Ram and A. G. Gray, Proceedings of the 17th ACM SIGKDD international conference on Knowledge discovery and data mining, pages 627-635
[4] A. A. Alves *et al.*, The LHCb detector at the LHC, JINST 3 (2008)
[5] J. de Favereau *et al.* [DELPHES 3 Collaboration], JHEP **1402** (2014) 057 doi:10.1007/JHEP02(2014)057 [arXiv:1307.6346 [hep-ex]].

Journal of Physics: Conference Series **762** (2016) 012043 doi:10.1088/1742-6596/762/1/012043

Development of Machine Learning Tools in ROOT

S. V. Gleyzer[1], L. Moneta[2], Omar A. Zapata[3]
[1] University of Florida
[2] CERN
[3] University of Antioquia and Metropolitan Institute of Technology

E-mail: Sergei.Gleyzer@cern.ch, Lorenzo.Moneta@cern.ch, Omar.Zapata@cern.ch

Abstract. ROOT is a framework for large-scale data analysis that provides basic and advanced statistical methods used by the LHC experiments. These include machine learning algorithms from the ROOT-integrated Toolkit for Multivariate Analysis (TMVA). We present several recent developments in TMVA, including a new modular design, new algorithms for variable importance and cross-validation , interfaces to other machine-learning software packages and integration of TMVA with Jupyter, making it accessible with a browser.

1. Introduction

ROOT is an object-oriented data analysis framework that provides statistical methods, visualization and storage libraries for data analysis of high-energy physics (HEP) experiments such as the Large Hadron Collider (LHC) in Geneva, Switzerland [1]. Although originally designed for HEP applications, ROOT is also widely used in other scientific fields outside of particle physics.

Machine learning algorithms have become integral to modern HEP analyses. The ROOT framework provides TMVA, the Toolkit for Multivariate Analysis [2],that contains widely-used machine learning algorithms in HEP such as:

- Fisher and Linear Discriminants (LD)
- Boosted Decision Trees (BDT)
- Decision Rule Ensembles
- k-Nearest Neighbor Classifier (KNN)
- Shallow Artificial Neural Networks (ANN)
- Deep Learning Neural Networks (DNN)
- Support Vector Machines (SVM)

Of these, most popular methods in HEP are boosted decision trees, neural networks and support vector machines. TMVA provides implementations of these popular methods applicable for machine-learning classification and regression. TMVA provides a way to compare the performance of these algorithms on the same dataset, allowing for an apples-to-apples comparison useful in choosing the optimal algorithm for a particular analysis task.

Recently, TMVA has been undergoing a significant makeover targeting greater flexibility, modular design and some new features and interfaces. As machine learning algorithms continue to evolve, the design of machine learning frameworks needs to be as flexible as possible.

2. New Algorithms and Features

Algorithms with useful information for the user have been added, such as cross-validation and variable importance. Also TMVA was integrated with Jupyter, allowing execution in a browser. In the next sections we describe some of the new functionality and features of TMVA.

2.1. Data Loader

One new design development in TMVA is the DataLoader class. This class creates greater flexibility and modularity in training different combinations of classifiers and variables. Previously, the choice of variables was defined once and could not be changed later. The new DataLoader class allows the user to combine different choices of variables and methods with the data. This design flexibility allows other useful algorithms, such as cross-validation and variable importance, described in the following sections, to be implemented.

Figures 1, 2 and 3 show the structure of the DataLoader class. New algorithm features require several DataLoaders at the same time. Various file formats, for example .root and comma-separated value (.csv) files are supported in the DataLoader class.

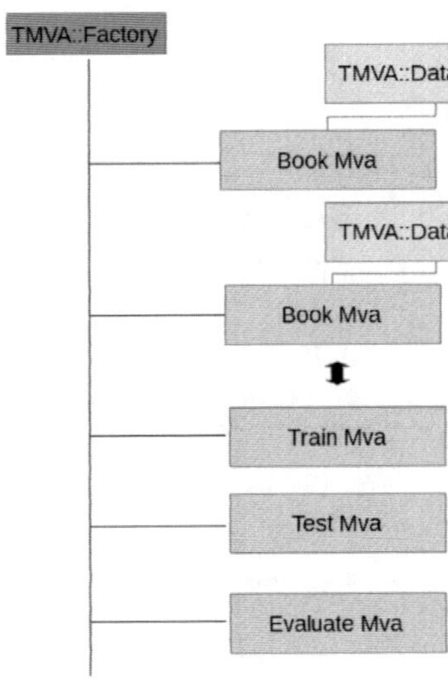

Figure 1. Booking methods with different dataloaders.

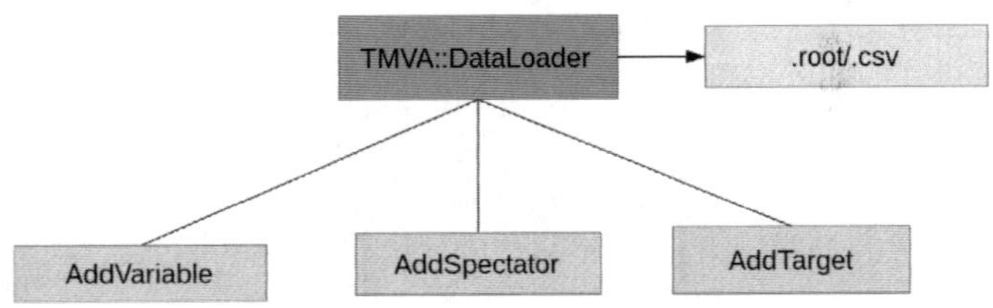

Figure 2. Loading data from files.

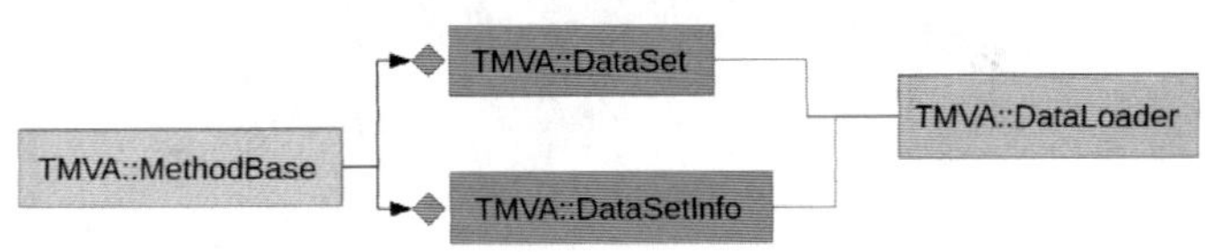

Figure 3. Storing data in objects in the class MethodBase.

2.2. Cross Validation

Cross-validation is a machine-learning model evaluation technique, important for model generalization to unseen data. During k-fold cross-validation, the dataset is partitioned into k folds or partitions. During one cross-validation round, k-1 folds are used for model training, and the remaining one for model testing. Several rounds of cross-validation are performed with different partitions and model performance results are averaged. One advantage of cross-validation over a simple split into training and testing set, is the use of the full dataset to validate the model. Performing cross-validation is known to reduce over-fitting of the data, leading to a more accurate estimate of the performance of the machine-learning model on unseen data [3].

Cross-validation in TMVA is done with a standalone CrossValidation class. Figure 5 illustrates five receiver-operating characteristic (ROC) curves for each cross-validation fold for a basic TMVA example. This example has four random variables with gaussian distributions plus several derived variables after applying basic mathematical operations to these variables.

2.3. Variable Importance

Currently, TMVA provides a number of method-specific variable importance algorithms. Each one is relevant only for the method chosen and is computed during construction. For example, for decision trees variable importance is derived by counting the number of splits for each variable weighted by the square of the information gained from the split, or for neural networks, as the sum of weights between the inputs and the hidden layer [2].

In addition to these, a new method-independent variable importance algorithm was added. This algorithm, described in [4], computes variable importance in the context of classifier performance. A number of seeds are randomly generated, each corresponding to a variable subspace. For each seed, individual variable contributions to classifier performance are measured as loss of classifier performance due to the removal of a variable. Figure 4 shows a sample variable importance plot for a basic example in TMVA.

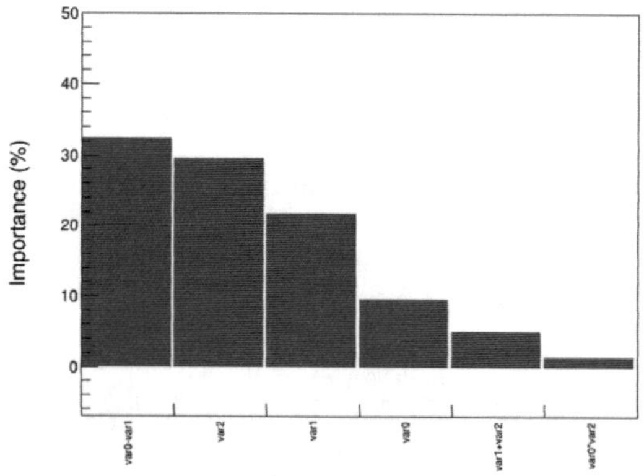

Figure 4. Histogram ranking the variables.

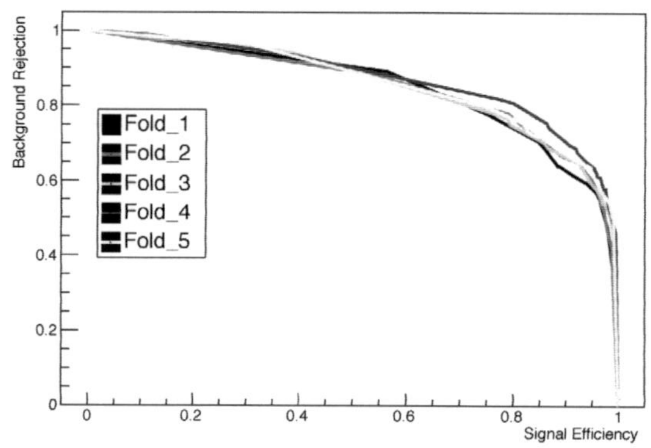

Figure 5. Cross validation ROC curves.

2.4. TMVA and Jupyter Notebooks

Another new feature in TMVA is the integration of TMVA and Jupyter Notebooks [5]. Jupyter is a web application that combines live code, rich text, links and formula in a user-friendly format. With the creation of a ROOT Jupyter kernel, additional integration of machine-learning tools in ROOT with Jupyter notebooks became possible. Currently, all the functionality of TMVA is available in Jupyter notebooks, requiring only access to a browser to run and execute TMVA.

3. Interfaces to external machine learning tools

Another useful functionality added to TMVA is the interface to external machine-learning tools written in R and Python languages. Figure 6 shows the relationship between ROOT, TMVA and other statistical packages.

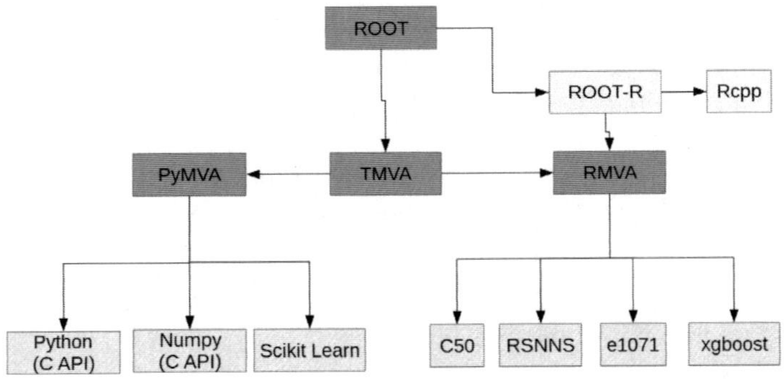

Figure 6. Interplay of Machine Learning Tools in ROOT.

3.1. ROOT-R Interface

R is a free software framework for statistical computing[6]. ROOT-R interface was developed to use R functions directly in ROOT. It opens a large set of statistical tools available in R, including machine-learning packages, for use within ROOT. The ROOT-R interface design is shown in Figure 7.

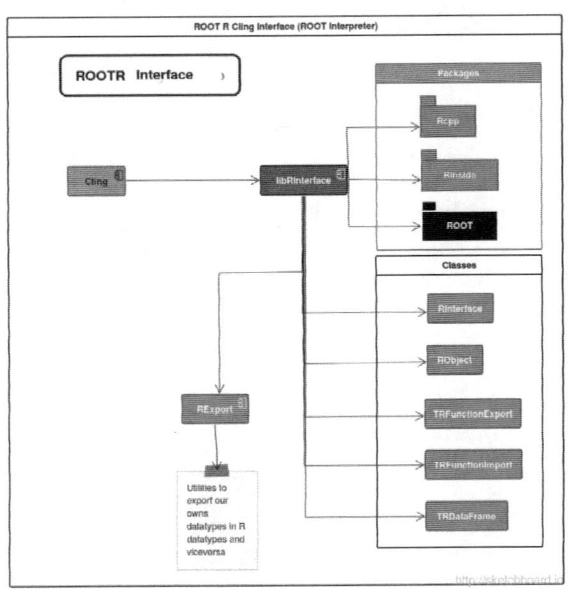

Figure 7. ROOT-R Interface design.

3.2. RMVA Interface

RMVA is a set of TMVA plugins based on the ROOT-R interface. It allows the use of machine-learning methods available in R directly from TMVA. The goal behind RMVA is not to replace the R package itself but to allow its direct comparison with existing tools in TMVA for a given problem, while adding more methods to TMVA users toolbox.

The RMethodBase class in TMVA starts the R environment using ROOT-R, imports the required modules and maps the DataLoader events in R data frame objects using the helper class ROOT::R::TRDataFrame. Each of the methods inherits from the base class RMethodBase as shown in Figure 8. Currently, the following machine-learning packages in R are supported:

- Decision trees and rule-based models (C50) [7].
- Stuttgart Neural Networks in R (SNNS)[8].
- Support Vector Machines in R (e1071)[9].
- eXtreme Gradient Boost (xgboost) An optimized general purpose gradient boosting library[10].

As shown in Figure 10 the above machine-learning methods in R can be executed within the TMVA framework for the basic example.

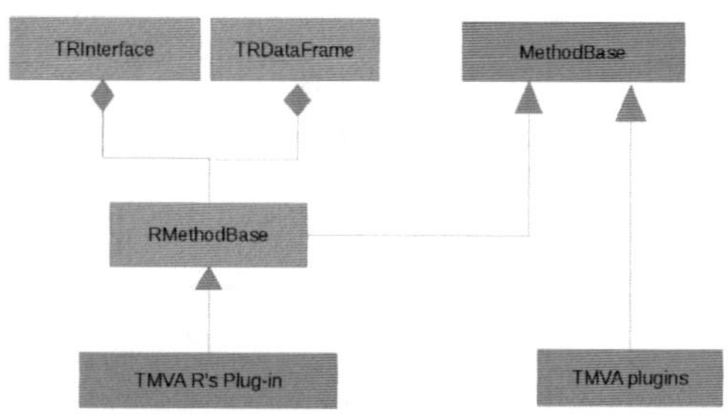

Figure 8. RMVA plugins system.

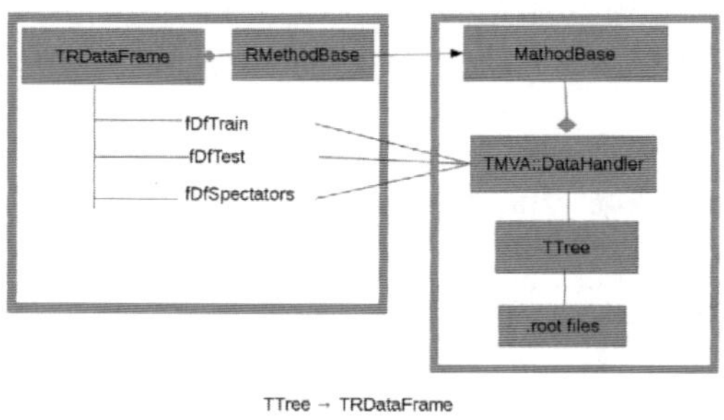

Figure 9. ROOT-R and TMVA data flow.

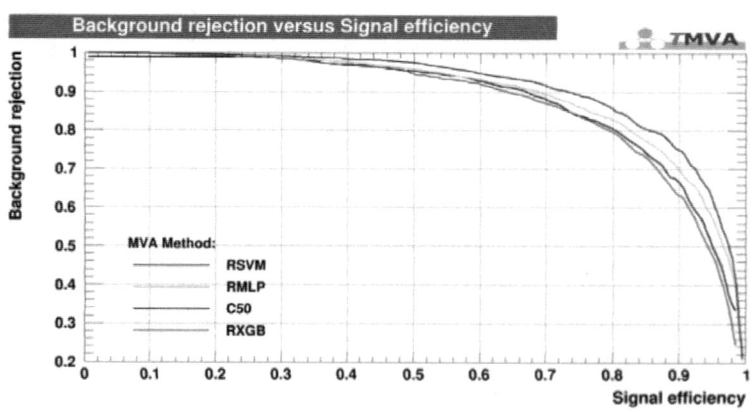

Figure 10. ROC Curves for RMVA methods.

3.3. Python with TMVA (PyMVA)

Similarly to RMVA, PyMVA is a set of TMVA plugins based on Python API that allows direct use of machine-learning methods written in Python from within TMVA. The goal, similarly to RMVA, is not to replace the original method, but to allow the comparison of the method with the other methods in TMVA using the same dataset, selections and performance metrics.
The PyMethodBase class in PyMVA initializes the Python environment, imports the required modules and maps the DataLoader events in numpy arrays using C-API. Each PyMVA method inherits from the base class PyMethodBase, as illustrated in Figure 11. Figure 12 shows how the dataset is mapped from ROOT trees to numpy arrays. The following Python based methods from the Scikit-learn software package [11] are currently available in TMVA:

- Random Forest (PyRandomForest)
- Gradient Boosted Regression Trees (PyGTB)
- Adaptive Boosting (PyAdaBoost)

Figure 13 shows the ROC curves of various PyMVA methods for a basic example.

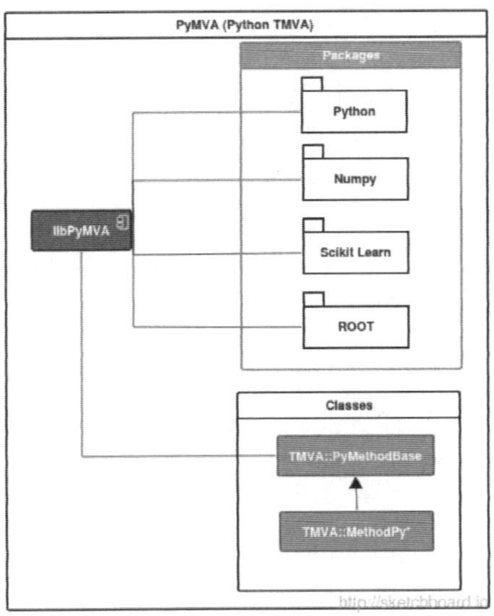

Figure 11. PyMVA plugins system.

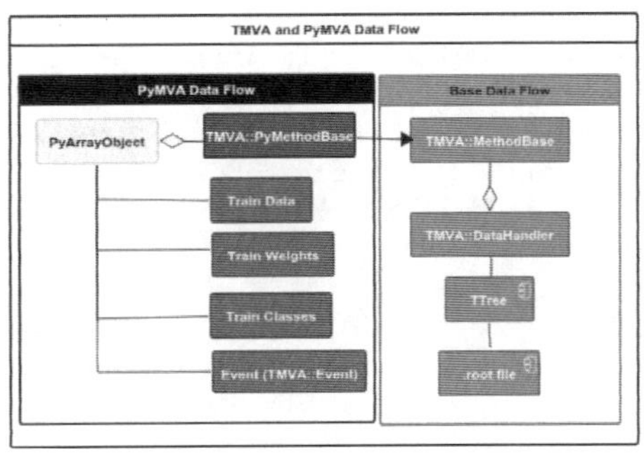

Figure 12. PyMVA data flow.

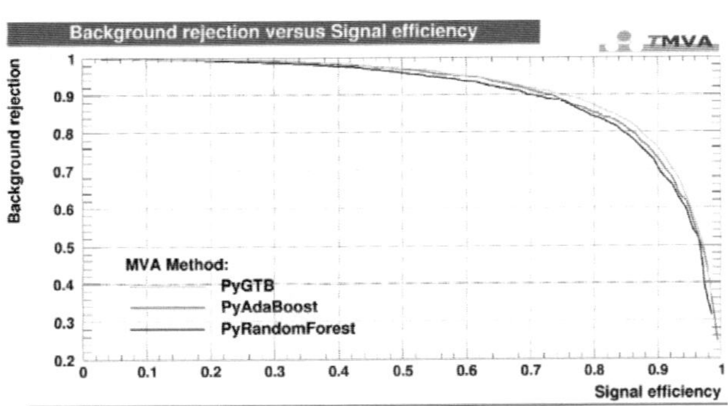

Figure 13. ROC Curves for PyMVA.

3.4. Conclusions

Machine learning tools in ROOT have undergone a significant makeover and upgrade. In particular, TMVA has a new design targeting greater flexibility and modularity, new features such as cross-validation, variable importance and interfaces to R and python-based machine-learning tools. In addition, TMVA is available in Jupyter notebooks, making it accessible in a web browser.

Acknowledgments

The work of O. A. Zapata was partially supported by Sostenibilidad-UdeA, UdeA/CODI grant IN361CE and COLCIENCIAS grant 111-556-934918.

References

[1] Antcheva I, Ballintijn M, Bellenot B, Biskup M, Brun R, Buncic N, Canal P, Casadei D, Couet O, Fine V, Franco L, Ganis G, Gheata A, Maline D G, Goto M, Iwaszkiewicz J, Kreshuk A, Segura D M, Maunder R, Moneta L, Naumann A, Offermann E, Onuchin V, Panacek S, Rademakers F, Russo P and Tadel M 2009 *Computer Physics Communications* **180** 2499 – 2512 ISSN 0010-4655 40 {YEARS} {OF} CPC: A celebratory issue focused on quality software for high performance, grid and novel computing architectures URL `http://www.sciencedirect.com/science/article/pii/S0010465509002550`

[2] Hoecker A, Speckmayer P, Stelzer J, Therhaag J, von Toerne E and Voss H 2007 (*Preprint* `physics/0703039`)

[3] Arlot S, Celisse A *et al.* 2010 *Statistics surveys* **4** 40–79

[4] Gleyzer S V and Prosper H B 2008 *PoS* 067

[5] Pérez F and Granger B E 2007 *Computing in Science and Engineering* **9** 21–29 ISSN 1521-9615 URL `http://ipython.org`

[6] R Core Team 2016 *R: A Language and Environment for Statistical Computing* R Foundation for Statistical Computing Vienna, Austria URL `https://www.R-project.org/`

[7] Kuhn M, Weston S, Coulter N and code for C50 by R Quinlan M C C 2015 *C50: C5.0 Decision Trees and Rule-Based Models* r package version 0.1.0-24 URL `https://CRAN.R-project.org/package=C50`

[8] Bergmeir C and Benítez J M 2012 *Journal of Statistical Software* **46** 1–26 URL `http://www.jstatsoft.org/v46/i07/`

[9] Meyer D, Dimitriadou E, Hornik K, Weingessel A and Leisch F 2015 *e1071: Misc Functions of the Department of Statistics, Probability Theory Group (Formerly: E1071), TU Wien* r package version 1.6-7 URL `https://CRAN.R-project.org/package=e1071`

[10] Chen T and He T 2015 *R package version 0.4-2*

[11] Pedregosa F, Varoquaux G, Gramfort A, Michel V, Thirion B, Grisel O, Blondel M, Prettenhofer P, Weiss R, Dubourg V *et al.* 2011 *Journal of Machine Learning Research* **12** 2825–2830

Journal of Physics: Conference Series **762** (2016) 012044 doi:10.1088/1742-6596/762/1/012044

GPUs for statistical data analysis in HEP: a performance study of *GooFit* on GPUs vs. *RooFit* on CPUs

Alexis Pompili and Adriano Di Florio
(on behalf of the CMS Collaboration)
Dipartimento Interateneo di Fisica di Bari and I.N.F.N.-Sezione di Bari,
via Amendola 173, I-70126 Bari, Italy

E-mail: `alexis.pompili@ba.infn.it`

Abstract. In order to test the computing capabilities of GPUs with respect to traditional CPU cores a high-statistics toy Monte Carlo technique has been implemented both in ROOT/RooFit and `GooFit` frameworks with the purpose to estimate the statistical significance of the structure observed by CMS close to the kinematical boundary of the $J/\psi\phi$ invariant mass in the three-body decay $B^+ \rightarrow J/\psi\phi K^+$. GooFit is a data analysis open tool under development that interfaces ROOT/RooFit to CUDA platform on nVidia GPU. The optimized GooFit application running on GPUs hosted by servers in the Bari Tier2 provides striking speed-up performances with respect to the RooFit application parallelised on multiple CPUs by means of PROOF-Lite tool. The considerably resulting speed-up, while comparing concurrent GooFit processes allowed by CUDA Multi Process Service and a RooFit/PROOF-Lite process with multiple CPU workers, is presented and discussed in detail. By means of GooFit it has also been possible to explore the behaviour of a likelihood ratio test statistic in different situations in which the Wilks Theorem may apply or does not apply because its regularity conditions are not satisfied.

1. Introduction to `GooFit`

The word 'GPU-accelerated computing' refers to an enhancement of application performances that can be obtained by offloading compute-intensive portions to the GPU, while the remaining code still runs on the CPUs. The computing capabilities are enhanced once a sequence of elementary arithmetic operations are performed in parallel on a huge amount of data. In the context of High Energy Physics (HEP) analysis application, `GooFit`[1] is an under development open source data analysis tool, used in applications for parameters' estimation, that interfaces ROOT[2]/RooFit[3] to the CUDA[4] parallel computing platform on nVidia's GPUs (it also supports OpenMP). `GooFit` acts as an interface between the MINUIT[5] minimization algorithm and a parallel processor which allows a Probability Density Function (PDF) to be evaluated in parallel. Fit parameters are estimated at each negative-log-likelihood (NLL) minimization step on the *host side* (CPU) while the PDF/NLL is evaluated on the *device side* (GPU)[6]. Description and details about `GooFit` can be found elsewhere[1]. In this study a comparison between RooFit and `GooFit` performances is presented when a huge amount of pseudo-experiments need to be fitted several times with the aim to estimate a p-value and thus the statistical significance of a new signal reconstructed from the data. The used hardware setup consists in two servers, one equipped with two nVidia TeslaK20 and 32 cores (16 + 16 by Hyper-Threading) and the other with one nVidia TeslaK40 and 40 (20 + 20) cores[7].

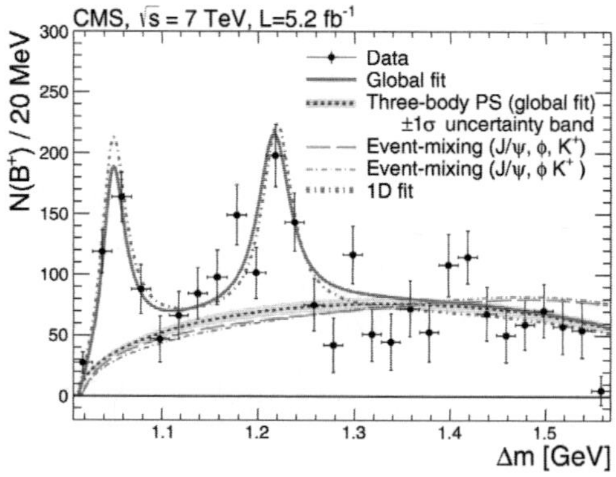

Figure 1. Fits to the background-subtracted $J/\psi\phi$ invariant mass in the $B^+ \to J/\psi\phi K^+$ decay[8]; the significance of the left peak is estimated in this study.

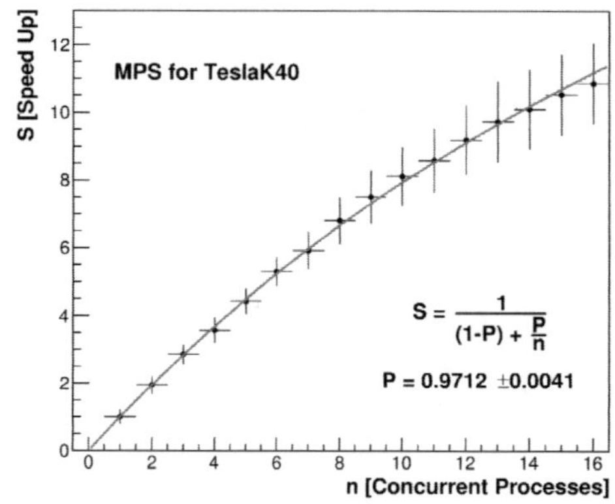

Figure 2. Amdhal's fit to speed-up perfomances of `GooFit`/MPS for a GPU TK40 and up to 16 CPUs.

2. Pseudo-experiments for p-value estimation and `GooFit` performances

In order to test the computing capabilities of GPUs with respect to CPU cores a high statistic toy MC technique has been implemented both in `GooFit` and RooFit frameworks to estimate the local statistical significance of the structure observed by CMS close to the kinematical boundary of the $J/\psi\phi$ invariant mass in the three-body decay $B^+ \to J/\psi\phi K^+$[8]. The found fit parameters (Fig. 1) are compatible with the state Y(4140) observed for the first time by CDF[9].

MC toys are used to estimate the probability (p-value) that background fluctuations would - alone - give rise to a signal as much significant as that seen in the data. A single toy fit cycle consists in the following sequence of steps:

1) generation of fluctuated background binned distribution according to the 3-body phase-space model (the number of entries are fixed to that in the data thus ignoring Poisson fluctuations);
2) a Binned Maximum Likelihood fit (BML) is done with phase-space model (null hypothesis);
3) 8 BML fits are performed by adding to the phase-space a voigtian model truncated to account for the kinematical threshold (alternative hypothesis); the gaussian resolution function has fixed width ($2MeV$) and the signal yield is constrained to be positive. For each bin the PDF value is estimated by integration over the bin since the signal is steep with respect to the bin size. The 8 fits differ by the starting values (2 masses and 4 widths) within the region of interest defined from CDF values[9] (no need to take the Look-Elsewhere-Effect (LEE) into account).
4) For each fit a $\Delta\chi^2$ value is calculated with respect to the null hypothesis fit and the best value is chosen among the 8 alternative fits. The $\Delta\chi^2$ (the test statistic) distribution is obtained over the whole sample of MC toys.

The speed-up for a `GooFit` single process with respect to a RooFit one is 64(48) for a TK40(20). The `GooFit` process does not exploit the whole computational power of a GPU ($\simeq 70\%$); the Multi Process Server (MPS) allows the execution of up to 16 simultaneous processes on the same GPU acting as a scheduler and allowing a balanced full use of GPU capabilities (each process uses one shared GPU and one exclusively assigned CPU). On the other hand PROOF-lite, a dedicated version of PROOF optimized for single multi-core machines[10], is used to efficiently run RooFit toys in parallel on the 72 CPUs available on the two servers. The speed-up of MPS (PROOF-Lite) with respect to one `GooFit` (RooFit) process as a function of the number of independent processes (workers) follows Amdhal's Law (Fig. 2); in both cases the parallelizable fraction is $\simeq 0.97$ i.e. the serial overhead is $\simeq 3\%$ of the application execution.

A first performance comparison can be obtained on the server hosting the two TeslaK20; when

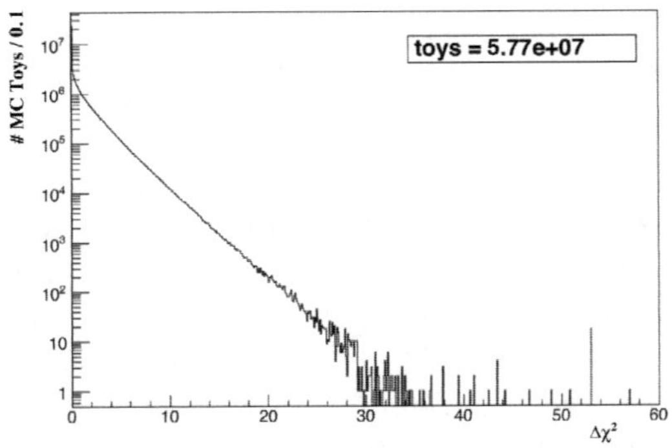

Figure 3. Final $\Delta\chi^2$ distribution with one toy (shown in Fig. 4) exceeding the value observed in the data ($\simeq$ 53.0) marked by a red tick.

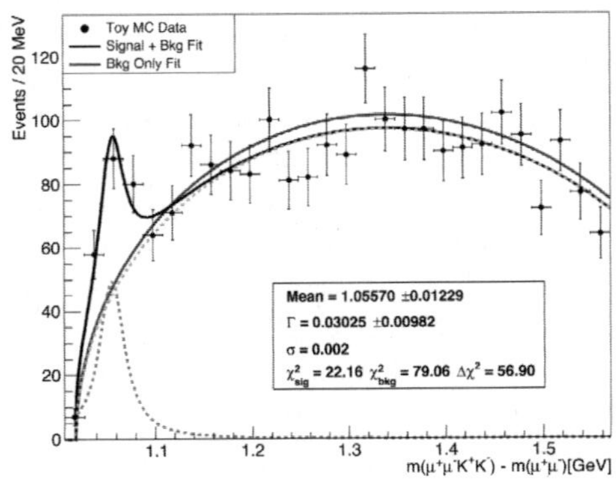

Figure 4. Bkg-only and bkg+signal fits to the MC toy characterized by a $\Delta\chi^2 \simeq 56.9$.

using 2 GooFit/MPS jobs on the 2 GPUs and 30 CPUs (each running 15 processes) against one PROOF-Lite job using 30 CPUs, the speed-up is $\simeq 45$ and pretty stable, as expected, with respect to a strongly varying number of MC toys.

A second comparison can be carried out on both the servers hosting both types of GPUs as a function of the number of produced toys, but limited to 16 independent processes because of the MPS limit for the single TK40. Specifically the comparison is between one RooFit/PROOF-Lite job using 16 workers (on 16 CPU cores) and one GooFit/MPS job running 16 simultaneous processes on a single TK40 or TK20. When using a TK40(TK20) the speed-up has been measured to be $\simeq 60$ (40); the speed-up representing the gain within the same micro-architecture (TK40 vs TK20) is $\simeq 1.5$. All speed-up values are stable with the number of toys.

A third performance comparison can be done from the end-user's point of view. Assuming the analyst has at his own disposal the 2 servers used in these studies equipped with 3 GPUs and 72 CPU cores, it has been measured that $1M$ of MC toys can be produced in ~ 11 days with RooFit/PROOF-Lite and ~ 6 hours with GooFit/MPS. To get a signal statistical significance $> 5\sigma$ a p-value $< 3 \cdot 10^{-}7$ is needed, namely at least $3.3M$ of MC toys are needed. However, as in the case under investigation, the significance estimation may require many more MC toys.

The final $\Delta\chi^2$ distribution, $f(\Delta\chi^2)$ is shown in Fig. 3; the MC toys production was stopped once a fluctuation with $\Delta\chi^2 > \Delta\chi^2_{obs}$ was found (Fig. 4). The p-value is estimated by:

$$P = \int_{\Delta\chi^2_{obs}}^{\infty} f(\Delta\chi^2) d(\Delta\chi^2) \simeq (57.7 \cdot 10^6)^{-1} \simeq 1.73 \cdot 10^{-8} \quad (1)$$

This corresponds to the statistical significance $Z\sigma = \Phi^{-1}(1-P)\sigma \simeq 5.52\sigma$, through the inverse function of the cumulative distribution of the standard gaussian, that is compatible with the lower limit of 5σ quoted in [8] on the basis of $50.5M$ of MC toys obtained by means of RooFit.

3. Exploring the applicability limits of Wilks theorem

By means of GooFit it has also been easier to explore the (asymptotic) behaviour of a likelihood ratio test statistic in different situations in which the Wilks theorem may apply or does not apply because its regularity conditions are not satisfied. The Wilks theorem[11] is often used to estimate the p-value associated to a new signal. Given two hypothesis, the *null* one, H_0, with ν_0 degrees of freedom (dof) and an *alternative* one, H_1, with ν_1 dof, any test statistic t, defined as a likelihood ratio $-2ln\lambda = -2ln(L_{H_0}/L_{H_1})$, or similarly (in the asymptotic limit) as a $\Delta\chi^2 = \chi^2_{H_0} - \chi^2_{H_1}$, approaches a χ^2 distribution with $\nu = \nu_1 - \nu_0$ dof, provided that the following regularity conditions hold: 1) H_0 and H_1 are nested (H_1 includes H_0), 2) while $H_1 \rightarrow H_0$, the H_1 parameters are well behaving (well defined and not approaching some limit), 3) asymptotic limit (namely in the enough large data sample regime). Once this theorem can be applied, the p-value associated to the signal is $p = \int_{t_{obs}}^{\infty} \chi^2_{\nu_1-\nu_0}(t)dt$ and the use of pseudo-experiments to estimate the p-value is not needed (even if still suggested).

When null hypothesis is background-only and the alternative is background plus signal, often the above conditions are not all satisfied, and the MC toys are mandatory. Indeed this is the case previously studied. The signal parameters in the model of H_1 hypothesis are: mass (m), width (Γ) and yield ($\mu \geq 0$); when $H_1 \rightarrow H_0$ the problem is that not only m and Γ are not well defined but also μ tends to the null limit. This explains the previous use of a MC toys technique.

In general the distributions of a test statistic are not predictible and can be extracted from pseudo-experiments. MC toys according to the previously discussed procedure and physics case have been generated for each of the following 4 cases: (1) m and Γ fixed, μ free; (2) m and Γ fixed, μ free but constrained to be positive; (3) m and Γ free, μ free; (4) m and Γ free, μ free but constrained to be positive. The $\Delta\chi^2$ distributions for the four cases are shown in Fig. 5. The fourth case was the one studied so far (with much higher statistics). The first two cases have also been studied and are going to be discussed hereafter.

3.1. *First case: m and Γ fixed, μ free (either positive or negative)*

Let us consider a likelihood ratio test statistic $t_\mu = -2ln\lambda(\mu)$, where μ is the `strength parameter`, as the basis of the statistical test. This can be a test of $\mu = 0$ for purposes of establishing the existence of a signal process.

In this first case, following [12], the PDF of the test statistic $f(t_\mu|\mu) = \frac{1}{\sqrt{2\pi}}\frac{1}{\sqrt{t_\mu}}e^{-t_\mu/2}$ asymptotically approaches a $\chi^2_{\nu=1}$ distribution; this is in agreement with the Wilks theorem, with the difference of dof being one and represented by μ.

A fit to the test statistic distribution with a χ^2_ν model has been performed, where the likelihood ratio distribution was obtained by the already discussed fit procedure but when fixing the values of mass and width parameters to the CMS estimates previously obtained (alternatively they could have been fixed to CDF values), while leaving μ free. The best estimate obtained for the number of dof is $\hat{\nu} \simeq 1.014 \pm 0.001$, thus close to the approximate theoretical prediction; the goodness of fit is checked using a chi-square test that returns a 11.8% probability.

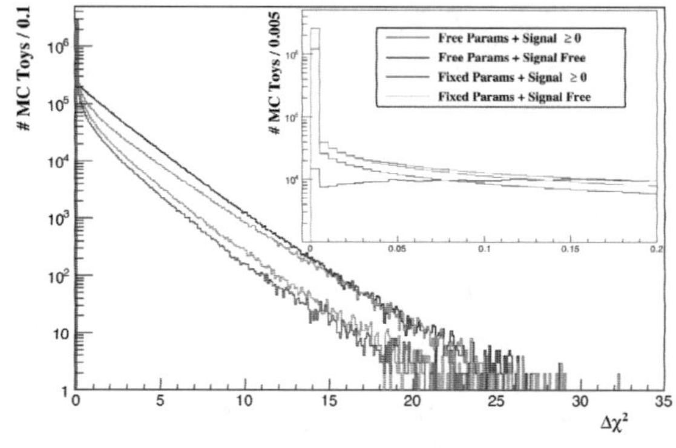

Figure 5. Different test statistic ($\Delta\chi^2$) distributions for the 4 cases discussed in the text, with the same number ($2M$) of MC toys.

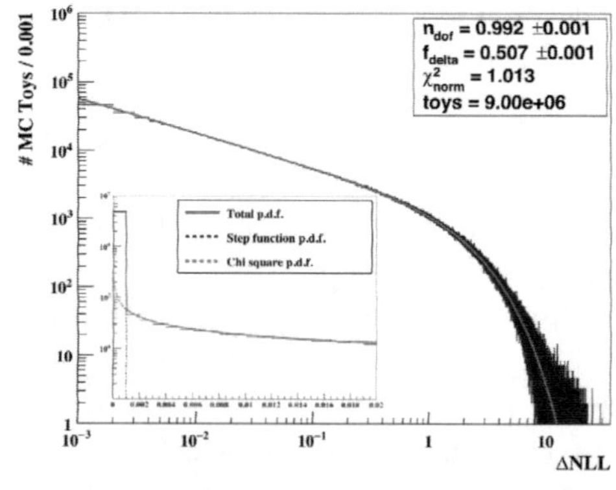

Figure 6. Fit to the ΔNLL distribution of case (2). Fit model has two components: a very narrow step function and a χ^2_ν.

3.2. Second case: m and Γ fixed, μ free but constrained to be positive

Let us consider the special case of the test statistic t_μ with the purpose to test $\mu = 0$ in a model where we assume $\mu > 0$; rejecting the null hypothesis ($\mu = 0$) leads to the discovery of a new signal. In this case, following [12], the test statistic is $q_0 = -2ln\lambda(0)$ if the estimated signal strength $\hat{\mu} \geq 0$ while is null otherwise, with $\lambda(0)$ being the profile likelihood ratio for $\mu = 0$. The authors of [12], derive analitically that an asymptotic approximation for the PDF of the statistic q_0 under assumption of the background-only ($\mu = 0$) hypothesis is an equal mixture of a delta function at 0 and a chi-square distribution for one dof: $f(q_0|0) = \frac{1}{2}\delta(q_0) + \frac{1}{2}\frac{1}{\sqrt{2\pi q_0}}e^{-q_0/2}$.

A fit to the test statistic distribution with a model consisting in a linear combination of a χ^2_ν function and a narrow step function at zero has been performed (Fig. 6), where the likelihood ratio distribution was obtained by the already discussed fit procedure but when fixing the values of mass and width parameters to the CMS estimates previously obtained, while leaving μ free. The best estimates obtained for the number of dof and the coefficient in front of the step function are $\hat{\nu} \simeq 0.992 \pm 0.001$ and $\hat{c} \simeq 0.507 \pm 0.001$, namely close to the approximate theoretical prediction. A chi-square test returns a 3.5% probability for this fit.

4. Future developments

The `GooFit` MC toys application run by means of the MPS provides a striking speed-up with respect to the RooFit application parallelized on multiple CPUs by means of PROOF-Lite. The method can be extended to situations with an unexpected signal and a global significance must be estimated: the LEE must be considered and a scanning technique must be implemented in order to consider relevant peaking behaviour with respect to the background model everywhere in the mass spectrum within the same fluctuation. This would increase the execution time of a cycle of fits performed on a single fluctuation. Different scan configurations should be tried to evaluate the systematic uncertainty associated to the scan. The RooFit-based approach would be unbearable for the long required processing time and the use of `GooFit` would be mandatory.

Aknowledgements

We are grateful for valuable suggestions and feedback from B.A.Hittle, M.D.Sokoloff and our CMS colleagues T.Dorigo, F.Pantaleo, L.Cristella and G.Donvito. Code development/testing and results were obtained on the IT resources of ReCas[13] hosted by the Bari Tier-2. Support was also provided by the italian *Project 20108T4XTM* of type `PRIN-2010/2011`.

References

[1] Andreassen R, Meadows B T, de Silva M, and Sokoloff M D 2014 *J. Phys.: Conf. Series* **513** 052003. The development of `GooFit` is supported by US NSF grants (*NSF-1005530, NSF-1414736*). `GooFit`'s source code lives in the repository `https://github.com/GooFit` that includes a manual and examples. A valuable tutorial can be found at http://indico.cern.ch/conferenceDisplay.py?confId-235992.
[2] Brun R and Rademakers F 1997 *Nucl. Instrum. Meth.* **A 389** 81-86
[3] Verkerke W and Kirkby D P 2003 *eConf* **C0303241** MOLT007 (*Preprint* physics/0306116)
[4] `https://docs.nvidia.com/`; for these studies CUDA version 6.5(7.0) were used for Tesla K20(40) boards.
[5] James F and Roos M 1989 CERN Program Library routine D506 (long write-up).
[6] An intense memory transfer takes place between CPU and GPU (checked by nVidia Visual Profiler (`nvvp`)).
[7] Tesla K20(K40) has 5(12)GB GDDRs; the 16(20) cores are E5-2640 v2 @ 2.0GHz with 64(256)GB of RAM.
[8] CMS Collaboration 2014 *Phys. Lett.* **B 734** 261
[9] CDF Collaboration 2009 *Phys. Rev. Lett.* **102** 242002 and 2011 (arXiv:1101.6058)
[10] Barbone L, Donvito G and Pompili A 2012 *J. Phys.: Conf. Series* **396** 042017
[11] Wilks S S 1938 *Ann. Math. Stat.* **9** 60-62
[12] Cowan G, Cranmer K, Gross E and Vitells O 2011 *Eur. Phys. J.* **C71** 1554
[13] ReCas is a project financed by the italian MIUR (*PONa3_00052, Avviso 254/Ric.*); its web page is `http://www.recas-bari.it/index.php/en/`.

Inclusive Flavour Tagging Algorithm

Tatiana Likhomanenko[1,2,3], **Denis Derkach**[1,2], **Alex Rogozhnikov**[1,2]

[1] National Research University Higher School of Economics (HSE), RU
[2] Yandex School of Data Analysis (YSDA), RU
[3] NRC "Kurchatov Institute", RU

E-mail: antares@yandex-team.ru

Abstract. Identifying the flavour of neutral B mesons production is one of the most important components needed in the study of time-dependent CP violation. The harsh environment of the Large Hadron Collider makes it particularly hard to succeed in this task. We present an inclusive flavour-tagging algorithm as an upgrade of the algorithms currently used by the LHCb experiment. Specifically, a probabilistic model which efficiently combines information from reconstructed vertices and tracks using machine learning is proposed. The algorithm does not use information about underlying physics process. It reduces the dependence on the performance of lower level identification capacities and thus increases the overall performance. The proposed inclusive flavour-tagging algorithm is applicable to tag the flavour of B mesons in any proton-proton experiment.

1. Introduction

B mesons contain either a b or a $\bar{b}$ quark, which defines their flavour. The flavour-tagging (FT) algorithms determine the flavour of a reconstructed signal B meson candidate at the production point in proton-proton collisions. The FT algorithms are used to measure differences in the behaviour of particles and antiparticles (e.g. measurements of flavour oscillations of $B^0_{(s)}$ mesons) and CP asymmetries to probe the validity of the Standard Model of particle physics.

The production of a B meson is usually accompanied by the production of another b hadron and other particles like kaons, pions, and protons (see Figure 1). At hadron collider experiments the FT algorithms are usually divided into two groups:

- opposite side (OS) taggers use the decay products of b hadrons that are produced together with the signal B (see [1]);
- same side (SS) taggers exploit light particles that evolve from the hadronisation process of the signal B meson like kaons, pions, and protons (see [2]).

The current version of the FT algorithm used by the LHCb [1, 2, 3], CMS [4], Atlas [5], CDF [6] and D0 [7] experiments tries to identify tracks and vertices produced on the OS/SS sides (SS tagging is done only by LHCb and CDF). It works as follows:

(i) the first step finds all tagging tracks and tagging vertices, where the latter is only used in OS taggers[1]. Both the OS and SS algorithms find one, maybe several, tagging track, while the OS algorithm also finds one tagging vertex.

[1] Note, a tagging track/vertex is a track/vertex whose charge is used to predict the flavour of the signal B. Ideal

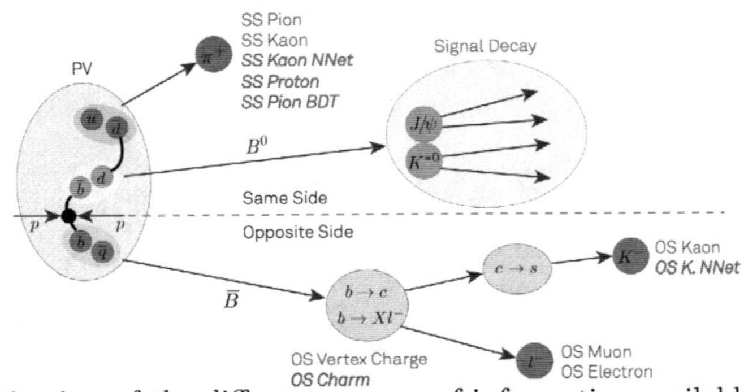

Figure 1. Schematic view of the different sources of information available to define the initial flavour of a signal B candidate.

(a) for the OS algorithm only lepton, kaon tracks and b hadron decay, secondary charm hadron vertices are considered;
(b) for the SS algorithm only pion, proton or kaon tracks are considered;
(c) other physically motivated selections are applied to leave only tracks/vertices which have the characteristics that help to define the flavour;
(d) if more than one track (vertex) is left after the previous steps for the OS or SS, a special rule is applied to select an appropriate track (vertex);

(ii) each of the OS/SS algorithms predicts a flavour based on the charge of the tagging track (vertex). Other characteristics of the track (vertex) are used to estimate the probability of incorrectly predicted flavour (i.e. misclassification, or mistag probability).

(iii) finally, the predictions of the OS and SS taggers are usually combined in the flavour-tagged analyses.

The first step naturally follows our physics intuition, but requires setting ad-hoc conditions, which require a deep understanding of the physics processes. From an analysis point of view, this pipeline causes some disadvantages:

- the algorithm relies heavily on the particle identification and reconstructed variables during the selection of the tagging tracks (vertices);
- the process of selecting a tagging particle is based on physics assumptions. This prevents the use of complex selection rules;
- a lot of information is lost since only a couple of tracks (vertices) are selected.

This paper describes a new approach to define the signal B flavour that exploits all available information in an event without using information about the underlying physics processes, like a tagging track (vertex) search.

2. Inclusive Probabilistic Model

The algorithm starts using an inclusive probabilistic model, which combines information from all tracks and vertices for each selected event containing a B candidate to tag. It uses an assumption similar to a naive Bayes model. Specifically, it assumes a strong independence of the tagging information available in the tracks and vertices[2].

choices for a tagging track are pion, kaon or proton tracks involved in the SS tagger and lepton or kaon tracks coming from a b hadron decay in the OS tagger. Ideal choice of a tagging vertex for the OS is a b hadron decay vertex or charm hadron vertex coming from a b hadron decay.

[2] Note that the use of a varying number of multipliers for each event is atypical for a naive Bayes approach in machine learning.

Let "components" refer to both tracks and vertices. Additionally, let s_p be the charge sign of a component ($+1$ or -1) and s_b be the flavour of the signal B ($+1$ for $\bar{b}$ and -1 for b). Then, assume the following:

$$\frac{P(\bar{b})}{P(b)} = \prod_{\text{components}} \frac{P(\bar{b}|B, \text{component}, s_p)}{P(b|B, \text{component}, s_p)} = \prod_{\text{components}} \left(\frac{P(s_b \cdot s_p > 0|B, \text{component})}{P(s_b \cdot s_p < 0|B, \text{component})} \right)^{s_p}$$

The last equality assumes that the spurious asymmetries introduced by different detection efficiencies for particles and antiparticles in the different regions of the detectors are negligible.

The usage of this formula, however, requires estimating probabilities $P(s_b \cdot s_p > 0|B, \text{component})$ and $P(s_b \cdot s_p < 0|B, \text{component})$. Note that these probabilities are established using different parameters of the signal B meson and a track/vertex, but not using their charges.

This approach has several key properties:

- it combines all available information from the components of the events under the naive probabilistic model;
- it implicitly determines the tagging tracks and vertices by the value of the ratio of the probabilities. Most of the particles will have a very small contribution;
- it does not depend on the tagging particle type (i.e. pion, kaon, electron, muon, proton) and it is not split into OS and SS tagging algorithms;
- it is symmetric with respect to matter/antimatter due to model definition.

Thus, the proposed FT algorithm is an inclusive model.

3. Inclusive Training

Charged B meson can be tagged using the charge of its decay components. Thus, the flavour of the meson can be defined ($\bar{b}$ for B^+ and b for B^-). $B^\pm \to J/\psi[\mu^+\mu^-]K^\pm$ decays are used for training. The charge of the kaon in the signal decay is used to independently infer the flavour of the B meson at production: $P(\bar{b}) = P(B^+), P(b) = P(B^-)$. The inclusive model is applied to the LHCb data samples that contain reconstructed signal decays $B^\pm \to J/\psi[\mu^+\mu^-]K^\pm$. The set of all tracks with the low probability to be ghost (fake track) and vertices for all events form the tracks and vertices datasets. Note the tracks and vertices forming the reconstructed signal decay are excluded.

In the probabilistic model conditional probabilities $P(s_b \cdot s_p > 0|B, \text{component})$ and $P(s_b \cdot s_p < 0|B, \text{component})$ are unknown. We can estimate them using a classification model. The target for this classification model is:

$$\text{target} = \begin{cases} 1, & \text{if } s_b \cdot s_p > 0, \\ 0, & \text{if } s_b \cdot s_p < 0. \end{cases}$$

Two gradient boosted decision tree (GBDT) algorithms are trained to predict the conditional probability $P(s_b \cdot s_p > 0|B, \text{component})$ for tracks and vertices. Kinematic properties of the tracks, vertices and signal B meson, information from the particle identification algorithm based on machine learning methods and track quality criteria are used as input observables. For the B meson the following features are used: transverse momentum, polar angle, impact parameter with respect to the primary interaction, pseudorapidity. For tracks, the particle identification algorithm output, polar angle, momentum, transverse momentum are used. Finally, for vertices, the number of tracks forming the vertex, mean of tracks impact parameters and mean of their transverse momenta, mass and momentum, which are calculated assuming pion mass for the incoming tracks, lifetime, angle between the signal B and the vertex are used.

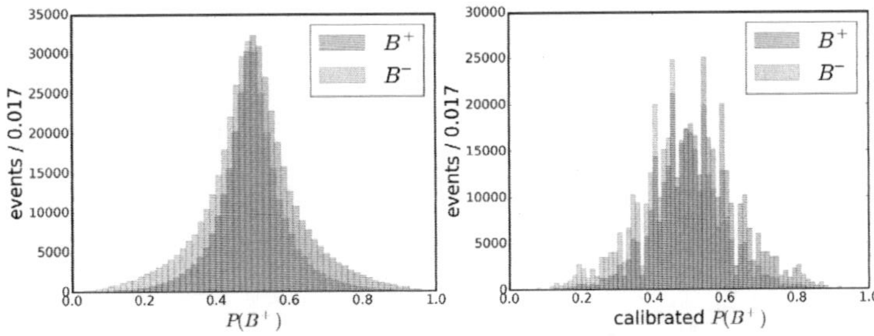

Figure 2. $P(B^+)$ distribution for B^+ (blue) and B^- (green) before calibration (left) and after isotonic regression calibration (right).

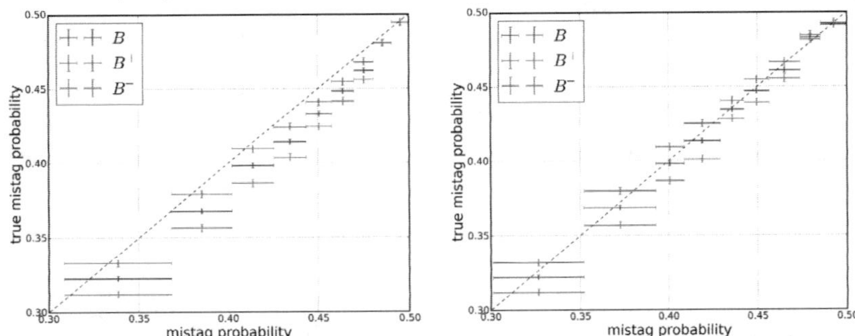

Figure 3. Calibration and symmetry check before (left) and after (right) isotonic regression. Estimation of the frequency based probability is used as a true probability in each bin for percentile 10 binning.

4. Symmetric Calibration

The conditional probability $P(s_b \cdot s_p > 0|B, \text{component})$ predicted by a classification model (i.e. by the GBDT) may be biased (see [8], [9]). Additionally, $P(\bar{b})$ and $P(b)$ computed by the probabilistic model may not be true probabilities due to the naive Bayes assumption. To compensate for these biases, the classifier output must be calibrated. Furthermore, the model should have the same behaviour for particles and antiparticles except small asymmetry of the production and detectors. This means that distributions for $P(B^+)$ and $P(B^-)$ should be symmetric around 0.5.

To calibrate the GBDT output, Platt scaling [10] and isotonic regression [11] were used. Platt scaling provides better results than isotonic regression and the uncalibrated probabilities. When calibrating $P(B^+)$ and $P(B^-)$, symmetric isotonic regression is used to preserve symmetry in the distributions. The calibration rule, f, is required to be symmetric, i.e. $f(1-x) = 1 - f(x)$, where x is $P(B^+)$. Figure 2 shows distributions for B^+ and B^- before and after the isotonic regression calibration. The comparison between the probability obtained from the inclusive model and frequency based estimation of the true probability is shown in Figure 3 before and after the calibration procedure. Distributions for $P(B^+)$ and $P(B^-)$ are checked to be symmetric around 0.5 after calibration (see Figure 4). After the calibration the inclusive model has improved Brier and logarithmic scores (see the scoring rules [12]), while the Platt scaling gives worse scores than isotonic regression.

5. Quality Metric

The figure of merit of a FT algorithm is the effective efficiency (see [1, 2, 3]) since the overall statistical power of the flavour-tagged sample is proportional to it. As a proxy metric of the effective efficiency the ROC curve is used in the analysis to optimize the FT algorithm. After the effective efficiency is checked to have increased value with respect to the previous results.

We analyze the ROC curves and check that the new version has a higher ROC curve at each

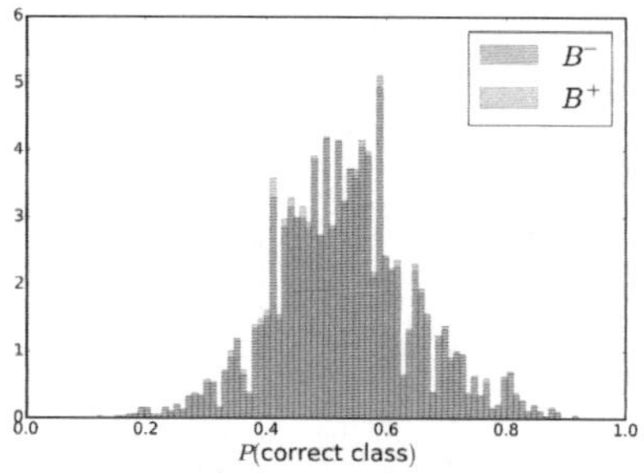

Figure 4. $P(B^+)$ for B^+ (blue) and $P(B^-)$ for B^- (green) distributions. Kolmogorov-Smirnov distance is 0.0163.

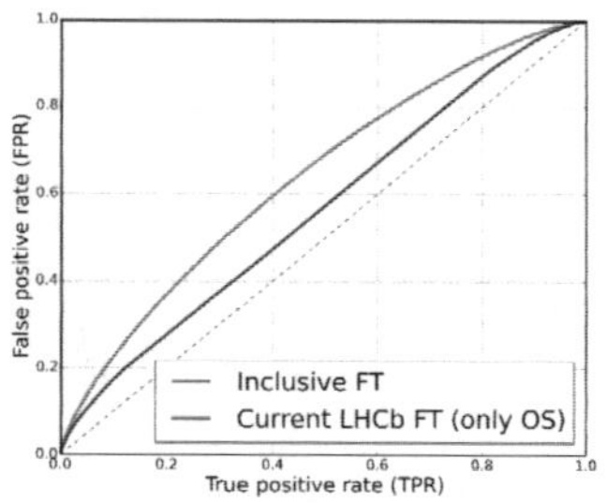

Figure 5. ROC curves for the current LHCb OS FT algorithm (blue) and proposed inclusive model (red).

point. The comparison of the ROC curve is shown in Figure 5 and the AUC (area under the ROC curve) values are 0.566 for the current OS FT. The AUC for the proposed inclusive model is 0.641. The ROC curve and AUC values were computed for all events including untagged events to compute the overall quality of the algorithm[3].

6. Conclusion

We proposed a simple flavour tagging technique, which efficiently combines information from vertices and tracks using a machine learning approach. The inclusive flavour tagging algorithm does not use information about underlying physics process and it is applicable to FT of B mesons in proton-proton experiments. The results demonstrate significant improvement in LHCb data, as seen in the ROC AUC score improvement from 0.566 to 0.641.

References

[1] LHCb Collaboration. Opposite-side flavour tagging of B mesons at the LHCb experiment. The European Physical Journal C 72.6 (2012): 1-16.
[2] LHCb Collaboration. A new algorithm for identifying the flavour of B_s^0 mesons at LHCb. Submitted to J. Instrum. arXiv:1602.07252 [hep-ex] (2016).
[3] LHCb Collaboration. B flavour tagging using charm decays at the LHCb experiment. JINST 10 (2015) P10005.
[4] CMS Collaboration. Measurement of the CP-violating weak phase ϕ_s and the decay width difference $\Delta\Gamma_s$ using the $B_s \to J/\psi\ \phi(1020)$ decay channel in pp collisions at $\sqrt{s} = 8$ TeV. Submitted to Physics Letters B. arXiv:1507.07527 [hep-ex] (2015).
[5] ATLAS Collaboration. Flavor tagged time-dependent angular analysis of the $B_s \to J/\psi\phi$ decay and extraction of $\Delta\Gamma_s$ and the weak phase ϕ_s in ATLAS. Physical Review D 90.5 (2014): 052007.
[6] CDF Collaboration. Measurement of the bottom-strange meson mixing phase in the full CDF data set. Physical review letters 109.17 (2012): 171802.
[7] D0 Collaboration. Measurement of the CP-violating phase $\phi_S^{J/\psi\phi}$ using the flavor-tagged decay $B_s^0 \to J/\psi\phi$ in 8 fb^{-1} of $p\bar{p}$-collisions. Physical Review D 85.3 (2012): 032006.
[8] Winkler, R. L., Murphy, A. H. (1968). Good probability assessors. Journal of applied Meteorology, 7(5), 751-758.
[9] Niculescu-Mizil, A., Caruana, R. (2005). Predicting good probabilities with supervised learning. In Proceedings of the 22nd international conference on Machine learning (pp. 625-632). ACM.
[10] Platt, J. (1999). Probabilistic outputs for support vector machines and comparisons to regularized likelihood methods. Advances in large margin classifiers, 10(3), 61-74.
[11] Zadrozny, B., Elkan, C. (2002). Transforming classifier scores into accurate multiclass probability estimates. In Proceedings of the eighth ACM SIGKDD international conference on Knowledge discovery and data mining (pp. 694-699). ACM.
[12] Bickel, J. E. (2007). Some comparisons among quadratic, spherical, and logarithmic scoring rules. Decision Analysis, 4(2), 49-6.

[3] Untagged events are those events for which all tracks and vertices did not pass selections; for them probabilities are set $P(B^+) = P(B^-) = 0.5$.

Journal of Physics: Conference Series **762** (2016) 012046 doi:10.1088/1742-6596/762/1/012046

Novel real-time alignment and calibration of LHCb detector for Run II and tracking for the upgrade.

Renato Quagliani[1,2] on behalf of the LHCb collaboration

[1]Laboratoire de l'Accelerateur Lineare, Orsay, France
[2]Bristol University, Bristol, United Kingdom

E-mail: rquaglia@cern.ch

Abstract. LHCb has introduced a novel real-time detector alignment and calibration strategy for LHC Run II. Data collected at the start of the fill is processed in a few minutes and used to update the alignment, while the calibration constants are evaluated for each run. The procedure aims to improve the quality of the online selection and performance stability. The required computing time constraints are met thanks to a new dedicated framework using the multi-core farm infrastructure for the trigger. A similar scheme is planned to be used for Run III foreseen to start in 2020. At that time LHCb will run at an instantaneous luminosity of 2×10^{33} cm^{-2} s^{-1} and a fully software based trigger strategy will be used. The new running conditions and the tighter timing constraints in the software trigger (only 13 ms per event are available) represent a big challenge for track reconstruction. The new software based trigger strategy implies a full detector read-out at the collision rate of 40 MHz. High performance and timing constraints are ensured by a new tracking system and a fast and efficient track reconstruction strategy.

1. Introduction

The LHCb detector [1] is a single-arm forward spectrometer covering a pseudorapidity range $1.9 < \eta < 4.9$ designed to study CP-violating processes and decay modes of hadrons containing b and c quarks which are highly suppressed in the Standard Model. The physics goals of LHCb are achieved thanks to an excellent tracking and particle identification (PID) system [2].

The LHCb tracking system is composed by three main detectors: a silicon-strip vertex detector (VELO) surrounding the *pp* interaction aiming to identify primary (PV) and secondary (SV) vertices, a large-area silicon strip detector (TT) consisting of four layers placed upstream the dipole magnet (4 Tm bending power) and three stations (four layers per station) of silicon-strip detectors (IT, in the inner region[1]) and straw drift tubes (OT, in the outer region) [4] downstream the dipole magnet.

Particle identification is achieved thanks to two ring-imaging Cherenkov (RICH1 and RICH2) detectors [5] placed upstream and downstream the magnet. Electrons, photons and hadrons are identified by the electromagnetic (ECAL) and hadronic (HCAL) calorimeter system [6]. Muons are identified by five muon tracking stations made of multi-wire proportional chambers alternated with layers of iron [7].

A sketch of the LHCb trigger strategy for Run I, Run II and the one foreseen for the upgrade phase in 2020 are summarized in Fig 1. Events are selected by a trigger consisting of a hardware

[1] Closer to the beam pipe

Table 1: Run I and Run II timing budget comparison in the HLT.

Trigger step	Output rate (Run I $\rightarrow$ Run II)	Time budget (Run I $\rightarrow$ Run II)
HLT1	(80 $\rightarrow$ 150) kHz	(20 $\rightarrow$ 35) ms/event
HLT2	(5 $\rightarrow$ 12.5) kHz	(150 $\rightarrow$ 650) ms/event

stage (L0) aiming to reduce the event rate to the maximal read-out rate of tracking sub-detectors of 1 MHz using solely information from muon stations and calorimeters, followed by a software stage (HLT)[2]. HLT is divided in two steps, HLT1 and HLT2 performing a partial and a full event reconstruction, respectively.

The larger timing budget for both the HLT stages (see Table 1) available in Run II allows the introduction of the real-time detector alignment and calibration procedure taking advantage of the multi-core infrastructure of the Event Filter Farm (EFF). Furthermore, the full offline event reconstruction (used in Run I) becomes available in the HLT2 without any performance losses. Those factors combined together boost the physics performance of the apparatus, minimizing the online and offline selection performance. It also allows to apply exclusive selections at trigger level (mainly taking advantage of the full calibration of the PID system) reducing the output rate. Furthermore, by using directly the trigger output to perform physics analysis, large amount of disk space ($\sim 90\%$) is saved. LHC Run III will be characterized by a centre-of-mass energy of colliding protons equal of 14 TeV (7 and 8 in Run I, 13 in Run II), with a bunch spacing of 25 ns (was 50 ns for Run I) and the LHCb instantaneous luminosity will be $2 \times 10^{33}\,\mathrm{cm}^{-2}\ \mathrm{s}^{-1}$, almost a factor 5 higher with respect to Run I and Run II. This will improve the current sensitivity of the key measurements, e.g. $2\beta_s$,γ, $\mathcal{B}(B_s \rightarrow \mu^+\mu^-)$, charm CPV search and photon polarisation in exclusive $b \rightarrow s\gamma^{(*)}$, by almost a factor 10 [8].

In Run III, a higher detector occupancy, track multiplicity and pile-up[3] is expected together with a more complex event topology which requires an efficient and more sophisticated trigger selection, especially for hadronic decay modes. If the current (Run I and Run II) read-out limit of the tracking system of 1 MHz would be kept, it would imply tighter cuts in the L0 trigger, reducing the yield for hadronic modes. For such reasons, the LHCb detector will be upgraded: the entire detector will be read-out at 40 MHz and a fully software-based trigger [9] performing tracking at the collision rate will be used. The tracking strategy for the upgrade will benefit from the upgraded tracking system described in Sec. 3. Critically, the full software-based trigger implies tighter timing constraints (estimated to be $\sim$ 13 ms per event) which are met by the upgrade tracking strategy described in Sec. 3.1.

2. Real-time alignment and calibration at LHCb

Sub-detector alignment and calibration constants are evaluated at regular intervals, *i.e.*, per fill, per run or even less frequently. A dedicated data sample for each sub-detector calibration and alignment task is collected with specific trigger selections from the HLT1 and the updated alignment and calibration constants become available in a few minutes.

If the automatic procedure finds a significant variation of the constants with respect to the ones used previously (each run has a baseline set of constants), a new run is triggered and the updated constants are used in the trigger (HLT1 and HLT2) and for the further offline processing. Otherwise, the events are processed with the baseline ones.

[2] The hardware stage will be removed for the upgrade.

[3] The expected pile-up in Run III is 7.6, three times larger than Run I and Run II.

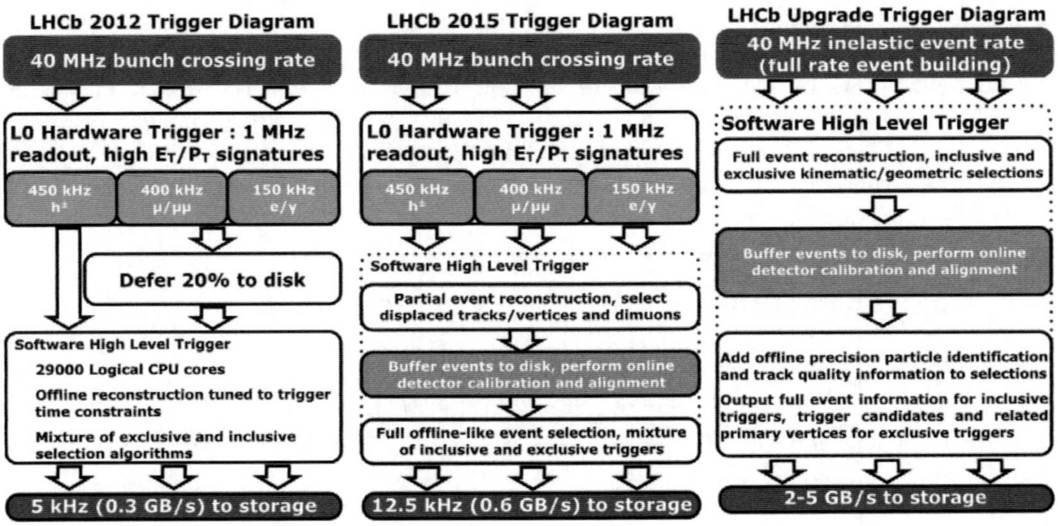

Figure 1: Diagrams of the LHCb trigger data-flow in Run I (left), Run II (center) and the one foreseen for Run III (right).

2.1. Framework

Two main activities are defined: alignment and calibration. The former benefits from the multi-core infrastructure of the EFF while the latter are mainly executed on a single node of the EFF (exception is done for the π^0 calorimeter calibration). The coordination of the different activities is guaranteed by a dedicated framework which exploits the multi-core infrastructure computing power. The framework is composed by the analyser, performing a massive parallelized (on ~ 1700 nodes) track reconstruction based on the alignment constants provided by the iterator. The iterator, running on a single node collects the analyser output and evaluate the constants for the next iteration. Depending on the task, the constants are evaluated in different ways: minimizing a Kalman filter fit χ^2 calculated from the residuals of reconstructed tracks for the tracking system alignment, and/or fitting monitoring histograms as it will be discussed later. The iterative procedure is interrupted when the variation of the χ^2 and/or the fit quality of the monitoring histograms are below a given threshold.

2.2. VELO, trackers and muon system alignment

The tracking system alignment is evaluated every fill. It is based on the minimization of a Kalman filter fit χ^2 calculated from the residual of already reconstructed tracks [10] [11]. The Kalman filter fit allows to include multiple scattering, energy loss and magnetic field effects and to provide multiple sub-detectors alignment at once. Moreover, mass and vertex constraints can be used to avoid global distorsion of the resulting alignment constants. Each sub-detector element constant is constrained to the nominal, surveyed or previously aligned position and the iterative alignment procedure to compute the updated alignment constants are stopped when the χ^2 variation is below threshold.

The VELO consists of two movable detector halves and it is the sub-detector closest to the interaction point (7.0 mm from beam pipe). During the LHC beam injection it is placed at a safe distance from the beam moving the two halves. Therefore, VELO alignment constants (3 translation and 3 rotation constants for each of the 86 detector elements) can change every fill.

The precision of the track based alignment procedure is $\mathcal{O}(2\mu\text{m})$ and the average observed misalignment of the VELO during Run I was 4μm with a maximum variation of $\mathcal{O}(10\mu\text{m})$. In

Fig. 2 one can see the stability of the VELO alignment during the data taking period in 2015 for the x and y translation constants (no relevant z translation is expected during data-taking).

The alignment constants are updated and a new Run is triggered only when the variation between the initial alignment constants and the ones evaluated by the automatic procedure is above threshold (black dashed line in Fig. 2). The full (empty) markers in Fig. 2 correspond to fills where a new run is (not) triggered.

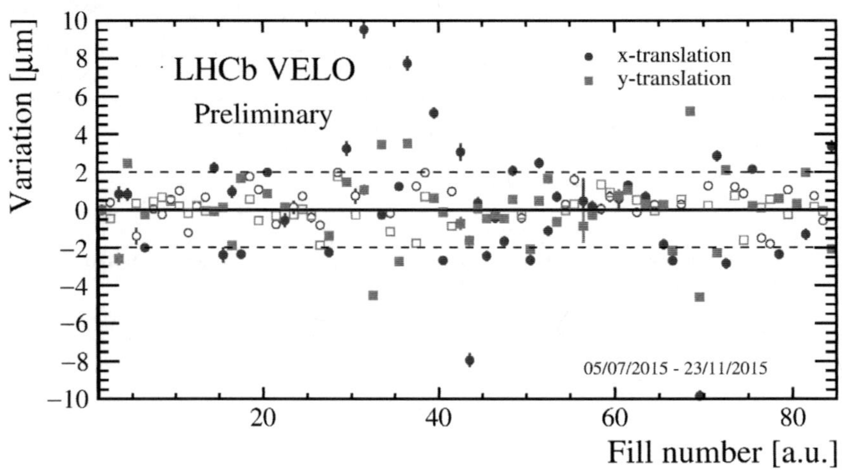

Figure 2: Stability of the alignment for translations in x and y direction of the VELO halves during Run II fills. On the y axis the difference between the initial alignment constant and the one computed by the automatic procedure.

Alignment of the trackers (upstream and downstream ones) is executed every fill after the VELO alignment. The constants might change during data taking due to magnet polarity switches, thus updates of the constants are expected every few weeks. Alignment constants for the muon stations are not expected to vary in time and the alignment procedure output is used as monitoring.

2.3. RICH mirror alignment

The RICH mirror alignment is executed every one or two fills and it is used as monitoring. It relies on the parallelized track reconstruction and the Cherenkov rings reconstruction on the Hybrid Photon Detector (HPD) plane. Both the tasks are performed by the analyser while the fit to the analyser output histograms to evaluate the alignment constants is performed by the iterator.

The Cherenkov light produced by particles entering the RICH radiators is reflected by a spherical and a flat mirror to the HPD plane. Therefore, the Cherenkov light cone produced is detected as a ring (with radius equal to the Cherenkov opening angle θ_{Ch}) centered at the intersection point of the incoming track with radiator material. Misalignment of the mirrors implies the observed Cherenkov ring center to be shifted by a quantity $\Delta\theta_{Ch}$ with respect to the expected position from tracking. The projected reconstructed track coordinate will then not correspond to the observed Cherenkov ring center. Therefore, the Cherenkov opening angle varies as a function of the azimuthal angle (ϕ) in the HPD plane. Misalignment of mirrors are encoded in the values of θ_x and θ_y which can be fitted for using $\Delta\theta_{Ch} = \theta_{Ch} - \theta_0 = \theta_x \cdot \cos\phi + \theta_y \cdot \sin\phi$, where θ_0 is the Cherenkov angle calculated from the momentum of the selected tracks of a given unambiguous PID assignment from other sub-detectors and from the refractive index of the radiator.

Fig. 3 shows $\Delta\theta_{Ch}(\phi)$ before and after the alignment procedure together with the stability of the Cherenkov angle resolution for RICH during 2015 data taking period.

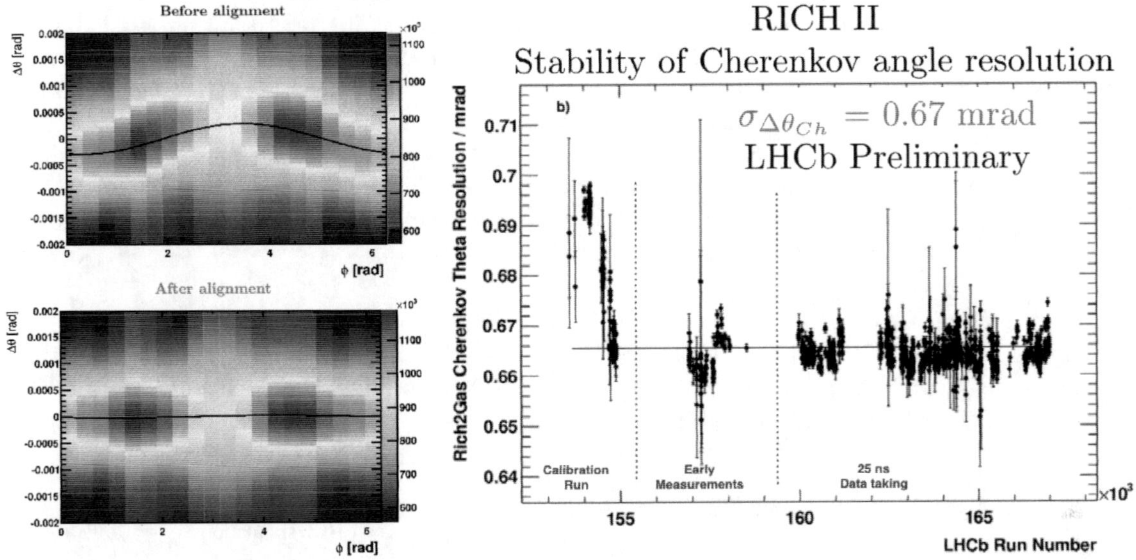

Figure 3: On the left the distribution of $\Delta\theta_{Ch}(\phi)$ used for the RICH mirrors alignment. On the right the stability of the Cherenkov angle resolution for RICH2 during 2015 data taking period.

2.4. Online calibration for RICH, OT and calorimeters

The HLT1 output of the online reconstruction and monitoring histograms is used to calibrate the RICH, OT timing offset t_0 and the calorimeter systems.

PID performance depends on the variation of the gas admixture, temperature and pressure of RICH radiators since they imply a variation of the refractive index of the radiators. At each run a fit for the difference between the reconstructed and the expected Cherenkov angle is performed and scale factors for the refractive index corrections are extracted.

The distance between the reconstructed track position and the wire position in the OT is proportional to the known drift time. Therefore a global offset of the drift time residual distribution for OT hits is used to correct the time-offset t_0 of read-out electronics with respect to the LHCb clock (relative to the time of collision). The calibration is performed every run and the t_0 offset constant is updated if the observed shift is above a given threshold.

Calorimeter calibration is achieved through relative and absolute methods consisting of adjustments of the photomultipliers (PMTs) HV settings. The relative method consists of a per-fill LED monitoring system aiming to detect aging of PMTs and it is performed for both ECAL and HCAL. The absolute method uses a reconstructed di-photons invariant mass spectrum peaking around the π^0 nominal value. Per-cell calibration coefficients are computed in order to shift as close as possible the π^0 reconstructed invariant mass to its nominal value.

Every technical stop a ^{137}Cs source scan is also performed to provide an absolute calibration of the HV settings for the PMTs.

3. LHCb upgrade

The whole tracking system will be upgraded for Run III. The current VELO will be substituted by a new design, consisting of hybrid pixel sensors placed at 5.1 mm from the beam-pipe [12].

From tracking point of view, the single hybrid pixel sensor has the benefit to provide directly an (x, y, z) measurement, leading to a more efficient, faster and simpler track reconstruction in the VELO. In fact, the magnetic field is negligible in the VELO region, thus VELO tracks are straight lines which are used as input for other pattern recognition algorithm. The trackers upstream and downstream the magnet will be substituted [13]. The upstream one will consists of four layers of large area silicon detector with a higher granularity and covering a larger acceptance in the central region. The downstream tracker, currently made of 3 stations (4 layers each for a total of 12 layers) of composite technologies will be fully replaced by a homogeneous scintillating fiber detector (SciFi) made of 2.5 m long vertically oriented fibers read-out by silicon photomultipliers.[4]. This technology solution guarantees a fast read-out of the detector at 40 MHz, a single hit position resolution in all the SciFi acceptance of $\sigma_{x-z} \sim 100\mu$m and a simpler and faster detector treatment from a tracking point of view.

3.1. *Fast track reconstruction strategy in LHCb 2020 HLT*

The online tracking in the first step of the HLT needs to match the timing requirements and efficiently select interesting events. The main steps for the online tracking sequence in LHCb 2020 HLT trigger [9] are summarized in Fig. 4.

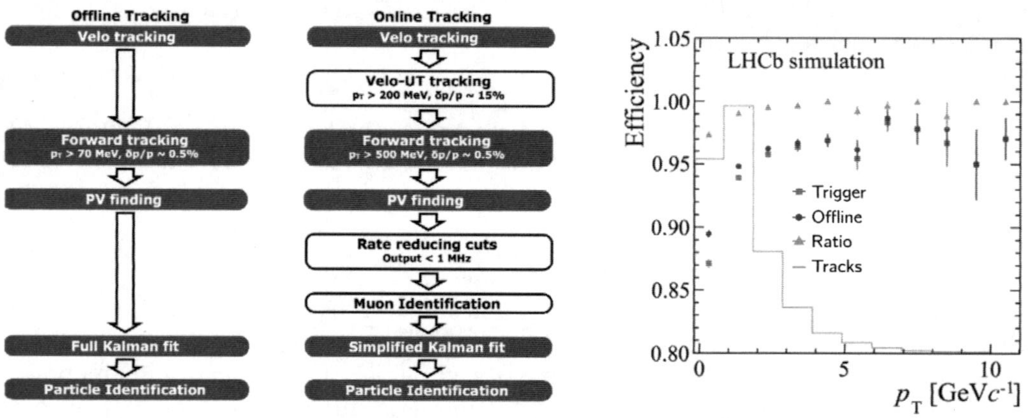

Figure 4: Online and offline tracking sequence comparison for the LHCb upgrade trigger (left). Online and offline tracking performance comparison as a function of the transverse momentum p_T of tracks (right). On the right also the transverse momentum distribution of tracks (in arbitrary scale) in a typical LHCb event is shown (gray line).

The main track reconstruction steps are:

(i) Velo tracking : segments in the VELO are looked for.

(ii) Velo-UT tracking : each segment is matched with a list of hits in the upstream tracker.

(iii) Forward tracking : VELO segments combined with the hits in the upstream tracker are used to look for matching segments in the SciFi.

(iv) PV finding: using the reconstructed tracks, PVs are searched for.

Velo-UT tracking allows to perform a momentum estimation (with $\frac{\delta p}{p} \sim 15\%$) and charge assignment which can be used to extrapolate the track in the downstream tracker and reduce the search windows, and thus the timing. The Forward tracking uses the momentum estimation and the charge information from VELO-UT tracking output to find matching segments in the

[4] Fibers are mirrored at $y = 0$, where y is the direction of the $\overrightarrow{B}$ field lines and z is the beam pipe direction.

SciFi. Each VELO segment is extrapolated to the SciFi stations and in each one of the 12 layers a search window is defined in order to collect the compatible hits, achieved thanks to a fast Hough transformation procedure [14]. The fast track reconstruction sequence combined with the upgraded tracking system is expected to find 98.7 % of the tracks which would be found by the slower offline sequence where no momentum requirements are added to the pattern recognition algorithms. The timing budget of 13 ms per event per single CPU is also satisfied, since the tracking sequence (VELO, VELO-UT, forward tracking and PV finding) takes 5.4 ms per event leaving the remaining timing budget for the leftover tasks. Events selected in this first fast partial event reconstruction will be used to feed the online alignment and calibration and the full event reconstruction at the reduced rate will be performed in a similar way as for Run II.

4. Summary

For Run II LHCb has implemented a fully automatic tracking system alignment and PID system alignment and calibration. Moreover, track and event reconstruction is fully performed in the online system relying on a fully calibrated and aligned system. This is possible thanks to the described dedicated framework developed for real-time alignment and calibration and the tracking timing improvements which allow to move the offline track reconstruction into the online system. In few minutes LHCb evaluates the updated calibration and alignment constants and the trigger output can be directly used to perform physics analysis.

For Run III data-taking, LHCb will upgrade the entire tracking system. The whole apparatus will be read-out at the collision rate and track reconstruction will be performed in real-time. The new tracking system and the fast tracking strategy described is expected to fit in the timing budget and guarantee high performance for the upgrade phase. Work is ongoing to improve the timing by exploiting possible new hardware architectures and to further improve the performances of tracking algorithms.

References

[1] Alves Jr A A *et al.* (LHCb) 2008 *JINST* **3** S08005
[2] Aaij R *et al.* (LHCb) 2015 *Int. J. Mod. Phys.* **A30** 1530022 (*Preprint* 1412.6352)
[3] Aaij R *et al.* (LHCb) 2014 *JINST* **9** 09007 (*Preprint* 1405.7808)
[4] Arink R *et al.* (LHCb Outer Tracker Group) 2014 *JINST* **9** P01002 (*Preprint* 1311.3893)
[5] Adinolfi, M and others (LHCb RICH Group) 2013 *Eur. Phys. J.* **C73** 2431 (*Preprint* 1211.6759)
[6] LHCb Collaboration 2000 LHCb calorimeters: Technical design report, *CERN-LHCC-2000-036*
[7] Alves, Jr, A A and others 2013 *JINST* **8** P02022 (*Preprint* 1211.1346)
[8] LHCb Collaboration 2011 Letter of Intent for the LHCb Upgrade, *CERN-LHCC-2011-001*
[9] LHCb Collaboration 2014 LHCb Trigger and Online Upgrade Technical Design Report, *CERN-LHCC-2014-016*
[10] Hulsbergen, Wouter 2009 *Nucl. Instrum. Meth.* **A600** 471–477 (*Preprint* 0810.2241)
[11] Amoraal J *et al.* 2013 *Nucl. Instrum. Meth.* **A712** 48–55 (*Preprint* 1207.4756)
[12] LHCb Collaboration (LHCb) 2013 LHCb VELO Upgrade Technical Design Report, *CERN-LHCC-2013-021*
[13] LHCb Collaboration 2014 LHCb Tracker Upgrade Technical Design Report, *CERN-LHCC-2014-001*
[14] Amhis, Y and Callot, O and De Cian, M and Nikodem, T 2014 Description and performance studies of the Forward Tracking for a scintilating fibre detector at LHCb, *LHCb-PUB-2014-001*

Parallel 4-Dimensional Cellular Automaton Track Finder for the CBM Experiment

Valentina Akishina and Ivan Kisel

Goethe University, Frankfurt am Main 60325, Germany, and Frankfurt Institute for Advanced Studies (FIAS), Frankfurt am Main 60438, Germany

E-mail: V.Akishina@gsi.de, I.Kisel@compeng.uni-frankfurt.de

Abstract. The CBM experiment (FAIR/GSI, Darmstadt, Germany) will focus on the measurement of rare probes at interaction rates up to 10 MHz with data flow of up to 1 TB/s. It requires a novel read-out and data-acquisition concept with self-triggered electronics and free-streaming data. In this case resolving different collisions is not a trivial task and event building must be performed in software online. That requires full online event reconstruction and selection not only in space, but also in time, so-called 4D event building and selection. This is a task of the First-Level Event Selection (FLES).

The FLES reconstruction and selection package consists of several modules: track finding, track fitting, short-lived particles finding, event building and event selection. The Cellular Automaton (CA) track finder algorithm was adapted towards time-slice-based reconstruction and included into the CBMROOT framework. In this article, we describe the modification done to the algorithm, as well as the performance of the developed time-based approach.

1. Introduction

The goal of the CBM experiment research program [1] is to explore the QCD phase diagram in the region of high baryon densities using high-energy nucleus-nucleus collisions. The CBM detector is designed to measure rare diagnostic probes such as multi-strange hyperons, charmed particles and vector mesons decaying into lepton pairs with unprecedented precision and statistics. In order to perform measurements with the required precision, the experiment is being designed to operate under interaction rates of up to 10 MHz, which puts strong constraints on the data processing (Fig. 1). This requires very fast and radiation-hard detectors, and a novel data read-out and analysis concept based on free streaming front-end electronics and a high-performance computing cluster for online event selection.

Thus, the traditional latency-limited trigger architectures, typical for conventional experiments with a hardware trigger, are inapplicable for the case of CBM. Instead, the experiment will ship and collect time-stamped data into a readout buffer in a form of a time-slice of a certain time length, not in the form of isolated collisions, and deliver it to a large computer farm, where online event reconstruction and selection will be performed. In this case a fraction of collisions inside a time-slice potentially may overlap in time with each other. The reconstruction algorithms in this case should be modified in order to process not event-associated data. The association of the hit information with physical events must be performed in software and requires full online event reconstruction not only in space, but also in time, so-called 4-dimensional track reconstruction. In order to study the problem of event association and to

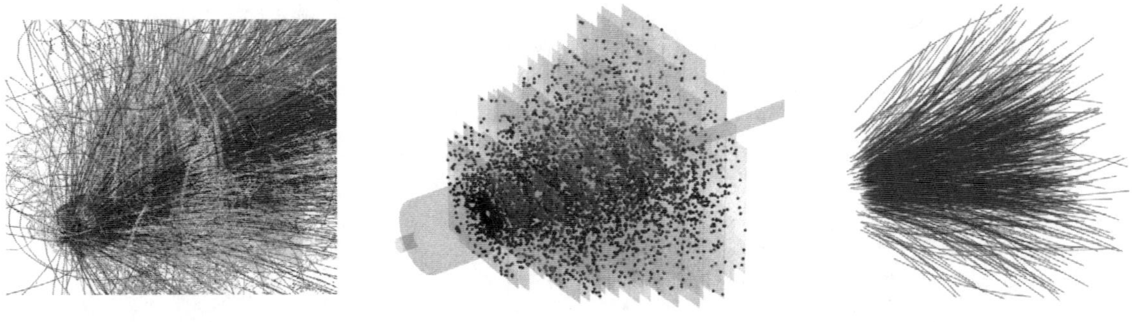

Figure 1. An illustration of complexity of the track finding problem (from left to right): particle trajectories in a simulated central Au+Au collision at 25 *A*GeV energy in the CBM experiment, only hits of the main tracking detector as the input information for track finding, and the reconstructed tracks found by the CA track finder.

develop proper algorithms, simulations must be performed which go beyond the traditional event-by-event processing as available from most experimental simulation frameworks.

2. Time-slice concept

In order to switch from the traditional event-by-event processing to realistic time-based simulations, a concept of time-slice should be introduced to the reconstruction chain. The beam at FAIR will constitute a free stream of particles without a bunch structure. Neglecting possible fluctuations in the beam intensity and, thus, non-constant average collision rate, the CBM collision distribution in time can be described as a Poisson process.

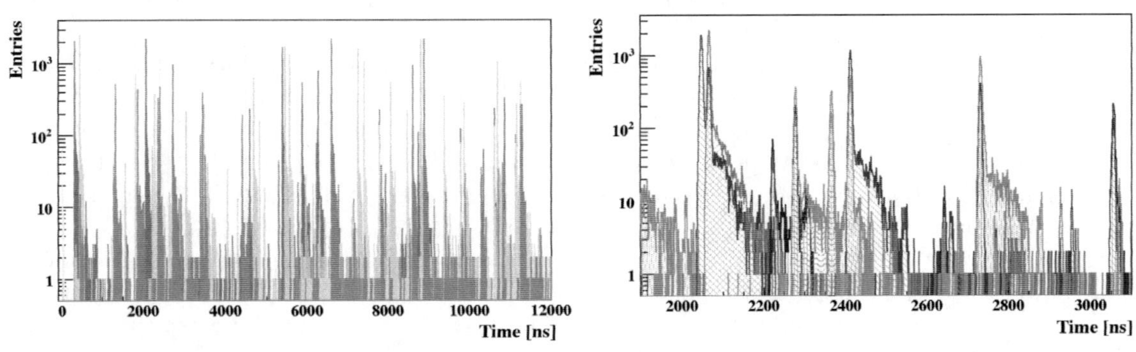

Figure 2. The distribution of hit time measurements for 100 minimum bias Au+Au collisions at 25 *A*GeV in the main tracking system of CBM, obtained assuming an average interaction rate of 10 MHz and Poisson distribution of time between subsequent events (left), the same distribution shown on a larger scale (right). Different collisions are shown with different color and filling.

A dedicated procedure was introduced to group the Monte-Carlo hits delivered by the event-by-event transport code into time-slices. At first, a group of 100 simulated UrQMD Au+Au minimum bias events at 25 *A*GeV[1] was taken. In future the number of events in a time-slice can be changed easily and should be adjusted to the power of a compute node.

The start time of each collision was obtained with the Poisson distribution, assuming the average interaction rate of 10^7 Hz. A time stamp, assigned to each hit in the main tracking detector Silicon Tracking System (STS) was calculated as a sum of the start time of the event, in which it was produced, and the time shift due to the time of flight from the collision point to a hit

[1] GeV per nucleon

position. In order to obtain from this absolutely precise time stamp the hit time measurement in the STS, the time stamp was smeared according to the Gaussian distribution with a σ-value of 5 ns, thus simulating the expected STS time resolution.

As a direct outcome of this procedure, the distribution of hit time measurements in the STS detector for a time-slice of 100 minimum bias events at 10 MHz interaction rate is presented in Fig. 2. The distribution clearly shows that at this interaction rate events do significantly overlap each other. The association of hits with events is no longer trivial. Space-time correlations must be employed, so that the tracking procedure takes into account not just three spacial dimensions, but in addition the fourth dimension — time.

3. 4-D CA track finder performance

The features of the Cellular Automaton (CA) method, namely the intrinsic parallelism and the ability to suppress large combinatorial enumeration, made the algorithm an appropriate solution for the track reconstruction in the Silicon Tracking System of CBM [2]. The CA track finder proposes a solid solution for the combinatorial search optimisation.

The method benefits from drastic suppression of combinatorial enumeration by introducing a phase of building up short track segments, namely triplets, at an early stage before going into the main search. Triplet is defined as a group of 3 hit measurements on adjacent detector stations, potentially produced by the same particle. After that the method does not work with the hits any more but instead with the created track segments. It puts neighbor relations between the segments according to the track model here and then one estimates for each segment its possible position on a track. After this process a set of tree connections of possible track candidates appears. Then one starts with the segments with the largest position counter and follows the continuous connection tree of neighbors to collect the track segments into track candidates. In the last step one sorts the track candidates according to their length and χ^2-values and then selects among them the best tracks.

In addition to that, the method is intrinsically local with respect to data processing and, thus, can be run in parallel on modern many-core CPU/GPU computer architectures, which is particularly suitable for CBM. The CA track finder has been significantly modified to be both vectorized (using SIMD instructions) and parallelized (between CPU cores). The algorithm shows now strong scalability on many-core systems. The speed-up factor of 10.6 was achieved on a Intel Xeon E7-4860 CPU with 10 hyper-threaded physical cores [3].

Since resolving different physical events in the case of CBM is a non-trivial task, the CA track finder has been further modified to take into account time information. The algorithm was adjusted to take as an input time-slices, containing STS hits. Each STS hit contains two spacial coordinates x and y, measured at a certain z-position of the detector plane, as well as a time measurement t.

The STS hit time measurement information was used in the CA track finder algorithm to improve the speed and performance. Since the triplets are to be build of three hits, potentially produced by the same particle, these hits should correlate not only in space, but also in time. Neglecting the time of flight between stations, the hits, belonging to the same track, should coincide in time measurement within the detector time precision. All the combinations of hits, whose time measurements differ from each other more than expected time of flight plus STS time precision should be rejected.

The time-based CA track finder version was tested to reproduce the results of the event-by-event analysis. The resulting reconstruction efficiency and the speed of the event-by-event analysis as well as for the 4D CA track finder in case of reconstructing time-slices, produced out of one Au+Au minimum bias event at $25A$ GeV, are presented in Tab. 1. For evaluation purposes a reconstructed track is assigned to a generated particle, if at least 70% of its hits have been caused by this particle. A generated particle is regarded as found, if it has been assigned

Table 1. Track reconstruction efficiency, clone and ghost levels (%), and timing performance (ms/event) for 100 minimum bias Au+Au collisions at 25 *A*GeV in case of event-by-event reconstruction, time-slice-based reconstruction at 0.1 MHz, 1 MHz and 10 MHz interaction rate, presented for different track categories.

Tracking efficiency, %	E-by-E	10^5Hz	10^6Hz	10^7Hz	CBMROOT
All tracks	92.1	92.6	92.6	92.2	91.3
Primary high-p	97.9	98.2	98.2	97.9	99.1
Primary low-p	93.6	94.1	94.1	93.5	93.6
Secondary high-p	92.0	92.7	92.7	92.0	88.9
Secondary low-p	65.7	66.7	66.6	65.9	56.8
Clone level	2.8	0.3	0.3	3.1	3.7
Ghost level	4.9	3.5	3.5	4.2	1.9
MC tracks found	145	146	146	145	88
Reconstruction time, ms/ev	11.7	11.97	11.92	13.60	17.30

to at least one reconstructed track. If the particle is found more than once, all additionally reconstructed tracks are regarded as clones. A reconstructed track is called a ghost, if it is not assigned to any generated particle according to the 70% criterion.

As one can see including the time and optimizing the 3D CA algorithm towards 4D reconstruction have made it possible to achieve the speed comparable to the case of the event-by-event analysis. Moreover, the track reconstruction efficiency has been improved after taking into account the STS time measurement, while comparing to event-based performance. The efficiency improvement is present even at the extreme case of 10 MHz interaction rate and can be explained by the presence of slow particles, which create random combinations of hits in case of event-based approach. These random combinations are rejected in the case of time-slices due to the hit time measurement cut, thus improving the performance.

The algorithm was included into the CBMROOT framework [4]. The simulation of detector response in the framework provides a time measurement, taking into account the anticipated behavior of the detector. For instance, a fake hit in this case is produced not only for the strips, which were accidentally fired simultaneously within a single event, but within a certain time interval. Thus, it puts the track finder in a more challenging condition.

The performance for the algorithm included into CBMROOT framework is presented in the last column of Tab. 1. It is comparable to the case of event-based analysis. The sightly higher clone level may be explained by the time measurement cut, which may be too strict for this case and may require additional tuning.

Residuals of the track parameters are determined as a difference between the reconstructed parameters and their true Monte-Carlo values. The normalized residuals (pulls) are determined as the residuals normalized by the estimated errors of the track parameters. In the ideal case these should be unbiased and Gaussian distributed with width of 1.0. Thus the pull distributions provide a measure of the track fit quality.

The residuals and the pulls for all track parameters are calculated at the first hit of each track. The distributions of residuals and the pulls for all track parameters in the CBM experiment together with their Gaussian fits are shown on Fig. 3. All distributions are not biased with pulls widths close to 1.0 indicating correctness of the fitting procedure.

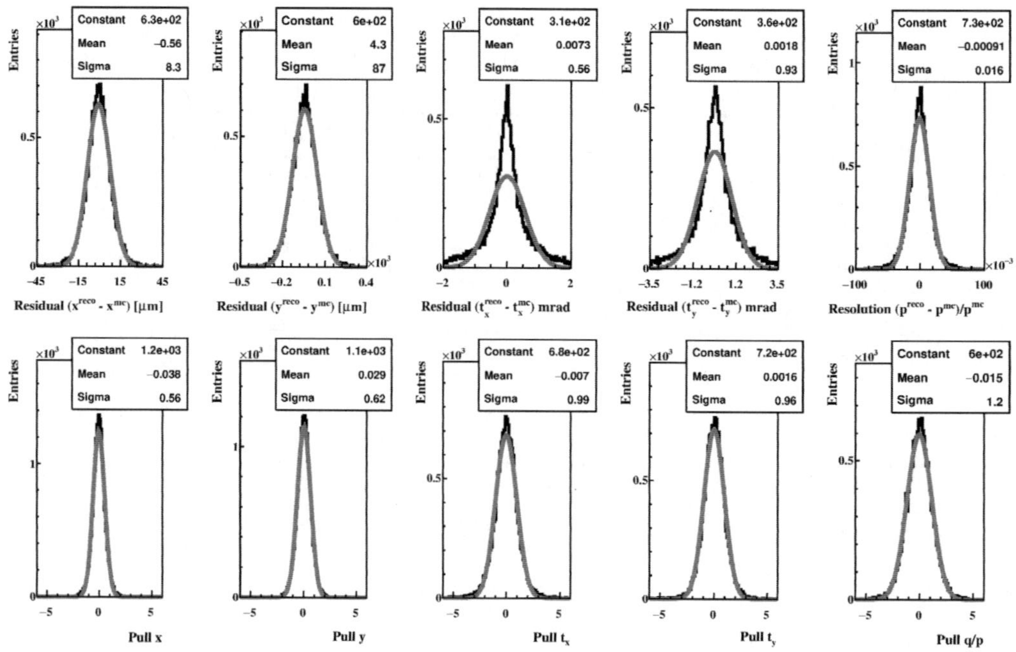

Figure 3. Residuals and pulls distributions for all parameters of tracks, reconstructed with the 4D CA track finder algorithm inside the CBMROOT framework.

4. Conclusion

The First-Level Event Selection (FLES) package for the CBM experiment contains all reconstruction stages: track finding, track fitting, short-lived particles finding, event building and event selection. For the most time-consuming part of the reconstruction procedure the Cellular Automaton track finder is used. In order to process large time-slices the CA track finder was vectorized using SIMD instructions and parallelized between CPU cores. The CA algorithm was further adapted towards time-slice-based reconstruction, which is a requirement in case of CBM for the event building. The 4D CA track finder algorithm is now able to achieve the efficiency and timing performance comparable with the event-by-event analysis even at the extreme interaction rate of 10 MHz.

Thus, the FLES package fulfills the experimental requirements and is ready for time-slice based 4D reconstruction in the CBM experiment.

Acknowledgments

The authors would like to thank I. Rostovtseva and M. Zyzak for their valuable advices and assistance during the development the 4D CA track finder.

The work is supported by BMBF, HIC for FAIR and HGS-HIRe for FAIR.

References

[1] The CBM Collaboration, "Compressed Baryonic Matter Experiment", Tech. Stat. Rep., GSI, Darmstadt, 2005; 2006 update.

[2] I. Kisel, I. Kulakov and M. Zyzak, Standalone First Level Event Selection package for the CBM experiment, IEEE Trans. Nucl. Sci. (2013), vol. 60, no. 5, 3703–3708.

[3] V. Akishina, I. Kisel, 4-Dimensional Event Building in the First-Level Event Selection of the CBM Experiment, J. Phys. Conf. Ser. 664 (2015) no. 7, 072027.

[4] http://cbmroot.gsi.de

Predicting dataset popularity for the CMS experiment

V. Kuznetsov[1], T. Li[1], L. Giommi[2], D. Bonacorsi[2], T. Wildish[3]

[1] Cornell University, Ithaca, NY 14850, USA
[2] University of Bologna, INFN-Bologna, Italy
[3] Princeton University, New Jersey, NJ 08544, USA

E-mail: vkuznet@gmail.com

Abstract. The CMS experiment at the LHC accelerator at CERN relies on its computing infrastructure to stay at the frontier of High Energy Physics, searching for new phenomena and making discoveries. Even though computing plays a significant role in physics analysis we rarely use its data to predict the system behavior itself. A basic information about computing resources, user activities and site utilization can be really useful for improving the throughput of the system and its management. In this paper, we discuss a first CMS analysis of dataset popularity based on CMS meta-data which can be used as a model for dynamic data placement and provide the foundation of data-driven approach for the CMS computing infrastructure.

1. Introduction

The Large Hadron Collider at Geneva, Switzerland, was designed to uncover mysterious parts of Universe building blocks. The tremendous efforts of physicists all around the globe were rewarded by Higgs discovery through analysis of Run-I data from both ATLAS and CMS experiments [1, 2]. The great success of Run-I was strongly dependent on reliable computing infrastructure. In CMS, the computing model is quite complex. The data collected from the detector are streamed to the HLT farm and organized into trigger streams. Later they are archived at the Tier-0 center at CERN and distributed to CMS Analysis facility (CAF) at CERN and to Tier-1 centers around the globe. Many Tier-2 centers worldwide share a portion of the data for further processing and Monte-Carlo generation. Finally, the Tier-3 centers (mostly at University levels) are used for various analysis tasks. Such distributed Computing Model is demonstrated to be reliable and flexible enough during Run-I operations, see [3].

During Run-I, CMS collected and processed more than 10B data events and simulated more than 15B events. The data transfer throughput peaks at few PB per week placing the data across hundreds of GRID sites for various user activities, ranging from data operations tasks to individual physics analysis. Even though CMS is able to successfully handle the data through Run-I operations [4] the experiment is looking at different ways to improve the system and reduce its operation cost and manual labor.

In this paper, we present a first attempt to understand different aspects of the CMS dataset popularity by the analysis of the CMS computing model. We start with the problem statement in Section 2, followed by discussions of various streams of information at our disposal from CMS data management systems. Section 3 provides a description of the architecture and implementation of different components used by Machine Learning (ML) algorithms to model and predict CMS dataset popularity. In particular, we discuss time series (referred to as "rolling") analysis for future predictions, operational cost of results obtained by ML algorithms and seasonality aspect of the data.

2. Problem statement

The CMS dataset popularity is a metric that quantifies user activities in CMS [5]. The data produced by the detector are logically organized into hierarchical structure, such as runs, files, blocks and datasets. The data coming from the detector are grouped into runs which represent a unit of the data-taking process. These runs are stored into files which are suitable for data processing by reconstruction or analysis programs. The files themselves are grouped into blocks which are used by data operations teams as atomic units for data transferring among the sites. Blocks are grouped into datasets which represent a processing chain of specific physics processes. Once data is produced and become available for end-users they use datasets as a common unit to submit their processing or analysis jobs. Every job will resolve a given dataset into series of files and process them at once. The outcome will be either stored in other files and a new dataset will be formed or it can be converted into n-tuples, histograms or any other form used by physicists in their analysis. Therefore it is possible to measure user activities in terms of dataset popularity, i.e. those datasets which are used in analysis jobs more often than others can be considered as popular.

During Run-I we collected more than 1TB of meta-data stored in various databases. More than 400K datasets, 4.5M blocks and 100M files were recorded in data bookkeeping system (DBS), see [4]. Only a portion of these datasets were actively accessed by end-users. The other parts were transferred once and had a very low access rate. The latter can lead to sub-optimal utilization of CMS storage space as well as overhead in the data management system with unnecessary data transfers. On the other hand, the reduction of dataset access latencies can be very useful for load balancing and scheduling jobs among the sites. This becomes especially important during rush periods when results are prepared for major events such as conferences. Therefore it is desirable to understand the nature of CMS dataset popularity in order to improve our computing model and reduce its operational cost. Among different possibilities to tackle these issues we decided to model and predict CMS dataset popularity via ML algorithms based on historical usage of datasets via various meta-data information available to us.

The CMS data management system is composed by several systems [4, 5]: PhEDEx (dataset replication system), DBS (data bookkeeping catalog), Dashboard (global monitoring system which keeps tracks of user jobs), SiteDB (site and user database), PopDB (CMS datasets popularity database), DDM (Dynamic Data Management system) and others[1]. The PopDB database keeps historical information of datasets usage in various metrics, e.g. number of total and user accesses per dataset, as well as total CPU time spent on each datasets during processing or analysis jobs. This information is used by DDM to replicate popular datasets to other sites. But the latter happens post-factum and requires sufficient amount of historical data to be accumulated up-front to trigger the replication process. To fill this gap we proposed to investigate a possibility to predict CMS dataset popularity based on existing meta-data for newly created datasets and use these predictions as a seeding mechanism for DDM machinery.

3. Architecture & Implementation

We started with evaluation of CMS data services and their underlying back-ends as a possible sources of information for CMS dataset popularity problem. We divided them into three categories: structured data which can be obtained from CMS data-services such as DBS,

[1] Only the systems which are relevant to and involved in the CMS datasets popularity are listed here.

PhEDEx, Dashboard, SiteDB, PopDB; the semi-structured data which can be found from CMS web logs, calendar systems and un-structured data, such as HyperNews forums, CMS twikies, etc. Obviously obtaining information from structured and semi-structured sources is much easier than un-structured ones. The latter will require significant pre-processing effort which is far beyond our abilities at the moment[2]. Therefore, we concentrate our work only on first source of information, i.e. structured data extracted from CMS data-services.

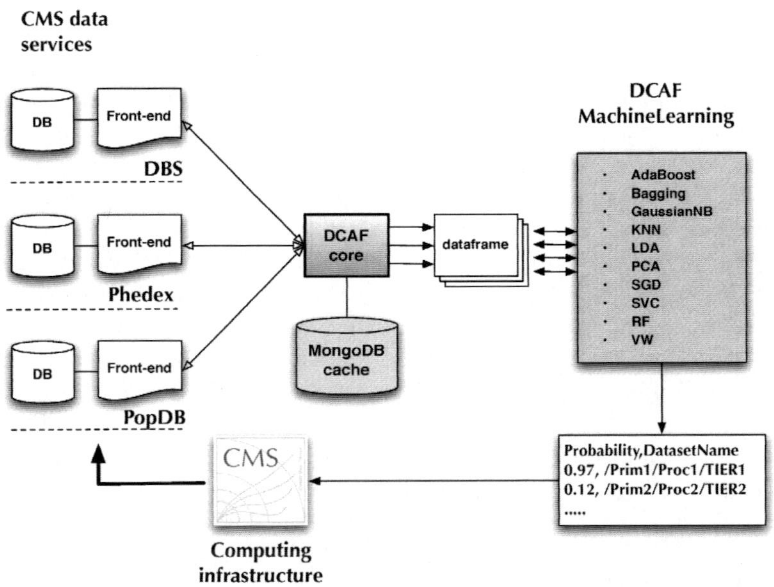

Figure 1. Flow diagram of DCAFPilot framework which collects and pre-processes information from various CMS data-services, anonymizes information within internal cache and passes them on to underlying ML algorithms.

We developed DCAFPilot (Data and Computing Analysis Framework Pilot) [7] framework in order to understand the metrics, the analysis workflow and the necessary tools (with possible technology choices) needed to tackle this problem. Its overall architecture is shown in Figure 1. It collects information from various CMS data-services, passes it to a pre-processing step which ensures that information stays anonymous and later to external ML algorithms for model building. The predictions are represented in a form of probabilities and dataset names which later can be fed back into the CMS computing infrastructure, e.g. used by DDM system as initial seeds for dataset replication.

The DCAFPilot framework was implemented in python and integrated with the scikit-learn ML library [8]. We also supplement it by Yahoo Vowal Wabbit [9] and Distributed Gradient Boosting xgboost ML libraries [10]. As a baseline we used five algorithms: RandomForest [11], LinearSVC [12], SGDClassifier [13], VW [9], and xgboost [10] to establish automated machinery to build and train the model, predict CMS dataset popularity in up-coming week and understand the cost of the modeling procedure.

[2] In fact, we foresee the usage of un-structured data but such effort should be justified at later stage of the project.

3.1. Data collection

We used five CMS data-services (DBS, PhEDEx, SiteDB, Dashboard, PopDB) to collect all required meta-data information suitable for ML analysis. In addition, the DBS services consists of four different instances used to publish globally accessible datasets and physics analysis ones and we used all of them during data collection phase. The data from CMS services were available via data-service APIs. In total, we utilized about a dozen of APIs from aforementioned CMS data-services, places 800K queries about 200K datasets, 900 software releases, 500 site entries and 5K user entries. The obtained data were stored in an internal cache, see Figure 1, and anonymized either by using internal database ids or via internal hash function. The data collection procedure was automated via UNIX cronjobs. The data were organized by weeks with total of 52 data files per year, where each file has 82 attributes. These attributes represent meta-data information obtained from various CMS data-services such as dataset name, number of files per dataset, etc. The total list of attributes can be found in Appendix A. The datasets were extracted from PopDB and complemented by datasets from the DBS system. The former datasets were used to define dataset popularity metric, while latter ones represented "unpopular" data. One year of data was roughly translated into the 82×600000 dataframe[3]. In total, we collected 130 dataframe files (52 files per year), worth of $\sim$ 1.5M measurements for different CMS datasets spawning over the years 2013-2015. Separately, we also generated conference count dataframe based on the CMS CINCO [14] database. It contains counters for up-coming conferences where CMS presents its results. The counters collect total number of conferences in 1,2,4,6,10 weeks followed by 5 week intervals up-to 70 weeks in the future. This dataframe was used as a complement to dataset meta-data to study the influence of conferences on the CMS dataset popularity (see Section 3.6 for discussion).

The collected data presents a good representation of various user activities through 2013-2015 years. These years cover Run-I era, preparation for Run-II and future detector upgrade studies. And, we expects that such activities will only grow through the Run-II phase and beyond.

3.2. Dataset popularity

Our first task was to define dataset popularity metric. As simple as it sounds it had many dimensions. The popularity DB provides us six different metrics: the number of accesses to a given dataset, the number of users/day recorded for a dataset, the total number of CPU hours spent on a given dataset, along with normalized values of those attributes over full number of datasets.

Figure 2 shows the behavior of some metrics recorded in popularity DB. Even though the shape of distributions of different years looks alike, we found that data themselves required further clean-up and understanding. For instance, number of users at low access patterns had gaps which represent the fact that our users were picky about certain datasets. The overall value of different metrics in 2015 were lower with respect to 2013/2014 years which reflects the usage of 2015 data sample (we only used half of the year available at the time of our studies). And, we have significant correlations among some attributes, e.g. number of dataset accesses is highly correlated with number of users used those datasets in their jobs, see Figure 2.

In addition, we were dealing with datasets composed out of 59 different data tiers within our sample. Each data-tier represents a specific interests within physics community, e.g. the raw data, processing data, the analysis data etc. The further work was only concentrated on five different data-tiers: AOD - Analysis Object Data, AODSIM - AOD with simulated information, MINIAOD - reduced AOD objects, MINIAODSIM - MINIAOD with simulated information and USER data-tier to denote user-based datasets. These five data-tiers represent the main interests

[3] Here and further we explicitly distinguish a CMS dataset vs a dataset suitable for ML algorithm which we call a dataframe.

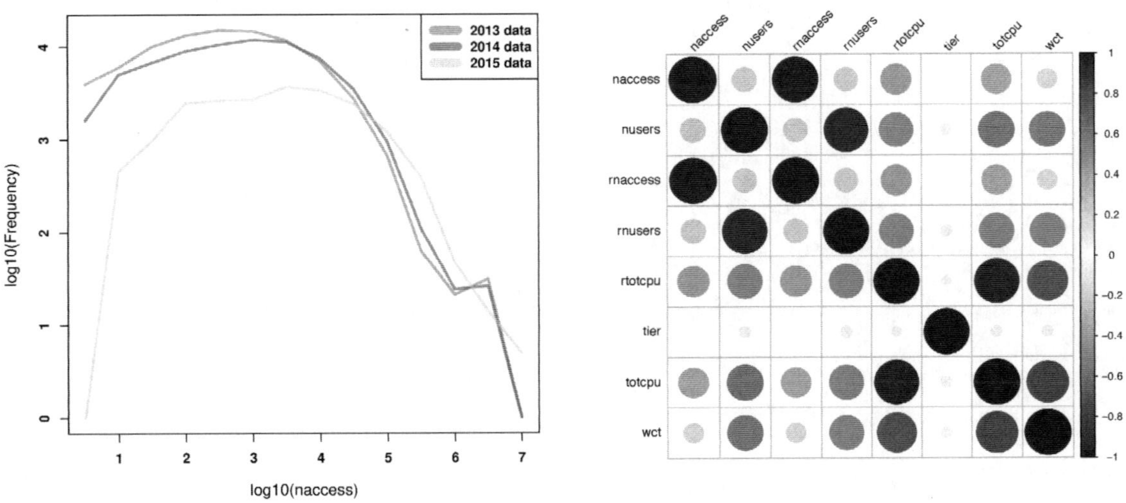

Figure 2. The left plot shows number of accesses per dataset and, on right hand side, we show a correlation plot for sub-set of available attributes in 2013-2015 data sets. The naccess/nusers attributes are the number of accesses and users per dataset, totcpu is a total cpu time spent by a user on a given dataset, tier is a dataset data-tier (e.g. RAW, AOD, etc.), and wct is a wall clock time of job processed this dataset. The right hand side plot shows the correlation plot between sub-set of dataset parameters. As we can see almost all attributes are correlated with each other to some degree except a tier attribute which shows no correlation with other parameters.

for dataset popularity since they correspond mostly to main user activities, such as analysis and data-processing, rather than daily activities of various data operations teams within the CMS experiment. The latter can be considered as "routine" tasks and obviously are well-defined within the experiment.

Table 1. Effects of applied cuts on the different data tiers.

Data tier/Cut	No cut	naccess>10	log(nuser)>2
AOD	4924 (7.25%)	4687 (8%)	1285 (35%)
AODSIM	21090 (31%)	18825 (32%)	1547 (42%)
MINIAOD	7 (0.01%)	6 (0.01%)	0
MINIAODSIM	1083 (1.5%)	792 (1.3%)	28 (0.8%)
USER	34127 (50%)	28777 (49%)	380 (10%)
ALL	67892	59222	3683

We collected the data for 2013, 2014 and first half of 2015 data and analyzed data-tiers breakdown within these samples. Table 1 shows the total number of datasets recorded in PopDB along with cuts for different data-tiers within 2014 dataframe. Further, we performed analysis of various cuts based on yield of TP (true positive), FP (false positive), TN (true negative) and FN (false negative) rates using Random Forest classifier. The Figure 3 shows model behavior under different *naccess* and *totcpu* cuts. Based on this studies we found that $naccess > 10$ and $totcpu > 10$ or $naccess > 10$ cuts provide the most the most promising TP/FP/TN/FN rates. For further studies we opted for simple threshold, $naccess > 10$.

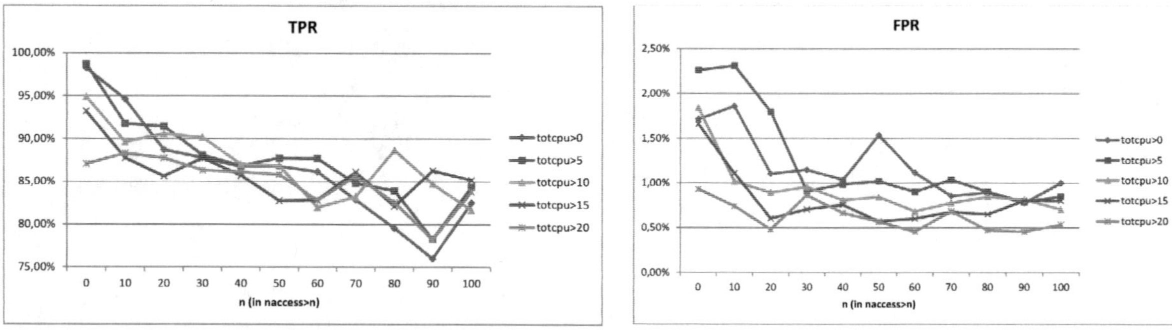

Figure 3. True positive and False positive rates for different values of *naccess* and *totcpu* parameters using Random Forest classifier over 2014 datasets.

3.3. Workflow

Our workflow was organized in the following way. The data was collected on a weekly basis via periodic cronjob which requested dataset information from PopDB for a given week. For every dataset in this set we obtained meta-data information from DBS, PhEDEx, SiteDB and Dashboard systems. This data was complemented by using random datasets from DBS system. The average time to collect 1 week of data took several hours and fade out gradually upon filling out internal cache. It was possible to run these jobs within single VM without requiring for additional resources. As mentioned in Section 3.1 we added to each dataframe the "unpopular" datasets meta-data with 1:10 ratio. This ratio was used to supplement signal data from PopDB with noise one from DBS system, i.e. we used random datasets from DBS which had no records in PopDB for given time period. All data attributes were anonymized via internal cache hashing mechanism and checked for correlations. Even though we were able to get 82 attributes representing each dataset we dropped 12 attributes due to their bias towards the popularity metric[4]. We used additional transformation step to convert obtained dataframes into classification problem, for which we used a cut discussed in Section 3.2. At this step the data were ready to be fed into any ML library.

3.4. Processing

Once the data were transformed into suitable format for ML we used the "rolling" approach which consists of the following steps:

- obtain a set of data up to the week in question (usually we kept one year worth of data) and train a specific classifier;
- obtain new data and make a prediction for selected data-tiers (we are only interested in user based data and AOD datasets);
- use the prediction as a seeding for DDM service who allocates replicas of popular dataset at various sites;

[4] For example, number of sites which hosted a given dataset certainly is biased towards popularity predictions since it represents posterior information.

- once data for the week in question becomes available in PopDB, compare the predictions to the actual dataset access patterns. This step can be used to adjust the model in the future.

The former two steps were performed with five ML classifiers: RandomForest, LinearSVC, SGDClassifier, VW and xgboost (see Section 3), and standard 70/30 split was used to test the model via cross-validation. We measured the outcome of True Positive (TP), True Negative (TN) rates for different data-tiers and estimated the operational cost of False Positive (FP) and False Negative (FN) rates. In the following, the results obtained from these studies are presented; the last two steps in the previous list are under implementation in the production system.

3.5. *Classification results*

We used 2013 data sample as the initial dataframe and gradually added weekly data to measure TP/TN/FP/FN rates for data through 2014. Among all used classifiers we found that SGDClassifier performs best, giving less than 10% error for TP/TN rates for the majority of the weeks.

From the results of various classifiers we found that prediction errors are larger during "vacation" periods, i.e. first and last weeks of the year, when CMS users are less active. During normal operation during the year all classifiers shown that it was feasible to keep TP/TN error below 10-15% using their default parameters. The fine tuning of algorithms were not performed since we mostly concentrated on setting up machinery for our task. The large deviations of errors were caused by different sensitivity of underlying algorithm to specific data tiers. Some data tiers, e.g. MINIAOD, were introduced during 2014 and did not gain too much access pattern among our users and therefore had larger errors. While others, e.g. AOD, AODSIM, and USER, had smaller errors throughout the entire year. Table 2 summarizes the mean and standard deviations for different statistics metrics grouped by data-tiers.

We found that only AOD, AODSIM and USER data-tiers gave reasonable predictions and their errors are stable through the entire year. The MINIAOD and MINIAODSIM datasets were less sensitive to algorithm due to the fact that they only become available for end-user through 2014 and did not attract user interest. In fact, they represented only 1.5% of entire data sample used in this analysis.

Table 2. Summary of TPR (true positive rate, i.e. sensitivity), TNR (true negative rate, i.e. specificity), FP (false positive) and FN (false negative) values for different data tiers.

Data tier	TPR	TNR	FP	FN
AOD	0.97 ± 0.05	0.99 ± 0.02	0.005 ± 0.011	0.015 ± 0.029
AODSIM	0.93 ± 0.13	0.99 ± 0.02	0.008 ± 0.016	0.021 ± 0.045
MINIAOD	0.11 ± 0.32	0.99 ± 0.02	0.014 ± 0.026	0.001 ± 0.007
MINIAODSIM	0.49 ± 0.48	0.99 ± 0.02	0.009 ± 0.016	0.007 ± 0.031
USER	0.93 ± 0.15	0.98 ± 0.02	0.014 ± 0.021	0.011 ± 0.023

The obtained errors on FP rates can be easily translated into data-transfer and extra disk space overhead we may face. For example, an average size of CMS dataset is around 2TB. If we have 1% of FP rate, see Table 2, and take on average 500 datasets created in CMS[5] this will translate into extra 10TB of data-transfer per week out of few PB of data transfer CMS is doing on weekly basis. Therefore the overhead would be $\sim$ 1% of total transfer rate. Such

[5] We looked at distribution of newly created datasets through 2014 and found that it has 565 ± 300 datasets on average.

overhead can be easily accounted by data transfer team. We found that the cost of FN rate can be considered as "irrelevant" from operational point of view since it will correspond to current mode of operations in CMS, since data placement is done manually via current snapshot of site occupancy. But in the ideal world where computing resources will be driven by such data popularity predictions the FN rate will translate into longer latencies of analysis user jobs. On average we have 1.5M jobs submitted every day[6]. The 1% of FN rate will correspond that $\sim$ 15K jobs will compete for resources of a site where popular dataset may resides. On another hand, the data driven approach would reduced overhead of site utilization. So far, our sites are populated with data which we put over there rather the data which will become popular. For example, we found that majority of datasets are accessed only by a few users. Therefore a better data placement will directly translate into site utilization savings, further studies of which are underway.

3.6. Prediction from information beyond dataset features

Even though we used historical information to predict popularity of datasets in a future we also considered that future events, such as up-coming conferences, group meetings and discussion within various physics groups can stimulate access patterns to certain datasets. Such studies will require data-mining of user based discussion groups, meetings agendas and up-comping conferences. As a starting point we used the CINCO CMS conference database [14] where physics conferences are registered and physicists are assigned to present their work on physics analysis.

Naively, we thought that some datasets which represent certain physics channels may become popular before up-coming conferences where physicist present their results. To test this idea we evaluated if the future conference weekly counts - if used together with the features of the datasets - could improve the classification of whether a dataset is popular or unpopular, with respect to using only the features of the dataset. This was done by comparing the classification accuracies, comparing the feature importance scores given by random forest classifiers for individual features, and comparing the p-values of the χ^2-independence test and the ANOVA F test between individual features and the popularity outcome. The experiment has not shown improvement yet, in part because the classification performance was already nearly perfect for the limited time period when the conference data was available. This also lead us to explore the influence of the conference data from alternative perspectives.

For instance, we calculated the cross correlation between the weekly count time series of conferences and the one of the accesses to each dataset at different time lags, and found that the time lag at which the cross correlation peaks is roughly 75 weeks for most datasets. We also found that for some datasets cross correlation is peaking at a positive lag implying that future conference schedules can affect the current dataset access, while for others it peaks at a negative lag meaning that past conferences can still have residual influence on the current dataset access. Figure 4 shows that such lags have significant cross correlations.

Finally, we tried to find out if each time series contains significant seasonality, i.e. periodicity. This was done by calculating the Discrete Fourier Transform (DFT) on the time series using the Fast-Fourier Transform (FFT) algorithm and searched for the most significant spike that represents the frequency of seasonality. We found a yearly seasonality effect for the conferences, a shorter-period seasonality for some datasets (for example, 15-week period), and no seasonality for other datasets (see Figure 5).

These studies demonstrate that the conference count and some CMS datasets access counts can show some seasonality, and there is significant correlation between conference count and most

[6] We used Dashboard CMS data-services and found that on average CMS has 1520669 ± 450069 completed jobs per week.

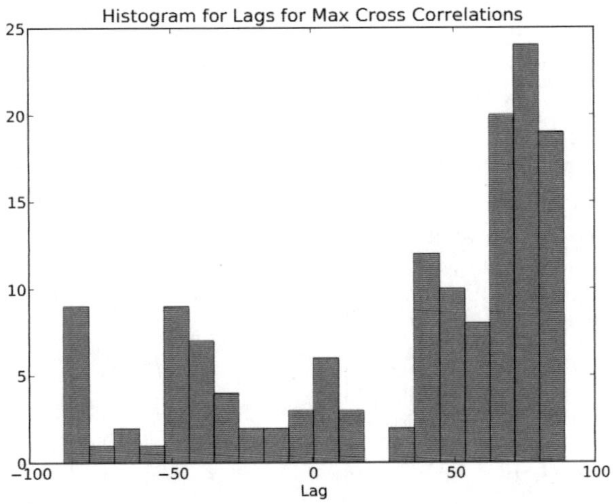

Figure 4. Histograms of lags with maximum cross correlation (see text for more details).

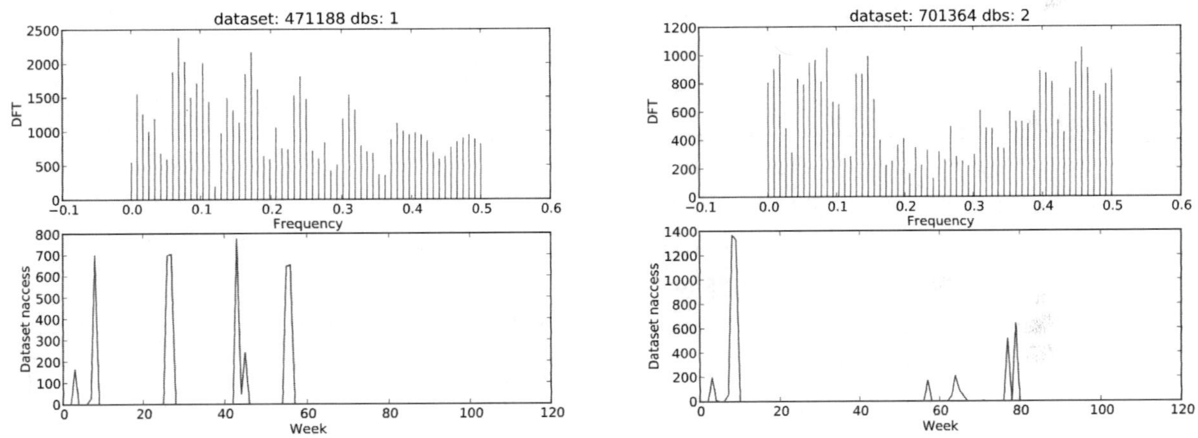

Figure 5. Seasonality effect of random datasets from 2014 data sample. Both plots show DFT of access count per week time series of a dataset. The length of time series is 116 weeks. The highest spike of top plot for dataset ID=471188 occurs at the 8-th frequency and corresponds to a time period of $116/8 = 15$ weeks. The DFT plot of another dataset ID=701364 does not show seasonality in the dataset access series.

dataset access counts. But further studies are required to see if such observations can improve the overall prediction for dataset popularity, and it is especially interesting in the context of specific data-tiers.

4. Summary

The results of this pilot project provides a solid foundation and a working pilot framework for future R&D work on data mining of several aspects of the CMS computing model. We demonstrated that we are able to successfully predict CMS dataset popularity based on collected

CMS meta-data. The obtained results shows that small FP/FN rates are possible to achieve and can be interpreted as overhead in data-transfer and job latencies. We estimated that obtained FP rate can introduce only 1% overhead to existing data transfer rate and can be easily managed by operational team. The small FN rate can lead to better sites utilization and provide operational savings in the long term. We also performed a time-series analysis of dataset popularity based on conference count information. Although we didn't observed a significant gain from this type of information we considered it as an important studies which requires further investigation. For example, we found a short-period seasonality of some datasets while others had no such effect. This fact can represent a data-ops or physics group activities which were performed on a regular basis, e.g. background studies for various physics analyses. Such type of datasets can be treated differently by DDM system, e.g. used with specific weight with respect to non-seasonal datasets.

We are evaluating to turn this pilot project into production and use dataset popularity predictions as initial seeding for the DDM [5] system. In addition, we are interested in further investigations of seasonality effects for specific data tiers. The relatively small errors for the few discussed data-tiers can be turned into better data placement and reduced latencies of analysis jobs during first week of data appearance. This will lead to better analysis throughput and user satisfaction with their data processing needs, more awareness in dynamic data placement tactics across different sites, as well as better usage of site resources in the CMS GRID infrastructure.

Acknowledgment
This work was partially supported by the National Science Foundation, contract No. PHY-1120138, and the Department of Energy of the United States of America.

Appendix A
Dataframe description:

- id: unique id
- cpu: CPU time reported by Dashboard data-service for given dataset
- creator: anonymized DN of the user who created given dataset, reported by DBS
- dataset: DBS dataset id (comes from DBS APIs/database back-end)
- dbs: DBS instance id
- dtype: anonymized DBS data type (e.g. data, mc)
- era: anonymized DBS acquisition era name associated with given dataset
- nblk: number of blocks in given dataset, reported by DBS
- nevt: number of events in given dataset, reported by DBS
- nfiles: number of files in given dataset, reported by DBS
- nlumis: number of lumi sections in given dataset, reported by DBS
- nrel: number of releases associated with given dataset, reported by DBS
- nsites: number of sites associated with given dataset, reported by PhEDEx
- parent: parent id of given dataset, reported by DBS
- primds: anonymized primary dataset name, reported by DBS
- proc_evts: number of processed events, reported by Dashboard
- procds: anonymized processed dataset name, reported by DBS
- rel1_N: defines N-th series CMSSW release associated with given dataset
- rel2_N: defines N-th major CMSSW release associated with given dataset
- rel3_N: defines N-th minor CMSSW release associated with given dataset

- s_X: PhEDEx site counter, i.e. number of Tier sites holding this dataset replica
- size: size of the dataset, reported by DBS and normalized to GB metric
- tier: anonymized tier name, reported by DBS
- wct: wall clock counter for given dataset, reported by Dashboard
- naccess: number of accesses to a dataset, reported by PopularityDB
- nusers: number of users*days to a dataset, reported by PopularityDB
- totcpu: number of cpu hours to accessed dataset, reported by PopularityDB
- rnaccess: naccess(dataset)/SUM_i naccess(i), reported by PopularityDB
- rnusers: nusers(dataset)/SUM_i nusers(i), reported by PopularityDB
- rtotcpu: totcpu(dataset)/SUM_i totcpu(i), reported by PopularityDB

The target variable was evaluated from naccess/nusers/totcpu attributes and final choice was done in favor of naccess> 10 cut. The rel1_N, rel2_N, rel3_N had various values for N parameter, e.g. rel2_5 defines a major 5th release of CMSSW used to process a given dataset. Please note, that CMSSW releases contain three components, e.g. CMSSW_X_Y_Z The first value in series (X) indicates a major release cycle, the second value (Y) corresponds to a major release within a cycle where new features were introduced with respect to the preceeding release, and the third value (Z) represents a release with some updates and bug fixes to the preceeding release. Here is a final set of attributes used in these studies:

```
id,cpu,creator,dataset,dbs,dtype,era,naccess,nblk,nevt,nfiles,nlumis,
nrel,nsites,nusers,parent,primds,proc_evts,procds,rel1_0,rel1_1,rel1_2,
rel1_3,rel1_4,rel1_5,rel1_6,rel1_7,rel2_0,rel2_1,rel2_10,rel2_11,rel2_2,
rel2_3,rel2_4,rel2_5,rel2_6,rel2_7,rel2_8,rel2_9,rel3_0,rel3_1,rel3_10,
rel3_11,rel3_12,rel3_13,rel3_14,rel3_15,rel3_16,rel3_17,rel3_18,rel3_19,
rel3_2,rel3_20,rel3_21,rel3_22,rel3_23,rel3_24,rel3_25,rel3_26,rel3_27,
rel3_3,rel3_4,rel3_5,rel3_6,rel3_7,rel3_8,rel3_9,relt_0,relt_1,relt_2,
rnaccess,rnusers,rtotcpu,s_0,s_1,s_2,s_3,s_4,size,tier,totcpu,wct
```

References

[1] ATLAS experiment, *Observation of a new particle in the search for the Standard Model Higgs boson with the ATLAS detector at the LHC*, Physics Letters B, Volume 716, Issue 1, pp 1-29, 2012.
[2] CMS experiment, *Observation of a new boson at a mass of 125 GeV with the CMS experiment at the LHC*, Physics Letters B, Volume 716, Issue 1, pp 30-61, 2012.
[3] J. Adelman et al, *CMS Computing Operations during Run 1*, Journal of Physics: Conference Series 513 (2014) 032040.
[4] M. Giffels, Y. Guo, V. Kuznetsov, N. Magini and T. Wildish *The CMS Data Management System* J. Phys.: Conf. Ser. 513 042052 doi:10.1088/1742-6596/513/4/042052
[5] F.H. Barrerio Megino et al, *Implementing data placement strategies for the CMS experiment based on a popularity model* J. Phys.: Conf. Ser. 396 032047 doi:10.1088/1742-6596/396/3/032047
[6] V. Kuznetsov, T. Wildish, L. Giommi, D. Bonacorsi *Exploring patterns and correlations in CMS Computing operations data with Big Data analytics techniques* International Symposium on Grids and Clouds (ISGC) 2015, Academia Sinica, Taipei, Taiwan
[7] https://github.com/dmwm/DMWMAnalytics/
[8] http://scikit-learn.org/stable/
[9] https://github.com/JohnLangford/vowpal_wabbit/wiki
[10] https://github.com/dmlc/xgboost
[11] L. Breiman, *Random Forests*, Machine Learning 45 (1) 532. doi:10.1023/A:1010933404324
[12] C. Cortes, V. Vapnik, *Support-vector networks*, Machine Learning, 20, 273-297 (1995),
[13] *Stochastic Gradient Descent*, Website, 2010. http://leon.bottou.org/projects/sgd
[14] Cms INformation on COnferences (CINCO) data-service, https://cms-mgt-conferences.web.cern.ch/

Ring-shaped Calorimetry Information for a Neural Egamma Identification with ATLAS Detector

João Victor da Fonseca Pinto[1], **on behalf of ATLAS Collaboration**
Federal University of Rio de Janeiro, Rio de Janeiro, Brazil
E-mail: joao.victor.da.fonseca.pinto@cern.ch

Abstract. This article presents the identification process of electrons based only on calorimeter information. It is based on the usage of ring-shaped description for a region of interest of the calorimeter which explores the shower shape propagation throughout the ATLAS calorimeters. This information is fed into a multivariate discriminator, currently an artificial neural network, responsible for hypothesis testing. The concept is evaluated for online selection (trigger), used for reducing storage rate into viable levels while preserving collision events containing desired signals. Preliminary results from Monte Carlo simulation data indicate that the background rejection can be reduced by as much as 50 % over the current method used in the High-Level Trigger, allowing for high-latency reconstruction algorithms such as tracking to run at a later stage of the trigger.

1. Introduction

Succeeding the discovery of the Higgs boson in 2012 during Run 1, the LHC has started a new data-taking period (Run 2). During the run start-up in 2015, it successfully increased the energy in the center-of-mass of proton-proton collision physics to 13 TeV and shrank bunch spacing to 25 ns. Also, an increase of the luminosity is foreseen, targeting $1.6 \times 10^{34} cm^{-2} s^{-1}$, which will result in pile-up conditions of 44 mean inelastic interactions per bunch crossing. These changes in experimental conditions bring a proper environment for possible new physics discoveries and measurements.

ATLAS [1] is the largest LHC detector and was designed for a large spectrum of physics studies. Many physics channels have electrons in their final states. For electron identification, the calorimeter system plays a major role. In ATLAS, the calorimeter system comprises both electromagnetic (EMCal) and hadronic (HCal) sections, where the latter envelopes the electromagnetic counter-part. In the central region, the EMCal Barrel and both EMCal and HCal End-Cap employ a Liquid Argon and absorber technology. The central HCal is based on steel absorber with plastic scintillators.

These calorimeters consists of, for most of the η[2] acceptance, three samplings along longitudinal (depth) segmentation. Additionally, each sampling layer has its own lateral ($\eta \times \phi$)

[1] Speaker.
[2] ATLAS uses a right-handed coordinate system with its origin at the nominal interaction point (IP) in the centre of the detector and the z-axis along the beam pipe. The x-axis points from the IP to the centre of the LHC ring, and the y-axis points upward. Cylindrical coordinates (r, ϕ) are used in the transverse plane, ϕ being the azimuthal angle around the z-axis. The pseudorapidity is defined in terms of the polar angle θ as $\eta = -\ln \tan(\theta/2)$.

granularity, which may vary with η. The EMCal is complemented by a presampler (PS) within $|\eta| < 1.8$.

The ATLAS trigger system [2] aims at providing online selection of promising bunch crossing events into a viable output rate for recording and further offline processing. A set of improvements and upgrades have been made to address the new circumstances in the Run 2 (cf. Ref. [3] for trigger related upgrades).

The trigger operates applying selection chains (referred to as chains for simplicity), in which each selection level increases in latency and physics reconstruction depth. The first level (L1) is the start of the selection chain and it is hardware-based in order to achieve a latency of less than 2.5 μs meanwhile scaling down the readout rate to, at most, 100 kHz. Thereafter operates the High-Level Trigger (HLT), a software-based selection with average target latency of 550 ms and average output rate of 1 kHz. The physics selection menu consists of about $\sim$1000 chains out of which about $\sim$100 aims at selecting electron and photons signatures (referred to as EM particles).

2. Baseline Algorithm

The Egamma chain (Figure 1) is based on the Region-of-Interest (RoI) concept. The following short description emphasizes the particle identification (ID) task, which consists of disentangling the EM particles (signal) from collimated hadronic jets (background).

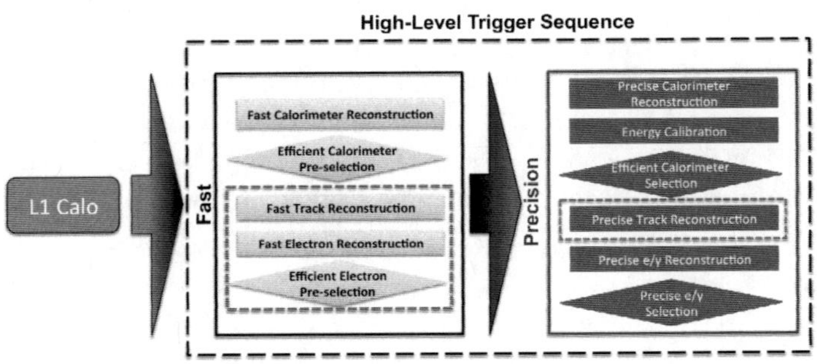

Figure 1: Flow diagram for the algorithm sequence of the Egamma chain. Steps surrounded by blue dashed lines only apply to electron chains.

A sliding window algorithm is used to build the L1 RoI, which is used to select EM objects by applying η-dependent E_{T} threshold and hadronic isolation requirements.

In the sequence, already in the HLT, fast reconstruction is applied to provide early rate reduction. Particularly for electron chains, the fast calorimeter and fast track reconstructions are combined to provide an efficient electron pre-selection. Calorimeter driven pre-selection is less resource intensive and, therefore, provides latency reduction. This reasoning is duplicated to the precision step where offline-like algorithms are applied.

The current algorithm used by Egamma ID makes extensive use of calorimeter shower shapes and widths, energy ratios, tracking quantities and track-cluster matching [4]. Offline reconstructed distributions of these discriminating quantities over Run 2 data can be found in Ref. [5].

Two standard selection variants are available. In cut-based ID, these quantities are individually compared to the respective optimized threshold. The likelihood (LH) ID variant, by contrast, uses a multivariate analysis technique, where a Naïve Bayes-like approach is applied over the very quantities (cf. Ref. [4] for more details on these variations). The LH ID variant is now the baseline method used on the precise calorimeter pre-selection and final selection for electrons as part of the Run 2 updates. Nevertheless, the fast stage still uses selection-based ID in which three, most discriminating, quantities are used by the calorimeter pre-selection.

3. Proposed Algorithm

The proposed algorithm (cf. Figure 2) aims at exploring the conic geometry of the shower shapes by using the cells of the calorimeter to build energy quantities describing the amount of energy available in each concentric ring around the cell of maximum energy deposition. This process preserves the lateral and longitudinal information of the shower shape and, thus, the physics interpretation, while reducing the amount of information with respect to using all cells.

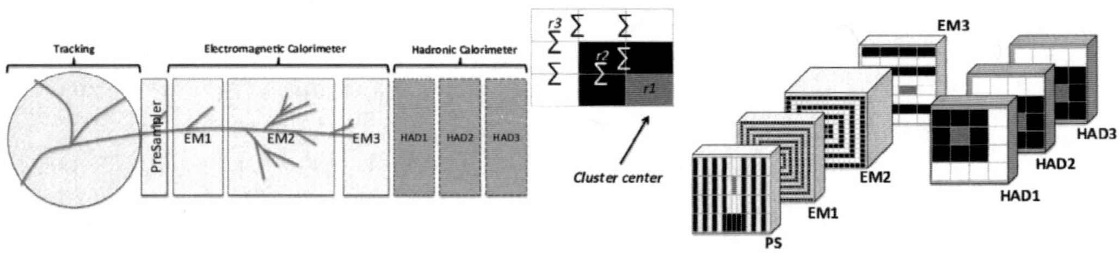

Figure 2: Sketches to illustrate the Ringer algorithm. Left: An EM shower propagation throughout the sampling layers of the calorimeters. Right: The calorimeter cells used to build the rings description over the calorimeters layers. The cell of first (central) ring is displayed in red, whereas the cells of the consecutive rings are alternated in black and white.

For the fast calorimeter pre-selection step, the Ringer RoI covers a region of the calorimeter, centered in the L1 provided RoI, of 0.4×0.4 in the $\eta \times \phi$ plane. The algorithm then starts on the second EM sampling layer (EM2)[3], where position (given in $\eta \times \phi$ plane) of the cell with highest E_{T} on the Ringer RoI is taken as the center of the cascade interaction with the calorimeter. The seed cell position is propagated to other calorimeter layers in order to define the axis of the rings in that layer ($c_{a,l}$). A ring $R_{n,l}$ contains all cells in a calorimeter layer l which are n cells from the axis. Formally,

$$R_{n,l} = \left\{ c_{n,l} \mid n = \left\lfloor \max\left(\frac{|\eta_{i,l} - \eta_{a,l}|}{h_{\eta,l}}, \frac{|\phi_{i,l} - \phi_{a,l}|}{h_{\phi,l}} \right) \right\rfloor, \forall c_{i,l} \in \Theta_{RoI,l} \right\}, \tag{1}$$

where (analogous to ϕ when suitable): $\eta_{i,l}$ and $\eta_{a,l}$ are respectively the $c_{i,l}$ and $c_{a,l}$ cells position in η; $h_{\eta,l}$ is the lth layer granularity in η; $\Theta_{RoI,l}$ is the set of cells in the lth layer which are within the Ringer RoI; $l \in \{\mathrm{PS, EM1, EM2, EM3, HAD1, HAD2, HAD3}\}$; and $n \in \{0, \ldots, N_l \mid N_{PS} = 7;\ N_{EM1} = 63;\ N_{EM2} = 7;\ N_{EM3} = 7;\ N_{HAD1} = 3;\ N_{HAD2} = 3;\ N_{HAD3} = 3\}$. If no cell is available for a given ring, the correspondent ring quantity is set to 0.

Sum of energies of cells in the ring over the sum of the cells energies in all rings form a vector of discriminating quantities. They are ordered outwards and from the innermost layer of the calorimeter and provided to a multivariate technique.

These discriminating quantities are fed into neural network (NN) [6] which performs the selection task. The architecture employed is a single hidden layer fully connected feedforward network (MLP — Multilayer Perceptron). The hidden and output layer activation function is set to $\tanh(\cdot)$. The data is divided into two datasets: the training set (contains 60% of available data), in which the NN weights are adjusted; and the test set (remaining 40%), used to verify that NN specialization does not occur and to evaluate performance. The NN weights are initialized by the Nguyen-Widrow [7] algorithm and optimized by applying the resilient back-propagation algorithm [8]. In order to avoid convergence to local optima, a total of 100 initializations was employed. The figure of merit used to evaluate performance was the threshold applied to the NN output which maximizes the SP-index,

[3] This convention is used for all sampling layers, thus HAD1 stands for the first hadronic layer, HAD2 the second hadronic layer etc.

$$SP(\%) = \sqrt{\sqrt{P_D \cdot (1 - FR)} \cdot \left(\frac{P_D + (1 - FR)}{2}\right)} \times 100\%, \tag{2}$$

where: P_D is probability of detection; and FR is the background rejection or fake rate.

The thresholds were adjusted using the signal efficiency of the benchmark for regions of η and E_{T}. Statistical fluctuations from the dataset are estimated by employing the cross-validation technique [6] with 50 sorts over 10 uniformly spread subsets of both signal and background datasets. The number of neurons in the hidden layer is chosen by the application of both cross-validation performance and generalization capability as criteria.

The operation points and fluctuations, respectively in terms of P_D and F_R, of the Ringer algorithm for each η region, when $E_{\mathrm{T}} < 30$ GeV, are of: 96.9 ± 0.1 and 4.6 ± 0.2 $(0 \leq |\eta| < 0.8)$; 96.9 ± 0.1 and 4.6 ± 0.2 $(0.8 \leq |\eta| < 1.37)$; 90.7 ± 0.1 and 6.2 ± 1.2 $(1.37 \leq |\eta| < 1.54)$; 95.1 ± 0.1 and 10.9 ± 1.2 $(1.54 \leq |\eta| < 2.5)$. The region $1.37 \leq |\eta| < 1.54$ is the transition between the barrel and end-cap calorimeters, where deterioration of the response of the detector occurs.

When $30 \leq E_{\mathrm{T}} < 50$, these values are the following: 99.0 ± 0.1 and 5.5 ± 0.7 $(0 \leq |\eta| < 0.8)$; 99.1 ± 0.1 and 6.3 ± 0.9 $(0.8 \leq |\eta| < 1.37)$; 95.9 ± 0.1 and 9.5 ± 2.1 $(1.37 \leq |\eta| < 1.54)$; 97.5 ± 0.1 and 11.2 ± 0.9 $(1.54 \leq |\eta| < 2.5)$.

Finally, for higher energies ($E_{\mathrm{T}} \geq 50$ GeV): 99.4 ± 0.1 and 2.0 ± 0.4 $(0 \leq |\eta| < 0.8)$; 99.7 ± 0.1 and 5.2 ± 1.0 $(0.8 \leq |\eta| < 1.37)$; 96.7 ± 0.1 and 2.7 ± 1.9 $(1.37 \leq |\eta| < 1.54)$; 98.4 ± 0.1 and 4.7 ± 0.9 $(1.54 \leq |\eta| < 2.5)$.

4. Results

The proposed algorithm, based upon ring-shaped calorimetry description, was compared to the baseline algorithm executed in the fast calorimeter pre-selection step. The selection chain of the trigger aims to pick events containing at least one electron with $E_{\mathrm{T}} > 24$ GeV. Prompt electrons from $Z \rightarrow ee$ decays simulated using Monte Carlo were used as signal input and, for the background, a dataset contained simulated hadronic jets with $E_{\mathrm{T}} > 17$ GeV. Both datasets have pile-up emulation of 20 mean inelastic interactions per bunch crossing.

Only electrons reconstructed in the full offline processing were considered in the efficiency computation. Analysis is done only for the fast calorimeter pre-selection decision. Thus, if a candidate is discarded in advance by the L1 algorithm, it is simply not taken into consideration and does not count for efficiency measurement. Likewise, the behavior of the remaining selection of the electron chain was not evaluated.

Apart from meeting above conditions, the training data were selected as follow: signal data are particles from $Z \rightarrow ee$ dataset considered as electrons by the most exigent criteria of the offline LH ID; and background data are particles from hadronic jets dataset, where the residual contamination of prompt electrons is removed by using simulation predictions.

The preliminary results in Figure 3 show that the proposed algorithm can operate with similar performance over both the full η acceptance and over the range of the pile-up conditions. Efficiency measurements use tag-and-probe[4] (T&P) technique to allow comparison with other studies which also consider collision data efficiencies. Notwithstanding, the efficiency measurements demonstrate the capacity of a factor of 2 reduction in the background efficiency: the signal efficiency of the ringer approach remains almost unchanged (97.6% → 97.7%) while the fraction of background surviving selection changes from 12.7% → 5.3%. Therefore, these results provide indication that the proposed algorithm is able to further improve Egamma HLT chain by upgrading the fast calorimeter pre-selection step.

[4] The tag-and-probe is a method to measure signal efficiency. $Z \rightarrow ee$ events with one electron passing default trigger and second (probe) reconstructed electron by offline are selected and used to check if probe electron has also passed trigger.

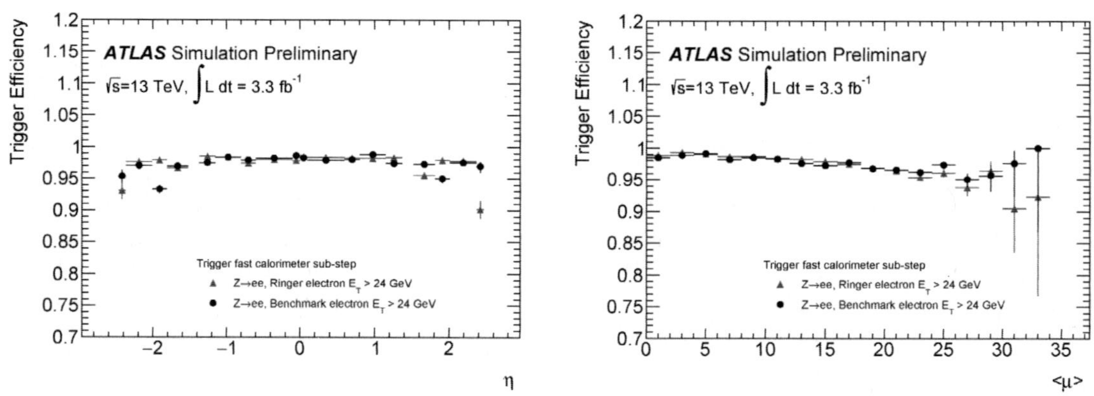

Figure 3: Electron T&P efficiency over $Z \rightarrow ee$ simulation data for the trigger fast calorimeter pre-selection. Left: as a function of the offline reconstructed electron candidates pseudorapidity (η). Right: as a function of the number of pile-up events (μ). Taken from Ref. [9].

5. Conclusions

A selection method for electrons using ring-shaped calorimetry information was described. It is an alternative for the fast calorimeter pre-selection stage of HLT trigger. Preliminary results from simulation data demonstrate that the algorithm is able to provide similar signal efficiency while improving background rejection by a factor of 2. Likewise, this gain also means a latency reduction due to background removal in early stages of the trigger chain. Also, the execution time of the proposed algorithm is equivalent to the baseline one. Hence, it may contribute to further improve the Egamma High-Level Trigger for ATLAS experiment. Finally, a version of the algorithm is being analysed on the offline reconstruction and it may also be applied to photon identification which relies on calorimeter information only.

6. Acknowledgements

We would like to thank CNPQ, CAPES, RENAFAE, FAPERJ (Brazil) and EU through the E-Planet project.

References

[1] ATLAS Collaboration 2008 *JINST* **3** S08003
[2] ATLAS Collaboration 2012 *Eur. Phys. J. C* **72** 1849 (*Preprint* `1110.1530`)
[3] ATLAS Collaboration 2013 Technical Design Report for the Phase-I Upgrade of the ATLAS TDAQ System Tech. Rep. CERN-LHCC-2013-018. ATLAS-TDR-023 CERN Geneva URL `https://cds.cern.ch/record/1602235`
[4] ATLAS Collaboration 2014 Electron efficiency measurements with the ATLAS detector using the 2012 LHC proton–proton collision data ATLAS-CONF-2014-032 URL `http://cdsweb.cern.ch/record/1706245`
[5] ATLAS Collaboration Electron shower shapes, tracking, isolation and invariant mass distributions from $Z \rightarrow ee$ and $J/\Psi \rightarrow ee$ events URL `https://atlas.web.cern.ch/Atlas/GROUPS/PHYSICS/PLOTS/EGAM-2015-003`
[6] Haykin S 1999 *Neural Networks: A Comprehensive Foundation* (Prentice Hall) ISBN 9780780334946
[7] Nguyen D and Widrow B 1990 *Proceeding of the International Joint Conference on Neural Networks* vol 3 (San Diego, USA) pp 21–26
[8] Riedmiller M and Braun H 1993 *Proceedings of the IEEE International Conference on Neural Networks* (San Francisco, USA) pp 586–591
[9] ATLAS Collaboration Expected performance of the ringer algorithm URL `https://twiki.cern.ch/twiki/bin/view/AtlasPublic/EgammaTriggerPublicResults`

Segmentation of HER2 protein overexpression in immunohistochemically stained breast cancer images using Support Vector Machines

Raquel Pezoa[1,2], Luis Salinas[1,2], Claudio Torres[1,2], Steffen Härtel[3], Cristián Maureira-Fredes[4] and Paola Arce[1]

[1] Department of Informatics, Universidad Técnica Federico Santa María, Valparaíso, Chile
[2] Centro Científico Tecnológico de Valparaíso, Universidad Técnica Federico Santa María, Valparaíso, Chile
[3] SCIAN-Lab, ICBM, BNI, University of Chile, Chile
4 Max Planck Institut für Gravitationsphysik (Albert-Einstein-Institut), D-14476 Potsdam, Germany

E-mail: raquel.pezoa@usm.cl

Abstract. Breast cancer is one of the most common cancers in women worldwide. Patient therapy is widely supported by analysis of immunohistochemically (IHC) stained tissue sections. In particular, the analysis of HER2 overexpression by immunohistochemistry helps to determine when patients are suitable to HER2-targeted treatment. Computational HER2 overexpression analysis is still an open problem and a challenging task principally because of the variability of immunohistochemistry tissue samples and the subjectivity of the specialists to assess the samples. In addition, the immunohistochemistry process can produce diverse artifacts that difficult the HER2 overexpression assessment.

In this paper we study the segmentation of HER2 overexpression in IHC stained breast cancer tissue images using a support vector machine (SVM) classifier. We asses the SVM performance using diverse color and texture pixel-level features including the RGB, CMYK, HSV, CIE L*a*b* color spaces, color deconvolution filter and Haralick features. We measure classification performance for three datasets containing a total of 153 IHC images that were previously labeled by a pathologist.

1. Introduction

A common practice for evaluating the progression of diverse types of cancers is the analysis of immunohistochemistry-stained tissue sections. Immunohistochemistry (IHC) is a method for detecting specific antigens in tissues or cells based on antigen-antibody recognition. This technique takes advantage of the specificity provided by the binding of an antibody with its antigen shown at a microscopic level [1]. The antigen-antibody interaction is visualized through a color-producing reaction in various cellular structures such as nucleus, cytoplasm or membrane, for instance.

A typical biomarker observed on the cell membrane is the Human Epidermal growth factor Receptor 2 (HER2). This is a protein normally expressed on the cell membrane surface of various organs epithelia such as lung, bladder, pancreas, breast, stomach and prostate [2]. When HER2 is present in normal amounts, the protein plays an important role in normal cell growth development. However, when the HER2 protein is overexpressed, cells multiply and grow more rapidly than normal, promoting carcinogenesis in diverse types of cancer [3].

The HER2 protein has become an important biomarker for invasive breast cancer. Around 15%-25% of breast cancer overexpress HER2 protein. Its measurement helps to guide the therapeutic decision process.

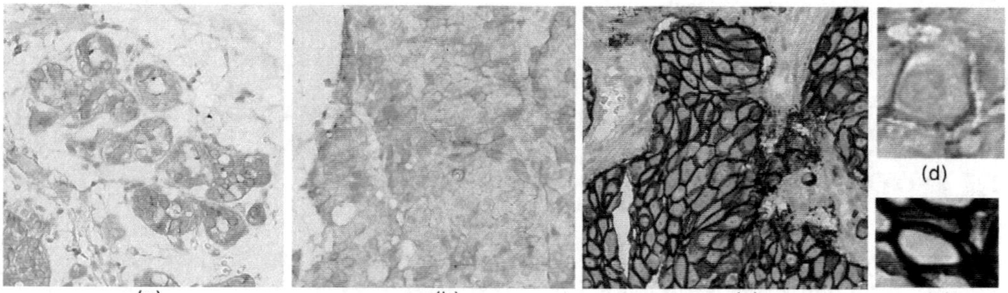

Figure 1. IHC-stained breast tissue images with score (a) 1+, (b) 2+ and (c) 3+ (40x magnification and 1000 × 1000 pixels). The cell structures of interest are cell membranes with HER2 overexpresion which are stained in brown color. Nuclei are stained in a blue color, to give contrast. (d) and (e) show two examples of cells with different levels of HER2 overexpression.

When IHC is used to assess the amount of HER2 overexpression, two main chemical compounds (commonly referred to as *stains*) are used to visualize cell nuclei and membranes: hematoxylin and diaminobenzidine (DAB). The hematoxylin is used as a counterstain and produces a blue-violet staining of nuclear elements. The diaminobenzidine produces a dark brown reaction on the cell membrane (only when there is HER2 overexpression).

In general, pathologists perform a qualitative assessment of HER2 overexpression through a visual inspection and classify IHC tissues according to the ASCO/CAP guidelines [4]. They should estimate two main parameters associated to HER2 overexpression: *intensity* and *completeness*. The former corresponds to the intensity of brown staining (from weak to intense) and the latter corresponds to the percentage of staining completeness with respect to the membrane length.

According to the two estimated parameters the observed specimen is classified into four categories: 0, 1+, 2+ and 3+. The classification system proposed in [4] indicates: 0 means no staining was observed, 1+ barely perceptible, 2+ weak to moderate and 3+ means strong intensity was observed in more than 10% of tumor cells (see Fig. 1).

Thus, the classification (and the whole analysis) is mainly based on the strength of the staining, whose evaluation clearly depends on the experience of the pathologist. This subjective analysis can produce inconsistent results. Hence, the importance of a quantitative and objective analysis is recognized by researchers in both, the image analysis and pathology fields [5].

Several cell segmentation methods have been proposed. However HER2 overexpression segmentation is still and open problem. This is due to the non-standardization of tissue samples and also due to the undesired brown staining over the cytoplasm. Regarding non-standardization, immunohistochemical tissue samples can be quite variable due to the use of different whole slide imaging (WSI) scanners, immunohistochemical kit systems or staining procedures [6], [7], [8]. This tissue variability makes difficult the finding of a general tissue pattern. On the other hand, HER2 overexpression is visualized through the brown color reaction produced on the membrane, but it is quite frequent that brown color spreads over the cytoplasm, which is an artefact and should not be considered as HER2 overexpression.

In this study we address the problem of segmentation of cell membranes with HER2 overexpression. This is a previous but crucial step before being able to estimate intensity and completeness parameters of the HER2 overexpression. We proposed a method for the segmentation of HER2 overexpression using a support vector machine (SVM). We asses the SVM performance using diverse color and texture pixel-level features including the RGB, CMYK, HSV, CIE L*a*b* color spaces, color deconvolution filter and Haralick features.

The outline of this paper is described as follows. Section 2 describes the proposed method, including the computation of color and texture features and SVM pixel classification. In section 3 we measure the performance of SVM classifiers on three different set of features and three datasets including images 1+ in data set one, images 2+ in data set two and images 3+ in data set three. Finally, Section 4 presents the conclusions and perspectives of this work.

2. Segmentation using SVM

The proposed method addresses the segmentation of HER2 overexpression as a binary classification problem and each pixel belongs to one of the two classes: overexpressed and non-overexpressed. The method uses a SVM as binary pixel classifier. As mentioned before, IHC images can be highly variable which makes difficult to find a common tissue pattern. Color variation is one of the most important issues, and this is produced for a number of reasons such as: the variable histochemical stains kits coming from different manufacturers, the staining procedure (which in many cases is performed in a manual way), or the use of different whole slide imaging (WSI) scanners. In fact, there is a lack of standardization between IHC clinical laboratories. Current developments in this field strongly suggest the need for the standardization of this process [9, 8, 7, 6]. Thus, considering this variable scenario, we decide to build one SVM classifier per IHC image.

The proposed method includes three main stages: pixel-level feature computation, training data generation, and SVM pixel classification (see Fig. 2). The aim of feature computation is to represent the IHC image pixels by means of a descriptive set of features that captures the complex characteristics of HER2 overexpression. In this paper, we compute a set of 210 color and texture features. Training data generation is a fundamental step of any supervised machine learning method. In this study, training pixels are selected from a small sub-image of the IHC image. Then, we build the SVM classifier in order to predict the label of the remaining pixels. Notice that the selection of the sub-image is an important issue. The proposed method needs that the expert provides training data to build the SVM classifier. This training data should be a small but representative set of the whole IHC image. The point of doing this is that we want to include the pathologist knowledge in order to guide the segmentation, due to the complexity of the IHC images, and pathologist are able to label small portions of the whole IHC-image. In this paper, we have the gold standard of the whole IHC image, thus, we are able to simulate when the pathologist generate the training data.

2.1. Pixel-level Feature Computation

In this work, we calculate a total of 210 color and texture pixel-level features (see Table 1). We use color features that correspond to different space color channels and a color deconvolution filter. We use four color spaces: RGB, CMYK, HSV and CIEL*a*b*, together with the grayscale image. In addition, we use a color deconvolution filter for stain separation in IHC stained images. This deconvolution is achieved by means of a transformation of the original RGB IHC image, which is based on the Lambert-Beer law and the optical density. In particular, we use the DAB channel obtained with the color deconvolution filter [10].

Texture features includes the calculation of Gray Level Co-occurrence Matrices (GLCMs) in order to obtain GLCM mean, GLCM standard deviation and eleven Haralick texture features [11] (a total of thirteen features). This calculation has a high computational cost, because we

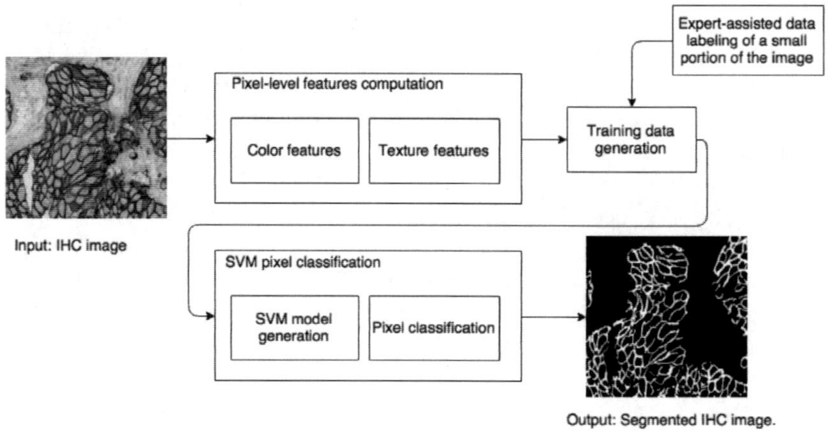

Figure 2. Overview of the main stages of the proposed method. The input is an RGB image obtained from a WSI scanner. The output is a binary image, where white pixels indicates overexpressed class, and black pixels indicate non-overexpressed class.

Type	**Feature**	
Color	RGB	3 channels, R, G and B
	CMYK	4 channels, C, M, Y and K
	HSV	3 channels, H, S and V
	CIEL*a*b*	3 channels, L*, a* and b*
	Grayscale image	1 channel
	Color deconvolution	1 DAB channel
Texture	Haralick features	11 Haralick features calculated for each the 15 color channels (which give a total of 165)
	GLCM mean	Calculated for each the 15 color channels
	GLCM standar deviation	Calculated for each the 15 color channels

Table 1. Color and texture pixel-level features.

are computing GLCM matrices in a pixel-level. For this calculation we define a neighborhood for each pixel and a GLCM matrix is calculated for each neighborhood. Then, from each GLCM matrix, we calculate thirteen texture features for each pixel. Due to the nature of the algorithmic process, we implemented a set of parallel routines in C++ code, using the MPI library, in order to speed up the whole process.

2.2. SVM pixel classification
Support Vector Machines (SVM) are supervised learning algorithms widely used for binary classification. SVM were introduced by Vladimir Vapnik et al. [12] and the general idea is to find an optimal separating hyperplane that maximizes the minimum distance between any data point in order to separate data into two classes. We use SVM as a binary pixel classifier, thus,

the segmentation of HER2 overexpression is handled as a classification problem. Each pixel is classified into one of the following classes: overexpressed or non-overexpressed. We build a SVM model for each IHC image using the pixels of a sub-image (training data). Then we predict the class labels of the remaining pixels (testing data).

3. Experiments

3.1. Data

A total of 153 images were used for the experiments, including around 750 cells. The images are grouped into three datasets. Dataset 1, dataset 2 and dataset 3 contain images of category 1+, 2+ and 3+, respectively. The images are regions of interest (ROIs) extracted from paraffin embedded breast cancer tissue slides obtained from the archives of the Department of Pathology at the Hospital Clínico Universidad de Chile. The tissues were scanned using the NanoZoomer-XR C12000 whole slide imaging scanner (Hamamatsu Photonics K.K.) at a magnification of 40x and a resolution of 0.23 μm/pixel. All images were manually labeled by a pathologist in order to generate training data and gold standard for the evaluation of the proposed method.

3.2. Experiment Setup

The goal of this experiment is to analyze the SVM performance using three set of features: the whole set of 210 features; (ii) a subset of 4 features corresponding to the channels of CMYK color space; (iii) and another subset of 14 features corresponding to the texture features applied to the gray channel (see Table 1).

For building the SVM classifieres, we select the Gaussian Radial Basis Function (RBF) as the SVM kernel function and we performe 5-fold cross-validation. The F1 metric [13] is used to measure performance in the cross-validation stage. We select the F1 metric because we are dealing with an unbalanced classification problem (there are more pixels belonging to the non-overexpressed class than the overexpressed class) [14]. Notice that the most commonly metric used in the machine learning literature is accuracy [13], however it is not well suited in this context.

In this experiment, for each IHC image I, we generate N sub-images s_k, $k = 1, \dots, N$. All the pixels of the sub-image s_k are used to train SVM classifiers svm_k, $k = 1, \dots, N$. For each k, the svm_k classifier is then used to classify the remaining pixels of the IHC image. Since we have the gold standard of the IHC image, we can calculate the performance η_k of the svm_k classifier. From the N sub-images s_k, we select the sub-image which gives the best SVM performance. We call this sub-image the "best sub-image", s^* (see Fig. 3).

This selection is repetead for each IHC image of our dataset. The performance presented in this paper correspond to the performance of the best sub-images of each IHC image. In addition, we group them into the 1+,2+ and 3+ categories and into the three different set of features (see box plots of Fig. 4).

Regarding the computational resources, we used a DELL Studio server to run the experiments. This server has 8 cores (Intel(R) Core(TM) i7 CPU 960 @3.20GHz) 24 GB RAM, and the Archlinux operating system. The framework was mainly programmed in Python, using the scikit-learn library [15] for machine learning algorithms. In particular, the Haralick features were calculated in the `datura` cluster (Max-Planck-Institut für Gravitationsphysik, Albert-Einstein-Institut, Germany), which has a total of 165 nodes (we used 128), and each node has 12 cores (CPU: Intel(R) Xeon(R) CPU X5650 @ 2.67GHz), 24 GB RAM, Scientific Linux 6.0 operating system, and InfiniBand for nodes communication.

3.3. Performance Evaluation

The SVM pixel classification performance is evaluated using five different metrics: ROC AUC, accuracy, precision, recall, and F1 [13]. The results can be seen in the box plots of Figure 4 and

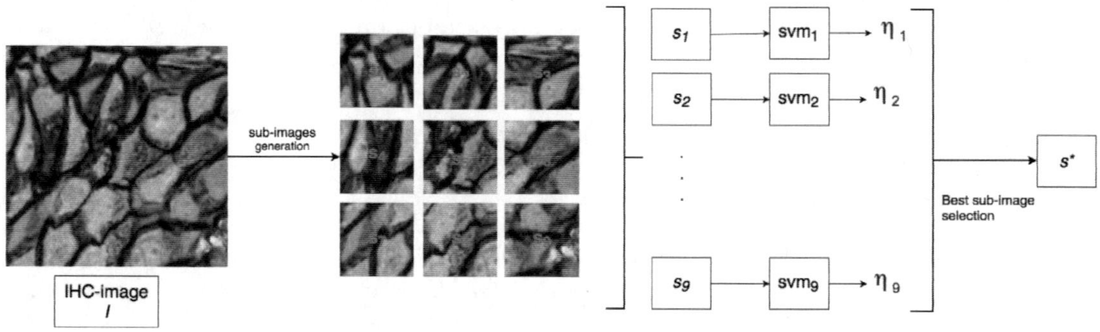

Figure 3. Selection of the best sub-image. We divide the image into N sub-images ($N = 9$ in this example) and we rank the sub-images according to the performance that is measured on the testing data (the whole IHC without the training data). The sub-image with the best performance, is called the *best sub-image.*

a summary of these is shown in Table 2.

In most of the cases, the CMYK feature set produced the better SVM performance. We may expect that using the full feature set (210 features) should produce the best performance, but the experiments with high dimensionality do not show this trend. Recall that the performance is measured using a gold standard manually labeled by one expert pathologist. This is a critical point, because the intrinsic subjectivity and variability of the gold standard can affect our results. Ficarra et al [16] clearly describe this situation.

In addition, the experiments show that the best sub-image is not always the same. We observed that when we change the feature set and the performance metric, the best sub-image also changes.

Dataset-Features	ROC AUC	Acc.	Prec.	Rec.	F1
1-210	0.75(0.08)	0.75(0.08)	**0.97(0.03)**	0.51(0.16)	0.65(0.13)
1-CMYK	**0.79(0.06)**	**0.79(0.06)**	0.95(0.03)	**0.61(0.14)**	**0.73(0.12)**
1-Haralick	0.73(0.06)	0.73(0.06)	0.95(0.03)	0.49(0.13)	0.64(0.12)
2-210	0.72(0.07)	0.72(0.07)	**0.98(0.01)**	0.44(0.14)	0.6(0.12)
2-CMYK	**0.79(0.07)**	**0.79(0.07)**	0.97(0.02)	**0.59(0.14)**	**0.72(0.12)**
2-Haralick	0.73(0.08)	0.73(0.08)	**0.98(0.01)**	0.47(0.16)	0.62(0.15)
3-210	0.79(0.08)	0.79(0.08)	**0.94(0.04)**	0.61(0.17)	0.73(0.13)
3-CMYK	**0.84(0.07)**	**0.84(0.07)**	0.93(0.03)	**0.73(0.14)**	**0.81(0.1)**
3-Haralick	0.80(0.07)	0.80(0.07)	0.93(0.04)	0.65(0.16)	0.75(0.11)

Table 2. Mean values (and standard deviation) of performance metrics obtained in each data set considering different sets of features.

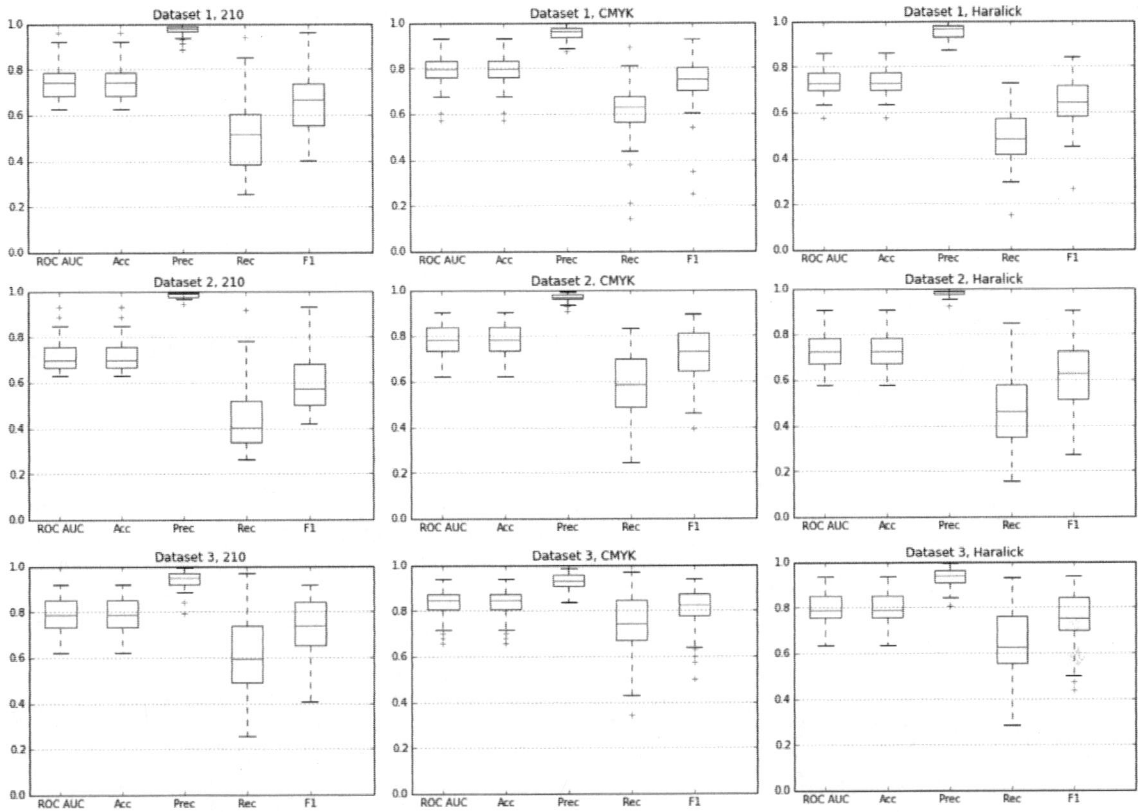

Figure 4. Box plots of SVM performance. We measured 5 performance metrics (ROC AUC, accuracy, precision, recall and F1). The SVM classifiers were built using the pixels of the best sub-image as training data. From left to rigth, columns correspond to 210, CMYK and Haralick features sets. From top to down, rows represent dataset 1 (42, 1+), datset 2 (50, 2+) and dataset 3 (61 images, 3+).

4. Conclusions

We have presented a method to segment HER2 overexpression in IHC stained breast cancer tissue images. We have described the main stages of the method, including the computation of 210 color and texture features, training data generation and SVM pixel classification; and performed experiments to evaluate the contribution of different sets of features to the SVM performance.

The experimental results have shown that using the best sub-image for training data selection and increasing the dimensionality of the feature set do not necessarily produce a better performance (considering our SVM configuration). The results showed that the set of 4 features corresponding to the CMYK color space produced the better SVM performace in most of the cases. This suggests that the aggregation of additional features, i.e., increasing dimensionality, does not contribute to a better segmentation. However, this is not enough evidence to remove those extra dimensions.

Two important aspects of our study have to be highlighted. First, we used a gold standard that was manually labeled by a pathologist. Thus, the interpretation of our results must consider the intrinsic subjectivity of the gold standard. Second, we were able to rank the sub-images because we have a gold standard, however this is precisely what is missing for a real case. Therefore, our next goals will be to continue analyzing the best set of features (using a gold standard generated for more than one pathologist), and developing methods able to detect the best sub-image, but now, from unlabeled set of images.

4.1. Acknowledgments

The authors acknowledge the assistance in tissue digitization performed at SCIAN Laboratory (University of Chile) and Hamamatsu TIGA Center (Heidelberg University). The authors thanks the Max-Planck-Institut für Gravitationsphysik (Albert-Einstein-Institut), Potsdam-Golm, Germany, for providing access to `datura` cluster for feature calculation.

Partial financial support for this research is acknowledged from CCTVAL-FB0821, FONDEF D11I1096 and FONDECYT 1150810.

5. References

[1] Taylor C R, Shi S R and Barr N J 2011 *Diagnostic Immunohistochemistry (THIRD EDITION)* ed Dabbs D J (Philadelphia: W.B. Saunders) pp 1 – 41

[2] Alvarez R, Cortés J, Mattos-Arruda L, Falzon M, Fasolo A, Gandy M, Gianni L, Harbeck N, Piccart M, Zambelli S *et al.* 2013 *Handbook of HER2-targeted agents in breast cancer* (Springer Healthcare Limited)

[3] Di Fiore P, Pierce J, Kraus M, Segatto O, King C and Aaronson S 1987 *Science* **237** 178–182

[4] Wolff A C, Hammond M E, Hicks D G, Dowsett M, McShane L M, Allison K H, Allred D C, Bartlett J M, Bilous M, Fitzgibbons P, Hanna W, Jenkins R B, Mangu P B, Paik S, Perez E A, Press M F, Spears P A, Vance G H, Viale G and Hayes D 2013 *Journal of Clinical Oncology* **31** 3997–4013

[5] Gurcan M, Boucheron L, Can A, Madabhushi A, Rajpoot N and Yener B 2009 *Biomedical Engineering, IEEE Reviews in* **2** 147–171

[6] Khan A, Rajpoot N, Treanor D and Magee D 2014 *Biomedical Engineering, IEEE Transactions on* **61** 1729–1738 ISSN 0018-9294

[7] Li X and Plataniotis K N 2015 *IEEE Transactions on Biomedical Engineering* **62** 1862–1873

[8] Bejnordi B E, Litjens G, Timofeeva N, Otte-Hller I, Homeyer A, Karssemeijer N and van der Laak J A 2016 *IEEE Transactions on Medical Imaging* **35** 404–415

[9] Lin F and Chen Z 2014 *Archives of Pathology & Laboratory Medicine* **138** 1564–1577

[10] Ruifrok A C and Johnston D A 2001 *Analytical and quantitative cytology and histology / the International Academy of Cytology [and] American Society of Cytology* **23** 291–299

[11] Haralick R M, Shanmugam K and Dinstein I 1973 *Systems, Man and Cybernetics, IEEE Transactions on* **SMC-3** 610–621

[12] Boser B E, Guyon I M and Vapnik V N 1992 *Proceedings of the Fifth Annual Workshop on Computational Learning Theory* COLT '92 (New York, NY, USA: ACM) pp 144–152

[13] Powers D M W 2011 *Journal of Machine Learning Technologies* **2** 37–63

[14] He H and Garcia E A 2009 *IEEE Transactions on Knowledge and Data Engineering* **21** 1263–1284 ISSN 1041-4347

[15] Pedregosa F, Varoquaux G, Gramfort A, Michel V, Thirion B, Grisel O, Blondel M, Prettenhofer P, Weiss R, Dubourg V, Vanderplas J, Passos A, Cournapeau D, Brucher M, Perrot M and Duchesnay E 2011 *Journal of Machine Learning Research* **12** 2825–2830

[16] Ficarra E, Cataldo S D, Acquaviva A and Macii E 2011 *IEEE Trans. Biomed. Engineering* **58** 1421–1429

Delphes 3: Latest Developments

Michele Selvaggi

Centre for Cosmology, Particle Physics and Phenomenology CP3, Université Catholique de Louvain, Chemin du Cyclotron, 1348 Louvain la Neuve, Belgium

E-mail: michele.selvaggi@uclouvain.be

Abstract. Latest developments in Delphes 3, such as the implementation of photons conversions into electron positron pairs, as well as a renovated energy-flow algorithm, which has been specifically designed to reconstruct complex boosted objects using all the available detector information, are described in this note.

1. Introduction

Delphes is a C++ framework [1, 2, 3], performing a fast multipurpose detector response simulation. The simulation includes a tracking system, embedded into a magnetic field, calorimeters and a muon system. The framework is interfaced to standard file formats [4, 5, 6] and outputs observables such as isolated leptons, missing transverse energy and collection of jets which can be used for dedicated analyses. The simulation of the detector response takes into account the effect of magnetic field, the granularity of the calorimeters and sub-detector resolutions.

Latest developments have to scope of extending the reach of Delphes. Complex event reconstruction algorithms are needed in the context of novel detector configurations, such as prototypes for future colliders experiments. Such detectors will operate at extreme conditions, and will be required to precisely reconstruct particle of large momenta in an environment of higher average occupancy compared to present detectors. An improved version of the energy-flow algorithm has been developed to be able to cope with such conditions and will be presented in section 2. Another recent development deals with the reconstruction of photons which has been made more realistic by including the simulation of photon conversions. Such improvement will be presented in section 3.

2. Energy-flow algorithm

Future detector environments will feature large instantaneous luminosities which typically result in high particle occupancy in the detector. Specific reconstruction algorithms [7] are needed to be able to reject particle originating from secondary scattering (pile-up interactions). In addition, largely "boosted" jets typically display the presence of high momentum hadrons and photons separated by small angles. Given that such boosted objects might indicate the presence of new physics at high scales, specific procedures to reconstruct them accurately are needed.

The detector model in Delphes is simple: charged and neutral particles originating from the interaction point propagate within a cylindrical volume embedded in a magnetic field until they reach the calorimeters. Calorimeters are described by a two-dimensional surface in the (η, ϕ)

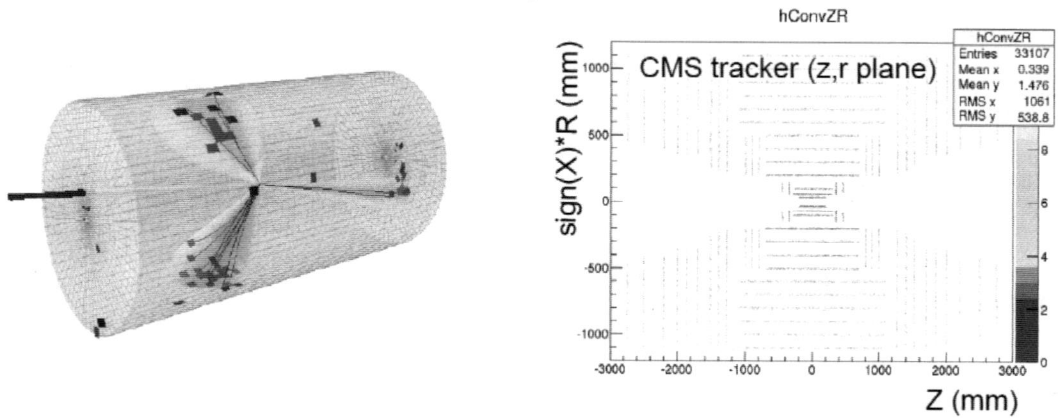

Figure 1. (Left) Delphes event display, showing the inner tracking volume surrounded by the calorimeter. The segmentation in (η, ϕ) is taken as (0.1×0.1), and matches the typical angular resolution from the hadronic calorimeters of the ATLAS and CMS experiments. (Right) Distribution of conversion vertices obtained with the material density map of the CMS tracker.

plane surrounding the propagation volume (see figure 1, left). This surface is subdivided in elementary cells that are called *towers*.

The main goal of the energy-flow algorithm in Delphes is to make use of all the available energy deposit information to optimally reconstruct objects such as tracks and tower deposits. These objects are later processed to form high-resolution complex objects such as jets, missing energy, or well identified particles. More specifically one wants to be able to exploit the difference in the performance provided in various energy-momentum range by the two main subdetectors: the tracker and the calorimeters. It is well known that tracking algorithms take advantage of the curvature of charged particles to be able to reconstruct their momenta. Therefore the momentum resolution of tracks is optimal for low transverse momenta , e.g. $p_T \leq 20$ GeV for the ATLAS [8] or CMS [9] detectors. Conversely, large energy deposits are well reconstructed by calorimeters.

We will now describe in detail the algorithm. Each calorimeter tower is characterised by its edges, which delimit its size (and control the angular resolution of the calorimeter) and by its total energy deposit, E^{cal}. For each tower, we count the number of tracks N that propagate into it, and separate them into two distinct categories:

- N_{good} tracks, such that $\sigma_{trk} < \sigma_{cal}$, depositing total charged energy E^{trk}_{good} ,
- N_{bad} tracks, such that $\sigma_{trk} > \sigma_{cal}$, depositing total charged energy E^{trk}_{bad} ,

where σ_{trk} and σ_{cal} are respectively the momentum resolution obtained with the tracking and the energy resolution obtained with the calorimeter. Good tracks have typically low transverse momentum, since they are reconstructed with high precision by the tracking system, whereas bad tracks have high momenta and are better reconstructed by calorimeters.

We define the excess neutral energy as:

$$\Delta E \equiv \max(E^{cal} - E^{trk}_{good} \;, \;\; 0)\,. \qquad (1)$$

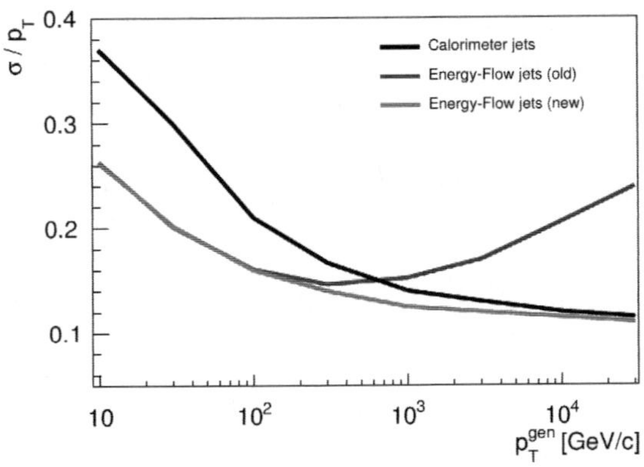

Figure 2. Jet momentum resolution obtained with the old and new versions of the energy-flow algorithm compared to jets built out of pure calorimeter towers.

According to the number N_{bad} of *bad* tracks that are found within a given tower, the algorithm proceeds in the following way:

(i) if no bad tracks are found, $N_{bad} = 0$:

If the excess energy ΔE is significant, that is, if:

$$\begin{cases} \Delta E & > E_{min} \,, \\ \frac{\Delta E}{\sigma_{cal}(\Delta E)} & > S_{min} \,, \end{cases} \tag{2}$$

where E_{min} and S_{min} are two configurable parameters, that respectively represent the minimal energy and the minimum significance for a deposit to be retained. If the above conditions are satisfied, a single neutral *energy-flow tower* object is produced, with energy $E = E^{cal} - E^{trk}_{good}$. The original (good) tracks are kept unchanged, and are simply re-labeled as *energy-flow tracks*.

(ii) if at least one bad track is found, $N_{bad} > 0$:

In this case there is no way to assert whether the excess ΔE is really originating from a neutral particle hitting the calorimeter, or if it is simply the result of poor tracking resolution producing an *over-smeared* track. We then use the superior calorimeter resolution to assess to momenta of *bad* tracks. The momenta of bad tracks are rescaled by a factor:

$$r = \frac{E^{cal} - E^{trk}_{good}}{E^{trk}_{bad}} \,. \tag{3}$$

In this case no *energy-flow tower* are created, good tracks are kept unchanged and re-labeled as *energy-flow tracks* as in the previous case, and bad tracks have their momenta rescaled by r. This approach ensures total energy conservation, since by construction we now have, after rescaling:

$$\sum E^{trk}_{bad} = E^{cal} - E^{trk}_{good} \,. \tag{4}$$

Case (i) is easily obtained in high-pile-up scenarios where a significant number of low momentum tracks randomly populate the detector. Case (ii) is instead very frequent when reconstructing highly energetic jets.

To resume, in both cases, for each calorimeter elementary cell, the energy-flow algorithm produces a set of objects called respectively *energy-flow towers* and *energy-flow tracks*. These *energy-flow* objects can be used as input to jet clustering algorithms, or to compute isolation variables or missing energy. The algorithm described in this note has been included starting from Delphes version 3.3.0.

The performance of the energy-flow procedure is shown in figure 2. In older Delphes versions, the simplistic assumption was made that $\sigma_{trk} < \sigma_{cal}$ regardless of the charged particle momentum, which resulted in poor jet resolution performance at high momenta, i.e. $p_T > 1$ TeV. Instead, the algorithm presented above dynamically accounts for the the two different momentum regimes, and gives the most precise measurement in all the considered energy range. Both versions of the energy-flow jets are compared to the calorimeter jets (i.e. reconstructed with pure calorimeter information). At low energies, the energy-flow jets outperforms calorimeter jets thanks to the superior tracking resolution, whereas at high energies both give comparable performance as expected.

3. Photon conversions

About 50% of the total number of long-lived particles produced in a collider experiment are photons originating from π^0 decays. Accurately reconstructing photons is therefore crucial. For incoming photon energies $E_\gamma \geq 1$ MeV, the presence of material in detectors is responsible for photon conversions into electron positron pairs. The pair of resulting charged particles can be reconstructed as tracks. This phenomenon can be used to measure photons more precisely, especially at moderate energies where the tracking provides excellent angular and momentum resolutions.

The number of photons "surviving" after travelling a distance Δx in a given material is:

$$N_\gamma(x) = N_0 e^{-\frac{\Delta x}{\lambda}}, \tag{5}$$

where λ is the *mean free path* and N_0 is the number of photons present in the incident beam. Hence, the probability for a photon of converting after travelling a distance x is given by:

$$P(x) = 1 - e^{-\frac{\Delta x}{\lambda}}. \tag{6}$$

A basic algorithm can therefore be built upon two simple external inputs:

- the step length, Δx, which controls the accuracy of the photon conversion simulation,
- the material budget map, $\lambda^{-1}(r, z, \phi)$, the average conversion rate per unit length. [1]

Also, we consider only a regime such that $E_\gamma \gg m_e$, we can therefore assume that the two resulting particles are emitted collinear to the photon. Within such massless approximation, one can simply generate the energy fraction x_F that gets attributed to each electron, according to the following formula [10]:

$$\frac{d\sigma}{dx_F} = 1 - \frac{4}{3} x_F (1 - x_F)\,. \tag{7}$$

The conversion algorithm proceeds as follows:

[1] In terms of well known quantities, $\lambda^{-1} = \frac{7\rho}{9X_0}$, where ρ is the material density and X_0 is the radiation length of the material.

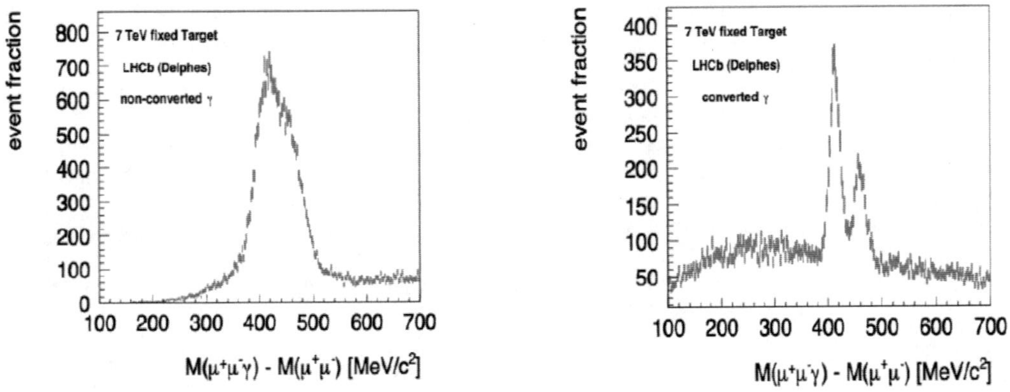

Figure 3. Distribution of the mass difference $M_{\mu^+\mu^-\gamma} - M_{\mu^+\mu^-}$ for χ_{c1} and χ_{c2} using calorimeter photons (left) and converted photons (right).

(i) the initial photon at x_0 is propagated for a distance Δx into position x_1,
(ii) the values $\lambda^{-1}(x_1)$ and $P(x_1)$ are computed,
(iii) a random number p between 0 and 1 is thrown,
(iv) if $p > P(x_1)$, keep propagating the photons from step (i). Otherwise the photon is converted according to formula 7 and the resulting e^+ and e^- are propagated from until they reach the calorimeters.

For sake of illustration, we show in figure 1 (right) a longitudinal projection of the distribution of conversion vertices obtained after specifying the material density map of the CMS tracker. The material distribution can be exactly reconstructed.

A possible application of photons conversions is the reconstruction of narrow resonances decaying into low energy photons [11]. For instance, χ_{c1} and χ_{c2} charmonia states decay to $J/\psi\gamma$ and have almost identical masses. Relying simply on calorimeter photons to reconstruct such resonances can result in not being able to distinguish the two distinct states figure 3 (left). Using converted photons instead, thanks the superior angular and momentum resolution of the tracking system at low energies, can help in resolving the two distinct states.

4. Conclusion

Recent developments such as an extension of the energy flow algorithm and the implementation of photon conversions in Delphes have been presented. Further developments on the energy flow algorithm include the possibility of parameterising efficiency loss as a function of high local particle occupancy. In order to account for the difficulty to reconstruct tracks originating from very displaced vertices, a parameterisation of the track reconstruction efficiency as a function of the transverse impact parameter might be beneficial as a possible extension of the photon conversion algorithm.

References

[1] de Favereau J, Delaere C, Demin P, Giammanco A, Lematre V, Mertens A and Selvaggi M (DELPHES 3) 2014 *JHEP* **02** 057 (*Preprint* 1307.6346)
[2] Selvaggi M 2014 *J. Phys. Conf. Ser.* **523** 012033
[3] Mertens A 2015 *J. Phys. Conf. Ser.* **608** 012045
[4] Alwall J *et al.* 2007 *Comput. Phys. Commun.* **176** 300–304 (*Preprint* hep-ph/0609017)

[5] Dobbs M and Hansen J B 2001 *Comput. Phys. Commun.* **134** 41–46
[6] Chekanov S V 2013 *Community Summer Study 2013: Snowmass on the Mississippi (CSS2013) Minneapolis, MN, USA, July 29-August 6, 2013* (*Preprint* 1306.6675) URL https://inspirehep.net/record/1240597/files/arXiv:1306.6675.pdf
[7] Bayatian G *et al.* (CMS) 2009 Particle-Flow Event Reconstruction in CMS and Performance for Jets, Taus, and MET Tech. Rep. CMS-PAS-PFT-09-001 CERN 2009. Geneva URL https://cds.cern.ch/record/1194487
[8] Aad *et al.* (ATLAS Collaboration) 2009 *Expected performance of the ATLAS experiment: detector, trigger and physics* (Geneva: CERN) URL https://cds.cern.ch/record/1125884
[9] Bayatian G *et al.* (CMS) 2006 CMS physics: Technical design report Tech. Rep. CERN-LHCC-2006-001, CMS-TDR-008-1
[10] Tsai Y S 1974 *Rev. Mod. Phys.* **46**(4) 815–851 URL http://link.aps.org/doi/10.1103/RevModPhys.46.815
[11] Aaij R *et al.* (LHCb) 2013 *JHEP* **10** 115 (*Preprint* 1307.4285)

Support Vector Machines and Generalisation in HEP

A. Bethani[1], A. J. Bevan[2], J. Hays[2] and T. J. Stevenson[2]

[1]LPSC, Grenoble, France. [2]Queen Mary University of London, London, United Kingdom.

E-mail: a.j.bevan@qmul.ac.uk, t.j.stevenson@qmul.ac.uk

Abstract. We review the concept of support vector machines (SVMs) and discuss examples of their use. One of the benefits of SVM algorithms, compared with neural networks and decision trees is that they can be less susceptible to over fitting than those other algorithms are to over training. This issue is related to the generalisation of a multivariate algorithm (MVA); a problem that has often been overlooked in particle physics. We discuss cross validation and how this can be used to improve the generalisation of a MVA in the context of High Energy Physics analyses. The examples presented use the Toolkit for Multivariate Analysis (TMVA) based on ROOT and describe our improvements to the SVM functionality and new tools introduced for cross validation within this framework.

1. Introduction

Machine learning methods are used widely within High Energy Physics (HEP). One promising approach, used extensively outside of HEP for applications such as handwriting recognition, is that of Support Vector Machines (SVMs), a supervised learning model used with associated learning algorithms for multivariate analysis (MVA). Developed originally in the 1960s, with the current standard version proposed in 1995 [1], SVMs aim to classify data points using a maximal margin hyperplane mapped from a linear classification problem to a possibly infinite dimensional hyperspace. However this means SVMs, like other MVA classifiers, have a number of free parameters which need to be tuned on a case by case basis. This motivates a number methods for ensuring the classifier is sufficiently generalised such that when used on an unseen dataset the performance can be accurately predicted. In this paper a brief overview of SVMs is given in Section 2, with an example using SVMs shown in Section 2.4. Generalisation is discussed in Section 3 with an illustrative example of how this can improve performance given in Section 3.3.

2. Support Vector Machines

2.1. Hard Margin SVM

Consider the problem of linear classification with the SVM where the training set, $\mathbf{x}$, is linearly separable. We define a separating hyperplane given by $\langle \mathbf{w} \cdot \mathbf{x} \rangle + b = 0$, where $\mathbf{w}$, the weight vector, is perpendicular to the hyperplane, and b, the bias, determines the distance of the hyperplane from the origin (Fig. 1(a)). A decision function defined by $y_i = \mathrm{sign}(\langle \mathbf{w} \cdot \mathbf{x_i} \rangle + b)$ is used to positively and negatively classify $\mathbf{x}_i$, the points in the training set. Without further constraint the choice of separating hyperplane is arbitrary and the geometric margin, γ, is introduced with the aim to maximise this. The functional margin, γ_i, for $\mathbf{x}_i$ is defined as

$$\gamma_i = y_i(\langle \mathbf{w} \cdot \mathbf{x}_i \rangle + b) \geq 1, \tag{1}$$

to impose the condition that no points lie within the margins. Considering two points that lie closest to the hyperplane referred to as support vectors (SVs), $\mathbf{x}^+$ and $\mathbf{x}^-$, with functional margins of $+1$ and -1 respectively, the geometric margin γ can be defined as $\gamma = \frac{1}{2}\left(\left\langle \frac{\mathbf{w}}{\|\mathbf{w}\|} \cdot \mathbf{x}^+ \right\rangle - \left\langle \frac{\mathbf{w}}{\|\mathbf{w}\|} \cdot \mathbf{x}^- \right\rangle\right) = \frac{1}{\|\mathbf{w}\|}$. The problem is solved in dual space by minimising the Lagrangian

$$\widetilde{L}(\alpha) = \sum_{i=1}^{n} \alpha_i - \frac{1}{2}\sum_{i,j} \alpha_i \alpha_j y_i y_j \langle \mathbf{x}_i \cdot \mathbf{x}_j \rangle\,, \tag{2}$$

for the parameters α_i (Lagrange multipliers) subject to $\alpha_i \geq 0$, where α_i is only non-zero for SVs, and $\sum_{i=1}^{n} \alpha_i y_i = 0$, providing the constraint equation. The full derivation leading to this minimisation and further details can be found in [2].

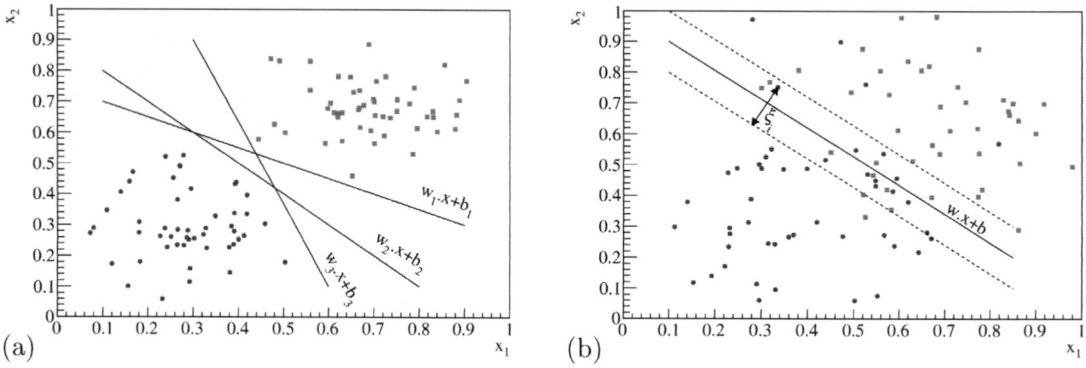

Figure 1: Illustrations of the (a) hard margin SVM where the solid line is the separating hyperplane, described by $\mathbf{w}$ and b, and the (b) soft margin SVM. The slack parameters, ξ_i, allow for misclassification of points weighted by the tunable cost parameter, C. Dotted lines are the margins, with the SVM finding the maximum geometric margin, γ.

2.2. Soft Margin SVM

The hard margin SVM is an important SVM algorithm, however it is not applicable in most real world cases. Issues arise if the data are not linearly separable due to being noisy, with data having small variations between different independent samples. This leads to the training becoming susceptible to over-fitting due to a few points in the training set, so some classification error must be allowed. To achieve this the hard margin constraint is relaxed and points are allowed on the incorrect side of the decision boundary, with the misclassification described by two additional parameters, ξ_i (slack) and C (cost), where ξ_i defines the distance from the correct margin to the ith incorrectly classified SV, and C is a tunable weight parameter multiplying the sum of slack variables penalising the misclassified points (Fig. 1(b)). This reduces to the same dual space Lagrangian problem as for the hard margin SVM, Eq. (2), however the constraint on the Lagrange multipliers is now $0 \leq \alpha_i \leq C$, while the constraint equation remains, $\sum_{i=1}^{n} \alpha_i y_i = 0$.

2.3. Kernel Functions and Feature Spaces

To this point the discussion of the SVM algorithm has been considered a linear decision boundary, however in most cases this is sub-optimal. In order to overcome this we can replace the inner product found in Eqns (1) and (2) with a kernel function (KF) $K(\mathbf{x}, \mathbf{y}) = \langle \phi(\mathbf{x}) \cdot \phi(\mathbf{y}) \rangle$ that maps the problem from the input space X to a potentially higher dimensionality, implicit feature space $F = \{\phi(\mathbf{x}) | \mathbf{x} \in X\}$, where the data may then be separable. In this process neither the feature map, $\phi(\mathbf{x})$, or the feature space, F, need to be known explicitly (known as the "Kernel Trick" [2]). This leads to specific properties required to define proper kernel functions in order for them to be used in this way. The kernel function should satisfy $K(\mathbf{x}, \mathbf{y}) = K(\mathbf{y}, \mathbf{x})$ and

$K(\mathbf{x}, \mathbf{y})^2 \leq K(\mathbf{x}, \mathbf{x})K(\mathbf{y}, \mathbf{y})$. These are not strong enough conditions to guarantee the existence of F and so a KF is also required to meet Mercer's condition [3].

An example of a kernel meeting these requirements is the polynomial KF of Eq. (3) with two tunable parameters, the order d and offset c. Another common choice of KF is the Radial Basis Function (RBF), Eq. (4), where the distance between two SVs is computed and used as the input to a Gaussian with one tunable parameter $\Gamma = 1/2\sigma^2$. The RBF does not take into account potential bandwidth differences between dimensions in the input data which may cause tension and the norm between SVs may result in loss of information. It can be extended to include a tunable parameter, Γ_i for each dimension in the input space referred to as the multi-Gaussian KF, Eq. (5). The KFs implemented in TMVA (along with products and sums thereof) are:

$$K(\mathbf{x}, \mathbf{y}) = (\mathbf{x} \cdot \mathbf{y} + c)^d = \left(\sum_{i=1}^{\dim(X)} \mathbf{x}_i \mathbf{y}_i + c \right)^d, \tag{3}$$

$$K(\mathbf{x}, \mathbf{y}) = e^{-\Gamma \|\mathbf{x}-\mathbf{y}\|^2}, \tag{4}$$

$$K(\mathbf{x}, \mathbf{y}) = \prod_{i=1}^{\dim(X)} e^{-\Gamma_i (\mathbf{x}_i - \mathbf{y}_i)^2}. \tag{5}$$

2.4. Higgs Boson Example: $H \to \tau^+\tau^-$

The Higgs Machine Learning Challenge dataset [4], with 6500 signal and 10000 background points is used to train SVMs with the KFs of Eqns (3-5), as well as a Boosted Decision Tree (BDT) [5]. The six variables offering the most discriminating power were used; MMC, $m_T(E_T^{miss}, l)$, $\Delta R(\tau, l)$, p_T^l/p_T^τ, ϕ Centrality and E_T^{miss}, see [4] for definition. The classifiers are optimised, trained and tested with TMVA [6] using the hold-out validation method, discussed in Sec. 3.1. Considering the receiver operating characteristic (ROC) curves, signal efficiency versus background rejection shown in Fig. 2(b), the classifiers all have similar performance for this problem. However it is not clear whether these solutions are fine tuned or optimal and so a further step is required to confirm if the results are generalised.

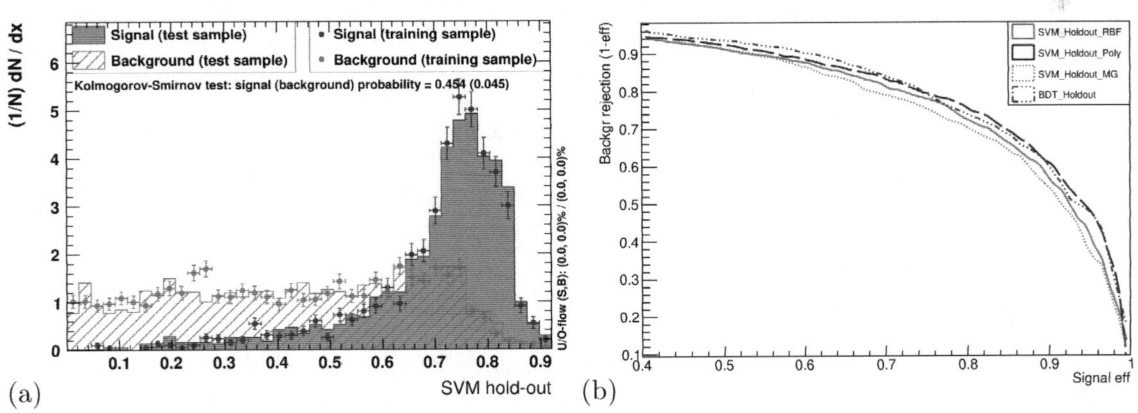

Figure 2: (a) TMVA outputs for SVM with RBF kernel optimised, trained and tested on the Higgs Machine Learning Challenge dataset with hold-out validation method. (b) ROC curves for SVMs with polynomial, RBF and multi-Gaussian kernel functions and a BDT.

3. Generalisation

Due to the often non-transparent nature of multivariate techniques one requires confidence that the trained MVA is robust against over-fitting, such that when used on an unseen data sample the performance of the MVA is accurately predicted and therefore the MVA is generalised. In order to ensure this, validation techniques are required for both model selection, as most MVAs have at least one free hyper-parameter (HP), and for performance estimation, measured with a figure of merit such as the misclassification error. For an unlimited dataset these issues are trivial, as one could simply iterate through the HP space and models using a different subset of the data each time, evaluating the performance and choosing the model and configuration with the lowest error rate. In reality datasets are often much smaller than required. Naïvely the entire dataset could be used to select and train the classifier and to estimate the error, however this leads to over-fitting or over-training as the classifier learns specific fluctuations in the dataset and subsequently performs worse on unseen data. This is more distinct in classifiers with a large number of tunable HPs. This also gives an overly optimistic estimation of the error.

3.1. Hold-out Validation

A natural extension and potential way to overcome these issues is to split the dataset into two subsamples, one for training and one for testing. This is referred to as hold-out validation [7]. The training sample is used to select optimal HPs, using a method such as back propagation for the Multilayer Perceptron [8], and train the classifier. The testing sample is used to evaluate the performance. This method is not ideal as the performance differs depending on how the data is split, which may lead to misleading error estimates.

3.2. k-fold Cross-validation

In circumstances where a relatively small dataset is available it may not be possible to reserve a large portion of data for testing and so the hold-out method may not be viable. Instead the approach of k-fold cross-validation (CV) [7] can be used, performed as follows:

(i) Split the data into k equally sized, randomly sampled, independent datasets (or folds);

(ii) Train the classifier with the data from $k-1$ of the folds;

(iii) Test the classifier using the remaining fold;

(iv) Repeat steps (ii) and (iii) k times for all permutations.

The overall performance for the classifier is then given by the average error rate, $E = \frac{1}{k}\sum_{i=1}^{k} E_i$, where E_i is the misclassification rate for each training. This method has the advantage of using the entire dataset for testing and training, offering potentially improved performance over the previous methods.

However it raises the question of the number of folds to be used. A large number of folds will give good estimation of the average error rate, as the bias of the estimator is small, but the variance is large and the computational time is large, whilst the reverse is true for a small number of folds. In reality the choice of number of folds should be motivated by the size of the dataset available for the problem, for example a sparse dataset may require the extreme of leave-p-out CV [7] in order to train on as much data as possible.

Building on the k-fold method, ideally the data should be split into three statistically independent, randomly sampled datasets for training, testing and validation, with each set split into k-folds. The purpose of the training set is to choose the model and to optimise the HPs for this model through iterative k-fold CV, with one set of HPs for each fold. These HPs are then tested via k-fold CV with the validation set to obtain their estimated performance. The training and validation samples are then combined and used to train the final classifier with the best performing HPs. The test sample, which has been reserved until this point, is then used to assess the performance of this final fully trained classifier.

Taking the best performing classifier from the optimisation process may not be the optimal solution or give the desired output, due to potential pathologies in the distributions. This also involves discarding a large number of trained classifiers from the process, so instead of using all HPs sets from the optimisation step, k classifiers can be trained and their performance estimated. These can then be used to give an averaged output on the testing sample which can help reduce fluctuations in the final distributions, improving agreement between the performance on the training and testing samples.

3.3. Higgs Boson Example: $H \to \tau^+\tau^-$

Using the Higgs Machine Learning Challenge dataset, discussed in Sec. 2.4, SVMs with RBF KFs were optimised, trained and tested using 5-fold CV to obtain the best performing and averaged SVMs, as well as using the hold-out validation method for comparison. The ROC curve comparing the different methods, Fig. 3(b), shows improvement for the k-fold CV classifiers over the hold-out method, both in the performance and shape of the curves. It is also worth noting that for the best performing and averaged classifiers from the k-fold CV, Fig. 3(a), show improved agreement between the training and test samples than for the hold-out validated method, Fig. 2(a), illustrating how this can lead to a more generalised result.

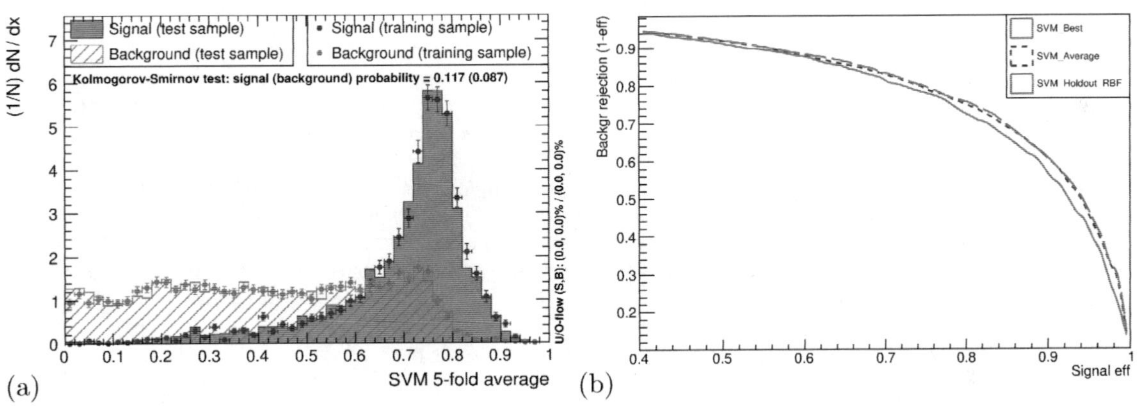

Figure 3: (a) TMVA output for a SVM with RBF kernel optimised, trained and tested on the Higgs Machine Learning Challenge dataset using 5-fold validation method averaging the classifiers from each fold. (b) ROC curves showing improved performance of the best and averaged 5-fold classifiers over the hold-out method classifier.

4. Summary

A brief overview of Support Vector Machines was presented, with an example showing similar performance to that of a BDT. However it is not clear without further checks as to whether the MVAs are sufficiently generalised. Hence a multistage cross-validation procedure has been outlined, which for the same example shows better performance as well as better agreement between the training and testing samples in the output distributions. This functionality is now available in the github version of ROOT.

References

[1] C. Cortes and V. Vapnik, Machine Learning Volume **20** Issue 3 (1995).
[2] J. Shawe-Taylor and N. Cristianini, Support Vector Machines and other kernel-based learning methods, Cambridge University Press (2000).
[3] J. Mercer, Philosophical Transactions of the Royal Society A **209** Issue 441–458 (1909).
[4] C. Adam-Bourdarios, G. Cowan, C. Germain-Renaud, I. Guyon, B. Kgl and D. Rousseau, "The Higgs Machine Learning Challenge," J. Phys. Conf. Ser. **664** (2015) No.7, 072015. See also URL: http://higgsml.lal.in2p3.fr/documentation/ (2014).
[5] L. Breiman, J. Friedman, R. Olshen, and C. Stone, Classification and Regression Trees, Wadsworth International Group (1984).
[6] A. Hoecker et al., PoS ACAT 040 (2007) 040.
[7] S. Arlot and A. Celisse, Statistics Surveys Vol. **4** 4079 (2010).
[8] R. Rojas, Neural Networks: A Systematic Introduction, Springer Science & Business Media (1996).

Journal of Physics: Conference Series **762** (2016) 012053 doi:10.1088/1742-6596/762/1/012053

The Matrix Element Method at the LHC: status and prospects for Run II

Sébastien Wertz

Centre for Cosmology, Particle Physics and Phenomenology - CP3,
Université catholique de Louvain, 2 Chemin du Cyclotron - Box L7.01.05,
B-1348 Louvain-la-Neuve, Belgium

E-mail: `sebastien.wertz@uclouvain.be`

Abstract. The Matrix Element Method (MEM) is a powerful multivariate method allowing to maximally exploit the experimental and theoretical information available to an analysis. Applications of the MEM at LHC experiments are discussed, such as searches for rare processes and measurements of properties of the Standard Model Higgs boson. The MadWeight software, allowing for a fast and automated computation of MEM weights for any user-specified process, is briefly reviewed. A new implementation of the MEM in the C++ language, MoMEMta, is presented. Building on MadWeight's tricks to accelerate the calculations, it aims at a much improved modularity and maintainability. Examples of this modularity are discussed: the possibility to compute several weights in parallel (propagation of systematic uncertainties), the Differential MEM (DMEM), and a novel way to search for non-resonant New Physics.

1. Introduction

The LHC has restarted last year at the record center-of-mass energy of 13 TeV and has already delivered an integrated luminosity of more than 2 fb^{-1}. Over the next two years, it is expected that the ATLAS and CMS experiments will collect about 50 times more data. As during LHC Run I, physics signals can be very hard to extract out of the huge backgrounds, requiring the use of advanced multivariate techniques. While Machine-Learning (ML) techniques such as Boosted Decision Trees (BDT) and Artificial Neural Networks (ANN) have become increasingly popular, the Matrix Element Method (MEM), pioneered at the Tevatron [1, 2, 3], represents a powerful alternative. Its main advantage is that the achieved discrimination is not limited by the amount of available simulated Monte-Carlo events, which can be an issue in searches for rare processes.

2. The Matrix Element Method

Being able to compute the likelihood $P(\mathbf{x}|\alpha)$ to observe an event $\mathbf{x}$ in a detector under a certain theoretical hypothesis α would give direct access to powerful tools such as:

- The Maximum likelihood fit of the sample likelihood $\mathcal{L} = \prod_{i \in \text{data}} P(\mathbf{x}_i|\alpha)$, allowing to measure a physical parameter in the model.
- The Neyman-Pearson discriminant $P(\mathbf{x}|\alpha)/P(\mathbf{x}|\beta)$, the most powerful discriminant for hypothesis α against hypothesis β [4], which can be used either for discrete hypothesis testing or as a discriminator in searches for rare processes.

Journal of Physics: Conference Series **762** (2016) 012053 doi:10.1088/1742-6596/762/1/012053

This likelihood[1] can actually be computed, as is obtained through the convolution of the theoretical likelihood for a partonic configuration y with so-called "transfer functions" and efficiency terms (defined below) linking y to the event $\mathbf{x}$:

$$P(\mathbf{x}|\alpha) = \frac{1}{A_\alpha \sigma_\alpha} \int d\Phi(y) \frac{dx_1 dx_2}{x_1 x_2 s} f(x_1) f(x_2) \left|\mathcal{M}_\alpha(y, x_1, x_2)\right|^2 W(\mathbf{x}|y)\, \epsilon_\alpha(y) \quad (1)$$

where:

- $\mathcal{M}_\alpha$ is the matrix element (usually LO) under hypothesis α.
- The transfer function $W(\mathbf{x}|y)$ gives the probability density that the selected event y ends up as the measured event $\mathbf{x}$. It describes parton shower, hadronisation, and detector and recontruction effects. It is normalised as $\int d\mathbf{x}\, W(\mathbf{x}|y) = 1$. The probability that y ends up as a selected event is included in the efficiency term $\epsilon_\alpha(y)$.
- f and x_i are the Parton Distribution Functions (PDFs) and Björken-x; $d\Phi(y)$ is the phase-space density term; s is the hadronic center-of-mass energy.
- The denominator $A_\alpha \sigma_\alpha$, including the total cross-section of the process α and the overall acceptance $A_\alpha = \langle \epsilon_\alpha(y) \rangle$, ensures that P is correctly normalised as a likelihood [5].

3. Use of the MEM at the LHC

Several analyses carried out by the ATLAS and CMS experiments during LHC Run I have successfully used the MEM and are briefly summarised here.

Searches for the $t\bar{t}h$ process

ATLAS and CMS have performed multiple searches for this process, in different final states, using either ML and cut-and-count techniques [6, 7, 8], the MEM [9], or a combination of ML and MEM [10]. Both MEM analyses concerned the $h \to b\bar{b}$ and semi-/fully-leptonic $t\bar{t}$ final state. Dedicated implementations of the MEM computation were used.

Search for s-channel single Top quark production

This rare process at the LHC has been searched for using BDTs, and upper limits on its cross section have been set by ATLAS [11] and CMS [12]. ATLAS has repeated the search [13] for this process by re-analysing its 8 TeV dataset using updated calibrations and the MEM. ATLAS was able to increase the expected/observed significance for the process in the background-only hypothesis, up from $1.4/1.3\sigma$ in [11] to $3.9/3.3\sigma$. Again, an in-house implementation of the MEM was used.

Measurement of $t\bar{t}$ spin correlation

CMS has measured [14] the correlation between pair-produced Top quarks in the μ+jets channel using a MEM, yielding the most precise measurement of this quantity in this channel. Additionnaly, the MEM was used to perform hypothesis testing, measuring the compatibility of the data with the correlated or uncorrelated spins hypotheses. The MadWeight software was used for the computation of the likelihoods.

Discovery of the Higgs boson and spin-parity measurements

ME techniques ("MELA") were used by CMS in the discovery of the SM Higgs boson, to increase the sensitivity of the search for the $h \to ZZ^* \to 4l$ process [15]. Similar tools were used by both ATLAS and CMS to constrain the spin-parity properties of the Higgs boson [16, 17]. Contrary to the three previous use cases presented this section, there is no need to compute integrals such as (1) here, but only to evalute the matrix element itself on the measured events.

[1] Called "weight" in general, to include use cases where there is no need for proper normalisation.

4. Implementation of the MEM

Any implementation of the MEM must include a common set of building blocks: matrix element evaluation, transfer functions, Monte-Carlo (MC) integrator, calls to PDFs, phase-space parametrisation, and input/output. While most of these are fairly simple to implement, it is in general non-trivial to parametrise the phase-space in an efficient way, since the integrand of (1) includes sharp peaks (propagators in the matrix element, transfer functions) and represents a challenge for adaptive MC integration algorithm.

As can be seen from the previous section, several dedicated implementations, suited to the analyses' specific needs (input/output, reduced set of processes, ...), have been developed. On the other hand, there also exists a fully general code, able to compute weights under almost any user-specified hypothesis: MadWeight.

MadWeight

MadWeight [18] is part of the MadGraph5_aMC@NLO [19] software suite. It relies on Vegas [20] to compute the integrals and builds on smart phase-space mappings allowing to align the integrand's peaks with the coordinate axes and analytically remove some of the peaks, greatly improving the convergence rate of the calculations (since Vegas assumes a separable weight function). While being extremely general and fully automatic, it is difficult to use in large-scale experimental analyses, due to inefficient input/output (text files), lack of active development, use of old Fortran syntax, and full automation preventing the user to implement useful tweaks or approximations. This observation has spawned us to launch the MoMEMta project.

MoMEMta

MoMEMta (Modular Matrix Element Method implementation) aims at a C++ framework allowing to compute weights under any desired hypothesis, keeping MadWeight's highly efficient parametrisations while featuring a modular architecture enabling easy tailoring to the user's specific needs (new features, interface to analysis code, ...). A set of modules (phase-space mappings, transfer functions, ...), as well as the possibility to quickly develop additional ones, will be provided. The code relies on Root[2], Cuba [21] (Monte-Carlo integration library) and LHAPDF [22], while the configuration of the modular structure is achieved using the Lua[3] scripting language. Modules are implemented as C++ classes. Linking of the modules is achieved using a memory pool built at startup, yielding minimal overhead during the computation compared to the non-modular code. The present-day status of the project can be summarised as follows:

- A non-modular prototype has been written to ensure our understanding of the computation in a particular topology (fully-leptonic $t\bar{t}$ production) and was validated against MadWeight.
- The modular framework is being actively developed and the code (with a working proof-of-principle) is already freely available[4].
- A new MadGraph matrix element export module in the C++11 language has been created. It allows easier linking of the output code to a larger project, and the resulting code is significantly faster than the default C++ exporter's. However, we would like to stress that MoMEMta is being designed to allow inclusion of any matrix element, not solely restricted to MadGraph's format.

[2] https://root.cern.ch

[3] http://lua.org

[4] The project can be retrieved at https://github.com/MoMEMta/MoMEMta

5. Example applications

This section discusses uses of the MEM that have not yet been applied to data analyses, and are targeted to be easily realised using the modular framework. The exploratory results presented here have been obtained using the non-modular prototype of MoMEMta (while possible using MadWeight, it requires significant input by the authors).

Parallel weights & systematic uncertainties

One of the main challegens when relying on the MEM for an analysis is the evaluation of systematic uncertainties. While some of them can be handled through Monte-Carlo event reweighting (scale and PDF uncertainties, scale-factors, ...), others require re-evaluations of the weights on slightly modified event contents (chiefly among them Jet Energy Scale and Resolution uncertainties - JES/JER), which affects the computing time needed for the analysis. However, for most events the object selection does not change, and evaluating these uncertainties only amounts to slight variations $E^{\pm}_{rec}$ of the event's reconstructed jet energies E_{rec}. It is therefore possible to define a *vectorial* integrand,

$$\begin{bmatrix} P^+ \\ P \\ P^- \end{bmatrix} = \int \mathrm{d}E_{gen} \, |\mathcal{M}(E_{gen})|^2 \times \begin{bmatrix} W(E^+_{rec}|E_{gen}) \\ W(E_{rec}|E_{gen}) \\ W(E^-_{rec}|E_{gen}) \end{bmatrix} \times \ldots, \tag{2}$$

where the Vegas grid remains nearly optimal for all components (variations being small). Since only the transfer functions need to be re-evaluated on each phase-space point in (2), not the matrix element (by far the slowest part), this procedure can bring a significant speed-up.

Differential MEM

While (1) provides a way to obtain a per-event likelihood, it is also possible to compute a per-event *differential* likelihood w.r.t. a variable Z, as was first proposed in [23]:

$$\frac{1}{P}\frac{\partial P(\mathbf{x})}{\partial Z}\bigg|_{Z_0} = \int \mathrm{d}\Phi(y) \, \frac{\mathrm{d}x_1 \mathrm{d}x_2}{x_1 x_2 s} \, f(x_1) f(x_2) \, |\mathcal{M}(y)|^2 \, W(\mathbf{x}|y) \, \delta(Z(y) - Z_0) \tag{3}$$

In practice, this amounts to filling a histogram, for each event, with the values of the variable $Z(y)$ during the computation of the weights P. Stacking these histograms yields a distribution of variable Z. The procedure can prove useful in cases where ambiguities prevent proper reconstruction of kinematical variables, such as in the fully-leptonic decay of a $t\bar{t}$ pair.

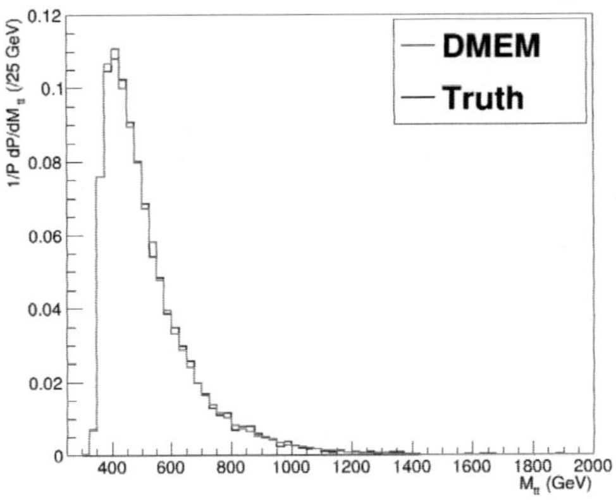

Figure 1. Distribution of $m_{t\bar{t}}$ obtained on a simulated sample of fully-leptonic $t\bar{t}$ events. Blue: true distribution. Red: distribution obtained using DMEM.

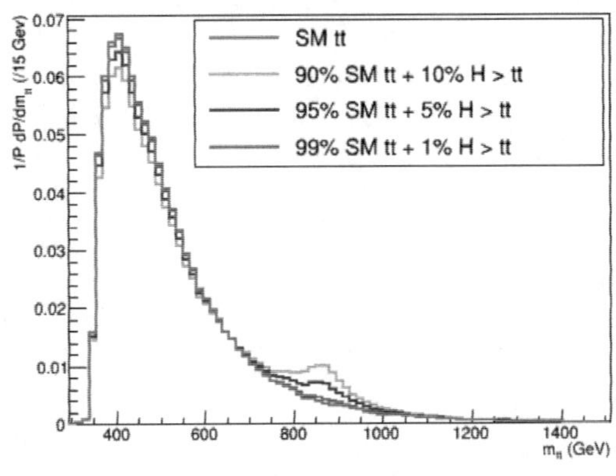

Figure 2. Distributions obtained through DMEM when adding mixtures of $H \to t\bar{t}$ events to a sample of SM $t\bar{t}$ events ($m_H = 875$ GeV).

Search for higher-dimensional operators
If new physics is too heavy to be produced directly at the LHC, its effects may be parametrised through so-called *effective operators* giving rise to new couplings amongst SM particles [24, 25, 26, 27]. In the case of $t\bar{t}$ production, considered here, the dominant effect is expected from the interference of dimension 6 operators with SM amplitudes. The experimental goal is then to extract a global fit of these operators' couplings from data. While possible using cross-section and asymmetry measurements, we are investigating the potential of the MEM. Denoting by $\tilde{P}_i$ the "weight" obtained by integrating as in (1) the interference between operator i and the SM amplitude, and $P_{t\bar{t}}$ the weight under the SM hypothesis, the variable

$$D_i = (\arctan(\log(|\tilde{P}_i|/P_{t\bar{t}})) + \pi/2)/\pi \tag{4}$$

is equivalent to a Neyman-Pearson discriminant between the hypotheses "Operator i (with a certain coupling) interfering with $t\bar{t}$ in the SM" and "SM $t\bar{t}$ production", while being conveniently bound between 0 and 1, and avoiding the need to normalise the weights as likelihoods.

Preliminary studies (in the fully-leptonic final state) were performed based on a set of 7 operators defined and implemented in [28, 29], using MADGRAPH5_AMC@NLO, PYTHIA8 [30, 31] and DELPHES [32]. Event samples were generated by considering only the interference between the SM and one operator at a time, yielding a set of "signals" whose contributions are linear in the operators' couplings, avoiding the need to rely on techniques such as clustering to reduce the number of samples. The constraining power of the D_i variables can then be probed by template-fitting the distributions obtained with the generated samples. While some gain over using simple kinematical variables is already visible when considering single operators in the fit, we are still in the process of investigating the use of these variables in a global fit.

6. Conclusions

The Matrix Element Method has been presented and its successful uses at the LHC so far have been reviewed. While the MADWEIGHT software is able to efficiently and automatically compute weights under almost any hypothesis, it is no longer in development and is difficult to use in a large scale analysis. We have therefore started the MOMEMTA project, with the aim to build a modular framework retaining MADWEIGHT's generality and effectiveness while being more maintainable and adaptable to the user's needs. Finally, novel use cases of the MEM, expected to be easily realised using MOMEMTA, have been presented.

Acknowledgments

We warmly thank the organisers of ACAT2016 for giving us the opportunity to present this work. Our gratitude goes to Alexandre Mertens and Sébastien Brochet for their substantial contributions, and to Pierre Artoisenet and Vincent Lemaître for their useful feedback. This work is supported through a FRIA grant by the FNRS.

References

[1] Kondo K 1988 *Journal of the Physical Society of Japan* **57** 4126–4140
[2] Kondo K 1991 *Journal of the Physical Society of Japan* **60** 836–844
[3] Abazov V M *et al.* (D0) 2004 *Nature* **429** 638–642 (*Preprint* hep-ex/0406031)
[4] Neyman J and Pearson E S 1933 *Phil. Trans. R. Soc. Lond.* A **231** 289–337 ISSN 0264-3952
[5] Volobouev I 2011 (*Preprint* data-an/1101.2259)
[6] Aad G *et al.* (ATLAS) 2015 *Phys. Lett.* **B749** 519–541 (*Preprint* hep-ex/1506.05988)
[7] Aad G *et al.* (ATLAS) 2015 *Phys. Lett.* **B740** 222–242 (*Preprint* hep-ex/1409.3122)
[8] Khachatryan V *et al.* (CMS) 2014 *JHEP* **09** 087 [Erratum: JHEP10,106(2014)] (*Preprint* hep-ex/1408.1682)
[9] Khachatryan V *et al.* (CMS) 2015 *Eur. Phys. J.* **C75** 251 (*Preprint* hep-ex/1502.02485)
[10] Aad G *et al.* (ATLAS) 2015 *Eur. Phys. J.* **C75** 349 (*Preprint* hep-ex/1503.05066)
[11] Aad G *et al.* (ATLAS) 2015 *Phys. Lett.* **B740** 118–136 (*Preprint* hep-ex/1410.0647)

[12] Khachatryan V *et al.* (CMS) 2013 CMS-PAS-TOP-13-009
[13] Aad G *et al.* (ATLAS) 2015 Submitted to *Phys. Lett.* B (*Preprint* hep-ex/1511.05980)
[14] Khachatryan V *et al.* (CMS) 2015 Submitted to *Phys. Lett.* B (*Preprint* hep-ex/1511.06170)
[15] Chatrchyan S *et al.* (CMS) 2012 *Phys. Lett.* **B716** 30–61 (*Preprint* hep-ex/1207.7235)
[16] Aad G *et al.* (ATLAS) 2015 *Phys. Rev.* **D91** 012006 (*Preprint* hep-ex/1408.5191)
[17] Chatrchyan S *et al.* (CMS) 2014 *Phys. Rev.* **D89** 092007 (*Preprint* hep-ex/1312.5353)
[18] Artoisenet P, Lemaître V, Maltoni F and Mattelaer O 2010 *J. High Energy Phys.* **2010** (*Preprint* hep-ph/1007.3300)
[19] Alwall J *et al.* 2014 *JHEP* **07** 079 (*Preprint* hep-ph/1405.0301)
[20] Lepage G P 1978 *Journal of Computational Physics* **27** 192
[21] Hahn T 2005 *Comput. Phys. Commun.* **168** 78–95 (*Preprint* hep-ph/0404043)
[22] Buckley A *et al.* 2015 *Eur. Phys. J.* **C75** 132 (*Preprint* 1412.7420)
[23] Pin A and Mattelaer O 2012 *Nuovo Cim.* **C035N3** 229 TOP2011 Conf. Rep.
[24] Weinberg S 1979 *Phys. Rev. Lett.* **43**(21) 1566–1570
[25] Buchmuller W and Wyler D 1986 *Nucl. Phys.* **B268** 621–653
[26] Leung C N, Love S T and Rao S 1986 *Zeitschrift für Physik C Particles and Fields* **31** 433–437 ISSN 1431-5858
[27] Grzadkowski B, Iskrzynski M, Misiak M and Rosiek J 2010 *JHEP* **10** 085 (*Preprint* hep-ph/1008.4884)
[28] Zhang C and Willenbrock S 2011 *Phys. Rev.* **D83** 034006 (*Preprint* hep-ph/1008.3869)
[29] Degrande C, Gerard J M, Grojean C, Maltoni F and Servant G 2011 *JHEP* **03** 125 (*Preprint* hep-ph/1010.6304)
[30] Sjostrand T, Mrenna S and Skands P Z 2006 *JHEP* **05** 026 (*Preprint* hep-ph/0603175)
[31] Sjostrand T, Mrenna S and Skands P Z 2008 *Comput. Phys. Commun.* **178** 852–867 (*Preprint* hep-ph/0710.3820)
[32] de Favereau J, Delaere C, Demin P, Giammanco A, Lemaître V, Mertens A and Selvaggi M 2014 *J. High Energy Phys.* **2014** (*Preprint* hep-ex/1307.6346)

Upgrading the ATLAS fast calorimeter simulation

Z Hubacek[1], on behalf of the ATLAS Collaboration

[1] Czech Technical University in Prague, FNSPE, Brehova 7, 115 19, Praha 1, Czech Republic

E-mail: zdenek.hubacek@cern.ch

Abstract. Many physics and performance studies with the ATLAS detector at the Large Hadron Collider require very large samples of simulated events, and producing these using the full Geant4 detector simulation is highly CPU intensive. Often, a very detailed detector simulation is not needed, and in these cases fast simulation tools can be used to reduce the calorimeter simulation time. In ATLAS, a fast simulation of the calorimeter systems was developed, called Fast Calorimeter Simulation (FastCaloSim). It provides a parametrized simulation of the particle energy response at the calorimeter read-out cell level. It is interfaced to the standard ATLAS digitization and reconstruction software and can be tuned to data more easily than Geant4. An improved parametrization is being developed, to eventually address shortcomings of the original version. It makes use of statistical techniques such as principal component analysis and a neural network parametrization to optimise the amount of information to store in the ATLAS simulation infrastructure.

1. Introduction

The success of the physics program of the ATLAS Experiment [1] at the Large Hadron Collider relies heavily on the large number of Monte Carlo (MC) simulated events. The MC production takes a significant part of the available grid CPU and disk space resources, therefore every possibility for improvement is studied. One of the possibilities is the use of fast simulation tools. The ATLAS Monte Carlo production chain and the Integrated Simulation Framework (ISF) are briefly described in Section 2, followed by the description of the main features of the Fast Calorimeter Simulation (FastCaloSim) in Section 3. The development of the new FastCaloSim version is described in Section 4.

2. ATLAS Monte Carlo Production Chain and ISF Framework

The typical ATLAS MC production chain [2] consists of several steps. The event generation simulates the truth-level particle collision with the possibility to choose from common Monte Carlo generators. The event information is stored in the EVNT file format, based on HepMC. The detector simulation computes the sensitive detector energy deposits (HITS) caused by the final particles from the EVNT files. Here either the full detector simulator (Geant4 [3]) and/or fast ATLAS specific simulators are used. Next, the digitization converts the HITS into raw data objects. During the digitization, in-time and out-of-time pileup effects can be merged into the signal event. The reconstruction finds and identifies physics objects (photons, leptons, jets, ...). With respect to the reconstructed physics objects, the output files contents are equivalent between the simulated and recorded data. In the final step, a conversion to physics analysis format is usually applied to allow fast ROOT [4] access to the reconstructed objects used in the physics analyses.

To reduce the CPU utilization, a number of developments are undertaken to speed up the ATLAS detector simulation. The Geant4 based simulation of the detector is taken as the most detailed and the most accurate detector simulation. Precomputed (so-called frozen showers) libraries of electromagnetic showers in the forward calorimeters are used by default also in the full ATLAS simulation. Additional fast simulators available consist of Fatras [5], a fast ATLAS tracking simulation, and of FastCaloSim [6], a parametrized ATLAS calorimeter response simulation.

The Integrated Simulation Framework (ISF) [7] was developed to allow combining various simulators (full and fast) for different particles within each event. In this way, one could combine the full and precise simulator for particles of interest and fast simulators for the rest of the event. The ISF is now fully integrated in the ATLAS Athena framework [8]. An example of a speedup of the simulation for different simulation setups is presented in Table 1.

ISF Simulation Setup	Speedup	Accuracy
Full Geant4	1	best possible
Geant4 + FastCaloSim	≈ 25	approximated calorimeter
Fatras + FastCaloSim	≈ 750	all subdetectors approximated
Fatras + FastCaloSim only simulating particles inside cones around photons	≈ 3000	all subdetectors approximated only partial event simulation

Table 1. Comparison of various ISF simulation setups in the $gg \to H \to \gamma\gamma$ events without pileup contributions.

3. FastCaloSim

The simulation of the particle shower evolution takes a large amount of time. In the FastCaloSim simulation, time is saved by considering a simplified detector geometry and replacing the simulation by parametrizations. Only three types of particles are parametrized and used by the FastCaloSim: photons, electrons and charged pions. The charged pions are used for all hadrons (both neutral and charged). In order to reflect the chosen simulation priorities, the parametrization model reproduces the longitudinal shower properties, including fluctuations and correlations, but only the average transverse shower shapes are used. The properties of the shower developments are stored in histogram form for each particle type, each energy and η points.

The FastCaloSim was validated and used in the Run 1. The average lateral shower shapes were tuned to data, but some shortcomings of the simplified model were observed, particularly in shower sub-structure (Fig. 1). Because of the limited computing resources, only the average lateral shower shapes could be saved. In the context of the new FastCaloSim development, a random shower shape fluctuations model has also been developed to improve the situation [9]. Overall, a good agreement between the FastCaloSim and the full Geant4 simulation was observed, and the fast simulation reduced the simulation time in the ATLAS calorimeter from several minutes to a few seconds per event.

4. Upgrade of Fast Calorimeter Simulation

A new version of the fast calorimeter simulation is currently being developed. The new parametrization is based on the latest Geant4 version used by the ATLAS experiment. The development is split again into the simulation of the calorimeter response of single particles, the

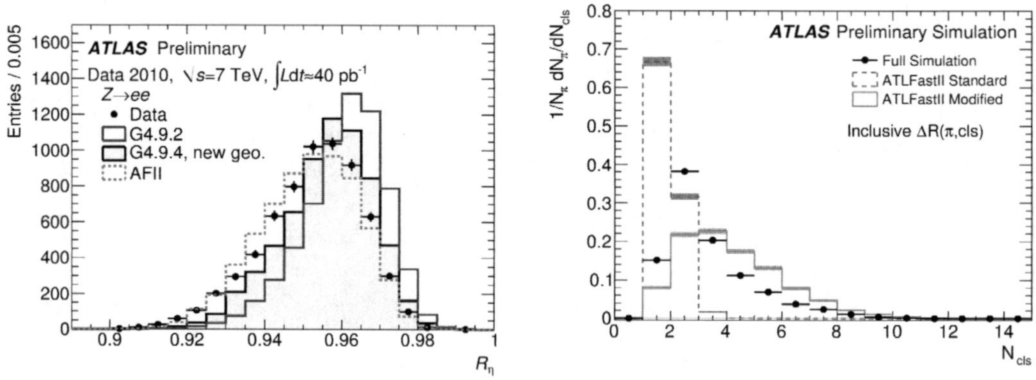

Figure 1. Second layer of the electromagnetic calorimeter R_η variable is the ratio of energy in a $\Delta\eta \times \Delta\phi = 3 \times 7$ cell cluster with respect to a 7×7 cluster. Two different Geant4 versions together with the FastCaloSim (denoted as AFII) setups are compared to the data collected in 2010 [10] (left) and a number of recontructed calorimeter clusters in a single pion sample (right). Black points correspond to the full Geant4 simulation. The red histogram is the original FastCaloSim model. The blue histogram is a model which introduced additional random shower shape fluctuations on top of the average model.

parametrization of the longitudinal energy deposition, the fitting of the average lateral shower shapes and implementing the shower shape fluctuations.

Single particle inputs of photons, electrons and pions are used as in the original FastCaloSim. The detector response to these particles is simulated with the latest Geant4 version and latest ATLAS geometry and conditions. The detailed spatial position of each energy deposit is saved. The samples are generated in grids of energy (covering single particle energies from 100 MeV up to about 4 TeV) and η bins (bin width of 0.05). The impact angle of the particle on a calorimeter cell surface is also considered.

Energy Parametrization describes the longitudinal shower development. It utilizes the Principal Component Analysis (PCA) to convert the set of correlated energy fractions into a linearly uncorrelated set of principle components. A TMVA [11] neural network (NN) regression is used to approximate cumulative histograms inputs (Fig. 2) in order to reduce the amount of stored information. The simulation procedure will start with the generation of uncorrelated random numbers and will invert the PCA matrix in order to simulate the energy fraction deposited in each calorimeter layer. An example output of the simulation procedure is shown in Fig. 2.

Lateral Shower Shape describes the average shower shape for each particle type in each calorimeter layer, and bins of the leading principle component variable. A NN is trained to describe the average shower shape instead of saving a full histogram information. A radial symmetric binning is optimized so that the number of hits remains roughly the same in each bin and a finer binning is chosen in the center of the shower as shown in Fig. 3. The NN is going to be trained for each particle type, each energy and η point and in bins of the leading PCA component from the energy parametrization. An illustrative example of the NN training is shown in Fig. 4.

Typical NN settings for both Energy Parametrization and Lateral Shower Shape steps are shown in Table 2.

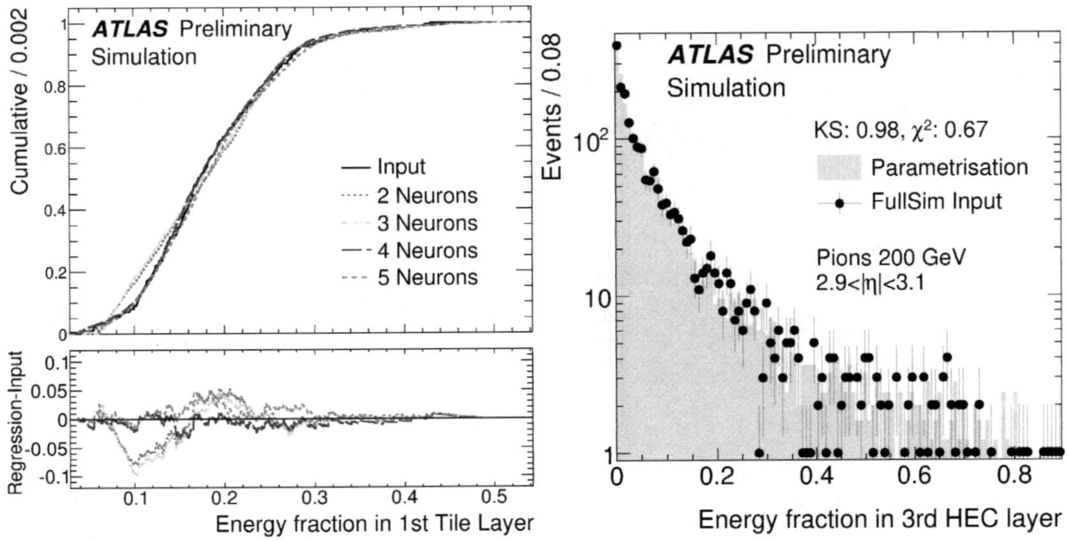

Figure 2. Illustration of the iterative regression with an increasing number of neurons for one of the cumulative histograms used in the longitudinal energy parametrization of charged pions. The example here shows the energy fraction deposited in the first layer of the Tile calorimeter (left) and energy fraction deposited in the 3rd layer of the Hadronic Endcap calorimeter by charged pions. The black points show the Geant4 inputs, and the result of the longitudinal energy parametrization is shown in light blue. A good agreement is observed. The results of a Kolmogorov (KS) and χ^2 test are displayed as well (right).

Parameter	Default value	Value used
HiddenLayers	N	2–10
TestRate	10	6
NeuronType	sigmoid	tanh
TrainingMethod	BP	BFGS
ConvergenceImprove	1e-30	1e-6
ConvergenceTests	-1	15

Table 2. TMVA NN parameter settings used in the Energy Parametrization and Lateral Shower Shape steps.

5. Summary

A development of a new ATLAS fast calorimeter simulation is underway. It is aiming at improving the current FastCaloSim speed and precision. The new fast calorimeter simulation is going to use PCA methods for the energy parametrization and NN regression for the fitting of the average shower shapes.

Acknowledgments

The author gratefully acknowledges the financial support from MSMT of the Czech Republic under Grant No. RVO 68407700.

References

[1] The ATLAS Collaboration 2008 The ATLAS Experiment at the Large Hadron Collider *JINST* **3** S08003

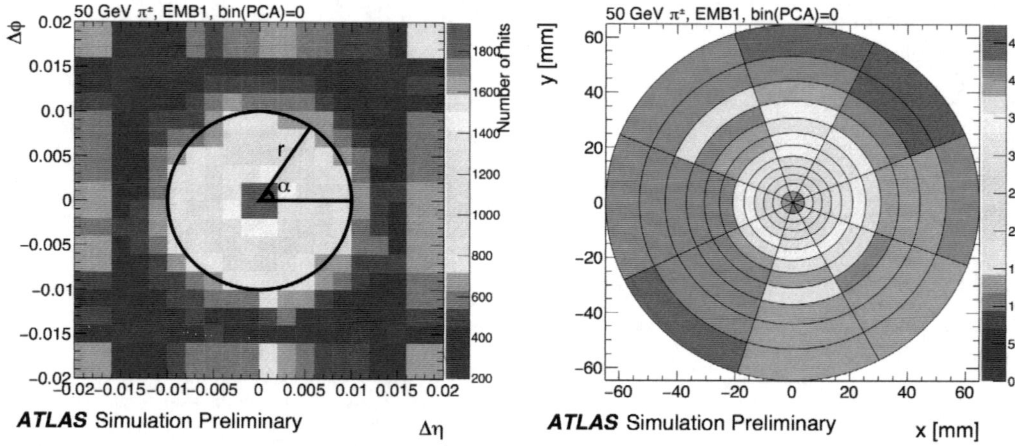

Figure 3. Illustration of the definition in terms of η and ϕ of the hits of the alpha and r variables, which are used to optimise the binning and perform the NN fit. This example is for 50 GeV central ($0.20 < |\eta| < 0.25$) pions in the first layer of the electromagnetic calorimeter (EMB1) and corresponds to events included in the first bin of the PCA energy parametrization (left) and of the number of hits per bin used for the NN fit. The number of hits per bin is kept as constant as possible for the NN fit stability. At the very center, a more detailed description of the shower shape is needed, therefore the bins are sub-divided there, hence the decrease observed in the number of hits per bin (right).

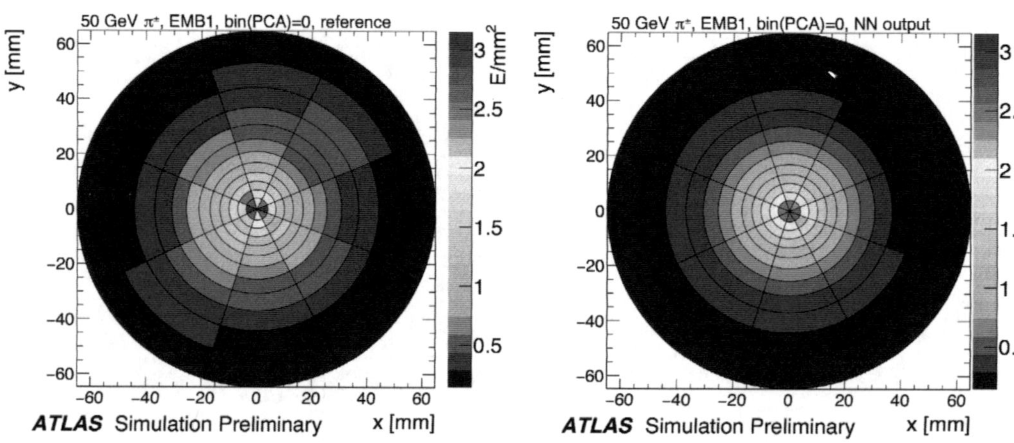

Figure 4. Illustration of the energy normalized per bin area used as input to the NN fit. This example is for 50 GeV central ($0.20 < |\eta| < 0.25$) pions in the EMB1 layer and corresponds to events included in the first bin of the PCA energy parametrization (left) and of the output of the NN parametrization of the input (right).

[2] The ATLAS Collaboration 2010 The ATLAS Simulation Infrastructure *Eur. Phys. J.* C **70** 823-874
[3] Agostinelli S et al. 2003 Geant4 a simulation toolkit *Nucl. Instrum. Methods Phys. Res.* A **506** 250303
Allison J et al. 2006 Geant4 developments and applications *IEEE Trans. Nucl. Sci.* **53** 270-278
[4] Brun R and Rademakers F 1996 ROOT - An Object Oriented Data Analysis Framework, Proceedings AIHENP'96 Workshop, Lausanne, Sep. 1996, *Nucl. Inst. & Meth. in Phys. Res.* A **389** 81-86

Antcheva I et al. 2011 ROOT A C++ framework for petabyte data storage, statistical analysis and visualization *Comput. Phys. Commun.* **182** 1384-1385

[5] Edmonds K et al. 2008 *The Fast ATLAS Track Simulation (FATRAS)* (Geneva: CERN) ATL-SOFT-PUB-2008-001

[6] Yamanaka T (for the ATLAS Collaboration) 2011 The ATLAS calorimeter simulation FastCaloSim *J. Phys.: Conf. Ser.* **331** 032053

[7] Ritsch E (on behalf of the ATLAS Collaboration) 2014 Concepts and Plans towards fast large scale Monte Carlo production for the ATLAS Experiment *J. Phys.: Conf. Ser.* **523** 012035

[8] The ATLAS Collaboration 2005 ATLAS Computing Technical Design Report ATLAS-TDR-017, CERNLHCC-2005-022

[9] The ATLAS Collaboration 2014 *Performance of the Fast ATLAS Tracking Simulation (FATRAS) and the ATLAS Fast Calorimeter Simulation (FastCaloSim) with single particles* (Geneva: CERN) ATL-SOFT-PUB-2014-001

[10] `http://atlas.web.cern.ch/Atlas/GROUPS/PHYSICS/EGAMMA/PublicPlots/20111005/ATL-COM-PHYS-2011-1299/`

[11] Hoecker A, Speckmayer P, Stelzer J, Therhaag J, von Torne E and Voss H 2007 TMVA: Toolkit for Multivariate Data Analysis *Proc. of Science* PoS ACAT (2007) 040

Vertex finding by sparse model-based clustering

R Frühwirth[1], K Eckstein[1] and S Frühwirth-Schnatter[2]

[1] Institute of High Energy Physics, Austrian Academy of Sciences, Vienna

[2] Institute for Statistics and Mathematics, Vienna University of Economics and Business

E-mail: rudolf.fruehwirth@oeaw.ac.at

Abstract. The application of sparse model-based clustering to the problem of primary vertex finding is discussed. The observed z-positions of the charged primary tracks in a bunch crossing are modeled by a Gaussian mixture. The mixture parameters are estimated via Markov Chain Monte Carlo (MCMC). Sparsity is achieved by an appropriate prior on the mixture weights. The results are shown and compared to clustering by the expectation-maximization (EM) algorithm.

1. Introduction

Primary vertex finding is an important step in the data analysis of collider experiments. In the case of the LHC experiments, there is in each bunch crossing a large background of so-called pile-up events which are superimposed to the signal event, i.e. the one that has triggered the recording of the bunch crossing. Finding the correct assignment of the charged tracks to their respective production vertices is crucial for the subsequent steps of vertex fitting and kinematic fitting. The distribution of the primary vertices is very narrow in the transverse directions, but much wider in the direction along the beams, which by convention is called the z direction. Finding the primary vertices can therefore be formulated as a one-dimensional clustering problem, namely finding clusters of primary tracks in the z direction. In a perfect clustering, each found cluster of tracks corresponds to a single primary vertex. In reality, a cluster produced by the clustering may contain tracks from several vertices, for instance when the clustering fails to resolve two close vertices and merges them into a single cluster.

In this paper two methods are compared: sparse model-based clustering [1] and the EM algorithm [2], an iterative implementation of maximum-likelihood estimation. In both methods the observations are assumed to be drawn from a mixture model. Model-based clustering has the advantage that it can include additional available information via prior densities. More specifically, prior knowledge on the following quantities can be used: the number of clusters or vertices; the cluster size, i.e. the number of tracks per vertex; and the spread of a cluster in the z direction. In the sparse version clusters are allowed, even encouraged, to be empty. The EM algorithm is less flexible, as the number of clusters is fixed, although clusters can effectively merge. The distribution used to model the clusters is in principle arbitrary; in the case of vertices a normal model is both adequate and easier to implement than other distributions.

2. Data

Proton-proton collisions were simulated using PYTHIA 8 [3], an event simulation engine for high-energy physics. The center of mass energy was set to 13 TeV, the one of the LHC run in 2015. The detailed settings can be obtained from the authors on request.

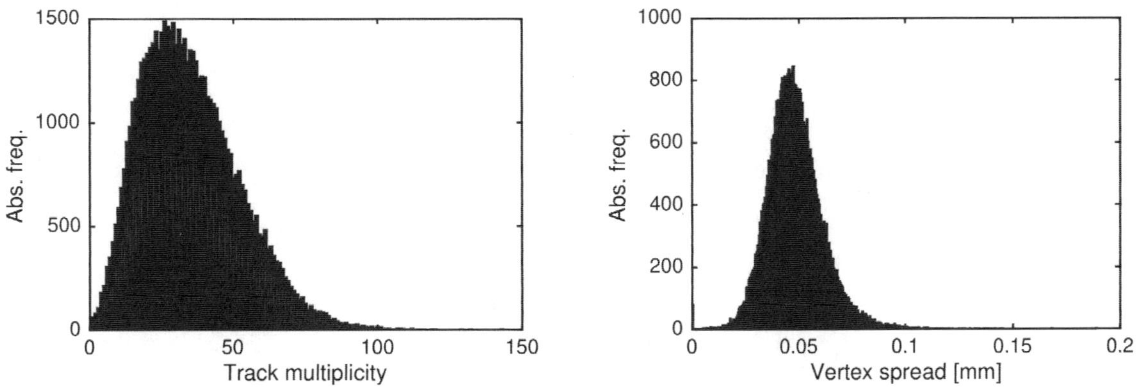

Figure 1. Left: distribution of the charged track multiplicity $M_{\rm v}$ per vertex, summed over all simulated vertices; right: distribution of the spread $s_{\rm v}$ of all simulated vertices.

Each collision produces a vertex with a number of particles attached to it. Only charged particles in the final state with a transverse momentum of at least $700\,\mathrm{MeV}/c$ are retained. In addition, their polar angle θ is restricted to the interval $9° \leq \theta \leq 171°$, roughly corresponding to the acceptance of a typical vertex detector. The multiplicity of the remaining charged tracks per vertex is denoted by $M_{\rm v}$. Its empirical distribution is shown in the left panel of Figure 1. In order to make use of the distribution in the clustering, it is smoothed by a kernel estimator.

The z-positions of the vertices are generated according to a normal distribution centered at the origin with standard deviation $\sigma = 65\,\mathrm{mm}$, corresponding to the luminous region in the LHC (see [4]). This standard deviation has been determined from $\sqrt{s} = 7\,\mathrm{TeV}$ data, but is about the same for $\sqrt{s} = 13\,\mathrm{TeV}$ collisions. The z-positions of the tracks belonging to a vertex are smeared by an additional statistical error due to multiple scattering in the beam tube plus the position and the extrapolation error from the innermost pixel layer. For each vertex, its spread $s_{\rm v}$ is computed as the r.m.s. of the z-positions of the tracks belonging to the vertex. The empirical distribution of $s_{\rm v}$ is shown in the right panel of Figure 1 and is characterized by its mean $\mu_s = 0.048\,\mathrm{mm}$ and its standard deviation $\sigma_s = 0.013\,\mathrm{mm}$.

To simulate an amount of pile-up similar to the one expected in the Phase I upgrade of LHC, the vertices are grouped into individual bunch crossings. Each bunch crossing consists of L superimposed vertices, where L is drawn from a Poisson distribution with mean $\lambda = 100$. As the average track multiplicity $M_{\rm v}$ per vertex is about 35, the average number of tracks per bunch crossing ist about 3500.

3. Clustering algorithms

3.1. Data preparation for clustering

Clustering produces groups of tracks, called clusters, that are compatible with the hypothesis that they belong to the same collision vertex. The input is the set of the z-positions z_i of all tracks in a bunch crossing. For a better performance, however, the clustering is not applied to all tracks in a bunch crossing, but rather to sections in z with boundaries that depend on the data. To find suitable boundaries, the z-positions of the tracks in a bunch crossing are filled into a 1D histogram with a bin width of $h = 1\,\mathrm{mm}$. Then the following three preparation steps are executed.

1. First, boundaries of *basic sections* (BS) are defined by empty bins. These BS usually contain between 1 and 3 vertices. Tracks belonging to the same vertex are always contained within a single BS, as there is a gap of at least 1 mm between two BS.

2. Next, as a trade-off between accuracy and speed, ten adjacent BS are merged into one *final section* (FS).

3. For each FS, the three most likely numbers of vertices K_1, K_2, K_3 are estimated. The likelihood of a value K is given by the probability of the vertex multiplicity $M = N/K$ in the empirical distribution $g(M_\text{v})$, where N is the number of tracks in the FS.

The cluster finding then proceeds independently in each FS. This can be done in parallel, although this was not exploited in the present study.

3.2. Sparse model-based clustering

The sparse model-based clustering algorithm and its implementation is explained in detail in [1]. The basic idea is to start with a large number of clusters and let the algorithm find out how many clusters are needed in the most economical clustering.

The input to the clustering are the z-positions z_i, $i = 1, \dots, N$, of the N tracks in the FS. The initial number of clusters K is the largest of $K_1 + K_0, K_2 + K_0, K_3 + K_0$ with $K_0 = 5$. The observations z_i are assumed to be drawn from a Gaussian mixture with K components and the following probability density function (pdf):

$$f(z_i|\boldsymbol{\theta}_1, \dots, \boldsymbol{\theta}_K, \boldsymbol{\eta}) = \sum_{k=1}^{K} \eta_k \, \varphi_k(z_i|\boldsymbol{\theta}_k), \tag{1}$$

where $\varphi_k(z_i|\boldsymbol{\theta}_k)$ is the Gaussian pdf with component specific parameters $\boldsymbol{\theta}_k = (\mu_k, \sigma_k^2)$, and η_k is the weight of component k. Solutions that are sparse with respect to K are obtained by choosing an appropriate prior distribution for the weights $\boldsymbol{\eta} = (\eta_1, \dots, \eta_K)$. In this study the prior is a symmetric Dirichlet distribution with concentration parameter e_0 and the following pdf:

$$p(\eta_1, \dots, \eta_K|e_0) = \frac{\Gamma(Ke_0)}{\Gamma(e_0)^K} \prod_{k=1}^{K} \eta_k^{e_0 - 1}. \tag{2}$$

Smaller values of e_0 tend to produce fewer clusters than larger values. The prior distribution of the component means is a Gaussian centered at the mean b of the observations. Its width depends on a hyperparameter N_0 that controls the amount of information in the prior relative to the information in the observations. The prior of the component variances is the inverse Gamma distribution $\mathcal{IG}(c, C)$, where the hyperparameters c and C are chosen such that the mean is μ_s and the variance is ${\sigma_s}^2$ (see Section 2 and Figure 1).

The clustering proceeds via data augmentation. For each observation z_i, a latent allocation variable S_i is created that takes integer values in the set $\{1, \dots, K\}$. If $S_i = k$, z_i is assigned to cluster k, and the pdf of z_i conditional on the assignment is given by:

$$f(z_i|\boldsymbol{\theta}_1, \dots, \boldsymbol{\theta}_K, S_i = k) = \varphi(z_i|\mu_k, \sigma_k^2), \quad \Pr(S_i = k|\boldsymbol{\eta}) = \eta_k. \tag{3}$$

Initial values of the allocation variables $\boldsymbol{S}$ are obtained from a tentative clustering by the simple and familiar k-means algorithm [5]. Estimation of the allocation variables is done via MCMC. A Gibbs sampler [6] generates a random sample from the posterior distribution of $\boldsymbol{S}$, given the data and all priors. It iterates the following steps:

(i) Simulate the weights η_k and the component specific parameters $\boldsymbol{\theta}_k = (\mu_k, \sigma_k^2)$ conditional on the current allocations, for $k = 1, \dots, K$.

(ii) Sample the allocation variables $S_i, i = 1, \dots, N$ from the discrete distribution on $\{1, \dots, K\}$ defined by:

$$\Pr(S_i = k|z_i, \eta_k, \mu_k, \sigma_k^2) \propto \eta_k \varphi(z_i|\mu_k, \sigma_k^2). \tag{4}$$

(iii) Update the hyperparameters b, N_0, c, C according to the current cluster sizes.

The full details of the sampler including the updating formulas are spelled out in [1]. After generating the Markov chain, the clusters are identified by choosing the configuration of $\boldsymbol{S}$ with the largest posterior probability. Note that the number of clusters that are actually populated is likely to be smaller than the initial number K.

3.3. EM clustering

The EM algorithm [2] is an iterative method for finding the maximum likelihood estimates of the weights η_k and the component parameters $\boldsymbol{\theta}_k$ of the mixture in Eq. (1) for given K. In contrast to the sparse model-based clustering, the EM algorithm is not adaptive. To compensate for this, multiple runs with different numbers of components were performed, using the previously estimated most likely numbers of vertices for the algorithm, namely $K_1, K_2, K_3, \ldots, K_3 + 5$. Finally, the clustering with the smallest Bayesian information criterion (BIC) was selected.

4. Results

The two methods were compared by means of a simulation study with 600 bunch crossings, containing about 60 000 vertices and about 2 million tracks. The following runs were performed:

- Run MB1: Model-based clustering, $e_0 = 0.1$, long Markov chain (5000 draws)
- Run MB2: Model-based clustering, $e_0 = 1.0$, long Markov chain (5000 draws)
- Run MB3: Model-based clustering, $e_0 = 1.0$, short Markov chain (1000 draws)
- Run EM: EM algorithm with BIC

The performance of the clustering methods can be assessed by the difference $\Delta K = K_{\text{est}} - K_{\text{true}}$ of the estimated and the true number of clusters in a FS, and by the cluster purity, defined as the largest fraction of tracks in the cluster belonging to the same vertex. The frequency distribution of ΔK is shown in Fig. 2 for the four runs. The numerical results are summarized in Table 1. It shows the average, the standard deviation and the root of the mean squared error (r.m.s.e.) of ΔK, the average purity and the fraction of perfect clusters with 100% purity, i.e. clusters without contamination from tracks belonging to other vertices.

The run with the smallest r.m.s.e. is MB2, and the average number of clusters is very close to the true one. The shorter chain in MB3 still gives the correct average, but a slightly larger standard deviation and r.m.s.e. The standard deviation of ΔK in run MB1 is smaller than in run MB2, but there are too few clusters on average. The EM algorithm combined with the BIC overestimates the number of clusters, a feature described in the literature [7]. In all runs, more than 80% of the clusters are perfect; in nearly all of the other cases the cluster purity is between 50% and 100%, indicating the fusion of two very close primary vertices into a single cluster. This problem can be resolved to some extent in a subsequent adaptive vertex fit [8].

Table 1. Summary of the results.

	$\Delta K = K_{\text{est}} - K_{\text{true}}$			Purity	
Run	average	std. dev.	r.m.s.e.	average	perfect
MB1	−0.98	1.43	1.73	0.95	0.83
MB2	−0.06	1.52	1.52	0.95	0.83
MB3	0.03	1.68	1.69	0.95	0.82
EM	1.47	3.15	3.47	0.93	0.81

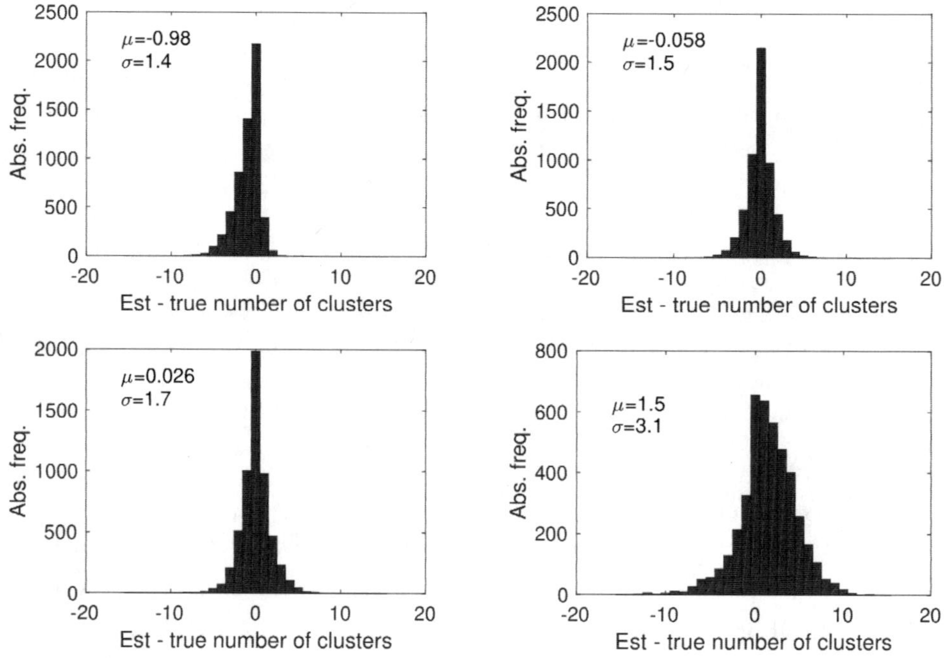

Figure 2. Estimated minus true number of clusters for runs MB1 (top left), MB2 (top right), MB3 (bottom left), and EM (bottom right).

5. Discussion and outlook

Model-based clustering has a number of attractive features: prior information on the number of tracks per vertex and the vertex spread is used; this information can be extracted from runs with low luminosity, where vertex finding is easy; only the expected (average) number of vertices depends on the luminosity; with a mimimal amount of tuning the correct average number of vertices is obtained, in contrast to the EM algorithm; and finer tuning of the parameters of the priors can be expected to give further improvements in the performance.

The main drawback is that sampling by MCMC is slow by HEP standards, so that model-based clustering is probably not suitable for standard vertex finding. Also, fine tuning of all hyperparameters is computationally intensive. The speed of the method is, however, less relevant in applications with fewer observations and fewer clusters, for instance ring finding in RICH detectors or cluster finding in calorimeters, where a lot of prior information on the cluster shape is available and can be incorporated in a natural manner. Further studies will have to concentrate on the optimal settings of the prior parameters and also on the robustness of the clustering against additional noise.

References

[1] Malsiner-Walli G, Frühwirth-Schnatter S and Grün B 2016 *Statistics and Computing* **26** 303–324
[2] Dempster A, Laird N and Rubin D 1977 *Journal of the Royal Statistical Society B* **39** 1–38
[3] Sjöstrand T, Mrenna S and Skands P 2008 *Computer Physics Communications* **178** 852–867
[4] Bartoldus R (for the ATLAS Collaboration) 2012 *Physics Procedia* **37** 2080–2088
[5] Anderberg M 1973 *Cluster Analysis for Applications* (New York: Academic Press)
[6] Tierney L 1994 *The Annals of Statistics* **22** 1701–1728
[7] Baudry J et al 2010 *Journal of Computational and Graphical Statistics* **9** 332–353
[8] Waltenberger W, Frühwirth R and Vanlaer P 2007 *Journal of Physics G: Nuclear and Particle Physics* **34** N343–N356

Alternative method of Reduction of the Feynman Diagrams to a set of Master Integrals

Julio Borja and Igor Kondrashuk
Grupo de Física de Altas Energías, Departamento de Ciencias Básicas,
Universidad del Bío-Bío, Campus Fernando May, Casilla 447, Chillán, Chile
E-mail: `igor.kondrashuk@gmail.com`

Abstract. We propose a new set of Master Integrals which can be used as a basis for certain multiloop calculations in massless gauge field theories. In these theories we consider three-point Feynman diagrams with arbitrary number of loops. The corresponding multiloop integrals may be decomposed in terms of this set of the Master Integrals. We construct a new reduction procedure which we apply to perform this decomposition.

1. Introduction

We describe the idea of an alternative method to calculate three-point vertex in a massless gauge field theory. At the first step, the method includes an algorithm to perform Lorentz algebra in the position space. This algorithm may be applied to the double ghost vertex of $\mathcal{N}=4$ SYM in Landau gauge. The result of such an application will be a decomposition of this vertex in terms of basis elements. We describe these basis elements. The proposed algorithm is simple for programming. At the next step, we show how the proposed basis elements may be represented in terms of the integrals corresponding to the triangle ladder diagrams in $d=4-2\varepsilon$ dimensions.

In Refs. [1, 2, 3, 4] the calculation of the two-loop correction to this auxiliary double ghost vertex for $\mathcal{N}=4$ super-Yang-Mills (SYM) theory has been done in the limit $\varepsilon \to 0$. The contributing diagrams are shown in Fig.1. This vertex is finite in $\mathcal{N}=4$ SYM in the Landau

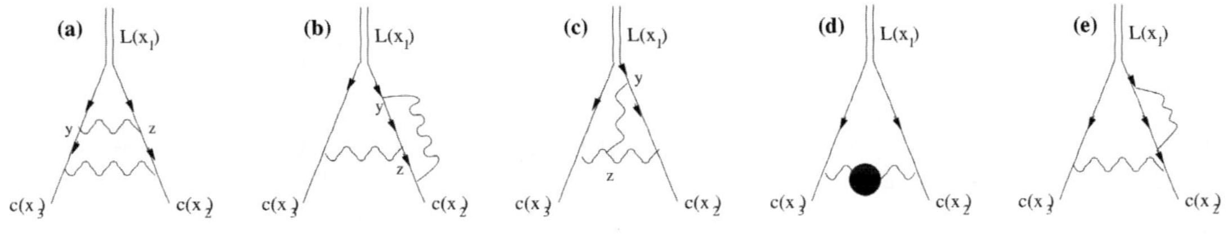

Figure 1. Two-loop planar corrections to the double ghost vertex. Wiggly lines correspond to gluon propagators, straight lines correspond to ghost propagators. The black disc in (d) is for one-loop contribution in the renormalization of the vector propagator from scalar, spinor and ghost fields.

gauge at any loop order, it does not have poles in ε in $d=4-2\varepsilon$ dimensions. The superficial divergence disappears due to property of the transversality of the gluon propagator in the Landau gauge and divergences in subgraphs disappear due to $\mathcal{N}=4$ supersymmetry.

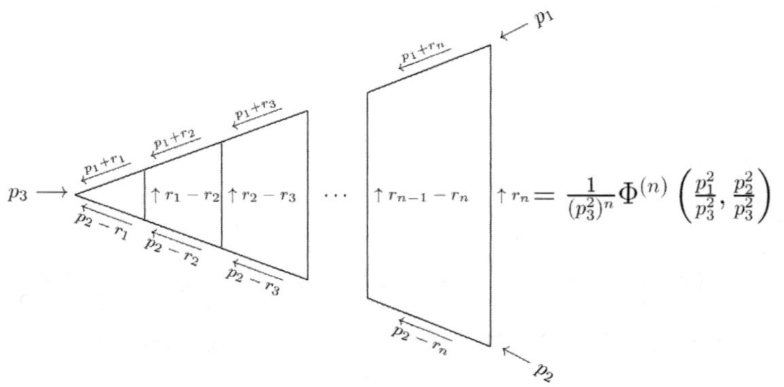

Figure 2. Triangle ladder diagrams for an arbitrary number n of loops in the momentum space. Each line corresponds to certain power of scalar propagator in the momentum space $1/(p^2)^\alpha$, α is the index of the line. If the index is not written, it is equal to 1. Vertex has no structure.

We show that all these contributions may be decomposed in terms of the integrals corresponding to the triangle ladder diagrams even in $d = 4 - 2$ dimensions. This may help to analyse the planar limit of $\mathcal{N} = 4$ super-Yang-Mills theory. The triangle ladder diagrams have been studied in $d = 4$ dimensions in Refs. [5, 6, 7] in the momentum space. These triangle ladders with an arbitrary number of loops n are shown in Fig. 2 which is reproduced from Ref. [8] with the same notation.

2. Mellin-Barnes transforms of the momentum integrals of the triangle ladders

It has been shown in Refs. [9, 10] at the diagrammatic level that the functions $\Phi^{(n)}(x\ y)$ obtained in Refs. [6, 7] for $d = 4$,

$$\Phi^{(n)}(x\ y) = -\frac{1}{n!\lambda}\ {}^{2n}_{j=n}\ \frac{(-1)^j j! \ln^{2n-j}(y\ x)}{(j-n)!(2n-j)!}\left[\mathrm{Li_j}\left(-\frac{1}{\rho x}\right) - \mathrm{Li}_j(-\rho y)\right]$$

$$\rho = \frac{2}{1-x-y+\lambda} \qquad \lambda = \overline{(1-x-y)^2 - 4xy}$$

of two variables that appear on the right hand side of Fig.2 possess the property of invariance with respect to Fourier transformation. In Ref. [11] the property of such an invariance of these functions has been formulated as

$$\frac{1}{[31]^2}\Phi^{(n)}\left(\frac{[12]}{[31]}\ \frac{[23]}{[31]}\right) = \frac{1}{(2\pi)^4}\quad d^4p_1 d^4p_2 d^4p_3\ \delta(p_1+p_2+p_3)\ \times$$
$$\times e^{ip_2x_2}e^{ip_1x_1}e^{ip_3x_3}\frac{1}{(p_2^2)^2}\Phi^{(n)}\left(\frac{p_1^2}{p_2^2}\ \frac{p_3^2}{p_2^2}\right) \qquad (1)$$

where the notation notation $[12] = (x_1 - x_2)^2$ of Ref. [2] is used. This property has been proven via Mellin-Barnes transformation, that is,

$$\Phi^{(n)}(x\ y) = \quad dz_2 dz_3 x^{z_2} y^{z_3}\mathcal{M}^{(n)}(z_2\ z_3) \qquad \frac{1}{[31]^2}\Phi^{(n)}\left(\frac{[12]}{[31]}\ \frac{[23]}{[31]}\right) =$$
$$\frac{1}{(2\pi)^8}\quad d^4p_1 d^4p_2 d^4p_3 d^4x_5 e^{ip_2(x_2-x_5)}e^{ip_1(x_1-x_5)}e^{ip_3(x_3-x_5)}\frac{1}{(p_2^2)^2}\Phi^{(n)}\left(\frac{p_1^2}{p_2^2}\ \frac{p_3^2}{p_2^2}\right) =$$
$$\frac{1}{(2\pi)^8}\quad d^4p_1 d^4p_2 d^4p_3 d^4x_5\quad dz_2 dz_3\frac{e^{ip_2(x_2-x_5)}e^{ip_1(x_1-x_5)}e^{ip_3(x_3-x_5)}}{(p_2^2)^{2+z_2+z_3}(p_1^2)^{-z_2}(p_3^2)^{-z_3}}\mathcal{M}^{(n)}(z_2\ z_3) =$$

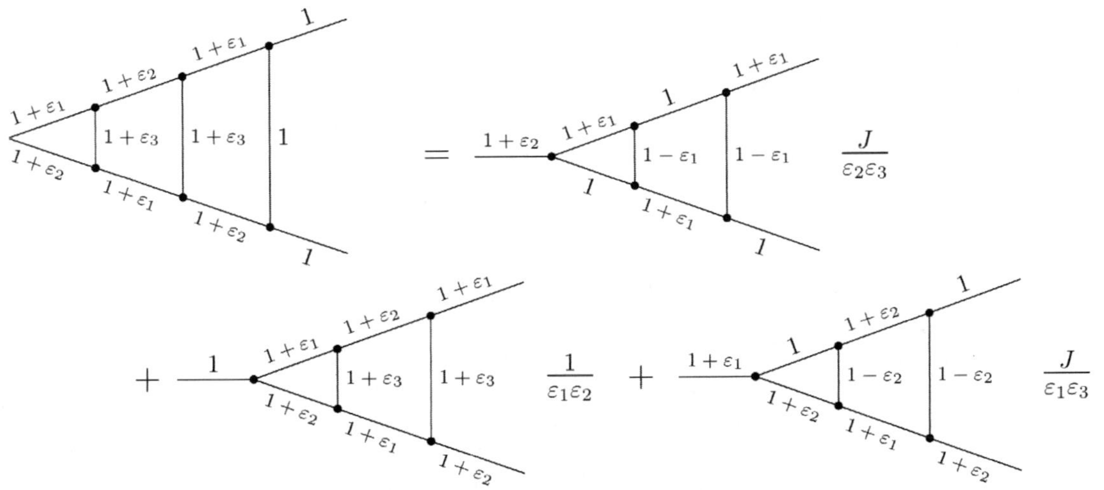

Figure 3. Loop reduction in $d = 4$ dimensions. Each line corresponds to certain power of scalar propagator in the position space $1/((y-z)^2)^\alpha$, α is the index of the line. The internal vertices y_μ and z_μ are integration points in the position space, they are depicted in boldface. The formula depicted is valid in $d = 4$ for $\varepsilon_1 + \varepsilon_2 + \varepsilon_3 = 0$, with complex valued $\varepsilon_1, \varepsilon_2, \varepsilon_3$. Factor J is defined in Eq.(3).

$$= \frac{(4\pi)^6}{(2\pi)^8} \quad d^4x_5 \quad dz_2dz_3 \frac{\Gamma(-z_2-z_3)}{\Gamma(2+z_2+z_3)} \frac{\Gamma(2+z_2)}{\Gamma(-z_2)} \frac{\Gamma(2+z_3)}{\Gamma(-z_3)}$$

$$\times \frac{2^{2z_2+2z_3-2(2+z_2+z_3)}\mathcal{M}^{(n)}(z_2 \ z_3)}{(x_2-x_5)^{-z_2-z_3}(x_1-x_5)^{2+z_2}(x_3-x_5)^{2+z_3}} = \quad dz_2dz_3 \frac{\mathcal{M}^{(n)}(z_2 \ z_3)}{[12]^{-z_3}[23]^{-z_2}[31]^{2+z_2+z_3}}$$

Such a simple proof suggests that the Mellin-Barnes transformation may be a helpful trick to study ladder diagrams in a non-integer dimension too. The Mellin-Barnes transforms of the momentum integrals corresponding to the ladder diagrams have been studied in Refs. [12, 13] for the ladders in $d = 4$ dimensions and in Refs. [15, 14] in $d = 4 - 2$ dimensions with indices $1 -$ on the rungs of the ladders. These integral transforms have been studied with a help of the diagrammatic relations of Ref. [5], for the review of these relations one can see Ref. [12]. This is a first known example when a three-point diagram with n loops was represented as linear combination of diagrams with $n-1$ loops. As it has been shown in Ref. [12], the diagrammatic relation shown in Fig. 3 corresponds to the integral relation

$$_C \ dz_2dz_3 \ D^{(u \ v)}[1+ \ _1 - z_3 \ 1+ \ _2 - z_2 \ 1+ \ _3]D^{(z_2 \ z_3)}[1+ \ _2 \ 1+ \ _1 \ 1+ \ _3] =$$

$$J \ \frac{D^{(u \ v- \ 2)}[1- \ _1]}{_2 \ _3} + \frac{D^{(u \ v)}[1+ \ _3]}{_1 \ _2} + \frac{D^{(u- \ 1 \ v)}[1- \ _2]}{_1 \ _3} \quad (2)$$

where

$$D^{(z_2 \ z_3)}[\nu_1 \ \nu_2 \ \nu_3] = \frac{\Gamma(-z_2)\Gamma(-z_3)\Gamma(-z_2-\nu_2-\nu_3+d \ 2)\Gamma(-z_3-\nu_1-\nu_3+d \ 2)}{\Pi_i\Gamma(\nu_i)\Gamma(d-\Sigma_i\nu_i)} \times$$
$$\times\Gamma(z_2+z_3+\nu_3)\Gamma(\Sigma\nu_i - d \ 2 + z_3 + z_2)$$

and where we have

$$_1 + \ _2 + \ _3 = 0 \quad D^{(u \ v)}[1+\nu] \equiv D^{(u \ v)}[1 \ 1 \ 1+\nu] \quad J = \frac{\Gamma(1- \ _1)\Gamma(1- \ _2)\Gamma(1- \ _3)}{\Gamma(1+ \ _1)\Gamma(1+ \ _2)\Gamma(1+ \ _3)} \quad (3)$$

Integral relation (2) has been proven via Barnes lemmas in [16]. Going back from such a type of integral relations to diagrammatic relations we may construct the diagrammatic relations similar to the relation depicted in Fig. 3 in $d = 4 - 2$ dimensions. This observation makes the MB transformation to be a powerful trick to calculate ladder diagrams in $d = 4 - 2$ exactly without expansion in Laurent series in [16].

3. Algorithm

The result for diagram (c) of Fig.1 may be found in Ref. [4]. This diagram contains gluon self-interaction, for example, in $d = 4$ the term

$$\frac{(31)_\nu}{[31]^2} \quad Dy \, \frac{(2y)_\sigma}{[2y]^2}\frac{(1y)_\lambda}{[1y]^2} \quad Dz \left({}^{(z)}_{\mu}\Pi_{\rho\nu}(z3) \right) \Pi_{\rho\lambda}(zy)\Pi_{\mu\sigma}(z2) \tag{4}$$

where the gluon propagator in $d = 4 - 2$ in the position space is

$$\Pi_{\rho\lambda}(zy) = \frac{g_{\rho\lambda}}{[yz]^{1-\epsilon}} + 2(1-\epsilon)\frac{(yz)_\rho (yz)_\lambda}{[yz]^{2-\epsilon}}$$

and the notation is $[yz] \equiv (y-z)^2$ $\quad [y1] \equiv (y - x_1)^2$ $\quad (yz)_\nu \equiv (y-z)_\nu$ $\quad (31)_\nu \equiv (x_3 - x_1)_\nu$ and $Dx \equiv \pi^{-\frac{d}{2}} d^d x$ is a measure of the integration in the position space. This measure is useful in massless theories [2]. The result for the diagram (c) from Ref. [4] in the limit $\to 0$ is

$$\frac{13\ 2}{[12]^2[23]^2} + \frac{-33\ 2}{[12]^2[31]^2} + \frac{15\ 2}{[23]^2[31]^2} + \frac{11}{[12][23][31]^2} + \frac{-14}{[12][23]^2[31]} + \frac{10}{[12]^2[23][31]} + \Bigg[\frac{6}{[12][23]^2}$$
$$+ \frac{-6}{[12][31]^2} + \frac{[12]}{[23]^2[31]^2} + \frac{4[23]}{[12]^2[31]^2} + \frac{-2}{[12]^2[31]} + \frac{-5}{[23]^2[31]} + \frac{-2[31]}{[12]^2[23]^2}\Bigg] \frac{1}{[31]}\Phi^{(1)}\left(\frac{[12]}{[31]}, \frac{[23]}{[31]}\right)$$
$$+ \left[\frac{-2}{[12]^2[23]^2} + \frac{6}{[12]^2[31]^2} + \frac{4}{[23]^2[31]^2} + \frac{1}{[12][23][31]^2} + \frac{-1}{[12][23]^2[31]} + \frac{-4}{[12]^2[23][31]}\right] \ln[12]$$
$$+ \left[\frac{-1\ 2}{[12]^2[23]^2} + \frac{-5}{[12]^2[31]^2} + \frac{-3\ 2}{[23]^2[31]^2} + \frac{-7\ 2}{[12][23][31]^2} + \frac{2}{[12][23]^2[31]} + \frac{7\ 2}{[12]^2[23][31]}\right] \ln[23]$$
$$+ \left[\frac{5\ 2}{[12]^2[23]^2} + \frac{-1}{[12]^2[31]^2} + \frac{-5\ 2}{[23]^2[31]^2} + \frac{5\ 2}{[12][23][31]^2} + \frac{-1}{[12][23]^2[31]} + \frac{1\ 2}{[12]^2[23][31]}\right] \ln[31]$$

The gluon self-interaction term of Eq.(4) may be decomposed in a sum of simpler terms which contain derivatives acting on the scalar propagator, for example,

$$Dy\ Dz \, \frac{1}{[31]^{1-\epsilon}}\frac{1}{[2y]^{1-\epsilon}}\frac{1}{[3z]^{1-\epsilon}}\frac{1}{[z2]^{1-\epsilon}} {}^{(y)}_{\mu}\frac{1}{[1y]^{1-\epsilon}} {}^{(y)}_{\mu}\frac{1}{[yz]^{1-\epsilon}} \tag{5}$$

There are other terms with higher number of pairs of the contracted derivatives. We call the terms of such a type basis elements.

Now we show how the basis elements may be decomposed in terms of the triangle ladder diagrams. The algorithm has already been described in Ref. [10]. Here we describe the idea at level of formulas. The trick of the reduction of the basis elements to integrals corresponding to the triangle ladder diagrams may be shown for the element of Eq. (5)

$$Dy\ Dz \, \frac{1}{[31]^{1-\epsilon}}\frac{1}{[2y]^{1-\epsilon}}\frac{1}{[3z]^{1-\epsilon}}\frac{1}{[z2]^{1-\epsilon}} {}^{(y)}_{\mu}\frac{1}{[1y]^{1-\epsilon}} {}^{(y)}_{\mu}\frac{1}{[yz]^{1-\epsilon}} =$$
$$\frac{1}{2} \quad Dy\ Dz \, \frac{1}{[31]^{1-\epsilon}}\frac{1}{[2y]^{1-\epsilon}}\frac{1}{[3z]^{1-\epsilon}}\frac{1}{[z2]^{1-\epsilon}} \times$$
$$\times \left[{}^{2}_{(y)}\left(\frac{1}{[1y]^{1-\epsilon}}\frac{1}{[yz]^{1-\epsilon}}\right) - \left({}^{2}_{(y)}\frac{1}{[1y]^{1-\epsilon}}\right)\frac{1}{[yz]^{1-\epsilon}} - \frac{1}{[1y]^{1-\epsilon}}\left({}^{2}_{(y)}\frac{1}{[yz]^{1-\epsilon}}\right)\right] \tag{6}$$

D Alembert operator acting on the scalar propagator generates Dirac δ-function. This means that the double integration disappears and the result reduces to the simple integration of three propagators in each of the terms on the right hand side of Eq. (6). In more complicate cases the d Alembertian at the end of these transformations may act on one of the external points of the triangle ladder diagram. As we have mentioned in the previous section, these integrals may be calculated exactly even in $d = 4 - 2$ dimensions via the MB transformation.

4. Conjecture

We make the conjecture that the diagrammatic relations in Fig. 1 of Ref. [10] established in $d = 4$ dimensions will maintain themselves in $d = 4 - 2$ dimensions, too. This may be proved via Mellin-Barnes transformation with help of the trick presented in talk [16]. On the other hand, the integrals on the right hand side of these relations are more easy to calculate than the ladder diagrams. Such a calculation again may be done by using the MB transformation. The result of this chain of transformations may be considered as an independent crosscheck of the method described in Section 2. Thus, the algorithm we have proposed in Section 3 gives us the decomposition of the three-point double ghost vertex in terms of the ladder diagrams on which the d Alembert operator acts as it is shown on Fig. 1 of Ref. [10], and these constructions may be calculated explicitly in $d = 4 - 2$ dimensions by two different methods. The first method is described in Section 2 and the second method is described in this Section.

5. Conclusion

The advantage of this approach is that the proposed algorithm is easy to program. Algebraically, it reduces the integration to the integrals in the momentum space corresponding to the triangle ladder diagrams. We have explained in the previous Sections why such diagrams may be calculated explicitly via the Mellin-Barnes transformation in $d = 4 - 2$ dimensions. The method may serve in any massless field theory, not necessarily in $\mathcal{N} = 4$ SYM only.

Acknowledgments

These results are based on the papers written in collaboration with Pedro Allendes (UdeC), Gorazd Cvetic (UTFSM), Ivan Gonzalez (UV), Bernd Kniehl (DESY), Anatoly Kotikov (JINR), Eduardo Notte-Cuello (ULS), Ivan Parra-Ferrada (U de Talca), Marko Rojas-Medar (UBB), Ivan Schmidt (UTFSM), Alvaro Vergara (UBB). I.K. is very grateful to all of them. This research is supported in part by Fondecyt (Chile) Grants Nos. 1040368, 1050512 and 1121030, by DIUBB (Chile) Grant Nos. 125009, GI 153209/C and GI 152606/VC.

References

[1] Cvetič G, Kondrashuk I and Schmidt I 2006 Effective action of dressed mean fields for $\mathcal{N} = 4$ super-Yang-Mills theory *Mod. Phys. Lett.* A **21** 1127 (*Preprint* hep-th/0407251)

[2] Cvetič G, Kondrashuk I, Kotikov A and Schmidt I 2007 Towards the two-loop Lcc vertex in Landau gauge *Int. J. Mod. Phys.* A **22** 1905 (*Preprint* hep-th/0604112)

[3] Cvetič G and Kondrashuk I 2008 Further results for the two-loop Lcc vertex in the Landau gauge *J. High Energy Phys.* JHEP02(2008)023 (*Preprint* hep-th/0703138)

[4] Cvetič G and Kondrashuk I 2008 Gluon self-interaction in the position space in Landau gauge *Int. J. Mod. Phys.* A **23** 4145 (*Preprint* arXiv:0710.5762 [hep-th])

[5] Belokurov V V and Usyukina N I 1983 Calculation Of Ladder Diagrams In Arbitrary Order *J. Phys.* A **16** 2811.

[6] Usyukina N I and Davydychev A I 1993 An Approach to the evaluation of three and four point ladder diagrams *Phys. Lett.* B **298** 363.

[7] Usyukina N I and Davydychev A I 1993 Exact results for three and four point ladder diagrams with an arbitrary number of rungs *Phys. Lett.* B **305** 136.

[8] Kondrashuk I and Vergara A 2010 Transformations of triangle ladder diagrams *J. High Energy Phys.* JHEP03(2010)051 (*Preprint* arXiv:0911.1979 [hep-th])

Journal of Physics: Conference Series **762** (2016) 012056 doi:10.1088/1742-6596/762/1/012056

[9] Kondrashuk I and Kotikov A 2009 Fourier transforms of UD integrals *Analysis and Mathematical Physics* Birkhäuser Book Series Trends in Mathematics, ed Gustafsson B and Vasil'ev A, (Basel Switzerland: Birkhäuser) pp 337-348 (*Preprint* arXiv:0802.3468 [hep-th])

[10] Kondrashuk I and Kotikov A 2008 Triangle UD integrals in the position space *J. High Energy Phys.* JHEP08(2008)106 (*Preprint* arXiv:0803.3420 [hep-th])

[11] Allendes P, Guerrero N, Kondrashuk I and Notte-Cuello E A 2010 New four-dimensional integrals by Mellin-Barnes transform *J. Math. Phys.* **51** 052304 (*Preprint* arXiv:0910.4805 [hep-th])

[12] Allendes P, Kniehl B, Kondrashuk I, Notte-Cuello E A, Rojas-Medar M 2013 Solution to Bethe-Salpeter equation via Mellin-Barnes transform *Nucl. Phys.* B **870** 243 (*Preprint* arXiv:1205.6257 [hep-th])

[13] Kniehl B, Kondrashuk I, Notte-Cuello E A, Parra-Ferrada I, Rojas-Medar M 2013 Two-fold Mellin-Barnes transforms of Usyukina-Davydychev functions *Nucl. Phys.* B **876** 322 (*Preprint* arXiv:1304.3004 [hep-th])

[14] Gonzalez I and Kondrashuk I 2013 Belokurov-Usyukina loop reduction in non-integer dimension *Phys.Part.Nucl.* **44** 268 (*Preprint* arXiv:1206.4763 [hep-th])

[15] Gonzalez I and Kondrashuk I 2013 Box ladders in non-integer dimension *Theor. Math. Phys.* **177** 1515 (*Preprint* arXiv:1210.2243 [hep-th])

[16] Kondrashuk I 2015 Talk at XVIII International Congress on Mathematical Physics, Santiago, Chile, 24 July - 1 August 2015, unpublished

Journal of Physics: Conference Series **762** (2016) 012057 doi:10.1088/1742-6596/762/1/012057

Automating QCD amplitudes with on-shell methods

Simon Badger

Higgs Centre for Theoretical Physics, School of Physics and Astronomy, The University of Edinburgh, Edinburgh EH9 3JZ, Scotland, UK.

E-mail: `sbadger@staffmail.ed.ac.uk`

Abstract. We review some of the modern approaches to scattering amplitude computations in QCD and their application to precision LHC phenomenology.

1. Introduction

New understanding of the structure of scattering amplitudes has led to efficient algorithms for high multiplicity final states which are now common in next-to-leading order simulations of LHC collisions. To keep up with the experimental accuracy expected during Run II, new theoretical methods are required to look at high multiplicity final states at higher accuracy. In this contribution we take a brief look at some new on-shell methods attempting to tackle this problem. As a specific example we consider applications of momentum twistor technology for the parameterisation of multi-leg kinematics.

Over the last ten years or so an increasingly large number of theoretical physicists have turned their attention to understanding the underlying structure of amplitudes in gauge, gravity and string theory. Much of this can be traced to Witten's 2003 paper on $\mathcal{N} = 4$ super-symmetric Yang-Mills theory as a string theory in twistor space [1] though on-shell unitarity methods [2, 3] have been hard at work in collider physics applications for much longer. Perhaps the most remarkable example of on-shell simplicity was the realisation of Parke and Taylor [4] that tree-level amplitudes for multi-gluon scattering amplitudes with specific external helicity states can be written in a compact form,

$$A_n^{(0)}(1^+, 2^+, 3^+, \ldots, n^+) = 0, \tag{1}$$

$$A_n^{(0)}(1^-, 2^+, 3^+, \ldots, n^+) = 0, \tag{2}$$

$$A_n^{(0)}(1^-, 2^-, 3^+, \ldots, n^+) = \frac{i\langle 12\rangle^4}{\langle 12\rangle\langle 23\rangle\langle 34\rangle\cdots\langle n1\rangle}. \tag{3}$$

This famous representation for the maximal helicity violating configuration at the amplitude level, due to later work by Mangano, Parke and Xu [5] and Berends and Giele [6], resulted from understanding the power of both colour ordering to exploit the $SU(N_c)$ group structure and the spinor-helicity formalism to simplify the on-shell kinematics. The motivation for studying these high multiplicity tree-level expressions came from collider physics applications where multi-jet backgrounds were still impossible even at leading order in perturbation theory.

Though the number of Feynman diagrams involved in a given computation are sometimes used as measure of the complexity, with a high degree of automation they can easily be applicable to relevant processes. Obtaining compact and efficient amplitude representations with this method is fairly hopeless however. In Table 1 we compare the set of Feynman graphs for the full colour amplitude $\mathcal{A}_n^{(0)}$ with the reduced colour ordered set of graphs which contribute to the partial amplitudes $A_n^{(0)}$ defined by,

$$\mathcal{A}_n^{(0)} = \sum_{\sigma \in S_{n-2}} \tilde{f}^{a_1\sigma(a_2)\sigma(a_3)} \ldots \tilde{f}^{\sigma(a_{n-2})\sigma(a_{n-1})a_n} A_n^{(0)}(1, \sigma(2), \ldots, \sigma(3), n) \tag{4}$$

Here we have used the compact colour decomposition into adjoint traces [8] which incorporates additional colour symmetry coming from Kleiss-Kuijf identities [9] and runs over $(n-2)!$ permutations. $\tilde{f}^{abc} = \sqrt{2} f^{abc}$ are the usual structure constants in $SU(N_c)$. By exploiting the symmetry in colour space the amplitude can be constructed from a minimal set of gauge invariant kinematic building blocks. The number of diagrams $N_{\mathcal{A}}(n)$ in the full colour amplitudes can be derived from a simple algorithm using differential operators [10]. If we let t be the number of three-gluon vertices and g be the number of internal and external gluons then,

$$N_{\mathcal{A}}(n) = \left(t\, g^3 \frac{\partial}{\partial g} + g \frac{\partial}{\partial t} \right)^{n-3} t\, g^3 \Bigg|_{t=g=1}, \tag{5}$$

where the first derivative adds a gluon to each gluon line via a three vertex and the second derivative adds a gluon to each three vertex via a four-point vertex. The colour ordered diagrams can be quickly computed using the recursive approach of Berends and Giele [6].

$$N_A(n) = \sum_{k=1}^{n-2} N_A(k+1) N_A(n-k) + \sum_{k=1}^{n-3} \sum_{l=k+1}^{n-3} N_A(k+1) N_A(l-k+1) N(n-l) \tag{6}$$

A more modern technique for the evaluation of tree-level amplitudes purely from on-shell building blocks is the recursive method of Britto, Cachazo, Feng and Witten [11, 12]. By counting the number of on-shell BCFW diagrams that contribute to a given helicity configuration, one can see that there is considerable simplification from the ordered diagrams to a fully on-shell amplitude. Automated packages which solve the on-shell recursion relation to obtain compact expression for all tree amplitudes in massless QCD have been developed [13, 7]. The complexity of helicity amplitudes then becomes clear with the addition of more and more negative helicity gluons.

For applications in collider phenomenology an individual helicity amplitude is unfortunately of little use and both sums of the helicity and colour configurations of the ordered partial amplitudes must be performed efficiently. Given the precision for the LHC experiments, leading order predictions in QCD are insufficient for four or even five particle final states, so it is necessary to perform this task at least to one-loop order.

n	3	4	5	6	7	8	9	10
Feynman diagrams	1	4	25	220	2485	34300	559405	10525900
Ordered diagrams	1	3	10	38	154	654	2871	12925
MHV	1	1	1	1	1	1	1	1
BCFW NMHV	-	-	-	3	6	10	15	21
BCFW N^{2}MHV	-	-	-	-	-	20	50	105
BCFW N^{3}MHV	-	-	-	-	-	-	-	175

Table 1. The number of graphs appearing in tree-level n-gluon scattering using different approaches. The number of terms in BCFW are chosen for the most complicated alternating helicity case and were obtained using Bourjaily's `bcfw` package [7].

Nevertheless a host of new algorithms have been developed in recent years which exploit this tree-level simplicity at the loop level and can provide the necessary information to the Monte Carlo event generators like SHERPA [14], MADGRAPH5_aMC@NLO [15] and HERWIG7 [16] used in experimental analyses. A few important insights were necessary to make this possible some of which I will expand on the next sections.

2. Unitarity, generalised unitarity and integrand reduction

One of the key insights which led to the automation of one-loop amplitudes is that a purely algebraic algorithm would kinematic algebra to be performed numerically and avoid a traditional bottleneck in multi-leg amplitude computations.

The starting point for this approach is the well known decomposition of one-loop amplitudes into a basis of known box, triangle and bubble integrals,

$$A_n^{(1)} = \sum_i c_i I_i^{4-2\epsilon} + \text{ rational} + \mathcal{O}(\epsilon). \tag{7}$$

In this formula for one-loop primitive (i.e. colour ordered) amplitudes c_i are rational functions while $I_i^{4-2\epsilon}$ are loop integrals in the dimensional regularisation parameter ϵ. The generalised unitarity algorithm uses complex momenta to systematically solve the multiple cut conditions and isolate the coefficients c_i from products of tree-level amplitudes [17, 18, 19]. The remaining rational term can be extracted from modified BCFW recursion relations [20, 21, 22, 23, 24] or D-dimensional unitarity cuts [25].

Another approach developed at the same time was the integrand reduction method of Ossola, Papadopoulos and Pittau [26]. Here the loop momentum dependence of the integrand is completely parameterised in terms of the scalar integral basis and additional 'spurious' terms which integrate to zero. The sum over topologies remains the same while the numerator is decomposed into monomials of irreducible scalar products $\Delta_i(k)$:

$$A_n^{(1)} = \int d^{4-2\epsilon} \sum_i \frac{\Delta_i(k)}{\prod_{\alpha\in i} D_\alpha(k)}, \tag{8}$$

where $D_\alpha(k)$ are the propagators for the ith topology. The integrand reduction procedure is compatible with D-dimensional generalised unitarity cuts [27, 28]. A complete reconstruction of the integrand, including spurious terms, is important to allow the algorithm to be implemented efficiently using fixed precision numerics. Further details on these algorithms can be found in reference [29].

Since the information required for any loop amplitude has now been parameterised in terms of rational functions, the generation of the input to these algorithms can then be performed using tree-level recursive techniques. An important development in this area was the extension of off shell recursive techniques to tensor integrands using the OPENLOOPS method [30]. Variations of these techniques have subsequently been implemented into computer codes such as GOSAM [31], OPENLOOPS [30], MADLOOP[1] [32] and RECOLA [33] that can handle a wide variety of complicated processes including QCD and Electro-weak corrections GOSAM [31] combines efficient Feynman diagram generation together with the integrand reduction algorithm NINJA [34].

The BLACKHAT [19] and NJET [35] one-loop amplitude providers are more specialised but use a fully on-shell approach that is currently able to attack higher multiplicity processes than other algorithms. These codes have been interfaced with the SHERPAMonte-Carlo to produce NLO

[1] MADLOOP is part of the MADGRAPH5_aMC@NLOMonte Carlo event generator which offers fully automated NLO simulations interfaced with parton showers.

predictions for high multiplicity processes like $pp \to 5j$ [36], $pp \to WW{+}3j$ [37], $pp \to \gamma\gamma{+}3j$ [38] and $pp \to W + 5j$ [39]. The bottleneck in these extreme configurations is no longer in the virtual corrections since most of the event generation time is spent in the real radiation phase-space integration. Nevertheless, without the advanced recursive tree amplitude algorithms in Comix [40] and Sherpa's automated dipole subtraction method [41] such predicitions wouldn't be possible at all.

3. D-dimensional generalised unitarity for multi-loop integrands

Extending the current multi-loop methods to higher multiplicity represents a serious challenge. The D-dimensional generalised unitarity cut algorithm has recently been extended to multi-loop integrands using integrand reduction [26] and elements of computational algebraic geometry [42, 43, 44, 45, 46, 47, 48, 49]. In contrast to the one-loop case, the basis of integrals obtained through this method is not currently known analytically and is much larger than the set of basis functions defined by standard integration-by-parts(IBP) identities. Nevertheless, new results in non-supersymmetric theories have been obtained for five gluon scattering amplitudes with all positive helicities [49, 50]. The maximal unitarity method [51], which incorporates IBP identites, has been applied to a variety of two-loop examples in four dimensions [52, 53, 54, 55, 56]. This approach can be seen as an extension of the generalised unitarity methods of Britto, Cachazo and Feng [17] and Forde [18]. Efficient algorithms to generate unitarity compatible IBPs are a key ingredient in both approaches and has been the focus of on-going investigations [57, 58, 59, 60].

Constructing loop amplitudes from trees for non-supersymmetric theories like QCD require the identification of the additional components of the dimensionally regulated loop momenta. At two-loops this information can be extracted from six-dimensional trees which are computed efficiently using the six-dimensional spinor-helicity formalism [61]. These cuts can then be dimensionally reduced to $4 - 2\epsilon$ in an analogous way to the one-loop approaches considered previously [62, 63].

Breaking loop amplitudes into trees also has benefits for colour decompositions and representations of the non-planer sector since tree-level amplitude relations such as Kliess-Kuijf [9] and Bern-Carrasco-Johansson [64] can be applied at the level of the cut amplitude. This technique has been shown to be a powerful tool in $\mathcal{N} = 4$ Yang-Mills and $\mathcal{N} = 8$ super-gravity computations [65], but recently been exploited in the computation of the complete five gluon all-plus amplitude at two loops in pure Yang-Mills [50]. Combining D-dimensional generalised unitarity cuts, BCJ relations and integrand reduction a remarkably compact form was obtained,

$$\mathcal{A}^{(2)}(1^+, 2^+, 3^+, 4^+, 5^+) =$$

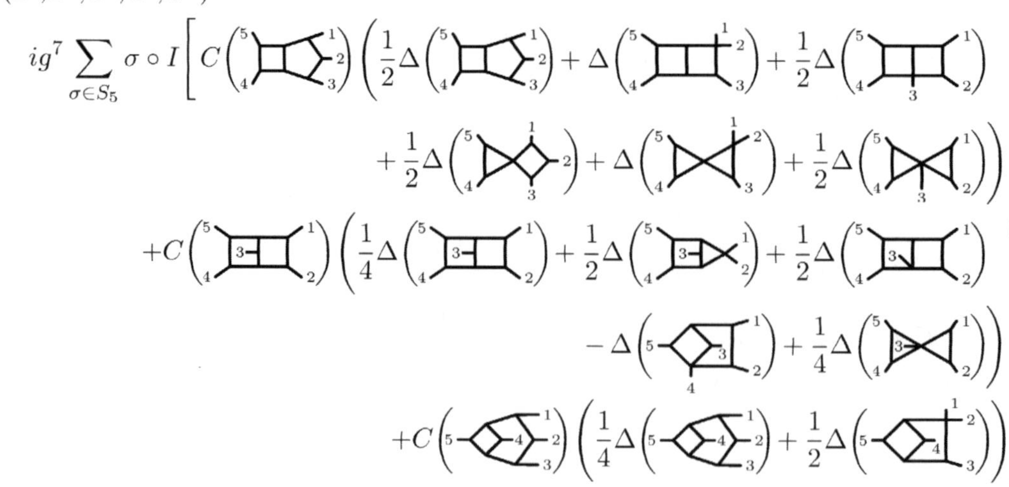

(9)

where the C functions are the colour factors in terms of adjoint $\tilde{f}$ structure constants and Δ are the irreducible numerators. Further details can be found in reference [50].

4. Momentum Twistors

In this section we collect some useful formulae for parameterising external kinematics using Hodges' momentum twistor formalism [66]. Momentum twistors are extremely useful for studying the geometric properties of amplitudes but they also have a practical property that can be used in general amplitude computations: these variable linearise the momentum conservation conditions and give a rational parameterisation of the phase-space.

The topic is somewhat more specialised than the techniques reviewed in the previous sections. The connection is that rational parameterisation of the kinematics can allow analytic computations to be performed in exactly the same way as the numerical algorithms described above and hence is a useful tool when combined with unitarity cuts and integrand reduction.

The notation used in this section relies on the spinor-helicity formalism of which good introductions can be found in references [67, 68, 69, 70]. The first observation stems from the fact that massless momenta can be decomposed in two component Weyl spinors,

$$p^2 = 0 \Rightarrow (\sigma \cdot p)^{\alpha\dot{\alpha}} = \lambda^{\alpha}(p)\tilde{\lambda}^{\dot{\alpha}}(p). \tag{10}$$

Hence the on-shell conditions are manifest when using the spinor-helicity formalism. Momentum conservation,

$$\sum_i p_i^{\mu} = 0, \tag{11}$$

still impose some non-trivial conditions on the amplitudes. Hodges [66] introduced momentum twistors as a natural extension of Penrose's twistor formalism. We being by defining dual momentum co-ordinates, x_i^{μ},

$$p_i^{\mu} = x_i^{\mu} - x_{i-1}^{\mu} \tag{12}$$

which can be inverted up to a fixed point, x_0^{μ},

$$x_i^{\mu} = x_0^{\mu} + \sum_{k=1}^{i} p_k^{\mu}. \tag{13}$$

The momentum twistor is then constructed from these dual co-ordinates and the holomorphic Weyl spinors:

$$Z_{iA} = \left(\lambda_{\alpha}(i), \mu^{\dot{\alpha}}(i) = \lambda_{\alpha}(i)(\sigma \cdot x_i)^{\alpha\dot{\alpha}}\right), \tag{14}$$

The two component object $\mu^{\dot{\alpha}}(i)$ is used instead of the $\tilde{\lambda}^{\dot{\alpha}}(i)$ spinor to define the kinematics for the n-particle system $i = 1, n$. The $\tilde{\lambda}(i)_{\dot{\alpha}}$ spinor is defined through the dual twistor,

$$W_i^A = (\tilde{\mu}_{\alpha}(i), \tilde{\lambda}^{\dot{\alpha}}(i)) = \frac{\varepsilon^{ABCD} Z_{(i-1)B} Z_{iC} Z_{(i+1)D}}{\langle i-1 i\rangle\langle i i+1\rangle} \tag{15}$$

from which we can find the definition of the anti-holomorphic spinor,

$$\tilde{\lambda}(i)^{\dot{\alpha}} = \frac{\langle i-1 i\rangle\mu^{\dot{\alpha}}(i+1) + \langle i+1 i-1\rangle\mu^{\dot{\alpha}}(i) + \langle i i+1\rangle\mu^{\dot{\alpha}}(i-1)}{\langle i-1 i\rangle\langle i i+1\rangle}. \tag{16}$$

By use of the Schouten identity it is easy to show that any $\tilde{\lambda}(i)_{\dot{\alpha}}$ defined through the dual twistor will automatically satisfy momentum conservation. Amplitudes can then be written in terms of holomorphic spinor products and momentum twistor 4-brackets:

$$\langle ijkl\rangle = \varepsilon^{ABCD} Z_{iA} Z_{jB} Z_{kC} Z_{lD}. \tag{17}$$

The $4 \times n$ matrix Z_{iA} has $3n - 10$ independent entries and we can generate rational phase-space points (for complex momenta) by filling the matrix with random integers. We can even pick rational functions to fill the momentum twistor matrix, which amounts to picking a special frame to evaluate general kinematics. A convenient choice is,

$$Z_i = \begin{pmatrix} \frac{\Sigma_i}{s_{12}} \\ 1 - \delta_{1i} \\ \frac{\langle 123i\rangle\langle 34\rangle[23]}{\langle 1234\rangle\langle 1i\rangle[12]} \\ \frac{-\langle 13\rangle\langle 124i\rangle + \langle 14\rangle\langle 123i\rangle}{\langle 1234\rangle\langle 1i\rangle} \end{pmatrix} \tag{18}$$

where

$$\frac{\Sigma_i}{s_{12}} = \begin{cases} 0 & i = 1 \\ \frac{\langle 13\rangle\langle 2i\rangle}{\langle 23\rangle\langle 1i\rangle} & \text{otherwise} \end{cases} \tag{19}$$

One additional subtlety in using this technique is that phase information ensures parity invariance is discarded in favour of rational functions. This can be easily restored as a pre-factor. In the explicit parameterisation above this is:

$$\Phi_n(1^{h_1}, \cdots, n^{h_n}) = \left(\frac{\langle 13\rangle}{[12]\langle 23\rangle}\right)^{-h_1} \prod_{i=2}^{n} \left(\frac{\langle 1i\rangle^2[12]\langle 23\rangle}{\langle 13\rangle}\right)^{-h_i} \tag{20}$$

where h_i are the helicities of the external gluons. As an example the MHV amplitude in this language would read,

$$\frac{A_n^{(0)}(1^-, 2^-, 3^+, \ldots, n^+)}{\Phi_n(1^-, 2^-, 3^+, \ldots, n^+)} = \frac{i\, s_{12}^{n-2}}{(\Sigma_4 - 1)\prod_{i=4}^{n-1}(\Sigma_i - \Sigma_{i+1})} \tag{21}$$

The power of this representation is that the above expression can be obtained by simply plugging in the parameterisation (18) into the thousands of ordered Feynman diagrams - assuming your computer has enough memory to perform the multi-variate factorisation. The MHV is a a special case of course since the being a single term the factorisation is guaranteed to land on the compact expression.

5. Conclusions

In these proceedings we have given a very brief overview of the on-shell methods used in automated loop amplitude computations. The success of this approach to the one-loop problem has motivated activity in applying the methods at higher loops and some useful progress has been made. Some subtleties of the integrand reduction method beyond one-loop have been identified in particular the non-minimal basis of integrals compared to those generated by traditional IBP reduction to master integrals.

We highlighted one of the new techniques used in modern high multiplicity loop computations. Though momentum twistors were originally applied in the context of maximally super-symmetric Yang-Mills theory these computations show they can be applied just as easily to general gauge theories.

References

[1] Witten E 2004 *Commun. Math. Phys.* **252** 189–258 (*Preprint* hep-th/0312171)
[2] Bern Z, Dixon L J, Dunbar D C and Kosower D A 1994 *Nucl. Phys.* **B425** 217–260 (*Preprint* hep-ph/9403226)
[3] Bern Z, Dixon L J, Dunbar D C and Kosower D A 1995 *Nucl. Phys.* **B435** 59–101 (*Preprint* hep-ph/9409265)
[4] Parke S J and Taylor T R 1986 *Phys. Rev. Lett.* **56** 2459

[5] Mangano M L, Parke S J and Xu Z 1988 *Nucl. Phys.* **B298** 653
[6] Berends F A and Giele W T 1988 *Nucl. Phys.* **B306** 759
[7] Bourjaily J L 2010 (*Preprint* 1011.2447)
[8] Del Duca V, Dixon L J and Maltoni F 2000 *Nucl. Phys.* **B571** 51–70 (*Preprint* hep-ph/9910563)
[9] Kleiss R and Kuijf H 1989 *Nucl. Phys.* **B312** 616
[10] Caravaglios F, Mangano M L, Moretti M and Pittau R 1999 *Nucl. Phys.* **B539** 215–232 (*Preprint* hep-ph/9807570)
[11] Britto R, Cachazo F and Feng B 2005 *Nucl. Phys.* **B715** 499–522 (*Preprint* hep-th/0412308)
[12] Britto R, Cachazo F, Feng B and Witten E 2005 *Phys. Rev. Lett.* **94** 181602 (*Preprint* hep-th/0501052)
[13] Dixon L J, Henn J M, Plefka J and Schuster T 2011 *JHEP* **01** 035 (*Preprint* 1010.3991)
[14] Gleisberg T, Hoeche S, Krauss F, Schonherr M, Schumann S, Siegert F and Winter J 2009 *JHEP* **02** 007 (*Preprint* 0811.4622)
[15] Alwall J, Frederix R, Frixione S, Hirschi V, Maltoni F, Mattelaer O, Shao H S, Stelzer T, Torrielli P and Zaro M 2014 *JHEP* **07** 079 (*Preprint* 1405.0301)
[16] Bellm J *et al.* 2016 *Eur. Phys. J.* **C76** 196 (*Preprint* 1512.01178)
[17] Britto R, Cachazo F and Feng B 2005 *Nucl. Phys.* **B725** 275–305 (*Preprint* hep-th/0412103)
[18] Forde D 2007 *Phys. Rev.* **D75** 125019 (*Preprint* 0704.1835)
[19] Berger C F, Bern Z, Dixon L J, Febres Cordero F, Forde D, Ita H, Kosower D A and Maitre D 2008 *Phys. Rev.* **D78** 036003 (*Preprint* 0803.4180)
[20] Bern Z, Dixon L J and Kosower D A 2005 *Phys. Rev.* **D71** 105013 (*Preprint* hep-th/0501240)
[21] Bern Z, Dixon L J and Kosower D A 2005 *Phys. Rev.* **D72** 125003 (*Preprint* hep-ph/0505055)
[22] Bern Z, Dixon L J and Kosower D A 2006 *Phys. Rev.* **D73** 065013 (*Preprint* hep-ph/0507005)
[23] Berger C F, Bern Z, Dixon L J, Forde D and Kosower D A 2006 *Phys. Rev.* **D74** 036009 (*Preprint* hep-ph/0604195)
[24] Berger C F, Bern Z, Dixon L J, Forde D and Kosower D A 2007 *Phys. Rev.* **D75** 016006 (*Preprint* hep-ph/0607014)
[25] Badger S D 2009 *JHEP* **01** 049 (*Preprint* 0806.4600)
[26] Ossola G, Papadopoulos C G and Pittau R 2007 *Nucl. Phys.* **B763** 147–169 (*Preprint* hep-ph/0609007)
[27] Ellis R K, Giele W T and Kunszt Z 2008 *JHEP* **03** 003 (*Preprint* 0708.2398)
[28] Giele W T, Kunszt Z and Melnikov K 2008 *JHEP* **04** 049 (*Preprint* 0801.2237)
[29] Ellis R K, Kunszt Z, Melnikov K and Zanderighi G 2012 *Phys. Rept.* **518** 141–250 (*Preprint* 1105.4319)
[30] Cascioli F, Maierhofer P and Pozzorini S 2012 *Phys. Rev. Lett.* **108** 111601 (*Preprint* 1111.5206)
[31] Cullen G *et al.* 2014 *Eur. Phys. J.* **C74** 3001 (*Preprint* 1404.7096)
[32] Hirschi V, Frederix R, Frixione S, Garzelli M V, Maltoni F and Pittau R 2011 *JHEP* **05** 044 (*Preprint* 1103.0621)
[33] Actis S, Denner A, Hofer L, Scharf A and Uccirati S 2013 *JHEP* **04** 037 (*Preprint* 1211.6316)
[34] Peraro T 2014 *Comput. Phys. Commun.* **185** 2771–2797 (*Preprint* 1403.1229)
[35] Badger S, Biedermann B, Uwer P and Yundin V 2013 *Comput. Phys. Commun.* **184** 1981–1998 (*Preprint* 1209.0100)
[36] Badger S, Biedermann B, Uwer P and Yundin V 2014 *Phys. Rev.* **D89** 034019 (*Preprint* 1309.6585)
[37] Cordero F F, Hofmann P and Ita H 2015 (*Preprint* 1512.07591)
[38] Badger S, Guffanti A and Yundin V 2014 *JHEP* **03** 122 (*Preprint* 1312.5927)
[39] Bern Z, Dixon L J, Febres Cordero F, Hche S, Ita H, Kosower D A, Matre D and Ozeren K J 2013 *Phys. Rev.* **D88** 014025 (*Preprint* 1304.1253)
[40] Gleisberg T and Hoeche S 2008 *JHEP* **12** 039 (*Preprint* 0808.3674)
[41] Gleisberg T and Krauss F 2008 *Eur. Phys. J.* **C53** 501–523 (*Preprint* 0709.2881)
[42] Mastrolia P and Ossola G 2011 *JHEP* **11** 014 (*Preprint* 1107.6041)
[43] Badger S, Frellesvig H and Zhang Y 2012 *JHEP* **04** 055 (*Preprint* 1202.2019)
[44] Zhang Y 2012 *JHEP* **09** 042 (*Preprint* 1205.5707)
[45] Kleiss R H P, Malamos I, Papadopoulos C G and Verheyen R 2012 *JHEP* **12** 038 (*Preprint* 1206.4180)
[46] Feng B and Huang R 2013 *JHEP* **02** 117 (*Preprint* 1209.3747)
[47] Mastrolia P, Mirabella E, Ossola G and Peraro T 2012 *Phys. Lett.* **B718** 173–177 (*Preprint* 1205.7087)
[48] Mastrolia P, Mirabella E, Ossola G and Peraro T 2013 *Phys. Lett.* **B727** 532–535 (*Preprint* 1307.5832)
[49] Badger S, Frellesvig H and Zhang Y 2013 *JHEP* **12** 045 (*Preprint* 1310.1051)
[50] Badger S, Mogull G, Ochirov A and O'Connell D 2015 *JHEP* **10** 064 (*Preprint* 1507.08797)
[51] Kosower D A and Larsen K J 2012 *Phys. Rev.* **D85** 045017 (*Preprint* 1108.1180)
[52] Larsen K J 2012 *Phys. Rev.* **D86** 085032 (*Preprint* 1205.0297)
[53] Caron-Huot S and Larsen K J 2012 *JHEP* **10** 026 (*Preprint* 1205.0801)
[54] Johansson H, Kosower D A and Larsen K J 2013 *Phys. Rev.* **D87** 025030 (*Preprint* 1208.1754)

[55] Johansson H, Kosower D A and Larsen K J 2014 *Phys. Rev.* **D89** 125010 (*Preprint* 1308.4632)
[56] Johansson H, Kosower D A, Larsen K J and Sgaard M 2015 *Phys. Rev.* **D92** 025015 (*Preprint* 1503.06711)
[57] Gluza J, Kajda K and Kosower D A 2011 *Phys. Rev.* **D83** 045012 (*Preprint* 1009.0472)
[58] Schabinger R M 2012 *JHEP* **01** 077 (*Preprint* 1111.4220)
[59] Ita H 2015 (*Preprint* 1510.05626)
[60] Larsen K J and Zhang Y 2016 *Phys. Rev.* **D93** 041701 (*Preprint* 1511.01071)
[61] Cheung C and O'Connell D 2009 *JHEP* **07** 075 (*Preprint* 0902.0981)
[62] Bern Z, Carrasco J J, Dennen T, Huang Y t and Ita H 2011 *Phys. Rev.* **D83** 085022 (*Preprint* 1010.0494)
[63] Davies S 2011 *Phys. Rev.* **D84** 094016 (*Preprint* 1108.0398)
[64] Bern Z, Carrasco J J M and Johansson H 2008 *Phys. Rev.* **D78** 085011 (*Preprint* 0805.3993)
[65] Bern Z, Carrasco J J M, Dixon L J, Johansson H and Roiban R 2012 *Phys. Rev.* **D85** 105014 (*Preprint* 1201.5366)
[66] Hodges A 2013 *JHEP* **05** 135 (*Preprint* 0905.1473)
[67] Elvang H and Huang Y t 2013 (*Preprint* 1308.1697)
[68] Henn J M and Plefka J C 2014 *Lect. Notes Phys.* **883** 1–195
[69] Dixon L J 1996 *QCD and beyond. Proceedings, Theoretical Advanced Study Institute in Elementary Particle Physics, TASI-95, Boulder, USA, June 4-30, 1995* (*Preprint* hep-ph/9601359) URL http://www-public.slac.stanford.edu/sciDoc/docMeta.aspx?slacPubNumber=SLAC-PUB-7106
[70] Mangano M L and Parke S J 1991 *Phys. Rept.* **200** 301–367 (*Preprint* hep-th/0509223)

ACAT2016 IOP Publishing
Journal of Physics: Conference Series **762** (2016) 012058 doi:10.1088/1742-6596/762/1/012058

Automation of analytical calculations in high energy physics with Redberry CAS

Stanislav Poslavsky

Institute for High Energy Physics NRC "Kurchatov Institute", 142281 Protvino, Russia
Institute for Theoretical and Experimental Physics NRC "Kurchatov Institute", 117218 Moscow, Russia

E-mail: stvlpos@mail.ru

Abstract. Here we present a brief report on recent developments of the `Redberry` computer algebra system through two practical examples: derivation of Feynman rules from arbitrary Lagrangians and computing NLO processes with heavy quarkonia.

1. Introduction

`Redberry` is a high performance computer algebra system [1, 2] which is specifically focused on applications in high energy physics (HEP). The main distinguishing feature of `Redberry` is that it considers all mathematical objects arising in HEP (scalars, spinors, matrices and in general arbitrary tensors) in a uniform way. This means that e.g. indices of tensors are automatically considered as patterns, all algebraic operations and functions are defined for objects with indices and the user need not worry about such things as dummy index clashes, relabelling of indices in substitutions etc. While `Redberry` comes with an extensive set of programming features which makes it easily extendable, out of the box it provides a wide range of functions for performing standard computations in HEP. Through the recent developments of the system, it is worth to mention the improved d-dimensional Dirac algebra for dimensional regularization, features for doing Fourier transform of the Lagrangian, a set of newly introduced transformations and language features and major performance improvements.

In the following two sections we will briefly illustrate the new features of `Redberry` by two practical examples: derivation of Feynman rules from the Lagrangian and computing next-to-leading order (NLO) contributions to processes with heavy quarkonia.

2. Deriving Feynman rules

For the first illustration let's consider an example which illustrates the derivation of Feynman rules for the gravitational field with the following action:

$$\mathcal{L} = \sqrt{-g}\left(-\frac{c}{\kappa^2}\mathcal{R} - b\,\mathcal{R}^2 + a\,\mathcal{R}_{ab}\mathcal{R}^{ab} + \gamma\,\mathcal{R}\phi^2 + g^{ab}\partial_a\phi\partial_b\phi - m^2\phi^2 + \mathcal{L}_{g.f.}\right),$$

where ϕ is a scalar matter field and the standard definition of the Riemann tensor $\mathcal{R}^{\lambda}{}_{\rho\alpha\beta} = \partial_\alpha\Gamma^\lambda{}_{\rho\beta} - \partial_\beta\Gamma^\lambda{}_{\rho\alpha} + \Gamma^\lambda{}_{\alpha\sigma}\Gamma^\sigma{}_{\rho\beta} - \Gamma^\lambda{}_{\beta\sigma}\Gamma^\sigma{}_{\rho\alpha}$ is implied. The first step towards the derivation of Feynman rules is to perform a weak field expansion:

$$g_{ab}(x) = \eta_{ab} + h_{ab}(x), \quad g^{ab}(x) = \eta^{ab} - h^{ab}(x) + h^a{}_c(x)h^{cb}(x) + \ldots, \quad \sqrt{-g} = 1 + h^a_a + \ldots,$$

where η_{ab} is the Minkowski metric and $h_{ab}(x)$ is considered as a quantum perturbation (assumed to be also symmetric). With this definition, the gauge fixing term $\mathcal{L}_{g.f.}$ in the initial Lagrangian is chosen as $1/(2f)(\partial_a h^{ab}(x) - \partial^b h^a_a/2)^2$.

The following boilerplate code inputs all required definitions into `Redberry`:

```
setSymmetric 'h_ab[x_a]'
L = 'det * (Lg + Lf + Li + ...)'.t //total Lagrangian
//Lg - gravity, Lf - matter, Li - interaction
Lg = 'Lg = -c*R/k**2 - b*R**2 + a*R_ab*R_cd*m^ac[x_a]*m^bd[x_a]'.t
Lf = 'Lf = m^ab[x_a]*f~(1)_a[x_a]*f~(1)_b[x_a] - m**2*f[x_a]**2'.t
Li = 'Li  = g*R*f[x_a]**2'.t
//Ri - Riemann, Chr - Christoffel, Ric - Ricci
Ri = 'R^a_bcd = G~(1)^{a}_{db c}[x_a] - ...'.t
Chr = ... ; Ric = ...; R = ...;

//WF - weak field, DT - sqrt(-g)
WF = 'm^ab[x_a] = g^ab - h^ab[x_a] + h^a_d[x_a]*h^db[x_a] + ...'.t
DT = 'det = 1 + h^a_a[x_a] + ...'.t

//subtitute all into the total Lagrangian
L <<= Lg & Lf & Li & R & Ric & Ri & Chr & WF & DT & ...
```

Here in the first line we specify that $h_{ab}(x)$ is a symmetric field; this will be automatically taken into account in all simplifications. The syntax `'...'.t` translates string expression into the computer object. For the original metric field $g_{ab}(x)$ the definition `m_ab[x_a]` is used while for the Minkowski metric there is a built-in definition `g_ab`. Partial derivatives are specified using special syntax `G~(1)^a_dbc[x_a]` which is treated as $\partial_c \Gamma^a_{db}(x)$. Finally, the join operator `&` used in the last line allows to join several substitutions into a single function and `expr<<=func` is a shorthand for the `expr = func(expr)`.

At the next step we need to translate the Lagrangian into momentum space. At this point we can select which terms to include into the expansion: we will keep terms with not higher than cubic interaction (i.e. $\sim hhh$, $\sim hh\phi$, $\sim h\phi\phi$).

```
$Degree = { expr -> Count(expr, 1, ['h_ab[x_a]', 'f[x_a]'].t, true) }
$SelectCubic = { expr ->
    (expr.class == Product && $Degree(expr) > 3) ? 0.t : expr
}

//expand all brackets and drop higher interaction terms
L <<= ExpandAndEliminate[$SelectCubic]
//perform Fourier transform
L <<= LagrangeFourier
```

Here we define a new function `$SelectCubic` which takes `expr` and in the case if it is a product containing more than three field multipliers it returns zero, otherwise it returns `expr` as is. The transformation *ExpandAndEliminate* just expands out all brackets and simplifies contractions with Minkowski metric. Additionally it takes an optional argument – a function (in our case `$SelectCubic`) which will be applied at each intermediate step of the expand procedure; thus all higher interactions will be dropped out at the earliest possible step, which reasonably speeds up the calculation. After applying *LagrangeFourier*, the expression will look like

```
println L

▷ h_{ab}[k_a]*h_{cd}[-k_a]*k^a*k^c*g^bd...
      + h_{ab}[k_a]*f[p_a]*f[-p_a-k_a]*p^a*k^b... + ...
```

In order to find a propagator of e.g. a graviton, we need to select quadratic part of the Lagrangian. This can be easily done using functional derivatives, e.g.:

$$P_{\mathrm{hh}}^{(-1)\,abcd}(p) = \frac{\delta}{\delta h_{ab}(p)}\frac{\delta}{\delta h_{cd}(-p)}\,\mathcal{L}$$

```
$hh = L
$hh <<= Differentiate['h_ab[p_a]', 'h_cd[-p_a]']
$hh <<= 'h_ab[p_a] = 0'.t & 'f[p_a] = 0'.t
$hh <<= ApplyDiracDeltas & 'DiracDelta[0] = 1'.t
//bind inverse propagator
$iP = 'iP^abcd[p_a]'.eq $hh
```

Differentiate will take into account symmetries and function arguments, so e.g.:

$$\frac{\delta\texttt{h\_ab[k\_a]}}{\delta\texttt{h\^{}mn[p\_a]}} = \frac{1}{2}\texttt{*(g\_am*g\_bn + g\_an*g\_bm)*DiracDelta[k\_a, p\_a]}$$

ApplyDiracDeltas just removes delta-functions in an appropriate way:

```
DiracDelta[k_a, p_a]*f[k_a]*... = f[p_a]*...
```

Having the inverse propagator, the propagator is determined from the equation:

$$P_{\mathrm{hh}}^{(-1)}{}_{mn\,ab}\,P_{\mathrm{hh}}^{mn\,cd} = \left(\delta_a^c\delta_b^d + \delta_a^d\delta_b^c\right)/2\,.$$

This tensor equation can be solved in `Redberry` using the *Reduce* function:

```
//specify symmetries of propagator
addSymmetries 'P_abcd[k_a]', [[0,1]].p, [[0,2], [1,3]].p
$eq = 'iP^abcd[k_a]*P_abpq[k_a] = (d^c_p*d^d_q + d^c_q*d^d_p)/2'.t
$eq <<= iP //substitute the inverse propagator in equation

$opts= [Transformations: 'd^n_n = 4'.t,
        ExternalSolver: [Solver: 'Mathematica', Path: '/usr/bin/']]
$hhPropagator = Reduce([$eq], ['P_abcd[k_a]'], $opts)
println $hhPropagator
```

```
▷ P_{abcd}[k_{a}] = 2*(-c+(-6*k**2*b+2*a*k**2)*k_{g}*k^{g}...
```

Our unknown variable in the equation for the propagator is `P_abcd[k_a]`. First we should specify symmetries which this unknown variable should satisfy. Then we call *Reduce* with the list of equations, list of unknown variables and additional options. It will make an ansatz for the unknown variable and reduce the equations to the system of scalar equations. Additional options specify that the produced scalar system should be solved with `Mathematica` (without this *Reduce* will return just a system of scalar equations). Note that the spacetime dimension is controlled everywhere just by substituting `d^n_n = 4`. The graviton propagator finally reads:

$$\begin{aligned}P_{abcd}(k) = {} & \frac{a\kappa^2k^2 + c - 2\kappa^2 f^2}{2f^2k^4(a\kappa^2k^2+c)}\left(g_{ac}k_bk_d + g_{bc}k_ak_d + g_{bd}k_ak_c + g_{ad}k_bk_c\right)\\ & + \frac{(g_{ac}g_{bd} + g_{ad}g_{bc})}{(a\kappa^2k^2+c)\kappa^2k^2} + \frac{\kappa^2(-\kappa^2ak^2 + c + 4b\kappa^2k^2)g_{ab}g_{cd}}{(\kappa^2ak^2+c)(2\kappa^2(a-3b)k^2-c)k^2}\\ & + \frac{2\kappa^4(a-2b)\left(g_{cd}k_ak_b + g_{ab}k_ck_d + 2k_ak_bk_ck_d/k^2\right)}{(a\kappa^2k^2+c)(2\kappa^2(a-3b)k^2-c)k^2}\end{aligned}$$

All other Feynman rules can be found in the same way. For example, take the $h\phi\phi$ vertex:

```
$hff = L
$hff <<= Differentiate['h_ab[k_a]', 'f[p_a]', 'f[q_a]']
$hff <<= 'h_ab[p_a] = 0'.t & 'f[p_a] = 0'.t
$hff <<= ApplyDiracDeltas

//bind vertex
$v3 = 'V3^ab[p_a, q_a, k_a]'.eq $hff
println $v3
```

```
▷ V3^ab[p_a, q_a, k_a] = (4*g-1)*p^{l}*q_{l}*g^{ab}-m**2...
```

which corresponds to

$$V^{ab}(p,q) =$$

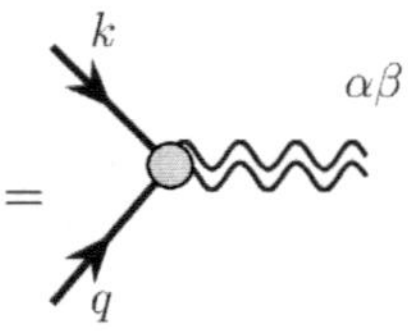

$$= -2g\left(q^a q^b + p^a p^b\right) + (1-2g)\left(q^a p^b + q^b p^a\right) + \\ + \left(2g(q^2+p^2) + (4g-1)q_l p^l - m^2\right) g^{ab}$$

3. NLO processes with heavy quarkonia

Let us now turn to applications in QCD and consider the process $e^+e^- \to \gamma^* \to J/\psi + \eta_c$ at the one-loop level. In total, there are 210 Feynman diagrams which could be generated using e.g. FeynArts[1] [3]. Assuming that we already have the diagrams in some format, let's focus on the part which can be easily done with Redberry.

The basic boilerplate setup for QCD in Redberry looks like:

```
setAntiSymmetric 'e_abcd', 'f_ABC'
setSymmetric 'd_ABC'
defineMatrices 'T_A', Matrix2.matrix, //unitary matrices
    'G_a', 'G5', Matrix1.matrix, //Dirac matrices
    'v[p_a]', Matrix1.vector, Matrix2.vector, //final-state quark
    'ubar[p_a]', Matrix1.covector, Matrix2.covector, //final-state antiquark
    ...
```

The first lines just sets up symmetries for the Levi-Civita tensor and the SU(N) structure constants. Next, we specify rules for non-commuting objects: spinors, SU(N) and Dirac matrices. The definition `Matrix1.matrix` tells nothing but that `G_a` has two additional (invisible) matrix indices etc. In this sense e.g. the line $\bar{u}(p)T_A\gamma_a v(k)$ will be treated internally as

```
ubar[p_a]*T_A*G_a*v[k_a] -> ubar_{a'A'}[p_a]*T^{A'}_{B'A}*G^{a'}_{b'a}*v^{b'B'}[k_a].
```

The particular setup for our process of interest is then:

```
$proj = 'v[pPsi2_a]*ubar[pPsi1_a] =
        (G^a*pPsi_a/2-MC)*G_m*epsPsi^m*(G^b*pPsi_b/2+MC)'.t & ...//onia projectors

$scalars  = setMandelstam([k1_a: 0, k2_a: 0, pPsi_a: '2*MC', pEta_a: '2*MC'])
$qScalars = 'q_a*pPsi^a = qp'.t & 'q_a*pEta^a = qe'.t & 'q_a*q^a = qq'.t

$PaVe  = PassarinoVeltman([2, 1], 'q_a', ['pPsi_a', 'pEta_a'])
$dSimplify = DiracSimplify[[Dimension: 'd', TraceOfOne: 4]]
$dTrace = DiracTrace[[Dimension: 'd', TraceOfOne: 4]]
```

[1] The details of the conversion from FeynArts can be found on the Redberry website.

In the first line the projectors $v(p_1)\bar{u}(p_2) = (\hat{P}/2 - m_c)\,\hat{S}\,(\hat{P}/2 + m_c)$ of charmed quarks on the quantum numbers of the final quarkonia are specified, where $\hat{S} = \gamma_5$ for η_c and $\hat{S} = \hat{\varepsilon}_\psi$ for the J/ψ and P is the total momentum of the quarkonia. Next, the function `setMandelstam` generates a list of replacements for the scalar products of momenta. The variable `$qScalars` holds auxiliary replacements of dot products with loop momenta, while `PassarinoVeltman(i, in, ext)` returns a list of replacements for loop momenta `in` and list of external momenta `ext`, e.g.:

```
println PassarinoVeltman(2, 'q_a', ['pPsi_a', 'pEta_a'])
  ▷ q_a*q_b = (...)*pPsi_a*pEta_b + (...)*g_ab + ...
```

Finally, *DiracSimplify* and *DiracTrace* are used with additional options in order to perform all algebra in dimensional regularization (Larin's scheme [4] for γ_5 is implied automatically).

After these preparations, the main calculation loop can be the following:

```
$amps = __CallFeynArts__().t //import diagrams from FeynArts
$answNLO = 0.t
for($amp in $amps){
    $amp <<= $proj //apply quarkonia projectors
    $amp <<= UnitaryTrace & UnitarySimplify // SU(N) algebra
    $amp <<= $qScalars & $PaVe & ExpandAndEliminate // PaVe reduction
    $amp <<= $dSimplify & $dTrace & LeviCivitaSimplify // dimensional regularization
    $answNLO += Factor($amp) //sum amplitudes
}
```

We just iterate through all diagrams and sequentially do: substitutions of projectors, color algebra, Passarino-Veltman reduction and dimensional regularization. After all, the amplitudes `$answNLO` will be in the form:

```
( ... ) * e_{abcd}*pPsi^b*pEta^c*epsPsi^d + (scalar) * (tensor) + ...,
```

where all same Lorentz structures will be collected automatically and each scalar prefactor is a sum of scalar loop integrals that should be either directly calculated numerically or reduced to a set of master integrals. Presently, the latter step is deferred to external tools. In the case of the above example, it is easy to export processed amplitudes in e.g. `Mathematica` format with

```
new File('amps_nlo.m') << $answNLO.toString(OutputFormat.WolframMathematica)
```

and reduce the scalar integrals with e.g. the `$Apart` package [5] and `FIRE` [6]. The whole reduction of the above 210 diagrams with `Redberry` takes just about two minutes on a standard laptop.

4. Conclusions

In this paper we have illustrated some new important features of `Redberry` system, which facilitates automatization of calculations of many types required in HEP. More examples with many details can be found on the `Redberry` website http://redberry.cc [1].

Acknowledgments

I would like to thank Andrei Kataev and Andrei Onishchenko for stimulating discussions. The work was supported by the RFBR grant #16-32-60017 and by the FAIR-Russia Research Center.

References

[1] Bolotin D A and Poslavsky S V 2013 (*Preprint* 1302.1219) URL http://redberry.cc
[2] Poslavsky S and Bolotin D 2015 *J. Phys. Conf. Ser.* **608** 012060
[3] Hahn T 2001 *Comput. Phys. Commun.* **140** 418–431 (*Preprint* hep-ph/0012260)
[4] Larin S A 1993 *Phys. Lett.* **B303** 113–118 (*Preprint* hep-ph/9302240)
[5] Feng F 2012 *Comput. Phys. Commun.* **183** 2158–2164 (*Preprint* 1204.2314)
[6] Smirnov A V 2008 *JHEP* **10** 107 (*Preprint* 0807.3243)

Journal of Physics: Conference Series **762** (2016) 012059 doi:10.1088/1742-6596/762/1/012059

Automation of NLO processes and decays and POWHEG matching in WHIZARD

Jürgen Reuter[1], Bijan Chokoufé[1], André Hoang[2,3], Wolfgang Kilian[4], Maximilian Stahlhofen[5,1], Thomas Teubner[6], Christian Weiss[1,4]

[1] DESY Theory Group, Notkestr. 85, 22607 Hamburg, Germany
[2] University of Vienna, Faculty of Physics, Boltzmanngasse 5, 1090 Vienna, Austria
[3] Erwin Schrödinger International Institute for Mathematical Physics, University of Vienna, Boltzmanngasse 9, 1090 Vienna, Austria
[4] University of Siegen, Emmy-Noether Campus, Walter-Flex-Str. 3, 57068 Siegen, Germany
[5] PRISMA Cluster of Excellence, Institute of Physics, Johannes Gutenberg University, 55128 Mainz, Germany
[6] University of Liverpool, Department of Mathematical Sciences, Liverpool L69 3BX, U.K.

E-mail: juergen.reuter@desy.de, bijan.chokoufe@desy.de, andre.hoang@univie.ac.at, kilian@physik.uni-siegen.de mastahlh@uni-mainz.de, thomas.teubner@liverpool.ac.uk, christian.weiss@desy.de

Abstract. We give a status report on the automation of next-to-leading order processes within the Monte Carlo event generator WHIZARD, using GoSam and OpenLoops as provider for one-loop matrix elements. To deal with divergences, WHIZARD uses automated FKS subtraction, and the phase space for singular regions is generated automatically. NLO examples for both scattering and decay processes with a focus on e^+e^- processes are shown. Also, first NLO-studies of observables for collisions of polarized leptons beams, e.g. at the ILC, will be presented. Furthermore, the automatic matching of the fixed-order NLO amplitudes with emissions from the parton shower within the Powheg formalism inside WHIZARD will be discussed. We also present results for top pairs at threshold in lepton collisions, including matching between a resummed threshold calculation and fixed-order NLO. This allows the investigation of more exclusive differential observables.

1. The WHIZARD Event Generator

WHIZARD [1] is a multi-purpose event generator for in principle arbitrary processes at hadron and lepton colliders. Being able to treat beam-spectra and initial-state photon radiation, it is especially suited for lepton collider physics studies. Moreover, its built-in scripting language SINDARIN allows for the analysis in the same framework. WHIZARD is written in Fortran2003. Its structure is strictly object-oriented, so that a modular setup enables the convenient interface to numerous other programs. The main sub-components of WHIZARD are O'Mega [2], VAMP [3] and CIRCE [4]: O'Mega provides multi-leg tree-level matrix elements using the helicity formalism. VAMP is used for Monte-Carlo integration and grid sampling. CIRCE creates and evaluates lepton beam spectra. Color factors are evaluated using the color-flow formalism [5]. WHIZARD can be used for event generation on parton level as well as for the subsequent shower simulation. For this purpose, it has its own analytical and k_T-ordered parton shower [6] as well as a built-in interface to Pythia6 [7]. An automated interface to Pythia8 [8] or HERWIG++ [9] is not yet present, but planned. Though WHIZARD spearheaded many BSM phenomenological studies [10–16], these proceedings focus on the automation of SM QCD NLO corrections and non-relativistic top threshold resummation.

2. Automated NLO Calculations in WHIZARD

Next-to-leading order (NLO) calculations have become standard for the prediction of most observables during the last decade. Substantial progress has been made in the automation of loop matrix-element computations as well as in NLO parton shower matching [17, 18]. At the LHC, NLO simulations are routinely employed. Computer programs in this field cover a range from dedicated, single-purpose codes to automated multi-purpose event generators.

With the ILC approaching its possible approval phase, ILC studies become more specific [19], thereby increasing the need for easy to use NLO programs. Here, we want to combine the expertise of WHIZARD in the field of lepton collisions with the improved accuracy of NLO predictions. There have been earlier works on NLO QED extensions to WHIZARD for certain supersymmetric processes [20, 21] as well as on NLO QCD corrections for $pp \to b\bar{b}b\bar{b}$ [22, 23] using Catani-Seymour subtraction. However, the fully generic NLO framework is a recent development.

Automated NLO programs deal with the infrared divergences of real and virtual matrix elements by using subtraction procedures. These rely on the extraction of the soft and collinear limit in its most generic form. This is possible because the $\mathcal{O}(\alpha_s)$-corrections factorize in the corresponding limits. These subtraction terms $d\sigma^{\mathrm{S}}$ are then added and subtracted, such that

$$d\sigma^{\mathrm{NLO}} = d\sigma^{\mathrm{LO}} + \underbrace{\int_{n+1} \left(d\sigma^{\mathrm{R}} - d\sigma^{\mathrm{S}}\right)}_{\text{finite}} + \underbrace{\int_{n+1} d\sigma^{\mathrm{S}} + \int_{n} d\sigma^{\mathrm{V}}}_{\text{finite}} . \qquad (1)$$

The explicit form of $d\sigma^{\mathrm{S}}$ is arbitrary. The two most common schemes employed are Catani-Seymour subtraction [24] and the FKS (Frixione-Kunszt-Signer) scheme [25]. WHIZARD uses the latter. The program automatically finds the singular regions involved in the desired process and computes the subtraction terms to the real matrix element correspondingly.

NLO computations require virtual loop matrix elements. However, O'Mega can only generate tree-level matrix elements. So, to obtain virtual matrix elements, external programs, so-called One-Loop Providers (OLP), can be interfaced to WHIZARD. Up to now, there are working interfaces to GoSam [26] and OpenLoops [27]. They use the Binoth Les Houches Accord (BLHA) [28], which unifies the interface between Monte-Carlo generator and OLPs. The BLHA interface in WHIZARD is generic, meaning that every other matrix element generator which complies to the BLHA standard can be interfaced very easily in the same fashion.

The NLO functionality of WHIZARD has first been presented in ref. [29], where we have focussed our discussion on fixed-order off-shell top production. In the following, we want to present the newest developments in WHIZARD for NLO.

2.1. Treatment of Resonances

A recent development in WHIZARD is the special treatment of radiative corrections occuring from particles which originate from a common resonance. This issue has been extensively discussed in ref. [30]. We have automated the scheme discussed therein in WHIZARD.

Subtraction methods like FKS lose accuracy when after the construction of the real phase space, the resonance momentum deviates significantly from its value in the Born phase space. This affects the propagator both in the real and the Born matrix element, which is part of the subtraction terms. The real matrix element and its approximation therefore do not match perfectly, impairing the convergence of the integration and distorting the invariant mass

Table 1. Comparison of integration histories for the standard and resonance-aware subtraction at $\sqrt{s} = 800\,\text{GeV}$. The Coulomb singularity which arises from $\gamma \to \mu^+\mu^-$ has been cut by imposing $m_{\mu\mu} > 20\,\text{GeV}$. Each iteration uses 12000 calls.

n_{it}	$\sigma[fb]$, standard	Error[%]	$\sigma[fb]$, resonances	Error[%]
1	9.68118	66.30	2.90570	2.87
2	2.85397	8.25	2.85920	1.82
3	2.49076	26.25	2.92779	1.40
4	2.76956	34.91	2.85123	1.40
5	2.43462	19.80	2.88554	1.34
5	2.75391	7.15	2.88420	0.71

Table 2. Real-subtracted integration component and, in the case of resonance-aware sutbraction, soft mismatch, for $\Gamma_H = 1000\,\text{GeV}$ at $\sqrt{s} = 800\,\text{GeV}$.

Method	$\sigma_{\text{soft}}[fb]$	$\sigma_{\text{mism}}[fb]$	n_{calls}
standard	-3.31997 $\pm$ 0.62%	0	5 $\times$ 100000
resonances	-1.62098 $\pm$ 0.29%	-1.70388 $\pm$ 0.56%	5 $\times$ 20000 (soft) + 5 $\times$ 20000 (mismatch)

spectrum in event generation [30].
This problem is solved by employing a phase space mapping that generates the real phase space in such a way that the resonance momentum is conserved. Thus the real matrix element is evaluated at the same point on the resonance's Breit-Wigner curve. However, this approach comes with the drawback of increased complexity in the implementation, since now the calculation has to be aware of the additional singular regions belonging to different resonance histories. Moreover, the so-called soft mismatch (eqn. (3.59) in [30]) enters the calculation as an additional new component of the integration. This is due to the fact that the sum over all singular regions does not reproduce the full real matrix element in this approach, because for each singular region, the FKS mapping is evaluated in the associated resonance's frame of reference. However, usual FKS requires all subtraction terms to be evaluated in the same frame of reference. This property can be restored, yielding the soft mismatch term.

The implementation of the resonance-aware subtraction has been applied to the process $e^+e^- \to b\bar{b}\mu^+\mu^-$, where there is one resonance topology with two associated resonance histories, namely $Z/H \to b\bar{b}$. This process is also of interest in studies of Higgsstrahlung processes at lepton colliders. We have set $m_b = 4.2\,\text{GeV}$, so that collinear divergences do not occur. Without resonance mappings, we observe that the integration does not perform well, due to the very narrow Higgs resonance ($\Gamma_H \sim 4\,\text{MeV}$). Even a relatively soft gluon emission is enough to alter the resonance momentum in such a way that the soft approximation does not fit well enough with the real matrix element. Using the resonance-aware subtraction we obtain significantly better results, as can be clearly seen in Table 1. Taking the large-width limit yields a good integration also for the non-resonant case, as desired. Table 2 shows that both calculations give the same result in this limit. However, the non-resonant approach needs significantly more integration calls to reach the same accuracy as its resonance-aware counterpart. Figure 1 shows a scan of the total cross section. The generation of fixed-order NLO events has been modified for the use

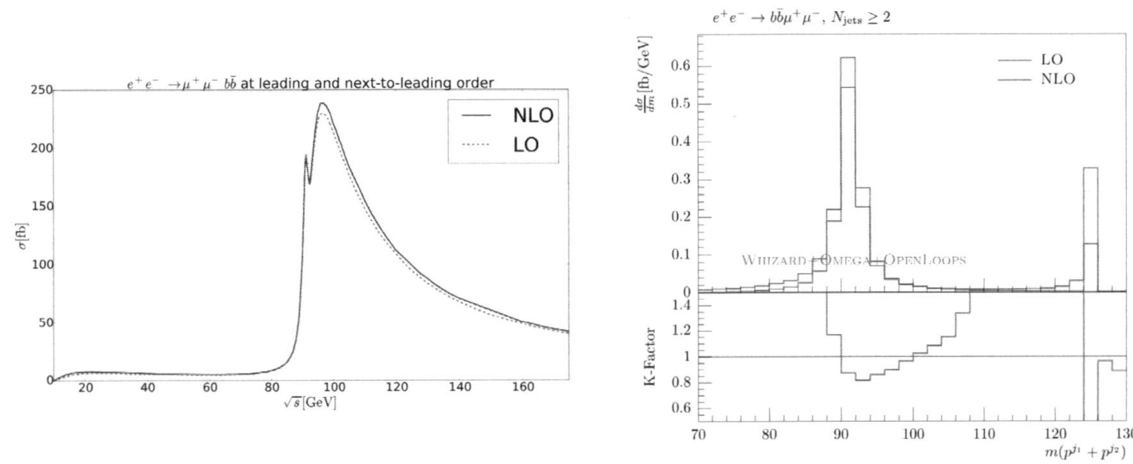

Figure 1. Left: σ_{tot} at the $Z\mu\mu$ and Zbb resonance. Right: Invariant mass distribution of the two hardest jets at $\sqrt{s} = 800\,\text{GeV}$. Jets have been clustered using `FastJet` [31]. The Z and especially the H peak are being washed out by gluon emissions.

with resonance-aware subtraction. In the earlier version of `WHIZARD`, weighted N-particle events were produced with weight $\mathcal{B} + \mathcal{V} + \mathcal{S}$ and associated with $N + 1$-particle events with weight $\mathcal{R}_{\alpha_r}$ for each singular region α_r. The events are saved in `HepMC`-files. If resonances are included, different α_r can yield the same phase space if they only differ by the underlying resonance mass. Therefore, to save disk space, the number of real events now is the number of different possible phase spaces. Their weight is the sum of all $\mathcal{R}_{\alpha_r}$ whose α_r belong to the same phase space. Figure 1 shows an example of NLO event generation in `WHIZARD`.

2.2. Other features

`WHIZARD` is able to produce events according to the POWHEG scheme, which can be used as input to parton showers and preserve the NLO accuracy of the subsequent simulations. This has been discussed in [32] and will not be elaborated further here.

Apart from the major NLO functionalities above, `WHIZARD` also can now deal with several minor aspects of NLO calculations. First, `WHIZARD` can also compute decay widths at next-to-leading order and generate fixed-order events accordingly. The first use case here is the top decay $t \to bW$. The computation takes into account both the initial-state gluon emissions from the top quark and final-state emissions from the bottom quark. The results of this computation play a crucial role in the study of the top threshold described in the next section, where it is important to keep consistency between the input parameters and the top width used.

Second, NLO calculations can be performed using polarized lepton beams. This is a desirable feature for future studies at the ILC or CLIC, which will operate with polarized leptons. For this purpose, `WHIZARD` uses a modified BLHA interface to `OpenLoops`, which allows it to provide the virtual matrix elements individually for each helicity configuration of the colliding lepton pair.

3. Top Threshold Resummation and Matching to the Continuum

High-energy lepton colliders operating at the top threshold or close to it allow for the most precise method to measure the mass of the top quark known to date. However, at threshold the top quark pair is non-relativistic and can thus form a toponium quasi-bound state, due to the attractive top quark potential. This potential relies on the exchange of an arbitrary number

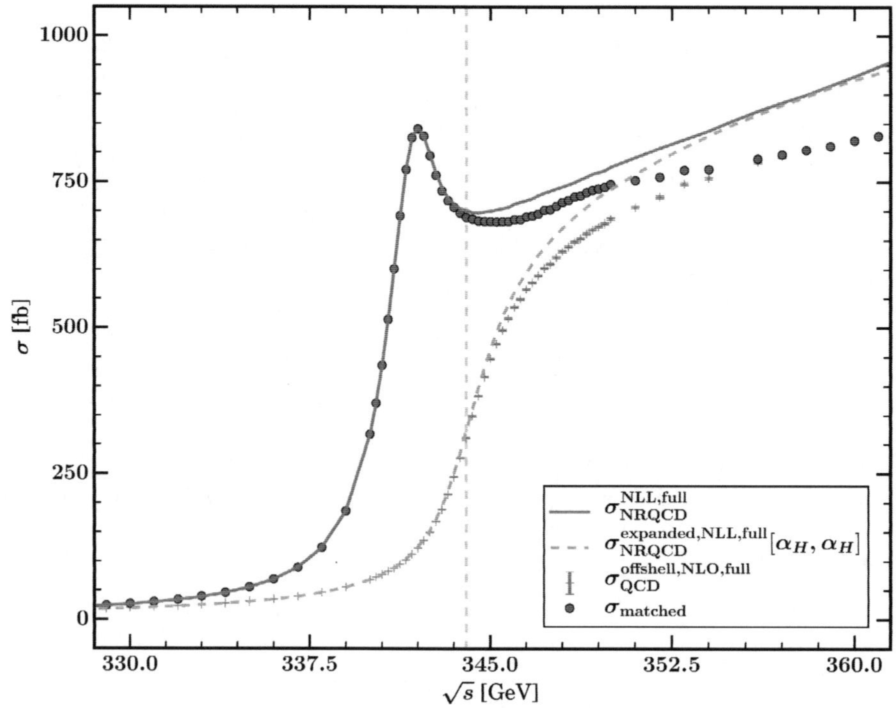

Figure 2. Matching the NLL resummed threshold prediction to the fixed-order NLO QCD continuum for the total cross section of the process $e^+e^- \to W^+bW^-b$. The blue dots show the result obtained by the (preliminary) matching prescription between vNRQCD and the continuum. The solid green line corresponds to the insertion of the NLL form factor from `TOPPIK` into the Born process. The dashed orange curve shows the same result with the form factor expanded to first order in α_s. The red crosses represent the full relativistic fixed-order NLO result.

of virtual gluons between the tops. The bound-state effects give rise to Coulomb singularities $(\alpha_s/v)^n$ and large logarithms $\ln^n v$ in the perturbative expansion of the cross section. Here, $v \sim \alpha_s \sim 0.1$ is the relative velocity of the non-relativistic top quarks. The Resummation of the threshold enhanced terms can be carried out in the vNRQCD framework [33–37]. The R-ration then takes the schematic form [38]

$$R = \frac{\sigma_{t\bar{t}}}{\sigma_{\mu\mu}} = v \sum_k \left(\frac{\alpha_s}{v}\right)^k \sum_l (\alpha_s \ln v)^l \times \left\{ \underbrace{1(\text{LL}); \alpha_s, v(\text{NLL});}_{\text{effective ttV vertex}} \alpha_s^2, \alpha_s v, v^2(\text{NNLL}); \ldots \right\}. \tag{2}$$

In **WHIZARD**, the resummation is achieved by replacing the $tt\gamma$ and ttZ vertices by non-relativistic form factors at NLL accuracy using the external code **TOPPIK** [39]. This interface was first discussed in [40]. A recent development is the matching of the vNRQCD-approximation to the relavistic continuum, complementing the accuracy of the two approximations. The matching between the resummed prediction at threshold and the relativistic NLO result, which is the reliable result for $\sqrt{s} \gg 2m_t$, is based on two main concepts. First, we have to subtract the first order expansion of the resummed computation when we want to add the NLO contributions. This has to be done in a way that respects the relevant scales of the vNRQCD calculation (hard

m_t, soft $m_t v$ and ultra-soft $m_t v^2$). In Figure 2 we can see how this expansion evaluated at the hard scale ($\alpha_H = \alpha_s m_t$), shown as orange dashes, reproduces the full NLO result, represented by red crosses, as it contains the dominant terms close to threshold. On the other hand, we face the problem that the resummed prediction keeps growing arbitrarily with $\sqrt{s}$, which is an artifact of the assumption that the computation is performed close to threshold and is seen by the rise of the green curve in Fig. 2. This is cured by multiplying the relevant scales with a switch-off function that smoothly approaches zero as one moves away from threshold. Combining these concepts gives a nice physical prediction in form of the blue dots. Note that while we have concentrated on the inclusive cross section here, the implementation in `WHIZARD` will eventually allow for the first time for abritrary differential distributions with NLO+NLL precision together with their uncertainties based on scale variation.

Acknowledgements
We want to thank Jonas Lindert for many useful discussions and especially the fast `OpenLoops` support. We also thank Tomáš Ježo for his advice about the resonance-aware FKS subtraction. JRR wants to thank the ACAT 2016 organizers for the beautiful venue.

References

[1] Kilian W, Ohl T and Reuter J 2011 *Eur. Phys. J. C* **71** 1742 (*Preprint* 0708.4233)
[2] Moretti M, Ohl T and Reuter J 2001 29 (*Preprint* hep-ph/0102195)
[3] Ohl T 1999 *Comput.Phys.Commun.* **120** 13–19 (*Preprint* hep-ph/9806432)
[4] Ohl T 1997 *Comput.Phys.Commun.* **101** 269–288 (*Preprint* hep-ph/9607454)
[5] Kilian W, Ohl T, Reuter J and Speckner C 2012 *JHEP* **2012** 22 (*Preprint* 1206.3700)
[6] Kilian W, Reuter J, Schmidt S and Wiesler D 2012 *JHEP* **2012** 13 (*Preprint* 1112.1039)
[7] Sjöstrand T, Mrenna S and Skands P 2006 *JHEP* **2006** (*Preprint* hep-ph/0603175)
[8] Sjöstrand T, Ask S, Christiansen J R, Corke R *et al.* 2015 *Comput. Phys. Commun.* **191** 159–177 (*Preprint* 1410.3012)
[9] Bähr M, Gieseke S, Gigg M A, Grellscheid D *et al.* 2008 *Eur. Phys. J. C* **58** 639–707 (*Preprint* 0803.0883)
[10] Kilian W, Rainwater D and Reuter J 2005 *Physical Review D* **71** 015008 (*Preprint* hep-ph/0411213)
[11] Hagiwara K, Kilian W, Krauss F, Ohl T *et al.* 2006 *Physical Review D* **73** 055005 (*Preprint* hep-ph/0512260)
[12] Beyer M, Kilian W, Krstonošić P, Mönig K *et al.* 2006 *Eur. Phys. J. C* **48** 353–388 (*Preprint* hep-ph/0604048)
[13] Alboteanu A, Kilian W and Reuter J 2008 *JHEP* **2008** 010 (*Preprint* 0806.4145)
[14] Kalinowski J, Kilian W, Reuter J, Robens T and Rolbiecki K 2008 *JHEP* **2008** 090 (*Preprint* 0809.3997)
[15] Kilian W, Sekulla M, Ohl T and Reuter J 2015 *Physical Review D* **91** 096007 (*Preprint* 1408.6207)
[16] Kilian W, Ohl T, Reuter J and Sekulla M 2015 (*Preprint* 1511.00022)
[17] Nason P 2004 *JHEP* **2004** 040 (*Preprint* hep-ph/0409146)
[18] Frixione S, Nason P and Oleari C 2007 *JHEP* **2007** 70 (*Preprint* 0709.2092)
[19] Baer H, Barklow T, Fujii K, Gao Y, Hoang A *et al.* 2013 *ILC-REPORT-2013* (*Preprint* 1306.6352)
[20] Kilian W, Reuter J and Robens T 2006 *Eur. Phys. J. C* **48** 389–400 (*Preprint* hep-ph/0607127)
[21] Kalinowski J, Kilian W, Reuter J, Robens T and Rolbiecki K 2008 *Acta Phys. Polon. B* **39** 1705–1714 (*Preprint* 0803.4161)
[22] Binoth T, Greiner N, Guffanti A, Reuter J *et al.* 2010 *Physics Letters B* **685** 293–296 (*Preprint* 0910.4379)
[23] Greiner N, Guffanti A, Reiter T and Reuter J 2011 *Physical review letters* **107** 102002 (*Preprint* 1105.3624)
[24] Catani S and Seymour M 1997 *Nuclear Physics B* **485** 291–419 (*Preprint* hep-ph/9605323)
[25] Frederix R, Frixione S, Maltoni F and Stelzer T 2009 *JHEP* **2009** 003 (*Preprint* 0908.4272)
[26] Cullen G, Greiner N, Heinrich G, Luisoni G *et al.* 2012 *Eur. Phys. J. C* **72** 1889 (*Preprint* 1111.2034)
[27] Cascioli F, Maierhöfer P and Pozzorini S 2012 *Physical review letters* **108** 1–5 (*Preprint* 1111.5206)
[28] Alioli S, Badger S, Bellm J, Biedermann *et al.* 2014 *Comput. Phys. Commun.* **185** 560–571 (*Preprint* 1308.3462)
[29] Weiss C, Chokoufé B, Kilian W and Reuter J 2015 *PoS* **EPS-HEP2015** 466 (*Preprint* 1510.02666)
[30] Ježo T and Nason P 2015 *JHEP* **2015** 1–47 (*Preprint* 1509.09071)
[31] Cacciari M, Salam G P and Soyez G 2012 *Eur. Phys. J. C* **72** 1896 (*Preprint* 1111.6097)
[32] Chokoufé B, Kilian W, Reuter J and Weiss C 2015 *PoS* **EPS-HEP2015** 317 (*Preprint* 1510.02739)
[33] Luke M E, Manohar A V and Rothstein I Z 2000 *Physical Review D* **61** 074025 (*Preprint* hep-ph/9910209)
[34] Pineda A 2002 *Physical Review D* **66** 054022 (*Preprint* hep-ph/0110216)
[35] Hoang A H and Ruiz-Femenía P 2006 *Physical Review D* **74** 114016 (*Preprint* hep-ph/0609151)

[36] Hoang A H, Reißer C J and Ruiz-Femenía P 2010 *Physical Review D* **82** 014005 (*Preprint* 1002.3223)
[37] Hoang A H and Stahlhofen M 2011 *JHEP* **2011** 88 (*Preprint* 1102.0269)
[38] Hoang A H and Stahlhofen M 2014 *JHEP* **2014** 121 (*Preprint* 1309.6323)
[39] Hoang A H and Teubner T 1999 *Physical Review D* **60** 114027 (*Preprint* hep-ph/9904468)
[40] Bach F and Stahlhofen M 2014 (*Preprint* 1411.7318)

Calculating four-loop massless propagators with Forcer

T Ueda[1], B Ruijl[1,2] and J A M Vermaseren[1]

[1] Nikhef Theory Group, Science Park 105, 1098 XG Amsterdam, The Netherlands
[2] Leiden University, Niels Bohrweg 1, 2333 CA Leiden, The Netherlands

E-mail: `tueda@nikhef.nl`, `benrl@nikhef.nl`, `t68@nikhef.nl`

Abstract. We present `Forcer`, a new `FORM` program for the calculation of four-loop massless propagators. The basic framework is similar to that of the `Mincer` program for three-loop massless propagators: the program reduces Feynman integrals to a set of master integrals in a parametric way. To overcome an ineludible complexity of the program structure at the four-loop level, most of the code was automatically generated or made with computer-assisted derivations. Correctness of the program has been checked with the recomputation of some quantities in the literature.

1. Introduction

Calculating massless propagator-type Feynman integrals is one of the basic building blocks in higher-order perturbative calculations in QCD (see Ref. [1] for a recent review). Up to the three-loop level, it turned out that dimensionally regularized [2, 3] massless propagator-type Feynman integrals can be reduced by the so-called triangle rule [4, 5, 6] obtained from integration-by-parts (IBP) identities [4, 5] and successive one-loop integrations with G-functions [7, 5] except for two special topologies, which are again reduced by manually solving IBP identities leading to two master integrals [5]. Together with the results of the master integrals [7], this led to a program `Mincer` [8], implemented in `SCHOONSCHIP` [9] and later reprogrammed [10] in `FORM` [11]. `Mincer` can analytically evaluate arbitrary scalar massless propagator-type Feynman integrals up to the three-loop level as Laurent series expansions in ϵ, where $D = 4 - 2\epsilon$ is the number of space-time dimensions.

On the other hand, at the four-loop level, people have used more systematic or generic ways to perform the IBP reductions: Laporta's algorithm [12] (for public implementations see [13, 14, 15, 16, 17, 18]), Baikov's method [19, 20] and heuristic approaches [21, 22]. The values of the master integrals at the four-loop level are known up to enough orders for practical applications [23, 24]. Then one might think that the `Mincer` program could be extended to the four-loop level, giving specialized (and hopefully optimized) routines for massless propagator-type integrals, and this would be more efficient than general-purpose IBP solvers. This approach has not been pursued, however, to the best of our knowledge. A reason may be that there are too many topologies at the four-loop level, and therefore it is impractical to program treatment of all topologies by hand as it was done for `Mincer`.

The aim of this work is to develop a new `FORM` program `Forcer` [25], which can be considered as an extension of `Mincer` to the four-loop level. This requires automatized code generation because

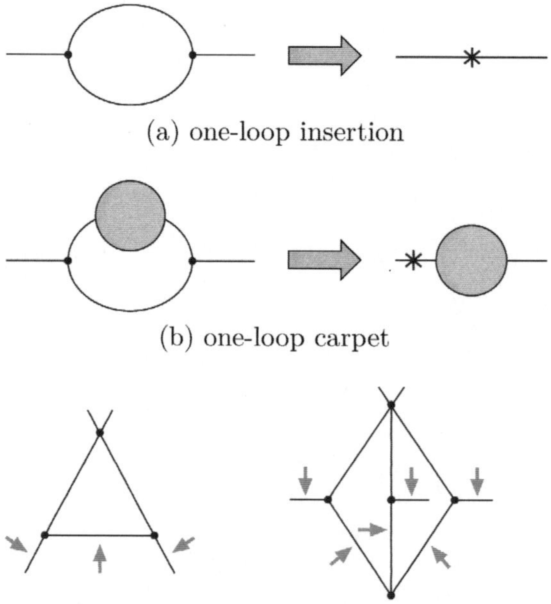

(a) one-loop insertion

(b) one-loop carpet

(c) triangle (d) diamond

Figure 1. Substructures of topologies that lead to reduction to simpler topologies. (a) One-loop insertion. A line with a non-integer power is indicated with "$*$". (b) One-loop carpet. (c) Triangle. The triangle rule removes one of the three lines indicated by arrows. (d) Two-loop diamond. The diamond rule removes one of the six lines indicated by arrows.

the large number of topologies makes the program structure too complicated to program by hand. Such a program is promising when one considers extremely time-consuming computations with existing approaches, for example, higher moments of four-loop splitting functions.

2. Reduction for each topology

Here we briefly summarize how one can reduce integrals belonging to a topology into those in simpler topologies from a diagrammatical point of view. Techniques of 2.1, 2.2 and 2.3 were used in the three-loop `Mincer` program. The technique of 2.4 is new at the four-loop level.

2.1. One-loop insertion integral

When a topology contains a massless one-loop insertion graph as its substructure and both propagators are massless, one can integrate out the corresponding loop momentum and it becomes a line but gets a non-integer power $1/(Q^2)^\epsilon$ in the resultant momentum Q (figure 1a). The powers of the two propagators can be arbitrary numbers. It is also allowed to have polynomic tensorial numerators with respect to the loop momentum.

2.2. One-loop carpet integral

This is another type of massless one-loop integrals that we can perform. When a subgraph is inserted to a one-loop graph, the outer loop can be integrated out first (figure 1b). Similar to the one-loop insertion integrals, the powers of the two propagators of the outer loop can be arbitrary numbers. Numerators are also allowed.

2.3. Triangle rule

When a topology contains a one-loop triangle graph depicted in figure 1c as a subgraph, recursive usage of an IBP identity leads to a sum of integrals corresponding to simpler topologies with one of the three lines indicated by arrows removed. This is possible when the powers of the three propagators are integer; hence this rule cannot be applied if any of the three lines has taken non-integer powers from one-loop integrals explained above. Irreducible numerators in the topology can always be chosen such that they do not interfere with the rule.

2.4. Diamond rule

The triangle rule can be extended for multi-loop diamond-shaped subgraphs [26]. Figure 1d shows the simplest diamond structure, in which one of six lines is removed in the end of the recursion.

2.5. Manual rules

If none of the above substructures is available in a topology, one needs to manually solve IBP identities such that recursive use of rules removes one of the propagators, or at least simplify integrals as much as possible. In the latter case, irreducible integrals are considered as master integrals of the problem. This part has not been fully automatized yet, though we used computer-assisted derivations of reduction rules.

Starting from a set of IBP identities obtained in a normal way, one can generate sets of identities by raising or lowering one of (or combinations of) powers of propagators and irreducible numerators in all possible ways. They are combined into a new set of identities, and then complicated integrals are eliminated by means of Gaussian elimination. Powers of the propagators remain parametric. We construct an reduction scheme out of the resulted identities.

3. Code generation for all topologies

In order to handle many topologies and transitions among them at the four-loop level, we need to automatize code generation for them.

First, we represent a topology as an undirected graph in graph theory. This makes it easy to detect the substructures explained in the previous section by pattern matchings of connections of vertices and edges. We implemented such routines in `Python` with a graph library `igraph` [27].

Then, starting from the top-level topologies, we consider to remove a line from each topology in all possible ways. If removing a line gives a massless tadpole, then the generated topology is immediately discarded because it is zero. Some of the generated topologies are actually identical and such graph isomorphisms are efficiently detected by the graph library. We keep track of all mappings of momenta between topologies before and after the transition. They are needed to rewrite propagators and irreducible numerators in a topology into those in another topology.

For each topology, the next action is determined from its substructures. Irreducible numerators are chosen such that they do not interfere with the next action. An adequate subroutine must be called for the next action in generated code. Symmetries in each topology are detected as graph automorphisms, which are helpful to reduce the number of terms we need to process in actual calculations.

Repeating this procedure until all topologies are reduced into a trivial Born graph, we obtain a tree of all possible topologies (figure 2). From 11 top-level topologies at the four-loop level, we obtained 437 non-trivial topologies in total. The numbers of topologies that have one-loop insertions, one-loop carpets, triangles and diamonds as their substructures are 335, 24, 53 and 4, respectively, and 21 topologies need construction of manual rules. From the topology tree, we generated `FORM` code for reductions in all topologies, and rewriting propagators and irreducible numerators at all topology transitions.

Journal of Physics: Conference Series **762** (2016) 012060 doi:10.1088/1742-6596/762/1/012060

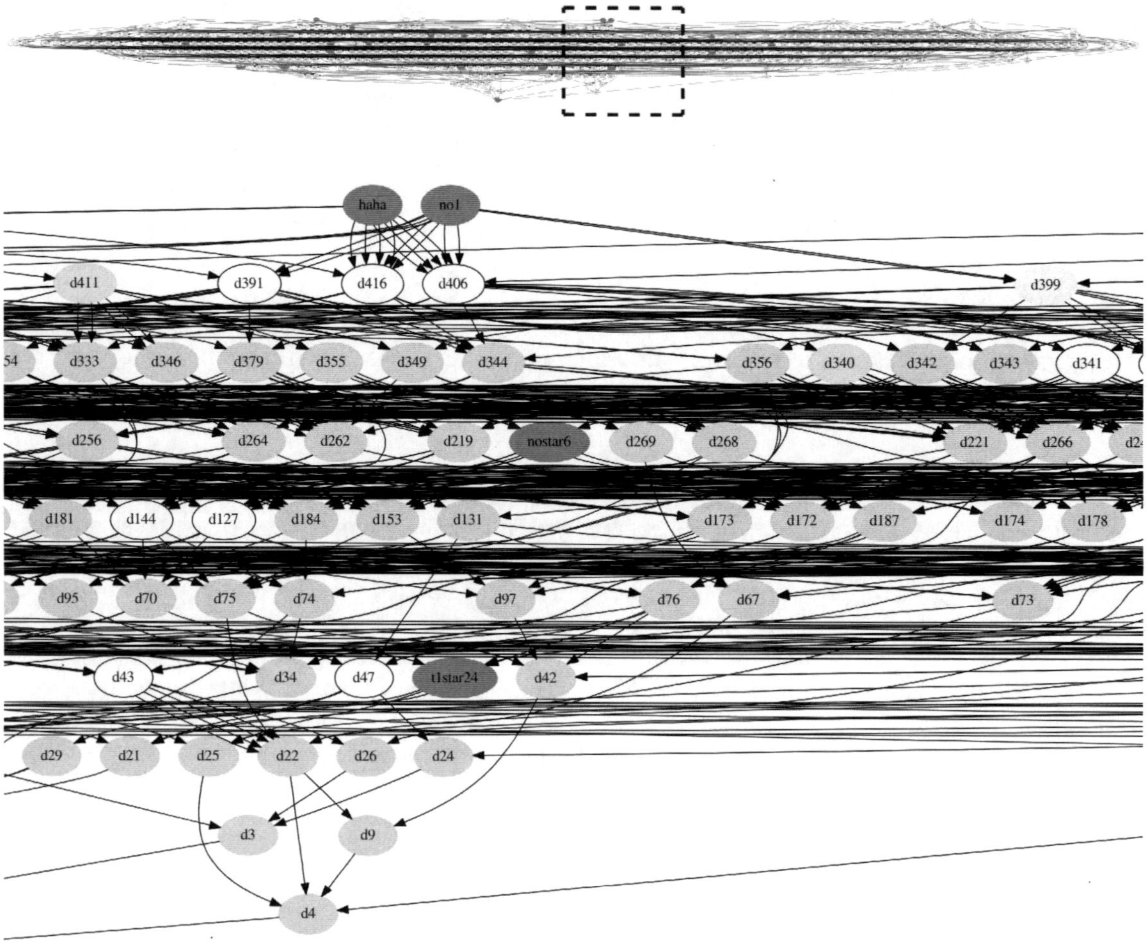

Figure 2. The tree-like graph structure of topology transitions at the four-loop level. The lower figure shows the enlarged view of the boxed region in the upper one. Green, cyan, white, yellow and red circles represent topologies in which one-loop insertion integral, carpet integral, triangle rule, diamond rule and none of them are available, respectively.

4. Conclusion

We have constructed a `FORM` program `Forcer`, which analytically computes massless propagator-type Feynman integrals up to the four-loop level. The complicated program structure made us opt to go for automatization of code generation. The problem that there are many topologies at this level was solved by representing all topology transitions as a topology tree and generate code in an automatized way. It is difficult to automatize deriving manual rules for topologies where none of known substructures is available. However, we observed that it is helpful to employ computer-assisted derivations and especially Gaussian elimination with a ordering of integrals by their complexities, even for parametric reduction rules.

In order to check correctness of the program, we performed recomputations of several known results in the literature. We recomputed the four-loop QCD β-function [28, 29] and checked the gauge-invariance of the result. Moreover, we computed the same quantity in the background field method and obtained the same result. Another set of recomputations were done for four-loop anomalous dimensions of fixed N-moments of the non-singlet twist-2 operator. We reproduced $N = 2$ [30, 31], $N = 3$ and $N = 4$ results [32, 33]. More physics results obtained by `Forcer` will be reported elsewhere [34].

Acknowledgments
We would like to thank Andreas Vogt for discussions and collaboration for physics applications. This work is supported by the ERC Advanced Grant no. 320651, "HEPGAME".

References

[1] Baikov P A, Chetyrkin K G and Kühn J H 2015 *Nucl. Part. Phys. Proc.* **261-262** 3–18 (*Preprint* `1501.06739`)
[2] Bollini C G and Giambiagi J J 1972 *Nuovo Cim.* **B12** 20–6
[3] 't Hooft G and Veltman M J G 1972 *Nucl. Phys.* **B44** 189–213
[4] Tkachov F V 1981 *Phys. Lett.* **B100** 65–8
[5] Chetyrkin K G and Tkachov F V 1981 *Nucl. Phys.* **B192** 159–204
[6] Tkachov F V 1983 *Theor. Math. Phys.* **56** 866–70
[7] Chetyrkin K G, Kataev A L and Tkachov F V 1980 *Nucl. Phys.* **B174** 345–77
[8] Gorishny S G, Larin S A, Surguladze L R and Tkachov F V 1989 *Comput. Phys. Commun.* **55** 381–408
[9] Strubbe H 1974 *Comput. Phys. Commun.* **8** 1–30
[10] Larin S A, Tkachov F V and Vermaseren J A M 1991 *Preprint* NIKHEF-H-91-18
[11] Kuipers J, Ueda T, Vermaseren J A M and Vollinga J 2013 *Comput. Phys. Commun.* **184** 1453–67 (*Preprint* `1203.6543`)
[12] Laporta S 2000 *Int. J. Mod. Phys.* **A15** 5087–159 (*Preprint* `hep-ph/0102033`)
[13] Anastasiou C and Lazopoulos A 2004 *JHEP* **07** 046 (*Preprint* `hep-ph/0404258`)
[14] Smirnov A V 2008 *JHEP* **10** 107 (*Preprint* `0807.3243`)
[15] Smirnov A V and Smirnov V A 2013 *Comput. Phys. Commun.* **184** 2820–7 (*Preprint* `1302.5885`)
[16] Smirnov A V 2014 *Comput. Phys. Commun.* **189** 182–91 (*Preprint* `1408.2372`)
[17] Studerus C 2010 *Comput. Phys. Commun.* **181** 1293–300 (*Preprint* `0912.2546`)
[18] von Manteuffel A and Studerus C 2012 *Preprint* `1201.4330`
[19] Baikov P A 1996 *Phys. Lett.* **B385** 404–10 (*Preprint* `hep-ph/9603267`)
[20] Baikov P A 2006 *Phys. Lett.* **B634** 325–9 (*Preprint* `hep-ph/0507053`)
[21] Lee R N 2012 *Preprint* `1212.2685`
[22] Lee R N 2014 *J. Phys. Conf. Ser.* **523** 012059 (*Preprint* `1310.1145`)
[23] Baikov P A and Chetyrkin K G 2010 *Nucl. Phys.* **B837** 186–220 (*Preprint* `1004.1153`)
[24] Lee R N, Smirnov A V and Smirnov V A 2012 *Nucl. Phys.* **B856** 95–110 (*Preprint* `1108.0732`)
[25] Ruijl B, Ueda T and Vermaseren J A M *in preparation*
[26] Ruijl B, Ueda T and Vermaseren J 2015 *Phys. Lett.* **B746** 347–50 (*Preprint* `1504.08258`)
[27] Csardi G and Nepusz T 2006 *InterJournal* **Complex Systems** 1695 `http://igraph.org`
[28] van Ritbergen T, Vermaseren J A M and Larin S A 1997 *Phys. Lett.* **B400** 379–84 (*Preprint* `hep-ph/9701390`)
[29] Czakon M 2005 *Nucl. Phys.* **B710** 485–98 (*Preprint* `hep-ph/0411261`)
[30] Baikov P A and Chetyrkin K G 2006 *Nucl. Phys. Proc. Suppl.* **160** 76–9
[31] Velizhanin V N 2012 *Nucl. Phys.* **B860** 288–94 (*Preprint* `1112.3954`)
[32] Velizhanin V N 2014 *Preprint* `1411.1331`
[33] Baikov P A, Chetyrkin K G and Kühn J H *results up to $N = 4$ in QCD were presented on the 19th Meeting of SFB/TR9 "Computational Particle Physics", 19 March 2013, Aachen*
[34] Ruijl B, Ueda T, Vermaseren J A M and Vogt A *in preparation*

Journal of Physics: Conference Series **762** (2016) 012061 doi:10.1088/1742-6596/762/1/012061

The Mathematica package TopoID and its application to the Higgs boson production cross section

Jens Hoff

Deutsches Elektronen-Synchrotron (DESY), Platanenallee 6, D-15738 Zeuthen, DE

E-mail: jens.hoff@desy.de

Abstract. We present the Mathematica package TopoID which aims at the automation of several steps in multiloop calculations. The algorithm which lies at the very core of the package is described and illustrated with an example. The main features of TopoID are stated and some of them are briefly demonstrated for NLO or NNLO Higgs boson production.

1. Introduction

With the LHC running in the coming years at 14 TeV, the experimental precision for observables will further improve. Likewise, the theoretical uncertainty needs to be decreased which means higher orders in the perturbation series need to be computed. Some quantities under consideration presently are for example the Higgs boson production cross section at N^3LO [1, 2], the four-loop $\overline{\text{MS}}$-OS relation [3], the four-loop cusp anomalous dimension [4] or the five-loop β-function [5]. All these calculations are based on the evaluation of Feynman diagrams which usually suffers from a factorial growth of complexity with higher orders.

For most of the necessary steps in such a calculation various program packages are already available publically. Color, Lorentz and Dirac algebra (tensorial structures) can be handled, e.g., with FORM [6], FeynCalc [7] or HEPMath [8]. The remaining scalar integrals can then be classified in so-called Feynman integral families or "topologies". TopoID [9, 10] is a package dedicated especially to this task (and others). The integrals within each topology are subject to integration-by-parts identities (IBPs) [11] which allow for a systematic reduction to "master integrals". That is, a large set of integrals can be expressed as linear combinations of only relatively few basis integrals. Public reduction programs include MINCER [12] for three-loop massless propagtors, MATAD [13] for three-loop massive tadpoles or AIR [14], FIRE [15] and Reduze [16] implementing Laporta's approach [17] for generic topologies.

TopoID aims at taming the diagrammatic or "topologic" complexity that arises in a calculation. The complexity involved in algebraic operations on the coefficients of integrals in reduction relations or in the function classes for master integrals, however, is a different and still open subject.

1.1. Example: Higgs boson production at NLO

Let us illustrate the concepts of topology and integral reduction by the single topology needed for the NLO corrections to Higgs boson production due to real radiation:

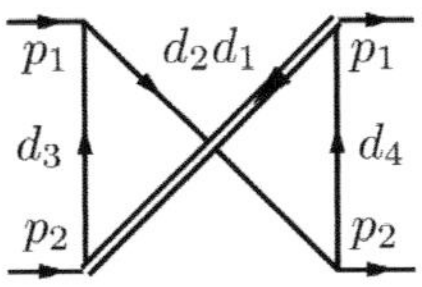

$$= \int \mathrm{d}k_1{}^D \, \frac{1}{d_1^{a_1} d_2^{a_2} d_3^{a_3} d_4^{a_4}} = T(a_1, a_2, a_3, a_4) \quad \text{with} \quad \begin{array}{l} d_1 = m_H^2 + k_1^2, \\ d_2 = (p_1 + p_2 + k_1)^2, \\ d_3 = (p_2 + k_1)^2, \\ d_4 = (p_1 + k_1)^2. \end{array} \tag{1}$$

The non-planar box topology T in Eq. (1) consists of four scalar propagators $1/d_1, \ldots, 1/d_4$ raised to generic integer powers $a_1, \ldots, a_4$ ($a_i = 0$ means contraction of line i). Note, this topology is defined in forward scattering kinematics to faciliate the use of the optical theorem (for a detailed discussion see Ref. [10]). The double line indicates the Higgs boson, arrows show the momentum flow. Due to the linear dependence of $d_1, \ldots, d_4$ (with respect to the scalar products $p_1 \cdot k_1$, $p_2 \cdot k_1$ and k_1^2) and since integrals with d_1 or d_2 eliminated do not contribute via the optical theorem, partial fractioning allows to eliminate either d_3 or d_4 in all appearing expressions. This leaves us with two symmetric triangle topologies $T(a_1, a_2, a_3, 0) = T(a_1, a_2, 0, a_3)$. Gauss' theorem in D dimensions gives in this case rise to three IBPs, one for the contraction with either the loop momentum k_1 or one of the external momenta p_1 and p_2:

$$0 = \int \mathrm{d}k_1{}^D \, \frac{\partial}{\partial k_1^\mu} \left\{ k_1^\mu, p_1^\mu, p_2^\mu \right\} \frac{1}{d_1^{a_1} d_2^{a_2} d_3^{a_3}}. \tag{2}$$

The relations obtained from Eq. (2) by fixing $a_1, \ldots, a_3$ to integer values then involve integrals with differing denominator powers. Generating a set of such relations for all the needed integrals of a family gives a linear system which can be solved in terms of master integrals. In this case only a single master integral (a bubble with $a_1 = a_2 = 1$, $a_3 = a_4 = 0$) emerges in the end.

2. Canonical ordering

The most important capability of `TopoID` is the fast and efficient **ID**entification of isomorphic **Topo**logies. This is essential since the routing of loop momenta and the order of propagators in the definition of a topology are ambiguous. The basic idea is to use the sum $\mathcal{U} + \mathcal{F}$ of the Symanzik polynomials appearing in the Feynman representation, which is independent of loop momenta, as an identifier. Therefore, the Feynman parameters $\{\alpha_j\}$ (corresponding to scalar propagators) are permuted in a unique algorithmic way. See also Ref. [18].

The task of the algorithm can be stated without reference to Feynman integrals: bring the polynomial P with m terms into a unique form $\hat{P}$ by renaming the n variables $\{x_j\}$. This can be achieved by the following steps:

(i) Convert P into a $m \times (n+1)$ matrix $M^{(0)}$. Each row corresponds to a term, the first column to constant coefficients and the remaining columns to powers of the $\{x_j\}$ in the monomial.

(ii) Start with the above $M^{(0)}$ and in the second column ($k = 1$).

(iii) Compute for all considered matrices $M^{(k),\sigma}$ all transpositions of columns k and $k+1, \ldots, n$ where the index σ collects all applied permutations.

(iv) Sort rows in each matrix lexicographically by the first k columns.

(v) Extract the lexicographically largest vector from columns k of all matrices.

(vi) Keep only matrices with this maximal vector. If $k < n-1$, increment k and go to step (iii).

(vii) Each remaining matrix encodes the same unique $\hat{P}_\sigma$ and a corresponding permutation of variables σ.

Journal of Physics: Conference Series **762** (2016) 012061 doi:10.1088/1742-6596/762/1/012061

2.1. Example: a simple polynomial in two variables

We demonstrate the above algorithm in a simple example where equation labels indicate the respective step and iteration:

$$P = x_1^2 + 2x_1x_2 + x_2^2 + x_3^2 \;\to\; M^{(0)} = \begin{pmatrix} 1 & 2 & 0 & 0 \\ 2 & 1 & 1 & 0 \\ 1 & 0 & 2 & 0 \\ 1 & 0 & 0 & 2 \end{pmatrix}. \tag{i}$$

$$S^{(1)} = \left\{ M^{(0)(123)} = M^{(0)} \right\}, \; k = 1. \tag{ii}$$

$$S'^{(1)} : M'^{(1)(123)} = \left(\begin{array}{cc|cc} 1 & 2 & 0 & 0 \\ 2 & 1 & 1 & 0 \\ 1 & 0 & 2 & 0 \\ 1 & 0 & 0 & 2 \end{array}\right), \; M'^{(1)(213)} = \left(\begin{array}{cc|cc} 1 & 0 & 2 & 0 \\ 2 & 1 & 1 & 0 \\ 1 & 2 & 0 & 0 \\ 1 & 0 & 0 & 2 \end{array}\right), \; M'^{(1)(321)} = \left(\begin{array}{cc|cc} 1 & 0 & 0 & 2 \\ 2 & 0 & 1 & 1 \\ 1 & 0 & 2 & 0 \\ 1 & 2 & 0 & 0 \end{array}\right). \tag{iii-1}$$

$$S''^{(1)} : M''^{(1)(123)} = \begin{pmatrix} 1 & \mathbf{0} & 0 & 2 \\ 1 & \mathbf{0} & 2 & 0 \\ 1 & \mathbf{2} & 0 & 0 \\ 2 & \mathbf{1} & 1 & 0 \end{pmatrix}, \; M''^{(1)(213)} = \begin{pmatrix} 1 & \mathbf{0} & 2 & 0 \\ 1 & \mathbf{0} & 0 & 2 \\ 1 & \mathbf{2} & 0 & 0 \\ 2 & \mathbf{1} & 1 & 0 \end{pmatrix}, \; M''^{(1)(321)} = \begin{pmatrix} 1 & 0 & 0 & 2 \\ 1 & 0 & 2 & 0 \\ 1 & 2 & 0 & 0 \\ 2 & 0 & 1 & 1 \end{pmatrix}. \tag{iv-1}$$

$$\hat{M}''^{(1)} = (0, 0, 2, 1)^{\mathrm{T}}. \tag{v-1}$$

$$S^{(2)} = \left\{ M''^{(1)(123)}, M''^{(1)(213)} \right\}, \; k = 2. \tag{vi-1}$$

$$\begin{aligned} S'^{(2)} : M'^{(2)(123)} &= \left(\begin{array}{ccc|c} 1 & 0 & 0 & 2 \\ 1 & 0 & 2 & 0 \\ 1 & 2 & 0 & 0 \\ 2 & 1 & 1 & 0 \end{array}\right), \; M'^{(2)(132)} = \left(\begin{array}{ccc|c} 1 & 0 & 2 & 0 \\ 1 & 0 & 0 & 2 \\ 1 & 2 & 0 & 0 \\ 2 & 1 & 0 & 1 \end{array}\right), \\ M'^{(2)(213)} &= \left(\begin{array}{ccc|c} 1 & 0 & 2 & 0 \\ 1 & 0 & 0 & 2 \\ 1 & 2 & 0 & 0 \\ 2 & 1 & 1 & 0 \end{array}\right), \; M'^{(2)(231)} = \left(\begin{array}{ccc|c} 1 & 0 & 0 & 2 \\ 1 & 0 & 2 & 0 \\ 1 & 2 & 0 & 0 \\ 2 & 1 & 0 & 1 \end{array}\right). \end{aligned} \tag{iii-2}$$

$$\begin{aligned} S''^{(2)} : M''^{(2)(123)} &= \begin{pmatrix} 1 & 0 & \mathbf{0} & 2 \\ 1 & 0 & \mathbf{2} & 0 \\ 1 & 2 & \mathbf{0} & 0 \\ 2 & 1 & \mathbf{1} & 0 \end{pmatrix}, \; M''^{(2)(132)} = \begin{pmatrix} 1 & 0 & 0 & 2 \\ 1 & 0 & 2 & 0 \\ 1 & 2 & 0 & 0 \\ 2 & 1 & 0 & 1 \end{pmatrix}, \\ M''^{(2)(213)} &= \begin{pmatrix} 1 & 0 & \mathbf{0} & 2 \\ 1 & 0 & \mathbf{2} & 0 \\ 1 & 2 & \mathbf{0} & 0 \\ 2 & 1 & \mathbf{1} & 0 \end{pmatrix}, \; M''^{(2)(231)} = \begin{pmatrix} 1 & 0 & 0 & 2 \\ 1 & 0 & 2 & 0 \\ 1 & 2 & 0 & 0 \\ 2 & 1 & 0 & 1 \end{pmatrix}. \end{aligned} \tag{iv-2}$$

$$\hat{M}''^{(2)} = (0, 2, 0, 1)^{\mathrm{T}}. \tag{v-2}$$

$$S^{(2)} = \left\{ M''^{(2)(123)}, M''^{(2)(213)} \right\}. \tag{vi-2}$$

$$\hat{P} = P = x_3^2 + x_2^2 + x_1^2 + 2x_1x_2, \; \hat{\sigma} = \{(123), (213)\}. \tag{vii}$$

We observe, P was already in its unique form $\hat{P}$ (apart from the order of terms) and the algorithm resulted in two permutations (123) and (213) which correspond to the apparent symmetry in exchanging x_1 and x_2. In the context of Feynman integrals this means the algorithm gives not just a unique identifier $\hat{\mathcal{U}} + \hat{\mathcal{F}}$ but also its symmetries.

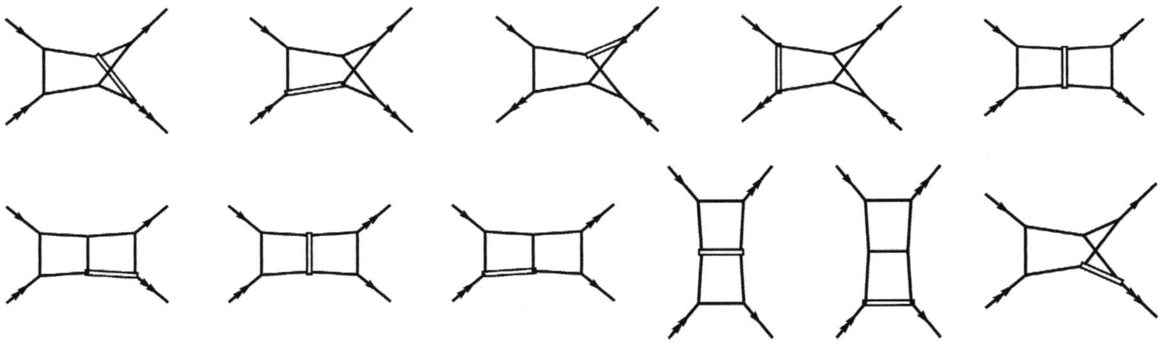

Figure 1. Minimal set of topologies needed for the real corrections to the Higgs boson cross section at NNLO. This set is sufficient for the calculation of all 2 946 appearing diagrams.

3. Features of `TopoID`

`TopoID` is written in `Mathematica` and assists in obtaining results for the amplitude of a given process in terms of a minimal set of master integrals. However, for the computationally expensive parts of a calculation, much faster `FORM` code is generated that does not demand for a `Mathematica` license. The main input for this task are the Feynman diagrams which can be generated, e.g., by `QGRAF` [19].

Starting from the diagrams, `TopoID` identifies a minimal set of topologies. That is, by contraction and permutation of propagators all scalar integrals arising from the diagrams can be mapped to at least one of the topologies in the minimal set. See Fig. 1 for a minimal set sufficient for the real contributions to NNLO Higgs boson production.

In case the topologies identified in a first step expose linearly dependent propagators, see Subs. 1.1, linear independent subtopologies are detected and partial fractioning relations are generated automatically. The algorithm described in Ref. [18] which employs Gröbner bases is used. For Eq. 1 the following set of rules is generated whose repeated application is guaranteed to terminate:

$$\begin{aligned} d_4 &\to -m_H^2 + s + d_1 + d_2 - d_3, \\ d_3/d_4 &\to \left(-m_H^2 + s + d_1 + d_2 - d_4\right)/d_4, \\ d_2/\left(d_3 d_4\right) &\to \left(m_H^2 - s - d_1 + d_3 + d_4\right)/\left(d_3 d_4\right), \\ d_1/\left(d_2 d_3 d_4\right) &\to \left(m_H^2 - s - d_2 + d_3 + d_4\right)/\left(d_2 d_3 d_4\right), \\ 1/\left(d_1 d_2 d_3 d_4\right) &\to \left(d_1 + d_2 - d_3 - d_4\right)/\left(\left(m_H^2 - s\right) d_1 d_2 d_3 d_4\right). \end{aligned} \tag{3}$$

It is of course understood that `TopoID` is capable of dealing with various properties of topolgies, such as completeness with respect to scalar products involving loop momenta, distinct and scaleless subtopologies and symmetries. Also graphs corresponding to a set of propagators can be reconstructed and unitarity cuts (used extensively in connection with the optical theorem and forward scattering) can be revealed in a very efficient way.

Finally, let us mention that also relations among master integrals (emerging from different topologies) can be found using `TopoID`. These relations can simplify a calculation tremendously or are very useful cross-checks. An example for such a relation from NNLO Higgs boson production can be sketched as follows:

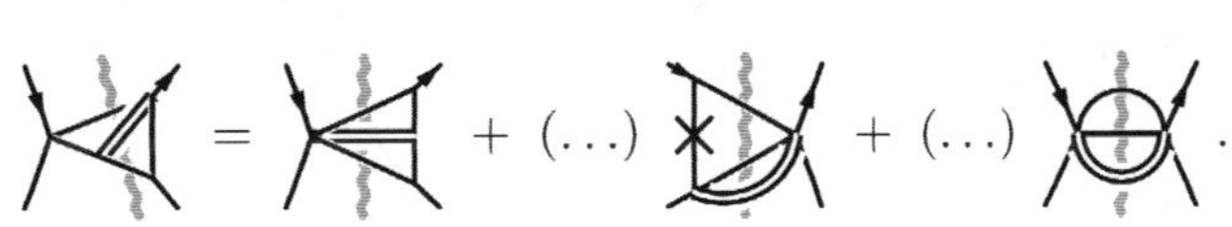

(4)

Here, the integral from the left-hand side stems from the fitfth and the first integral on the right-hand side from the ninth topology in Fig. 1, the coefficients $(\ldots)$ are rational functions in the kinematic invariants m_H^2 and s and the space-time dimension D.

4. Conclusion

The package `TopoID` is a generic, process independent tool for multiloop calulations, especially in case many topologies are involved. Until now it has been applied successfully to the qq'-channel in Higgs boson production at N^{3}LO [2] and to the NNLO soft-virtual corrections to Higgs boson pair production [20]. It can also handle the classification of the massless five-loop propagator topologies, including all their symmetries. Along with this proceedings contribution, a first public version can be obtained from the web page `http://topoid.hepforge.org/` where also future versions, a detailed documentation and additional material will be provided.

5. Acknowledgments

We would like to thank Alexey Pak who is the author of the initial version of `TopoID` and Matthias Steinhauser for his help related to debugging the package. This work was supported partially by the European Commission through contract PITN-GA-2012-316704 (HIGGSTOOLS). The attendence at ACAT was enabled by the generous support from the DAAD (German Academic Exchange Service).

References

[1] C. Anastasiou, C. Duhr, F. Dulat, E. Furlan, T. Gehrmann, F. Herzog and B. Mistlberger, JHEP **1503**, 091 (2015) doi:10.1007/JHEP03(2015)091 [arXiv:1411.3584 [hep-ph]].
[2] C. Anzai, A. Hasselhuhn, M. Höschele, J. Hoff, W. Kilgore, M. Steinhauser and T. Ueda, JHEP **1507**, 140 (2015) doi:10.1007/JHEP07(2015)140 [arXiv:1506.02674 [hep-ph]].
[3] P. Marquard, A. V. Smirnov, V. A. Smirnov and M. Steinhauser, arXiv:1601.03748 [hep-ph].
[4] A. von Manteuffel, E. Panzer and R. M. Schabinger, arXiv:1510.06758 [hep-ph].
[5] K. G. Chetyrkin, P. A. Balkov and J. H. Kühn, PoS RADCOR **2013**, 056 (2013) [arXiv:1402.6606 [hep-ph]].
[6] J. Kuipers, T. Ueda, J. A. M. Vermaseren and J. Vollinga, Comput. Phys. Commun. **184**, 1453 (2013) doi:10.1016/j.cpc.2012.12.028 [arXiv:1203.6543 [cs.SC]].
[7] V. Shtabovenko, R. Mertig and F. Orellana, arXiv:1601.01167 [hep-ph].
[8] M. Wiebusch, Comput. Phys. Commun. **195**, 172 (2015) doi:10.1016/j.cpc.2015.04.022 [arXiv:1412.6102 [hep-ph]].
[9] J. Grigo and J. Hoff, PoS LL **2014**, 030 (2014) [arXiv:1407.1617 [hep-ph]].
[10] J. Hoff, "Methods for multiloop calculations and Higgs boson production at the LHC", Dissertation, KIT, 2015.
[11] K. G. Chetyrkin and F. V. Tkachov, Nucl. Phys. B **192**, 159 (1981). doi:10.1016/0550-3213(81)90199-1
[12] S. G. Gorishnii, S. A. Larin, L. R. Surguladze and F. V. Tkachov, Comput. Phys. Commun. **55**, 381 (1989). doi:10.1016/0010-4655(89)90134-3
[13] M. Steinhauser, Comput. Phys. Commun. **134**, 335 (2001) doi:10.1016/S0010-4655(00)00204-6 [hep-ph/0009029].
[14] C. Anastasiou and A. Lazopoulos, JHEP **0407**, 046 (2004) doi:10.1088/1126-6708/2004/07/046 [hep-ph/0404258].
[15] A. V. Smirnov, Comput. Phys. Commun. **189**, 182 (2014) doi:10.1016/j.cpc.2014.11.024 [arXiv:1408.2372 [hep-ph]].
[16] A. von Manteuffel and C. Studerus, arXiv:1201.4330 [hep-ph].
[17] S. Laporta, Int. J. Mod. Phys. A **15**, 5087 (2000) doi:10.1016/S0217-751X(00)00215-7 [hep-ph/0102033].
[18] A. Pak, J. Phys. Conf. Ser. **368**, 012049 (2012) doi:10.1088/1742-6596/368/1/012049 [arXiv:1111.0868 [hep-ph]].
[19] P. Nogueira, J. Comput. Phys. **105**, 279 (1993). doi:10.1006/jcph.1993.1074
[20] J. Grigo, J. Hoff and M. Steinhauser, Nucl. Phys. B **900**, 412 (2015) doi:10.1016/j.nuclphysb.2015.09.012 [arXiv:1508.00909 [hep-ph]].

Computer system SANC: its development and applications

A Arbuzov[1], D Bardin[1], S Bondarenko[1], P Christova[1], L Kalinovskaya[1], R Sadykov[1], A Sapronov[1], T Riemann[2]

[1]Joint Institute for Nuclear Research, Dubna, 141980, Russia
[2]Königs Wusterhausen, 15711, Germany

E-mail: arbuzov@theor.jinr.ru

Abstract. The SANC system is used for systematic calculations of various processes within the Standard Model in the one-loop approximation. QED, electroweak, and QCD corrections are computed to a number of processes being of interest for modern and future high-energy experiments. Several applications for the LHC physics program are presented. Development of the system and the general problems and perspectives for future improvement of the theoretical precision are discussed.

1. Introduction

The main problem of particle physics today is to define the energy domain of the Standard Model (SM) applicability. Precision tests of the Standard Model (SM) at LHC nowadays become more and more important. The accuracy of the corresponding experimental studies grows continuously with collected statistics, improved detector calibration, elaboration of analysis techniques *etc.* The experimental data should be confronted to high-precision theoretical predictions obtained within the SM. Searches for new physical phenomena require accurate SM theoretical predictions as well. The absence of clear signals of SUSY and other new physics scenarios at LHC makes SM studies more and more actual. On the other hand, Tevatron has proved that a hadron collider can make precision measurements of electroweak (EW) processes and verify the SM predictions. Remind that Tevatron reached the precision of the w boson mass measurement better than LEP. For high-precision theoretical predictions we need to take into account several effects of different kind. The predictions should be presented in a form suitable to be used in the data analysis. All that leads to new requirements on the accuracy of theoretical predictions challenged by the experimental data.

The subject of the present report is the description of the SANC computer system which is a tool for systematic derivation of high-precision theoretical predictions within the Standard Model for various processes studied at modern high-energy colliders. Several examples of SANC applications will be given.

2. The SANC Project

SANC is a project to Support Analytic and Numeric calculations for experiments at Colliders. The development of the SANC project started at JINR in 2001. The roots of the project are the computer programs ZFITTER [1] and HECTOR [2] which were developed earlier with participation of the SANC team members. These programs were extensively used in the analysis of LEP and HERA experimental data.

The SANC system implements calculations of complete (real and virtual) NLO QCD and EW corrections for the Drell–Yan process, associative Higgs and gauge boson production, single top production and many other SM processes. All calculations are performed within the on-mass-shell renormalization scheme in the R_ξ gauge which allows an explicit control of the gauge invariance by examining cancellation of the gauge parameters in the analytical expression for the matrix element. The theoretical basis of the system is provided by the book [3].

During the first phase (2001–2005), the computer system SANC–v1.10 for semi-automatic calculations at the one-loop precision level (EW and QCD) of realistic and pseudo-observables (event distributions and decay rates) has been created [4, 5]. The SANC system was made publicly available for external users at `http://sanc.jinr.ru`. During its second phase (2006–2009) the work on the SANC integrated development environment was suspended, and instead the concept of SSFM (Standard SANC FORM/FORTRAN Modules), aimed for usage in physical applications, was realized [6]. The third phase (2010–present time) was mostly devoted to physical applications of SANC Monte Carlo Integrators and Generators based on the above mentioned modules. Meantime, modules for several more processes were implemented into the SANC framework: top quark decays, QCD corrections to Drell–Yan processes, 4-boson processes, and single top quark production.

The SANC framework features advanced computing methods for the calculation of a number of processes in the next-to-leading order approximation in QCD and EW theory. It has been extensively verified and cross checked against similar instruments and was proven to provide reliable advanced predictions at both parton and hadron levels.

The scheme of the SANC framework is shown in figure 1. Analytical expressions are obtained for the form factors and amplitudes of generalized processes $ffbb \to 0$ and $4f \to 0$ and stored as FORM expressions. The latter are translated to Fortran modules for specific parton level processes with NLO QCD and EW corrections. The modules are utilizing LoopTools and SANClib packages for loop integral evaluations. To build a Monte Carlo code one convolutes the partonic cross sections from the modules with the parton density functions and feeds the result as an integrand to any Monte Carlo algorithm implementation, e.g. FOAM or Cuba. The module's procedures for partonic cross sections are significantly unified and allow to calculate fully differential hadronic cross sections.

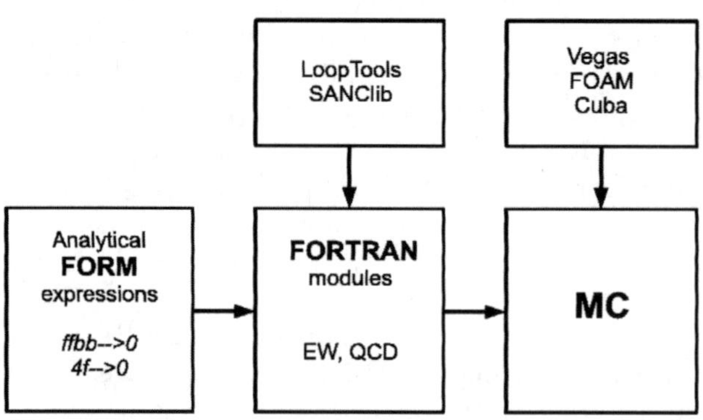

Figure 1. The SANC framework scheme.

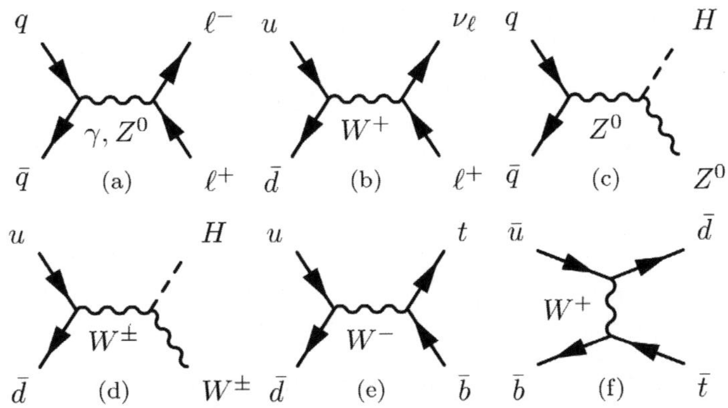

Figure 2. Feynman graphs for tree level Drell–Yan processes (neutral (a) and charged (b) currents), Higgs and gauge boson production (neutral (c) and charged (d) currents), and single top-quark production (s-channel (e) and t-channel (f)), implemented in the MCSANC integrator.

Table 1. NC and CC DY processes. LO, NLO EW, NLO QCD cross sections are given in picobarns and compared with corresponding values obtained with the program MCFM. The correction factors δ are shown in %.

$pp \to$	$Z^0(\mu^+\mu^-)$	$W^+(\mu^+\nu_\mu)$	$W^-(\mu^-\bar{\nu}_\mu)$
LO	3338(1)	10696(1)	7981(1)
LO MCFM	3338(1)	10696(1)	7981(1)
NLO QCD	3388(2)	12263(4)	9045(4)
NLO MCFM	3382(1)	12260(1)	9041(5)
NLO EW	3345(1)	10564(1)	7861(1)
δ_{QCD}, %	1.49(3)	14.66(1)	13.35(3)
δ_{EW}, %	0.22(1)	-1.23(1)	-1.49(1)

2.1. MCSANC integrator

The list of processes implemented in the recently updated version of the MCSANC integrator [7, 8] includes Drell–Yan processes (inclusive), associated Higgs and gauge boson production and single-top quark production in s and t channels, see figure 2. The MCSANC integrator is suited for simulation of realistic distributions taking into account LHC experimental conditions. The code allows to study how various radiative corrections affect observable distributions.

Below we provide numerical cross checks for the MCSANC integrator. The SANC DY NLO electroweak corrections were thoroughly verified with other calculations earlier during theoretical workshops on the subject and tête-á-tête comparisons, see e.g. Refs. [9, 10, 11]. The newer QCD results are validated using the MCFM [12] program. table 1 contains results for integrated LO and NLO EW and QCD cross sections obtained with the MCSANC integrator. The LO and NLO QCD values are in agreement with the MCFM program within statistical errors.

In the MCSANC-v1.20 version of the Monte-Carlo tool based on the SANC modules, the inverse photon ($q\gamma$) and ($\gamma\gamma$) configurations in the initial pp state of beam, which contribute to Drell–Yan processes, are added. The MCSANC-v1.20 version was used recently to calculate the

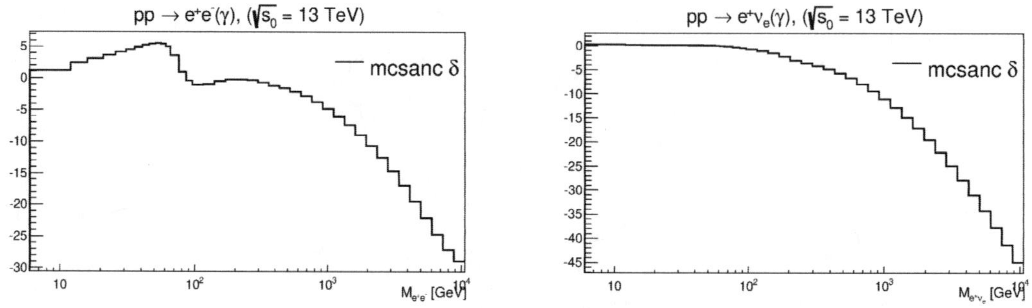

Figure 3. $\delta_{\rm missed}$ contributions $pp \to e^+e^-(\gamma)$ (left) and $pp \to e^+\nu_e(\gamma)$ right at $\sqrt{s} = 13$ TeV.

Process	LO, pb	NLO(qγ), pb	δ, %
pp→ e^+e^-	606.63(1)	606.20(1)	-0.071(1)
pp→ $\mu^+\mu^-$	586.49(1)	586.07(1)	-0.073(1)
pp→ $e^+\nu_e$	8.291(1)	8.378(1)	1.06(1)
pp→ $\mu^+\nu_\mu$	5.476(1)	5.540(1)	1.17(1)
pp→ $e^-\nu_e$	3298.46(3)	3299.51(3)	0.032(1)
pp→ $\mu^-\nu_\mu$	2610.50(2)	2611.25(2)	0.029(1)
Process	LO, pb	NLO($\gamma\gamma$), pb	δ, %
pp→ e^+e^-	606.63(1)	607.38(1)	0.124(1)
pp→ $\mu^+\mu^-$	586.49(1)	587.22(1)	0.124(1)

Table 2. Contributions of the (qγ) and ($\gamma\gamma$) configuration in the initial pp state

following corrections to the Drell–Yan processes at $\sqrt{s} = 13$ TeV: 1) the so-called missed, *i.e.* pure weak, initial and interference QED one-loop contributions to the M_{inv} distribution; 2) the inverse photon contributions for typical ATLAS fiducial cuts. The obtained results were used by the ATLAS Standard Model group for the Drell–Yan data analysis.

The estimation of four-boson background processes is very important for new physics searches. For example, according to the SM calculations of $\gamma\gamma \to \gamma Z$ the number of background events is about 5×10^3 to 3×10^4 corresponding to clear $\gamma\gamma \to H \to \gamma\gamma$ signal of about 45 to 70 events per 500 fb^{-1} for the 160 to 320 GeV energy range of a $\gamma\gamma$ mode of a linear e^+e^- machine. To study the Higgs boson properties one needs to have angular distributions for each helicity amplitude of these processes. We implemented into the four-boson sector of the SANC system processes like: $\gamma\gamma \to \gamma\gamma$, $\gamma\gamma \to \gamma Z$, $Z \to \gamma\gamma\gamma$, $\gamma\gamma \to ZZ$ in the SM at the one-loop level in the R_ξ gauge with taking into account of all masses (Z boson and internal ones) [13]. The corresponding package for numeric calculations is presented on the project web-site. In order to cross-check the results we made a comparison with other calculations.

3. SANC Applications

3.1. Electroweak corrections for Drell–Yan analysis

The Drell–Yan-like processes, *i.e.* single Z and W boson production with consequent decay into a lepton pair, provide at LHC the ultimate benchmark in the experimental precision. The theoretical description of these processes within the SM is constructed taking into account various possible effects including radiative corrections, PDF uncertainties and scale variations. Tuned comparisons between independent results of several research groups show a perfect agreement in description of QCD and electroweak radiative corrections in the one-loop (NLO) approximation. Higher order effects are also shown to be important to provide the required

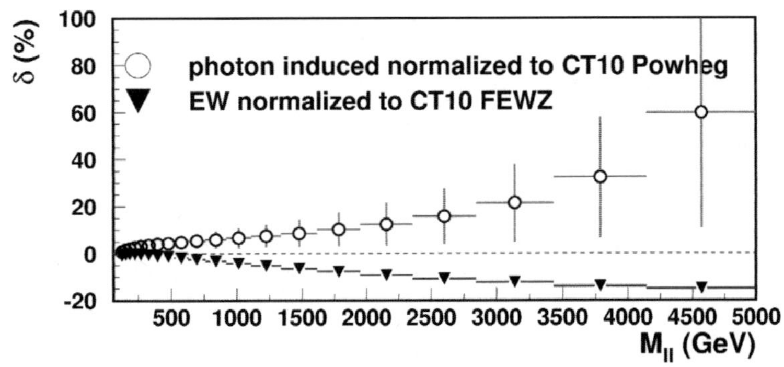

Figure 4. Photon induced background contributions calculated with $p_{T,\ell} > 25$ GeV and $\eta_\ell < 2.5$, MRST2004QED photon PDF and $\sqrt{s} = 8$ TeV over a wide invariant mass range. Calculations are performed with `MCSANC` [7].

accuracy level.

Monte Carlo simulations in an ATLAS analysis are typically based on the NLO QCD hard process event generators like `POWHEG++` or `MC@NLO` complemented with `PHOTOS` to generate final state electromagnetic radiation. This approach does not take into account a set of higher order electroweak corrections (HO EW) when considering Drell–Yan processes: pure weak (PW) contributions, initial–final QED interference (IFI) and what remains from initial state radiation (ISR) after subtraction of collinear divergences. These corrections are sometimes referred in this text under the term "missed".

At present the best fixed order approximation of the Drell–Yan cross section is NNLO QCD and NLO EW. However, their proper combination requires also the calculation of $\mathcal{O}(\alpha\alpha_S)$ contributions, which is currently not available in complete form. In view of it's importance for the data analysis, two methods of combination of HO EW and QCD corrections in the theoretical predictions were compared in [14].

LHC data provide access to invariant mass regions where the photon induced contribution to the Drell–Yan process becomes substantial relative to quark-antiquark annihilation. A more accurate estimation of this background for high mass resonance searches requires the inclusion of $\gamma\gamma \to \ell^+\ell^-$ into the theory predictions. The predictions were obtained using an implementation of this process in the `MCSANC` integrator with MRST2004QED PDF, which albeit deprecated, was the only set containing a photon density at that time. The results of this estimation are presented in figure 4 together with HO EW (except QED FSR) corrections, which appear to have opposite sign. The corrections were included as a systematic uncertainty in the ATLAS searches for high mass dilepton resonances [15].

Calculation of HO EW (except QED FSR) corrections and their NNLO QCD combination methodology was routinely used in the ATLAS Standard Model and BSM analysis:

- Measurements of the Drell–Yan differential cross sections at $\sqrt{s} = 7$ TeV in the e and μ channels for invariant masses between 26 and 66 GeV using an integrated luminosity of 1.6 fb^{-1} collected in 2011 [16]. Theory comparisons show that fixed order next-to-next-to-leading-order QCD predictions provide a significantly better description of the results than next-to-leading-order QCD calculations, see figure 5.

- Measurements of the high-mass Drell–Yan differential cross sections at $\sqrt{s} = 7$ TeV in the e^+e^- channel based on integrated luminosity of 4.9 fb^{-1}. Invariant mass of the electron pair, covered by the measurement, is $116 < M_{ee} < 1500$ GeV [17].
- Measurement of the inclusive $W^{\pm}$ and Z/γ cross sections in the electron and muon decay channels in pp collisions at $\sqrt{s} = 7\text{TeV}$ [18].
- A QCD analysis performed with the ATLAS data of inclusive W and Z boson production, jointly with ep deep inelastic scattering data from HERA. The ATLAS data exhibited sensitivity to the light quark sea composition and magnitude at Bjorken $x \sim 0.01$. Specifically, the data supported the hypothesis of a symmetric composition of the light quark sea at low x. The ratio of the strange-to-down sea quark distributions is determined to be 1.00(+0.25-0.28) at absolute four-momentum transfer squared $Q^2 = 1.9$ GeV^2 and x = 0.023 [19].
- Search for high-mass dilepton resonances in 20 fb^{-1} of pp collisions at $\sqrt(s) = 8$ TeV. Results are presented from an analysis of proton-proton collisions at a center-of-mass energy of 8 TeV corresponding to an integrated luminosity of 20.3 fb^{-1} in the e^+e^- channel and 20.5 fb^{-1} in the $\mu^+\mu^-$ channel. A narrow resonance with Standard Model Z couplings to fermions is excluded at the 95% confidence level for masses less than 2.79 TeV in the e^+e^- channel, 2.53 TeV in the $\mu^+\mu^-$ channel, and 2.90 TeV in the two channels combined [15].

3.2. QCD analysis and HERAFitter development

A precise determination of PDFs as a function of x and Q^2 requires large amounts of experimental data that cover a wide kinematic region and that are sensitive to partonic densities of different kinds. Measurements of inclusive Neutral Current (NC) and Charge Current (CC) Deep Inelastic Scattering (DIS) at the lepton-proton (ep) collider HERA provided crucial information for determining the PDFs. Different processes in proton-proton (pp) and proton-antiproton ($p\bar{p}$) collisions at the LHC and the Tevatron, respectively, provide complementary information on the DIS measurements.

The PDFs are determined from χ^2 fits of the theoretical predictions to the data. The xFitter (formerly HERAFitter) [20] project is developed for the QCD analysis of experimental data, was launched by our colleagues from DESY and initially was meant for the analysis of DIS results obtained at HERA experiments. With participation of SANC group members it was subsequently extended to include proton-(anti)proton collisions measured at the LHC experiments [20, 21, 22].

There are also examples of SANC system applications to calculations of radiative corrections to neutrino deep inelastic scattering [23] and to the process of high-energy lepton bremsstrahlung on heavy nuclei [24].

4. Outlook

The plans of the SANC project for the nearest future are listed below.

• The implementation of new processes into the SANC Monte Carlo generators and integrators at the hadron level. To obtain relevant theoretical predictions for the cross sections at the NLO level of the processes with LHC conditions and other experiments at accelerators: $ud \to W\gamma$ (EW&QCD); $ff \to W^+W^-$ (QCD); $b\gamma \to tW$ (QCD), $bg \to tW$ (QCD).

• Study of QED radiative correction contributions in the initial and final states of Drell–Yan type processes in order to coordinate them with standard programs for the analysis of experimental data.

• Research under development of the measurement of effective parameters of the Standard Model such as weak mixing angle, gauge couplings, and rho-parameter using recent LHC data. Having access to and knowledge of the specifics in the higher order electroweak corrections

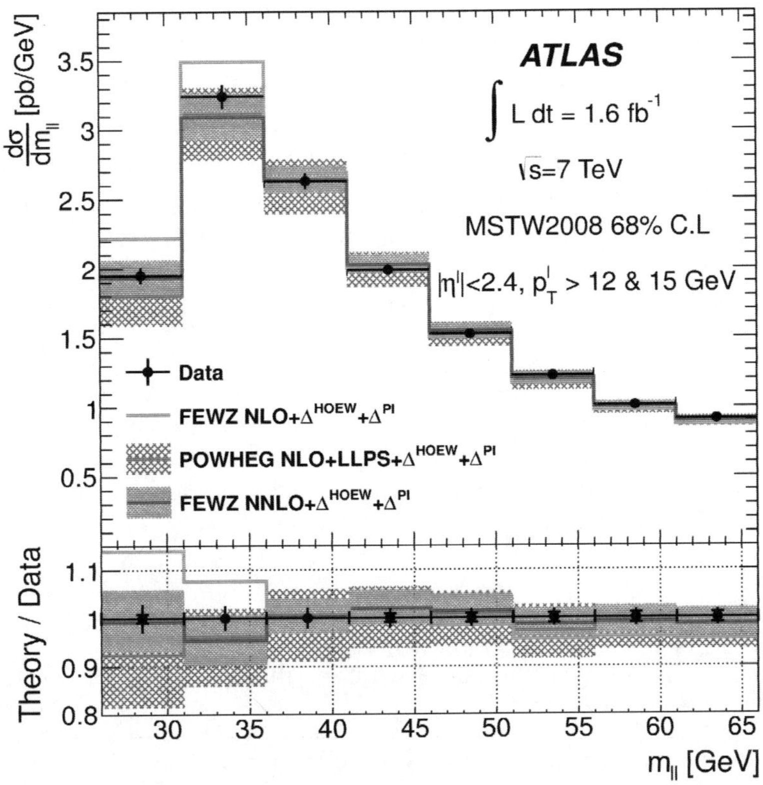

Figure 5. The measured fiducial differential cross section, $d\sigma/dm_{\ell\ell}$ for the nominal analysis as a function of the invariant mass $m_{\ell\ell}$ (solid points) compared to NLO predictions from FEWZ, NLO+LLPS predictions from POWHEG and NNLO predictions from FEWZ (all including higher-order electroweak and photon induced corrections). The predictions are calculated using MSTW2008 PDF sets with the appropriate order of perturbative QCD. The uncertainty bands include the PDF and α_S variations at 68% CL, scale variations between 0.5 and 2 times the nominal scales, and the uncertainty in the PI correction. The ratios of all three theoretical predictions (solid lines) to the data are shown in the lower panels. The data (solid points) are displayed at unity with statistical (inner) and total (outer) measurement uncertainties.

implementation in the `MCSANC` code, it is possible to set up an analysis chain to measure the mentioned variables. A sensitivity of the effective Weinberg angle to ATLAS data has been observed and the possibility to extract its value will be studied.

• Analysis of Drell–Yan type processes in the context of QCD. The purpose of this research is to tune the functions of parton distributions based on experimental data of proton-proton collisions. Application of HERAFitter to RUN-I data showed additional information on the densities of the momentum distributions of s-quark at small values of x and of gluons at large x. There is a need to continue these studies in RUN-II with high kinematic ranges and higher statistics.

• Collaboration with DESY (Hamburg) on the project ZeeD. The project ZeeD (analysis of $Z \to ee$ events at DESY) is aimed to study Drell–Yan processes. Now it was developed a technique and software for the analysis of RUN-I data. It's necessary to adapt the program interface to the new RUN-II data format and expand its functionality.

Working in this direction we try to approach several general problems which are relevant for increasing the precision of theoretical predictions. The key problems are:

- Matching specific higher order corrections with general tools like PYTHIA or HERWIG without double counting of effects
- Estimation of the theoretical uncertainty. We have a considerable scheme and scale dependence for EW corrections and make the optimal choice.
- Interplay of EW and QCD RC to DY: implementation of $\mathcal{O}(\alpha\alpha_s)$ RC with matching to existing results.
- NLO treatment of QED evolution in PDFs should be done for a consistent treatment of QED initial state radiation (ISR).

In conclusion, we would like to underline that precision tests of the SM at LHC are of ultimate importance. Many effects of different nature should be taken into account for these tests. The SANC project contributes to studies of EW observables. But only common efforts of different groups guarantee the construction of reliable theoretical predictions. The forthcoming analysis of the LHC Run2 data provide new challenges for theoreticians.

References

[1] Bardin D Yu, Christova P, Jack M, Kalinovskaya L, Olchevski A, Riemann S and Riemann T 2001 *Comput. Phys. Commun.* **133** 229–395 (*Preprint* hep-ph/9908433)
[2] Arbuzov A, Bardin D Yu, Blumlein J, Kalinovskaya L and Riemann T 1996 *Comput. Phys. Commun.* **94** 128–184 (*Preprint* hep-ph/9511434)
[3] Bardin D and Passarino G 1999 *The standard model in the making: Precision study of the electroweak interactions* International series of monographs on physics (Oxford University Press)
[4] Andonov A, Arbuzov A, Bardin D, Bondarenko S, Christova P *et al.* 2006 *Comput. Phys. Commun.* **174** 481–517 (*Preprint* hep-ph/0411186)
[5] Bardin D, Bondarenko S, Kalinovskaya L, Nanava G, Rumyantsev L *et al.* 2007 *Comput. Phys. Commun.* **177** 738–756 (*Preprint* hep-ph/0506120)
[6] Andonov A, Arbuzov A, Bardin D, Bondarenko S, Christova P *et al.* 2010 *Comput. Phys. Commun.* **181** 305–312 (*Preprint* 0812.4207)
[7] Bondarenko S G and Sapronov A A 2013 *Comput. Phys. Commun.* **184** 2343–2350 (*Preprint* 1301.3687)
[8] Arbuzov A, Bardin D, Bondarenko S, Christova P, Kalinovskaya L, Klein U, Kolesnikov V, Sadykov R, Sapronov A and Uskov F 2015 (*Preprint* 1509.03052)
[9] Buttar C *et al.* 2006 *Physics at TeV colliders. Proceedings, Workshop, Les Houches, France, May 2-20, 2005* (*Preprint* hep-ph/0604120)
[10] Buttar C *et al.* 2008 *Physics at TeV colliders, Les Houches 2007 : 11-29 June 2007* pp 121–214 (*Preprint* 0803.0678)
[11] Bardin D, Bondarenko S, Jadach S, Kalinovskaya L and Placzek W 2009 *Acta Phys. Polon.* **B40** 75–92 (*Preprint* 0806.3822)
[12] Campbell J M and Ellis R K 2010 *Nucl. Phys. Proc. Suppl.* **205-206** 10–15 (*Preprint* 1007.3492)
[13] Bardin D Y, Kalinovskaya L and Uglov E 2013 *Phys. Atom. Nucl.* **76** 1339–1344 (*Preprint* 1212.3105)
[14] Butterworth J, Dissertori G, Dittmaier S, de Florian D, Glover N *et al.* 2014 (*Preprint* 1405.1067)
[15] Aad G *et al.* (ATLAS) 2014 *Phys. Rev.* **D90** 052005 (*Preprint* 1405.4123)
[16] Aad G *et al.* (ATLAS) 2014 *JHEP* **1406** 112 (*Preprint* 1404.1212)
[17] Aad G *et al.* (ATLAS) 2013 *Phys. Lett.* **B725** 223–242 (*Preprint* 1305.4192)
[18] Aad G *et al.* (ATLAS) 2012 *Phys. Rev.* **D85** 072004 (*Preprint* 1109.5141)
[19] Aad G *et al.* (ATLAS) 2012 *Phys. Rev. Lett.* **109** 012001 (*Preprint* 1203.4051)
[20] Alekhin S, Behnke O, Belov P, Borroni S, Botje M *et al.* 2014 (*Preprint* 1410.4412)
[21] Belov P *et al.* (HERAFitter developers' team) 2014 *Eur. Phys. J.* **C74** 3039 (*Preprint* 1404.4234)
[22] Camarda S *et al.* (HERAFitter developers' Team) 2015 (*Preprint* 1503.05221)
[23] Arbuzov A B, Bardin D Yu and Kalinovskaya L V 2005 *JHEP* **06** 078 (*Preprint* hep-ph/0407203)
[24] Arbuzov A B 2008 *JHEP* **01** 031 (*Preprint* 0710.3639)

Journal of Physics: Conference Series **762** (2016) 012063 doi:10.1088/1742-6596/762/1/012063

Mass of the bottom quark from Upsilon(1S) at NNNLO: an update

César Ayala[1], Gorazd Cvetič[2], Antonio Pineda[3]

[1]Department of Theoretical Physics, IFIC, University of Valencia - CSIC, Valencia, Spain
[2]Physics Department, Universidad Técnica Federico Santa María, Valparaíso, Chile
[3]Physics Department and IFAE, Universitat Autònoma de Barcelona, Barcelona, Spain

Abstract. We update our perturbative determination of $\overline{\rm MS}$ mass $\overline{m}_b(\overline{m}_b)$, by including the recently obtained four-loop coefficient in the relation between the pole and $\overline{\rm MS}$ mass. First the renormalon subtracted (RS or RS') mass is determined from the known mass of the $\Upsilon(1S)$ meson, where we use the renormalon residue N_m obtained from the asymptotic behavior of the coefficient of the 3-loop static singlet potential. $\overline{\rm MS}$ mass is then obtained using the 4-loop renormalon-free relation between the RS (RS') and $\overline{\rm MS}$ mass. We argue that the effects of the charm quark mass are accounted for by effectively using $N_f = 3$ in the mass relations. The extracted value is $\overline{m}_b(\overline{m}_b) = 4222(40)$ MeV, where the uncertainty is dominated by the renormalization scale dependence.

1. Introduction

The ($\overline{\rm MS}$) mass of the bottom (b) quark, $\overline{m}_b \equiv \overline{m}_b(\overline{m}_b)$, is an important quantity in particle physics, free of renormalon ambiguities, and appears in many physical observables. Since it is relatively high, ~ 4 GeV, perturbative QCD methods are suitable for its extraction. The mass of the ground state of the $b\bar{b}$ quarkonium, $\Upsilon(1S)$, is one of the best quantities for such an extraction, $M^{(\rm th)}_{\Upsilon(1S)} = 2m_b + E_{\Upsilon(1S)} = 9.460$ GeV, where m_b is the pole mass of the bottom quark, and $E_{\Upsilon(1S)}$ is the binding energy. We use the available perturbative expansions of $2m_b/\overline{m}_b$ and of $E_{\Upsilon(1S)}/\overline{m}_b$ in powers of QCD coupling $a(\mu) \equiv \alpha_s(\mu)/\pi$ and thus extract the value of $\overline{m}_b$. In the extraction, we use the fact that the leading infrared (IR) renormalon ambiguity of $2m_b$ cancels out with that of $E_{\Upsilon(1S)}$ [1, 2, 3].

These proceedings are a brief review of our previous work [4], which we update by including in the analysis the recently calculated [5] four-loop coefficient of the relation between the pole mass and the $\overline{\rm MS}$ mass. Here we outline: (1) The correct treatment of charm quark mass effects in the perturbation expansion of $m_b/\overline{m}_b$; (2) Asymptotic expressions for the coefficients in the perturbation expansion of the ratio $m_b/\overline{m}_b$ and of the static singlet potential $V(r)$, and the extraction of the renormalon residue N_m; (3) The construction of the (modified) renormalon-subtracted mass $m_{b,{\rm RS}(')}$ (using N_m), and the renormalon-free relation between $m_{b,{\rm RS}(')}$ and $\overline{m}_b$; (4) Renormalon-free perturbation expansion for $M^{(\rm th)}_{\Upsilon(1S)}$ in terms of $m_{b,{\rm RS}(')}$, and extraction, from $M^{(\rm th)}_{\Upsilon(1S)} = 9.460$ GeV, of the values of $m_{b,{\rm RS}(')}$ ($\Rightarrow \overline{m}_b$).

2. Charm mass effects in the bottom pole mass

The pole mass m_b and the $\overline{\text{MS}}$ mass $\overline{m}_b$ are related:

$$m_b = \overline{m}_b \left(1 + S(N_f)\right) + \delta m_c^{(+)} \;, \tag{1}$$

$$\text{where} \qquad S(N_f) = \frac{4}{3} a_+(\mu) \left[1 + r_1^{(+)}(\mu) a_+(\mu) + r_2^{(+)}(\mu) a_+^2(\mu) + r_3^{(+)}(\mu) a_+^3(\mu) + \mathcal{O}(a_+^4)\right] \tag{2}$$

and the evaluation is usually performed in QCD with $N_f = N_l + 1 = 4$ active flavors: $r_j^{(+)}(\mu) \equiv r_j(\mu; N_f)$, $a_+(\mu) = a(\mu; N_f)$. The coefficients $R_0 = 4/3$ and r_j $(j = 1, 2)$ were obtained in Refs. [6], [7], [8, 9], respectively. Recently, numerical values of the 4-loop coefficient r_3 were obtained [5], and we incorporate them here in the form given in [10].

These coefficients have a specific dependence on the renormalization scale μ, dictated by μ-independence of $S(N_f)$

$$r_1(\mu; N_f) = r_1(N_f) + \beta_0 L_m(\mu) \;, \text{ etc.} \tag{3}$$

where $L_m(\mu) = \ln(\mu^2/\overline{m}_b^2)$, and we maintain, for simplicity, the notation $r_j \equiv r_j(\overline{m}_b)$. We will use the notations $\beta_0 = (1/4)(11 - 2N_f/3)$ and $\beta_1 = c_1\beta_0 = (102 - 38N_f/3)/16$ for the first two coefficients of the RGE of $a(\mu)$

$$\frac{da(Q)}{d\ln Q^2} = -\beta_0 a^2(Q) \left(1 + c_1 a(Q) + c_2 a^2(Q) + c_3 a^3(Q) + \cdots\right) \;. \tag{4}$$

Finite-mass charm quark effects are incorporated in

$$\delta m_c^{(+)} = \delta m_{(c,+)}^{(1)} a_+^2(\overline{m}_b) + \delta m_{(c,+)}^{(2)} a_+^3(\overline{m}_b) + \mathcal{O}(a_+^4) \,, \tag{5}$$

which vanishes in the $m_c \to 0$ limit. We have

$$\delta m_{(c,+)}^{(1)} = \frac{4}{3} \overline{m}_b \Delta[\overline{m}_c/\overline{m}_b] = 1.9058 \text{ MeV [7]}, \quad \delta m_{(c,+)}^{(2)} = 48.6793 \text{ MeV [11]}, \tag{6}$$

$$\Rightarrow \quad \delta m_{(c,+)}^{(1)} a_+^2(\overline{m}_b) = 9.3 \text{ MeV}, \qquad \delta m_{(c,+)}^{(2)} a_+^3(\overline{m}_b) = 18.1 \text{ MeV}, \tag{7}$$

so $\delta m_c^{(+)}$ is badly divergent. Why? At loop order n, the natural scale of the loop integral for m_b is $m_b e^{-n}$ [12], which for n large enough is: $m_b e^{-n} < m_c$. Therefore, for large n (> 2) charm quark appears as very heavy (decoupled), leading to the effective number of flavors being $N_l = 3$ and not $N_f = N_l + 1 = 4$. Therefore, it is convenient to rewrite the relation between the pole and the $\overline{\text{MS}}$ mass in terms of $a_-(\mu) = a(\mu; N_l)$ and $r_j^{(-)}(\mu) \equiv r_j(\mu; N_l)$ $[N_l = 3]$

$$m_b = \overline{m}_b \left(1 + S(N_l)\right) + \delta m_c \;, \tag{8}$$

$$\text{where} \quad S(N_l) \;=\; \frac{4}{3} a_-(\mu) \left[1 + r_1^{(-)}(\mu) a_-(\mu) + r_2^{(-)}(\mu) a_-^2(\mu) + r_3^{(-)}(\mu) a_-^3(\mu) + \mathcal{O}(a_-^4)\right], \tag{9}$$

and $r_j^{(-)}(\overline{m}_b) = 7.74,\ 87.2,\ 1265.3 \pm 16.1$, for $j = 1, 2, 3$. The effects of the decoupling of S $(N_f \mapsto N_l = 3)$ are absorbed in the new δm_c

$$\delta m_c \;=\; \left[\delta m_{(c,+)}^{(1)} + \delta m_{(c,\text{dec.})}^{(1)}\right] a_-^2(\overline{m}_b) + \left[\delta m_{(c,+)}^{(2)} + \delta m_{(c,dec.)}^{(2)}\right] a_-^3(\overline{m}_b) + \mathcal{O}(a_-^4) \;, \tag{10}$$

where $\delta m^{(j)}_{(c,\mathrm{dec.})}$ are generated by this decoupling and read

$$\delta m^{(1)}_{(c,\mathrm{dec.})} = \frac{2}{9}\overline{m}_b \left(\ln\left(\frac{\overline{m}_b^2}{\overline{m}_c^2}\right) - \frac{71}{32} - \frac{\pi^2}{4}\right) \tag{11}$$

and $\delta m^{(2)}_{(c,\mathrm{dec.})}$ can be found in Ref. [4].

Numerical evaluation gives for $\left[\delta m^{(1)}_{(c,+)} + \delta m^{(1)}_{(c,\mathrm{dec.})}\right] a_-^2(\overline{m}_b) = -1.6$ MeV and $\left[\delta m^{(2)}_{(c,+)} + \delta m^{(2)}_{(c,\mathrm{dec.})}\right] a_-^3(\overline{m}_b) = -0.3$ MeV. This means that the previous divergent series (in $\mathrm{QCD}_{N_f=4}$) $\delta m_c^{(+)} = (9.3 + 18.1 + \ldots)$ MeV [Eq. (7)] now tranforms (in $\mathrm{QCD}_{N_l=3}$) to

$$\delta m_c = (-1.6 - 0.3 + \ldots) \text{ MeV}. \tag{12}$$

The series for δm_c in $\mathrm{QCD}_{N_l=3}$ formulation is convergent, strong cancellation takes place between $\delta m^{(j)}_{(c,+)}$ and $\delta m^{(j)}_{(c,\mathrm{dec.})}$, as expected.

3. Leading renormalon of the pole mass

The asymptotic behaviour of r_N is determined by the leading IR renormalon:

$$\frac{4}{3} r_N^{\mathrm{asym}}(\mu) \simeq \pi N_m \frac{\mu}{\overline{m}_b}(2\beta_0)^N \frac{\Gamma(\nu+N+1)}{\Gamma(\nu+1)}\left[1 + \sum_{s=1}^{3}\frac{\nu\cdots(\nu-s+1)}{(N+\nu)\cdots(N+\nu-s+1)}\widetilde{c}_s + \mathcal{O}\left(N^{-4}\right)\right]. \tag{13}$$

$$\frac{4}{3} r_N(\mu) = \pi N_m \frac{\mu}{\overline{m}_b}(2\beta_0)^N \frac{\Gamma(\nu+N+1)}{\Gamma(\nu+1)}\left[1 + \sum_{s\geq 0}\frac{\nu\cdots(\nu-s+1)}{(N+\nu)\cdots(N+\nu-s+1)}\widetilde{c}_s\right] + h_N(\mu), \tag{14}$$

where h_N is dominated by subleading renormalons, and the coefficients $\widetilde{c}_s$ ($s = 1,2,3$) are given in [13, 14, 15, 4] ($\widetilde{c}_0 = 1$ by convention).

Determining the pole mass from $\Upsilon(1S)$ mass has large uncertainties due to the pole mass renormalon ambiguity $\delta m_b \sim \Lambda_{\mathrm{QCD}}$ [13]. In order to avoid this problem, we work with the renormalon-subtracted (RS) bottom mass $m_{b,\mathrm{RS}}$ instead [14]. Then, $\overline{m}_b$ is obtained from its stable (renormalon-free) relation with the m_{RS} mass.

The use of m_{RS} in the theoretical evaluation of the $\Upsilon(1S)$ mass is convenient because it has no leading IR renormalon ambiguity, and the renormalon cancellation in the quarkonium mass $M_{\Upsilon(1S)} = 2m_b + E_{\Upsilon(1S)}$ is implemented automatically and explicitly.

4. Determination of the renormalon residue N_m and N_V

The asymptotic behavior of the coefficients $v_N(\mu)$ of the static singlet potential,

$$V(r) = -\frac{4\pi}{3}\frac{1}{r} a_-(\mu)\left[1 + v_1(\mu)a_-(\mu) + v_2 a_-(\mu)^2 + v_3 a_-(\mu)^3 + \ldots\right], \tag{15}$$

can be determined in complete analogy with those of r_N

$$-\frac{4}{3} v_N(\mu) = N_V \mu r (2\beta_0)^N \sum_{s\geq 0}\widetilde{c}_s \frac{\Gamma(\nu+N+1-s)}{\Gamma(\nu+1-s)} + d_N(\mu) \quad \Rightarrow \tag{16}$$

$$-\frac{4}{3} v_N^{\mathrm{asym}}(\mu) \approx N_V \mu r (2\beta_0)^N \frac{\Gamma(\nu+N+1)}{\Gamma(\nu+1)}\left[1 + \sum_{s=1}^{3}\frac{\nu\cdots(\nu-s+1)}{(N+\nu)\cdots(N+\nu-s+1)}\widetilde{c}_s\right], \tag{17}$$

where in Eq. (17) $d_N = 0$ was taken. We can determine the "strength", N_V, of the leading IR renormalon by approximating the asymptotic $v_N^{\rm asym}(\mu)$ with the exact $v_N(\mu)$ ($N = 0, 1, 2, 3$): $v_N^{\rm asym}(\mu) \approx v_N(\mu) \Rightarrow$

$$N_V \approx -\frac{4}{3} v_N(\mu) \Bigg/ \left\{ \mu r (2\beta_0)^N \frac{\Gamma(\nu + N + 1)}{\Gamma(\nu + 1)} \left[1 + \sum_{s=1}^{3} \frac{\nu \cdots (\nu - s + 1)}{(N + \nu) \cdots (N + \nu - s + 1)} \widetilde{c}_s \right] \right\}. \quad (18)$$

The result for N_V should be the best for the highest available N ($N = 3$) and should also have reduced spurious μ-dependence. At present, the v_j are known up to N^{3}LO (v_3) [16, 17, 18, 19, 20, 21, 22, 23, 24, 25].

In the sum $2m_b + V(r)$ the leading IR renormalon gets cancelled. N_V is then related with N_m by the renormalon cancellation of the sum $2m_b + V(r)$: $2N_m + N_V = 0$. Determining N_m via N_V gives us the value that we use [4]

$$N_m = -N_V/2 = 0.56255(260) \; (N_l = 3). \quad (19)$$

5. Renormalon-subtracted (RS, RS') mass of bottom

The RS mass is defined by subtracting the leading IR renormalon singularity from the pole mass [14]:

$$m_{b,\rm RS}(\nu_f) = m_b - N_m \pi \nu_f \sum_{N=0}^{\infty} a_-^{N+1}(\nu_f)(2\beta_0)^N \sum_{s \geq 0} \widetilde{c}_s \frac{\Gamma(\nu + N + 1 - s)}{\Gamma(\nu + 1 - s)} \quad (20)$$

Equation (20) is still formal. In practice, one rewrites m in terms of $\overline{m}$ using Eqs. (8)-(9)

$$m_b = \overline{m}_b(1 + (4/3)a_-(\nu_f) + \ldots), \quad (21)$$

and reexpands the perturbation series in Eq. (20) around the same coupling $a_-(\mu)$, at fixed but otherwise arbitrary scale μ:

$$m_{b,\rm RS}(\nu_f) = \overline{m}_b \left[1 + \sum_{N=0}^{\infty} h_N(\nu_f) a_-^{N+1}(\nu_f) \right] \Rightarrow m_{b,\rm RS}(\nu_f) = \overline{m}_b \left[1 + \sum_{N=0}^{\infty} \widetilde{h}_N(\nu_f; \mu) a_-^{N+1}(\mu) \right], \quad (22)$$

where $h_N(\nu_f)$ is determined from Eq. (14) (with $\mu = \nu_f$ and with the sum truncated at $\widetilde{c}_3$) for $N = 0, 1, 2, 3$. For $N \geq 4$ we take $h_N(\overline{m}_b) = 0$. The coefficients $\widetilde{h}_N(\nu_f; \mu)$ in Eq. (22) are obtained by expanding $a_-(\nu_f)$ in the expansion in powers of $a_-(\mu)$. Note that $m_{b,\rm RS}(\nu_f)$ will only marginally depend on μ when we truncate the infinite sum in Eq. (22). On the other hand, the coefficients h_N are functions of ν_f, μ, and $\overline{m}_b$, and are much smaller than $r_N(\mu)$.

A variant of the RS mass is the modified renormalon-subtracted (RS') mass $m_{b,\rm RS'}$, where subtractions start at $\sim a^2$ [14]. Specifically in this case, Eqs. (20) and (22) are repeated, with the replacements $m_{b,\rm RS}(\nu_f) \mapsto m_{b,\rm RS'}(\nu_f)$ and $\sum_{N=0}^{\infty} \mapsto \sum_{N=1}^{\infty}$.

6. Bottom mass from heavy quarkonium

The perturbation expansion of $M_{\Upsilon(1S)}^{(th)}$ is presently known up to $\mathcal{O}(m_b a^5)$ [19, 20, 21, 22]:

$$\begin{aligned} M_{\Upsilon(1S)}^{(th)} &= 2m_b - \frac{4\pi^2}{9} m_b a_-^2(\mu) \Bigg\{ 1 + a_-(\mu)\left[K_{1,0} + K_{1,1} L_p(\mu)\right] + a_-^2(\mu) \sum_{j=0}^{2} K_{2,j} L_p(\mu)^j \\ &\quad + a_-^3(\mu) \left[K_{3,0,0} + K_{3,0,1} \ln a_-(\mu) + \sum_{j=1}^{3} K_{3,j} L_p(\mu)^j \right] + \mathcal{O}(a_-^4) \Bigg\}, \end{aligned} \quad (23)$$

μ is the renormalization scale, $L_p(\mu) = \ln(\mu/\mu_b)$ where $\mu_b = (4\pi/3)m_b a_-(\mu)$. $K_{i,j}(N_f)$ and $K_{3,0,j}$ are given, e.g., in [4]. We then rewrite m_b in terms of $m_{b,\rm RS}$ to implement the leading IR renormalon cancellation. This gives

$$\frac{M^{(th)}_{\Upsilon(1S)}}{m_{b,\rm RS}(\nu_f)} = 2 + \left[2\pi N_m b a \mathcal{K}_0 - \frac{4\pi^2}{9}a^2\right] + \left[2\pi N_m b a^2 \left(\mathcal{K}_1 + z_1 \mathcal{K}_0\right) - \frac{4\pi^2}{9}a^3 \left(K_{1,0} + K_{1,1} L_{\rm RS}\right)\right]$$
$$+ \left[2\pi N_m b a^3 \left(\mathcal{K}_2 + 2z_1\mathcal{K}_1 + z_2\mathcal{K}_0\right) - \frac{4\pi^2}{9}\left(a^4 \sum_{j=0}^{2} K_{2,j} L^j_{\rm RS} + b a^3 \pi N_m \mathcal{K}_0\right)\right] + \mathcal{O}(ba^4, a^5). \quad (24)$$

The terms $\mathcal{O}(ba^4, a^5)$ have a similar structure and were written in [4]. The notations are

$$a \equiv a_-(\mu) = a(\mu, N_f = 3) \; ; \quad b \equiv b(\nu_f) = \nu_f / m_{b,\rm RS}(\nu_f) \, , \quad N_m = N_m(N_l = 3) \, , \quad (25a)$$

$$L_{\rm RS} \equiv L_{\rm RS}(\mu) = \ln\left(\frac{\mu}{(4\pi/3) m_{b,\rm RS}(\nu_f) a_-(\mu)}\right) , \; \mathcal{K}_N = (2\beta_0)^N \sum_{s=0}^{3} \widetilde{c}_s \frac{\Gamma(\nu + N + 1 - s)}{\Gamma(\nu + 1 - s)} \, . \quad (25b)$$

In the expression (24) for $M_{\Upsilon(1S)}$, the terms of the same order $(\nu_f/m_{b,\rm RS})a^n$ and a^{n+1} were combined in common brackets [...], in order to account for the renormalon cancellation.

If using the RS' mass in our approach instead, the above expressions are valid without changes, except that $m_{b,\rm RS} \mapsto m_{b,\rm RS'}$ and $\mathcal{K}_0 \mapsto 0$ (and: $h_0(\mu) \mapsto 4/3$).

We note that we take $N_l = 3$ active flavours, as the charm quark mass effects in the binding energy $E_{\Upsilon(1S)}$ are negligible [26].

We extract the bottom masses from the condition $M^{(th)}_{\Upsilon(1S)} = M^{(exp)}_{\Upsilon(1S)} (= 9.460$ GeV). The error estimates are made assuming $\mu = 2.5^{+1.5}_{-1.0}$ GeV [we varied μ in Eq. (24) but not in Eq. (22)], $\nu_f = 2 \pm 1$ GeV, $\alpha_s(M_z) = 0.1184(7)$ (and decoupling at $\overline{m}_b = 4.2$ GeV and at $\overline{m}_c = 1.27$ GeV), $N_m = 0.56255(260)$, and $(4/3) r_3(\overline{m}_b; N_l) = 1687.1 \pm 21.5$ [10].

In RS and RS' approaches we extract, in MeV, respectively

$$m_{b,\rm RS}(2\rm GeV) = 4\,437^{-11}_{+43}(\mu)^{-3}_{+5}(\nu_f)^{-2}_{+2}(\alpha_s)^{-41}_{+41}(N_m)^{-0}_{+0}(r_3); \quad (26a)$$
$$\Rightarrow \quad \overline{m}_b = 4\,217^{-10}_{+39}(\mu)^{-3}_{+5}(\nu_f)^{-5}_{+5}(\alpha_s)^{-1}_{+1}(N_m)^{-4}_{+4}(r_3). \quad (26b)$$
$$m_{b,\rm RS'}(2\ \rm GeV) = 4\,761^{-16}_{+41}(\mu)^{-3}_{+5}(\nu_f)^{+4}_{-3}(\alpha_s)^{-26}_{+26}(N_m)^{-0}_{+0}(r_3); \quad (26c)$$
$$\Rightarrow \quad \overline{m}_b = 4\,223^{-14}_{+36}(\mu)^{-2}_{+4}(\nu_f)^{-4}_{+4}(\alpha_s)^{-1}_{+1}(N_m)^{-4}_{+4}(r_3). \quad (26d)$$

The uncertainties in $\overline{m}_b$ are dominated by the variation of the renormalization scale μ.

The renormalon cancellations are reflected numerically in Eq. (24) [we take $\mu = 2.5$ GeV]:

$$\mathrm{RS}: M_{\Upsilon(1S)} = (8874 + 431 + 167 + 18 - 30)\ \mathrm{MeV}\, , \quad (27a)$$
$$\mathrm{RS}': M_{\Upsilon(1S)} = (9521 - 150 + 112 + 8 - 31)\ \mathrm{MeV}\, , \quad (27b)$$

The convergence is good; except for the last (NNNLO) term $\mathcal{O}(a^5, ba^4)$, where the factorization scale dependence becomes stronger, which may signal the importance of ultrasoft effects.

The relations between RS (RS') mass and $\overline{\rm MS}$ mass are reasonably convergent:

$$m_{b,\rm RS}(2\ \mathrm{GeV}) = (4217 + 191 + 36 + 12 - 19)\ \mathrm{MeV}\ , \quad (28a)$$
$$m_{b,\rm RS'}(2\ \mathrm{GeV}) = (4223 + 478 + 60 + 18 - 17)\ \mathrm{MeV}\ , \quad (28b)$$

where the expansion parameter is taken to be $a(2.5$ GeV). A bigger value for the renormalization scale, closer to the bottom quark mass, makes the last term smaller.

Until now we have approximated $\delta m_c = 0$ in Eq. (20). However, $\delta m_c \approx -2$ MeV, Eq.(12). Hence, we have to add 2 MeV to the values of $\bar{m}_b$ obtained in Eqs. (26b) and (26b) (in Ref. [4] it was incorrectly subtracted), leading to the final average of the RS and RS' extractions

$$\overline{m}_b = 4222(40)\ \text{MeV}\ . \tag{29}$$

where we have rounded the $\pm$ variation of each parameter to the maximum and added them in quadrature.

7. Conclusions

(i) We presented strong numerical indications that the charm quark decouples in the relation between m_b and $\overline{m}_b$ ($\Rightarrow N_l = 3$).

(ii) An improved determination of the residue of the leading renormalon for the bottom pole mass (and static potential with $N_l = 3$) was performed: $N_m = 0.56255(260)$.

(iii) Use of the 3-loop ($\sim a^5 \overline{m}_b$) corection to the $\Upsilon(1S)$ binding energy, and 4-loop relation between m_b and $\overline{m}_b$, allowed us to perform extraction of $m_{b,\mathrm{RS}^{(\prime)}}$ and $\overline{m}_b$ to NNNLO, with the resulting values Eq. (29). The uncertainties are dominated by the variation of the renormalization scale.

Acknowledgments

This work was supported in part by FONDECYT (Chile) Grant No. 1130599 (GC); Spanish Government and ERDF funds (EU Commission) FPA2014-53631-C2-1-P and by CONICYT "Becas Chile" Grant No. 74150052 (CA); and Spanish Government FPA2014-55613-P, FPA2013-43425-P (AP).

References

[1] Pineda A 1998 "Heavy quarkonium and nonrelativistic effective field theories". *PhD. thesis*
[2] Hoang A H, Smith M C, Stelzer T and Willenbrock S 1999 *Phys. Rev.* D **59** 114014 (arXiv:hep-ph/9804227)
[3] Beneke M 1998 *Phys. Lett.* B **434** 115 (arXiv:hep-ph/9804241)
[4] Ayala C, Cvetič G and and Pineda A 2014 *JHEP* **1409** 045 (arXiv:1407.2128 [hep-ph])
[5] Marquard P, Smirnov A V, Smirnov V A and Steinhauser M 2015 *Phys. Rev. Lett.* **114** 142002 (arXiv:1502.01030 [hep-ph])
[6] Tarrach R 1981 *Nucl. Phys.* B **183** 384
[7] Gray N, Broadhurst D J, Grafe W and Schilcher K 1990 *Z. Phys.* C **48** 673
[8] K. G. Chetyrkin K G and M. Steinhauser M 1999 *Phys. Rev. Lett.* **83** 4001 (arXiv:hep-ph/9907509)
[9] Melnikov K and van Ritbergen T 2000 *Phys. Lett.* B **482** 99 (arXiv:hep-ph/9912391)
[10] Kiyo Y, Mishima G and Sumino Y 2015 *JHEP* **1511** 084 (arXiv:1506.06542 [hep-ph])
[11] Bekavac S, Grozin A, Seidel D and Steinhauser M 2007 *JHEP* **0710** 006 (arXiv:0708.1729 [hep-ph])
[12] Ball P, Beneke M and Braun V M 1995 *Nucl. Phys.* B **452** 563 (arXiv:hep-ph/9502300)
[13] Beneke M 1995 *Phys. Lett.* B **344** 341 (arXiv:hep-ph/9408380)
[14] Pineda A 2001 *JHEP* **0106** 022 (arXiv:hep-ph/0105008)
[15] Contreras C, Cvetič G and Gaete P 2004 *Phys. Rev.* D **70** 034008 (arXiv:hep-ph/0311202)
[16] Fischler W 1977 *Nucl. Phys.* B **129** 157
[17] Billoire A 1980 *Phys. Lett.* B **92** 343
[18] Schröder Y 1999 *Phys. Lett.* B **447** 321 (arXiv:hep-ph/9812205)
[19] Pineda A and Yndurain F J 1998 *Phys. Rev.* D **58** 094022 (arXiv:hep-ph/9711287)
[20] Brambilla N, Pineda A, Soto J and Vairo A 1999 *Phys. Lett.* B **470** 215 (arXiv:hep-ph/9910238)
[21] Kniehl B A, Penin A A, Smirnov V A and Steinhauser M 2002 *Nucl. Phys.* B **635** 357 (arXiv:hep-ph/0203166)
[22] Penin A A and Steinhauser M 2002 *Phys. Lett.* B **538** 335 (arXiv:hep-ph/0204290)
[23] Smirnov A V, Smirnov V A and Steinhauser M 2008 *Phys. Lett.* B **668** 293 (arXiv:0809.1927 [hep-ph])
[24] Anzai C, Kiyo Y and Sumino Y 2010 *Phys. Rev. Lett.* **104** 112003 (arXiv:0911.4335 [hep-ph])
[25] Smirnov A V, Smirnov V A and Steinhauser M 2010 *Phys. Rev. Lett.* **104** 112002 (arXiv:0911.4742 [hep-ph])
[26] Brambilla N, Sumino Y and Vairo A 2002 *Phys. Rev.* D **65** 034001 (arXiv:hep-ph/0108084)

FeynCalc 9

Vladyslav Shtabovenko
Technische Universität München, Physik-Department T30f, James-Franck-Str. 1, 85747 Garching, Germany
E-mail: v.shtabovenko@tum.de

Abstract. We report on a new version of FEYNCALC, a well-known MATHEMATICA package for symbolic computations in quantum field theory and provide some explicit examples for using the software in different types of calculations.

1. Introduction

Modern perturbative calculations in quantum field theory (QFT) often heavily rely on different software tools that are used to automatize both the algebraic and the numerical stages of the evaluation.

In the last decades, much effort was invested to develop packages (e.g. FORMCALC [1], GOSAM [2], GRACE [3]) that are able to perform all the steps needed to arrive to physical observables (e.g. decay rates or cross-sections) in a highly automatized way. The level of the automation in such packages is so advanced, that in many cases the user just has to specify the incoming and outgoing particles, while the underlying code will take care of everything else and produce final results.

A complementary approach is represented by tools (e.g. HEPMATH [4], PACKAGE-X [5]) that do not attempt to automatize everything bur rather provide a collection of easy to use instruments for accomplishing most common tasks in QFT calculations, like contractions of Lorentz indices, calculation of Dirac traces or tensor decomposition of loop integrals. Such semi-automatic packages give the user much flexibility but also require him or her to carefully handle every step of the calculation. This bears a strong resemblance to calculations done by pen and paper, with the difference that one can proceed in a much faster way .

While both philosophies have their supporters in the high energy physics community, one should of course understand that they are not competing with each other but rather cover different use cases. On the one hand, many tree-level and 1-loop calculations in Standard Model and its popular (e.g. supersymmetric) extensions can be nowadays carried out in a highly automatic fashion, since the essential steps of these calculations are sufficiently well understood. It is therefore natural to use suitable packages like FORMCALC or GOSAM to study such processes.

On the other hand, multi-loop calculations, determination of matching coefficients in effective field theories or evaluation of Feynman diagrams in non-relativistic QFTs are good examples for tasks that are too complex or non-standard to be automatized in full generality. Although for such projects people often prefer to write their own code from scratch (e.g. implemented in FORM [6] or MATHEMATICA), it can also be useful to use different instruments provided by publicly available semi-automatic packages instead of or in addition to the private codes.

FEYNCALC belongs to the oldest semi-automatic packages that are still actively developed and used in research. The first version of the program was published almost 25 years ago [7] and followed the idea to provide a flexible and easy to use MATHEMATICA package for symbolic calculations and for the evaluation of Feynman diagrams in QFT at tree-level and at 1-loop. In this context, "symbolic calculations" means that FEYNCALC can deal not only with Feynman diagrams but also standalone QFT expressions like $\int \frac{d^D q}{(2\pi)^D} \frac{l^\mu l^\nu}{l^2-m^2}$ or $\text{Tr}(\gamma^\mu \gamma^\nu \gamma^\rho \gamma^\sigma)$. The users of FEYNCALC are therefore not forced to follow a particular workflow (e.g. first enter the Lagrangian, then generate Feynman diagrams and finally evaluate them) but are free to use the software as some sort of "calculator" for QFT expressions. This design philosophy has been preserved over the years and made FEYNCALC a valuable tool for many QFT practitioners.

This note is organized in the following way. In Sec. 2 we briefly describe the most interesting new features of FEYNCALC 9 and provide some examples for purposes of illustration. Sec. 3 shows the calculation of the gluon self-energy at 1-loop in QCD, where the final result will be presented in terms of the Passarino-Veltman scalar functions A_0 and B_0. In Sec. 4 we summarize and mention main directions for the future development of FEYNCALC.

2. New features in FeynCalc 9

FEYNCALC 9 was released in January 2016 [8]. While the development of the program between 2001 and 2014 was mostly limited to providing fixes for the discovered bugs, this version also introduces some new features that are mostly related to the evaluation of 1-loop and multi-loop integrals. Version 9 requires at least MATHEMATICA 8 and can be installed by just evaluating

```
Import["https://raw.githubusercontent.com/FeynCalc/feyncalc/master/install.m"]
InstallFeynCalc[]
```

in a MATHEMATICA notebook. The automatic installer also handles the setup of FEYNARTS [9], a MATHEMATICA package that can generate Feynman diagrams, which can be then (after some conversion) evaluated with FEYNCALC. An example for the conversion from FEYNARTS to FEYNCALC will be described in Sec. 3.

Tensor decomposition of 1-loop integrals using the `TID` routine has received a lot of improvements in FEYNCALC 9 and can in principle handle 1-loop tensor integrals of arbitrary rank and multiplicity. Provided that there are no vanishing Gram determinants, one can choose (via the option `UsePaVeBasis`) between the full decomposition into Passarino-Veltman scalar functions

```
In[1]:= int = FCI[GAD[μ].(m + GSD[q]).GAD[μ] FAD[{q, m},{q−p}]]
```

$$\text{Out[1]:=} \frac{\gamma^\mu.(m+\gamma\cdot q).\gamma^\mu}{(q^2-m^2).(q-p)^2}$$

```
In[2]:= TID[int, q]//ToPaVe[#, q]&
```

$$\text{Out[2]=} \frac{i\pi^2(D-2)\text{A}_0(m^2)\gamma\cdot p}{2p^2} - \frac{i\pi^2\text{B}_0(p^2,0,m^2)(Dm^2\gamma\cdot p-2Dmp^2+Dp^2\gamma\cdot p-2m^2\gamma\cdot p-2p^2\gamma\cdot p)}{2p^2}$$

and the decomposition into Passarino-Veltman coefficient functions

```
In[3]:= TID[int, q, UsePaVeBasis -> True, PaVeAutoReduce -> False] //
        ToPaVe[#, q] &
```

$$\text{Out[3]:=} i\pi^2\text{B}_0(p^2,0,m^2)(Dm - D\gamma\cdot p + 2\gamma\cdot p) - i\pi^2(D-2)\gamma\cdot p\text{B}_1(p^2,0,m^2)$$

Zero Gram determinants are automatically recognized and force the algorithm to avoid the full decomposition

```
In[4]:= FCClearScalarProducts[]
        SPD[p1, p2] = x;
        SPD[p1, p1] = x;
        SPD[p2, p2] = x;
        FCI[FVD[l, μ] FVD[l, ν] FVD[l, ρ] FAD[{l, m0}, {l + p1, m1}, {l + p2, m2}]]
```

$$\text{Out[4]:=}\ \frac{l^{\mu}l^{\nu}l^{\rho}}{\left(l^2-\text{m0}^2\right).\left((l+\text{p1})^2-\text{m1}^2\right).\left((l+\text{p2})^2-\text{m2}^2\right)}$$

```
In[5]:= TID[(1/(I Pi^ 2)) %, l]
```

$$\begin{aligned}\text{Out[6]:=}\ &(\text{p1}^{\rho}g^{\mu\nu}+\text{p1}^{\nu}g^{\mu\rho}+\text{p1}^{\mu}g^{\nu\rho})\,\text{C}_{001}\left(x,0,x,\text{m0}^2,\text{m1}^2,\text{m2}^2\right)\\&+(\text{p2}^{\rho}g^{\mu\nu}+\text{p2}^{\nu}g^{\mu\rho}+\text{p2}^{\mu}g^{\nu\rho})\,\text{C}_{002}\left(x,0,x,\text{m0}^2,\text{m1}^2,\text{m2}^2\right)\\&+\text{p1}^{\mu}\text{p1}^{\nu}\text{p1}^{\rho}\text{C}_{111}\left(x,0,x,\text{m0}^2,\text{m1}^2,\text{m2}^2\right)\\&+(\text{p1}^{\nu}\text{p1}^{\rho}\text{p2}^{\mu}+\text{p1}^{\mu}\text{p1}^{\rho}\text{p2}^{\nu}+\text{p1}^{\mu}\text{p1}^{\nu}\text{p2}^{\rho})\,\text{C}_{112}\left(x,0,x,\text{m0}^2,\text{m1}^2,\text{m2}^2\right)\\&+(\text{p1}^{\rho}\text{p2}^{\mu}\text{p2}^{\nu}+\text{p1}^{\nu}\text{p2}^{\mu}\text{p2}^{\rho}+\text{p1}^{\mu}\text{p2}^{\nu}\text{p2}^{\rho})\,\text{C}_{122}\left(x,0,x,\text{m0}^2,\text{m1}^2,\text{m2}^2\right)\\&+\text{p2}^{\mu}\text{p2}^{\nu}\text{p2}^{\rho}\text{C}_{222}\left(x,0,x,\text{m0}^2,\text{m1}^2,\text{m2}^2\right)\end{aligned}$$

where we multiplied the result by $\frac{1}{i\pi^2}$ to cancel the normalization factor that appears in the conversion of loop integrals to Passarino-Veltman functions.

With the new function `FCMultiLoopTID` it is now also possible to compute tensor decompositions of multi-loop integrals, although in this case a special treatment of integrals with zero Gram determinants is not available yet.

```
In[7]:= FCI[FVD[q1, μ] FVD[q2, ν] FAD[q1, q2, {q1 - p1}, {q2 - p1}, {q1 - q2}]]
```

$$\text{Out[7]:=}\ \frac{\text{q1}^{\mu}\text{q2}^{\nu}}{\text{q1}^2.\text{q2}^2.(\text{q1}-\text{p1})^2.(\text{q2}-\text{p1})^2.(\text{q1}-\text{q2})^2}$$

```
In[8]:= FCMultiLoopTID[%, {q1, q2}]
```

$$\begin{aligned}\text{Out[8]:=}\ &\frac{D\text{p1}^{\mu}\text{p1}^{\nu}-\text{p1}^2g^{\mu\nu}}{4(D-1)\text{q2}^2.\text{q1}^2.(\text{q2}-\text{p1})^2.(\text{q1}-\text{q2})^2.(\text{q1}-\text{p1})^2}-\\&\frac{\text{p1}^2g^{\mu\nu}-\text{p1}^{\mu}\text{p1}^{\nu}}{2(D-1)\text{p1}^2\text{q2}^2.\text{q1}^2.(\text{q2}-\text{p1})^2.(\text{q1}-\text{p1})^2}+\frac{\text{p1}^2g^{\mu\nu}-\text{p1}^{\mu}\text{p1}^{\nu}}{(D-1)\text{p1}^2\text{q2}^2.\text{q1}^2.(\text{q1}-\text{q2})^2.(\text{q1}-\text{p1})^2}-\frac{D\text{p1}^{\mu}\text{p1}^{\nu}-\text{p1}^2g^{\mu\nu}}{2(D-1)\text{p1}^4\text{q1}^2.(\text{q2}-\text{p1})^2.(\text{q1}-\text{q2})^2}\end{aligned}$$

Simplification of scalar loop integrals with linearly dependent propagators is another feature of FeynCalc that was greatly improved in the version 9 by adopting the partial fractioning algorithm of F. Feng [10]. The corresponding function ApartFF is not only faster and more efficient than the old ScalarProductCancel (which is now considered legacy), but is also capable to handle multi-loop integrals

```
In[9]:= FCI[SPD[p, q1] ^2 SPD[p, q2] FAD[{q1, m}, {q2, m}, q1 - p, q2 - p, q1 - q2]]
```

$$\text{Out[9]:=}\ \frac{(p\cdot\text{q2})(p\cdot\text{q1})^2}{\left(\text{q1}^2-m^2\right).\left(\text{q2}^2-m^2\right).(\text{q1}-p)^2.(\text{q2}-p)^2.(\text{q1}-\text{q2})^2}$$

```
In[10]:= ApartFF[%,{q1,q2}]
```

$$\begin{aligned}\text{Out[10]:=}\ &\frac{\left(m^2+p^2\right)^3}{8\left(\text{q1}^2-m^2\right).\left(\text{q2}^2-m^2\right).(\text{q2}-p)^2.(\text{q1}-\text{q2})^2.(\text{q1}-p)^2}-\frac{\left(m^2+p^2\right)^2}{4\left(\text{q1}^2-m^2\right).\left(\text{q2}^2-m^2\right).(\text{q1}-\text{q2})^2.(\text{q1}-p)^2}\\&+\frac{\left(m^2+p^2\right)^2}{4\text{q2}^2.\text{q1}^2.\left((\text{q1}-p)^2-m^2\right).(\text{q1}-\text{q2})^2}+\frac{\left(m^2+p^2\right)\left(p^2-p\cdot\text{q1}\right)}{4\text{q2}^2.\text{q1}^2.(\text{q1}-\text{q2})^2.\left((\text{q2}-p)^2-m^2\right)}\\&-\frac{\left(m^2+p^2\right)(p\cdot\text{q1})}{4\left(\text{q1}^2-m^2\right).\left(\text{q2}^2-m^2\right).(\text{q2}-p)^2.(\text{q1}-\text{q2})^2}-\frac{p\cdot\text{q1}}{4\left(\text{q2}^2-m^2\right).(\text{q1}-\text{q2})^2.(\text{q1}-p)^2}-\frac{m^2+p\cdot\text{q1}+p^2}{4\left(\text{q1}^2-m^2\right).(\text{q2}-p)^2.(\text{q1}-\text{q2})^2}\\&+\frac{m^2+2(p\cdot\text{q1})+p^2}{8\left(\text{q1}^2-m^2\right).\left(\text{q2}^2-m^2\right).(\text{q2}-\text{q1})^2}\end{aligned}$$

A more detailed description of these and other new and improved functions can be found in [8].

Journal of Physics: Conference Series **762** (2016) 012064 doi:10.1088/1742-6596/762/1/012064

3. Gluon self-enery at 1-loop in QCD

To make some connection to more realistic examples, let us look at the 1-loop gluon self-energy in QCD. While this is, of course, a standard textbook calculation that can be done by pen and paper with comparably little effort, it shows how FeynCalc can be used to evaluate Feynman diagrams generated by FeynArts.

The first step is always to load FeynCalc with the global option `$LoadFeynArts` set to `True`, which ensures that the patched copy of FeynArts, that usually resides inside the FeynCalc installation, will we loaded in a proper way. It is also convenient to set the option `$FAVerbose` to `0`, which prevents FeynArts from displaying too much extra information.

```
In[1]:= $LoadFeynArts = True;
<< FeynCalc`
$FAVerbose = 0;
```

Then we can use FeynArts' functions `CreateToplogies` and `InsertFields` [1] to generate the diagrams that we want to evaluate.

```
In[2]:= diags = InsertFields[CreateTopologies[1, 1 -> 1 , ExcludeTopologies -> {Tadpoles}],
    {V[5]} -> {V[5]}, InsertionLevel -> {Classes}, Model -> "SMQCD",
    ExcludeParticles -> {S[_], V[2 | 3], U[1 | 2 | 3 | 4], F[4]}];
```

With the following code

```
In[3]:= amps = FCFAConvert[CreateFeynAmp[diags, Truncated -> True, PreFactor -> -1],
   IncomingMomenta -> {p}, OutgoingMomenta -> {p}, LoopMomenta -> {q},
   DropSumOver -> True, ChangeDimension -> D, UndoChiralSplittings -> True,
  LorentzIndexNames -> {μ, ν}] /. {MQU[Index[Generation, 3]] -> m};
```

the amplitudes generated by FeynArts are converted into a valid FeynCalc input. Here the output of the standard FeynArts function `CreateFeynAmp` is passed to `FCFAConvert`, a new FeynCalc routine that handles the conversion according to the options submitted by the user. The resulting expression `amps` is a list with four entries that correspond to the four diagrams displayed in Fig. 1, where for convenience the FeynArts expression for the up-type quark mass `MQU[Index[Generation, 3]]` was replaced by `m`.

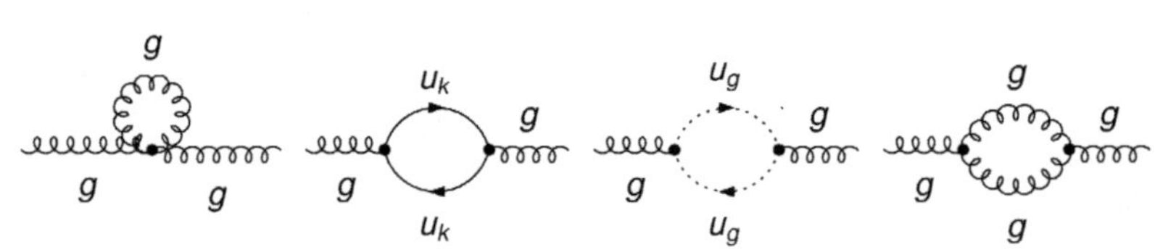

Figure 1. 1-loop contributions to the gluon self-energy in QCD

These four diagrams can be evaluated with

```
In[4]:= amps2 = TID[#, q] & /@ SUNSimplify /@ Contract /@ amps;
```

where we apply the FeynCalc functions `Contract` (contraction of Lorentz indices), `SUNSimplify` (simplification of the $SU(N)$ algebra) and `TID` (tensor reduction of 1-loop integrals) to each amplitude. We known that the first amplitude in Fig. 1 vanishes in dimensional regularization, as it is proportional to a scaleless loop integral. FeynCalc confirms that this is indeed the case

```
In[5]:= amps2[[1]]
Out[5]:= 0
```

[1] The syntax for these and other FeynArts functions can be found in the "FeynArts User's Guide", available from `www.feynarts.de`

Summing up the contributions from all the diagrams (for simplicity we consider here only one quark flavor) we obtain

In[6]:= ampsTotal = **Total**[amps2] // **Simplify**

Out[6]:= $$\frac{g_s^2 \delta^{\text{Glu1Glu2}} \left(p^\mu p^\nu - p^2 g^{\mu\nu}\right) \left(\frac{(3D-2)p^2 C_A}{q^2.(q-p)^2} - \frac{2\left((D-2)p^2+4m^2\right)}{\left(q^2-m^2\right).\left((q-p)^2-m^2\right)} + \frac{4(D-2)}{q^2-m^2} \right)}{2(D-1)p^2}$$

For convenience, we may also rewrite this result in terms of the Passarino-Veltman scalar functions A_0 and B_0 by applying the function `ToPaVe` that is available since FeynCalc 9

In[7]:= res = ampsTotal // **ToPaVe**[#, q] & // **Simplify**

Out[7]:= $$\frac{i\pi^2 g_s^2 \delta^{\text{Glu1Glu2}} \left(p^2 g^{\mu\nu} - p^\mu p^\nu\right) \left((2-3D)p^2 C_A \text{B}_0\left(p^2,0,0\right) + 2\left((D-2)p^2+4m^2\right) \text{B}_0\left(p^2,m^2,m^2\right) - 4(D-2) \text{A}_0\left(m^2\right)\right)}{2(D-1)p^2}$$

Such an output is the simplest possible form that can be achieved with FeynCalc. However, with the aid of some auxiliary tools one can in principle obtain even more explicit results with the 1-loop master integrals evaluated analytically. An add-on called FeynHelpers that connects FeynCalc to such tools is currently in development, with the early version of the code already available for testing [2]

4. Summary

In summary, we described some of the new developments in FeynCalc 9 and provided an explicit example for using the combination of FeynCalc and FeynArts to compute the 1-loop gluon self-energy in QCD. While better support for multi-loop calculations is one of the important goals in the future development of FeynCalc, we are also looking into possibilities to combine FeynCalc with other useful software tools by developing the corresponding interfaces.

Acknowledgments

This work has been supported by the DFG and the NSFC through funds provided to the Sino-German CRC 110 "Symmetries and the Emergence of Structure in QCD", and by the DFG cluster of excellence "Origin and structure of the universe" (`www.universe-cluster.de`). This report has the preprint number TUM-EFT 78/16.

References

[1] Hahn T and Perez-Victoria M 1999 *Comput. Phys. Commun.* **118** 153–165 (*Preprint* `hep-ph/9807565`)

[2] Cullen G, van Deurzen H, Greiner N, Heinrich G, Luisoni G, Mastrolia P, Mirabella E, Ossola G, Peraro T, Schlenk J, von Soden-Fraunhofen J F and Tramontano F 2014 *Eur. Phys. J. C* **74** 8, 3001 (*Preprint* `1404.7096`)

[3] Belanger G, Boudjema F, Fujimoto J, Ishikawa T, Kaneko T, Kato K and Shimizu Y 2006 *Phys. Rept.* **430** 117–209 (*Preprint* `hep-ph/0308080`)

[4] Wiebusch M 2014 *Comput. Phys. Commun.* **195** 172–190 (*Preprint* `1412.6102`)

[5] Patel H H 2015 *Comput. Phys. Commun.* **197** 276–290 (*Preprint* `1503.01469`)

[6] Vermaseren J A M 2007 (*Preprint* `math-ph/0010025`)

[7] Mertig R, Böhm M and Denner A 1991 *Comput. Phys. Commun.* **64** 345–359

[8] Shtabovenko V, Mertig R and Orellana F 2016 (*Preprint* `1601.01167`)

[9] Hahn T 2001 *Comput. Phys. Commun.* **140** 418–431 (*Preprint* `hep-ph/0012260`)

[10] Feng F 2012 *Comput. Phys. Commun.* **183** 2158–2164 (*Preprint* `1204.2314`)

[2] github.com/FeynCalc/feynhelpers

Journal of Physics: Conference Series **762** (2016) 012065 doi:10.1088/1742-6596/762/1/012065

FormCalc 9 and Extensions

T. Hahn[1], S. Paßehr[2], C. Schappacher[3]

[1] Max-Planck-Institut für Physik, Föhringer Ring 6, D–80805 Munich, Germany,
[2] DESY, Notkestr. 85, D–22607 Hamburg, Germany,
[3] Institut für Theoretische Physik, KIT, D–76128 Karlsruhe, Germany (former address).

E-mail: `hahn@mpp.mpg.de`, `sebastian.passehr@desy.de`, `schappacher@kabelbw.de`

Abstract. We present Version 9 of the Feynman-diagram calculator FormCalc and a flexible new suite of shell scripts and Mathematica packages based on FormCalc, which can be adapted and used as a template for calculations. Report DESY 2016-060, MPP-2016-63

1. Introduction

The Mathematica package FormCalc [1] simplifies Feynman diagrams generated by FeynArts [2] up to one-loop order. It provides the analytical results and can generate Fortran or C code for the numerical evaluation of the squared matrix element.

This note presents new features in FormCalc 9:

- Combination of processes with (almost) arbitrary kinematics.
- A wider range of cuts.
- Improved code generation and driver programs.
- Optional output of non-vectorized code.
- Various QCD functions (running, $\overline{\text{MS}} \to \overline{\text{DR}}$, etc.).
- Improvements in the makefiles and the build system.

Most of these improvements are the 'collected wisdom' from several nontrivial SUSY calculations [3, 4, 5, 6] and provide, as it were, 'extended support' for the MSSMCT model file [7].

Furthermore a new suite of shell scripts and Mathematica packages based on FormCalc is introduced. Presently used for the computation of the two-loop $\mathcal{O}(\alpha_t^2)$ Higgs-mass corrections in the MSSM, this code can easily be adapted and employed as a template for further calculations, and in this sense complements and generalizes what FormCalc does at one-loop.

2. New Features and Improvements in FormCalc 9

2.1. Combining Processes

FormCalc 9 allows to combine processes with (almost) arbitrary kinematics – only the phase-space dimension must match, i.e. the number of external legs must be the same. This is mostly used in hadronic computations, but one could do e.g. $e^+e^- \to \{e^+e^-, \mu^+\mu^-, \tau^+\tau^-\}$ in one go this way.

In a first step, the user generates code for all partonic processes needed with `WriteSquaredME` as usual, though with output to separate folders via the `Folder` option. The symbol `ProcName` may be used in the latter, as in:

```
WriteSquaredME[..., Folder -> {"squaredme", ProcName}]
```

and is substituted by the process name, abridged in such a way that it can safely be used as symbol or file name.

The file `partonic.h` determines which processes are combined and how. For each subprocess an entry of the following form must be added:

```
#define PID 1
#define PARTON1 i
#define PARTON2 j
#include "folder/specs.h"
#include "parton.h"
```

- `PID` is an identifier enumerating the partonic processes and runs from 1 to the number of processes (though not necessarily in ascending order),
- `PARTON1` and `PARTON2` give the PDG codes for the incoming partons, if the `lumi_hadron.F` module is used (they are ignored otherwise),
- *folder* is the folder into which code for the subprocess has been written,
- `parton.h` comes with FormCalc (same for all subprocesses) and does the processing of the partonic information given.

Optionally, the user can choose in `process.h` whether the results shall be combined at the differential or integrated level:

- `JOIN_PARTONIC` = 0, add differential partonic cross-sections, then integrate sum (default),
- `JOIN_PARTONIC` = 1, integrate each differential partonic cross-section, then add up.

Once `partonic.h` is complete it takes just the usual `./configure` and `make` to build the code.

2.2. Veto Cuts

Before FormCalc 9, only few one-particle cuts were available. The selection was restricted since the cuts were applied by actually restricting the integration bounds, hence solvability of the cut equations was a limiting factor.

While these direct cuts are still available (and likely more efficient, in particular if large parts of phase-space are removed), a much wider range of so-called 'veto cuts' has now been added. They operate through an indicator ('veto') function which sets the cross-section to zero whenever a phase-space point removed by the cuts is sampled. The following one- and two-particle cuts are currently available and it is straightforward to add more in FormCalc's `cuts.F`:

```
CUT_E(i)                  CUT_R(i,j)
CUT_k(i)                  CUT_rho(i,j)
CUT_ET(i)                 CUT_deltay(i,j)
CUT_kT(i)                 CUT_deltaeta(i,j)
CUT_y(i)                  CUT_yprod(i,j)
CUT_eta(i)                CUT_etaprod(i,j)
CUT_deltatheta(i)         CUT_deltaalpha(i,j)
CUT_cosdeltatheta(i)      CUT_cosdeltaalpha(i,j)
                          CUT_invmass(i,j)
```

The overall cut condition is assembled from the one- and two-particle cuts in `run.F`, in the definitions for `CUT1...20`, for example

```
#define CUT1 CUT_kT(3) .gt. 10
```

requires the transverse momentum of particle #3 to be greater than 10 GeV. The `CUT1...20` definitions together make up one single logical expression and may include logical operators, e.g.

```
#define CUT1 CUT_kT(3) .gt. 10
#define CUT2 .and. CUT_kT(4) .gt. 10
```

2.3. Non-vectorized Code

FormCalc's `WriteSquaredME` function generates code which is by default vectorized for the helicities of the external particles [8].

While vectorization is meanwhile almost universally available on all modern hardware, having this feature turned on for all code is not ideal, either. For example, if the user wishes to include the generated subroutines in own code, the vectorization constitutes an extra layer of complexity.

For this reason FormCalc 9 allows to remove all vector instructions and special macros by choosing the output language with the `"novec"` qualifier, viz.

```
SetLanguage["Fortran","novec"]
SetLanguage["C","novec"]
```

2.4. Debugging and Checking Instructions

FormCalc has for long allowed to add debugging statements to the generated code, as in:

```
var = expr
DEB("var", var)
```

with (e.g.)

```
#define DEB(tag,var) print *, tag, "=", var
```

This has been extended in Version 9 to include *checking statements* of the form

```
CHK_PRE(var)
var = expr
CHK_POST("var", var)
```

which might be implemented as

```
#define CHK_PRE(var) tmp = var
#define CHK_POST(tag,var) if(abs(var-tmp) .gt. 1D7) stop tag
```

The idea here is to catch enormous variations from one parameter point to the next, which often point to some problem of the calculation, e.g. an untreated resonance. Finding the error locus is much more difficult using regular debug statements due to their copious output, which is moreover difficult to compare using standard 'diff' tools because even results that are close (but not identical) numerically are usually highlighted in the diff output.

The user can individually choose the actual debugging statement (`$DebugCmd[i]`) as well as commands issued before (`$DebugPre[i]`) and after (`$DebugPost[i]`) it, for $i = 1$ (debug), 2 (pre-check), and -2 (post-check).

2.5. New/improved QCD Functions

Various QCD utility functions have been adapted, mostly from RunDec [9], and added to the 'util' library of FormCalc. They are used e.g. for translating between various quantities in the MSSMCT model file [7].

- subroutine `AlphaS`: $\alpha_s^{\overline{\text{MS}}}(Q)$ up to 4-loop order [9],
- subroutine `MqRun`: $m_q^{\overline{\text{MS}}}(Q_1) \to m_q^{\overline{\text{MS}}}(Q_2)$ up to 3-loop order [9],

- subroutine `MqMSbar2OS`: $m_q^{\overline{\rm MS}}(Q) \to m_q^{\rm OS}$ up to 3-loop order [9],
- subroutine `AsMSbar2DRbar`: $\alpha_s^{\overline{\rm MS}} \to \alpha_s^{\overline{\rm DR}}$ up to 2-loop order [10].
- subroutine `SetQCDPara`: set inputs ($\alpha_s(M_Z)$, thresholds) for the above routines.

2.6. More Renormalization Constants

The calculation of renormalization constants (RCs) has been extended to functions of arbitrary head, e.g. mass shifts. Output is written to a separate file for each head, for example:

`RenConst[.] := ...`	$\to$	`RenConst.F, .h`	*(so far)*
`MassShift[.] := ...`	$\to$	`MassShift.F, .h`	*(new)*

This effectively creates subsets of RC-like objects, i.e. results of a loop calculation but constant (external kinematics fixed). This is useful, for example, when computing the shifts used in the iterative determination of the Yukawa-resummation parameter Δb since not all RCs, only the mass shifts, have to be recomputed in each iteration.

Results for head h can either be requested explicitly by adding h to the argument list, e.g.

`CalcRenConst[`*expr*`, MassShift]`

or computed together with the others in the usual way (no extra arguments) by adding h to the `$RenConst` list of RC-heads, e.g.

```
$RenConst = {RenConst, MassShift}
```

A model file could for example request that `MassShift` be computed by adding it to `$RenConst`.

3. Extensions

FeynArts/FormCalc have been designed, as many other software packages – and not just in high-energy physics, to do a 'complete' job. That is, all the steps from the generation of Feynman diagrams to the numerical computation of a cross-section are executed from a single control program (e.g. a single Mathematica session) – that at least is how the demo programs insinuate usage. There is nothing inherently wrong with this 'monolithic' approach, it is maybe just not obvious how to extend it beyond the limitations of the package(s) used.

Sometimes one needs to use other packages for specific tasks, or there are special requirements or adaptations necessary, for example

- Resummations (e.g. the *hbb* Yukawa coupling in the MSSM),
- Approximations (e.g. the gaugeless limit),
- K-factors,
- Nontrivial renormalization.

In the following we shall introduce a suite of scripts and small Mathematica programs [11] made for the computation of the two-loop $\mathcal{O}(\alpha_t^2)$ corrections to the Higgs masses in the MSSM [12], with optimized output suitable for inclusion in FeynHiggs [13]. While each step is specific to the corrections computed here the code can be used as a template due to its modular structure and adapted with little effort for similar calculations.

3.1. Components of the Calculation

The MSSM Higgs masses receive leading two-loop (2L) corrections of $\mathcal{O}(\alpha_s\alpha_t)$ and $\mathcal{O}(\alpha_t^2)$. We are performing a diagrammatic calculation of the latter in the full complex MSSM in the gaugeless limit at $p^2 = 0$. The following ingredients are needed:

① The unrenormalized 2L self-energies $\Sigma^{(2)}_{hh}$, $\Sigma^{(2)}_{hH}$, $\Sigma^{(2)}_{hA}$, $\Sigma^{(2)}_{HH}$, $\Sigma^{(2)}_{HA}$, $\Sigma^{(2)}_{AA}$, $\Sigma^{(2)}_{H^+H^-}$ in gaugeless approximation at $p^2 = 0$ at leading order in $\mathcal{O}(\alpha_t^2)$.

② The 1L diagrams with insertions of 1L counterterms.

③ The 2L counterterms for ①.

④ The 2L tadpoles $T_h^{(2)}$, $T_H^{(2)}$, $T_A^{(2)}$ at $\mathcal{O}(\alpha_t^2)$ appearing in ③.

3.2. Template for Calculations

Our present code is based on several Mathematica Notebooks used in the former calculation [12]. From the aspect of software development these Notebooks had various shortcomings: duplicate code (e.g. gaugeless limit implemented multiply), parallel instructions poured all over, even the requirement to run with a particular Mathematica version – all signs that the use of different packages together with various special requirements was nontrivial and that the reorganization was helpful.

We broke the calculation into several steps and implemented each step as an independent (stand-alone) shell script. These scripts are run from the command line during development, while in the production run a makefile organizes the entire sequence. In lieu of *in vivo* debugging, e.g. setting breakpoints, which is not easily possible with a shell script we have set up detailed log files. The outcome is a template for 2L calculations with optimized output in a nontrivial model with nontrivial renormalization.

The scripts use several external packages: FeynArts for the diagram generation [2] using the MSSM model file with 1L counterterms MSSMCT.mod [7], FormCalc for the 1L tensor reduction and code generation [1], and TwoCalc for the 2L tensor reduction [14].

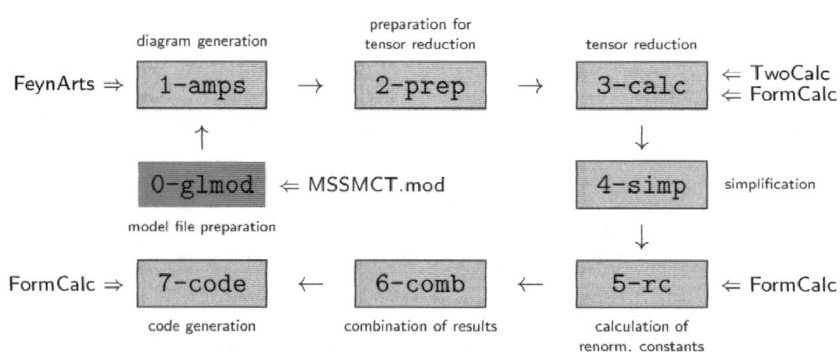

3.3. Script Structure

The shell scripts (`/bin/sh`) have several traits in common:

- Each script is run from the command line with up to two arguments, e.g.

  `scripts/1-amps` arg_1 arg_2

  where arg_1 = `h0h0`, `h0HH`, `h0A0`, `HHHH`, `HHA0`, `A0A0`, `HmHp` (self-energies), `h0`, `HH`, `A0` (tadpoles),
  arg_2 = `0` for virtual 2L diagrams, `1` for 1L diagrams with 1L counterterms.

  That is, each of the components listed in Sect. 3.1 is computed and stored individually. Arguments are checked upon entry and meaningless arguments are omitted (e.g. no arg_2 after combination of virtual and counterterm diagrams).

- Inputs and outputs are defined in the first few lines, e.g.

```
in=m/$1/2-prep.$2
out=m/$1/3-calc.$2
```

- Symbolic output including log files are written to the '`m`' subdirectory. The log file names are the output file names with `.log.gz` appended.
- Fortran code is written to the '`f`' subdirectory.
- The scripts use the shell's 'here documents' to run the Mathematica Kernel [15].

3.4. Script Tasks

In the following we shall briefly describe the main tasks of each script. The total running time for all scripts for all self-energies/tadpoles (i.e. the complete calculation) is about 15–20 min.

Step 0: Gaugeless Limit

We have to work in the gaugeless limit and also set $m_b = 0$ in order that the $\mathcal{O}(\alpha_t^2)$ corrections form a supersymmetric and gauge-invariant subset. The gaugeless limit consists of ① setting the gauge couplings $g, g' = 0$ and hence $M_W, M_Z = 0$ while keeping ② the weak mixing angle and ③ the quantities $\delta M_W^2/M_W^2$ and $\delta M_Z^2/M_Z^2$ finite.

No. ① and ② are most efficiently accomplished by modifying the Feynman rules, carried out by the `0-glmod` script:

- Load the `MSSMCT.mod` model file.
- Modify couplings and remove those that become zero.
- Write out an `MSSMCTgl.mod` model file.

Step 1: Diagram Generation The `1-amps` script generates the 2L virtual ($arg_2 = 0$) and 1L-plus-counterterm ($arg_2 = 1$) diagrams using wrappers around FeynArts functions provided by the `FASettings.m` package. The wrappers mainly simplify the diagram selection, e.g. the following list of filters, one for each topology, is composed from elementary functions like `htb[i]` which restrict the particles on the i-th propagator:

```
sel[0][S[_] -> S[_]] = {
  t[3] && htb[6],
  t[3] && tb[6],
  t[3] && tb[6],
  t[3] && t[4] && htb[5],
  t[3] && htb[5|6],
  t[3] && htb[5],
  t[3] && t[5],
  t[5] && ht[3|4],
  t[3|4|5] && ht[3|4|5] }
```

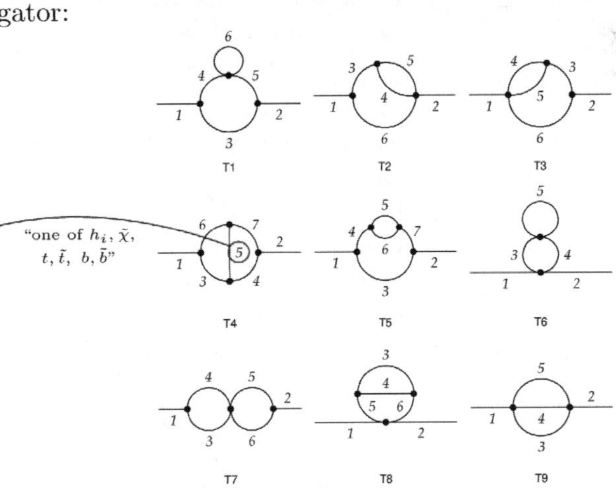

Step 2: Preparation for Tensor Reduction The tensor reduction is traditionally the step which increases the number of terms in the amplitude most, therefore we inserted Step 2 before and Step 4 after the tensor reduction to keep the number of terms as small as possible at all times.

Step 2, 'before', takes the $p^2 \to 0$ limit and simplifies the ubiquitous sfermion mixing matrices U_{ij}, mostly by exploiting unitarity ($\sim 50\%$ size reduction).

The unitarity relations of 2×2 matrices are easily written down, e.g. $U_{11}U_{11}^* + U_{12}U_{12}^* = 1$, but Mathematica rarely arranges expressions in just the way that they apply directly. This can be improved by adding definitions for single summands, e.g. $|U_{22}|^2 = |U_{11}|^2$. Because they apply *while* Mathematica ponders the simplification strategy they increase the incentive for `Simplify` to choose the unitarity-simplified version. The product of two matrix elements is too 'deep' for Mathematica to consider as an l.h.s., however, hence we must introduce intermediate symbols: we substitute `USf[1,`j`] USfC[1,`j`]` $\to$ `UCSf[1,`j`]` etc. and can then formulate unitarity as e.g. `UCSf[2,2]` $=$ `UCSf[1,1]`. The `USfSimplify.m` package contains the complete set of such rules.

Step 3: Tensor Reduction The actual tensor reduction is a relatively straightforward invocation of TwoCalc [14] for the 2L and FormCalc [1] for the 1L amplitudes. Noteworthy here is perhaps that TwoCalc and FormCalc cannot be loaded into one Mathematica session because of symbol conflicts; this is easily accomodated in the shell-script setup, however.

Step 4: Simplification The amplitudes are simplified again after tensor reduction using a largely empirical recipe. The following trick, lifted from [16], enhances simplification efficiency greatly: already during diagram generation a function of the form `DiagMark[`m_i`]` has been inserted by the FeynArts wrapper functions, where m_i are the masses running in the loop. Few simplifications can be made between parts with different `DiagMark`, hence simplification can be restricted to the prefactors of the `DiagMark`, rather than the entire expression, which is of course much faster.

Step 5: Calculation of Renormalization Constants The 1L RCs are computed with FormCalc. As mentioned in Step 0, $\delta M_V^2/M_V^2$ ($V = W, Z$) must remain finite in the gaugeless limit in which M_V vanishes. By substituting the explicit mass dependence: `dMVsq1` $\to$ `MV2 dMVsq1MV2` the masses cancel and we can take the limit safely.

Furthermore we insert expansions of the loop integrals in $\varepsilon = \frac{4-D}{2}$ (package `ExpandDel.m`) and collect the RCs w.r.t. powers of ε. For example, the 1L top-mass counterterm becomes

```
{ dMf1[3,3] → RC[-1, dMf1[-1,3,3]] + RC[0, dMf1[0,3,3]],
  {dMf1[-1,3,3] → ..., dMf1[0,3,3] → ...} }
```

where the first line denotes the expansion in ε (the interim function `RC` is used for power counting later) while the second gives the actual expressions for the coefficients. This enables us to write down just the ε^0-coefficient of the final result in Step 6 without substituting back all RCs.

Step 6: Combination of Results The virtual 2L and 1L-plus-counterterm amplitudes, computed separately so far, are combined with the genuine 2L counterterms (hand-coded, in `MSSMCT.rc2`) into one expression for each self-energy/tadpole. We insert the ε-expansions of the RCs from Step 5 and once more those of the loop integrals, and extract the coefficient of ε^0 (or of ε^{-1} or ε^{-2} if a debug flag is set). This expression is simplified again and forms then the final analytical result.

Step 7: Code Generation The renormalized self-energies are grouped into three subroutines for output, `TLsp_atat_c` for the ones needed for renormalization (`A0A0`, `HmHp`), `TLsp_atat_e` for the CP-even ones, and `TLsp_atat_o` for the CP-odd ones. These are the units in which FeynHiggs needs to compute the self-energies based on its flags.

Abbreviations are introduced for loop integrals and common subexpressions to shorten the overall expression ($\sim 50\%$ size reduction). Optimized Fortran code is then written out using FormCalc's `WriteExpr` and ancillary functions [17]. Finally, static code is added which embeds the generated routines into FeynHiggs. The Fortran output directory '`f`' totals about 350 kBytes in size and can directly be moved into the FeynHiggs source tree.

4. Résumé
FormCalc Version 9, available for download from `http://feynarts.de/formcalc`, introduces various new features, most adapted/generalized from real-life SUSY calculations:

- Combination of processes,
- More cuts,
- Improved code generation,
- New/improved driver and utility programs.

As a showcase of how to flexibly use FeynArts and FormCalc together with other packages, a suite of shell scripts and Mathematica packages was presented, originally made for the computation of the two-loop $\mathcal{O}(\alpha_t^2)$ Higgs-mass corrections, which may serve as template for similar calculations:

- Two-loop,
- Nontrivial model (MSSM) and renormalization,
- Specific approximations (gaugeless, $p^2 = 0$),
- Optimized, compact output.

The code is included in FeynHiggs 2.11.0 and above in the `gen/tlsp/` subdirectory.

[1] Hahn T, Pérez-Victoria M, 1999, *Comput. Phys. Commun.* **118** 153 [hep-ph/9807565].
[2] Hahn T, 2001, *Comput. Phys. Commun.* **140** 418 [hep-ph/0012260].
[3] Fritzsche T, Heinemeyer S, Rzehak H, Schappacher C, 2011, *Phys. Rev.* **D86** 035014 [arXiv:1111.7289].
[4] Heinemeyer S, Schappacher C, 2012, *Eur. Phys. J.* **C72** 1905 [arXiv:1112.2830].
[5] Heinemeyer S, Schappacher C, 2015, *Eur. Phys. J.* **C75** 198 [arXiv:1410.2787].
[6] Heinemeyer S, Schappacher C, arXiv:1511.06002.
[7] Fritzsche T, Hahn T, Heinemeyer S, von der Pahlen F, Rzehak H, Schappacher C, 2014, *Comput. Phys. Commun.* **185** 1529 [arXiv:1309.1692].
[8] Chokoufe Nejad B, Hahn T, Lang JN, Mirabella E, 2010, *J. Phys. Conf. Ser.* **523** 012050 [arXiv:1310.0274].
[9] Chetyrkin KG, Kühn JH, Steinhauser M, 2000, *Comput. Phys. Commun.* **133** 43 [hep-ph/0004189].
[10] Harlander R, Mihaila L, Steinhauser M, 2007, *Phys. Rev.* **D76** 055002 [arXiv:0706.2953].
[11] Hahn T, Paßehr S, arXiv:1508.00562.
[12] Hollik W, Paßehr S, 2014, *Phys. Lett.* **B733** 144 [arXiv:1401.8275].
[13] Frank M, Hahn T, Heinemeyer S, Hollik W, Rzehak H, Weiglein G, 2007, *JHEP* **0702** 047 [hep-ph/0611326].
[14] Weiglein G, Scharf R, Böhm M, 1994, *Nucl. Phys.* **B416** 606 [hep-ph/9310358],
Weiglein G, Mertig R, Scharf R, Böhm M, 1992, in: New computing techniques in physics research II, 617–623, La Londe-les-Maures.
[15] Hahn T, Illana J, 2006, *Nucl. Phys. Proc. Suppl.* **160** 101 [hep-ph/0607049].
[16] Heinemeyer S, Stöckinger D, Weiglein G, 2004, *Nucl. Phys.* **B690** 62 [hep-ph/0312264] and 2004, *Nucl. Phys.* **B699** 103 [hep-ph/0405255].
[17] Hahn T, 2010, PoS ACAT **2010** 078 [arXiv:1006.2231].

Fun with higher-loop Feynman diagrams

Thomas Luthe[1] **and York Schröder**[2]

[1] Institut für Theoretische Teilchenphysik, Karlsruhe Institute of Technology (KIT), Karlsruhe, Germany

[2] Grupo de Fisica de Altas Energias, Universidad del Bio-Bio, Casilla 447, Chillan, Chile

E-mail: thomas.luthe@kit.edu, yschroeder@ubiobio.cl

Abstract. We review recent progress that we have achieved in evaluating the class of fully massive vacuum integrals at five loops. After discussing topics that arise in classification, evaluation and algorithmic codification of this specific set of Feynman integrals, we present some selected new results for their expansions around $4-2\varepsilon$ dimensions.

1. Introduction

In high-energy physics experiments performed at current colliders such as the LHC, the flood of precision data requires matching theoretical efforts, in order extract the underlying event's structure. This is particularly relevant for Run II, where important processes are plagued with large QCD backgrounds, requiring high-precision theory calculations.

To this end, we will showcase a few techniques and results related to investigations of the structure of higher-loop Feynman integrals which provide one of the basic building blocks of high-precision perturbative calculations within elementary particle physics. We will discuss new results on the current (five-)loop frontier, which constitute basic building blocks for important quantities such as anomalous dimensions in gauge field theories such as QCD.

2. Classification

Let us consider fully massive 5-loop tadpoles, in Euclidean space-time, having the same mass m in all propagators $1/(q_i^2+m^2)$. Due to the absence of external legs there are no further scales, such that all integrals scale trivially with the mass (which we can hence set to unity), such that our problem corresponds to computing scale-free functions of the space-time dimension d only. The corresponding 4-loop problem has been solved already more than a decade ago in 4d [1] as well as 3d [2].

To fully define the 5-loop integral family, we need 15 propagators (lines), for which we choose

$$q_i \in \{k_1, k_2, k_3, k_4, k_5, k_{13}, k_{14}, k_{15}, k_{23}, k_{24}, k_{25}, k_{35}, k_{45}, k_{124}, k_{34}\}\ , \qquad (1)$$

where $k_{a\ldots bc} = k_a + \cdots + k_b - k_c$. Trivalent 5-loop graphs have 12 lines, and there are four independent such sectors (one of them corresponding to the last diagram of Fig. 1). We label different integral sectors by their binary representation (where a 1 at position j corresponds to a line with momentum q_j in the representation Eq. (1)), identify unique graphs, find all isometries and corresponding momentum shifts, and choose the largest representative from each class.

Figure 1. Some examples of 5-loop master integrals with 'dots' that we choose as elements of our basis. Each solid line corresponds to a massive propagator with unit power, while each dot represents an extra power. Note that we push all dots on a single line, which is the most natural choice with respect to the difference equation setup for their evaluation.

Furthermore, using the freedom of normalization of the integral measure, we find it convenient to divide out a factor of [1-loop tadpole]$^{\#loops}$ from each integral, achieving a convention-independent normalization of loop integrals. It is useful to recall that in 4d, the [1-loop tadpole]$\sim 1/\varepsilon$, while it is finite in 3d.

To show some 5-loop numerology: of the $2^{15} = 32768$ possible sectors that can be constructed from our list of momenta, 5151 do not correspond to a Feynman graph; 1941 of them are trivially zero (in dimensional regularization, since they do not depend on one of the loop momenta), while a further 3625 vanish non-trivially (since they can be shifted such that they become independent of one of the integration momenta); 21962 sectors can be transformed (by linear momentum shifts with unit Jacobian) to a larger representative, while 22 others never occur in a sane Feynman diagram labelling (since they would require a shift with Jacobian 1/2); leaving us with 48 unique irreducible 5-loop master sectors, plus 19 unique master sectors that correspond to products of lower-loop integrals. To compare, at 1/2/3/4-loop there were 1/1/3/10 unique sectors (plus 0/1/2/6 factorized ones).

While each of the 48+19 sectors contains at least one master integral (which can be chosen to be the 'corner integral' having all lines with power one), a (small) IBP reduction reveals that some sectors contain multiple master integrals. In particular, we need in 62 (+3 factorized ones) additional masters with 'dots', examples of which are shown in Fig. 1. Recall that at 1/2/3/4-loop there were only 0/0/0/3 masters with 'dots' [3].

3. Evaluation

Generalizing our set of master integrals to carry a symbolic power x on one of their lines, we can perform an IBP reduction on this modified set of masters and derive a hierarchical system of difference equations for all of them. Symbolically,

$$I(x) \equiv \int \frac{1}{D_1^x D_2 ... D_N} , \qquad \sum_{j=0}^{R} p_j(x) I(x+j) = F(x) \tag{2}$$

where the integer R is called order of the equation, the p_j are rational functions (in two variables, x and d), and the functions F on the right-hand sides contain integrals that are simpler (according to the ordering prescription chosen). Boundary conditions for these equations are simple, and can be derived by shrinking a line (i.e., at $x = 0$) or be related to lower-loop integrals (i.e., expanding the integrand for $x \gg 1$).

Typically, one finally wants to extract the value for $I(1)$. To this end, one needs to solve the difference equation; this is possible explicitly for $R = 1$, while higher-order equations call for a much more general setup [4]. Laporta's recipe suggests to first attempt a general solution in terms of factorial series, schematically:

$$I(x) = I_0(x) + \sum_{j=1}^{R} I_j(x) , \quad \text{where} \quad I_j(x) = \mu_j^x \sum_{s=0}^{\infty} a_j(s) \frac{\Gamma(x+1)}{\Gamma(x+1+s-K_j)} . \tag{3}$$

The difference equation then determines the special solution I_0, fixes the constants μ_j, $K_j(d)$ and implies a recurrence relation for the series coefficients $a_j(s)$ of the homogeneous solutions. Second, one needs a boundary condition for fixing, say, $a_j(0)$; it is useful to observe decoupling at large x, providing a particularly simple and automatable way to compute the necessary boundary values:

$$I(x) \;=\; \int_{k_1} \frac{g(k_1)}{(k_1^2+1)^x} \quad\Rightarrow\quad I(x) \sim (1)^x x^{-d/2} g(0)\;. \tag{4}$$

This is then to be compared with the large-x behavior of the factorial series, $\sum_j \mu_j^x x^{K_j} a_j(0)$.

Third, having all analytic ingredients for their solution in terms of infinite sums at hand, to obtain actual results for the master integrals it is most useful to resort to a numeric treatment. This involves truncating the sum in Eq. (3) at some $s_{\max}$, evaluating R consecutive values of $I(x)$ around some large x (where the factorial series converges well), and using the difference equation Eq. (2) to push their argument down to the required integrals, such as $I(1)$.

4. Implementation

The above method brings with it a number of advantages: it allows for a high level of automation; works well with divergent integrals; does not rely on specific function classes; gives high-precision results for arbitrary ε orders; allows for expansions around any dimension; provides simple but highly non-trivial cross-checks by putting x on different lines.

However, we have found that, following the program proposed by Stefano Laporta [4], one is faced with some problems and limitations when treating complex problems: the method is of limited use for multi-scale integrals; the complexity of coefficients in high-order equations can grow enormously; one typically obtains recurrence relations of high orders; in numerical evaluation, one often faces instabilities of the factorial series solutions.

To alleviate these problems and make feasible the computation of our set of 5-loop massive tadpoles, we had to considerably change the traditional setup. Much of our progress can be attributed to a number of fixes done to the original proposal. To name a few key ingredients [5]: we use coupled IBP equations, in order to tame the growth of complexity; we reduce (the orders of) recurrence relations, essentially by re-using the linear solver developed for the system of difference equations; our code predicts instability factors, in order to assign sufficient numerical precision.

5. Results

In order to be specific, let us focus on results in $d = 4 - 2\varepsilon$ here, recalling that we normalize all our integrals with the corresponding power of the massive 1-loop tadpole. Given our high-precision numerical results, one might wonder whether it is possible to determine some of the coefficients in analytic form. To this end, we employ the integer-relation finder `PSLQ` [6] together with an educated guess of a basis of irrationals (such as e.g. the well-classified set of multiple zeta values and/or alternating Euler sums), to guess some of these coefficients. It helps to keep the basis as compact as possible, so in order to absorb single powers of π as well as powers of $\ln 3$, we define

$$h_n \;\equiv\; \sum_{k=0}^{\infty} \frac{\Gamma(k+1/2)}{\Gamma(k+1)\Gamma(1/2)}\,\frac{(3/4)^k}{(2k+1)^n} = {}_{n+1}F_n[\underbrace{\{1/2,..,1/2\}}_{n+1},\underbrace{\{3/2,..,3/2\}}_{n},3/4]\;, \tag{5}$$

$$H_n \;\equiv\; h_n + h_1\,\mathrm{Coefficient}\Big[1-\frac{3^{\varepsilon/2}\Gamma(1-\varepsilon)}{\Gamma^2(1-\varepsilon/2)}+\mathcal{O}(\varepsilon^n),\varepsilon,n-1\Big]\;, \tag{6}$$

$$\Rightarrow\; H_1 = h_1 = \frac{2\pi}{3\sqrt{3}}\;,\quad H_2 = h_2 - \frac{1}{2}h_1\ln 3\;,\quad H_3 = h_3 - \frac{h_1}{8}(\ln^2 3 + 2\zeta_2)\;,\quad etc. \tag{7}$$

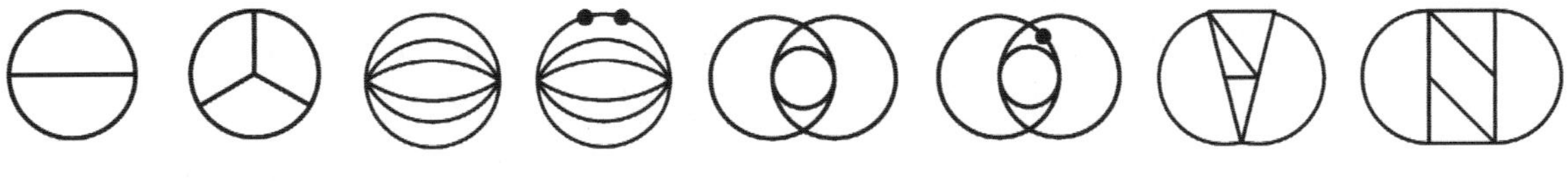

Figure 2. Sample integrals for which results are given in Eqs. (8), (12), (13-18).

The specific combinations H_n have actually been inspired by the 2-loop sunset (leftmost diagram of Fig. 2), which can be written to all orders in terms of the H_n, the first few of which read[1]

$$I_{7.1.1} = -\frac{3}{2} - \frac{3}{2}\,\varepsilon + (9H_2 - 3)\,\varepsilon^2 + (9H_2 - 18H_3 - 6)\,\varepsilon^3 + (18H_2 - 18H_3 + 36H_4 - 12)\,\varepsilon^4\,. \quad (8)$$

Furthermore, in order to absorb powers of $\ln 2$, let us define as elements of the MZV basis

$$a_n \equiv \sum_{k=0}^{\infty} \frac{1}{2^k\,k^n} = \mathrm{PolyLog}[n, 1/2] = \mathrm{Li}_n(\frac{1}{2})\,, \quad (9)$$

$$A_n \equiv a_n + (-1)^n \frac{\ln^n 2}{n!}\Big(1 - \frac{n(n-1)}{2}\,\frac{\zeta_2}{\ln^2 2}\Big)\,, \quad (10)$$

$$\Rightarrow\ A_4 = a_4 - \zeta_2 \frac{\ln^2 2}{2\cdot 2!} + \frac{\ln^4 2}{4!}\,, \quad A_5 = a_5 + \zeta_2 \frac{\ln^3 2}{2\cdot 3!} - \frac{\ln^5 2}{5!}\,, \quad etc., \quad (11)$$

such that e.g. the 3-loop mercedes-type integral (second diagram of Fig. 2) reads

$$I_{63.1.1} = +(0)\,\varepsilon^0 + (0)\,\varepsilon^1 + (-2\zeta_3)\,\varepsilon^2 + (-16A_4 + 27H_2^2 + \frac{34\zeta_2^2}{5})\,\varepsilon^3 + \dots\,. \quad (12)$$

Our numerical approach has passed a number of cross-checks with flying colors: placing the power x of each generalized master on topologically inequivalent lines, the values at $x = 1$ coincide; all known results up to 4-loop level [1, 7–9] have been reproduced or improved; all factorized sectors perfectly agree with the respective products of lower-loop integrals; and the few previously known 5-loop coefficients [10–12] have been reproduced. Let us finally show a couple of new results on the five-loop level (for the rightmost 6 graphs shown in Fig. 2), using our `PSLQ` fits for low ε-orders (the divergent ones), and truncating the constant term to 50 numerical digits for readability:

$$I_{28686.1.1} = +(-3)\,\varepsilon^0 + (-\frac{3}{2})\,\varepsilon^1 + (\frac{13}{24})\,\varepsilon^2 + (-\frac{1267}{1440})\,\varepsilon^3 + (-\frac{4193}{3456})\,\varepsilon^4 + {} +135.95072868792871461956492733702218574897992953584\,\varepsilon^5 + \dots\,, \quad (13)$$

$$I_{28686.1.3} = +(0)\,\varepsilon^0 + (\frac{3}{2})\,\varepsilon^1 + (-\frac{1}{2})\,\varepsilon^2 + (-\frac{443}{360})\,\varepsilon^3 + (\frac{95}{216})\,\varepsilon^4 + {} -38.292059175062436961881799538284449799148385376441\,\varepsilon^5 + \dots\,, \quad (14)$$

$$I_{30862.1.1} = +(-\frac{3}{5})\,\varepsilon^0 + (-\frac{27}{10})\,\varepsilon^1 + (-\frac{4\zeta_3}{5} - \frac{421}{60})\,\varepsilon^2 + (-\frac{12\zeta_2^2}{25} + \frac{24\zeta_3}{5} + \frac{211}{24})\,\varepsilon^3 + (\frac{72\zeta_2^2}{25} - 98\zeta_3 + \frac{32\zeta_5}{5} + \frac{12959}{48})\,\varepsilon^4 + {} +1143.1838307558764599466030303839590323268318605888\,\varepsilon^5 + \dots\,, \quad (15)$$

[1] Regarding our notation for integral labels: $I_{a,b,c}$ refers to the master integral with binary code a as explained in Sec. 2, having propagators with unit powers, except line number b, which carries exponent c (or $c-1$ 'dots').

$$\begin{aligned}
I_{30862.1.2} &= +(\frac{1}{5})\,\varepsilon^0 + (\frac{11}{30})\,\varepsilon^1 + (-\frac{1}{30})\,\varepsilon^2 + (-\frac{12\zeta_3}{5} - 9)\,\varepsilon^3 + (-\frac{36\zeta_2^2}{25} + \frac{548\zeta_3}{15} - \frac{1229}{15})\,\varepsilon^4 + \\
&\quad -102.42854342605587086319606311891941160276036031953\,\varepsilon^5 + \ldots\,, \qquad (16) \\
I_{30231.1.1} &= +(0)\,\varepsilon^0 + (0)\,\varepsilon^1 + (\frac{3\zeta_3}{5})\,\varepsilon^2 + (\frac{9\zeta_2^2}{25} + \frac{21\zeta_3}{5} + 3\zeta_5)\,\varepsilon^3 \\
&\quad +(-36H_2\zeta_3 + \frac{12\zeta_2^3}{7} + \frac{63\zeta_2^2}{25} - \frac{21\zeta_3^2}{5} + 27\zeta_3 - \frac{24\zeta_5}{5})\,\varepsilon^4 + \\
&\quad -531.32391547725635267943444561495368318398901378435\,\varepsilon^5 + \ldots\,, \qquad (17) \\
I_{31420.1.1} &= +(0)\,\varepsilon^0 + (0)\,\varepsilon^1 + (0)\,\varepsilon^2 + (0)\,\varepsilon^3 + (-\frac{36\zeta_3^2}{5})\,\varepsilon^4 + \\
&\quad +167.81535305918474061962120601112466233675898298296\,\varepsilon^5 + \ldots\,. \qquad (18)
\end{aligned}$$

6. Conclusions

We have studied the class of fully massive vacuum diagrams, which are essential building blocks for a number of phenomenological applications such as QCD thermodynamics, anomalous dimensions, or moments. A classification of all 5-loop integrals contained in this class reveals 48 independent sectors, with 48+62 master integrals (the latter having 'dots').

For the evaluation, we have chosen to employ a hierarchical system of difference equations and factorial series, amenable to automated numerical treatment. The setup required for deriving the difference equations, the recursion relations for factorial series coefficients, as well as the numerical solution has been implemented in the `C++` code `TIDE` [5], essential parts of which are parallelized. We use `Fermat` [13] for polynomial algebra in 2 variables, and manage to employ a substantial fine-tuning of the Laporta approach, without which these results could not have been derived on the computer hardware that we have at our disposal. At 5 loops, our code had to deal with difference equations of up to order 20, and recurrence relations of up to order 28.

To date, we have numerical results for 44 of the 48 non-factorized 5-loop master-sectors (including the 'dotted' integrals of those sectors), with about 300 digits accuracy, for at least 10 ε-orders around $d = 4 - 2\varepsilon$ and $d = 3 - 2\varepsilon$. These results include all divergent integrals, bringing 5-loop anomalous dimensions within reach (once the integral reduction has been performed).

Acknowledgments

The work of T.L. has been supported in parts by DFG grants GRK 881 and SCHR 993/2, and he wishes to thank the Bielefeld Graduate School in Theoretical Sciences for facilitating finalizing this project via a BGTS mobility grant. Y.S. acknowledges support from DFG grant SCHR 993/1, FONDECYT project 1151281 and UBB project GI-152609/VC.

References

[1] Laporta S 2002 *Phys. Lett.* **B549** 115–122 (*Preprint* `hep-ph/0210336`)
[2] Schroder Y and Vuorinen A 2003 (*Preprint* `hep-ph/0311323`)
[3] Schroder Y 2003 *Nucl. Phys. Proc. Suppl.* **116** 402–406 (*Preprint* `hep-ph/0211288`)
[4] Laporta S 2000 *Int. J. Mod. Phys.* **A15** 5087–5159 (*Preprint* `hep-ph/0102033`)
[5] Luthe T PhD thesis, Bielefeld University 2015
[6] Ferguson H R P, Bailey D H and Arno S 1999 *Math. Comp.* **68** 351–369
[7] Czakon M 2005 *Nucl. Phys.* **B710** 485–498 (*Preprint* `hep-ph/0411261`)
[8] Schroder Y and Vuorinen A 2005 *JHEP* **06** 051 (*Preprint* `hep-ph/0503209`)
[9] Chetyrkin K G, Faisst M, Sturm C and Tentyukov M 2006 *Nucl. Phys.* **B742** 208–229 (*Preprint* `hep-ph/0601165`)
[10] Groote S, Korner J G and Pivovarov A A 2007 *Annals Phys.* **322** 2374–2445 (*Preprint* `hep-ph/0506286`)
[11] Broadhurst D J 1985 *Phys. Lett.* **B164** 356
[12] Brown F and Schnetz O 2012 (*Preprint* `1208.1890`)
[13] Lewis R H URL `http://home.bway.net/lewis/`

Generalizations of polylogarithms for Feynman integrals

Christian Bogner
Institut für Physik, Humboldt-Universität zu Berlin, *D - 10099 Berlin, Germany*
E-mail: bogner@math.hu-berlin.de

Abstract. In this talk, we discuss recent progress in the application of generalizations of polylogarithms in the symbolic computation of multi-loop integrals. We briefly review the Maple program MPL which supports a certain approach for the computation of Feynman integrals in terms of multiple polylogarithms. Furthermore we discuss elliptic generalizations of polylogarithms which have shown to be useful in the computation of the massive two-loop sunrise integral.

1. Motivation: Multiple polylogarithms and Feynman integrals

Classical polylogarithms Li_n are obtained as a generalization of the logarithm function

$$\mathrm{Li}_1(z) = -\ln(1-z) = \sum_{j=1}^{\infty} \frac{z^j}{j},\ |z| < 1,$$

by allowing for higher integer powers of the summation variable in the denominator:

$$\mathrm{Li}_n(z) = \sum_{j=1}^{\infty} \frac{z^j}{j^n},\ |z| < 1.$$

These functions can be expressed in terms of integrals. For the dilogarithm, Leibniz [54] already found the identity

$$\mathrm{Li}_2(z) = -\int_0^z \frac{dx}{x} \ln(1-x).$$

In general, for weights $n \geq 2$, we have

$$\mathrm{Li}_n(z) = \int_0^z \frac{dx}{x} \mathrm{Li}_{n-1}(x). \tag{1}$$

If we write all integrations on the right-hand side of this equation explicitly, we obtain an iterated integral

$$\mathrm{Li}_n(z) = \int_0^z \frac{dx_n}{x_n} \dots \int_0^{x_3} \frac{dx_2}{x_2} \int_0^{x_2} \frac{dx_1}{1-x_1}. \tag{2}$$

In this talk, we denote iterated integrals by

$$[\omega_r|...|\omega_2|\omega_1] = \int_0^z \omega_r(x_r)... \int_0^{x_3} \omega_2(x_2) \int_0^{x_2} \omega_1(x_1)$$

where the ω_i are differential 1-forms in some given set. In eq. 2 we see that the set

$$\Omega_{\text{P}} = \left\{ \frac{dx}{x}, \frac{dx}{1-x} \right\}$$

suffices to construct the classical polylogarithms.

Generalizations of polylogarithms can be obtained by either generalizing the terms in the sum representation or by extending the set of differential 1-forms. In both ways, one arrives at multiple polylogarithms. They are defined as the series [44, 45]

$$\text{Li}_{n_1,...n_k}(z_1, ..., z_k) = \sum_{0<j_1<...<j_k} \frac{z_1^{j_1}...z_k^{j_k}}{j_1^{n_1}...j_k^{n_k}} \text{ for } |z_i| < 1$$

and they can be expressed in terms of iterated integrals known as hyperlogarithms [61, 52, 53]. These are obtained from differential 1-forms of the set

$$\Omega_{\text{Hyp}} = \left\{ \frac{dx}{x}, \frac{dx}{x-y_i} \middle| i = 1, ..., k \right\}. \tag{3}$$

Some powerful methods and computer programs for the analytical computation of Feynman integrals rely on either the sum representation or on an integral representation of multiple polylogarithms. In section 2 we review the computer program MPL which supports an approach based on iterated integrals. For Feynman integrals which can not be expressed in terms of multiple polylogarithms, we are in search of alternatives. In section 3 we briefly recall the concept of elliptic functions and in section 4 we discuss an elliptic generalization of polylogarithms which arises from the computation of the massive sunrise integral.

2. Iterated integrals and the program MPL

As an alternative to hyperlogarithms, we consider a class of iterated integrals over differential 1-forms in the set

$$\Omega_{\text{MPL}} = \left\{ \frac{dx_1}{x_1}, ..., \frac{dx_k}{x_k}, \frac{d(p_{a,b})}{p_{a,b}} \middle| 1 \le a \le b \le k \right\} \tag{4}$$

where

$$p_{a,b} = \prod_{a \le i \le b} x_i - 1. \tag{5}$$

In order to obtain a framework of well-defined functions of the k variables x_1, ..., x_k, we construct only iterated integrals which are homotopy invariant. In general, an iterated integral admits this property, if and only if it satisfies the condition [40]

$$\mathcal{D}[\omega_1|...|\omega_m] = 0 \tag{6}$$

where the operator $\mathcal{D}$ is defined by

$$\mathcal{D}[\omega_1|...|\omega_m] = \sum_{i=1}^{m} [\omega_1|...|\omega_{i-1}|d\omega_i|\omega_{i+1}|...\omega_m] + \sum_{i=1}^{m-1} [\omega_1|...|\omega_{i-1}|\omega_i \wedge \omega_{i+1}|...|\omega_m]. \tag{7}$$

For example, among the two integrals

$$I_1 = \left[\frac{dx_3}{x_3} + \frac{dx_2}{x_2} \middle| \frac{d(x_2x_3)}{x_2x_3 - 1} \right], \; I_2 = \left[\frac{dx_3}{x_3} \middle| \frac{d(x_2x_3)}{x_2x_3 - 1} \right],$$

only I_1 is homotopy invariant while I_2 fails eq. 6. For any number of variables k, we can apply algorithms described in [22, 23] to construct a basis of all homotopy invariant iterated integrals over 1-forms in Ω_{MPL}. Together with certain boundary conditions at a tangential basepoint (see [23]), this construction provides a $\mathbb{Q}$-vectorspace $V(\Omega_{\text{MPL}})$ of functions, including the class of multiple polylogarithms.

MPL [21] is a Maple program for computations with this class of functions. Its main algorithms [23] rely on the mathematical theory developed in [31]. One of the main purposes of the program is the computation of definite integrals of the type

$$I = \int_0^1 dx_n \frac{q}{\prod_j p_j^{a_j}} f \tag{8}$$

where $f \in V(\Omega_{\text{MPL}})$, q is some arbitrary polynomial in x_n, all $a_j \in \mathbb{N}$ and all p_j are polynomials of the type of eq. 5. For example, the program computes analytically

$$\int_0^1 dx_1 \int_0^1 dx_2 \int_0^1 dx_3 \frac{x_1^4(1-x_1)^4 x_2^9 (1-x_2)^4 x_3^4 (1-x_3)^4}{(1-x_1x_2)^5(1-x_2x_3)^5} = -\frac{11424695}{144} + 66002\zeta(3).$$

Such integrals appear in various contexts. Examples are given in [34, 21, 23].

The other main purpose of the program MPL is the analytical computation of a certain class of scalar Feynman integrals. For some Feynman graph G, consider the D-dimensional, scalar L-loop integral

$$I(\Lambda) = \Gamma(\nu - LD/2) \left(\prod_{i=1}^{N} \int_0^\infty \frac{dx_i x_i^{\nu_i - 1}}{\Gamma(\nu_i)} \right) \delta(H) \frac{\mathcal{U}^{\nu-(L+1)D/2}}{(\mathcal{F}(\Lambda))^{\nu - LD/2}}, \tag{9}$$

where N is the number of edges of G, ν_i are integer powers of the Feynman propagators, ν is the sum of all ν_i, Λ is a set of kinematical invariants and masses and $H = 1 - \sum_{i\in S} x_i$ for some choice of $S \subseteq \{1, ..., N\}$. The terms $\mathcal{U}$ and $\mathcal{F}$ are the Symanzik polynomials in the Feynman parameters $x_1, ..., x_N$ (see e.g. [25]). Applying the methods of [58, 56, 35] we can expand such a possibly divergent integral as a Laurent series in a parameter ϵ of dimensional regularisation,

$$I = \sum_{j=-2L}^{\infty} I_j \epsilon^j,$$

such that the integrals I_j are finite. The integrands of these I_j will involve Symanzik polynomials of G and of related graphs.

In general, Symanzik polynomials are more complicated than the polynomials of eq. 5. Therefore, the computation involves some additional steps. Before each integration, MPL attempts to express the integrand in the form of eq. 8 by an appropriate change of variables. Then the integral is computed and the result is mapped back to Feynman parameters, as a preparation of the next integration. In order for all Feynman parameters to be integrated out in this way, the (Symanzik) polynomials in the original integrand have to satisfy the condition of linear reducibility as discussed in [32, 33, 60]. This and two further conditions can be checked by the program. If they are satisfied, the integral can be computed automatically with MPL. Examples are given in [21] and in a manual obtained with the program.

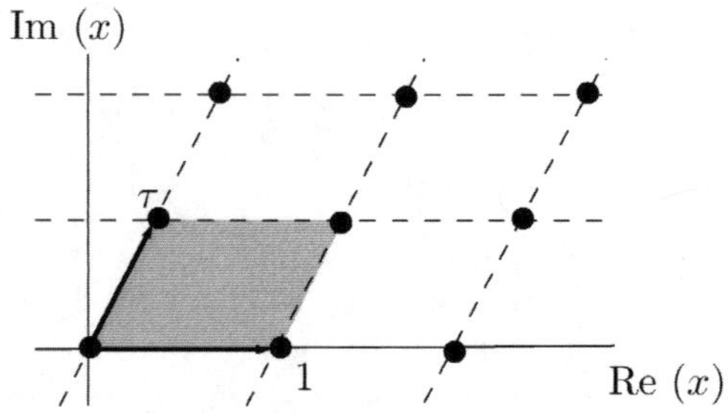

Figure 1. A lattice in the complex plane

A similar approach is followed by Panzer's program HyperInt [59], based on hyperlogarithms, which is publicly available as well, and related methods are applied by programs discussed in [1, 2, 5, 3, 4, 30].

What if a given Feynman integral does not satisfy the criterion of linear reducibility? In some cases, this problem is just an artefact of the parametrization and after some clever change of variables, the above approach can still be applied[1]. However, there are as well Feynman integrals, which can not be expressed in terms of multiple polylogarithms, no matter which parameters or classes of iterated integrals we try to apply. For such Feynman integrals, we have to turn to other frameworks of functions. The given success with multiple polylogarithms suggests to give further generalizations of polylogarithms a try.

3. Elliptic generalizations

Let us recall the basic concept of an elliptic function. In the complex plane of a variable $x \in \mathbb{C}$ we consider the lattice $L = \mathbb{Z} + \tau\mathbb{Z}$, where $\tau \in \mathbb{C}$ with $\text{Im}(\tau) > 0$ (the points in fig. 1). A function $f(x)$ is called elliptic with respect to L if

$$f(x) = f(x + \lambda) \text{ for } \lambda \in L. \tag{10}$$

It makes sense to consider such a function f only in one cell of the lattice (the grey area in fig. 1), as its behaviour in all other cells are just copies. If τ is the quotient of two periods ψ_1, ψ_2 of an elliptic curve E, this cell of the lattice is isomorphic to E and we can consider f as a function on the elliptic curve.

Now we introduce a change of variables, considering the function $f'(z)$ of $z \in \mathbb{C}^\star$ given by

$$f'\left(e^{2\pi i x}\right) = f\left(x\right).$$

Clearly, if f is elliptic with respect to L, then with respect to the new variable z, eq. 10 implies

$$f'\left(z\right) = f'\left(z \cdot q\right) \text{ where } q = e^{2\pi i \lambda} \text{ for } \lambda \in L. \tag{11}$$

Now there is a simple idea for the construction of such elliptic functions. If $f'(z)$ can be defined with the help of some other function $g(z)$ as

$$f'(z) = \sum_{n\in\mathbb{Z}} g\left(z \cdot q^n\right),$$

it satisfies eq. 11 by construction.

[1] An example for such a case is the graph found to be irreducible in [24] and later computed in [48, 58].

This concept can be applied to define elliptic generalizations of polylogarithms. A first version of an elliptic dilogarithm was defined in [17] for the single-valued Bloch-Wigner dilogarithm. Later the concept was generalized in various directions (see [66, 15, 55, 43]). Let us refer particularly to [36] where elliptic polylogarithms of the form

$$E_m(z) = \sum_{n\in\mathbb{Z}} u^n \mathrm{Li}_m\left(z\cdot q^n\right) \tag{12}$$

with a damping factor u are considered and where the concept is furthermore generalized to establish multiple elliptic polylogarithms.

In the following section, a related class of functions appears in the context of a Feynman integral.

4. The massive sunrise integral

The massive sunrise integral

$$S(D,\,t) = \int \frac{d^D k_1 d^D k_2}{\left(i\pi^{D/2}\right)^2} \frac{1}{\left(-k_1^2+m_1^2\right)\left(-k_2^2+m_2^2\right)\left(-\left(p-k_1-k_2\right)^2+m_3^2\right)}$$

is a Feynman integral which can not be expressed in terms of multiple polylogarithms. This integral was extensively considered in the literature [11, 12, 13, 14, 16, 27, 39, 37, 38, 46, 47, 51, 62, 63, 65, 64, 26, 50, 41, 42]. In a recent computation of the case $D = 2$ and equal masses, $m_1 = m_2 = m_3$, for the first time an elliptic polylogarithm was applied explicitly to express a Feynman integral [20]. Here we discuss further cases of the sunrise integral where elliptic generalizations of polylogarithms arise.

At first, let us consider the integral with three different particle masses as Laurent series at two and around four dimensions:

$$\begin{aligned} S(2-2\epsilon,\,t) &= S^{(0)}(2,\,t) + S^{(1)}(2,\,t)\epsilon + \mathcal{O}\left(\epsilon^2\right), & (13)\\ S(4-2\epsilon,\,t) &= S^{(-2)}(4,\,t)\epsilon^{-2} + S^{(-1)}(4,\,t)\epsilon^{-1} + S^{(0)}(4,\,t) + \mathcal{O}(\epsilon). & (14)\end{aligned}$$

Here we have used $t = p^2$. We begin with the result for exactly $D = 2$ dimensions, $S^{(0)}(2,\,t)$. In this case, the Feynman parametric representation (eq. 9) of the sunrise integral only involves the second Symanzik polynomial

$$\mathcal{F} = -x_1x_2x_3t + \left(x_1m_1^2 + x_2m_2^2 + x_3m_3^2\right)\left(x_1x_2 + x_2x_3 + x_1x_3\right)$$

whose zero-set intersects the integration domain at three points P_1, P_2, P_3. Together with each possible choice of one of these points as the origin, this zero-set defines an elliptic curve.

In [7] the following functions are introduced:

$$\mathrm{ELi}_{n;m}(x;y;q) = \sum_{j=1}^{\infty}\sum_{k=1}^{\infty} \frac{x^j}{j^n}\frac{y^k}{k^m} q^{jk} = \sum_{k=1}^{\infty} \frac{y^k}{k^m}\mathrm{Li}_n(q^k x), \tag{15}$$

$$\mathrm{E}_{n;m}(x;\,y;\,q) = \begin{cases} \frac{1}{i}\left(\frac{1}{2}\mathrm{Li}_n(x) - \frac{1}{2}\mathrm{Li}_n(x^{-1}) + \mathrm{ELi}_{n;m}(x;y;q) - \mathrm{ELi}_{n;m}(x^{-1};y^{-1};q)\right) & \text{for } n+m \text{ even,}\\ \frac{1}{2}\mathrm{Li}_n(x) + \frac{1}{2}\mathrm{Li}_n(x^{-1}) + \mathrm{ELi}_{n;m}(x;y;q) + \mathrm{ELi}_{n;m}(x^{-1};y^{-1};q) & \text{for } n+m \text{ odd.}\end{cases} \tag{16}$$

Notice that these definitions are related to the basic ideas recalled in section 3 but slightly differ[2] from the functions of eq. 12. By use of the differential equation of second order [57] for $S^{(0)}(2,\, t)$, we obtain

$$S^{(0)}\left(2,\, t\right) = \frac{\psi_1(q)}{\pi} \sum_{i=1}^{3} \mathrm{E}_{2;\,0}(w_i(q);\; -1;\; -q) \text{ where } q = e^{i\pi \frac{\psi_2}{\psi_1}}. \tag{17}$$

Here ψ_1 and ψ_2 are periods of the elliptic curve defined by $\mathcal{F}$, which are given by complete elliptic integrals of the first kind [6]. The three arguments $w_i(q)$, $i = 1, 2, 3$, are directly related to the three intersection points P_1, P_2, P_3 (see [7]).

While all terms in eq. 17 can be nicely related to the underlying elliptic curve, the situation becomes considerably more complicated for the higher coefficients of eqs. 13 and 14. Here the integrands under consideration depend on both Symanzik polynomials. However, the functions defined in eqs. 15 and 16 remain to be useful. Generalizing this concept, we furthermore introduce the multi-variable functions [10]

$$\begin{aligned} &\mathrm{ELi}_{n_1,\ldots,n_l;m_1,\ldots,m_l;2o_1,\ldots,2o_{l-1}}\left(x_1,\ldots,x_l;y_1,\ldots,y_l;q\right) \\ &= \sum_{j_1=1}^{\infty}\cdots\sum_{j_l=1}^{\infty}\sum_{k_1=1}^{\infty}\cdots\sum_{k_l=1}^{\infty} \frac{x_1^{j_1}}{j_1^{n_1}}\cdots\frac{x_l^{j_l}}{j_l^{n_l}}\frac{y_1^{k_1}}{k_1^{m_1}}\cdots\frac{y_l^{k_l}}{k_l^{m_l}} \frac{q^{j_1k_1+\ldots+j_lk_l}}{\prod_{i=1}^{l-1}\left(j_ik_i+\ldots+j_lk_l\right)^{o_i}}. \end{aligned} \tag{18}$$

By use of this set-up of functions, the coefficients $S^{(1)}(2,\, t)$ and $S^{(0)}(4,\, t)$ are computed in [8]. Furthermore it is shown in [10], that in the case of equal masses, every higher coefficient of the two-dimensional case $S(2-2\epsilon,\, t)$ can be expressed in terms of these functions as well. This result includes an explicit algorithm for the computation of these coefficients.

We want to point out that other elliptic generalizations of polylogarithms recently found further applications to the two- and the three-loop sunrise graph [18, 19] and to integrals arising in string theory [49, 28, 29].

5. Conclusions

With their double nature as nested sums and iterated integrals, multiple polylogarithms provide a very useful framework for the computation of Feynman integrals. The Maple program MPL serves for the computation of a certain class of Feynman integrals in terms of these functions. The program is publicly available and supports computations with a class of iterated integrals, which arise in other contexts as well.

The massive two-loop sunrise integral is a Feynman integral which can not be expressed in terms of multiple polylogarithms. For the computation of various cases of this integral, a class of elliptic generalizations of polylogarithms has shown to be useful. These and other appearances of elliptic generalizations give rise to the hope, that when we have to leave the realm of multiple polylogarithms, we might not have to dispense with all of its advantages.

References

[1] J Ablinger, *Preprint* arXiv:1011.1176 [math-ph].
[2] J Ablinger, *Preprint* arXiv:1305.0687 [math-ph].
[3] J Ablinger, J Blmlein, A Hasselhuhn, S Klein, C Schneider and F Wibrock, Nucl. Phys. **B864** (2012) 52, (*Preprint* arXiv:1206.2252 [hep-ph]).
[4] J Ablinger, J Blmlein, C G Raab and C Schneider, J. Math. Phys. **55** (2014) 112301, (*Preprint* arXiv:1407.1822 [hep-th]).

[2] For $n = 2$, the relation between these definitions is made more explicit in [9].

[5] J Ablinger, J Blmlein, C Raab, C Schneider and F Wissbrock, Nucl. Phys. **B885** (2014) 409, (*Preprint* arXiv:1403.1137 [hep-ph]).
[6] L Adams, C Bogner, and S Weinzierl, J. Math. Phys. **54**, 052303 (2013), arXiv:1302.7004 [hep-ph].
[7] L Adams, C Bogner, and S Weinzierl, J.Math.Phys. **55** (2014) no.10, 102301, (*Preprint* arXiv:1405.5640 [hep-ph]).
[8] L Adams, C Bogner, and S Weinzierl, J. Math. Phys. **56** (2015) no.7, 072303, (*Preprint* arXiv:1504.03255 [hep-ph]).
[9] L Adams, C Bogner, and S Weinzierl, talk at the Workshop on Multiple Zeta Values, Modular Forms and Elliptic Motives II, ICMAT Madrid, December 2014, (*Preprint* arXiv:1510.03883 [hep-ph]).
[10] L Adams, C Bogner, and S Weinzierl, J.Math.Phys. **57** (2016) no.3, 032304, (*Preprint* arXiv:1512.05630 [hep-ph]).
[11] D H Bailey, J M Borwein, D Broadhurst, and M L Glasser, J.Phys. **A41** (2008) 205203, (*Preprint* arXiv:0801.0891).
[12] S Bauberger, M Böhm, G Weiglein, F A Berends and M Buza, Nucl. Phys. Proc. Suppl. **37B**, 95 (1994), (*Preprint* arXiv:hep-ph/9406404).
[13] S Bauberger, F A Berends, M Böhm and M Buza, Nucl. Phys. **B434**, 383 (1995), (*Preprint* arXiv:hep-ph/9409388).
[14] S Bauberger and M Böhm, Nucl. Phys. **B445**, 25 (1995), (*Preprint* arXiv:hep-ph/9501201).
[15] A Beilinson and A Levin, Proceedings of Symposia in Pure Mathematics, Vol. **55** (1994), Part 2, pp. 126-196.
[16] F A Berends, M Buza, M Böhm and R Scharf, Z.Phys. **C63**, 227 (1994).
[17] S Bloch, *Higher regulators, algebraic K -theory and zeta-functions of elliptic curves*, lecture notes, U.C.Irvine, 1977.
[18] S Bloch, M Kerr and P Vanhove, Compos.Math. 151 (2015) 2329-2375, (*Preprint* arXiv:1406.2664).
[19] S Bloch, M Kerr and P Vanhove, *Preprint* arXiv:1601.08181 [hep-th].
[20] S Bloch and P Vanhove, Journal of Number Theory **148** (2015) pp. 328–364, (*Preprint* arXiv:1309.5865 [hep-th]).
[21] C Bogner, Comput.Phys.Commun. 203 (2016) 339-353, (*Preprint* arXiv:1510.04562 [physics.comp-ph]).
[22] C Bogner and F Brown, PoS LL2012 (2012) 053, (*Preprint* arXiv:1209.6524 [hep-ph]).
[23] C Bogner and F Brown, Commun. Num. Theor. Phys. **09** (2015) 189-238, (*Preprint* arXiv:1408.1862 [hep-th]).
[24] C Bogner and M Lüders, in L. Alvarez-Consul, J.I. Burgos-Gil, K. Ebrahimi-Fard (Eds.) *Feynman Amplitudes, Periods and Motives*, AMS Contemp. Math. 648, (*Preprint* arXiv:1302.6215 [hep-ph]).
[25] C Bogner and S Weinzierl, Int.J.Mod.Phys. **A25** (2010) 2585-2618, (*Preprint* arXiv:1002.3458 [hep-ph]).
[26] D Broadhurst, (2008), *Preprint* arXiv:0801.4813.
[27] D J Broadhurst, J Fleischer and O Tarasov, Z.Phys. **C60**, 287 (1993), (*Preprint* arXiv:hep-ph/9304303).
[28] J Broedel, C R Mafra, N Matthes and O Schlotterer, JHEP 1507 (2015) 112, (*Preprint* arXiv:1412.5535).
[29] J Broedel, N Matthes and O Schlotterer, J.Phys. **A49** (2016) no.15, 155203, (*Preprint* arXiv:1507.02254 [hep-th]).
[30] J Broedel, O Schlotterer and S Stieberger, Fortsch.Phys. **61** (2013) 812-870 (2013) , (*Preprint* arXiv:1304.7267 [hep-th]).
[31] F Brown, Ann. Sci. Ec. Norm. Sup'er. (4) 42 (2009), 371-489, (*Preprint* [math.AG/0606419]).
[32] F Brown, Commun. Math. Phys. 287(3):925-958, (2009).
[33] F Brown, *Preprint* math.AG/0910.0114.
[34] F Brown, *Preprint* arXiv:1412.6508 [math.NT].
[35] F Brown and D Kreimer, Lett.Math.Phys. 103 (2013) 933-1007, (*Preprint* arXiv:1112.1180 [hep-th]).
[36] F Brown and A Levin, *Preprint* arXiv:1110.6917[math.NT].
[37] M Caffo, H Czyz, S Laporta and E Remiddi, Nuovo Cim. **A111**, 365 (1998), (*Preprint* arXiv:hep-th/9805118).
[38] M Caffo, H Czyz, M Gunia and E Remiddi, Comput. Phys. Commun. **180**, 427 (2009), (*Preprint* arXiv:0807.1959).
[39] M Caffo, H Czyz and E Remiddi, Nucl. Phys. **B634**, 309 (2002), (*Preprint* arXiv:hep-ph/0203256).
[40] K T Chen, Bull. Amer. Math. Soc. **83**, (1977), 831-879.
[41] A I Davydychev and B Delbourgo, J. Math. Phys. **39**, 4299 (1998), (*Preprint* arXiv:hep-th/9709216).
[42] A I Davydychev and V A Smirnov, Threshold expansion of the sunset diagram, Nucl.Phys. B554 (1999) 391-414.
[43] H Gangl and D Zagier, in: The arithmetic and geometry of algebraic cycles Banff, AB, 1998, NATO Science Series C, Mathematical and Physical Sciences, vol 548, 561-615; Kluwer Academic Publishers, Dordrecht, 2000.
[44] A B Goncharov, Math Res. Letters **5**, (1998) 497-516, (*Preprint* arXiv:1105.2076 [math.AG]).
[45] A B Goncharov, *Preprint* arXiv:math.AG/0103059.

Journal of Physics: Conference Series **762** (2016) 012067 doi:10.1088/1742-6596/762/1/012067

[46] S Groote, J G Körner, and A A Pivovarov, Annals Phys. **322**, 2374 (2007), (*Preprint* arXiv:hep-ph/0506286).
[47] S Groote, J Körner, and A A Pivovarov, Eur.Phys.J. **C72**, 2085 (2012), (*Preprint* arXiv:1204.0694).
[48] J M Henn, A V Smirnov and V A Smirnov, JHEP 1403 (2014) 088, (*Preprint* arXiv:1312.2588 [hep-th]).
[49] E D'Hoker, M B Green, O Gurdogan and P Vanhove, (*Preprint* arXiv:1512.06779 [hep-th]).
[50] M Yu Kalmykov and B A Kniehl, Nucl. Phys. B809, 365 (2009), (*Preprint* arXiv:0807.0567).
[51] S Laporta and E Remiddi, Nucl. Phys. **B704**, 349 (2005), (*Preprint* hep-ph/0406160).
[52] J A Lappo-Danilevsky, Rec. Math. Moscou 34 (1927), no. 6 pp. 113–146.
[53] J A Lappo-Danilevsky, vol. I–III. Chelsea, 1953.
[54] G W Leibniz, Letter to Johann Bernoulli, dated 9.11.1696, in: Smtliche Schriften und Briefe, Dritte Reihe, Siebter Band, pp. 178-180, Akademie Verlag Berlin, 2011.
[55] A Levin, Compositio Math. 106 (1997), 267-282.
[56] A von Manteuffel, E Panzer and R M Schabinger, JHEP 1502 (2015) 120, (*Preprint* arXiv:1411.7392 [hep-ph]).
[57] S Mller-Stach, S Weinzierl and R Zayadeh, Commun. Num. Theor. Phys. **6** (2012) 203-222, (*Preprint* arXiv:1112.4360 [hep-ph]).
[58] E Panzer, JHEP 1403 (2014) 071, (*Preprint* arXiv:1401.4361 [hep-th]).
[59] E Panzer, Comput. Phys. Commun. **188** (2014) 148-166, (*Preprint* arXiv:1403.3385 [hep-th]).
[60] E Panzer, PhD thesis, Humboldt University, (*Preprint* arXiv:1506.07243) [math-ph].
[61] H Poincar, Acta Mathematica 4 (1884), no. 1 pp. 201–312.
[62] S Pozzorini and E Remiddi, Comput. Phys. Commun. **175**, 381 (2006), (*Preprint* arXiv:hep-ph/0505041).
[63] E Remiddi and L Tancredi, Nucl.Phys. **B880**, 343 (2014), (*Preprint* arXiv:1311.3342).
[64] E Remiddi and L Tancredi, (*Preprint* arXiv:1602.01481 [hep-ph]).
[65] N I Ussyukina and A I Davydychev, Phys. Lett. **B298**, 363 (1993).
[66] D Zagier, Math.Ann 286, 613-624 (1990).

ACAT2016 IOP Publishing
Journal of Physics: Conference Series **762** (2016) 012068 doi:10.1088/1742-6596/762/1/012068

Geometrical splitting and reduction of Feynman diagrams

ndrei a ydyc e

c lum erger F c lum erger ri e ugar and e as
nstitute for uclear ysics osco tate ni ersity osco ussia

mail da yd t eory si np msu ru

stract geometrical approac to t e calculation of point Feynman diagrams is re ie ed t is s o n t at t e geometrical splitting yields useful connections et een Feynman integrals it di erent momenta and masses t is demonstrated o t ese results can e used to reduce t e num er of aria les in t e occurring functions

ntroduction

geometrical interpretation of inematic in ariants and ot er uantities related to point Feynman diagrams s o n in gure elps us to understand t e analytical structure of t e results for t ese diagrams s an e ample singularities of t e general t ree point function can e descri ed pictorially t roug a tetra edron constructed out of t e e ternal momenta and internal masses uc a geometrical isuali ation can e used to deri e andau e uations de ning t e positions of possi le singularities see also in

n general t e one loop point diagrams as s o n in gure depend on - momentum in ariants $_l$ p p_l and masses of t e internal particles m_i ere and elo e follo t e notation used in in particular t e po ers of t e internal scalar propagators are denoted as $_i$ and t e space time dimension is denoted as n so t at e can also deal it t e dimensionally regulated integrals it n elo e ill mainly consider t e cases en all $_i$ n it as demonstrated o suc geometrical ideas could e used for an analytical calculation of one loop point diagrams For t e geometrical interpretation a asic simple in dimensional uclidean space is employed a triangle for a tetra edron for etc and t e o tained results can e e pressed in terms of an integral o er a dimensional

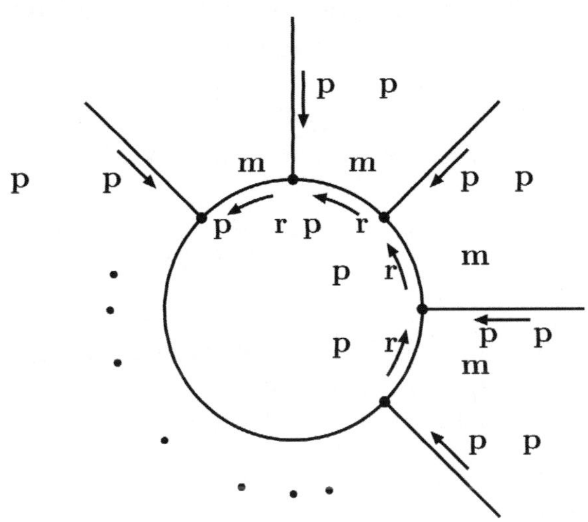

Figure point diagram

sp erical or yper olic simple ic corresponds to t e intersection of t e asic simple and t e unit ypersp ere or t e corresponding yper olic ypersurface it a eig t function depending on t e angular distance et een t e integration point and t e point corresponding to t e eig t of t e asic simple see in For n t is eig t function is e ual to and t e results simplify for t e case n see in and for t e case n see in t er interesting e amples of using t e geometrical approac can e found e g in

n t is paper e ill s o t at t e natural ay of splitting t e asic simple as prescri ed it in t e geometrical approac discussed a o e leads to a reduction of t e e ecti e num er of independent aria les in separate contri utions o tained as a result of suc splitting

o point function

For t e t o point function t ere is only one e ternal momentum in ariant and t e sides of t e corresponding asic triangle are m m and $\sqrt{}$ as s o n in gure a e angle et een t e sides m and m is de ned t roug cos c m m m m and in t e sp erical case t e integration goes o er t e arc of t e unit circle as s o n in gure

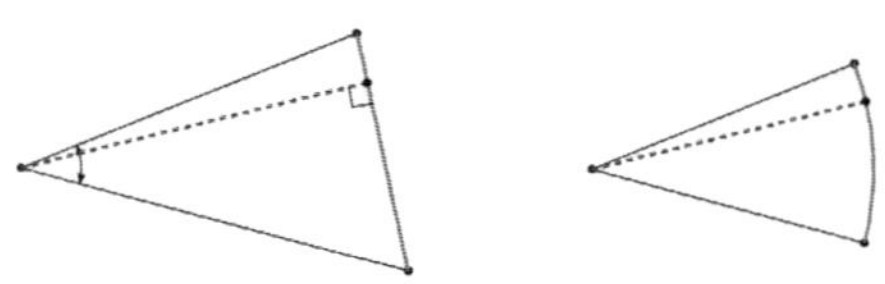

Figure o point case a t e asic triangle and t e arc

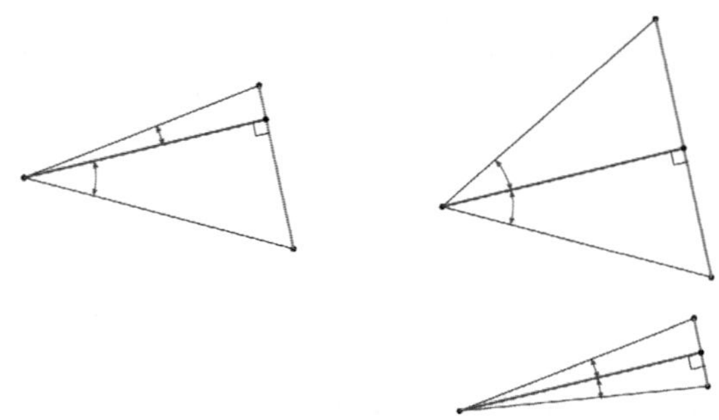

Figure o point function reduction to e ual mass integrals

For splitting e use t e eig t of t e asic triangle m and o tain t o triangles it t e sides m m $\sqrt{}$ and m m $\sqrt{}$ respecti ely ere m m m sin $\sqrt{}$ m m and m m note t at m m and m m ac of t e resulting integrals can e associated it a t o point function and e arri e at t e follo ing decomposition

$$\left(n \quad m \; m\right) \quad \text{—}\left\{ \quad m \quad m \quad \left(n \quad m \; m\right) \quad m \quad m \quad \left(n \quad m \; m\right)\right\}$$

is is an e ample of a functional relation et een integrals it di erent momenta and masses similar to t ose descri ed in oreo er using t e geometrical relation s o n in gure e can represent t e rig t and side in terms of t e e ual mass integrals

$$\left(n \quad m \; m\right) \quad \text{—}\left\{ \quad m \quad m \quad \left(n \quad m \; m\right) \quad m \quad m \quad \left(n \quad m \; m\right)\right\}$$

et us loo at t e num er of aria les n t e original integral (n m m) e a e t ree independent aria les t o masses and one momentum in ariant out of t em e can construct t o dimensionless aria les n t e integral (n m m) e a e one e tra condition on t e aria les m m so t at e get t o independent aria les i e one dimensionless aria le e same is alid for (n m m) ere t e e tra condition is due to t o e ual masses erefore t e result for t e t o point function in ar itrary dimension can e e pressed in terms of a com ination of functions of a single dimensionless aria le indeed e no t at it can e presented in terms of t e Gauss ypergeometric function F see e g in ose e pansion is no n to any order

ree point function

For t e t ree point function t ere are t ree e ternal momentum in ariants and and t e sides of t e corresponding asic tetra edron are m m m $\sqrt{\ }$ $\sqrt{\ }$ and $\sqrt{\ }$ as s o n in gure a e angles and et een t e sides m m and m are de ned t roug cos c m m m m and in t e sp erical case t e integration e tends o er t e sp erical triangle of t e unit sp ere see in gure

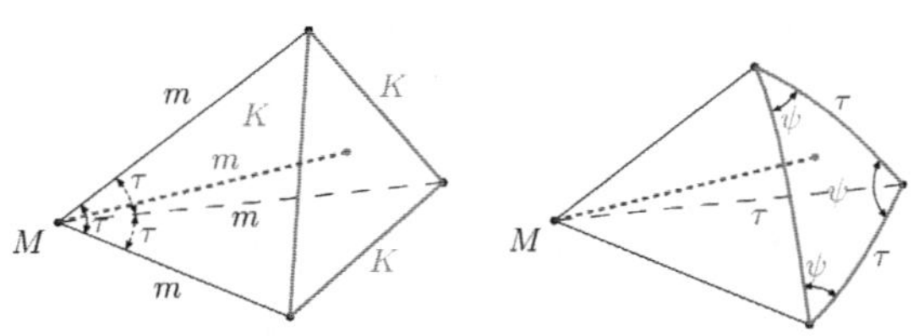

Figure ree point case a t e asic tetra edron and t e solid angle

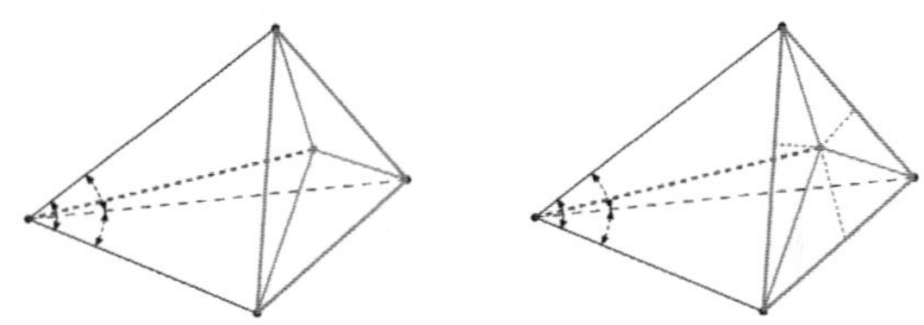

Figure a plitting t e asic tetra edron into t ree tetra edra and furt er splitting into si tetra edra

For t e splitting e use t e eig t of t e asic tetra edron m and o tain t ree tetra edra as s o n in gure a ne of t em as t e sides m m m $\sqrt{\ }$ $\sqrt{\ }$ and $\sqrt{\ }$ and t e sides for t e ot ers can e o tained y permutation of t e indices ere m m m m m m and m m m m $\sqrt{\ }$ ere - [ ] and det c is t e Gram determinant see in for more details ac of t e resulting integrals can e associated it a speci c t ree point function and e arri e at t e follo ing decomposition

$$\left(n \quad m \; m \; m\right) \quad \frac{m\,m\,m}{\ }\left\{\frac{F}{m} \quad \left(n \quad m \; m \; m\right)\right.$$
$$\frac{F}{m} \quad \left(n \quad m \; m \; m\right)$$
$$\left.\frac{F}{m} \quad \left(n \quad m \; m \; m\right)\right\}$$

it

$$F \quad \frac{\ }{m\,m}\Big[\ \Big(\quad m \quad m \quad m\Big) \quad m \quad m \ \Big(\quad \Big)\Big]$$

etc so t at $\sum_i$ F_i m_i m m m

Journal of Physics: Conference Series **762** (2016) 012068 doi:10.1088/1742-6596/762/1/012068

y dropping perpendiculars onto t e sides $\sqrt{\ }$ etc eac of t e resulting tetra edra can e split into t o so t at in total e get si irectangular tetra edra as s o n in gure Furt ermore for eac of t em e can use t e geometrical relation similar to one s o n in gure reducing t em to t e integrals it t o e ual masses

$$(n \quad m \quad m \quad m)$$

$$\text{—}\Big\{ \quad m \quad m \quad \Big(n \quad \frac{m \quad m}{\quad} \quad m \quad m \quad m\Big)$$

$$m \quad m \quad \Big(n \quad \frac{m \quad m}{\quad} \quad m \quad m \quad m\Big)\Big\}$$

et us analy e t e num er of aria les n t e integral $(n \quad m \quad m \quad m)$ e a e si independent aria les t ree masses and t ree momentum in ariants out of t em e can construct e dimensionless aria les n $(n \quad m \quad m \quad m)$ e a e t o e tra conditions on t e aria les m m and m m so t at e get four independent aria les i e t ree dimensionless aria les For t e integral $(n \quad m \quad m \quad m \quad m \quad m)$ e a e t ree relations it an additional condition due to t e t o e ual masses erefore t e result for t e t ree point function in ar itrary dimension s ould e e pressi le in terms of a com ination of functions of t o dimensionless aria les indeed e no t at it can e presented in terms of t e ppell ypergeometric function F see e g in

Four point function

For t e four point function t ere are si e ternal momentum in ariants ut of t em and are t e s uared momenta of t e e ternal legs ilst and correspond to t e andelstam aria les s and t e sides of t e corresponding asic four dimensional simple are m m m m and si additional sides $_1$ $\sqrt{\ _1}$ as s o n in gure a e si angles $_1$ et een t e corresponding sides m and m_l are de ned t roug cos $_1$ c $_1$ m m_l $_1$ m m_l and in t e sp erical case t e integration e tends o er t e sp erical tetra edron of t e unit ypersp ere as s o n in gure for t e yper olic case one can use analytic continuation

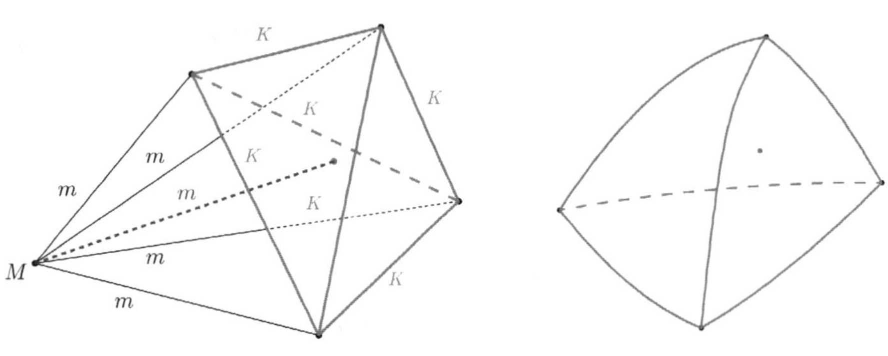

Figure Four point case a t e asic simple and t e sp erical tetra edron

For splitting e use t e eig t of t e asic simple m and o tain four simplices as s o n in gure a ne of t em as t e sides m m m m $\sqrt{\ }$ $\sqrt{\ }$

√ √ √ and √ and t e sides of t e ot ers can e o tained y permutation of t e indices s efore i mi m i ereas m m m m m √ ere det c 1 and det 1 see in for more details ac of t e four resulting integrals can e associated it a certain four point function t t e ne t step in eac of t e four tetra edra dra n in red e drop t e perpendiculars onto t e triangle sides as s o n in gure splitting eac of t em into t ree and t en di iding eac of t e resulting tetra edra into t o y dropping perpendiculars onto t e √ 1 sides as s o n in gure c s a result of t is splitting e get simplices e corresponding steps of splitting t e sp erical tetra edron are s o n in gure

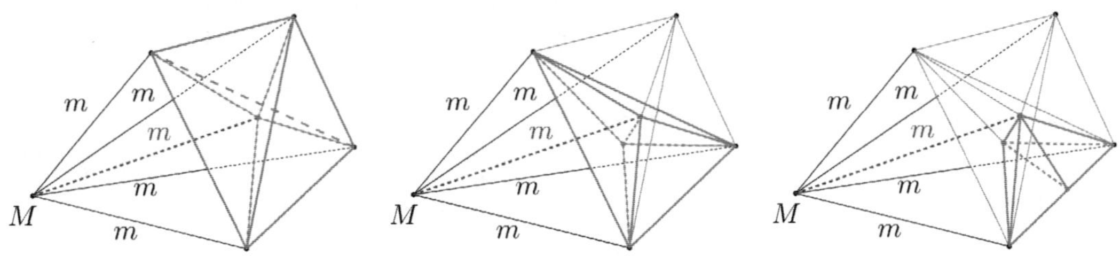

Figure Four point case splitting t e asic four dimensional simple

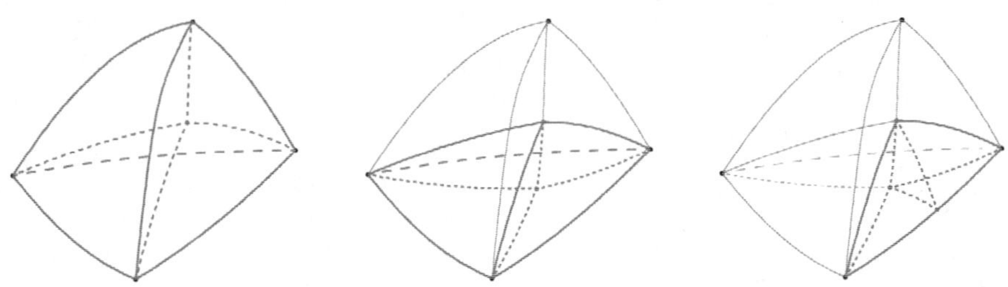

Figure Four point case splitting t e sp erical tetra edron

et us loo at t e num er of aria les n t e integral (n 1 mi) e a e ten independent aria les four masses and si momentum in ariants out of t em e can construct nine dimensionless aria les fter t e rst step gure a e a e t ree e tra conditions on t e aria les m m m m and m m so t at e get se en independent aria les i e si dimensionless aria les fter t e second step gure e get t o e tra conditions due to t e rig t triangles and after t e t ird step gure c e get one more condition s a result for eac of t e resulting four point functions e a e si relations so t at e end up it four independent aria les i e t ree dimensionless aria les erefore t e result for t e four point function in ar itrary dimension s ould e e pressi le in terms of a com ination of functions of t ree dimensionless aria les suc as e g auricella functions and t eir generali ations see e g in

General remar s and conclusions

sing a geometrical approac e can relate t e one loop point Feynman diagrams to certain olume integrals in non uclidean geometry Geometrical splitting pro ides a straig tfor ard ay of reducing general integrals to t ose it lesser num er of independent aria les n t is ay e can predict t e set and t e num er of t ese aria les in t e resulting integrals Furt ermore it allo s us to deri e functional relations et een integrals it di erent momenta and masses

um ers of dimensionless aria les in separate contri utions for point diagrams efore and after t e splitting are summari ed in t e ta le

a le um er of aria les efore and after t e splitting

	total of dimensionless aria les	of splitting pieces	reduced of aria les
ar itrary	–		

c no ledgements

am t an ful to el ourgo and u almy o it om started to or on t is su ect am grateful to t e organi ers of and t e ni ersity of o o for t eir support and ospitality

eferences

andau *Nucl. Phys.*
allen G and ig tman *Mat. Fys. Skr. Dan. Vid. Selsk.*
andelstam *Phys. Rev.*
ut os y *J. Math. Phys.*
aylor *Phys. Rev.*
a ydyc e *J. Math. Phys.*
a ydyc e *J. Math. Phys.*
t ooft G and eltman *Nucl. Phys.*
ollini G and Giam iagi *Nuovo Cim.*
s more F *Lett. Nuovo Cim.*
icuta G and ontaldi *Lett. Nuovo Cim.*
a ydyc e and el ourgo *J. Math. Phys.*
a ydyc e *Nucl. Instr. Meth.*
a ydyc e *Preprint* ep t
ic el G *J. Math. Phys.*
rtner and agner *Ann. Inst. Henri Poincaré (Phys. Théor.)*
agner *Indag. Math.*
a ydyc e and el ourgo *J. Phys.*
Gors y and i oedo *J. Phys.*
loc and reimer *Commun. Num. Theor. Phys.*
c net *Preprint* ar i
ason and inner *J. Phys.*
andan aulos F pradlin and olo ic *J. High Energy Phys.*
araso *Phys. Lett.*
nie l and araso *Nucl. Phys.*
erends F a ydyc e and mirno *Nucl. Phys.*
a ydyc e *Phys. Rev.*
a ydyc e and almy o u *Nucl. Phys. B (Proc. Suppl.)*
a ydyc e and almy o u *Nucl. Phys.*
araso *Nucl. Phys. B (Proc. Suppl.)*
Fleisc er egerle ner F and araso *Nucl. Phys.*
yte almy o u and oc *Comput. Phys. Commun.*

Journal of Physics: Conference Series **762** (2016) 012069 doi:10.1088/1742-6596/762/1/012069

Higgs Boson production in association with jets in gluon-gluon fusion

G Luisoni[1], N Greiner[2], S Höche[3], M Schönherr[2], J Winter[4] and V Yundin[5]

[1] Theoretical Physics Department, CERN, Geneva, Switzerland
[2] Physik-Institut, Universität Zürich, Wintherturerstrasse 190, CH-8057 Zürich, Switzerland
[3] SLAC National Accelerator Laboratory, Menlo Park, CA 94025, USA
[4] Department of Physics and Astronomy, Michigan State University, East Lansing, MI 48824, USA
[5] Max-Planck-Institut für Physik, Föhringer Ring 6, D-80805 München, Germany

E-mail: gionata.luisoni@cern.ch

Abstract. We present the computation of the NLO QCD corrections to the production of a Standard Model Higgs boson in association with up to three jets through gluon-gluon fusion. The computation is performed in the approximation of an infinitely heavy top quark. Before studying some phenomenologically important observables, we describe some technical detail about the computation.

1. Introduction

A major step after the discovery of the Higgs particle [1, 2] is the precise determination of its nature. Among the important properties, besides spin and parity, there are its couplings to fermions and other bosons. In order to discriminate between the electroweak symmetry breaking mechanism in the Standard Model and possible Beyond Standard Model scenarios, measurements have to be compared with very precise predictions.

The main Higgs boson production mechanism is given by the gluon fusion channel, where the Higgs boson is produced out of two initial state gluons via a loop of heavy quarks. The presence of two gluon in the initial state enhances the tendency of producing further QCD radiation in this production channel, where also higher order corrections are very large. For this reason they were recently computed to N^{3}LO [3]. The unavoidable presence of further jets makes the gluon fusion mechanism an interesting process on its own, in particular to study the Higgs boson in a very jetty environment. However, it is also an irreducible background to another very important production channel, namely vector boson fusion (VBF).

During Run II at the LHC, the VBF mechanism will play a leading role. In this production mode, a Higgs boson is created by annihilation of virtual W or Z bosons, radiated off the initial-state (anti-)quarks in a t-channel scattering process with no color exchange at leading order [4, 5]. This allows to directly probe the couplings between the Higgs boson and the electroweak gauge bosons providing at the same time a clean signature consisting of two forward jets and only little hadronic energy between these tagging jets.

In order to discriminate between these two production mechanisms it is crucial to be able to estimate the contamination of gluon fusion events in the VBF sample, and to devise the best

Journal of Physics: Conference Series **762** (2016) 012069 doi:10.1088/1742-6596/762/1/012069

selection strategies, which allow to keep untertainties under control providing at the same time the highest signal-to-background ratios.

In this talk we discuss the calculation and the phenomenlogy of the production of a Standard Model Higgs boson in association up to three jets, as described in detail in the references [6–8]. The calculation takes into account next-to-leading order QCD corrections and is carried out in the limit of an infinitely heavy top quark. We discuss two different sets of cuts: a set of basic cuts that are suitable for the gluon fusion contribution and a more restrictive set of cuts, which mimics the one adopted in VBF analyses. In addition we discuss different schemes for defining the tagging jets.

2. Computational setup

2.1. The codes

The computation is performed using the automated tools GoSam [9, 10] and Sherpa [11], linked via the interface defined in the Binoth Les Houches Accord [12, 13].

The one-loop amplitudes are generated with the new version 2.0 of GoSam, and are based on an algebraic generation of d-dimensional integrands using a Feynman diagrammatic approach. The expressions for the amplitudes are generated employing QGraf [14], Form [15, 16] and Spinney [17]. For the reduction of the tensor integrals at running time we used Ninja [18, 19], which is an automated package carrying out the integrand reduction via Laurent expansion [20], and OneLoop [21] for the evaluation of the scalar integrals. Unstable phase space points are detected automatically and reevaluated with the tensor integral library Golem95 [22–24]. Alternatively, one can use other reduction techniques such as the standard OPP method [25–27] as implemented in Samurai [28]. All the reduction programs had to be upgraded in order to be able to deal with the higher rank loop integrals generated by the presence of the effective gluon-gluon-Higgs boson vertex.

The calculation of tree-level matrix elements for the Born and the real emission contribution as well as the subtraction terms in the Catani-Seymour approach [29] have been done within Sherpa using the matrix element generator Comix [30]. These amplitudes were validated using a combination of MadGraph 4 [31, 32], MadDipole [33, 34] and MadEvent [35]

2.2. Root *Ntuples*

Because of the relatively slow NLO computation and the high statistics needed for such a large multiplicity final state, the Monte Carlo events are stored in the form of Root Ntuples. They are generated by Sherpa and were first used in the context of vector boson production in association with jets [36].

For this calculation, sets of Ntuples files with Born (B), virtual (V), integrate subtraction terms (I) and real minus subtraction term (RS) type of events have been generated for H + 1 jet, H + 2 jets and H + 3 jets at the center-of-mass energies of 8, 13, 14 and 100 TeV. The events were generated such that jets can be clustered using the k_T or anti-k_T algorithm [37, 38] as implemented in the FastJet package [39] and with radii that can vary between $R = 0.1$ and $R = 1$. On the jets a minimal generation cut was imposed by requiring $p_T > 25$ GeV, $|\eta| < 4.5$, which allow to post-process the events in every analysis with more inclusive cuts.

2.3. Analysis cuts and parameter settings

We present here results for two center of mass energies at 8 and 13 TeV. In both cases we have applied two sets of cuts: a baseline set with minimal cuts to render the cross section finite, and a more restrictive set of cuts, which is typically used in the context of VBF searches. In both cases jets are clustered using the anti-k_T algorithm [37, 38] as implemented in the FastJet package [39]. If not specified explicitly, the jet radius and PDF set have been set to $R = 0.4$ and CT10nlo [41], respectively. The baseline set consists of the following cuts: $p_T > 30$ GeV,

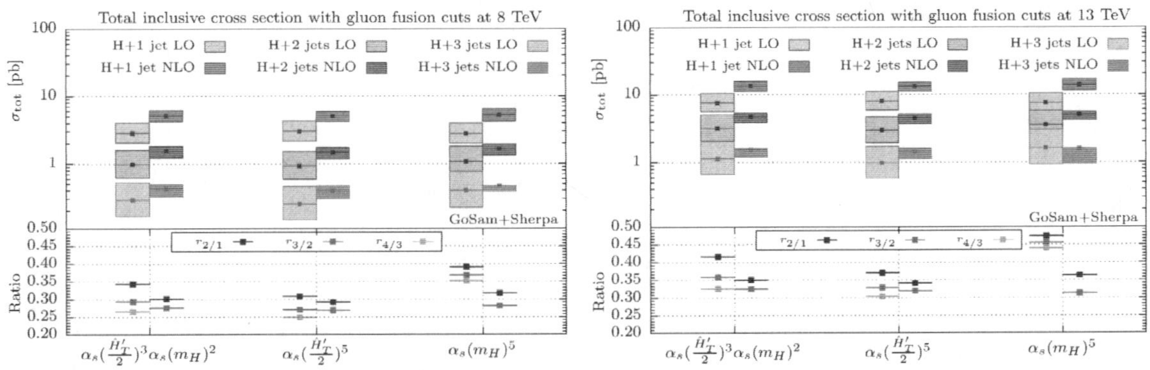

Figure 1: Total cross sections at LO (left side of each column) and NLO (right side of each column) for H + 1 jet (green), H + 2 jets (blue) and H + 3 jets (red) production using the three different scale choices as explained in text. In the lower part of the plots, the ratios $r_{2/1}$ (blue), $r_{3/2}$ (red) and $r_{4/3}$ (orange) are shown. Results have been obtained for 8 TeV and 13 TeV (left and right plot respectively).

$|\eta| < 4.4$. In the VBF case two additional cuts have been imposed, given by $m_{j_1 j_2} > 400$ GeV, $|\Delta y_{j_1,j_2}| > 2.8$. Here the two jets, j_1 and j_2, denote the tagging jets. Their selection is not unique and we study two different schemes, one where the two tagging jets are the two jets with the highest p_T (p_T-tagging), and one where the most forward and most backward jet (in rapidity) yield the two tagging jets (y-tagging).
The central scale for the renormalization and the factorization scale is chosen to be

$$\mu_F = \mu_R \equiv \frac{\hat{H}'_T}{2} = \frac{1}{2}\left(\sqrt{m_H^2 + p_{T,H}^2} + \sum_i |p_{T,i}|\right) . \tag{1}$$

However it is not obvious whether this dynamical scale is also a good choice to be used for the gluon-gluon-Higgs coupling. One might argue that the Higgs mass is the appropriate scale there. Therefore we consider three different scale choices, defined as

$$\mathrm{A}: \quad \alpha_s\left(x \cdot \frac{\hat{H}'_T}{2}\right)^3 \alpha_s\,(x \cdot m_H)^2 \,, \qquad \mathrm{B}: \quad \alpha_s\left(x \cdot \frac{\hat{H}'_T}{2}\right)^5 , \qquad \mathrm{C}: \quad \alpha_s\,(x \cdot m_H)^5 \,. \tag{2}$$

The presence of the factor x indicates that this scale is varied in the range $x \in [0.5, \ldots, 2]$.

3. Numerical results for gluon fusion setup

We start the discussion of the numerical results with the basic gluon fusion cuts. One of the most important observables is the total cross section. In figure 1 we report the total cross sections for both leading order and next-to-leading order for the processes H + 1 jet, H + 2 jets and H + 3 jets at $E_{\mathrm{cm}} = 8$ TeV (left plot) and $E_{\mathrm{cm}} = 13$ TeV (right plot). The results are shown for the three scale choices described above. On the level of total cross sections one only observes a mild dependence on the scale choice, in particular for the NLO results. For the fixed scale one observes an enhancement of the LO ratios. An increase of the center of mass energy from 8 to 13 TeV also only has a small influence on the overall pattern. The situation is different when one looks at differential distributions. In figure 2 we show the transverse momentum distribution for the Higgs boson for the H + 3 jets process at a center of mass energy of 8 TeV.

The subplot on the right shows the distribution for scale choice B, while Fig. 2a shows the results for the different scales normalized to the NLO result of scale A. The advantage of scale B is the flatness of the K-factor over the entire p_T range. This supports our choice to make scale

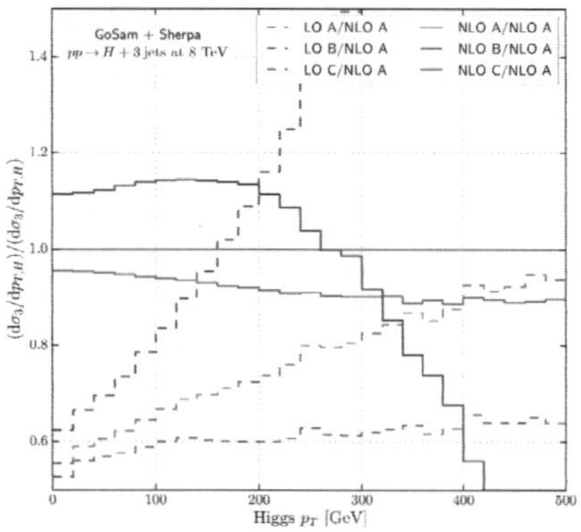

(a) Ratio

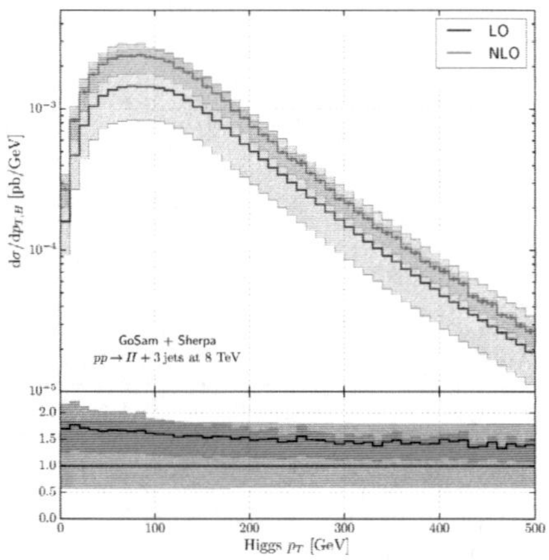

(b) Scale choice B

Figure 2: On the right: the p_T distribution of the Higgs boson in H + 3 jets production at the 8 TeV LHC presented for scale B. The subplot 2a shows the same central predictions normalized to the NLO result for scale A. Each ratio plot depicts the respective differential K-factors and their envelopes obtained from scale variations at LO and NLO.

B the default scale. For the lower p_T region up to $\sim$ 250 GeV, scale C seems to be a sensible choice as well. However, it completely breaks down for higher p_T, and the K-factor can even become negative. Further support for using scale B as the default choice comes from looking at the p_T distribution of the 'wimpiest' jet for each multiplicity. This means looking at the first jet for H + 1 jet, at the second jet for H + 2 jets, and at the third jet for H + 3 jets. This is illustrated in figure 3. The left hand side shows the p_T of the three jets for scale choice B, the right hand side shows scale choice A. The lower part of the plots shows the ratio of NLO versus LO for each jet. For better visibility the ratios are multiplied with a prefactor to avoid overlap of the bands of the different jets. As can be seen from the ratio plots, the purely dynamical scale choice B leads to flat K-factors, whereas scale choice A shows a decrease of the K-factor with increasing transverse momentum.

Another interesting question is how observables that are defined independently of a certain jet multiplicity, like the transverse momentum of the Higgs, change under the presence of additional QCD radiation. This is shown in figure 4. The upper plots show the NLO distributions for one, two and three jets (which we have obtained from the one-jet, two-jet and three-jet NLO calculations, respectively). Unless stated otherwise, the jet multiplicity is exclusive, labelled by 'excl', i.e. a veto on any additional jet activity is in place. The 1-jet and the 3-jet processes are shown twice, once for the exclusive case, and once for the inclusive case, labelled by 'incl'. The lower subpanels show each contribution normalized to the inclusive prediction of the core process, i.e. the most inclusive one, here given by the H + 1 jet process. The plots in the middle and lower panel are constructed following the same principle but using the NLO core process of increased jet multiplicity, namely H + 2 jets and H + 3 jets, respectively. The middle row of figure 4 hence depicts the same situation but without accounting for the H + 1 jet process; and, for the lower row, there are only two distributions left to show, the one for the exclusive and the one for the inclusive H + 3 jets process. One can see that the low energy region is dominated by the exclusive H + 1 jet contribution. The H + 2 jets contribution is negligible in that region, however starts to dominate already in a region above approx. 200 GeV. Going further up in

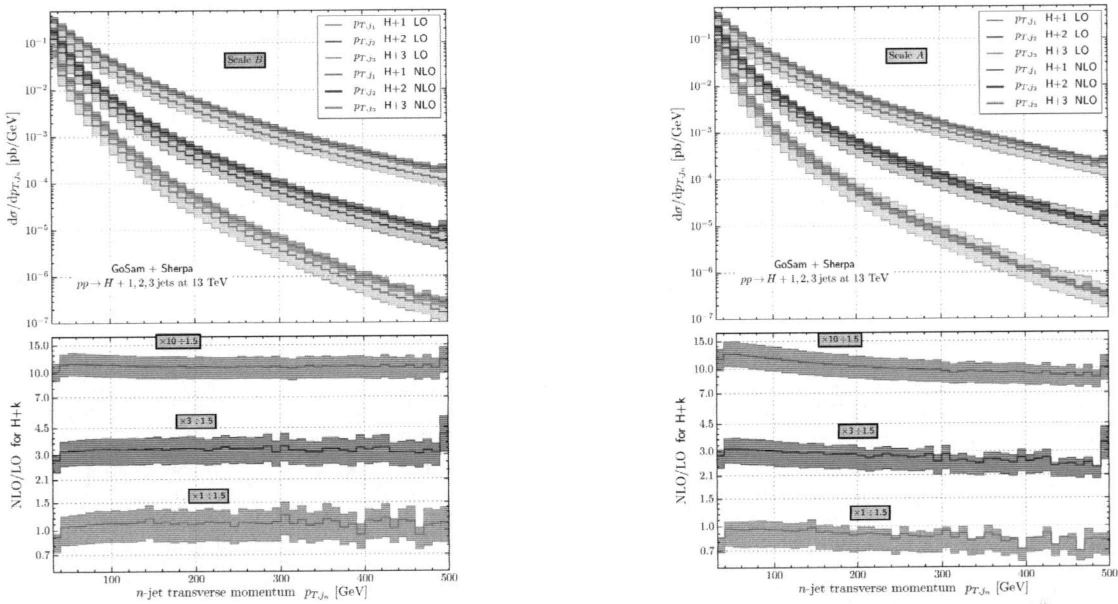

Figure 3: Transverse momentum distribution of the 'wimpiest' jet in $\mathrm{H}+n\,\mathrm{jets}$ production at the LHC. Using p_T ordering the first, second and third leading jet are shown in $\mathrm{H}+1\,\mathrm{jet}$, $\mathrm{H}+2\,\mathrm{jets}$ and $\mathrm{H}+3\,\mathrm{jets}$ at 13 TeV, respectively; on the left with the default scale choice B, on the right with the scale choice A.

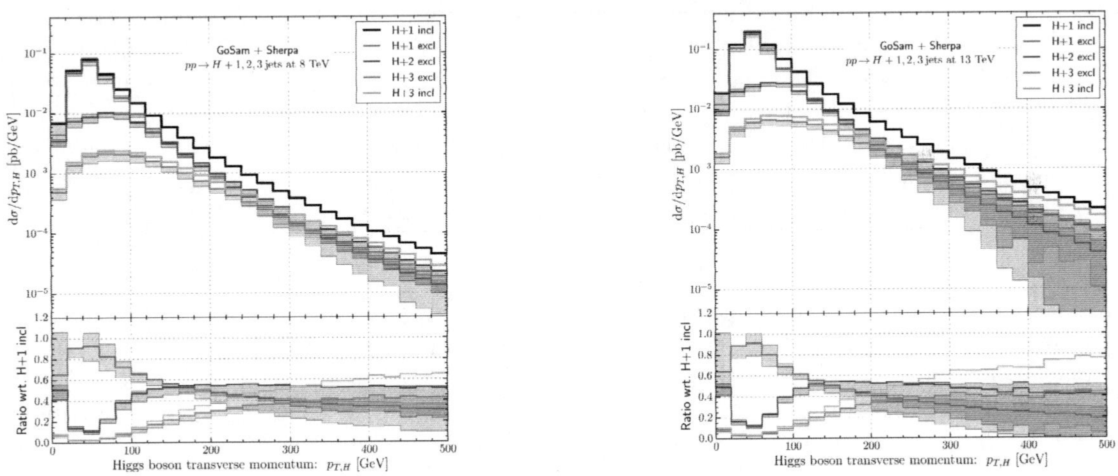

Figure 4: Transverse momentum distribution of the Higgs for the different jet multiplicities. L.h.s. shows the result for 8 TeV, r.h.s. shows the distribution for 13 TeV.

the spectrum increases the $\mathrm{H}+3\,\mathrm{jets}$ contribution which will eventually dominate the spectrum. In other words, at high enough energies it is more and more likely to produce further jets. This has to be kept in mind when comparing an inclusive measurement with a fixed order calculation for a given number of jets. For low multiplicities the description becomes inaccurate already at relatively low energies of around 200 GeV. A theoretical prediction that is based on a merged result of different multiplicities will yield a better description of the data.

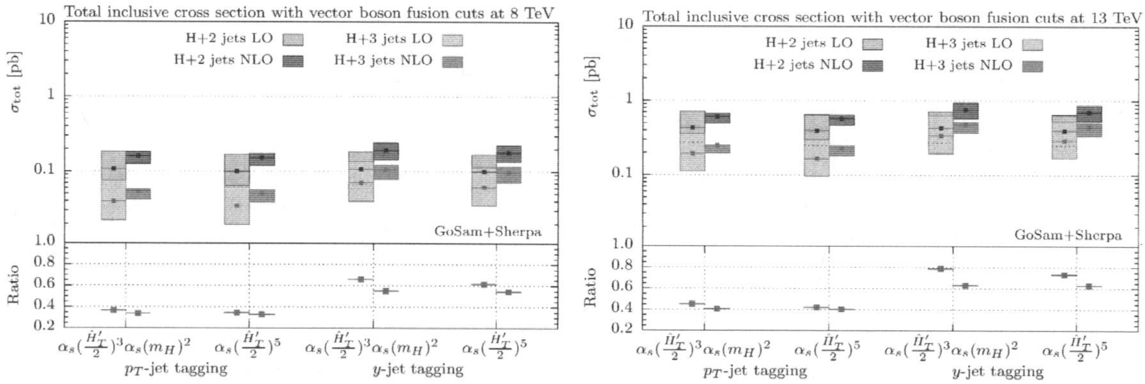

Figure 5: Total cross sections at LO and NLO for H + 2 jets (blue) and H + 3 jets (red) using VBF kinematical cuts and two different tagging jet definitions. Results are shown for the two scale choices A and B, as well as the two energies of 8 TeV (left plot) and 13 TeV (right plot). The lower part of each plot depicts the inclusive cross section ratios $r_{3/2}$ for the different scales and tag jet approaches.

4. Phenomenology with vector boson fusion cuts

As already mentioned, gluon fusion is an irreducible background to the VBF channel. A challenging task is therefore to provide a precise prediction of its rate compared to the signal. In this section we discuss the results obtained from the gluon fusion contribution under the presence of the additional VBF cuts described in section 2.3. Again we start the discussion with the total cross section, for the VBF selection we now also consider the differences between the two tagging schemes described above, p_T-tagging and y-tagging. The total cross section for the different energies, scales and tagging schemes is shown in figure 5. Having ruled out the fixed scale as a sensible choice we only show the result for the two scales A and B. Also in the case of the VBF selection, the differences between the scale choices are rather small. The tagging scheme has a much bigger impact. The y-tagging increases the ratios which means that it increases the fraction of the processes with higher multiplicity. This stresses the importance of the inclusion of NLO results with higher multiplicities into the theoretical prediction.

5. Conclusions

In this talk we presented the computation and some phenomenological results for the production of a Standard Model Higgs boson via the gluon fusion mechanism in the heavy top mass limit in association with up to three jets. We investigated the role of the scale choice as well as the effects of different sets of cuts, also allowing to assess the role of the gluon fusion contribution in VBF searches. Furthermore we discussed a variety of important observables allowing for a better discrimination between the gluon fusion and vector boson fusion contribution. Further improvements could certainly be achieved by providing a merged NLO result of the different jet multiplicities, but also through the inclusion of top-quark mass effects as well as the matching of the H + 3 jets NLO result with a parton shower.

Acknowledgments

We would like to thank the members of the GoSam collaboration for their help and effort. NG was supported by the Swiss National Science Foundation under contract PZ00P2_154829. This research used computing resources from the Rechenzentrum Garching.

References

[1] Aad G *et al.* (ATLAS Collaboration) 2012 *Phys.Lett.* **B716** 1–29 (*Preprint* 1207.7214)

[2] Chatrchyan S *et al.* (CMS Collaboration) 2012 *Phys.Lett.* **B716** 30–61 (*Preprint* 1207.7235)
[3] Anastasiou C, Duhr C, Dulat F, Furlan E, Gehrmann T, Herzog F, Lazopoulos A and Mistlberger B 2016 (*Preprint* 1602.00695)
[4] Cahn R and Dawson S 1984 *Phys.Lett.* **B136** 196
[5] Kane G L, Repko W and Rolnick W 1984 *Phys.Lett.* **B148** 367–372
[6] Greiner N, Hoeche S, Luisoni G, Schoenherr M, Winter J C and Yundin V 2016 *JHEP* **01** 169 (*Preprint* 1506.01016)
[7] Cullen G, van Deurzen H, Greiner N, Luisoni G, Mastrolia P *et al.* 2013 *Phys.Rev.Lett.* **111** 131801 (*Preprint* 1307.4737)
[8] van Deurzen H, Greiner N, Luisoni G, Mastrolia P, Mirabella E *et al.* 2013 *Phys.Lett.* **B721** 74–81 (*Preprint* 1301.0493)
[9] Cullen G, Greiner N, Heinrich G, Luisoni G, Mastrolia P *et al.* 2012 *Eur.Phys.J.* **C72** 1889 (*Preprint* 1111.2034)
[10] Cullen G, van Deurzen H, Greiner N, Heinrich G, Luisoni G *et al.* 2014 *Eur.Phys.J.* **C74** 3001 (*Preprint* 1404.7096)
[11] Gleisberg T, Höche S, Krauss F, Schönherr M, Schumann S *et al.* 2009 *JHEP* **0902** 007 (*Preprint* 0811.4622)
[12] Binoth T, Boudjema F, Dissertori G, Lazopoulos A, Denner A *et al.* 2010 *Comput.Phys.Commun.* **181** 1612–1622 (*Preprint* 1001.1307)
[13] Alioli S, Badger S, Bellm J, Biedermann B, Boudjema F *et al.* 2014 *Comput.Phys.Commun.* **185** 560–571 (*Preprint* 1308.3462)
[14] Nogueira P 1993 *J.Comput.Phys.* **105** 279–289
[15] Vermaseren J A M 2000 (*Preprint* math-ph/0010025)
[16] Kuipers J, Ueda T, Vermaseren J and Vollinga J 2013 *Comput.Phys.Commun.* **184** 1453–1467 (*Preprint* 1203.6543)
[17] Cullen G, Koch-Janusz M and Reiter T 2011 *Comput.Phys.Commun.* **182** 2368–2387 (*Preprint* 1008.0803)
[18] van Deurzen H, Luisoni G, Mastrolia P, Mirabella E, Ossola G *et al.* 2014 *JHEP* **1403** 115 (*Preprint* 1312.6678)
[19] Peraro T 2014 *Comput.Phys.Commun.* **185** 2771–2797 (*Preprint* 1403.1229)
[20] Mastrolia P, Mirabella E and Peraro T 2012 *JHEP* **1206** 095 (*Preprint* 1203.0291)
[21] van Hameren A 2011 *Comput.Phys.Commun.* **182** 2427–2438 (*Preprint* 1007.4716)
[22] Heinrich G, Ossola G, Reiter T and Tramontano F 2010 *JHEP* **1010** 105 (*Preprint* 1008.2441)
[23] Binoth T, Guillet J P, Heinrich G, Pilon E and Reiter T 2009 *Comput.Phys.Commun.* **180** 2317–2330 (*Preprint* 0810.0992)
[24] Cullen G, Guillet J P, Heinrich G, Kleinschmidt T, Pilon E *et al.* 2011 *Comput.Phys.Commun.* **182** 2276–2284 (*Preprint* 1101.5595)
[25] Ossola G, Papadopoulos C G and Pittau R 2007 *Nucl. Phys.* **B763** 147–169 (*Preprint* hep-ph/0609007)
[26] Mastrolia P, Ossola G, Papadopoulos C and Pittau R 2008 *JHEP* **0806** 030 (*Preprint* 0803.3964)
[27] Ossola G, Papadopoulos C G and Pittau R 2008 *JHEP* **0805** 004 (*Preprint* 0802.1876)
[28] Mastrolia P, Ossola G, Reiter T and Tramontano F 2010 *JHEP* **1008** 080 (*Preprint* 1006.0710)
[29] Catani S and Seymour M H 1997 *Nucl. Phys.* **B485** 291–419 (*Preprint* hep-ph/9605323)
[30] Gleisberg T and Höche S 2008 *JHEP* **0812** 039 (*Preprint* 0808.3674)
[31] Stelzer T and Long W F 1994 *Comput. Phys. Commun.* **81** 357–371 (*Preprint* hep-ph/9401258)
[32] Alwall J *et al.* 2007 *JHEP* **09** 028 (*Preprint* 0706.2334)
[33] Frederix R, Gehrmann T and Greiner N 2008 *JHEP* **09** 122 (*Preprint* 0808.2128)
[34] Frederix R, Gehrmann T and Greiner N 2010 *JHEP* **1006** 086 (*Preprint* 1004.2905)
[35] Maltoni F and Stelzer T 2003 *JHEP* **02** 027 (*Preprint* hep-ph/0208156)
[36] Bern Z, Dixon L J, Febres Cordero F, Hoeche S, Ita H, Kosower D A and Maitre D 2014 *Comput. Phys. Commun.* **185** 1443–1460 (*Preprint* 1310.7439)
[37] Cacciari M and Salam G P 2006 *Phys.Lett.* **B641** 57–61 (*Preprint* hep-ph/0512210)
[38] Cacciari M, Salam G P and Soyez G 2008 *JHEP* **04** 063 (*Preprint* 0802.1189)
[39] Cacciari M, Salam G P and Soyez G 2012 *Eur.Phys.J.* **C72** 1896 (*Preprint* 1111.6097)
[40] Hamilton K, Nason P and Zanderighi G 2012 *JHEP* **1210** 155 (*Preprint* 1206.3572)
[41] Lai H L, Guzzi M, Huston J, Li Z, Nadolsky P M *et al.* 2010 *Phys.Rev.* **D82** 074024 (*Preprint* 1007.2241)

High Performance and Increased Precision Techniques for Feynman Loop Integrals

K Kato[1], E de Doncker[2], T Ishikawa[3], J Kapenga[2], O Olagbemi[2] and F Yuasa[3]

[1] Department of Physics, Kogakuin University, Shinjuku, Tokyo 163-8677, Japan
[2] Department of Computer Science, Western Michigan University, 1903 W. Michigan Avenue, Kalamazoo, MI 49008, United States
[3] High Energy Accelerator Research Organization (KEK), 1-1 Oho Tsukuba, Ibaraki 305-0801, Japan

E-mail: kato@cc.kogakuin.ac.jp

Abstract. For the investigation of physics within and beyond the Standard Model, a precise evaluation of higher order corrections in perturbative quantum field theory is required. We have worked on the development of a computational method for Feynman loop integrals with a fully numerical approach. It is based on numerical integration and extrapolation techniques. In this paper, we describe the status and new developments in our techniques for the numerical computation of Feynman loop integrals.

Separation of ultra-violet divergences is important for the renormalization procedure. In our analyses, the separation can be done numerically. For 2-loop integrals we have performed the calculations for up to 4-point functions, and for 2-point functions we can handle up to 4-loop integrals. We report the status and accuracy of the computations with detailed numerical comparisons to results in the literature, in order to demonstrate that our method will evolve into an important component of automated systems for the study of higher-order radiative corrections.

1. Introduction

In high energy physics, we encounter the need for large-scale computations to obtain the theoretical prediction required to study experimental data. When the number of final particles is large, the matrix element and phase space becomes complicated and lengthy. When good experimental data are available, i.e., data are of high statistical quality, the higher-order radiative correction is mandatory. Such a large-scale computation is generally beyond man-power calculation, so that automated systems to calculate the perturbative expansion in quantum field theory (QFT) prove to be an essential tool for the present and future high-energy experiments.

Perturbation calculations in QFT follow a well established algorithm and many systems are already provided for practical use to analyze experimental data. For the radiative corrections, the one-loop order calculation is incorporated in several systems, GRACE [1], CompHEP [2], FeynArt/Calc [3], FDC [4], and so forth. Considering future accelerator experiments, the development of a system to perform computations beyond one-loop will be of great importance. One indispensable component of such a system is a library of multi-loop integrals. The library should be robust and general to treat any combination of internal masses and external momenta.

While the analytical calculation of multi-loop integrals is studied by many authors [5], the analytic expression for the most general form seems to be far from accomplished. Therefore we have started to solve the problem by numerical methods.

2. Computational method for loop integrals

The integral we study is

$$I = (-1)^N \frac{\Gamma(N - nL/2)}{(4\pi)^{nL/2}} \int \prod dx_r \frac{\delta(1 - \sum x_r)}{U^{n/2}(V - i\rho)^{N-nL/2}} \tag{1}$$

where n is the space-time dimension, N is the number of propagators, L is the number of loops and

$$V = M^2 - \frac{W}{U}, \qquad M^2 = \sum_r x_r m_r^2. \tag{2}$$

Here the functions U and W are polynomials in the integration variables x_r, and are derived from the structure of the graph. The coefficients of the W polynomial are determined by the external momenta and internal masses.

In the analytic method, one needs a sophisticated variable transformation of the x_r. However, we dare not exploit the transformation since the library is intended as a component of an automated system where the integrand will be supplied by the automatic diagram generator, so that the assignment of the x_r and other elements may not be under our control.

When a Feynman diagram includes fermions and/or gauge bosons, the integral has a momentum dependent numerator. Although the integral studied here is of scalar type, the introduction of a numerator does not cause a serious problem (although it may make the numerical integration more computationally intensive).

We introduce dimensional regularization to handle ultraviolet divergence, i.e., by setting

$$n = 4 - 2\varepsilon. \tag{3}$$

In the numerical evaluation of Eq.(1), we might encounter a singularity for $\rho \to 0$ and/or for $\varepsilon \to 0$. In the analytical treatment of loop integrals, the singularity for $\rho \to 0$ can be avoided by analytic continuation. However, in the numerical method, if we calculate Eq.(1) by naively taking $\rho = 0$, the result may be divergent or unstable. We have developed the direct computation method (DCM) to manage the problem [6, 7, 8, 9]. In DCM, we keep the numerical value of ρ non-zero. This value is not necessarily very small. When ρ is finite, Eq.(1) can be computed numerically as a function of ρ. For a sequence of $\rho = \rho_k$, $k = 1 \dots K$, we obtain a sequence of numerical values $I(\rho_k)$, $k = 1 \dots K$. From this sequence we can estimate the integral I either by a nonlinear extrapolation with Wynn's algorithm (ϵ algorithm) [12], or by using a linear solver. Note that a geometric sequence is needed when using the ϵ algorithm.

In this paper, we discuss the singularity as $\varepsilon \to 0$, originating from the ultra-violet (UV) singularity. The infrared singularity also appears as a pole in ε. As the latter is already studied in [8], we concentrate here on the UV-pole. The UV pole in Eq.(1) originates from the Gamma function and/or the integral.

Generally the integral I can be expanded as

$$I = \cdots + \frac{C_{-2}}{\varepsilon^2} + \frac{C_{-1}}{\varepsilon} + C_0 + C_1\varepsilon + C_2\varepsilon^2 + \cdots \tag{4}$$

and we extract the coefficients C_j numerically using the same methods as for the case of $\rho \to 0$, i.e., by linear or nonlinear extrapolation.

3. Numerical results

Using our numerical method, we have computed 2-loop diagrams with up to 4 external lines, and self-energy diagrams with up to 4 loops. The results are compared with those in the literature and show good agreement. As described in the sections below, we use various integration libraries, acceleration and hardware platforms, and check the efficiency and convergence behavior of the computation.

3.1. 4-loop self-energy integrals

We study the diagrams in Fig.1. The integral part is finite, and the Gamma factor is divergent for Fig.1(1) and (2). We consider massless internal particles and $p^2 = 1$ for comparison with the results in [15].

Results are shown in Table 1. The parallel execution time for Table 1(1) is 48s with ParInt [10], using 64 MPI [11] processes on four 16-core nodes of the *thor* cluster at Western Michigan University. The integration for Table 1(2) is performed in 4.8h elapsed time for a 64-thread computation using a double-exponential integration formula (DE) [13] on SR-16000. Fig.1(3) and (4) are also calculated with ParInt on the *thor* cluster.

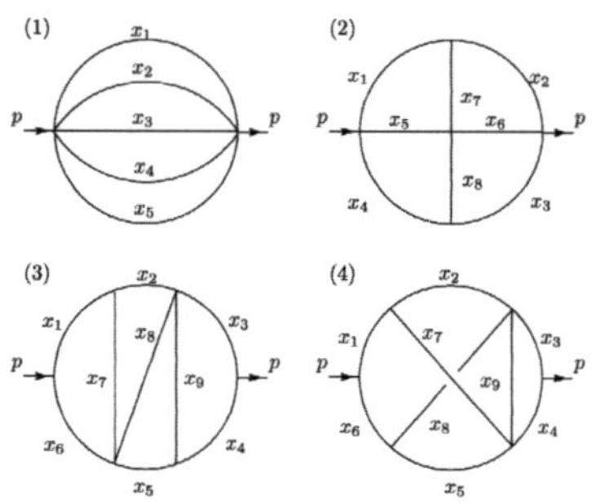

Figure 1. 4-loop self-energy diagrams

Table 1. 4-loop self-energy integrals for Fig.1. The rows with *BC* list results of [15].

	C_{-1}	C_0	C_1
Fig.1(1)	-0.001736111111109	-0.016927083381	-0.011842916
Fig.1(1) *BC*	-0.001736111111111	-0.016927083333	-0.011842930
Fig.1(2)	5.1846392	-2.582434	70.39877
Fig.1(2) *BC*	5.18463877	-2.5824360	70.399151
Fig.1(3)	–	55.585150	Not yet
Fig.1(3) *BC*	–	55.5852539	
Fig.1(4)	–	52.017714	Not yet
Fig.1(4) *BC*	–	52.017868	

3.2. 3-loop self-energy integrals

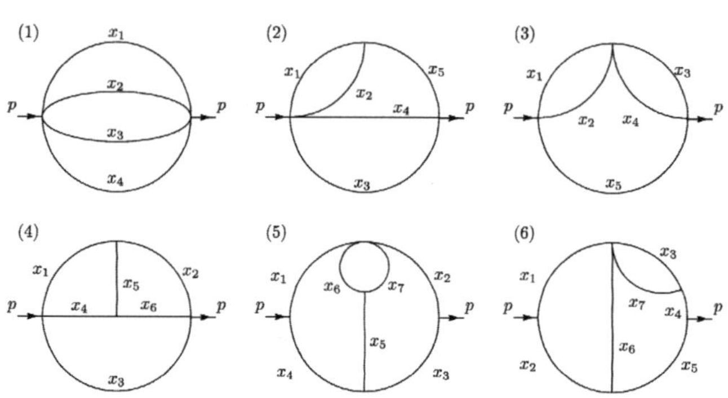

Figure 2. 3-loop self-energy diagrams

Table 2. 3-loop self-energy integral for Fig.2(5). The row marked *L* gives the results of [14].

	C_{-1}	C_0
Fig.2(5)	0.92370	-2.4201
Fig.2(5) *L*	0.923631827	-2.42349163

We deal with the diagrams in Fig.2 where maximally a $\frac{1}{\varepsilon^3}$ singularity appears, corresponding to a double-pole divergence ($\sim \frac{1}{\varepsilon^2}$).

As an example, we show the results for Fig.2(5), *ladybug*. To compare with [14], we let $m_r = 1$, $p^2 = 1$. The Gamma factor is finite and the integral is single-pole divergent ($\sim \frac{1}{\epsilon}$). This calculation is carried out with DE and performed in quadruple precision on (dual) E5-2687W v3 @ 3.10GHz using 40 threads. Extrapolated results obtained with Wynn's algorithm are shown for Fig.2(5) in Table 2. Computations for the other diagrams in Fig.2 result in a similar agreement with [14].

3.3. 2-loop vertex integrals

As an example, we consider the diagram in Fig.3(1), which has UV divergence from both the Gamma function and a single pole in the integral part. In total, it is divergent as $\frac{1}{\varepsilon^2}$.

For the numerical computation, we take $m_r = 1$, $p_1^2 = p_2^2 = p_3^2 = 1$, so that there is no singular threshold. In Table 3 we show the convergence behavior of the C coefficients, where a linear solver is used to estimate the coefficients. The bottom row of the table lists the values from [14].

The diagram of Fig.3(2) is also studied and the result for C_{-1} is in good agreement with [14].

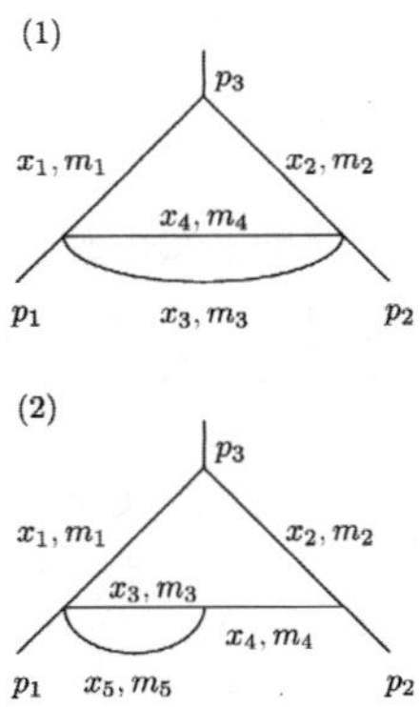

Figure 3. 2-loop vertex diagrams

Table 3. 2-loop vertex integral for Fig.3(1). An integration is performed for each $\varepsilon_j = 1/b_j$ in the first column, and the four coefficients are estimated by a linear extrapolation.

b_j	T [s]	C_{-2}	C_{-1}	C_0	C_1
3	1.4				
4	1.5	0.514890217	0.535680679		
6	1.8	0.505861627	0.598880809	−0.108343080	
8	5.2	0.501111606	0.660631086	−0.364844229	0.342001531
12	2.5	0.500159017	0.680635463	−0.515353351	0.822106580
16	4.2	0.500017647	0.685300679	−0.573315125	1.161395016
24	20.5	0.500001357	0.686098882	−0.588595022	1.307352233
32	19.1	0.500000078	0.686192225	−0.591298134	1.347594589
48	8.6	0.500000001	0.686200327	−0.591641476	1.355242242
[14]		0.5	0.686200636	−0.591666701	1.356196533

3.4. 2-loop box integrals

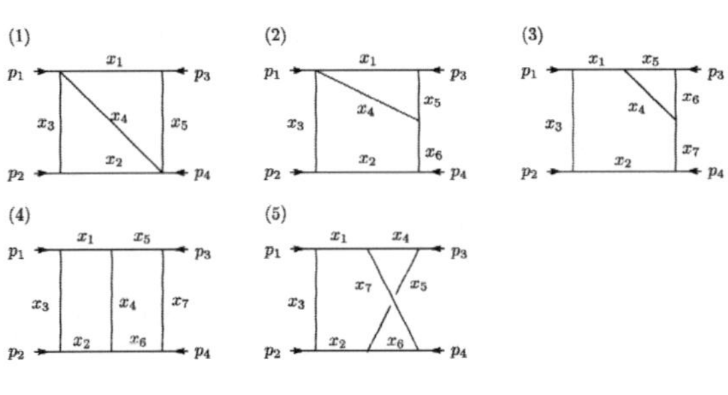

Figure 4. 2-loop box diagrams

Table 4. 2-loop box integrals for Fig.4. Listed are: integral dimension, error tolerance, maximum number of integral evaluations, sequential and parallel time (in seconds), and parallel speedup, $R = T_1/T_{64}$.

	D.	Tol.	Eval.	T_1 [s]	T_{64} [s]	R
Fig.4(1)	5	10^{-10}	400M	32.6	0.74	44.1
Fig.4(2)	6	10^{-9}	3B	213.6	5.06	42.2
Fig.4(3)	7	10^{-8}	5B	507.9	8.83	57.5
Fig.4(4)	7	10^{-8}	2B	189.9	4.33	43.9
Fig.4(5)	7	10^{-7}	300M	27.6	0.49	56.3
Fig.4(5)	7	10^{-9}	20B	1893	34.6	54.7

Finally, we show results for the box diagrams in Fig.4. These are not UV divergent; the computation is done with $\varepsilon = 0$. Furthermore, the test integral is calculated in the unphysical region, $m_r = 1$, $s = t = 1$, $p_j^2 = 1$, so that it is possible to take $\rho = 0$. The comparison with [14] was shown before in [7].

Here, we report the speedup of the parallel computation on MPI [11]. We use ParInt on the *thor* cluster at Western Michigan University. As shown in Table 4, we gain significant acceleration with 64 processes. The computation times with one process (T_1) and with 64 processes (T_{64}) are given in seconds (s), and the speedup is $R = T_1/T_{64}$.

4. Summary

Numerical methods are tested to calculate loop integrals with 2 to 4 loops, and up to 4-point functions. In parameter space, up to 8-dimensional integrals are computed. The DCM method works to estimate both UV-divergent and finite terms using dimensional regularization. As the approach is expected to be part of an automated system, no special handling of the integrand is utilized. Automatic decisions on the variable transformation might improve the status. Problems with large CPU-times can be targeted using suitable parallel software and platforms. The study should be extended to the general cases with physical masses and external momenta, including a numerator part in the integrand function.

Acknowledgments

We acknowledge the support from the National Science Foundation under Award Number 1126438. This work is further supported by Grant-in-Aid for Scientific Research (24540292 and 15H03668) of JSPS, and the Large Scale Simulation Program No.14/15-06 and No.15/16-06 of KEK.

References

[1] Khiem P H, Fujimoto J, Ishikawa T, Kaneko T, Kato K, Kurihara Y, Shimizu Y, Ueda T, Vermaseren J A M and Yasui Y 2013 Eur. Phys. J. C 73:2400
[2] Boos E, Bunichev V, Dubinin M, Dudko L, Edneral V, Ilyin V, Kryukov A, Savrin V, Semenov A and Sherstnev A 2008 PoS ACAT08:008
[3] Hahn T 2010 PoS ACAT2010:078
[4] Wang J-X 2004 Nucl. Instrum. Meth. A534 241-245
[5] See, for an instance, other papers in the proceedings of ACAT2016 in track 3.
[6] de Doncker E, Yuasa F and Assaf R 2013 Journal of Physics: Conf. Ser. 454 doi:10.1088/1742-6596/454/1/012082
[7] de Doncker E and Yuasa F 2014 XV Adv. Comp. and Anal. Tech. in Phys. Res., Journal of Physics: Conference Series (ACAT 2013) 523 doi:10.1088/1742-6596/523/1/012052
[8] de Doncker E, Fujimoto J, Hamaguchi N, Ishikawa T, Kurihara Y, Shimizu Y and Yuasa F 2011 Journal of Computational Science (JoCS) 3 102-112 doi:10.1016/j.jocs.2011.06.003
[9] Yuasa F, de Doncker E, Hamaguchi N, Ishikawa T, Kato K, Kurihara Y, Fujimoto J and Shimizu Y 2012 Computer Physics Communications 183 2136-2144
[10] ParInt `http://www.cs.wmich.edu/parint`
[11] Open-MPI `http://www.open-mpi.org`
[12] Wynn P 1956 Mathematical Tables and Aids to Computing 10 91-96
[13] Takahasi H and Mori M 1974 Publications of the Research Institute for Mathematical Sciences 9 721-741
[14] Laporta S 2000 Int. J. Mod. Phys. A 15 5087-5159 arXiv:hep-ph/0102033v1
[15] Baikov B A and Chetyrkin K G 2010 Nuclear Physics B 837 186-220

Journal of Physics: Conference Series **762** (2016) 012071 doi:10.1088/1742-6596/762/1/012071

Making extreme computations possible with virtual machines

J Reuter[1], B Chokoufe Nejad[1], T Ohl[2]

[1] DESY Theory Group, Notkestr. 85, D-22607 Hamburg
[2] University of Würzburg, Emil-Hilb-Weg 22, D-97074 Würzburg

E-mail: juergen.reuter@desy.de,bijan.chokoufe@desy.de,ohl@physik.uni-wuerzburg.de

Abstract. State-of-the-art algorithms generate scattering amplitudes for high-energy physics at leading order for high-multiplicity processes as compiled code (in Fortran, C or C++). For complicated processes the size of these libraries can become tremendous (many GiB). We show that amplitudes can be translated to byte-code instructions, which even reduce the size by one order of magnitude. The byte-code is interpreted by a Virtual Machine with runtimes comparable to compiled code and a better scaling with additional legs. We study the properties of this algorithm, as an extension of the Optimizing Matrix Element Generator (O'Mega). The bytecode matrix elements are available as alternative input for the event generator WHIZARD. The bytecode interpreter can be implemented very compactly, which will help with a future implementation on massively parallel GPUs.

1. Introduction

Meta-programming is a popular approach in high-energy physics to determine the expression of a cross section in a higher level programming language while the numerical evaluation is performed in high-performance languages. A problem, however, arises when the expression becomes so large that it is impossible to compile and link, and hence to evaluate numerically, due to the sheer size. In `Fortran`, which is known for its excellent numerical performance, we typically encounter this problem for source code of gigabyte sizes irrespective of the available memory. In these proceedings, we sketch the ideas of Ref. [1], how to completely circumvent the tedious compile step by using a virtual machine (VM). To avoid confusions, we note that a VM is in our context a compiled program, an *interpreter*, that is able to read instructions, in the form of *byte-code*, from disk and perform an arbitrary number of operations out of a finite *instruction set*. A VM allows the complexity of the computation to be only set by the available hardware and not limited by software design or intermediate steps. Furthermore, it is easy to implement and makes parallel evaluation obvious.

An important concern is of course whether the VM can still compete with compiled code in terms of speed. The instructions have to be translated by the VM to actual machine code, which is a potential overhead. However, in the computation of matrix elements a typical instruction will correspond to a product of scalar, spinor or vector currents which involves $O(10)$ arithmetical operations on complex numbers. This suggests that the overhead might be small, which has however to be proven by a concrete implementation. In fact, we will show that a VM can even be faster than compiled code for certain processes and compilers since the formulation in terms of a VM has also benefits, especially for large multiplicities, as is discussed in detail below. More importantly, the runtime is in general in the same order of magnitude as the compiled code and as such the VM is very usable for general purpose applications, where the clever use of Monte Carlo (MC) techniques can easily change the number of points needed for convergence by orders of magnitude.

2. General Virtual Machines

One might imagine a VM as a machine, which has a number of registers, and is given instructions, encoded in the byte-code, how to act on them. This picture is quite similar to a CPU, except that we are doing this on a higher level, i.e. our registers are arrays of e.g. wave functions or momenta and the instructions can encode scalar products or more complicated expressions. In order to use VMs for arbitrary processes, a dynamic construction is desirable. For such a construction, it is necessary to include a *header* in the byte-code, which contains the number of objects that have to be allocated. After this the body of instructions follows, whereby each line corresponds to a certain operation that the VM should perform on its registers. The first number of an instruction is the *operation code (opcode)* that specifies which operation will be performed. For illustration, consider the example `1 5 4 3`, which could be translated into `momentum(5) = momentum(4) + momentum(3)`, a typical operation to compute the s-channel momentum in a $2 \to 2$ scattering process. Depending on the context, set by the opcode, the following numbers have different meanings but are typically addresses, i.e. indices, of objects, or specify how exactly the function should act on the operands, by what numbers the result should be multiplied, etc. The interpreter is a very simple program that reads the byte-code into memory and then loops over the instruction block with a `decode` function, which is basically a `select/case` statement depending on the opcode. The instructions can be instantly translated to physical machine code, compared to the execution time of the relevant instructions, since the different types of operations are already compiled and only the memory locations of the objects have to be inserted. The byte-code file that is given to the interpreter completely dictates the specific problem, or process in the cross section context, that should be computed. Input data or external parameters are given as arguments to the function call of the VM. The generation of events for collider physics usually parallelizes trivially. Since an integral is in most cases needed, the same code is just evaluated multiple times with different input data. The situation can change, however, for an extreme computation that already uses all caches. Depending on the size of the caches and the scheduler, evaluating such code with multiple data at the same time, can run even slower than the single-threaded execution. Obviously, the computation is then so large, containing numerous objects, that it is worth trying to parallelize the execution with a single set of input data with shared memory.

The byte-code lends itself nicely to parallelization as it can be split into recursion *levels*, whereby in each level all *building blocks* are non-nested and may be computed in parallel. Different levels are separated by necessary synchronization points. It is clear that one should aim to keep the number of synchronization points to the inherent minimum of the computation for optimal performance. This parallelization is straightforward in `Fortran95/2003` and `OpenMP` and shown explicitly in Ref. [1].

As a side note, we mention that the sketched parallelization should be very well suited for an implementation on a graphics processing unit (GPU). A common problem, encountered when trying to do scientific computing on a GPU, is the finite kernel size problem. As noted e.g. in Ref. [2], large source code cannot be processed by the `CUDA` compiler, which is related to the fact that the numerous cores on a GPU are designed to execute simple operations very fast. Dividing an amplitude into smaller pieces, which are computed one by one, introduces more communication overhead and is no ultimate solution since the compilation can still fail for complex amplitudes [2]. The VM on the other hand is a fixed small kernel, no matter how complex the specific computation is. A potential bottleneck might be the availability of the instruction block to all threads, but this question has to be settled by an implementation and might have a quite hardware-dependent answer.

Finally, we note that the phase-space parallelization mentioned in the beginning of this subsection can still be applied. When considering heterogeneous cluster or grid environments, where each node is equipped with multi-core processors, a combination of distributed memory parallelization for the combination of different phase-space points and shared memory parallelization of a single point seems to be a quite natural and extremely potent combination.

3. O'Mega Virtual Machine

The concept of a VM can be easily applied to evaluate tree-level matrix elements of arbitrary multiplicity. The Optimizing Matrix Element Generator, O'MEGA [3], avoids the redundant representation of amplitudes in the form of Feynman diagrams by using one-particle off-shell wave functions (1POWs) recursively. O'MEGA tames herewith the computational growth with the number of external particles from a factorial to an exponential one.

O'MEGA uses the meta-programming ansatz mentioned in Section 1 whereby the symbolic representation is determined in `OCaml`. This abstract expression is then translated to valid `Fortran` code that is automatically compiled and used in WHIZARD [4, 5, 6] for event generation. We replace the last step now with an output module that produces byte-code instead of `Fortran` code.

The number of distinct operations that have to be performed in the computation of a cross section is related to the Feynman rules and therefore quite limited. As such, these operations are very good candidates for the translation to byte-code. In fact, this results in only about 80 different opcodes for the complete standard model (SM), which have been implemented in the O'Mega virtual machine (OVM). For the parallel execution, we identify the different levels by the number of external momenta a wave function is connected to or equivalently the number of summands in the momentum of the wave function.

4. Speed Benchmarks

All processes shown here, and various others, have been validated against the compiled versions where possible. We stress that every process is computed in its respective model (QCD or SM) to *full* tree-level order including all interferences and we have not restricted e.g. the Drell-Yan amplitudes to only one electroweak propagator. We use the term SM for the full electroweak theory together with QCD and a nontrivial Yukawa matrix but without higher dimensional couplings like $\mathrm{H} \to \mathrm{gg}$.

To investigate the compiler dependence of the results, we use two different compilers, `gfortran 4.7.1` and `ifort 14.0.3`. We do not claim that our results are necessarily representative for all `Fortran` compilers or even compiler versions, but they should still give a good impression of the expected variance in performance. The evaluation time measurements are performed on a computer with two Intel(R) Xeon(R) E5-2440 @ 2.40GHz central processing units (CPUs), having 16 MiB L3 cache on each socket, and 2x 32 GiB RAM running under Scientific Linux 6.5. In Figure 1, we show the measured CPU times for QCD and SM processes with two different optimization levels for the compiled code and the OVM using the GNU and Intel compiler. Since the evaluation times are highly reproducible, we use only three runs to obtain mean and standard deviation. The times are normalized for *each* process to `gfortran-O3`, which is why the times are not growing with the number of particles. For `gfortran`, we observe for most processes the fastest performance with `-O3` and for `ifort` with `-O2`, which is an effect commonly encountered. The fastest performance is given by the source code compiled with `ifort-O2` being roughly 0.75 times the time needed by `gfortran-O3`.

The crucial point, however, is that `ifort` fails to compile the $n = 7$ gluon and the $\mathrm{u\bar{u}} \to \mathrm{e^+e^-}6j$ Drell-Yan process while the OVM immediately starts computing. The GNU

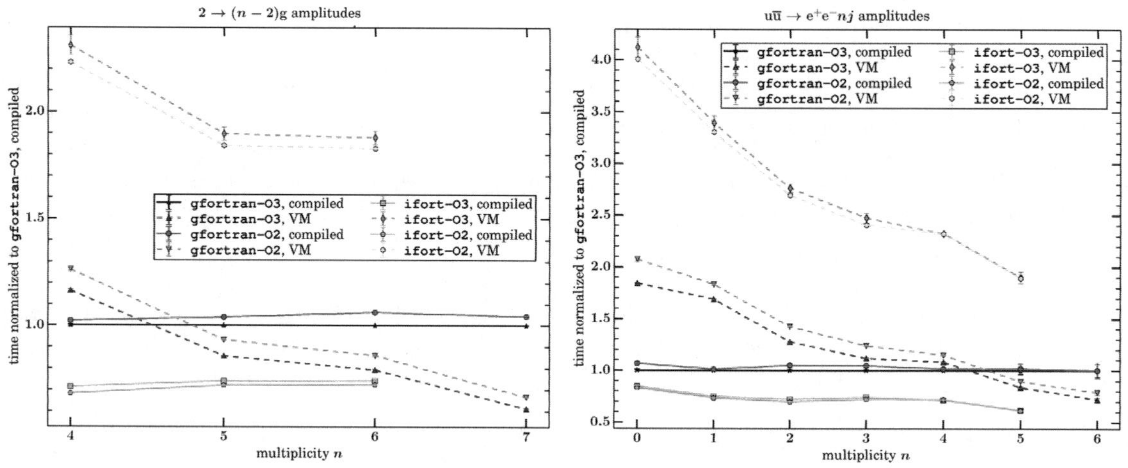

(a) Pure gluon amplitudes $2 \to (n-2)g$ (b) SM Drell-Yan process $\mathrm{u}\bar{\mathrm{u}} \to \mathrm{e}^+\mathrm{e}^- nj$

Figure 1: CPU times measured with the **Fortran** intrinsic `cpu_time` and normalized for each process to the compiled source code using `gfortran -O3`. Dashed (solid) lines represent the OVM (compiled source code).

compiler is usually able to compile one multiplicity higher compared to the Intel before breaking down. This fits together with the better performance of the compilable processes and longer compile times as `ifort` seems to apply more optimizations.

Another interesting observation is that the OVM gets faster compared to the compiled code with increasing multiplicity of external particles though this feature is more pronounced in SM and QCD processes. As discussed in Ref. [1], this is not due to costs from initialization or virtualization but likely caused by the explicit loop over the instructions in the VM. It gives a higher probability to keep the decode function in the instruction cache compared to the compiled code. We observe roughly the same effect for both compilers, but the OVM compiled with `ifort` is about a factor of two slower than the version with `gfortran`. This is probably solvable with a profile-guided optimization, which allows the compiler to use data instead of heuristics to decide what and how to optimize. Amdahl's idealized law [7] simply divides an algorithm into parallelizable parts p and strictly serial parts $1-p$. Therefore, the possible speedup s for a computation with n processors is

$$s(n) \equiv \frac{t(1)}{t(n)} = \frac{1}{(1-p)+\frac{p}{n}} \,. \tag{1}$$

Communication costs between processors $\mathcal{O}(n)$ have been neglected hereby in the denominator of eq. (1). This means that we have $\lim_{n\to\infty} s(n) = 1/(1-p)$ in the idealized case and $\lim_{n\to\infty} s(n) = 0$ including communication costs. In reality, we are interested in high speedups for finite n and also have to care about efficient cache usage.

In Figure 2, we show the speedup with multiple cores, N, by either using the parallelization procedure discussed above to compute one amplitude in parallel or by computing multiple amplitudes for multiple phase-space points in parallel. In Figure 2a, we can see that the $n=7$ and $n=8$ gluon amplitudes parallelize very well with both methods with parallelizable parts above 95 %. In the shared memory parallel evaluation of the amplitude (A), the impact of the hardware architecture is quite obvious. For $N=7$, i.e. when the second socket of the board is activated, we see a drop in efficiency. This relative drop is the stronger the higher the communication costs are compared to the calculation done in the individual threads and can

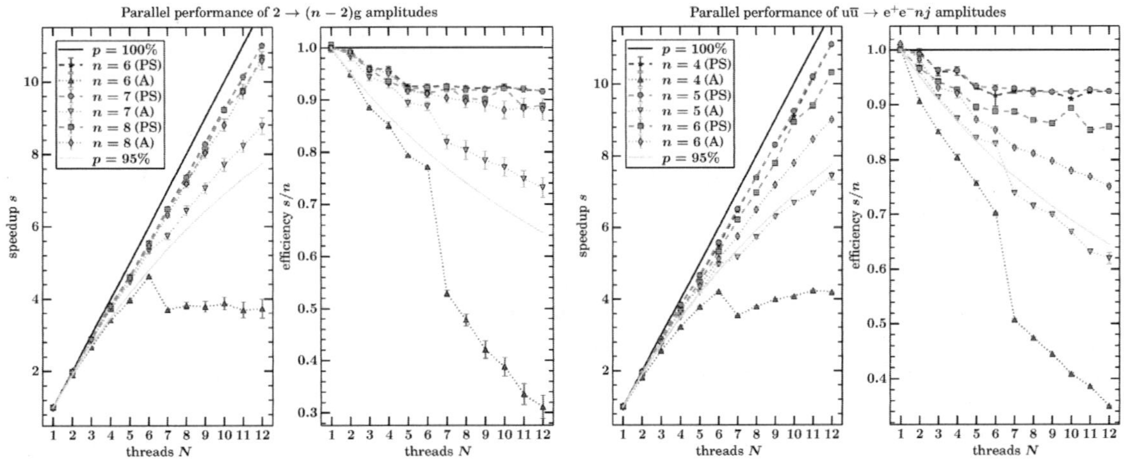

(a) Pure gluon amplitudes $2 \to (n-2)g$ (b) SM Drell-Yan process $\mathrm{u\bar{u}} \to \mathrm{e^+e^-}nj$

Figure 2: Speedup and efficiency to compute a fixed number of phase-space points for the parallel evaluation of multiple phase-space points (PS) and the parallel evaluation of the amplitude itself (A) are shown as dashed and dotted lines. The solid lines represent Amdahl's law for a fixed value of the parallelizable part p.

thus be seen most cleanly for the $n = 6$ gluon and the $n = 4$ Drell-Yan processes. This is a cache coherency issue due to the synchronization of the caches of the two CPUs. We observe that the byte-code for the OVM is about one order of magnitude smaller. Some processes together with their compile times are shown in Table 1. The smaller output format leads to less required RAM and time to produce it. Especially for many color flows, where the generation time of O'MEGA is dominated by the output procedure, we observe e.g. for $\mathrm{gg} \to 6\mathrm{g}$ a reduction in memory from 2.17 GiB to 1.34 GiB and in generation time from 11 min 52 s to 3 min 35 s, while staying roughly the same for small processes.

Table 1: Size of the byte-code (`BC`) compared to the `Fortran` source code together with the corresponding compile time with `gfortran`. The compile times were measured on a computer with an i7–2720QM CPU. The $2\mathrm{g} \to 6\mathrm{g}$ process fails to compile.

process	`BC` size	`Fortran` size	t_{compile}
$\mathrm{gg} \to \mathrm{gggggg}$	428 MiB	4.0 GiB	—
$\mathrm{gg} \to \mathrm{ggggg}$	9.4 MiB	85 MiB	483(18) s
$\mathrm{gg} \to q\bar{q}q'\bar{q}'q''\bar{q}''g$	3.2 MiB	27 MiB	166(15) s
$e^+e^- \to 5\,(e^+e^-)$	0.7 MiB	1.9 MiB	32.46(13) s

5. Summary and Outlook

A VM circumvents the compile and link problems that are associated with huge source code as it emerges from very complex algebraic expressions. VMs are indeed a viable option that is maintaining relatively high performance in the numerical evaluation of these expressions and allows to approach the hardware limits. In practice, a VM saves at least hours of compile time.

The concept has been successively applied to construct the OVM that is now an alternative method to compute tree-level matrix elements in the publicly available package O'MEGA and can be chosen in WHIZARD with a simple option since version 2.2.3. Any computation can in principle be performed with a VM though the benefits are clearly in the regime of extreme computations that run into compiler limits with the conventional method. Here, we have seen that VMs can even perform better than compiled code. Also the parallelization of the amplitude is for very complex processes close to the optimum.

It would be an interesting experiment to remove the virtualization overhead by using dedicated hardware that has the same instruction set as the OVM to compute matrix elements. The number of instructions corresponding to different wave function fusions and propagators is finite for renormalizable theories (including effective theories up to a fixed mass dimension) and implemented similarly in the various matrix element generators. If the authors can agree on a common set of instructions and conventions this machine could therefore be used by all those programs. Field programmable gate arrays (FPGAs) can serve as such a machine as they have comparable if not superior floating-point performance with respect to current microprocessors and the OVM and its instruction set is the first step to test the feasibility and potential gains of computing matrix elements in this environment.

Acknowledgments

JR wants to thank the organizers for the beautiful venue.

References

[1] Chokoufe Nejad B, Ohl T and Reuter J 2015 *Computer Physics Communications* **196** 58–69 (*Preprint* 1411.3834)
[2] Hagiwara K, Kanzaki J, Okamura N, Rainwater D and Stelzer T 2010 *The European Physical Journal C* **66** 477–492 (*Preprint* 0908.4403)
[3] Moretti M, Ohl T and Reuter J 2001 (*Preprint* hep-ph/0102195)
[4] Kilian W, Ohl T and Reuter J 2011 *The European Physical Journal C* **71** 1742 (*Preprint* 0708.4233)
[5] Kilian W, Ohl T, Reuter J and Speckner C 2012 *Journal of High Energy Physics* **2012** 22 (*Preprint* 1206.3700v2)
[6] Kilian W, Reuter J, Schmidt S and Wiesler D 2012 *Journal of High Energy Physics* **2012** 13 (*Preprint* 1112.1039)
[7] Amdahl G 1967 *AFIPS Conference Proceedings* **30** 483–485

Modelling the nuclear parton distributions

S A Kulagin

Institute for Nuclear Research of the Russian Academy of Sciences, Moscow 117312, Russia

E-mail: `sergey.kulagin@gmail.com`

Abstract. We review a semi-microscopic model of nuclear parton distributions, which takes into account a number of nuclear effects including Fermi motion and nuclear binding, nuclear meson-exchange currents and off-shell corrections to bound nucleon distributions as well as nuclear shadowing effect. We also discuss applications of the model to the lepton-nuclear deep-inelastic scattering, Drell-Yan process and neutrino total cross sections.

1. Introduction

The leading contribution to cross sections of various hard processes is determined by the parton distributions (PDFs). Thus, PDFs are universal process-independent characteristics of the target at high invariant momentum transfer Q. The nuclear parton distributions (NPDFs) are the subject of significant nuclear effects which span a wide region of Bjorken x as demonstrated by the deep inelastic scattering (DIS) experiments and the measurements of the Drell-Yan process (DY) (for a review see [1, 2]). A relative rate of nuclear effects is more than one order of magnitude larger than the ratio of the nuclear binding energy to the nucleon mass thus indicating that the nuclear environment plays an important role even at high energies and momenta.

A number of phenomenological approaches to NPDFs are available in literature [3, 4, 5, 6]. Although such studies are useful in constraining nuclear effects for different partons, they provide little information about the underlying physics responsible for the nuclear corrections. In this contribution we follow a different approach and present the results of a study of NPDFs using a model of Ref.[7] which takes into account a few different mechanisms of nuclear corrections. The model, which is briefly reviewed in Sec.2, explains the observed x, Q^2 and A dependencies of the nuclear structure functions in the deep inelastic scattering (DIS) for a wide range of nuclear targets from ^{2}H to ^{207}Pb [7, 8, 9]. In what follows we focus on the region of high Q^2 and review predictions [10] of nuclear effects for the valence and sea quark distributions (Sec.2), discuss applications to nuclear DIS, nuclear DY process (Sec.3) and the total neutrino-nuclear cross sections (Sec.4).

2. Nuclear PDFs and DIS

We will use the notation $q_{a/T}(x, Q^2)$ for the distribution of quarks of the flavor a in a target T. The parton distributions in a nucleus receive a number of contributions and can be summarized as [7, 10]

$$q_{a/A} = \left\langle q_{a/p} \right\rangle + \left\langle q_{a/n} \right\rangle + \delta q_a^{\mathrm{coh}} + \delta q_a^{\mathrm{MEC}}. \tag{1}$$

For brevity, we suppress explicit dependencies on x and Q^2 in Eq.(1). The first two terms on the right side are the contribution from the partons from bound protons and neutrons. The brackets stand for the averaging with the energy-momentum distribution of bound nucleons in a nucleus, as discussed in detail in Ref.[7, 10]. Note that the evaluation of these terms requires the proton and the neutron PDFs in off-mass shell region [11]. The off-shell correction together with the nucleon momentum distribution (Fermi motion) and the nuclear binding effect [12] plays an important role in the valence quark region [7, 9].

The correction δq^{coh} arises due to propagation effects of intermediate hadronic states of a virtual boson in nuclear environment. This term involves contributions from multiple scattering series and typically lead to a negative correction relevant at low x (nuclear shadowing effect, for a review see, e.g., Ref.[14]).

The last term in Eq.(1) is a contribution from the meson degrees of freedom in nuclei. We recall that the meson fields mediate the nucleon-nucleon interaction in nuclei and also generate an additional quark-gluon content in nuclei. This term is relevant for intermediate region of x and results in a some enhancement of the nuclear sea quark distribution. Also its contribution is important to balance the overall nuclear light-cone momentum. More details on the treatment of each term in Eq.(1) can be found in Refs.[7, 10].

In analysis of Ref.[10] we choose the PDFs with definite C-parity $q^{\pm} = q \pm \bar{q}$ as a basis. Note that the C-even combination xq^{+} contributes to the structure function F_2, for both the charged-lepton and neutrino deep-inelastic scattering, while the C-odd xq^{-} is the valence quark distribution which contributes to the neutrino structure function xF_3. For light quarks we also consider the isoscalar $q_0 = u+d$ and the isovector $q_1 = u-d$ combinations. Below we summarize our results on the nuclear valence and sea quark PDFs.

In order to illustrate the nuclear effects, in Fig.1 we show the ratios $R_a^A = q_{a/A}/(Zq_{a/p} + Nq_{a/n})$ for the valence quarks (left panel) and the antiquarks (right panel) in the lead nucleus. Here Z and N are the proton and the neutron number in the discussed nucleus with the total nucleon number $A = Z + N$, and $q_{a/p}$ and $q_{a/n}$ are the free proton and the neutron PDF, respectively. As can be seen from Fig.1 the resulting nuclear effect is different for the nuclear

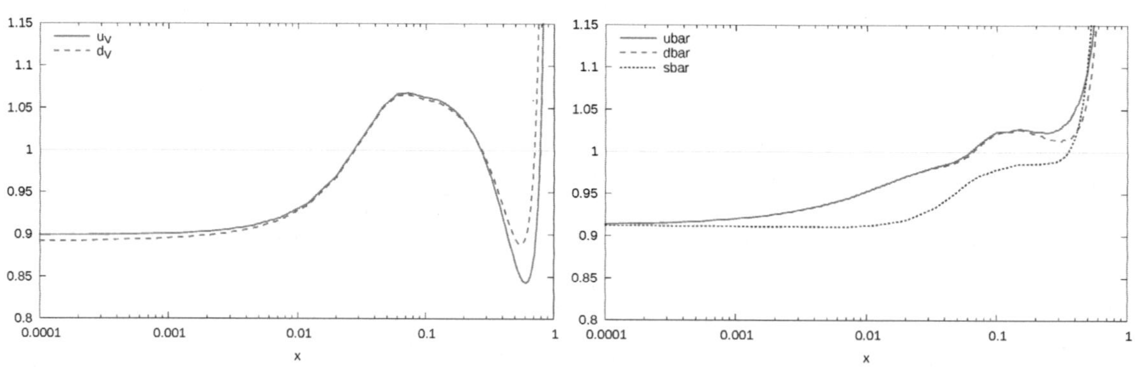

Figure 1. The nuclear ratios for the valence quarks (left panel) and the $\bar{u}$, $\bar{d}$, and $\bar{s}$ antiquarks (right panel) in ^{208}Pb calculated at $Q^2 = 25\,\mathrm{GeV}^2$ (see text for more detail).

valence and sea distribution and also depends on the PDF flavor. We briefly comment on characteristic features of the nuclear ratios for different PDFs, for a more detail discussion see Ref.[10].

In the region $x \ll 0.1$, all NPDFs are suppressed due to the nuclear *shadowing* effect (a negative δq^{coh} term). Note that the rate of this effect is not universal and differs for the valence and sea quark distribution [7, 10]. We also remark that the result of averaging of the nucleon PDF with the nuclear spectral function in the first two terms in the right side of Eq.(1) depends on the details of the x dependence of considered PDF. For this reason, this correction is positive for the valence quarks thus partially reducing the shadowing effect. For the sea quarks this correction is negative in this region. However, the MEC correction is positive and we also have a partial cancellation of different effects.

In the intermediate region of $x \sim 0.1$ (which is usually referred to as *antishadowing* region) we observe interplay between different nuclear corrections. For the valence quark PDF (C-odd q^- distribution) we find an enhancement which is due to constructive interference in the multiple scattering effect between the C-even and C-odd amplitudes in the $\delta q^{\rm coh}$ term. The antishadowing effect on the antiquark PDFs is somewhat weaker and shifted towards larger x because of a partial cancellation between different contributions.

At large $x > 0.2$ the nuclear PDFs are dominated by incoherent scattering from bound nucleons, i.e. by the first term on the right in Eq.(1). For the valence quarks, the averaging of the proton and neutron PDFs with the nuclear spectral function together with the off-shell correction produces a pronounced 'EMC-effect' shape at large x [7]. This effect was noticed long ago in [12, 13] and further discussed in Ref.[7]. The relative rate of this correction strongly depends on the particular x dependence of the given PDF and for this reason the ratios $R_{\rm val}$ and $R_{\rm sea}$ are quite different at large x.

It should be remarked that different nuclear effects are related through a number of sum rules which reflect different global symmetries. In particular, the conservation of the valence quark number links together the off-shell and the shadowing corrections to the valence quark distributions. In Ref.[7, 10] the normalization conditions for the isoscalar and the isovector valence quark distributions were used as equations to determine the unknown amplitudes controlling the coherent nuclear correction. We then solved these equations in terms of the off-shell correction to the corresponding distributions. The energy-momentum conservation causes the light-cone momentum sum rule at two different levels. At the hadronic level, the nuclear light-cone momentum is shared between nucleons and mesons. This requirement allows to constrain the MEC correction to NPDFs. At the partonic level, the light-cone momentum is balanced between quarks, antiquarks and gluons. The study of different contributions to the light-cone momentum sum rule can provide insights on modification of gluon distribution in nuclei.

A detailed analysis of data on the ratio of the DIS structure functions for different nuclear targets $R_2(A'/A) = F_2^{A'}/F_2^A$ (usually the ratio is taken with respect to the deuteron) with targets ranging from ^{4}He to ^{208}Pb shows a very good agreement of model predictions with observed x, Q^2 and nuclear dependence of data [7, 9]. The performance of discussed model is illustrated in Fig.2, in which we show the model predictions in comparison with data for a number of nuclei. It should be remarked that the model predictions in Fig.2 are given at a fixed $Q^2 = 5\,\mathrm{GeV}^2$ while the data from different measurements have essentially different average Q^2. Furthermore, each x bin of data has different average Q^2 value. For a detailed comparison with data and discussion of consistency of different data sets see Ref.[7, 9].

3. NPDFs and Drell-Yan process

The reaction of muon pair production in hadron-hadron collisions (Drell-Yan process) is an important source of information on the proton, pion and nuclear PDFs [20]. In the context of NPDFs, the use of DY data in combination with DIS data allows a separation of the nuclear valence and sea quark distributions. In the DY reaction with the proton beam by tuning the kinematics of the muon pair one can select a region in which the DY cross sections are driven by annihilation of valence quarks in the beam and antiquarks in the target. Then the ratio of the DY cross sections off different nuclear targets provide a tool to measure the nuclear dependence of antiquark PDFs

$$\frac{\sigma_A^{DY}}{\sigma_B^{DY}} \approx \frac{\bar{u}_A(x_T) + \bar{d}_A(x_T)}{\bar{u}_B(x_T) + \bar{d}_B(x_T)}, \tag{2}$$

Journal of Physics: Conference Series **762** (2016) 012072 doi:10.1088/1742-6596/762/1/012072

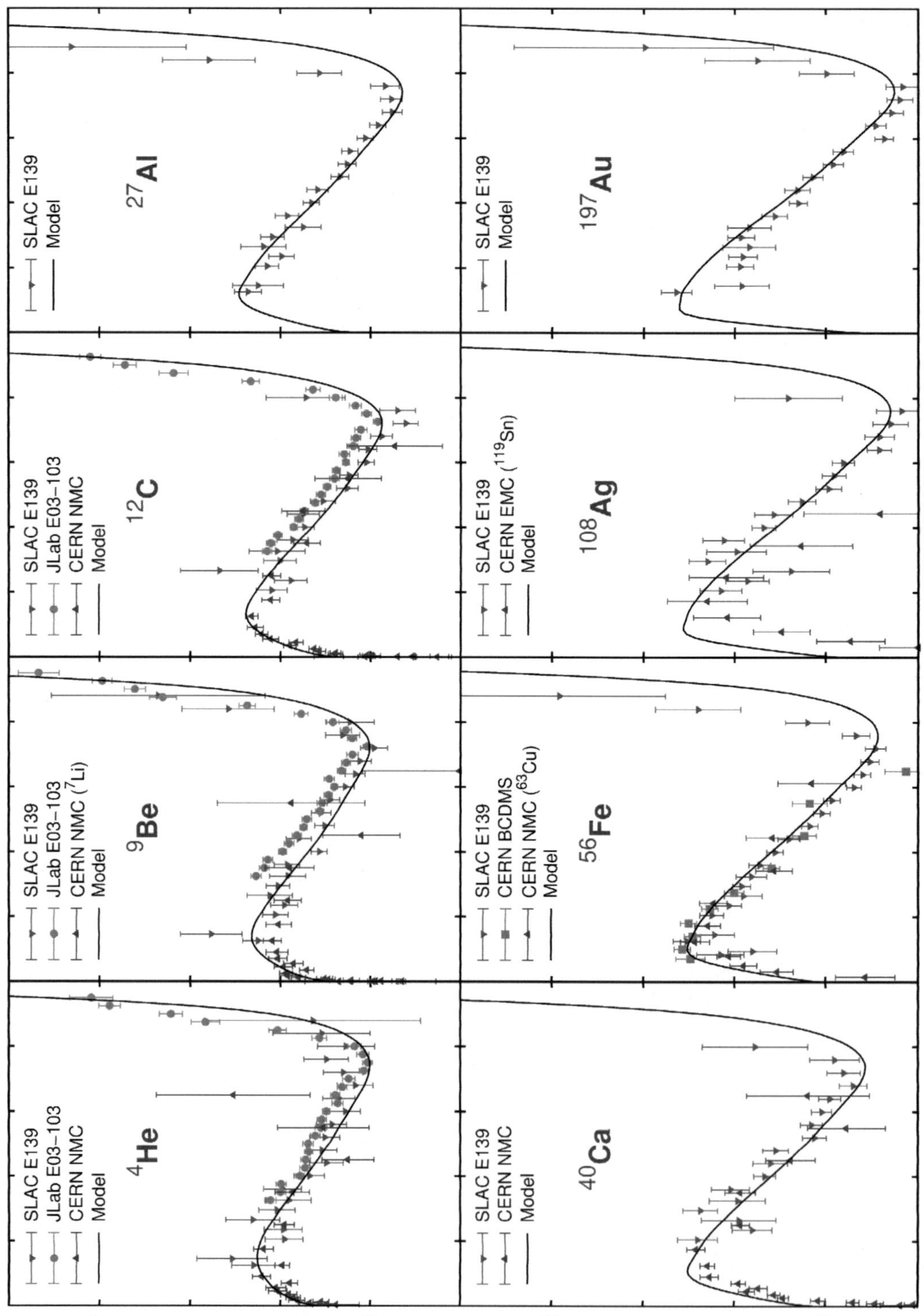

Figure 2. The ratio of a heavy nucleus to the deuteron structure function F_2^A/F_2^D. Data are from experiments at CERN (EMC [15], NMC [16] and BCDMC [17]), SLAC (E139 [18]), and JLab [19]. The model predictions are given for a fixed $Q^2 = 5\,\mathrm{GeV}^2$.

where x_T is the Bjorken variable for a nuclear target. This ratio was measured in the experiments E772 [21] and E866 [22] with the proton beam momentum 800 GeV/c at Fermilab in the region $x_T < 0.15$ for a number of nuclear targets. In contrast to DIS, the DY data show no *antishadowing* (i.e. enhancement of nuclear antiquark distributions) at $x_T \sim 0.1$ that was a long standing puzzle since the nuclear binding should result in an excess of nuclear mesons, which is expected to produce a marked enhancement in the nuclear anti-quark distributions [25].

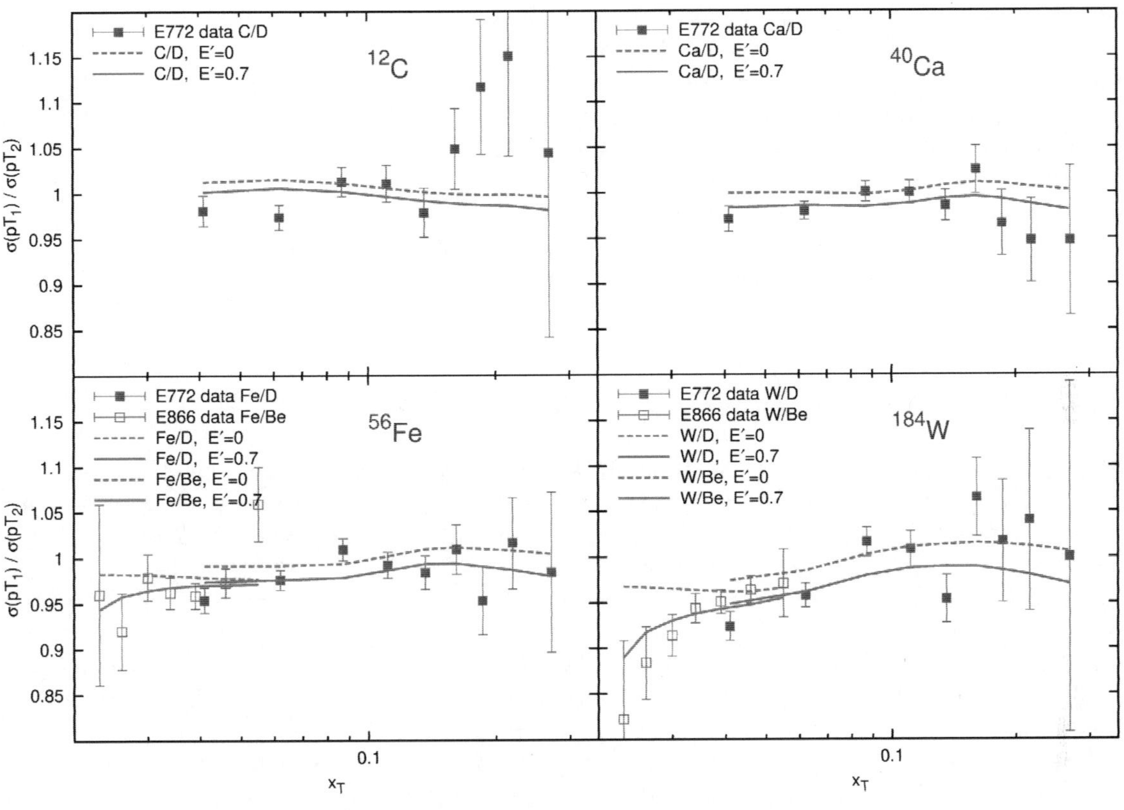

Figure 3. Ratio of the DY reaction cross sections for different nuclei vs. x_T. Data points are from the E772 experiment [21] (■) and E866 experiment [22] (□). Note that the ratio is normalized to one bound nucleon and taken relative to the deuteron for E772 and ^{9}Be for E866. The curves are the predictions of Ref.[10] for the deuteron (green) and beryllium (red) ratios with (——) and without (- - - -) the projectile energy loss effect (see text and legend).

As discussed above, the model of Ref.[7, 10] predicts a significant cancellation of different nuclear effects in the antiquark distribution in the region $x \sim 0.1$ in agreement with the nuclear DY data. In Fig.3 we show the data along with our predictions on the ratios of the DY cross sections [10]. We also remark that nuclear dependence of the DY process comes from two different sources: (i) the modification of the nuclear target PDFs, and (ii) the initial state interaction of the projectile particle (parton) within the nuclear environment that causes the parton energy loss [23] before annihilation into a dimuon pair. The rate of this effect is characterized by the parton energy loss in a nucleus per unit length E' and effectively results in a change of the projectile parton Bjorken x [24] which in turn modify the ratio in Eq.(2).

The data from the E866 experiment is shifted towards lower values of target's x_T and higher values of projectile's x_B with respect to E772 data thus falling into a region where both the shadowing and the energy loss effects become more prominent. Our analysis [10] indicates that the model of Ref.[7] is in a good agreement with data at a moderate energy loss effect. The solid curves in Fig.3 show our predictions with $E' = 0.7$ GeV/fm. Note also that the cross section ratios are taken relative to the deuterium for E772 and beryllium for E866. For this reason the corresponding curves in Fig.3 are not identical in the overlap region.

4. NPDFs and total neutrino cross sections

For the reason of statistics the neutrino scattering data have been collected on heavy nuclei. Thus the comparison of theory predictions with data on the neutrino cross sections serves as an important and independent test of both the proton and the nuclear PDFs. In the following we focus on the total cross sections as the available data on total cross sections have typically better accuracy than the corresponding measurements of differential cross sections and span a wide energy region.

We calculate the total cross sections by integrating the double differential inelastic cross sections $\mathrm{d}^2\sigma^{(\nu,\bar\nu)}/\mathrm{d}x\mathrm{d}y$ over allowed region of x and y. In the leading α_S order we have for the total cross sections off an isoscalar target (i.e. the combination of $p+n$ or a nucleus with equal number of protons and neutrons)

$$\frac{\sigma_{\rm tot}(E)}{E} = \frac{2G_F^2 M}{3\pi(1+Q^2/M_W^2)}\left(x_0^+ \pm x_0^-/2 + 3x_s\right), \tag{3}$$

where G_F is the Fermi coupling constant, M and M_W are the proton and the W-boson mass, $x_0^\pm = \int_0^1 \mathrm{d}x x q_0^\pm(x, Q^2)$ is the average light-cone momentum for the isoscalar C-even (q_0^+) and C-odd (q_0^-) quark distributions, and, similarly, x_s is the light-cone momentum carried by s quark (we assume $s = \bar s$ and also neglect a small contributions from heavy quarks). The $+/-$ sign in Eq.(3) corresponds to $\nu/\bar\nu$ scattering. The average invariant momentum transfer Q in Eq.(3) is difficult to calculate precisely, however by order of magnitude $Q^2 = ME/2$.

In Fig.4 we show the results on the total cross section for the isoscalar nucleon calculated by Eq.(3). We also go beyond the LO approximation and numerically compute the (anti)neutrino cross sections using the NNLO structure functions with the NNLO PDFs of Ref.[29].

A few comments are in order. The ratio $\sigma_{\rm tot}(E)/E$ is almost a constant in the considered energy region. A weak energy dependence comes through Q^2 dependence of the parton distributions (structure functions). A higher order α_S corrections on the structure functions result in a overall negative correction to the total cross section reducing significantly the LO cross section in the energy region up to a few hundred GeV. It is worth noting that this effect leads to a better agreement with data.

In Fig.4 we show the data from CCFR [26], NOMAD [27] and MINOS [28] neutrino experiments. Note, however, those data were taken on different nuclear targets, ^{56}Fe for CCFR and MINOS and ^{12}C for NOMAD. Therefore, in order to make a meaningful comparison, it is important to address the nuclear dependence of the (anti)neutrino cross sections. For this reason we discuss the normalized ratios of the nuclear and the proton and neutron neutrino cross sections $R_A^\nu = \sigma_A^\nu/(Z\sigma_p^\nu + N\sigma_n^\nu)$ and a similar ratio $R_A^{\bar\nu}$ for antineutrino. In Fig.5 (left panel) we show the ratios R_A^ν and $R_A^{\bar\nu}$ calculated for ^{12}C and ^{56}Fe nuclei using the NNLO structure functions of Ref.[29] within the approach outlined above (for more detail see Refs.[8, 10]). We observe that in considered energy region the rate of nuclear effects is about 2% at maximum for both the neutrino and the antineutrino and gradually decreasing with energy. Note also that the ratios R_A^ν and $R_A^{\bar\nu}$ show almost no nuclear dependence that justifies the comparison of total neutrino cross sections for different nuclear targets in Fig.4.

Figure 5 indicates a suppression of the neutrino-nuclear total cross section while the antineutrino cross section is somewhat enhanced in nuclei (the latter in turn further improves

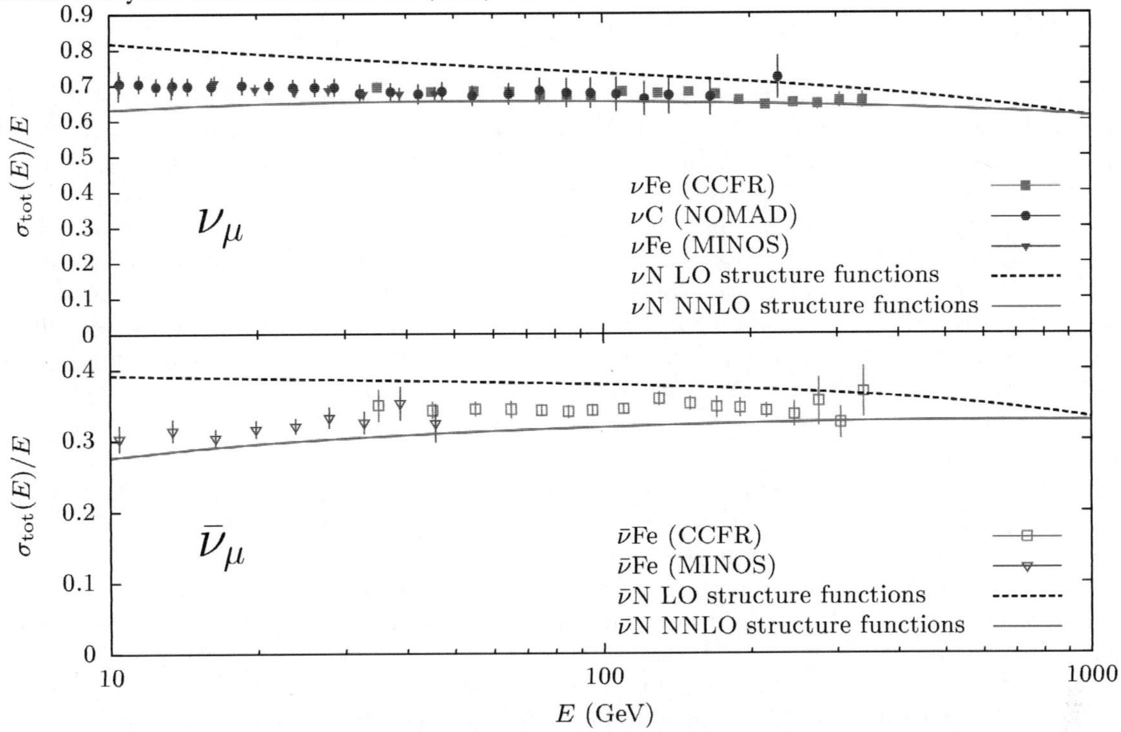

Figure 4. The neutrino (upper panel) and antineutrino (lower panel) total cross sections as a function of neutrino energy E. The units are $10^{-38}\,\mathrm{cm}^2/\mathrm{GeV}$. The data points are the cross sections per one bound nucleon from CCFR [26], NOMAD [27] and MINOS [28] experiments corrected for the neutron excess in iron. The curves are the LO (dashed) and the NNLO (full) predictions for ^{56}Fe target calculated with the PDFs of Ref.[29].

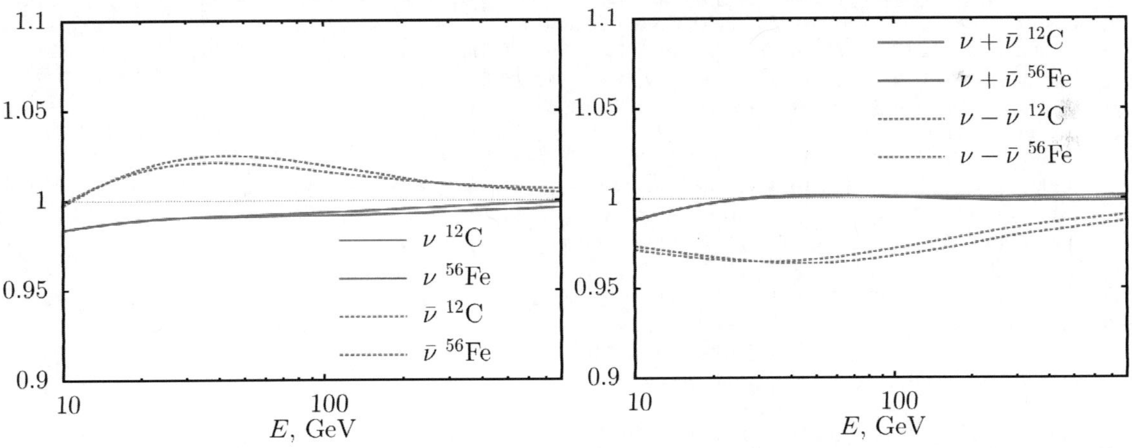

Figure 5. Left panel: The ratios R_A^ν (solid curve) and $R_A^{\bar{\nu}}$ (dashed curve) calculated for ^{12}C (red) and ^{56}Fe (green) nuclei as a function of neutrino energy E. Right panel: Similar ratios for the C-even $\nu+\bar{\nu}$ (solid) and C-odd $\nu-\bar{\nu}$ (dashed) combinations of the total cross sections.

agreement with $\bar{\nu}$ data in Fig.4). These two effects cancel each other for the C-even combination $\nu+\bar{\nu}$ of the cross sections as illustrated in the right panel of Fig.5. On the other hand, the nuclear correction to the C-odd combination $\nu-\bar{\nu}$ is somewhat enhanced.

It is instructive to compare the nuclear corrections for the total cross sections with those for PDFs (see Fig.1 and 2). We note in this context a direct relation between the total neutrino cross sections and average light-cone momenta of different quark PDFs in Eq.(3). The nuclear modification of average light-cone momenta have much smaller magnitude compared to a typical amplitude of nuclear effects for PDFs shown in Fig.1 and 2. This is because of a cancellation between nuclear corrections of different sign in different kinematic regions in the integral over the Bjorken x. The isoscalar $\nu+\bar{\nu}$ combination of the total cross sections is proportional to the total quark and antiquark light-cone momentum in the target. Thus a cancellation of nuclear effects in $\sigma(\nu+\bar{\nu})$ suggests that the overall light-cone momentum of quarks and antiquarks does not change in nuclei. Referring to the light-cone sum rule we also conclude that the average gluon light-cone momentum is not modified in nuclei. However, discussion of associated problems goes beyond of the scope of the present article and will be given elsewhere.

5. Summary

In summary, we reviewed a semi-microscopic model of nuclear PDFs addressing a few different mechanisms of modification of PDFs in nuclear environment. A number of applications were discussed including charged-lepton DIS, proton-nuclear DY reaction and calculation of total (anti)neutrino-nuclear cross sections in charged-current interaction.

I am grateful to R. Petti for fruitful collaboration on the topics of this paper and A. Kataev for motivating discussions. I would like to thank the Organizing Committee of ACAT 2016 Workshop for support and for creating a friendly and working environment during the meeting.

The work was supported by the Russian Science Foundation grant No. 14-22-00161.

References

[1] Arneodo M 1994 *Phys. Rept.* **240** 301 (1994)
[2] Norton P R 2003 *Rept. Prog. Phys.* **66** 1253 (2003)
[3] Eskola K J, Paukkunen H and Salgado C A 2009 *J. High Energy Phys.* JHEP0904(2009)065
[4] Hirai M, Kumano S and Nagai T-H 2007 *Phys. Rev.* C **76** 065207 (2007)
[5] de Florian D, Sassot R, Zurita P and Stratmann M 2012 *Phys. Rev.* D **85** 074028 (2012)
[6] Kovarik K *et al.* 2015 *Preprint* arXiv:1509.00792 [hep-ph].
[7] Kulagin S A and Petti R 2006 *Nucl. Phys.* A **765** 126 (2006)
[8] Kulagin S A and Petti R 2007 *Phys. Rev.* D **76** 094023 (2007)
[9] Kulagin S A and Petti R 2010 *Phys. Rev.* C **82** 054614 (2010)
[10] Kulagin S A and Petti R 2014 *Phys. Rev.* C **90** 045204 (2014)
[11] Kulagin S A, Piller G and Weise W 1994, *Phys. Rev.* C **50** 1154 (1994)
[12] Akulinichev S V, Kulagin S A and Vagradov G M 1985, *Phys. Lett.* B **158** (1985) 485
[13] Kulagin S A 1989, *Nucl. Phys.* A **500** 653 (1989)
[14] Piller G and Weise W 2000, *Phys. Rept.* **330** 1 (2000)
[15] Ashman J *et al.* 1993, *Z. Phys.* C **57** 211 (1993)
[16] Armeodo M *et al.* 1995, *Nucl. Phys.* B **441** 12 (1995); *Ibid.* 3 (1995)
[17] Bari G *et al.* 1985, *Phys. Lett.* B **163** 282 (1985)
[18] Gomez J *et al.* 1994, *Phys. Rev.* D **49** 4348 (1994)
[19] Seely J *et al.* 2009, *Phys. Rev. Lett.* **103** 202301 (2009)
[20] Peng J C and Qiu J W 2014, *Prog. Part. Nucl. Phys.* **76** 43 (2014)
[21] Alde D M *et al.* 1990, *Phys. Rev. Lett.* **64** 2479 (1990).
[22] Vasilev M A *et al.* 1999, *Phys. Rev. Lett.* **83** 2304 (1999).
[23] Bjorken J D 1982, FERMILAB-PUB-82-059-THY
[24] Garvey G T and Peng J C 2003, *Phys. Rev. Lett.* **90** 092302 (2003)
[25] Bickerstaff R P, Birse M C and Miller G A 1984, *Phys. Rev. Lett.* **53** 2532 (1984)
[26] Seligman W G 1997, FERMILAB-THESIS-1997-21
[27] Wu Q *et al.* 2008, *Phys. Lett.* B **660** 19 (2008)
[28] Adamson P *et al.* 2010, *Phys. Rev.* D **81** (2010) 072002
[29] Alekhin S, Blumlein J, and Moch S 2014, *Phys. Rev.* D **89** 054028 (2014)

Journal of Physics: Conference Series **762** (2016) 012073 doi:10.1088/1742-6596/762/1/012073

Numerical multi-loop calculations: tools and applications

S. Borowka[1], G. Heinrich[2], S. Jahn[2], S. P. Jones[2], M. Kerner[2], J. Schlenk[2], T. Zirke[2]

[1] Institute for Physics, University of Zürich, Winterthurerstr.190, 8057 Zürich, Switzerland
[2] Max Planck Institute for Physics, Föhringer Ring 6, 80805 München, Germany

E-mail: sborowka@physik.uzh.ch, gudrun@mpp.mpg.de, sjahn@mpp.mpg.de, sjones@mpp.mpg.de, kerner@mpp.mpg.de, jschlenk@mpp.mpg.de, zirke@mpp.mpg.de

Abstract. We report on the development of tools to calculate loop integrals and amplitudes beyond one loop. In particular, we review new features of the program SecDec which can be used for the numerical evaluation of parametric integrals like multi-loop integrals.

1. Introduction

Precision calculations for the LHC and future colliders are entering more and more the domain of predictions beyond the next-to-leading order (NLO). The rapid progress was mainly triggered by the development of better calculational methods and tools, but also by the development of more powerful computing resources. Both aspects lay the groundwork allowing to move towards the automation of calculations beyond NLO.

In these proceedings we will focus on the automation of two-loop calculations, in particular two-loop integrals with several mass scales. While the latter are very difficult to calculate analytically, numerical methods may offer a solution. However, the numerical integration can only succeed if the various types of singularities which may occur in a two-loop amplitude can be isolated or evaded beforehand, preferably in an automated way. The program SecDec [1–4] is able to perform this task, and to subsequently integrate the resulting finite parameter integrals numerically. In the following we will describe the program SecDec and some of its new features, as well as its embedding into a framework to generate two-loop amplitudes.

2. Amplitude generation at one- and two loops

The steps to be performed in order to generate a virtual amplitude (at in principle any loop order) in an approach based on Feynman diagrams are roughly the following:

(i) generation of the contributing diagrams in terms of algebraic expressions, based on the Feynman rules derived from a Lagrangian for a certain model,

(ii) casting the expression into a form "coefficients $\otimes$ (tensor) loop integrals",

(iii) reduction of the loop integrals to a set of master integrals (ideally, this set is a minimal basis in loop integral space) times coefficients, where the latter coefficients will depend on the dimension D of loop momentum space,

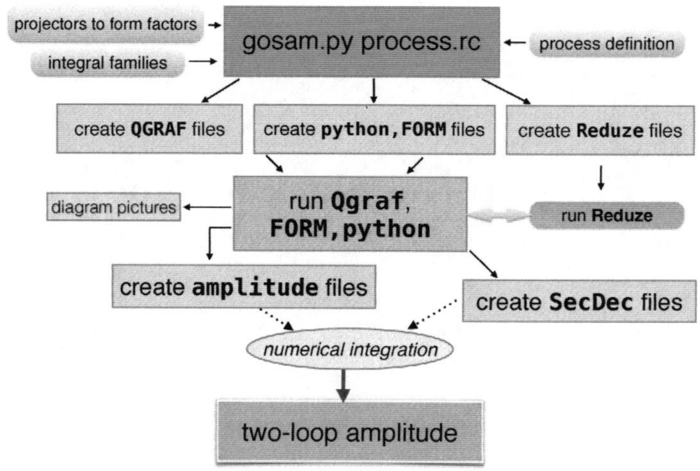

Figure 1. Building blocks of GoSam-2loop. Boxes with rounded corners denote user input. The reduction to master integrals is run separately.

(iv) evaluation of the master integrals, combination with the coefficients forming the amplitude.

We are constructing a program package which is designed to perform the steps listed above in an automated way. It partly builds on the program package GoSam [5, 6] for automated one-loop calculations; for example it also uses Qgraf [7] for the diagram generation. However, the reduction step and the evaluation of the master integrals are much more challenging than at one loop, as the integral basis is usually not known beforehand, and the master integrals may not be available in analytic form. Within the GoSam-2loop framework, we constructed an interface to the reduction program Reduze [8, 9]. The reduction to master integrals is then run separately, and the resulting database is imported back into GoSam-2loop. The latter also contains an interface to the program SecDec which is used to evaluate the integrals numerically. The flowchart for GoSam-2loop, shown in Fig. 1, illustrates the main steps.

3. The program SecDec

The program SecDec [1–4] is based on the sector decomposition algorithm described in [10, 11]. It can be applied to multi-loop integrals as well as more general parametric integrals which contain dimensionally regulated end-point singularities. Other public implementations of sector decomposition are described in [12–16].

3.1. Feynman Integrals

We consider scalar multi-loop integrals in order to simplify the notation. The case of tensor integrals is analogous, for details we refer to [11]. A scalar Feynman integral I in D dimensions at L loops with N propagators, where the propagators can have arbitrary, not necessarily integer powers ν_j, has the following representation in momentum space:

$$I = \frac{\mu^{4-D}}{i\pi^{\frac{D}{2}}} \int d^D k_l \, \frac{1}{\prod\limits_{j=1}^{N} P_j^{\nu_j}(\{k\},\{p\},m_j^2)} \,,$$

$$P_j(\{k\},\{p\},m_j^2) = q_j^2 - m_j^2 + i\delta \,, \qquad (1)$$

where the q_j are linear combinations of external momenta p_i and loop momenta k_l. Introducing Feynman parameters and carrying out the momentum integrations leads to

$$G = \frac{(-1)^{N_\nu}}{\prod_{j=1}^{N}\Gamma(\nu_j)}\Gamma(N_\nu - LD/2)\int_0^\infty \prod_{j=1}^{N} dx_j\, x_j^{\nu_j-1}\,\delta(1-\sum_{l=1}^{N} x_l)\frac{\mathcal{U}(\vec{x})^{N_\nu-(L+1)D/2}}{\mathcal{F}(\vec{x},\{s_{ij},m_i^2\})^{N_\nu-LD/2}}\,, \quad (2)$$

with $N_\nu = \sum_{j=1}^{N}\nu_j$. The graph polynomial $\mathcal{U}$ is a positive-semidefinite function of the Feynman parameters only, while $\mathcal{F}$ is a function of Feynman parameters and the kinematic invariants characterising the integral. In the region where all Lorentz invariants formed from external momenta are negative, which we call the *Euclidean region*, $\mathcal{F}$ is also a positive semi-definite function of the Feynman parameters x_j and the invariants. If some of the invariants (masses, squared external momenta) are zero, the vanishing of $\mathcal{F}$ may lead to an infrared (soft or collinear) divergence. Thus it depends on the *kinematics* and not only on the topology (as in the ultraviolet case) whether a zero of $\mathcal{F}$ leads to a divergence. If some of the kinematic invariants are positive, $\mathcal{F}$ can also vanish at points in Feynman parameter space which are not the endpoints of the parametric integration. To be able to deal with such thresholds (or pseudo-thresholds), a deformation of the integration contour into the complex plane has been implemented [2].

The program SecDec can construct the graph polynomials $\mathcal{U}$ and $\mathcal{F}$ automatically. Acting on polynomials raised to some power which usually contains the dimensional regulator ϵ, it will extract the poles in $1/\epsilon$ and evaluate their coefficients numerically, as explained in the next subsection.

3.2. Basic structure of the program

The program consists of two main parts, one being designed for loop integrals, to be found in a directory called `loop`, the other one for more general parametric integrals, located in a directory called `general`. The procedure to isolate the poles in the regulator ϵ and to do the subtractions and integrations is very similar in the two branches. However, only the `loop` part contains the possibility of contour deformation, because only for loop integrals the analytic continuation can be performed in an automated way, following Feynman's "$i\delta$" prescription. The basic flowchart of the program is shown in Fig. 2.

3.3. New features

SecDec-3.0 [4] has a number of new features, aiming more towards applications in which the program is used as part of the calculation of a full virtual amplitude, rather than serving only as a numerical check for master integrals which are calculated analytically. Among these features are:

- The usage on a cluster has been facilitated. This has been achieved by decoupling the parts where Mathematica is needed completely from the numerical part, and by creating a directory structure such that the files for the numerical integration created by SecDec can be transferred to a cluster, where they are compiled and run. Run scripts are provided for the PBS, LSF and `Condor` batch systems. (The scripts for the LSF system are only available in the most recent version 3.0.9.)
  The kinematics can be defined in a simple text file `kinem.input` containing the numerical values for the invariants occurring in the integral, where the program will loop over all kinematic points defined in `kinem.input`. This new way to define the kinematic points to be calculated is also available in single machine mode.
- Integrals with numerators can be calculated not only as contracted tensor integrals, but also by defining the numerators as inverse propagators. Further, the list of propagator powers

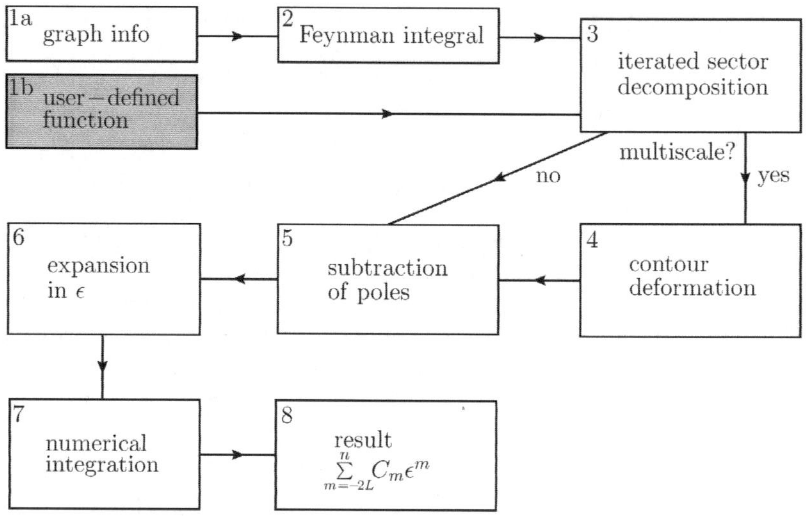

Figure 2. Flowchart showing the main steps the program performs to produce the numerical result as a Laurent series in ϵ.

can also contain zero entries. These features allow an easy interface to integral reduction programs as e.g. `Reduze` [9], where the propagator list defines the integral family, while the list of propagator powers defines the considered integral within this family.

- Two new decomposition strategies, based on algebraic geometry [17, 18], have been implemented. These strategies by construction cannot run into an infinite recursion at the iterated decomposition. Remarkably, the strategy `G2` often leads to a lower number of subsectors than the "default" strategy `X`. This new feature is described in detail in [19].
- The masses contained in the propagators can also be complex. For details we also refer to [19].
- Integrals with propagators which are only linear in the loop momentum can be calculated as well. If these integrals are non-Euclidean, care has to be taken to ensure the correct sign for the contour deformation. The program will assume the same $i\delta$ prescription as for Feynman propagators which are quadratic in the loop momenta.
- A new treatment of spurious poles which have a worse than logarithmic divergence behaviour (related e.g. to factors of $x^{-2-\epsilon}$ in the decomposed integrand) is available, which uses integration by parts (IBP) to remap these poles to logarithmic poles. This option is invoked if the parameter `IBPflag` is nonzero.
- The user can define thresholds below which the imaginary part of the integral is known to be zero; in this case SecDec will not attempt to calculate the imaginary part. This can speed up the calculation considerably.
- For the numerical integration, the user can also choose to use an integrator from Mathematica. For one-dimensional integrals, the program automatically switches to CQUAD [20], being faster and more accurate than Monte Carlo integration.
- The user interface has been facilitated, decoupling completely the following three types of input: (a) definition of the graph name, desired expansion order in ϵ and parameters for the numerical integration (`param.input`), (b) graph definition (`math.m`), (c) definition of the kinematics (`kinem.input`).

The next release of SecDec will also contain the possibility to create a library containing the sector functions produced by the program, such that the integrals can be linked to the calculation of a full amplitude. This will allow to adjust the precision to which an individual integral is calculated according to its importance within the full amplitude.

Finally, for SecDec-4.0, the algebraic part will be completely restructured, and will be implemented in `python` rather than `Mathematica`. This will be described in more detail in the following subsection.

3.4. Implementation in `python`

The forthcoming implementation of the algebraic part of SecDec in `python` has several advantages. One of them is certainly given by the fact that `python` is an open source software. In addition, the new implementation offers much more modularity. The various steps the program performs, like construction of the graph polynomials, iterated decomposition, subtraction, ϵ-expansion (see also Fig. 2), are available in the form of `python` modules and can be called separately. The decomposition can act on products of polynomials raised to some power, so it is not limited to a loop integral structure of the form $\mathcal{F}^{\text{expoF}}\mathcal{U}^{\text{expoU}}$ times a numerator (see Eq. (2)). Further, the new implementation allows to perform an expansion in several regulators, not only the dimensional regulator ϵ. The additional regulator could stem for example from analytic regularisation [21] or from parameter integrals containing the string tension [22].

An alpha-version of the python implementation can be found at `http://secdec.hepforge.org`, where a manual is also linked.

4. Applications: Two-loop four-point integrals with two mass scales

As an example for an integral where the analytic result is unknown, we give results for a non-planar two-loop four-point function (see Fig. 3) containing both internal masses (m_t) and massive external legs with a different mass (m_H). The result for this diagram is of the form $I = P_1/\epsilon + P_0$. Numerical results in the form of an interpolated scan over 20 values for s_{12}/m_t^2 times 20 values for s_{23}/m_t^2 for the real part of the finite term P_0 are shown in Fig. 4.

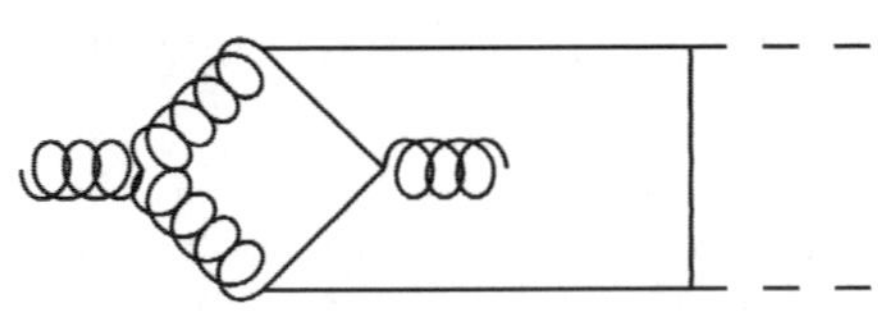

Figure 3. Example of a non-planar two-loop box diagram with two mass scales. Solid lines denote massive propagators (m_t). The dashed lines denote massive legs with a mass (m_H) different from the internal mass.

5. Conclusions

We have presented new features of the program SecDec, which is a program to factorise poles from dimensionally regulated parameter integrals, and to evaluate the pole coefficients numerically. The embedding of SecDec into GoSam-2loop, a general framework for the calculation of 2-loop amplitudes, has also been briefly described. The new features of SecDec and further developments of these programs aim at the usage of SecDec within the calculation of full two-loop virtual amplitudes in cases where the analytic result for most of the master integrals is unknown, rather than the calculation of just a few kinematic points to check an analytic calculation of individual integrals. This development is in line with the rapid developments in high performance computing.

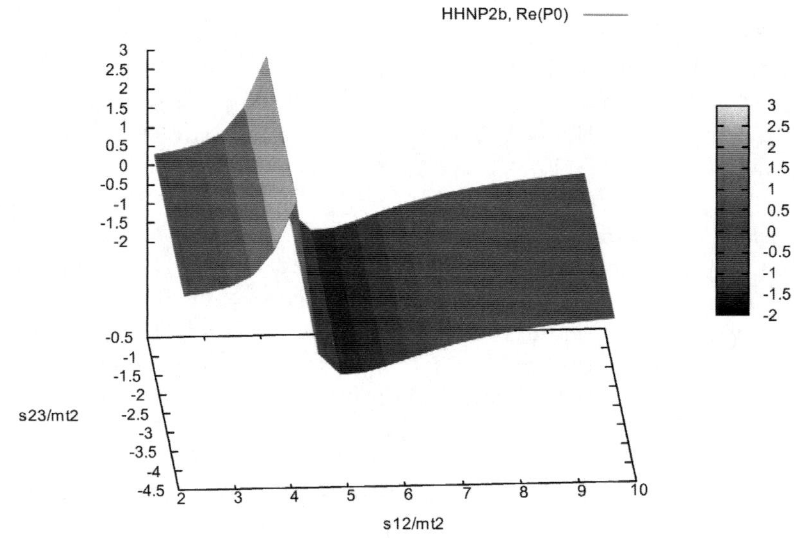

Figure 4. Real part of the finite part of the graph shown in Fig. 3, for the values m_t =173 GeV, m_H =125 GeV, $u = -s - t + 2m_H^2$.

Acknowledgments

GH would like to thank the organizers of ACAT2016 for the nice conference. This research was supported in part by the Research Executive Agency (REA) of the European Union under the Grant Agreement PITN-GA2012316704 (HiggsTools). S. Borowka gratefully acknowledges financial support by the ERC Advanced Grant MC@NNLO (340983).

References

[1] Carter J and Heinrich G 2011 *Comput.Phys.Commun.* **182** 1566–1581 (*Preprint* 1011.5493)
[2] Borowka S, Carter J and Heinrich G 2013 *Comput.Phys.Commun.* **184** 396–408 (*Preprint* 1204.4152)
[3] Borowka S and Heinrich G 2013 *Comput.Phys.Commun.* **184** 2552–2561 (*Preprint* 1303.1157)
[4] Borowka S, Heinrich G, Jones S P, Kerner M, Schlenk J and Zirke T 2015 *Comput. Phys. Commun.* **196** 470–491 (*Preprint* 1502.06595)
[5] Cullen G, Greiner N, Heinrich G, Luisoni G, Mastrolia P, Ossola G, Reiter T and Tramontano F 2012 *Eur. Phys. J.* **C72** 1889 (*Preprint* 1111.2034)
[6] Cullen G *et al.* 2014 *Eur. Phys. J.* **C74** 3001 (*Preprint* 1404.7096)
[7] Nogueira P 1993 *J. Comput. Phys.* **105** 279–289
[8] Studerus C 2010 *Comput. Phys. Commun.* **181** 1293–1300 (*Preprint* 0912.2546)
[9] von Manteuffel A and Studerus C 2012 (*Preprint* 1201.4330)
[10] Binoth T and Heinrich G 2000 *Nucl. Phys.* **B585** 741–759 (*Preprint* hep-ph/0004013)
[11] Heinrich G 2008 *Int. J. Mod. Phys.* **A23** 1457–1486 (*Preprint* 0803.4177)
[12] Bogner C and Weinzierl S 2008 *Comput. Phys. Commun.* **178** 596–610 (*Preprint* 0709.4092)
[13] Gluza J, Kajda K, Riemann T and Yundin V 2011 *Eur.Phys.J.* **C71** 1516 (*Preprint* 1010.1667)
[14] Smirnov A and Tentyukov M 2009 *Comput.Phys.Commun.* **180** 735–746 (*Preprint* 0807.4129)
[15] Smirnov A V 2014 *Comput.Phys.Commun.* **185** 2090–2100 (*Preprint* 1312.3186)
[16] Smirnov A V 2015 (*Preprint* 1511.03614)
[17] Kaneko T and Ueda T 2010 *Comput. Phys. Commun.* **181** 1352–1361 (*Preprint* 0908.2897)
[18] Kaneko T and Ueda T 2010 (*Preprint* 1004.5490)

[19] Schlenk J and Zirke T 2016 *Proceedings, 12th International Symposium on Radiative Corrections (Radcor 2015) and LoopFest XIV (Radiative Corrections for the LHC and Future Colliders)* (*Preprint* 1601.03982)
[20] Gonnet P 2010 *CoRR* **abs/1006.3962** URL http://arxiv.org/abs/1006.3962
[21] Becher T and Bell G 2012 *Phys. Lett.* **B713** 41–46 (*Preprint* 1112.3907)
[22] Puhlfürst G and Stieberger S 2016 *Nucl. Phys.* **B902** 186–245 (*Preprint* 1507.01582)

ACAT2016 IOP Publishing
Journal of Physics: Conference Series **762** (2016) 012074 doi:10.1088/1742-6596/762/1/012074

Search for Physics beyond Standard Model at the Precision Frontiers

A. Aleksejevs and S. Wu
Grenfell campus of Memorial University, Corner Brook, Canada

S. Barkanova
Acadia University, Wolfville, Canada

Y. Bystritskiy
Joint Institute for Nuclear Research, Dubna, Russia

V. Zykunov
Belarussian State University of Transport, Gomel, Belarus

Abstract. The article outlines the recent developments in the theoretical and computational approaches to the higher-order electroweak effects needed for the accurate interpretation of MOLLER and Belle II experimental data, and shows how new-physics particles enter at the one-loop level. By analyzing the effects of Z'-boson on the polarization asymmetry, we show how this hypothetical interaction carrier may influence the future experimental results.

1. Introduction

The recent availability of computer-algebra tools in particle physics research provides a unique opportunity to perform Next-to-Leading-Order (NLO) and Next-to-NLO (NNLO) Standard Model (SM) calculations with the high degree of precision requred by MOLLER and Belle II. Here, full SM calculations are required, with no approximations at the NLO level, and must include leading order NNLO contributions, which can only be achieved with some degree of automatization. We do this for both MOLLER ($e^- + e^- \to e^- + e^-$) and Belle II ($e^+ + e^- \to \mu^+ + \mu^-$), and compare the results of calculations performed with different sets of renormalization conditions using on-shell renormalization. This provides a straightforward test of gauge invariance for the polarization asymmetry. A discrepancy between SM predictions and experimental measurements would signal the physics beyond the SM. Since MOLLER and Belle II are the most sensitive to the parity-violating (PV) interaction, we include U(1)' extension of SM with a mass mixing scenario, which results in the extension of the SM by the parity-violating Z' boson. Our analysis for Z' extends to NLO level, giving us a refined set of constraints on the coupling and mass. First, we start with details on the NLO and NNLO (quadratic) calculations for MOLLER and then continue with Belle II. In the second part of the paper, we provide results and analysis of the polarization asymmetry with Z' boson present at LO and NLO orders.

2. SM Predictions for Polarization Asymmetry in MOLLER and Belle II

We consider two processes, $e^-_{k_1} + e^-_{k_2} \to e^-_{k_3} + e^-_{k_4}$, for MOLLER, and $e^+_{k_1} + e^-_{k_2} \to \mu^+_{k_3} + \mu^-_{k_4}$, for Belle II. For MOLLER, the most sensitive observable to PV new physics (aka Z') is the polarization asymmetry:

$$A_{LR} = \frac{\sigma_L - \sigma_R}{\sigma_L + \sigma_R} \simeq \frac{2\Re(M_\gamma M_Z^+ + M_\gamma M_{Z'}^+ + M_Z M_{Z'}^+)_{LR}}{\sigma_L + \sigma_R}. \quad (1)$$

In Eq.1, Z' will enter the numerator of the asymmetry through the interference term. For the $e^- + e^- \to e^- + e^-$ process, the asymmetry at LO order is given by the following expression:

$$A^0_{LR(MOLLER)} = \frac{s}{m_W^2}\frac{y(1-y)}{1+y^4+(1-y)^4}\frac{1-4s_W^2}{s_W^2},\ y = -t/s. \quad (2)$$

Here, $s = (k_1+k_2)^2$, $t = (k_3-k_1)^2$ and $u = (k_1-k_4)^2$ is the usual set of Mandelstam variables, and the Weinberg mixing angle is defined as $s_W^2 \equiv \sin^2\theta_W$. As one can see, the LO asymmetry is proportional to $1-4s_W^2$, which results in strong sensitivity to s_W^2. This provides an excellent opportunity for the precision measurements of s_W^2, or, accordingly, the measurement of the weak charge of electrons. Although PV asymmetry in Eq.2 is quite small, the accuracy of modern experiments exceed the accuracy of the theoretical result at LO order; the NLO order calculations have been completed by a number of authors [1, 2, 3]. We can express the perturbative expansion (up to NNLO) for differential scattering cross-section in orders of α as:

$$\frac{d\sigma}{d\cos\theta} = \frac{\pi^3}{2s}\Big|M_0 + M_1 + M_2\Big|^2 \simeq$$

$$\frac{\pi^3}{2s}\left(\alpha^2 M_0^{'} M_0^{'+} + \alpha^3 2\Re M_0^{'} M_1^{'+} + \alpha^4\left(M_1^{'} M_1^{'+} + 2\Re M_0^{'} M_2^{'+}\right)\right), \quad (3)$$

where matrix elements M_i are related to $M_i^{'}$ by $M_i = \alpha^{i+1}M_i^{'}$. The first term corresponds to LO, the second to NLO and the third forms the NNLO contribution, which is composed of a quadratic term ($\alpha^4 M_1^{'}M_1^{'+}$) and the two-loops ($\alpha^4 2\Re M_0^{'}M_2^{'+}$) contribution. The NLO contribution to LO asymmetry is rather big ($\sim 69\%$) [3], and in order to match 1% MOLLER uncertainty, we calculated a full set NNLO (quadratic) [4] and leading order NNLO (two-loops) contributions [5] (and references therein). The precision is essential, so we control it in two ways. First, we applied "on paper" on-shell calculations using the renormalization conditions of [6] and low energy approximations for $\frac{\{|t|,|u|\}}{m^2_{Z,W}} \ll 1$ for $\sqrt{s} \ll 30\,(GeV)$. Second, we performed semi-automated [7, 8, 9, 10, 11] calculations for the full set of Feynman diagrams without any approximations and using renormalization conditions of [12]. This approach was implemented for NLO and NNLO (quadratic) contributions. The semi-automated full two-loop calculations are yet to be completed which is our next goal. Let us demonstrate how these two approaches compare to each other. First, we introduce a correction to the asymmetry, as:

$$\delta_A^C = \frac{A_{LR}^C - A_{LR}^0}{A_{LR}^0}, \quad (4)$$

where A_{LR}^C stands for the NLO-corrected asymmetry. If we take $\alpha = 1/137.0359$, $m_W = 80.398\,(GeV)$, $m_Z = 91.1876\,(GeV)$ and kinematics relevant to MOLLER experiment ($E_{lab} = 11\,(GeV)$), we can see on Fig.1 (right plot) that results obtained in both approaches differ less than 0.1%. We find that the NNLO (quadratic) contribution, Fig.1(left plot),

$$\Delta_A = \frac{A_{LR}^{1-loop-Q} - A_{LR}^{1-loop}}{A_{LR}^0}, \quad (5)$$

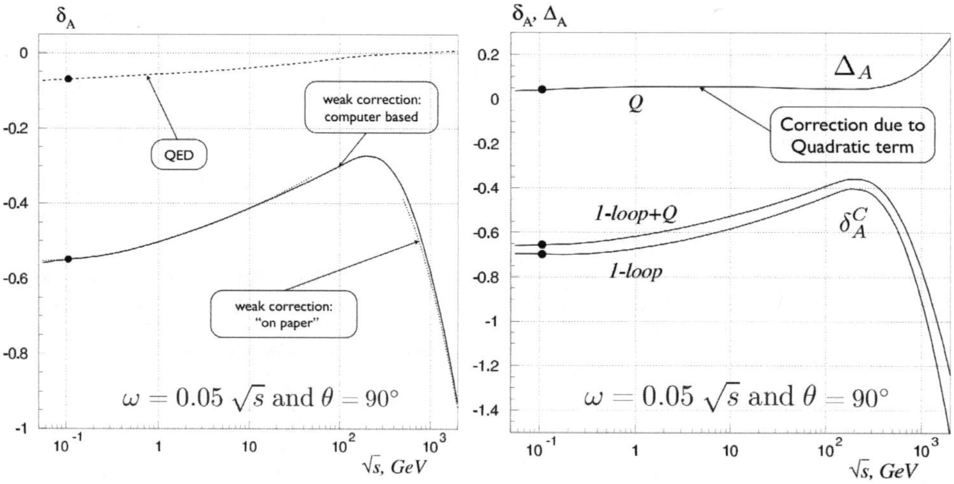

Figure 1. The graphs show corrections to the polarization asymmetry at the $\theta = 90^o$ and soft photon cut $\omega = 0.05\sqrt{s}$. On the left graph, the dashed line shows correction due to QED effects, the solid line represents weak correction based on semi-automated calculations, and the dotted line gives "on-paper" approximated calculations for the same correction. The graph on the right shows corrections for the pure quadratic ("Q") contribution (Δ_A), NLO ("1-loop") and NLO+NNLO ("1-loop-Q") results. The black dot on both graphs corresponds to $E_{lab} = 11\,(GeV)$.

θ^o	10	30	50	70	90	110	130	150	170
"on-paper"	0.0180	-0.0456	-0.0738	-0.0935	-0.1099	-0.1264	-0.1460	-0.1743	-0.2378
SA	0.0179	-0.0455	-0.0738	-0.0934	-0.1099	-0.1263	-0.1459	-0.1742	-0.2372

Table 1. Relative NLO correction to unpolarized cross section (δ_{00}^C) in the $e^+ + e^- \to \mu^+ + \mu^-$ process for the various angles in CM reference frame. Both approaches used the soft-photon approximation in the treatment of infrared divergences. Soft photon cut is $\omega = 0.05\sqrt{s}$.

is responsible for $\sim 5\%$ suppression of the total correction at $E_{lab} = 11\,(GeV)$. This is a clear signal that, in the light of proposed precision experiments, the NNLO contributions are very important. Similar to the Moller process, $e^+ + e^- \to \mu^+ + \mu^-$ polarization asymmetry (first addressed in [13]), also shows strong sensitivity to the s_W^2:

$$A^0_{LR(Belle\,II)} = -\frac{s}{4m_W^2}\frac{(y-1)^2}{2(y-1)y+1}\frac{1-4s_W^2}{s_W^2},\ y = -t/s. \tag{6}$$

We improve the precision by implementing the same two-way approach, for Belle II kinematics specifically, taking into account the full set of NLO electroweak corrections. For $\sqrt{s} = 10.57\,(GeV)$, Tabl.1 shows our results for the NLO relative correction to the unpolarized cross section ($\delta_{00}^C = \frac{\sigma_{00}^C - \sigma_{00}^0}{\sigma_{00}^0}$) computed using the semi-automated (SA) and "on-paper" calculation methods in the on-shell renormalization. Evidently, the difference in both approaches, for the broad range of scattering angles, is negligible. Although the NLO contribution (Fig.2) for MOLLER and Belle II experiments is significant, judging by the excellent agreement between the two approaches, the theoretical uncertainty at NLO level is at a sub-percent level. However, that result does not include the NNLO contribution, needed to interpret the ultra-precision measurements.

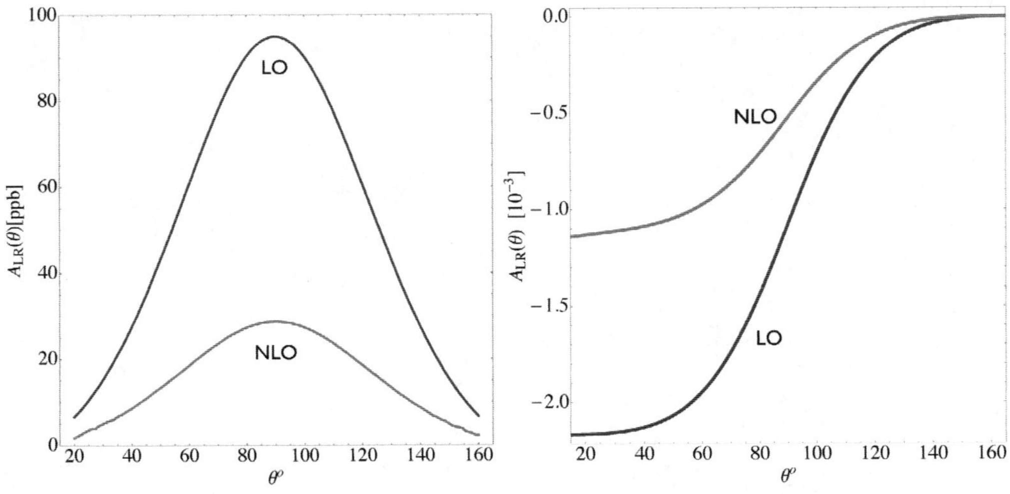

Figure 2. LO and NLO corrected polarization asymmetries for the MOLLER (left, $E_{lab} = 11\,(GeV)$) and Belle II (right, $\sqrt{s} = 10.57\,(GeV)$) experiments for the $20^o < \theta < 160^o$ in the CM reference frame.

3. Beyond the Standard Model Physics with Dark Vector

The discrepancy between experimental data and SM theoretical predictions would signal new interaction carriers. We use the simple U(1)' extension of SM proposed in [14], which uses kinetic type of mixing between dark vector (A'_μ) and hyper-charge (B_μ) fields:

$$\mathfrak{L}_{kin} \quad = \quad -\frac{1}{4}B_{\mu\nu}B^{\mu\nu} + \frac{1}{2}\frac{\epsilon}{\cos\theta_W}B_{\mu\nu}A'^{\mu\nu} - \frac{1}{4}A'_{\mu\nu}A'^{\mu\nu}, \tag{7}$$

where fields tensors are given by $\{A'_{\mu\nu},\, B_{\mu\nu}\} = \partial_\mu\{A'_\nu,\, B_\nu\} - \partial_\nu\{A'_\mu,\, B_\mu\}$, with $B_\mu = \cos\theta_W A_\mu - \sin\theta_W Z_\mu$ and ϵ is the $(B_\mu - A'_\mu)$ kinetic mixing parameter. With SM Higgs doublet plus the Higgs singlet (used for breaking the $U(1)'$ symmetry and giving mass to A'_μ), a Lagrangian describing the interaction between the SM fermions and the dark vector boson A'_μ is:

$$\mathfrak{L}_{int} = -\,eQ_f\bar{f}\gamma_\mu f\cdot(V^\mu + \epsilon A'_\mu) -$$

$$\frac{e}{\sin\theta_W\cos\theta_W}\bar{f}(c^f_V\gamma_\mu + c^f_A\gamma_\mu\gamma_5)f\cdot(Z^\mu + \epsilon_{Z'}A'_\mu). \tag{8}$$

Here, Q_f is the charge of the fermion in units of e, and the c^f_V and c^f_A constants are the usual SM vector and axial-vector coupling strengths, respectively. As we can see from Eq.8, the dark A'_μ couples to fermions through both parity-conserving and violating terms, which is similar to the weak Z_μ coupling. This type of dark A'_μ in [15] is called the dark Z'_μ boson and is derived from an additional mass mixing term characterized by mixing parameter $\epsilon_{Z'} = \frac{m_{z'}}{m_z}\delta$. Here, $m_{Z'}$ is the the mass of the dark Z'_μ boson and δ is an arbitrary model-dependent parameter. The fact that Z'_μ is represented as a superposition of mixings between dark vector with electromagnetic and Z-boson fields makes it possible to include Z'_μ at the NLO level. We include the Z'_μ at NLO in order to match NLO calculations with our SM predictions. Our results are shown in the form of exclusion plots in Fig.3 for MOLLER (left) and Belle II (right).

Fig.3 explores small mixing ϵ and small Z'_μ mass scenarios for MOLLER. We show up-to-1% deviation from the SM central prediction exclusion plots for Z'_μ, which is included at LO

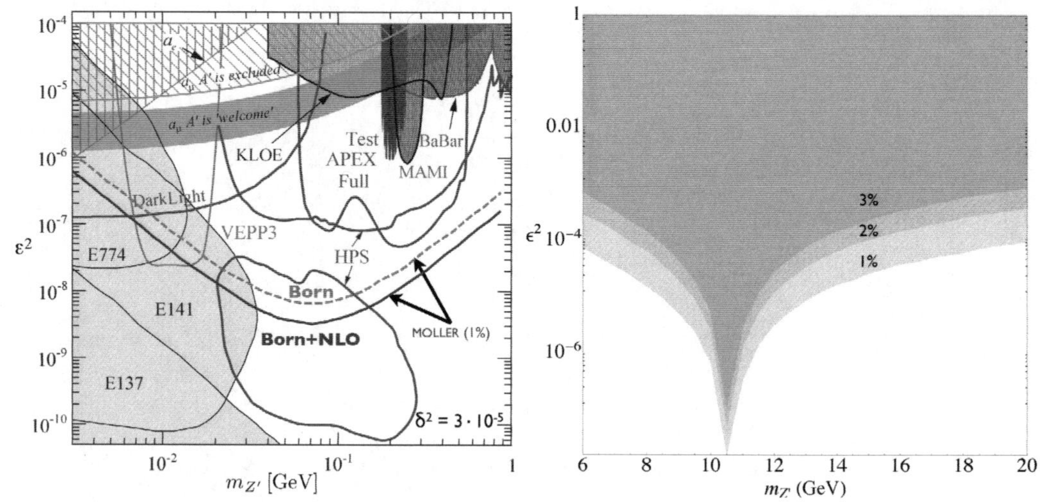

Figure 3. Exclusion plots for MOLLER (left) and Belle II (right). For MOLLER, we show exclusion for 1% deviation from SM prediction. Here, Z'_μ is included either at LO (Born) or up to NLO (Born+NLO) orders. Plot for Belle II shows up-to-NLO exclusion regions for 1%, 2% and 3% deviations from SM prediction. For both plots, we take $\delta^2 = 3 \cdot 10^{-5}$.

and NLO orders. Inclusion of the Z'_μ at the NLO order increases the exclusion region for ϵ for all masses of Z'_μ by about 25%. While this increase is not substantial, it could become an important factor in the determination of the Z'_μ mass and coupling if Z'_μ is discovered. According to [15, 16], if no Z'_μ is discovered, MOLLER will exclude the region where Z'_μ is used to explain the $(g-2)_\mu$ anomaly. For Belle II (right plot of Fig.3), we concentrate on the resonance region at $\sqrt{s} = 10.57\,(GeV)$, where the sensitivity of Belle II to the Z'_μ (up to NLO order) is the highest and is complimentary to the MOLLER experiment. We conclude that the inclusion of NLO and NNLO electroweak radiative corrections is essential for the search of new physics at the precision frontier, and that the computer-algebra tools are indispensable to this task.

Acknowledgments
This work has been supported by NSERC.

References

[1] Czarnecki A and Marciano W 1996 *Phys. Rev.* D **53** 1066
[2] Denner A and Pozzorini S 1999 *Eur. Phys. J.* C **7** 185
[3] Aleksejevs A, Barkanova S, Ilyichev A and Zykunov V 2010 *Phys. Rev.* D **82** 093013
[4] Aleksejevs A *et al.* 2012 *Phys. Rev.* D **85** 013007
[5] Aleksejevs A *et al.* 2012 (*Preprint* arXiv:1202.0378)
[6] Hollik W 1990 *Fortschr. Phys.* **38** 165
[7] Hahn T 2001 *Comput. Phys. Commun.* **140** 418
[8] Hahn T and Perez-Victoria M 1999 *Comput. Phys. Commun.* **118** 153
[9] Vermaseren J 2000 (*Preprint* math-ph/0010025)
[10] Shtabovenko V, Mertig R and Orellana F 2016 (*Preprint* arXiv:1601.01167)
[11] Mertig R, Böhm M and Denner A 1991 *Comput. Phys. Commun.* **64** 345
[12] Denner A 1993 *Fortschr. Phys.* **41** 307
[13] Böhm M and Hollik W 1982 *Nucl. Phys.* B **204** 45
[14] Holdom B 1986 *Phys. Lett.* B **166** 196
[15] Davoudiasl H, Lee H and Marciano W 2012 *Phys. Rev.* D **85** 115019
[16] Aleksejevs A, Wu S, Barkanova S and Zykunov V 2014 (*Preprint* arXiv:1410.6970)

Prediction of the higher-order terms based on Borel resummation with conformal mapping

M.V. Kompaniets

St. Petersburg State University, 7/9 Universitetskaya nab., St. Petersburg, 199034 Russia.

E-mail: m.kompaniets@spbu.ru

Abstract.
In this paper we discuss the method of the resummation of the asymptotic series suggested by Kazakov et al. in [1,2] and predictions of the higher order terms based on this approach. An application of this method to the φ^4 model is discussed.

1. Introduction

Motivated by the recent six loop computations of the field anomalous dimension in φ^4 model [3] (ε-expansion) and upcoming calculations of the six loop beta function in this model [4], we discuss in this paper the method of the resummation of the asymptotic series with factorially growing coefficients suggested by Kazakov et al. in [1,2]. This method is based on the Borel transform with a conformal mapping [1,2,5–7] and utilizes information about high order asymptotics (HOA) [8] as well as information about the asymptotic behavior at large values of the coupling constant.

2. General considerations

Let us consider a quantity $A(g) = \sum_n A_n g^n$ which is defined as an infinite series with factorially growing coefficients. The procedure of the Borel resummation has the following steps: first, one should find a Borel image $B(x) = \sum_n B_n x^n$ (1) (coefficients B_n will have no factorial growth), then sum it up and, at last, perform the inverse Borel transform (1) to get the resummed value $A^B(g)$. Usually, a series defined by coefficients B_n has finite radius of convergence, thus to define $B(x)$ on $x \in [0, \infty)$ one needs to perform an analytical continuation for $B(x)$ outside of the convergence radius.

The Borel transform can be defined as follows:

$$A^B(g) = \int_0^\infty dt\, e^{-t} t^{c_1 - c_2} \left(t \frac{\partial}{\partial t} \right)^{c_2} \sum_n B_n^{(c_1,c_2)} (gt)^n \,, \quad B_n^{(c_1,c_2)} \equiv \frac{A_n}{\Gamma(n+1+c_1-c_2) n^{c_2}} \,. \tag{1}$$

Usually the Borel transform with $c_2 = 0$ is used as the most simple one, but in some cases the transform with $c_2 \neq 0$ gives a better convergence (see [1,2])

Working within the perturbative approach, the quantity $A(g)$ is known only up to some finite order of the perturbation theory:

$$A^{(N)}(g) = \sum_{n=0}^{N} A_n g^n \,. \tag{2}$$

One can easily see that for a finite number of terms (2), the Borel resummation described above simply reproduces the initial series $A^{(N)}(g)$. Thus to get some nontrivial result we need to make some assumptions about the higher order terms (A_n, $n > N$). This is the most crucial step in the procedure of the Borel resummation, because a good choice (which will most accurately reproduce original (may be not known) series) will lead to a better convergence of the resummation procedure, while bad choice may lead to completely inconsistent results.

Knowing only first N terms of the series (2), continuation of the series to large N is a very ambiguous procedure, for example, one may use Pade approximants for that [9, 10]:

$$A^{Pade-Borel} = \int_0^\infty dt\, e^{-t} t^{c_1-c_2} \left(t\frac{\partial}{\partial t}\right)^{c_2} B(gt)\,, \quad B(x) = \frac{P_L(x)}{P_M(x)}\,. \tag{3}$$

Polynomials $P_L(x)$ and $P_M(x)$ are chosen in such a way that the initial terms of the series expansion of $B(x)$ reproduce the Borel image (1) of our series (2). For the series considered in φ^4 model and similar models, results obtained by Pade-Borel resummation are highly dependent on the choice of the approximant (L, M), and sometimes the approximant has a singularity in the integration domain, which prevents one from performing the inverse Borel transformation (see e.g. [11, 12]).

To make results more reliable one needs to incorporate into the reconstructed series all the information we know about it [13]. For example, for the φ^4 model (and some other models) we know high order asymptotics (HOA) for A_n [8]:

$$A_n \sim (-1)^n\, n!\, n^{b_0}\, a^n (1 + \mathcal{O}(1/n))\,. \tag{4}$$

Parameters a and b_0 are determined from the instanton analysis (see [8]). To incorporate this information into the resummation procedure the method of the Borel transform with conformal mapping was developed [1, 2, 5–7]. In this method function $B(x)$ is defined as follows:

$$B(x) = \sum_{n=0}^{N} C_n w(x)^n\,, \qquad \text{where } w(x) = \frac{\sqrt{1+ax}-1}{\sqrt{1+ax}+1}\,, \tag{5}$$

here a is the parameter of HOA (4), and b_0 fixes the value of $c_1 = b_0 + 3/2$ in (1) to fix HOA of the reconstructed series in accordance with (4). And again, expansion of the $B(x)$ must reproduce the initial coefficients of the Borel image of the original series (2). The variable transformation in (5) ensures that the reconstructed series has an analytical continuation in the whole integration domain. This resummation method gives more reliable results for the resummed values (see e.g. [14, 15]), but still has some disadvantages: despite the fact that the series reconstructed in such a way reproduces the initial part of the original series (2) and corresponding HOA (4), the terms of the series with $n > N$ are reproduced incorrectly. Actually this means that we resum a series different than the original one (of course if N is big enough this procedure must converge to a correct value, but given maximum 6 terms [3] the correct choice of the function $B(x)$ is very important).

The next step was performed in the papers of Kazakov et al. [1, 2]: the function $B(x)$ in (5) was modified in such a way that it reproduces not only initial terms and HOA, but also the resummed function has the same asymptotic behavior at $g \to \infty$ as the initial function $A(g)$:

$$B(x) = \left(\frac{x}{w}\right)^{\nu} \left(\sum_{n=0}^{N} C_n w(x)^n\right), \qquad \text{where } w(x) = \frac{\sqrt{1+ax}-1}{\sqrt{1+ax}+1}\,, \tag{6}$$

obviously, the parameter ν governs the behavior of the resummed function at $g \to \infty$ (in (1) $x = gt$)

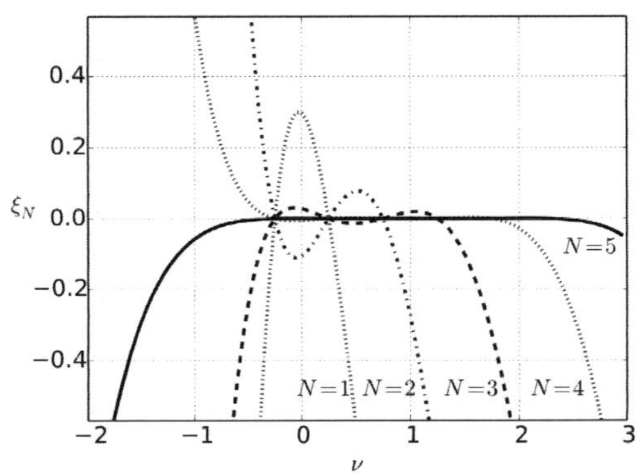

(a) Relative prediction error $\xi_N = (A_6^{(N)} - A_6)/A_6$ as a function of ν for different values of N.

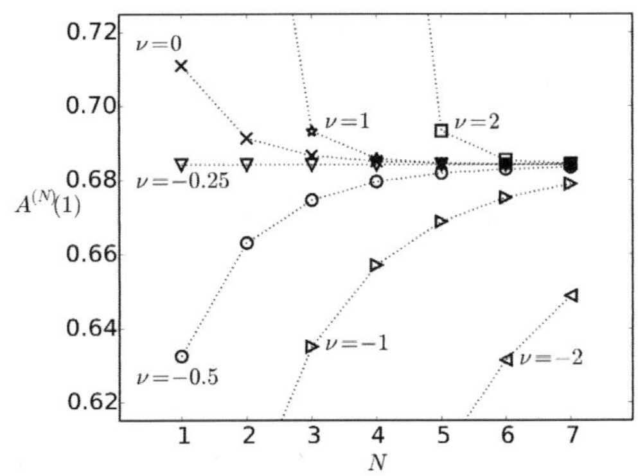

(b) The value of the resummed series $A^{(N)}$ at $g = 1$, for different N and ν.

Figure 1. Zero dimensional φ^4 model, see Eq. (7).

In [1,2] it was shown that for the series with known asymptotic behavior at $g \to \infty$ the most correct resummed values are obtained when the parameter ν is chosen close to the asymptotic. Moreover, in this case the expansion of the $B(x)$ reproduces the terms with $n > N$ with high accuracy. Thus we can state that with this approach we resum the series which is really close to the original one.

Let us illustrate this on the expansion of the following integral:

$$A(g) = \int_0^\infty dx\, e^{-x^2 - gx^4} = \sum_{n=0}^{\infty} A_n g^n\,, \quad \left(A_0 = \frac{\sqrt{\pi}}{2},\ A_n = \frac{(-1)^n \sqrt{\pi}(4n-1)!}{(4)^{2n} n!(2n-1)!},\ (n > 0) \right)\,. \quad (7)$$

High order asymptotic of this series is given by (4) with $a = 4$ and $b_0 = -1$ and $A(g) \sim g^{-1/4}$, when $g \to \infty$.

If we calculate the function $B^{(N)}(x)$ from (6) for the first N terms of the expansion of the integral (7), we can expand it back in x and reconstruct the series, which we are actually going to resum $B^{(N)}(x) = \sum A_n^{(N)} x^n$. From Fig.1 one can see that a better convergence of the resummation procedure (Fig.1(b)) appears for such values of ν where predictions of higher-order terms are most accurate (Fig.1(a)). From Fig.1(a) one can see that large negative and positive values of ν give us completely incorrect predictions, while starting from $\nu = -1/4$ we have an area with good predictions. One can easily see that the representation (6) with $\nu = -1/4$ exactly reproduces the coefficients of the initial series (7) thus giving a resummed result starting from $N = 0$. Of course, this fact can be explained by the simplicity of the example considered, and in more complicated cases (like φ^4 model) it is not the case. Any way, using this example we can investigate the stability of this resummation procedure and it's convergence.

3. Resummation of the φ^4 model in $D = 4 - 2\varepsilon$ space dimensions

The situation similar to the described above appears for the φ^4 model beta-function (see Fig.2). In this model the exact result is not known as well as coefficients of the beta-function for arbitrary number of loops. But we can verify the predictive power of this approach by checking predictions for the highest known term from lower ones as well as investigating the stability of the resummation procedure like in zero dimensional model.

For this model the plot with the relative prediction error $\xi_N^\beta = (\beta_M^{(N)} - \beta_M)/\beta_M$ (Fig.2(a)) has a similar form, of course there is no exact intersection as in Fig.1(a). One can see that

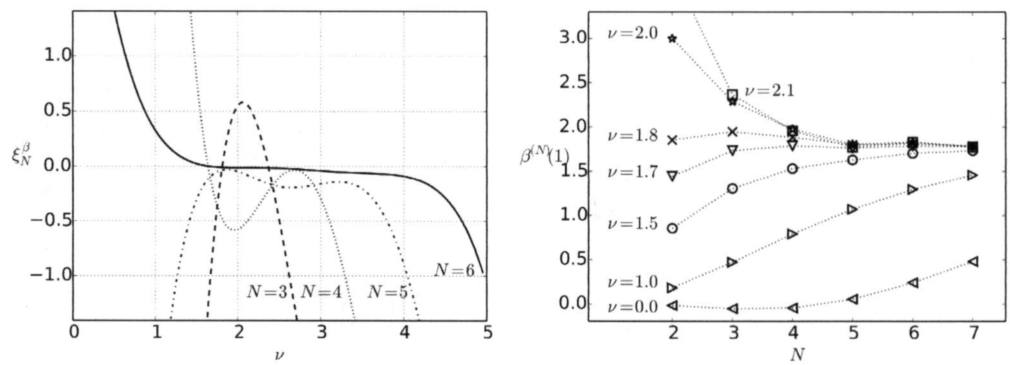

(a) The relative prediction error ξ_N^β as a function of ν, for the different values of N.

(b) The value of the resummed beta function $\beta^{(N)}$ at $g = 1$, for different N and ν.

Figure 2. φ^4 model ($D = 4 - 2\varepsilon$, MS-scheme).

the standard Borel transform with conformal mapping (5) (which corresponds to $\nu = 0$) gives completely incorrect predictions, while near $\nu = 2$ predictions are almost correct. In papers [1,2] the range $1.7 < \nu < 2.2$ was obtained for the beta-function and the value $\nu = 2$ was recommended to use for the resummation.

The same situation appears for the convergence of the resummation procedure: on Fig.2(b) the convergence of the resummation of the beta function $\beta^{(N)}(g)$ at $g = 1$ is shown for the different values of the parameter ν and the number of terms N taken into account (note that e.g. $N = 3$ corresponds to the two loop approximation; the preliminary value of the six loop term for the beta function is taken from [4]). The ν dependence of the resummed beta function $\beta^{(N)}(g = 1)$[1] has a form similar to the zero dimensional model (Fig. 1(b)). And again we can see that the value of ν with the most correct predictions ($\nu \simeq 1.8$) gives us the most stable resummed value of the beta-function, therefore by analogy with the zero dimensional model we should use this value in the resummation procedure. Also it should be noted that the resummation procedure for the field anomalous dimension for $N = 2$ and 3 leads to completely inconsistent results for any ν, but starting from the $N = 4$ shows the same properties (see Fig. 3), and according to this plot for the resummation of the field anomalous dimension the value of the parameter ν must be chosen as $\simeq 3$.

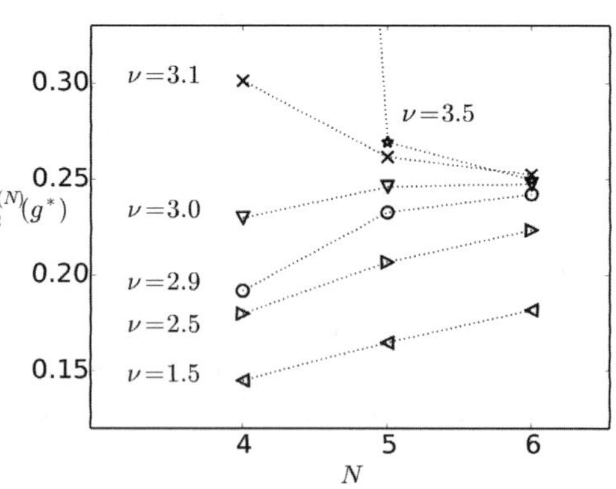

Figure 3. Resummed values of the Fisher exponent $\eta \equiv 2\gamma_\varphi(g^*)$ at $g^* = 1.155$ (fixed point at $D = 2$) for the different values of N and ν.

Performing the resummation analogous to the one made in [3] but with $\nu_\beta = 1.8$[2] and $\nu_\eta = 3$ we arrive to the resummed exponents shown in Table 1.

Comparing with the table in [3] one can see that the value $\nu_\beta = 1.8$ gives us better convergence of the resummation procedure as well as better predictions for the beta function coefficients. For a five loop term of the beta-function predicted

[1] The similar behavior appears for any reasonable(not very large) g.
[2] In [3] the value $\nu_\beta = 2$ was used.

Table 1. Resummation result ($\nu_\beta = 1.8$) for the Fisher exponent η for different number of loops taken into account. Expected values are 0.25 and 0.0364(5) [15], for $D = 2$ and $D = 3$.

Loops β/γ_φ	3/4	3/5	3/6	4/4	4/5	4/6	5/5	5/6
$D = 2$	0.1952	0.2079	0.2090	0.2233	0.2388	0.2403	0.2341	0.2354
$D = 3$	0.03351	0.03406	0.03410	0.03556	0.03623	0.03628	0.03599	0.03604

by a 4 loop beta function we have ($M = 6$, $N = 5$) $\beta_6^{(5)} = 2821.65$, while the exact value is $\beta_6 = 2848.57...$ which is only 1% greater than the prediction, in [1, 2, 16] value $\simeq 2808$ was obtained for $\beta_6^{(5)}$. For a six loop term on top of a five loop beta function we have ($M = 7$, $N = 6$) $\beta_7^{(6)} = -34393.3$. Of course, we didn't expect that a six loop term of the beta function will exactly fit this prediction, moreover it is expected that the absolute value of the six loop term is a bit greater than that of prediction.

4. Conclusions

Summarizing, we have shown that it is possible to choose such a form of the analytical continuation of the Borel image which will most accurately reproduce higher order terms of the series. This will provide a better convergence of the resummation procedure as well as predictions of the higher order terms. Despite the fact that some arguments for the choice of the $B(x)$ in Eq. (6) and the particular value of ν for φ^4 model are presented, this question is still open. It is still necessary to find more strict arguments for the choice of $B(x)$ and ν.

Acknowledgments

I am grateful to L Ts Adzhemyan, D I Kazakov, M Yu Nalimov, A I Sokolov and E Zerner-Käning for fruitful discussions, to ACAT'16 Organizing Committee, and in particular to I Kondrashuk (Univ. of Bio-Bio, Chillán), L Salinas (UTFSM, Valparaiso) and Y Schroder (Univ. of Bio-Bio, Chillán) for support and hospitality, to A L Kataev for the support in the process of the preparation of this talk. The work was supported by Saint-Petersburg State University (project 11.38.185.2014).

References

[1] Kazakov D, Shirkov D and Tarasov O 1979 *Theor. Math. Phys.* **38** 9
[2] Kazakov D and Shirkov D 1980 *Fortsch. Phys.* **28** 465
[3] Batkovich D, Chetyrkin K and Kompaniets M 2016 *Nuclear Physics B* **906** 147 – 167 (*Preprint* `hep-th/1601.01960`)
[4] Kompaniets M and Panzer E *in preparation*
[5] Loeffel J 1996 *Workshop on Pade approximants* ed Bessis D, Gilewicz J and Merry P (ACM Press)
[6] Le Guillou J C and Zinn-Justin J 1977 *Phys. Rev. Lett.* **39**(2) 95–98
[7] Le Guillou J C and Zinn-Justin J 1980 *Phys. Rev. B* **21**(9) 3976–3998
[8] Lipatov L 1977 *J. Exptl. Theoret. Phys.* **72** 411
[9] Baker G 1975 *Essentials of Pade Approximants* (Academic, New York)
[10] Baker G A, Nickel B G and Meiron D I 1978 *Phys. Rev. B* **17**(3) 1365–1374
[11] Antonenko S A and Sokolov A I 1995 *Phys. Rev. E* **51**(3) 1894–1898
[12] Orlov E and Sokolov A 2000 *Physics of the Solid State* **42** 2151–2158
[13] Nalimov M Y, Sergeev V A and Sladkoff L 2009 *Theor. Math. Phys.* **159** 499–508
[14] Zinn-Justin J 2002 *Quantum Field Theory and Critical Phenomena* (Oxford: Clarendon Press)
[15] Pelissetto A and Vicari E 2002 *Physics Reports* **368** 549–727 (*Preprint* `cond-mat/0012164`)
[16] Chetyrkin K G, Gorishny S G, Larin S A and Tkachev F V 1983 *Phys. Lett. B* **132** 351–354

Journal of Physics: Conference Series **762** (2016) 012076 doi:10.1088/1742-6596/762/1/012076

SModelS: A Tool for Making Systematic Use of Simplified Models Results

Wolfgang Waltenberger (on behalf of the SModelS group)

Institute for High Energy Physics, Austrian Academy of Sciences, Nikolsdorfergasse 18, 1050 Wien, Austria

E-mail: walten@teilchen.at

Abstract. We present an automated software tool "SModelS" to systematically confront theories Beyond the Standard Model (BSM) with experimental data. The tool consists of a general procedure to decompose such BSM theories into their Simplified Models Spectra (SMS). In addition, SModelS features a database containing the majority of the published SMS results of CMS and ATLAS. These results consist of the 95% confidence level upper limits on signal production cross sections. The two components together allow us to quickly confront any BSM model with LHC results. As a show-case example we will briefly discuss an application of our procedure to a specific supersymmetric model.

It is one of our ongoing efforts to extend the framework to include also efficiency maps produced either by the experimental collaborations, by efforts performed within the phenomenological groups, or possibly also by ourselves.

While the current implementation can handle null results only, it is our ultimate goal to build the Next Standard Model in a bottom-up fashion from both negative and positive results of several experiments. The implementation is open source, written in python, and available from http://smodels.hephy.at.

1. Introduction

The discovery of a Higgs-like boson [1, 2] has been announced with much fanfare in July 2012, marking the end of a fifty-year old quest to complete the Standard Model. And yet, many questions remain unanswered, notably the question about the particle nature of dark matter, or the puzzle of the stability of the electroweak scale. Many theoretical frameworks have been devised that aim at solving these questions. It is common to many of them that they predict partners of Standard Model particles at energies $\lesssim$ 2 TeV. No such particles have been discovered to date by the CMS and ATLAS experiments at the LHC, though a whole slew of searches has been conducted [3, 4]. So indeed, the Higgs boson is likely to have been found, but where is the physics that stabilizes its mass?

Simplified Models

It has become common practise to present one's results of searches for BSM physics in terms of SMSes, quantifying the result as an upper limit on the cross section of the produced BSM particles, as a function of the BSM particles' masses, see Fig. 1 (left). It is the aim of this work to make maximum usage of the published SMS results, enabling us to take an arbitrary BSM model and quickly compare it against an entire database of SMS results.

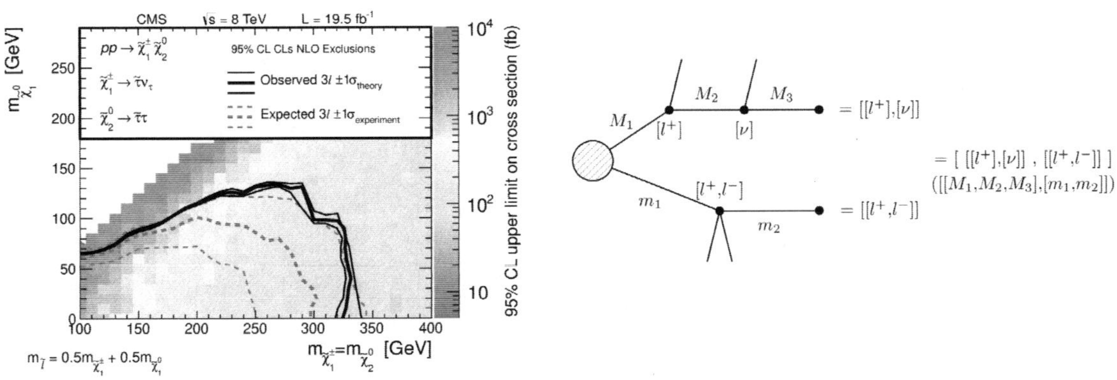

Figure 1. Anatomy of an SMS result, taken from Ref. [5] (left). The labeling scheme adopted (right).

2. The Procedure

The SModelS procedure can be split up into a few subtasks, such as:

- the decomposition of the full model into its SMS equivalent "elements",
- the potential compression of elements,
- the potential combination of elements to match experimental results,
- and finally the comparison of the theoretically predicted cross section with the experimentally given upper limit on that cross section.

In what follows, we shall discuss briefly each step of the procedure.

2.1. Decomposition

The main guiding principle of the SModelS framework is that the signal efficiencies of a given analysis depends mostly on the event kinematics and is not majorly affected by the idiosyncracies of a particular model. That way, it is possible for us to

- decompose a full model into its Simplified Models Spectrum,
- describe this spectrum in terms of SMS equivalent elements. These elements depend only on the final state SM particles plus a dark matter candidate (DMC), the masses of the BSM particles, and the element's production cross section times branching ratio.

Fig. 2 shows this description of an element. As a next step, a formalism has been devised that describes our SMS equivalent elements in a single, simple string, by making use of a square bracket notation to distinguish between branches and vertices, respectively, and assigning letters to the SM particles, as can been seen in Fig. 1 (right). In the current implementation, a fundamental $\mathbb{Z}_2$-symmetry is assumed, making it unnecessary to mention the dark matter particle explicitly in our formalism – it is expected to terminate the decay chains of every branch.

2.2. Compression

In case the difference in mass between two subsequent BSM particles in a given decay chain is small, it is assumed that the emergent SM particles are too soft to be detected by the LHC experiments. In that case these soft particles are omitted from the description of the

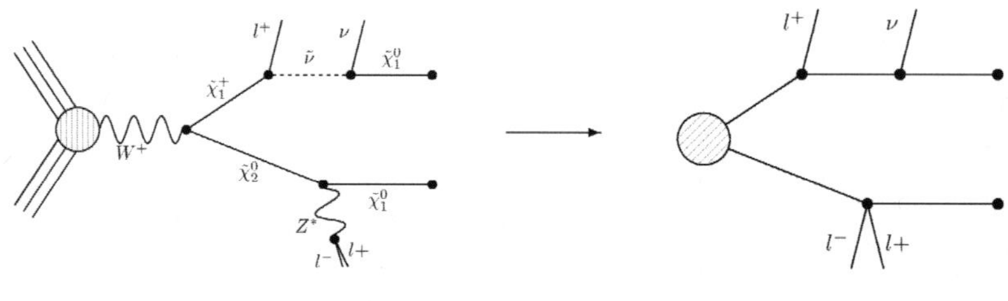

Figure 2. A full model diagram (left) and its SMS equivalent element (right).

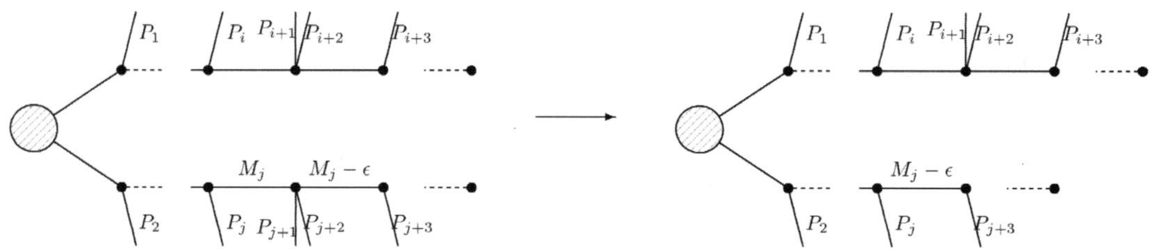

Figure 3. A signal topology with a compressed mass spectrum (left) and its mass compressed equivalent (right). M_j and $M_j - \epsilon$ represent the masses of the quasi-degenerate states.

topology: the topology is "compressed" – see Fig. 3. Likewise, in the case that multiple invisible particles appear at the end of a decay chain – such as a DMC and a neutrino – the particles are experimentally indistinguishable from a single invisible particle of a mass that is the equivalent of the invariant mass of the sum of all invisible particles. In this case, the topology is compressed accordingly: the invisible particles are replaced by a single invisible particle with the corresponding mass.

2.3. The Database of SMS Results

Another essential building block of the SModelS framework is its database of experimental results. Taking SMS results like Fig. 1 as the input data, its meta information such as the analysis' integrated luminosity, its center-of-mass energy, or the analysis "id" (e.g. ATLAS-CONF-2012-105, or SUS-12-005) is stored alongside with the efficiency maps and cross section upper limit maps published by the experiments, see Tab. 1. In addition, the efficiency and upper limit maps are described in the SModelS formalism in the aforementioned table, in a field coined "constraint", referring to the fact that the label defines what part of a full model is constrained by that experimental result. At times, additional conditions have to be met in order for an experimental result to be applicable to a specific full model. Such conditions are also described in terms of our SMS labeling formalism. The according entry in the database is called "fuzzycondition". See Ref. [6] for examples of such conditions, and also for a full list of the results that entered the database.

2.4. Putting it All Together

Once a full model is decomposed according to the procedure described in Sec. 2.1, the model is ready to be tested against the database of SMS results. In case an experimental result matches not a single but an entire set of elements (such as is the case for e.g. `[[1],[1],[1],[1]]` – it

```
id: CMS-SUS-13-006
sqrts: 8.00
lumi: 19.50
url: https://twiki.cern.ch/twiki/bin/view/CMSPublic/PhysicsResultsSUS13006
constraint: TChiChipmSlepStau -> [[['L'],['L']],[['nu'],['ta']]]
fuzzycondition: TChiChipmSlepStau ->
Cgtr([[['mu'],['mu']],[['nu'],['ta']]] , [[['e'],['e']],[['nu'],['ta']]])
Cgtr([[['e'],['e']],[['nu'],['ta']]] , [[['ta'],['ta']],[['nu'],['ta']]])
category: TChiChipmSlepStau -> eweakino
axes: TChiChipmSlepStau: M1 M0 005 - M1 M0 050 - M1 M0 095
```

Table 1. Metadata describing the SMS result in Fig. 1

matches both the case of electrons and muons), the two elements are combined, which is to say that their cross sections times branching ratios are added up – see Fig. 4. The experimental upper limit on the cross section can then be directly compared against the theoretical prediction, and the full model is either excluded by this specific analysis or not. Finally, the most important elements that are not covered by any analysis are identified; they are dubbed "missing topologies".

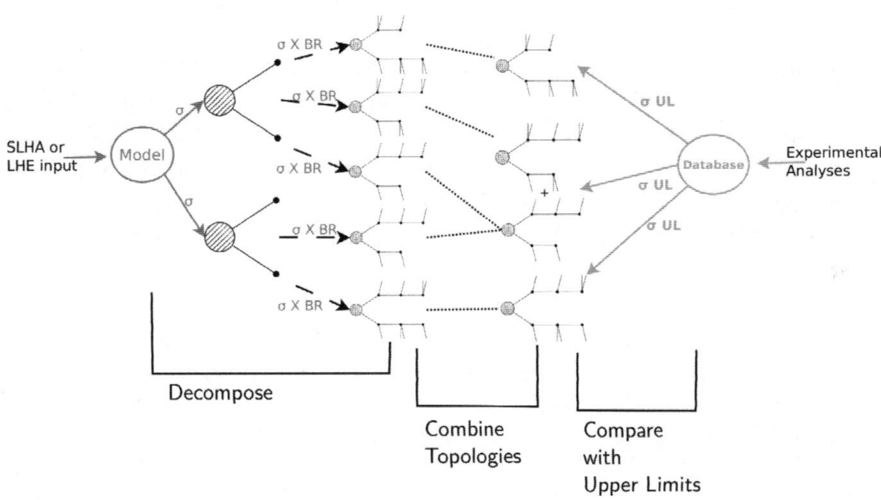

Figure 4. The working principle of SModelS

3. A Simple Application

SModelS v1.0 has already been applied to a variety of full models (see e.g. Refs. [7–12]); in this document we only exemplarily show one result of a study that targets a low-fine tuning scenario within the pMSSM, see Ref. [12]. The scenario itself has been defined in Ref. [13]. Fig. 5 shows the distribution of the missing topologies, as a function of $m_{\tilde{t}_1}$ and $m_{\tilde{\chi}_1^0}$ (leaving out weakino production): in order to constrain the stop mass in the low-fine tuning scenario, it can be seen that it is beneficial to consider pair-produced stop and sbottom particles with subsequent hadronic decays of SM vector bosons.

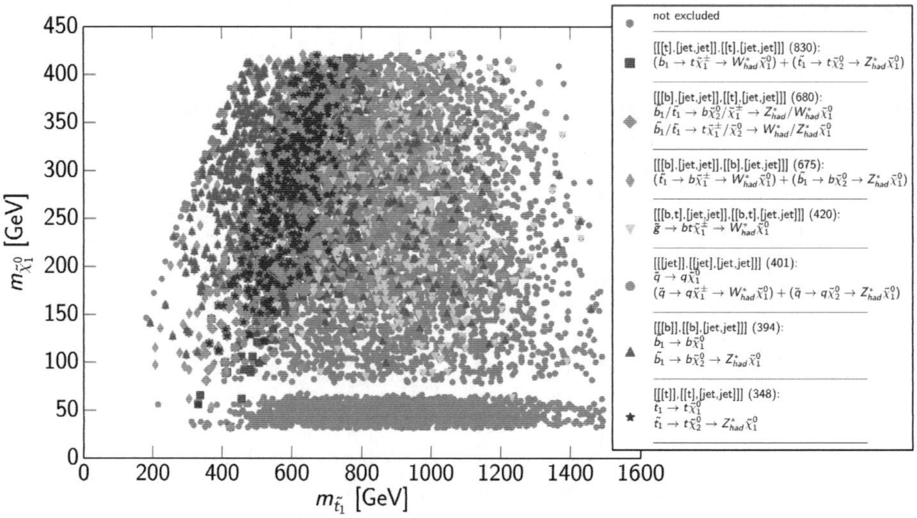

Figure 5. Most frequent missing topologies with $(\sigma \times BR) > 1$ fb and relevant decays only, as function of $m_{\tilde{t}_1}$ and $m_{\tilde{\chi}_1^0}$.

4. Conclusion

A tool has been presented that allows for an automatic comparison of a full model with SMS results. It is based on a method to decompose a given full model into its SMS topologies, a database of SMS results, a formalism to describe what a certain SMS result constrains, and a framework that finally matches the theoretical predictions of a given SMS topology with experimental results. Support for heavy stable charged particles has already been realized as a patch to SModelS [7]; it will eventually become a part of SModelS proper.

Acknowledgements

We thank the ATLAS and CMS SUSY groups for helpful discussions on their results, and in particular for providing (most of) the SMS cross section upper limits used here in digital format. This work is supported in part by the French ANR project DMAstroLHC, the "New Frontiers" program of the Austrian Academy of Sciences, the "Investissements d'avenir, Labex ENIGMASS", and by the "Fundação de Amparo à Pesquisa do Estado de São Paulo".

References

[1] Aad G *et al.* (ATLAS) 2012 *Phys.Lett.* **B716** 1–29 (*Preprint* 1207.7214)
[2] Chatrchyan S *et al.* (CMS) 2012 *Phys.Lett.* **B716** 30–61 (*Preprint* 1207.7235)
[3] https://twiki.cern.ch/twiki/bin/view/AtlasPublic/SupersymmetryPublicResults
[4] https://twiki.cern.ch/twiki/bin/view/CMSPublic/PhysicsResultsSUS
[5] Khachatryan V *et al.* (CMS) 2014 *Eur.Phys.J.* **C74** 3036 (*Preprint* 1405.7570)
[6] Kraml S, Kulkarni S, Laa U, Lessa A, Magerl W, Proschofsky-Spindler D and Waltenberger W 2014 *Eur. Phys. J.* **C74** 2868 (*Preprint* 1312.4175)
[7] Heisig J, Lessa A and Quertenmont L 2015 *JHEP* **12** 087 (*Preprint* 1509.00473)
[8] Bélanger G, Drieu La Rochelle G, Dumont B, Godbole R M, Kraml S and Kulkarni S 2013 *Phys. Lett.* **B726** 773–780 (*Preprint* 1308.3735)
[9] Grober R, Mühlleitner M, Popenda E and Wlotzka A 2015 *Phys. Lett.* **B747** 144–151 (*Preprint* 1502.05935)
[10] Belanger G, Ghosh D, Godbole R and Kulkarni S 2015 *JHEP* **09** 214 (*Preprint* 1506.00665)
[11] Barducci D, Bélanger G, Hugonie C and Pukhov A 2016 *JHEP* **01** 050 (*Preprint* 1510.00246)
[12] Magerl V 2015 Master's thesis TU Wien
[13] Cahill-Rowley M, Hewett J L, Ismail A and Rizzo T G 2013 *Community Summer Study 2013: Snowmass on the Mississippi (CSS2013) Minneapolis, MN, USA, July 29-August 6, 2013* (*Preprint* 1307.8444)

Journal of Physics: Conference Series **762** (2016) 012077 doi:10.1088/1742-6596/762/1/012077

Track 3: Computations in theoretical physics – techniques and methods

Gionata Luisoni[1], **Stanislav Poslavsky**[2] **and York Schröder**[3]

[1] Theoretical Physics Department, CERN, Geneva, Switzerland
[2] Institute for High Energy Physics NRC "Kurchatov Institute", 142281 Protvino, Russia
[3] Grupo de Fisica de Altas Energias, Universidad del Bio-Bio, Casilla 447, Chillan, Chile

E-mail: gionata.luisoni@cern.ch, stvlpos@mail.ru, yschroeder@ubiobio.cl

Abstract. Here, we attempt to summarize the activities of Track 3 of the 17th International Workshop on Advanced Computing and Analysis Techniques in Physics Research (ACAT 2016).

1. Introduction

Today's computations in theoretical high energy physics (HEP) are hardly feasible without the use of modern computer tools and advanced computational algorithms. This is due to the immense progress in both theoretical and experimental techniques that we have witnessed in recent decades, especially with the launch of powerful machines such as the LHC. As a matter of fact, the processes of interest that are accessible at present colliders (be it in Higgs physics, flavor physics or searches for physics beyond the Standard Model) are typically rather complicated; as such, from the theoretical side, one needs to deal with high multiplicities, and be able to go (at times far) beyond leading-order approximations in order to adequately match the high precision of experimental data.

As a consequence, within weak-coupling expansions, one has to consider processes involving hundreds or even thousands of Feynman diagrams, whose numerical evaluation sometimes requires using powerful computer clusters with thousands of CPUs over timescales of weeks. Given this situation, the development of both, computer tools for improving the automation of the computations as well as advanced algorithmic calculational methods, becomes highly relevant.

During this edition of ACAT, we have seen 18 parallel and 7 plenary talks devoted to computations in theoretical HEP. While most of the presented topics are closely interrelated (and in one way or the other related to computer tools), for the sake of this summary we decided to divide them into three main categories: computer algebra tools, algorithms for multi-loop computations and applications to physical processes. Given the nature of this summary, we will not attempt to do justice to all new developments and tools in the field, but only concentrate on the subset that has actually been discussed at the meeting.

2. Computer tools

During the workshop, ten new computer packages or new versions of existing packages have been presented. It is quite interesting to analyze which programming languages and frameworks

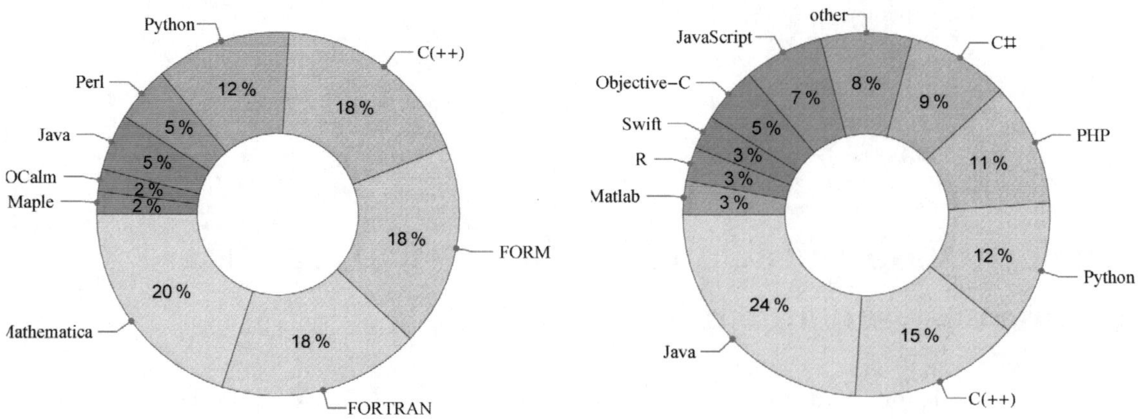

Figure 1. The distribution of programming languages and tools used in the theoretical high-energy physics community (left panel) and in the software industry (right panel). For the former, we have analyzed the packages presented and used by the participants of the session. For the latter, data from the PYPL index was used [3].

are used to code in theoretical high energy physics research. Looking through the different packages presented during the conference, one can find the distribution shown in Fig. 1 (left) (for comparison, the popularity of languages in the industry is shown in the right panel of Fig. 1).

We see that the `Mathematica` language [1] is the most popular. This is not surprising since many computations in theoretical HEP are based on algebraic manipulations, and `Mathematica` provides (besides the language itself) a wide set of computer algebra tools coming with a very convenient user interface.

For large-scale HEP problems, however, the performance of `Mathematica` leaves much to be desired since it was designed as a general-purpose computer algebra system (CAS) without any focus on specific problems arising in HEP. Another drawback of `Mathematica` is that it is proprietary software. These gaps are filled by `FORM` [2] – a free high performance CAS specifically focused on the needs of HEP. Actually, there is a consensus that today `FORM` has the highest performance regarding computer algebra applications in HEP. It is worth noting that, strictly speaking, `FORM` does not provide a programming language in the full sense, but it provides enough functionality for extension and implementation of specific routines required by the particular physical problems.

For low-level programming and, in particular, numerical computations, the standard languages in HEP are `C(++)` and `FORTRAN`. While `C(++)` shares the same place in the industry and in the physics computing community, the quite wide spread usage of `FORTRAN` is a peculiarity of the latter.

Finally, we note that `Python` fills the place of a compact high-level language for general-purpose programming, and as can be seen from Fig. 1 it plays a comparable role in the industry and physics research.

Among the computer tools that are used nowadays in HEP, `FeynCalc` [4] is perhaps one of the most popular and convenient-to-use packages written on top of `Mathematica`. The set of added features includes Dirac and SU(N) algebra, Feynman diagram calculation, tools for loop computations etc. The recent release of `FeynCalc9` [5] includes many important improvements: code quality improvements; further interoperability with external programs (including `FeynArts` [6] for diagrams generation, `FIRE` [7] for integration-by-parts reduction,

PackageX [8] for evaluating oneloop integrals etc.); and important new features including tools for partial fractioning (based on the $Apart package [9]), or multi-loop tensor decomposition. It is also worth mentioning that the new release shows considerable performance improvement.

Automation of the computational process plays an important role for large scale problems. While it is well known how to compute each individual ingredient of a next-to-leading order (NLO) calculation (and one can find plenty of computer packages for each single ingredient), it is still very convenient to have a universal tool, which aggregates all these ingredients into a single program. Such an approach makes the calculation more easily testable and less error-prone. FormCalc [10] is a well-known and powerful tool for automatic computation of one-loop processes. It uses FeynArts [6] to generate diagrams for the process of interest, then passes them to FORM, which performs the d-dimensional Dirac algebra and other things required for computing one-loop amplitudes, and finally it produces optimized FORTRAN code, which can be compiled and run in order to obtain the cross section and cross section distributions for the process. A new suite of shell scripts and Mathematica packages demonstrates how to flexibly use FeynArts and FormCalc beyond one-loop, and together with other packages. Originally constructed for the computation of the two-loop $\mathcal{O}(\alpha_t^2)$ Higgs-mass corrections in FeynHiggs [11], the code is relatively general and may serve as a template for similar calculations.

An important project towards the automation of higher-order processes is the computer system SANC [12], for which recent developments were presented in the plenary talk of Andrej Arbuzov. SANC has its origins in the ZFITTER program [13] for the calculation of fermion pair production and radiative corrections at high energy e^+e^- colliders, but it additionally supports QCD processes (including convolution with partonic distributions) and both QCD and EW corrections at the one-loop precision level. The list of processes implemented in SANC includes Drell-Yan processes (there was recently an important update [14] including a systematic treatment of the photon-induced contribution in proton-proton collisions and electroweak corrections beyond NLO approximation), associated Higgs and gauge boson production, and single-top quark production in s- and t-channels [15].

The new computer algebra system Redberry [16] is focused specifically on the applications in high energy physics and gravity. The main feature of this system is that it considers all mathematical objects that arise in HEP (ordinary scalars, vectors, spinors, matrices and in general arbitrary tensors) in a uniform and robust way. The graph-theoretical approach implemented in Redberry allows to achieve unmatched performance of operations involving symbolical tensors. While Redberry comes with an extensive set of programming features which make it easily extendable, out of the box it provides a range of common tools for performing computations in HEP such as Dirac & SU(N) algebra, simplifications with spinors, or calculation of one-loop counterterms in curved space-time, amongst others. Recent developments include derivation of Feynman rules from a given Lagrangian, and the computation of NLO processes.

As we have mentioned above, in order to match experimental precision it sometimes needed to consider high multiplicity processes, whose evaluation requires to run on large computer clusters for days or even weeks. For some important processes, the standard numerical approach quickly becomes infeasible. For example, the size of a file containing the matrix element for $2 \to 6$ gluon scattering written in a FORTRAN code is about 2 gigabytes. It is difficult to compile file which are that large in reasonable time. In such cases, some kind of virtual machine (VM) can prove very useful. Instead of writing the analytical expression for the amplitude in some language like C(++) or FORTRAN and compile it, one can translate it into a binary format, which then can be interpreted by the VM and calculated just-in-time (so-called JIT compilation) – this means that the actual translation to machine instructions done during execution of a program (at run time) rather than prior to execution. This paradigm is implemented as an extension [17] for the Optimized Matrix Element Generator O'Mega [18]. The implemented technique is very simple and can be naturally parallelized on several computational cores or even on GPUs. The performance shown using this approach is comparable to compiled code, and scales favorably for extremely high multiplicity processes.

The computation of NLO corrections for multi-leg processes requires a large degree of automation in the generation and computation of the amplitudes and the phase space. `Whizard` is one of the several tools able to perform this task. Despite being used mainly for lepton collider predictions, it can compute predictions also for proton-proton collisions. The new release (v.2.2.8) features an implementation of the Binoth Les Houches Accord (BLHA), to be able to interface with codes computing one-loop corrections, an automatic generation of subtraction terms based on the formalism developed by Kunszt, Frixione and Signer [19] and an implementation of the `POWHEG` matching [20] of fixed order NLO computations with parton shower algorithms. Furthermore a new phase space remapping was introduced for a correct treatment of resonances in NLO computations. The release of the new version was accompanied by validation examples based on several processes in electron-positron annihilation [21].

Another tool used for the computation of NLO corrections for several LHC processes is `GoSam` [22, 23]. A concrete example is mentioned later in Section 4. `GoSam` provides a framework for the automatic computation of one-loop amplitudes and is now being upgraded to be able to deal also with two and higher loop amplitudes. The basic idea is to use the interface to `QGRAF` [24] and `FORM`, already present in `GoSam`, to generate the amplitudes and reduce them to a linear combination of master integrals, which can than be computed with another external tool. One of these tools in the code `SecDec`, of which version 3.0 was recently released [25]. `SecDec` can compute multi-loop master integrals using the method of sector decomposition, by producing a Laurent expansion of the integral and computing the coefficients of the expansion numerically. The new version has a better user interface and can be run also on clusters. Furthermore, besides the several improvements in the possible reduction strategies, it allows to compute integrals containing propagators with zero or negative power and also complex masses. All these new features are currently extensively tested on practical examples and in particular for the computation of NLO QCD corrections to double Higgs boson production in gluon-gluon fusion [26].

3. Techniques for (multi-)loop calculations

One of the main motivations behind the most recent developments on advanced techniques for Feynman-diagrammatic computations is the need of precise QCD predictions. These are going to be crucial to make the best out of the large amount of data, which will be collected during the LHC Run2. As example we remind that, in some phase-space regions, NNLO QCD effects can be of the order of 20% (such as e.g. in diboson production). In this respect, increasing the precision of theory predictions by including more loops and/or more legs almost always leads to an improved accuracy. As reflected in many of the talks, we have witnessed a number of recent developments which pushed the boundaries of what can be done on the multi-loop front further, both analytically as well as numerically.

These developments are driven by many new ideas, of which we had interesting overviews in the plenaries. Some of the key advances include: the integration-by-parts (IBP) over finite fields, a broad attack on the bottlenecks of Feynman integral reduction [27], the employment of on-shell methods to automate high-multiplicity QCD amplitude computations [28]; the usage of hyperlogarithms in an attempt to classify the functional content of a wide class of integrals [29] and an elliptic generalization of the ever-present multiple polylogarithms (MPLs), to allow for insight into new function classes lurking behind Feynman diagrams [30].

As a general and uniting theme, most of the research programs devoted to multi-loop techniques are pushing towards highly algorithmic methods, due to the high level of automation needed in modern perturbative quantum field theory computation. To this end, many (public as well as private) codes, ideas and results have been discussed in the sessions. We will give a brief account of them below.

For long time one-loop calculations have been the bottleneck in the computation of NLO corrections. In the last decade, the development of systematic reduction algorithms and automated techniques for the computation of amplitudes allowed to make a big step forward. This led to tools which can automatically perform NLO QCD and Electroweak computations. The same ideas are now being applied and improved also beyond one-loop and many efforts are put in automation for NNLO. Besides the recent improvement in the computation of subtraction terms, the main bottleneck at NNLO is again given by the two loop amplitudes. Despite the several recent results on $2 \to 2$ processes, only few things are known about two loop amplitudes for $2 \to 3$ processes, which became recently a very active field of research [28]. A very nice overview of the latest development was given in the plenary talk by Simon Badger.

As has become clear in recent studies of the algebraic structure of Feynman integrals, many of them can be expressed in terms of the above-mentioned MPLs. There are, however, a number of well-known exceptions, most notably when studying electro-weak physics, QCD, or even massless N=4 Super-Yang-Mills theories. As Christian Bogner has reviewed in his plenary talk, the former type of integrals have by now been completely understood in terms of their iterated structure, a development that has led to software (such as `HyperInt` [31] and `MPL` [32]) which analytically solve a large class of Feynman integrals that respect a certain condition (the so-called linear reducibility of their associated graph polynomials). Going beyond linear reducibility (the two-loop massive sunset integral being a simple example), he furthermore presented a remarkable and new class of functions that generalize the MPLs, the elliptic polylogarithms [33], which harbor promising potential to become standard building blocks, once their structure is understood in more detail.

In order to predict the functional content of the special case of one-loop, N-leg Feynman integrals, it proves useful to switch to a geometric interpretation of their Feynman parameters, mapping integrals onto volumes. The minimal set of independent kinematic invariants and masses that the function depends on corresponds to objects such as a 2-dimensional triangle (for the 2-point function), the 3d tetrahedron (at 3pt), a 4d simplex (at 4pt) etc. [34]. From this (non-Euclidean) geometric intuition, it is possible to derive additional relations among the parameters, essentially reducing the number of variables from N^2 to N, which in facts severely constrains the functional forms. An all-N proof of this reduction of complexity, as well as a relation to other geometric constructions, such as the amplituhedron, is still open.

It is sometimes possible to use iterative structures of a specific set of Feynman integrals in order to obtain results for high (or even all) loop orders. As an example, 3-point ladder-type functions provide such a set, relevant in planar sectors of gauge theory. In particular, utilizing repeated transformations between momentum- and dual space when the propagator powers satisfy specific relations [35], one can systematically reduce the number of rungs, eventually mapping on sums of lower-loop integrals [36]. It would be interesting to check this method against the above-mentioned criterion of linear reducibility.

Looking at Feynman integrals from a more general perspective, one can view them as forming a linear vector space, since the well-known IBP relations form linear relationships between them. The problem of performing perturbation theory in an efficient way can then be regarded as choosing the most convenient basis. To this end, Andreas von Manteuffel highlighted three methods of different maturity level in his plenary talk. First, the master integrals required for the full NNLO QCD corrections to diboson production at LHC have been solved via a linear system of first order differential equations in invariants, mapping onto MPLs after a suitable basis change, which allow to construct optimized functions for fast and stable numerics by requiring absence of spurious denominators. Second, it was discussed how the problem of divergences in Feynman parameter integrals (which obstruct direct ε-expansions on the integrand level) can be overcome by changing to a basis of quasi-finite integrals and perform a direct expansion and integration circumventing sector decomposition or analytic regularization [37]. Third, observing that the polynomial GCD operations needed for the rational functions in large IBP reduction problems consume most of the CPU time, significant speedups may be obtained by switching to modular arithmetic and substituting variables with integers, reducing modulo finite prime fields, and using rational reconstruction (Hensel lifting) to reconstruct the original answer [27].

Turning towards the classification of Feynman integral families and topologies, a new `Mathematica` package `TopoID` has been presented [38]. It can be used in various ways, such as in preprocessing IBP identities or optimizing integral families, and works on the level of graph polynomials, providing a unique ordering that allows for unique and efficient topology mapping. As an application and first check, `TopoID` has been successfully used in evaluating N^3LO corrections to the Higgs boson production cross section [39], while ultimately aiming for e.g. optimizing the evaluation of 5-loop propagators, with applications to SM parameters such as the 5-loop QCD Beta function.

Pushing the loop frontier, a study of the class of five-loop fully massive vacuum diagrams has been presented [40]. This specific class finds applications in e.g. QCD thermodynamics and anomalous dimensions. Using IBP methods to identify the complete set of master integrals, those have been evaluated numerically to high precision, using IBP, difference equations, and factorial series. A parallelized (private) `C++` implementation employs `Fermat` for polynomial algebra, and implements substantial fine-tuning of the classic Laporta approach [41]. Due to the diverse nature of envisaged applications, the corresponding code `TIDE` is able to optimize large systems of difference and recursion relations for ε-expansions around 3 and 4 dimensions.

The loop-level record at this meeting, however, has been set by Mikhail Kompaniets who reported on analytic results for the six-loop beta function in scalar theory [42]. Massively profiting from treating φ^4 theory where each Feynman diagram corresponds to a single integral only (as opposed to orders of magnitudes more in gauge theories), the presented strategy restricts to the 3-vertex-free subset of 6-loop massless propagators without dots. Relying on IBP paired with a (private) R^* implementation for systematically disentangling UV- and IR-divergences, the 6-loop integrations can ultimately be performed in terms of hyperlogarithms (see above), to arrive at an analytic textbook-quality result in terms of multiple Zeta values.

For specific classes of Feynman integrals, it is possible to build a package that performs an exact reduction and solution of any given member, hence providing a method for exact evaluation of that class. For massless two-point functions, this has been achieved at the three-loop (`Mincer` [43]) and most recently on the four-loop level [44]. Due to the large tree of different integral sectors, the latter development has been made possible through a high level of automation ("write a small program that produces a large (41k lines) one"), incorporates new IBP tricks such as the exact diamond rule [45], and culminated in the `FORM` code `Forcer` that has already been successfully tested on a number of problems, and which will enable the authors to evaluate e.g. the $N = 6$ Mellin moments of 4-loop splitting functions. Looking ahead, the automated derivation of exact reduction rules might enable even a '5-loop Mincer' package.

Another way to compute loop integrals numerically, is to explicitly make use of the Feynman ϵ-prescription computing the integral for a finite value of ϵ and than extrapolate to the limit for $\epsilon \to 0$. This so called *direct computation method*, presented by K. Kato, needs high performance computing facilities, but allows to compute diagrams at several loops in a systematic way. Examples were shown for 2-, 3- and 4-loop diagrams.

4. Phenomenological applications

The complexity and variety of theoretical HEP computations is such that not all computations can directly be used for phenomenology. This was the case for several computations described in the previous sections, which are fundamental building blocks for future phenomenological applications, but whose outcome can currently not yet be compared with experimental data. In this section however we present some results which have direct impact on phenomenology.

The important problem of the relation between pole and running masses of heavy quarks is still a very active research field. To date, we have some known results for these relations up to four loops, part of which are analytical, while the complete result is known numerically only and has been obtained very recently [46]. This numerical result is known for a fixed number of flavors only. The authors of Ref. [47] used a least squares method in order to extract the full dependence of the relation between heavy quark pole- and running mass on the number of flavors. They show that results for pole masses of heavy quarks form an asymptotic series, which is a well-known feature of perturbative series expansions. For the bottom quark, the asymptotic nature arise only at the four-loop level, while for the top quark even up to four loops inclusive, the perturbation series converges. So, for the t quark it is natural to take into account these corrections. As of today, experimental uncertainties in the top mass are larger than the corresponding theory errors, but this situation might be reversed in the nearest future, in which case the presented results come to play an important role. Additionally, the least squares method used by the authors allows to estimate theoretical uncertainties which is quite informative when using numerical methods. These uncertainties turn out to be three times larger than the numerical errors obtained in [46]; however, the authors note that their uncertainties may be overestimated, an issue that can be resolved by the direct calculation of the full analytical result.

Another important aspect related to quark masses, is their experimental determination. The $\overline{\text{MS}}$ bottom-quark mass can for example be determined precisely from $\Upsilon(1S)$, which is the lowest energy $b\bar{b}$-quarkonium state. The mass of the quarkonium state can be written as a sum of the bottom quark pole mass and the binding energy. These quantities however suffer from renormalon ambiguity. This issue can be solved by expanding them in powers of the ratios over the $\overline{\text{MS}}$ bottom-quark mass and by subtracting the leading renormalon extracted from the static potential. With this technique it was recently possible to determine the value of the $\overline{\text{MS}}$ bottom-quark mass to percent accuracy [48].

The recent discovery of the long-sought-after Higgs boson by the ATLAS and CMS collaborations has ushered in an era of precision measurements to determine the nature of the new particle. The recent computation of Higgs boson production in gluon-gluon fusion at N^{3}LO accuracy [49], or the computation of the differential NNLO QCD corrections of H+1 jet [50, 51], Z+1 jet [52] or W+1 jet [53] are just few examples. The fact that it was recently possible to compute so many processes at NNLO accuracy in relatively short time relies on better automation, on improved algorithms (an example are the new subtraction methods developed recently [53, 54]), and on the smart combination of parallelization algorithms which allow to perform this calculations on large high performance computing facilities, obtaining the results within few hours of running.

Another important aspect to keep in mind when performing such CPU-expensive computations is the question of storing results in the most flexible way possible. A possibility which became popular recently for NLO computations, is to store events in form of `Root` N-Tuples [55]. This allows to perform the most computationally expensive part of the calculation only once and to re-run the phenomenological analysis varying cuts, parton distribution functions or scales several time in a fast way. An example of a possible study that can be performed with N-Tuples was presented during the conference for the NLO QCD corrections to H+1, 2, and 3 jets [56].

Hadronic parton distribution functions (PDFs) determine the leading contributions to the cross sections of various hard processes. This topic was covered in the plenary talk given by Sergei Kulagin. The PDFs of hadrons determine the leading contributions to the cross sections of various hard processes. The deep-inelastic scattering (DIS) and Drell-Yan (DY) experiments with nuclei demonstrated significant nuclear effects on the parton level. These observations

Figure 2. Typical keywords of Track 3 presentations.

rule out the naive picture of the nucleus as a system of quasi-free nucleons and indicate that the nuclear environment plays an important role even at energies and momenta much higher than those involved in typical nuclear ground state processes. Aiming at a better understanding of nuclear parton distributions, a detailed semi-microscopic model of nuclear PDFs has been developed. It includes the QCD treatment of nucleon structure functions, and addresses a number of nuclear effects such as shadowing, Fermi motion and nuclear binding, nuclear pion and off-shell corrections to bound nucleon structure functions [57–59]. Using this approach, the authors were able to reproduce data on nuclear effects in DIS and DY experiments. Predictions for the neutrino-nuclear cross sections have also been discussed.

Automation and precision is needed also in searches for physics Beyond the Standard Model (BSM). This is for example the case when performing very precise computations of Standard Model process to search for indirect deviations from the predictions. An example presented during the conference refers to the first computation of NNLO corrections, namely the interference of one-loop contributions, to the polarized Møller scattering [60]. It is also possible to consider the effect of considering a further Z' in the theoretical predictions and produce precise exclusion limits.

Exclusion limits are however also produced by several experimental analysis. The question can therefore also be reversed: how can the BSM exclusion limits determined by the LHC experiment be systematically exploited to discriminate between the many possible BSM extensions? A possible answer is given by the `SModels` project [61]. The key idea is to devise a formal language to describe the LHC results collected into a database, which can then be matched to the decomposed spectrum of each given model. This allows to quickly check if the experimental results already exclude a given model or whether there is still room for it to be realized in nature.

Rotational speed is an important physical parameter of stars: knowing the distribution of stellar rotational velocities is essential for understanding stellar evolution. However, rotational speed cannot be measured directly, but is given by the convolution between rotational speed and the sine of the inclination angle – $v\sin(\alpha)$. The problem of deconvolving stellar rotational speeds was discussed in the plenary by Michel Cure. The problem itself can be described via a Fredholm integral of the first kind. The new method [62] to deconvolve this inverse problem and to obtain the cumulative distribution function for stellar rotational velocities is based on the work of Chandrasekhar and Münch [63]. The proposed method can be also applied to the distribution of extrasolar planets and brown dwarfs (in binary systems) [64]. For stars in a cluster, where all members are gravitationally bounded, the standard assumption that rotational axes are uniformly distributed over the sphere is questionable. On the basis of the proposed techniques, a simple approach to model this anisotropy of rotational axes was developed, opening up the possibility to 'disentangle' simultaneously both the rotational speed distribution and the orientation of rotational axes.

5. Conclusions

We leave the reader with the word cloud presented in Fig. 2, which attempts to represent the main keywords of all Track 3 talks, as drawn from the actual presentations.

Acknowledgments

We would like to thank the organizers of the ACAT 2016 workshop, especially Luis Salinas and Claudio Torres, who took the major part of the organization. Our thanks extend to all involved staff of the Technical University Federico Santa Maria who provided a very comfortable atmosphere during the workshop. We would like to thank Andrei Kataev, Tord Riemann and Igor Kondrashuk for valuable discussions. The work of S.P. was supported by the Chilean Center for Mathematical Modeling (CMM) and partially by the RFBR grant #16-32-60017 and by the FAIR-Russia Research Center. Y.S. acknowledges support from DFG grant SCHR 993/1, FONDECYT project 1151281 and UBB project GI-152609/VC.

References

[1] Wolfram Mathematica, `http://www.wolfram.com`
[2] Vermaseren J A M 2000 (*Preprint* `math-ph/0010025`)
[3] PYPL Index, `http://pypl.github.io`
[4] Mertig R, Bohm M and Denner A 1991 *Comput. Phys. Commun.* **64** 345–359
[5] Shtabovenko V, Mertig R and Orellana F 2016 (*Preprint* `1601.01167`)
[6] Hahn T 2001 *Comput. Phys. Commun.* **140** 418–431 (*Preprint* `hep-ph/0012260`)
[7] Smirnov A V 2008 *JHEP* **10** 107 (*Preprint* `0807.3243`)
[8] Patel H H 2015 *Comput. Phys. Commun.* **197** 276–290 (*Preprint* `1503.01469`)
[9] Feng F 2012 *Comput. Phys. Commun.* **183** 2158–2164 (*Preprint* `1204.2314`)
[10] Hahn T and Perez-Victoria M 1999 *Comput. Phys. Commun.* **118** 153–165 (*Preprint* `hep-ph/9807565`)
[11] Hahn T and Paehr S 2015 (*Preprint* `1508.00562`)
[12] Andonov A, Arbuzov A, Bardin D, Bondarenko S, Christova P, Kalinovskaya L, Nanava G and von Schlippe W 2006 *Comput. Phys. Commun.* **174** 481–517 [Erratum: Comput. Phys. Commun.177,623(2007)] (*Preprint* `hep-ph/0411186`)
[13] Arbuzov A B, Awramik M, Czakon M, Freitas A, Grunewald M W, Monig K, Riemann S and Riemann T 2006 *Comput. Phys. Commun.* **174** 728–758 (*Preprint* `hep-ph/0507146`)
[14] Arbuzov A, Bardin D, Bondarenko S, Christova P, Kalinovskaya L, Klein U, Kolesnikov V, Sadykov R, Sapronov A and Uskov F 2015 (*Preprint* `1509.03052`)
[15] Bondarenko S G and Sapronov A A 2013 *Comput. Phys. Commun.* **184** 2343–2350 (*Preprint* `1301.3687`)
[16] Bolotin D A and Poslavsky S V 2013 (*Preprint* `1302.1219`)
[17] Chokoufe Nejad B, Ohl T and Reuter J 2015 *Comput. Phys. Commun.* **196** 58–69 (*Preprint* `1411.3834`)
[18] Moretti M, Ohl T and Reuter J 2001 1981–2009 (*Preprint* `hep-ph/0102195`)
[19] Frixione S, Kunszt Z and Signer A 1996 *Nucl. Phys.* **B467** 399–442 (*Preprint* `hep-ph/9512328`)
[20] Frixione S, Nason P and Oleari C 2007 *JHEP* **11** 070 (*Preprint* `0709.2092`)
[21] Reuter J, Chokoufe B, Hoang A, Kilian W, Stahlhofen M, Teubner T and Weiss C 2016 *these proceedings* (*Preprint* `1602.06270`)
[22] Cullen G, Greiner N, Heinrich G, Luisoni G, Mastrolia P, Ossola G, Reiter T and Tramontano F 2012 *Eur. Phys. J.* **C72** 1889 (*Preprint* `1111.2034`)
[23] Cullen G *et al.* 2014 *Eur. Phys. J.* **C74** 3001 (*Preprint* `1404.7096`)

[24] Nogueira P 1993 *J.Comput.Phys.* **105** 279–289
[25] Borowka S, Heinrich G, Jones S P, Kerner M, Schlenk J and Zirke T 2015 *Comput. Phys. Commun.* **196** 470–491 (*Preprint* 1502.06595)
[26] Borowka S, Heinrich G, Jahn S, Jones S P, Kerner M, Schlenk J and Zirke T 2016 *these proceedings* (*Preprint* 1604.00267)
[27] von Manteuffel A and Schabinger R M 2015 *Phys. Lett.* **B744** 101–104 (*Preprint* 1406.4513)
[28] Badger S, Mogull G, Ochirov A and O'Connell D 2015 *JHEP* **10** 064 (*Preprint* 1507.08797)
[29] Panzer E 2015 Ph.D. thesis Humboldt U., Berlin, Inst. Math. (*Preprint* 1506.07243)
[30] Adams L, Bogner C and Weinzierl S 2014 *J. Math. Phys.* **55** 102301 (*Preprint* 1405.5640)
[31] Panzer E 2014 *Comput. Phys. Commun.* **188** 148–166 (*Preprint* 1403.3385)
[32] Bogner C 2015 (*Preprint* 1510.04562)
[33] Bogner C 2016 *these proceedings* (*Preprint* 1603.00420)
[34] Davydychev A 2016 *these proceedings*
[35] Usyukina N I and Davydychev A I 1993 *Phys. Lett.* **B305** 136–143
[36] Borja J and Kondrashuk I 2016 *these proceedings* (*Preprint* 1604.01353)
[37] von Manteuffel A, Panzer E and Schabinger R M 2015 *JHEP* **02** 120 (*Preprint* 1411.7392)
[38] Hoff J 2016 *these proceedings*
[39] Anzai C, Hasselhuhn A, Hoschele M, Hoff J, Kilgore W, Steinhauser M and Ueda T 2015 *JHEP* **07** 140 (*Preprint* 1506.02674)
[40] Luthe T and Schroder Y 2016 *these proceedings* (*Preprint* 1604.01262)
[41] Laporta S 2000 *Int. J. Mod. Phys.* **A15** 5087–5159 (*Preprint* hep-ph/0102033)
[42] Batkovich D V, Chetyrkin K G and Kompaniets M V 2016 *Nucl. Phys.* **B906** 147–167 (*Preprint* 1601.01960)
[43] Gorishnii S G, Larin S A, Surguladze L R and Tkachov F V 1989 *Comput. Phys. Commun.* **55** 381–408
[44] Ueda T 2016 *these proceedings*
[45] Ruijl B, Ueda T and Vermaseren J 2015 *Phys. Lett.* **B746** 347–350 (*Preprint* 1504.08258)
[46] Marquard P, Smirnov A V, Smirnov V A and Steinhauser M 2015 *Phys. Rev. Lett.* **114** 142002 (*Preprint* 1502.01030)
[47] Kataev A L and Molokoedov V S 2015 (*Preprint* 1511.06898)
[48] Ayala C, Cvetic G and Pineda A 2014 *JHEP* **09** 045 (*Preprint* 1407.2128)
[49] Anastasiou C, Duhr C, Dulat F, Herzog F and Mistlberger B 2015 *Phys. Rev. Lett.* **114** 212001 (*Preprint* 1503.06056)
[50] Boughezal R, Caola F, Melnikov K, Petriello F and Schulze M 2015 *Phys. Rev. Lett.* **115** 082003 (*Preprint* 1504.07922)
[51] Boughezal R, Focke C, Giele W, Liu X and Petriello F 2015 *Phys. Lett.* **B748** 5–8 (*Preprint* 1505.03893)
[52] Boughezal R, Campbell J M, Ellis R K, Focke C, Giele W T, Liu X and Petriello F 2015 (*Preprint* 1512.01291)
[53] Boughezal R, Focke C, Liu X and Petriello F 2015 *Phys. Rev. Lett.* **115** 062002 (*Preprint* 1504.02131)
[54] Gaunt J, Stahlhofen M, Tackmann F J and Walsh J R 2015 *JHEP* **09** 058 (*Preprint* 1505.04794)
[55] Bern Z, Dixon L J, Febres Cordero F, Hoeche S, Ita H, Kosower D A and Maitre D 2014 *Comput. Phys. Commun.* **185** 1443–1460 (*Preprint* 1310.7439)
[56] Luisoni G 2016 *these proceedings*
[57] Kulagin S A and Petti R 2006 *Nucl. Phys.* **A765** 126–187 (*Preprint* hep-ph/0412425)
[58] Kulagin S A and Petti R 2010 *Phys. Rev.* **C82** 054614 (*Preprint* 1004.3062)
[59] Kulagin S A and Petti R 2014 *Phys. Rev.* **C90** 045204 (*Preprint* 1405.2529)
[60] Aleksejevs A, Wu S, Barkanova S, Bystritskiy Y and Zykunov V 2016 *these proceedings* (*Preprint* 1603.03006)
[61] Kraml S, Kulkarni S, Laa U, Lessa A, Magerl W, Proschofsky-Spindler D and Waltenberger W 2014 *Eur. Phys. J.* **C74** 2868 (*Preprint* 1312.4175)
[62] Curé M, Rial D F, Christen A and Cassetti J 2014 *Astronomy & Astrophysics* **565** A85 (*Preprint* 1401.1054)
[63] Chandrasekhar S and Münch G 1950 *Astrophysical Journal* **111** 142
[64] Curé M, Rial D F, Cassetti J, Christen A and Boffin H M J 2015 *Astronomy & Astrophysics* **573** A86 (*Preprint* 1412.0431)

Journal of Physics: Conference Series **762** (2016) 012078 doi:10.1088/1742-6596/762/1/012078

From perturbative calculations of the QCD static potential towards four-loop pole-running heavy quarks masses relation

A. L. Kataev[1,2], **V. S. Molokoedov**[1,2]

[1] Institute for Nuclear Research of the Russian Academy of Sciences (INR), 60th October Anniversary Prospect, 7a, 117312 Moscow, Russia

[2] Moscow Institute of Physics and Technology (MIPT), 141700, Dolgoprudny, Russia

E-mail: kataev@ms2.inr.ac.ru[1] , viktor_molokoedov@mail.ru[2]

Abstract. The summary of the available semi-analytical results for the three-loop corrections to the QCD static potential and for the $\mathcal{O}(\alpha_s^4)$ contributions to the ratio of the running and pole heavy quark masses are presented. The procedure of the determination of the dependence of the four-loop contribution to the pole-running heavy quarks mass ratio on the number of quarks flavours, based on application of the least squares method is described. The necessity of clarifying the reason of discrepancy between the numerical uncertainties of the α_s^4 coefficients in the mass ratio, obtained by this mathematical method by the direct numerical calculations is emphasised.

1. Introduction.

It is known that in the Standard Model it is possible to introduce several definitions of heavy quark masses. Of particular relevance are the potential subtracted masses $m_{PS,q}(\mu^2)$ [1], the $\overline{\rm MS}$-scheme running masses $\overline{m}_q(\mu^2)$ and the scale-independent pole masses M_q. The relations between these definitions of heavy quarks masses were studied in a number of theoretical works on the subject (see e.g. [1], [2], [3]). Since these masses are extracted at present from different rather precise experimental data, e.g. from the spectroscopy of heavy hadrons and from the measurable cross-sections of the productions of heavy quarks in different observable processes, it is important to know the relations between M_q, $\overline{m}_q(\mu^2)$ and $m_{PS,q}(\mu^2)$ with high enough precision, which is gained by evaluation of high order perturbative corrections to the relations between pole and $\overline{\rm MS}$-scheme running heavy quarks masses and to the QCD static potential as well. This talk is devoted to the summary of the available results of the determination of three-loop corrections to the QCD static potential and to the $\mathcal{O}(\alpha_s^4)$ relations between different definitions of heavy quarks masses. Special attention is paid to the discussions of theoretical errors in the existing numerical calculations of the $\mathcal{O}(\alpha_s^4)$ contributions to the static potential [4],[5],[6] and to the ratio of the running and pole heavy quark masses [3]. The determination of the dependence of the QCD correction on the number of lighter quarks flavours n_l by means of the ordinary least squares (OLS) mathematical method [7] is described.

2. The static potential in perturbative QCD.

The QCD static potential is strictly defined through the vacuum expectation value of the Wilson loop [8]:

$$\mathrm{V}(r) = -\lim_{T\to\infty}\frac{1}{iT}\ln\frac{\langle 0 \mid \mathrm{Tr}\, \mathrm{P}e^{ig\oint_C dx^\mu A^a_\mu t^a} \mid 0\rangle}{\langle 0 \mid \mathrm{Tr}1 \mid 0\rangle} = \int\frac{d^3\vec{q}}{(2\pi)^3}e^{i\vec{q}\vec{r}}\,\tilde{\mathrm{V}}(\vec{q}^{\,2}) \tag{1}$$

where C is a closed rectangular contour, T and r are the time and three-dimensional space variables, P is the path-ordering operator, A^a_μ is a gluon field and t^a are the generators of Lie algebra of the SU($\mathrm{N_c}$) group.

Perturbation theory expression for the Fourier transform $\tilde{\mathrm{V}}(\vec{q}^{\,2})$ of the static potential is known up to $\mathcal{O}(\alpha_s^4)$-corrections:

$$\tilde{\mathrm{V}}(q) = -\frac{4\pi C_F\alpha_s(\vec{q}^{\,2})}{\vec{q}^{\,2}}\left(1 + a_1 a_s(\vec{q}^{\,2}) + a_2 a_s^2(\vec{q}^{\,2}) + \left(a_3 + \frac{\pi^2 C_A^3}{8}\ln\frac{\mu^2}{\vec{q}^{\,2}}\right)a_s^3(\vec{q}^{\,2})\right) \tag{2}$$

where C_F and C_A are the Casimir operators and the QCD coupling constant $a_s(\vec{q}^{\,2}) = \alpha_s(\vec{q}^{\,2})/\pi$ is defined in the $\overline{\mathrm{MS}}$-scheme. The coefficients a_1 and a_2 in Eq. (2) were obtained by evaluating in the analytical form the corresponding one and two-loop Feynman diagrams (see [9],[10] and [11], [12] respectively). The additional term $\pi^2 C_A^3/8$, correctly evaluated in [13], arises due to the infrared (IR) divergences, which begin to manifest themselves in the static potential at the three-loop level. However, in the concrete applications of the effective non-relativistic QCD these IR-divergent terms are cancelling out.

The three-loop contribution a_3 to Eq. (2) can be presented as

$$a_3 = a_3^{(3)}n_l^3 + a_3^{(2)}n_l^2 + a_3^{(1)}n_l + a_3^{(0)} \tag{3}$$

where n_l are the number of quarks, which contribute to the QCD corrections to the static potential, responsible for the strong interactions of the heavy quark-antiquark pair with the flavour number $n_f = n_l + 1$. Note, that the masses of all lighter quarks with the flavour number n_l are usually neglected.

The renormalon-type n_l^3 contribution and the n_l^2-dependent term, which was evaluated analytically in [4], have the following form

$$a_3^{(3)} = -\frac{125}{729}T_F^3\,, \qquad a_3^{(2)} = \left(\frac{12541}{15552} + \frac{23}{12}\zeta(3) + \frac{\pi^4}{135}\right)C_A T_F^2 + \left(\frac{7001}{2592} - \frac{13}{6}\zeta(3)\right)C_F T_F^2 \tag{4}$$

Definite parts of the n_l-dependent coefficient $a_3^{(1)}$ and the overall expression for the constant term $a_3^{(0)}$ have not yet been computed analytically. The semi-analytical expression for $a_3^{(1)}$ was obtained in [4] and reads:

$$\begin{aligned} a_3^{(1)} &= -\frac{709.717}{64}C_A^2 T_F + \left(-\frac{71281}{10368} + \frac{33}{8}\zeta(3) + \frac{5}{4}\zeta(5)\right)C_A C_F T_F \\ &+ \left(\frac{143}{288} + \frac{37}{24}\zeta(3) - \frac{5}{2}\zeta(5)\right)C_F^2 T_F - \frac{56.83(1)}{64}\frac{d_F^{abcd}d_F^{abcd}}{N_A}\,. \end{aligned} \tag{5}$$

where $d_F^{abcd} = \mathrm{Tr}(t^a t^{(b}t^c t^{d)})/6$ and $d_A^{abcd} = \mathrm{Tr}(C^a C^{(b}C^c C^{d)})/6$ are totally symmetric tensors, $(C^a)_{bc} = -if^{abc}$ are the generators of the adjoint representation of the Lie algebra of the SU($\mathrm{N_c}$) group. Note that in the QED limit with C_A=0 the n_l-dependent terms, which are proportional to the powers of T_F in (4) and (5), are in agreement with the $\overline{\rm MS}$-scheme three-loop corrections to the photon vacuum polarisation function in QED, presented previously in [19]. The inaccuracy of the numerical evaluation of the $C_A^2 T_F$ coefficient was not indicated in [4], since it is much smaller than the uncertainty of the $d_F^{abcd}d_F^{abcd}$ coefficient in the presented in (5) expression for the $a_3^{(1)}$-term. The most precise calculation for $a_3^{(0)}$ was made in [5]. Its result

$$a_3^{(0)} = \frac{502.24(1)}{64}C_A^3 - \frac{136.39(12)}{64}\frac{d_F^{abcd}d_A^{abcd}}{N_A} \tag{6}$$

was published almost simultaneously with the following result of the independent calculation, performed in [6]:

$$a_3^{(0)} = \frac{502.22(12)}{64}C_A^3 - \frac{136.8(14)}{64}\frac{d_F^{abcd}d_A^{abcd}}{N_A} \tag{7}$$

Using the same computer code one of the authors of [6] improved recently the numerical precision of the second term in (7), which now reads[1]:

$$a_3^{(0)} = \frac{502.22(12)}{64}C_A^3 - \frac{136.6(2)}{64}\frac{d_F^{abcd}d_A^{abcd}}{N_A} \tag{8}$$

Within the error bars the results (6), (7) and (8) have a common intersection region. The obtained in [5] contributions to $a_3^{(0)}$ have the smallest uncertainties. However, it is highly desirable to understand what is the reason of the differences in the numerical results of (6), (7) and (8).

Indeed, in the process of calculations the authors of [5] and [6] used the same theoretical methods. Let us mention the basic steps of these calculations. First, the diagrams, contributing to the 3-loop correction to static potential are generated. Then they are reduced to about 2000 3-loop integrals. In [5] these integrals were reduced to a relatively small number of the concrete master integrals using the integration-by-parts (IBP) method [14] and the FIRE algorithm [17], which is based in part on the Laporta algorithm [16]. Almost all master integrals were evaluated analytically apart of *three* of them, which were evaluated numerically. To check the results of these analytical and numerical calculations the program FIESTA [17], [18] was used. The obtained in [5] and [4] numerical results in the expressions for $a_3^{(0)}$ and $a_3^{(1)}$ terms are related to the results of the numerical evaluation of *three* mentioned above master integrals.

In the independent work [6] the IBP method [14] and the Laporta algorithm were applied in another way. To evaluate the master integrals the different from FIESTA computer program, written by the authors of [6] was used. Therefore, it is difficult to relate the numerical uncertainties of the results, given in (7) and (8) with the ones, presented in (6), which are determined by the numerical evaluation of the known *three* master integrals. In view of this it may be of real interest to consider the possibility of the detailed comparison of the FIESTA computer program with the one, created by the authors of [6].

Note, that in [20] using the maximal transcendentality hypothesis the guess on the analytical structure of *four* numerically known coefficients in the the expression for the $\mathcal{O}(a_s^3)$ correction to the static potential was made. More definitely, in [20] we proposed to consider the reliability of the existence of the following decomposition

$$709.717 = R_1 + R_2\pi^2 + R_3\pi^4 + R_4\zeta(3) + R_5\pi^2\zeta(3) + R_6\zeta(5) \tag{9}$$
$$502.24(1) = R_7 + R_8\pi^2 + R_9\pi^4 + R_{10}\zeta(3) + R_{11}\pi^2\zeta(3) + R_{12}\zeta(5) \tag{10}$$
$$56.83(1) = R_{13} + R_{14}\pi^2 + R_{15}\pi^4 + R_{16}\zeta(3) \tag{11}$$
$$136.39(12) = R_{17} + R_{18}\pi^2 + R_{19}\pi^4 + R_{20}\zeta(3) \tag{12}$$

[1] We are grateful to Y. Sumino for informing us about this new unpublished result of his personal calculations.

where R_i are still unknown rational numbers. One should note that the concrete coefficients R_i in (9)-(12) may be zero. Considering carefully the analytical results of [4] in [20] we expressed the hope that R_5, R_{11}, R_{12} and R_{18} may be really zero. The status of the discussed above hypothesis of [20] will be clarified when the remaining numerically calculated in [4] *three* master integrals will be evaluated analytically. The work on the solution of this calculation problem is now in progress [2].

3. The relation between static potential and potential subtracted heavy quark masses.

The information on the QCD static potential and on the uncertainties of its perturbative calculations are important in the definitions of the heavy quarks masses and their values. This is possible to understand after consideration of introduced in [1] notion of heavy quark masses in the potential subtraction scheme. This definition is motivated by the studies of the behaviour of the cross-sections of the productions of heavy quarks near thresholds. It is based on the long-distance modifications of the perturbative static heavy quark potential $\tilde{\mathrm{V}}(\vec{q}^{\,2})$, oriented on the suppression of the IR renormalon contributions (for the consideration of the possible other power-suppressed non-perturbative contributions to the Coulomb-like potential for heavy quarks, which are not related to the IR renormalon effects, see e.g. [21]). In the static potential the IR effects can be eliminated by introducing boundary condition $|\vec{q}| > \mu_f$ to its Fourier transform, where the factorisation scale μ_f varies in the region $\Lambda_{QCD} < \mu_f < M_q v$, and v is the relative velocity of two heavy quarks, which is rather small near thresholds of their production. The equivalent way of the fixation of the factorisation scale is motivated by the uncertainty principle, which gives $\mu_f \sim 1/r \sim M_q v \sim M_q\alpha_s(M_q)$. The factorisation scale μ_f enters the definition of the subtracted potential $\mathrm{V}(r,\mu_f)$ as:

$$\mathrm{V}(r,\mu_f) = \mathrm{V}(r) + 2\delta m_q(\mu_f) \quad . \tag{13}$$

The residual heavy quark mass $\delta m_q(\mu_f)$ is determined in the following way

$$\delta m_q(\mu_f) = -\frac{1}{2}\int\limits_{|\vec{q}|<\mu_f} \frac{d^3\vec{q}}{(2\pi)^3}\tilde{\mathrm{V}}(\vec{q}^{\,2}) \tag{14}$$

where the perturbative expression for $\tilde{\mathrm{V}}(\vec{q}^{\,2})$ is presented above in (2). The defined by this way subtracted potential $\mathrm{V}(r,\mu_f)$ does not contain long-distance contributions. As a consequence, in the calculations of the cross sections of heavy quark production in the near threshold region the pole mass M_q (which contain long-distance IR renormalon contributions in its relation to the $\overline{\mathrm{MS}}$-scheme running mass [22], [23]) should be changed to the potential subtracted quark mass $m_{\mathrm{PS},q}(\mu_f)$, defined in [1] as

$$m_{\mathrm{PS},q}(\mu_f) = M_q - \delta m_q(\mu_f) \quad . \tag{15}$$

In the r.h.s. of (15) the high loop perturbative corrections to $\delta m_q(\mu_f)$, which generate IR renormalon contributions cancel exactly with long-distance contributions to the pole mass. Using now the expansion of the pole mass in terms of the $\overline{\mathrm{MS}}$-scheme running mass

$$M_q = \overline{m}_q(\overline{m}_q^2)(1 + \sum_{i\geq 1} l_i a_s^i(\overline{m}_q^2)) \tag{16}$$

and substituting it into (15) one can eliminate the dependence on the factorisation scale μ_f and change it to the $\overline{\mathrm{MS}}$-scheme running mass $\overline{m}_q(\overline{m}_q^2)$. To finalise this section we note that it is

[2] We are grateful to V.A.Smirnov for this information.

possible to obtain a link between the pole heavy quark mass and the $\overline{\text{MS}}$ running mass through the subtracted-potential mass. Therefore the explicit expressions for the coefficients $a_3^{(0)}$ and $a_3^{(1)}$ in (3) can be related with the contributions to the $\mathcal{O}(a_s^4)$ coefficients in the the ratio of the running and pole heavy quark masses and may be used in the future to check the results to be discussed in the next section.

4. The four-loop relation between $\overline{\text{MS}}$ running and pole heavy quark masses.

The perturbative relation between pole and $\overline{\text{MS}}$ running heavy quark masses contains the discovered in [22], [23] long-distance IR renormalon effects, which result from the asymptotic structure of the perturbative QCD series (for discussions of the asymptotic behaviour of the perturbative series in the renormalised quantum field theory models see [24]). However, to understand when the asymptotic nature of these expansions is starting to manifest itself and when the truncated perturbative series can be used it is necessary to study the concrete values of high-order corrections. To analyse these problems we consider the ratios of the defined in the $\overline{\text{MS}}$-scheme running heavy quark masses and the pole heavy quark masses, namely

$$\frac{\overline{m}_q(M_q^2)}{M_q} = 1 + \sum_{i=1}^{\infty} z_m^{(i)} a_s^i(M_q^2) \ . \tag{17}$$

The coefficients $z_m^{(i)}$ can be expanded in powers of n_l. All coefficients $z_m^{(i)}$ with $1 \leq i \leq 3$ were evaluated analytically in the case of SU(N_c) colour gauge group. The one-loop correction was found in [25], the two-loop contribution was analytically evaluated in [26], the three-loop term is known from the analytical and semi-analytical calculations, performed in [27] and [28] correspondingly. The coefficient $z_m^{(4)}$ can be presented in analogy to equation (3) as

$$z_m^{(4)} = z_m^{(40)} + z_m^{(41)} n_l + z_m^{(42)} n_l^2 + z_m^{(43)} n_l^3 \ . \tag{18}$$

The terms $z_m^{(43)}$ and $z_m^{(42)}$ are known in analytical form from the calculations of [29]. In the case of $\text{SU}_\text{c}(3)$ group of colour symmetry their expressions read:

$$z_m^{(43)} = \frac{42979}{1119744} + \frac{317\zeta_3}{2592} + \frac{89\pi^2}{3888} + \frac{71\pi^4}{25920} \ , \tag{19}$$

$$\begin{aligned} z_m^{(42)} = & -\frac{32420681}{4478976} - \frac{40531\zeta_3}{5184} - \frac{63059\pi^2}{31104} - \frac{103\pi^2\ln 2}{972} + \frac{11\pi^2\ln^2 2}{243} - \frac{2\pi^2\ln^3 2}{243} - \frac{5\pi^2\zeta_3}{48} + \\ & + \frac{241\zeta_5}{216} - \frac{30853\pi^4}{466560} - \frac{31\pi^4\ln 2}{9720} + \frac{11\ln^4 2}{486} - \frac{\ln^5 2}{405} + \frac{44}{81}\text{Li}_4\left(\frac{1}{2}\right) + \frac{8}{27}\text{Li}_5\left(\frac{1}{2}\right) \end{aligned} \tag{20}$$

The first two terms in (18) are not yet known analytically. However, the overall numerical expressions for $z_m^{(4)}$ at fixed n_l were calculated in [3] with the help of computer program FIESTA, created in [17], [18]. The obtained in [3] results read:

$$z_m^{(4)}\Big|_{n_l=3} = -1744.8 \pm 21.5, \quad z_m^{(4)}\Big|_{n_l=4} = -1267.0 \pm 21.5, \quad z_m^{(4)}\Big|_{n_l=5} = -859.96 \pm 21.5 \ . \tag{21}$$

They were recently used by us in [7] to get the numerical expressions for the analytically unknown terms $z_m^{(40)}$ and $z_m^{(41)}$ with the help of rigorous optimal least squares (OLS) mathematical method. Below we summarise definite steps of this work and discuss its main results. To get the values of *two* unknown parameters from the results of (21) we use the following overdetermined system of *three* linear equations:

$$z_m^{(40)} + 3z_m^{(41)} = -1371.77 \ , \quad z_m^{(40)} + 4z_m^{(41)} = -614.68 \ , \quad z_m^{(40)} + 5z_m^{(41)} = 142.32 \tag{22}$$

To solve this system by means of the OLS method the following function is introduced

$$\Phi(z_m^{(40)}, z_m^{(41)}) = \sum_{k=1}^{3} \Delta_{l_k}^2 = \sum_{k=1}^{3} (z_m^{(40)} + z_m^{(41)} n_{l_k} - y_{l_k})^2$$

where Δ_{l_k} are the squared deviations, $1 \le k \le 3$ labels the concrete equations in the system (22) and the expressions for y_{l_k} are fixed by the numbers on the r.h.s. of each equation from this system. Its solutions $(z_m^{(40)}, z_m^{(41)})$ are determined by the following requirements:

$$\frac{\partial \Phi}{\partial z_m^{(40)}} = 0, \quad \frac{\partial \Phi}{\partial z_m^{(41)}} = 0. \tag{23}$$

It is possible to show that their solution always exists and is unique due to the fact that the rank of the matrix, composed from the l.h.s. of the system, is equal to two and coincides with the number of unknowns parameters. This method allows to determine theoretical uncertainties of the solutions of Eq. (23) of the parameters. Theoretical errors were defined in [7] as

$$\Delta z_m^{(40)} = \frac{\sqrt{\sum\limits_{k=1}^{3} n_{l_k}^2}}{\sqrt{3\sum\limits_{k=1}^{3} n_{l_k}^2 - \left(\sum\limits_{k=1}^{3} n_{l_k}\right)^2}} \Delta y_l \ , \quad \Delta z_m^{(41)} = \frac{\sqrt{3}\Delta y_l}{\sqrt{3\sum\limits_{k=1}^{3} n_{l_k}^2 - \left(\sum\limits_{k=1}^{3} n_{l_k}\right)^2}} \tag{24}$$

where $\Delta y_l = \sigma = 21.5$ are the given in [3] uncertainties of the results of (21). Using equations (23), supplemented by (24), in [3] we found the following expressions for two parameters we are interested in :

$$z_m^{(40)}(M_q^2) = -3642.9 \pm 62.0, \quad z_m^{(41)}(M_q^2) = 757.05 \pm 15.2. \tag{25}$$

Similar expressions were also obtained in [30] with the help of the *fitting procedure* of the numerical results of [3] (see Eq. (21)), supplemented by the calculated in [29] analytical expressions for $z_m^{(43)}$ and $z_m^{(42)}$ and by renormalon-based large β_0-representation of the n_l-dependence for $z_m^{(4)}$, fixed in [31, 22]. This additional theoretical input was not used by us. However, our results from Eq. (25) agree very well with the numbers

$$z_m^{(40)}(M_q^2) = -3643 \pm 21.5, \quad z_m^{(41)}(M_q^2) = 757 \pm 21.5, \tag{26}$$

which follow from the ones obtained in [30] for the $\mathcal{O}(a_s^4)$ contributions to the inverse ratio $M_q/\overline{m}_q$, where the running mass was normalised at another scale $\mu = \overline{m}_q$. The essential difference is that in [30] theoretical errors of the outcomes of their fits were not analysed separately but fixed by the errors of the the numerical results, obtained in [3].

5. Discussions of the theoretical errors.

In our work of [7] we raised the question on the reason of the coincidence of the numerical errors in the presented in [3] results of the numerical calculations of the $\mathcal{O}(a_s^4)$ corrections to the relations between running and pole heavy masses of c, b and t-quarks (see Eq. (21). The proposed in [7] explanation of the flavour-independence of the numerical error is that this error corresponds to the uncertainty of the n_l-independent contribution to (18), namely of the coefficient $z_m^{(40)}$. If this explanation is correct, then the error of the numerical calculations of the coefficient $z_m^{(41)}$ should be negligibly small. Therefore, the given in [30] errors are not absolutely correct. Indeed, they do not satisfy the proposed in [7] explanation of the "paradox of the coincidence of the numerical errors". The determined in [7] OLS method error of $z_m^{(40)}$-term really dominates over the uncertainty of $z_m^{(41)}$ (see (25)) and it is 3 times larger than the numerical error of this coefficient, which follows from the results of [3]. In spite of this positive fact we understand that the errors obtained within the OLS method may be overestimated (in order to get more definite estimates it is necessary to have more than three equations). However, even these results indicate that it may be useful to clarify in more detail how the numerical errors of the results of [3] were obtained. This problem is really important. Indeed, the difference in the numerical errors discussed above leads to different theoretical estimates of the uncertainties of the top-quark pole mass value (in the case of applications of the results of [3] this error is 0.005 GeV, while if the results of [7] are used the error is 4 times larger, namely 0.023 GeV). The analytical evaluation of the coefficients $z_m^{(40)}$ and $z_m^{(41)}$ will remove this problem. It is highly desirable to get their analytical expressions. These calculations are realistic and already started in [32] from the creation of the special computer program.

Acknowledgements
We are grateful to V.A. Smirnov for discussions of the definite subjects discussed in this talk. One of us (ALK) would like to thank the Organisers of ACAT-2016 Workshop in Valparaiso, Chile (18-22 January 2016) for the invitation and partial financial support. This work of ALK was supported by Russian Science Foundation, grant No. 14-22-00161.

References

[1] Beneke M 1998 *Phys. Lett.* B **434** 115
[2] Ayala C, Cvetic G, and Pineda A 2014 *JHEP* **1409** 045 and talk at ACAT-2016 Int. Workshop.
[3] Marquard P, Smirnov A V, Smirnov V A and Steinhauser M 2015 *Phys. Rev. Lett.* **114** no.14 142002
[4] Smirnov A V, Smirnov V A and Steinhauser M 2008 *Phys. Lett.* B **668** 293
[5] Smirnov A V, Smirnov V A and Steinhauser M 2010 *Phys. Rev. Lett.* **104** 112002
[6] Anzai C, Kiyo Y and Sumino Y 2010 *Phys. Rev. Lett.* **104** 112003
[7] Kataev A L and Molokoedov V S 2015, Preprint INR-TH-2015-014, arXiv:1511.06898 [hep-ph]
[8] Brambilla N., Pineda A, Soto J and Vairo A 2005 *Rev. Mod. Phys.* **77** (2005) 1423
[9] Fischler W 1977 *Nucl. Phys.* B **129** 157
[10] Billoire A 1980 *Phys. Lett.* B **92** 343
[11] Peter M 1997 *Phys. Rev. Lett.* **78** 602
[12] Schroder Y 1999 *Phys. Lett.* B **447** 321
[13] Kniehl B A, Penin A A, Smirnov V A and Steinhauser M 2002 *Nucl. Phys.* B **635** 357
[14] Chetyrkin K G and Tkachov F V 1981 *Nucl. Phys.* B **192** 159
[15] Smirnov A V 2008 *JHEP* **0810** 107
[16] Laporta S 2000 *Int. J. Mod. Phys.* A **15** 5087
[17] Smirnov A V and Tentyukov M N 2009 *Comput. Phys. Commun.* **180** 735
[18] Smirnov A V 2014 *Comput. Phys. Commun.* **185** 2090
[19] Gorishny S G, Kataev A L and Larin S A 1991 *Phys. Lett.* B **273** 141; 1992 **275** 512(E)
[20] Kataev A L and Molokoedov V S 2015 *Phys. Rev.* D **92** no.5 054008
[21] Akhoury R and Zakharov V I 1998 *Phys. Lett.* B **438** 165
[22] Beneke M and Braun V M 1994 *Nucl. Phys.* B **426** 301
[23] Bigi I , Shifman M A, Uraltsev N G and Vainshtein A I 1994 *Phys. Rev.* D **50** 2234
[24] Kazakov D I and Shirkov D V 1980 *Fortsch. Phys.* **28** 465
[25] Tarrach R 1981 *Nucl. Phys.* B **183** 384
[26] Gray N, Broadhurst D J, Grafe W and Schilcher K 1990 *Z. Phys.* C **48** 673
[27] Melnikov K and van Ritbergen T 2000 *Phys. Lett.* B **482** 99
[28] Chetyrkin K G and Steinhauser M 2000 *Nucl. Phys.* B **573** 617
[29] Lee R, Marquard P, Smirnov A V, Smirnov V A and Steinhauser M 2013 *JHEP* **1303** 162
[30] Kiyo Y, Mishima G and Sumino Y 2015 *JHEP* **1511** 084
[31] Beneke M and Braun V M 1995 *Phys. Lett.* B **348** 513
[32] Lee R N and Mingulov K T arXiv:1507.04256 [hep-ph].